AF179434

ELIHU BURRITT LIBRARY
CENTRAL CONNECTICUT STATE UNIVERSITY
NEW BRITAIN, CONNECTICUT 06050

WITHDRAWN

NONLINEAR VISION:

Determination of Neural Receptive Fields, Function, and Networks

NONLINEAR VISION:

Determination of Neural Receptive Fields, Function, and Networks

Edited by

Robert B. Pinter, Ph.D.
Professor
Departments of Electrical Engineering and Zoology
University of Washington
Seattle, Washington

Bahram Nabet, Ph.D.
Assistant Professor
Department of Electrical and Computer Engineering
Drexel University
Philadelphia, Pennsylvania

CRC

CRC Press
Boca Raton Ann Arbor London Tokyo

11/10

QP
475
N66
1992

Library of Congress Cataloging-in-Publication Data

Nonlinear vision : determination of neural receptive fields, function,
and networks / editors, Robert B. Pinter and Bahram Nabet.
p. cm.
Includes bibliographical references and index.
ISBN 0-8493-4292-9
1. Vision—Mathematical models. 2. Neural networks (Computer
science) I. Pinter, Robert B. II. Nabet, Bahram.
QP475.N66 1992
591.1'823'0724—dc20 91-45482
CIP

This book represents information obtained from authentic and highly regarded sources. Reprinted material is quoted with permission, and sources are indicated. A wide variety of references are listed. Every reasonable effort has been made to give reliable data and information, but the authors and the publisher cannot assume responsibility for the validity of all materials or for the consequences of their use.

All rights reserved. This book, or any parts thereof, may not be reproduced in any form without written consent from the publisher.

Direct all inquiries to CRC Press, Inc., 2000 Corporate Blvd., N.W., Boca Raton, Florida, 33431.

© 1992 by CRC Press, Inc.

International Standard Book Number 0-8493-4292-9

Library of Congress Card Number 91-45482

Printed in the United States 1 2 3 4 5 6 7 8 9 0

Printed on acid-free paper

PREFACE

This text brings to vision research a treatment different from that often found in books on the subject in its emphasis on nonlinear aspects of vision, from human perception to eye cells of the fly. There is considerable emphasis on mathematics, which forms not only models but the algorithms for processing data. Accordingly, there are texts to be considered references, such as Volumes II through V and also X of *Vision and Visual Dysfunction* (Cronly-Dillon, J., Ed., CRC Press, 1991); *Visual Perception: The Neurophysiological Foundations* (Spillman, L. and Werner, J. S., Eds., Academic Press, 1990); *Vision: Coding and Efficiency* (C. Blakemore, Ed., Cambridge University Press, 1990); *Sensory Neural Networks: Lateral Inhibition* (Nabet, B. and Pinter, R. B., Eds., CRC Press, 1991); *Analysis of Physiological Systems* (Marmarelis, P. Z. and Marmarelis, V. Z., Eds., Plenum Press, 1978). From one or more of these texts the vision researcher, ophthalmologist, optometrist, physicist, mathematician, and engineer can draw the requisite background for any of the chapters herein, both from the mathematical and the physiological standpoint.

Nonlinear aspects of vision occur mostly above threshold. The ubiquitous linear analysis of visual system modulation transfer functions (MTF) and receptive fields is without doubt limited when suprathreshold contrasts or great changes in mean or ambient light level occur. For example, in the insect compound eye the physiological optics adapt via several processes in a significant manner (see Chapter 15). Photoreceptors and second order visual interneurons have adapting gain and dynamic properties and may rectify significantly. These kinds of nonlinearities also occur in the vertebrate retina, and thus there are many important nonlinearities in visual processing on levels more peripheral than processing for the "primal sketch" of David Marr.

Unfortunately, the mathematical world of nonlinear analysis demands a taxonomy too. There is not the unlimited generality of linear analysis within its narrow region of validity. But partially as a result of the kinds of visual stimuli utilized in quantitative experimental analysis of visual systems, there are a few dominant and specific categories of nonlinearity. They are in this text more limited than general statements of neural networks that learn, because identification of the subunits in such general models is not possible with present day approaches to experimental analysis. Thus, here the nonlinear aspects are quite basic and specific: parametric changes in gain, nonlinear gain transfer characteristics such as rectifiers of a general function type (Chapter 9), parametric changes in dynamics of time and space constants and receptive fields (Chapters 12, 15, 16, 17, 18, 19, and 21). All of these can be considered under a category of differential equation models of processes which have been identified quantitatively.

However, the general and specific nonlinear differential equation approach is limited in one sense by the identification problem. That is, if a nonlinear differential equation is assumed, how does one go about finding or identifying the coefficients of coupling and the nonlinear coupling functions themselves? There is no general solution to this problem, and if the specific form of the nonlinear differential equation is not assumed there is usually no method by which such a hypothesis can be tested. Fortunately this difficult approach in nonlinear systems is counterbalanced by the Volterra and Wiener kernel series methods, introduced clearly in Chapter 1 and in a quite different manner in Chapters 2 and 4. The Volterra series can be considered an assumed general solution to an unknown nonlinear integro-differential equation. Norbert Wiener took the important step of orthogonalizing the Volterra series. That result, the Wiener series, has a systematic identification method to find the deterministic kernels in the series. For broad-band noise inputs this can simply take the form of cross-correlation of the input and output, but the ramifications of other inputs are treated here in Chapters 5 and 6. A new set of assumptions about the underlying functions is explored in the fast orthogonal search method of Chapter 7. When motion of the visual stimulus is involved specifically, a reduction of space-time dimensionality is possible and a new motion kernel is defined and elucidated in Chapters 3 and 4.

Is there a link between the Volterra and Wiener series approach and the differential equation? It is possible to derive such a link via the method of perturbation analysis which is presented in Chapter 19. This, in turn, is linked to electronic implementations, in Chapter 18. Considerations about using the visual information and the problem of sensory-motor coordination are discussed in Chapter 20.

This text encompasses experimental subjects from the human to the fly. Each experimental study is recent and on the frontier of vision research. It may be true that the simpler the nervous system and experimental preparation the closer one gets to the underlying neuronal mechanism responsible for the nonlinearity, in the sense of requiring fewer linking or bridging hypotheses for a model. But the field of view is narrower as well. Form vision and VEP (visual evoked potential), psychophysical and behavioral response are treated in Chapters 8 through 13.

Ultimately neurons are responsible for the multipliers, rectifiers, and demodulators which can account for nonlinearity in the VEP, cortical simple and complex cells, and in the fly optic lobe visual interneurons. The neuronal properties of Nernst potentials, nonlinear conductances, presynaptic inhibition and postsynaptic conductance modulation are good candidate mechanisms for these nonlinearities. But the network connections are often very difficult to ascertain. Yet we have here a set of candidate mechanisms far more specific and useful than a Turing machine.

For example the dichotomy of even and odd functions representing receptive fields in models of human perception (Chapter 11) is also found in the second order cells (lamina, LMC) of the fly's retina. The spatio-temporal receptive field of the LMC cell is an odd function at bright light levels and becomes an even function at low light levels. This is corroborated in the consistent behavior of the fly walking towards edges at the same calibrated bright levels and towards dark blobs (scototaxis) at the same low light levels (Pinter et al., *Visual Neurosci.*, 4, 579–584, 1990; Srinivasan et al., *Proc. R. Soc. London, Ser. B*, 240, 279–293, 1990). Thus the even and odd function dichotomy appears already at the periphery in second order cells via a parametric adaptation. Conductance modulation in postsynaptic cell membrane in cell to cell or cell subunit to cell subunit coupling is a mechanism which can account for this adaptation (Chapters 16, 18, and 19). It is, however, not possible at present to anatomically pinpoint which synapses of a given type in the lamina are responsible for the modulation or multiplication and to ascertain the specific role of inherent cell membrane current-voltage nonlinearities.

In suprathreshold transmission of information many nonlinearities appear in the visual system. Whether these nonlinearities are in each case optimal in some strict sense is difficult to prove. However, many nonlinearities appear to be compensating for the changing visual environment as ambient light levels change or implementing motion detection, discrimination and quantification (Chapters 11, 15 to 19, and 21). The very nature of these nonlinearities is important because the organism has evolved with them or perhaps partially by means of them. Of course we are limited in our ability to describe and assign levels of importance by our mathematics and experimental apparatus, which is somewhat crude and brittle by comparison to biological capability and adaptations, on several time scales. This suggests that many more unknown nonlinear mechanisms exist in biological neural networks to accomplish this purposeful variability.

In the interaction of mathematical theory and experiment there is often an iteration: theory, experiment, theory . . . , and this text represents a strong iteration on behalf of nonlinear mathematical theory but with experiment. In neurobiology and the study of biological neural networks there is always the danger that the mathematics and theory overwhelm the physical and physiological fact. We have tried to avoid that by presenting all nonlinear mathematical analysis in the context of visual system experimental applications. The mathematics is there to organize the data—a synthesis.

R. B. Pinter
Vashon Island, Washington

THE EDITORS

Robert B. Pinter, Ph.D., is Professor of Electrical Engineering and Adjunct Professor of Zoology at the University of Washington, Seattle, Washington. Dr. Pinter received his M.S.E.E. in 1960 and Ph.D. in 1964 from Northwestern University, Evanston, Illinois. He joined the Electrical Engineering Department of the University of Washington in 1964 as Assistant Professor, and the Bioengineering Center as N.I.H. Fellow. He has also been a Visiting Professor of the Psychology Department of Dalhousie University, Halifax, Nova Scotia, Canada and a Visiting Fellow of the Centre for Visual Sciences at the Australian National University, Canberra, for continuing studies and research in vision physiology.

Dr. Pinter is a member of the Society for Neuroscience, the I.E.E.E., Sigma Xi, and the honorary societies Tau Beta Pi and Eta Kappa Nu. He is the technical chairman for Biological Cybernetics in the I.E.E.E. Systems, Man and Cybernetics Society, organizing workshops and meeting sessions. He is an Associate Editor of the I.E.E.E. Transactions on Systems, Man and Cybernetics.

Dr. Pinter has authored or co-authored more than sixty publications, including chapter contributions to four books. His major research interests focus on neurons and their networks including experimental neurophysiology of vision, especially in insects. His current research interests are nonlinear, adaptive models of networks of neurons and the physiology of detection and resolution of magnetic fields by neurons.

Bahram Nabet, Ph.D., was born in Iran in 1956. He received his B.S.E.E. from Purdue University in 1977, after completing an Honors Curriculum. He pursued graduate studies at the University of Washington in 1984 and was awarded M.S.E.E. and Ph.D. degrees in 1985 and 1989, respectively. He was a Postdoctoral Research Associate at the University of Washington before joining the Electrical and Computer Engineering faculty of Drexel University where he is presently an Assistant Professor. His research interests include neural networks, microelectronics, vision, and nonlinear systems. Dr. Nabet is co-author with R. B. Pinter of the book *Sensory Neural Networks: Lateral Inhibition* published by CRC Press in 1991. He is recipient of the Drexel University Research Scholar Award and the Engineering Foundation Research Award.

CONTRIBUTORS

A. B. Bonds
Department of Electrical Engineering
Vanderbilt University
Nashville, Tennessee

Abdesselam Bouzerdoum
University of Adelaide, S.A.
Australia

David C. Burr
Istituto di Neurofisiologia del CNR
Pisa, Italy

Mark C. Citron
Department of Neurology
Childrens Hospital
University of Southern California
Los Angeles, California

S. H. Courellis
University of Southern California
Los Angeles, California

Robert C. Emerson
University of Rochester
Rochester, New York

Stephen Grossberg
Boston University
Boston, Massachusetts

Andrew C. James
Australian National University
Canberra, Australia

M. E. Jernigan
University of Waterloo
Waterloo, Ontario
Canada

Stanley Klein
University of California
Berkeley, California

Michael Korenberg
Queen's University
Kingston, Ontario
Canada

Michael Kuperstein
Neurogen, Inc.
Brookline, Massachusetts

V. Z. Marmarelis
University of Southern California
Los Angeles, California

Peter D. McIntyre
University of New South Wales
Australian Defense Force Academy
Canberra, Australia

G. F. McLean
University of Victoria
Victoria, B.C.
Canada

M. Concetta Morrone
Scuola Normale Superiore
Pisa, Italy

Bahram Nabet
Drexel University
Philadelphia, Pennsylvania

D. Osorio
Australian National University
Canberra, Australia

Robert B. Pinter
University of Washington
Seattle, Washington

Keith Purpura
Cornell University Medical College
New York, New York

D. M. Regan
York University
North York, Ontario
Canada

M. P. Regan
York University
North York, Ontario
Canada

M. V. Srinivasan
Australian National University
Canberra, Australia

Erich Sutter
Smith-Kettlewell Eye Research Institute
San Francisco, California

Jonathan D. Victor
Cornell University Medical College
New York, New York

Eric J. Warrant
Department of Zoology
University of Lund
Lund, Sweden

Lonce Wyse
Boston University
Boston, Massachusetts

To Marie

R. B. P.

To Zohreh

B. N.

TABLE OF CONTENTS

Chapter 1
Nonlinear Systems Analysis in Vision: Overview of Kernel Methods 1
Jonathan D. Victor

Chapter 2
Nonlinear Operator Network Models of Processing in the Fly Lamina 39
Andrew C. James

Chapter 3
Linear and Nonlinear Mechanisms of Motion Selectivity in Simple
Cells of the Cat's Striate Cortex 75
Robert C. Emerson and Mark C. Citron

Chapter 4
Nonlinear Functional Representations for Motion Detection and Speed
Estimation Schemes 91
S. H. Courellis and V. Z. Marmarelis

Chapter 5
Optimizing the Estimation of Nonlinear Kernels 109
Stanley Klein

Chapter 6
A Deterministic Approach to Nonlinear Systems Analysis 171
Erich Sutter

Chapter 7
Applications to the Study of Visual Systems of the Fast
Orthogonal Search and Parallel Cascade Methods 221
Michael Korenberg

Chapter 8
Alterations of Form Vision at Isoluminance Identified Through the
Use of Visual Textures 243
Jonathan D. Victor and Keith Purpura

Chapter 9
A Frequency Domain Method for Isolating Specific Kinds of
Nonlinear Neural Processing and for Testing Nonlinear Multineuron
Models Against Data 265
M. P. Regan and D. M. Regan

Chapter 10
Nonlinearities in Psychophysical Models of the Processing of Spatial
Form and Motion 293
D. M. Regan

Chapter 11
A Nonlinear Model of Feature Detection 309
David C. Burr and M. Concetta Morrone

Chapter 12
Spatial and Temporal Nonlinearities in Receptive Fields of the Cat
Striate Cortex 329
A. B. Bonds

Chapter 13
How Flying Bees Compute Range from Optical Flow: Behavioral
Experiments and Neural Models 353
M. V. Srinivasan

Chapter 14
Retinotopic Vision in the Locust 377
D. Osorio

Chapter 15
The Trade-Off Between Resolution and Sensitivity in Compound Eyes 391
Eric J. Warrant and Peter D. McIntyre

Chapter 16
Nonlinear Lateral Inhibition Applied to Motion Detection in the Fly
Visual System 423
Abdesselem Bouzerdoum and Robert B. Pinter

Chapter 17
Lateral Inhibition and Image Processing 451
M. E. Jernigan and G. F. McLean

Chapter 18
Electronic Hardware for Vision Modeling 463
Bahram Nabet

Chapter 19
Multiplicative Inhibition and Volterra Series Expansion 475
Bahram Nabet and Robert B. Pinter

Chapter 20
INFANT Neural Controller for Adaptive Sensory-Motor Coordination 493
Michael Kuperstein

Chapter 21
A Neural Network Architecture for Figure-Ground Separation of
Connected Scenic Figures 515
Stephen Grossberg and Lonce Wyse

Index 545

Chapter 1

NONLINEAR SYSTEMS ANALYSIS IN VISION: OVERVIEW OF KERNEL METHODS

Jonathan D. Victor

TABLE OF CONTENTS

I. Introduction 2

II. Linear Systems 2
- A. The Superposition Property 2
- B. The Impulse Response 3
- C. The Transfer Function 4
- D. Systems with More Than One Input 7

III. The Kernel Approach to Nonlinear Systems Analysis: Theoretical Background 9
- A. Functional Expansions 9
- B. The Volterra Approach 9
- C. The Wiener Approach 11
- D. Comparison of Wiener and Volterra Representations 13

IV. Methods of Kernel Measurement 16
- A. Gaussian White Noise as a Stimulus 17
- B. Other Orthogonal Expansions 21
- C. Stochastic Systems 24
- D. The M-Sequence Approach 25
- E. Frequency Domain Methods 28
- F. Related Approaches 30
- G. Nonlinear Systems with More Than One Input 30

V. Closing Remarks: The Importance of Nonlinearities 34

Acknowledgments 34

References 34

I. INTRODUCTION

The aim of this chapter is to provide an overview of several related approaches to the analysis of nonlinear processes that occur in vision.

The term "systems analysis" denotes a theoretical framework for the characterization of the input-output properties of physical systems. As it is used in experimental vision research, the term often connotes techniques for investigating the internal structure of a system from measurement of its input-output properties. Thus, at one level, systems analysis is a technique of mathematical dissection.

However, mathematical dissection is not the only role of systems analysis in vision. A systems-analytic description of a neuron or collection of neurons is a concise representation of the operations performed by these elements, independent of hardware. As such, an accurate systems-analytic representation is a means to understand neural computations on an algorithmic level. These representations may serve both to test computational models of natural vision processes and to inspire the design of artificial vision systems.

Both arms of systems theory have important roles in vision: the analysis of the behavior of an empirical model (the "forward" problem) is key to understanding the algorithms of visual processing, while the deduction of internal structure from input-output data (the "inverse" problem) is an analytical technique which complements the techniques of biochemical or anatomical analysis available to the neuroscientist.

This survey will focus on principles and concepts. By avoiding mathematical details, I do not mean to imply that they are unimportant or dull. Rather, the interested reader is encouraged to consult original treatises (Volterra 1932, Wiener 1958), more recent texts (Marmarelis and Marmarelis 1978, Schetzen 1980, Rugh 1981), review articles (Billings 1980, Sakai et al. 1988, Korenberg and Hunter 1990), and primary papers (Barrett 1963, Palm and Poggio 1978, Victor and Knight 1979, Yasui 1979), in addition to the chapters in this book.

II. LINEAR SYSTEMS

To place nonlinear systems analysis in context, we begin by highlighting some important ideas from the analysis of linear systems. These ideas include the imposition of an algebraic structure on input and output, the relationship of time-domain and frequency-domain viewpoints, how systems-analytic representations can provide clues to internal structure, and how techniques applied to systems with a single input (in space) may be generalized to systems with spatially distributed inputs. With a firm grasp of the foundations of linear systems analysis in hand, the consequences of departures from linearity will be more readily apparent.

In both the linear and nonlinear cases, we will restrict attention to systems which are "stationary" and of "finite memory". A stationary system is one whose intrinsic input-output properties do not change with time. A system with finite memory is one whose input-output properties depend on only a finite amount of its stimulation history. These properties guarantee that a sequence of responses to a particular stimulus repeated at sufficiently separated times represents independent observations of the same system behavior, altered neither by the passage of time nor by the stimuli themselves. Although it is clear that these properties represent ideals which a real biological system can at best approximate, it is equally clear that these (or similar) assumptions are required for an input-output analysis to begin.

A. THE SUPERPOSITION PROPERTY

The defining property of a linear system is that the response to two stimuli combined can be predicted by summing up the responses to each of the components. That is, if the stimulus S_1 leads to the response R_1 and if the stimulus S_2 leads to the response R_2, then the stimulus $S_1 + S_2$ must lead to the response $R_1 + R_2$.

Closely related to the property of superposition is that of scaling. Multiplication of the input by a scale factor must lead to multiplication of the output by the same scale factor. That is, if the stimulus S_1 leads to the response R_1, then the stimulus $b \cdot S_1$ must lead to the response $b \cdot R_1$. Scaling is usually cited as a separate property of linear systems, independent of superposition. However, scaling (for rational scale factors) is a straightforward consequence of superposition, and more general scaling follows from technical arguments concerning continuity of response as a function of input.

Intrinsic to the notion of superposition is the notion of algebraic manipulation (addition and scaling) of stimuli and responses. In vision, a stimulus is typically a pattern of luminance values in time. The numerical value associated with a particular luminance is the difference between that luminance and a reference luminance. The reference luminance is typically the mean luminance of the background, and this difference is then known as the contrast. Addition of two stimuli means addition of their contrasts at corresponding points in time. With intracellular studies, the usual response measure is the time series of fluctuations of intracellular voltages about a mean potential. Addition of two responses corresponds to addition of voltage fluctuations at corresponding points in time. Note that measures of stimuli and responses were defined as deviations from a mean. In this way, the zero stimulus (background illumination) automatically elicits a zero response (mean intracellular voltage). This is a prerequisite for superposition.

Fundamentally, the quantification of stimuli and responses and the notion of addition represent the imposition of an algebraic structure on the experimental situation. In principle, one could have defined stimuli, responses, and their addition in a different way (for example, in terms of luminance, rather than contrast), but then, superposition could not be anticipated to hold. Because of the far-reaching consequences of the superposition property, it is usually advantageous to impose this algebraic structure in a way that superposition is likely to hold.

Derivation of a response measure from extracellular recordings is less straightforward, since a spike train represents a sequence of all-or-none voltage events, for which superposition cannot be anticipated. The usual strategy is to derive a measure of the "instantaneous firing rate" from the impulse train. For repetitive stimuli, the "instantaneous firing rate" is essentially the quantity measured by a histogram of impulse occurrences following many repeats of a stimulus. For nonrepetitive stimuli often used in the analysis of nonlinear systems, more sophisticated strategies are required (Brodie et al. 1978a,b).

B. THE IMPULSE RESPONSE

The superposition property leads to an exhaustive but concise characterization of the input-output behavior of a linear system. We consider first a system with a single input, and later discuss how the ideas extend to systems with two or more spatially distributed inputs. For analytical purposes, we restrict observations to closely spaced discrete times. An input signal S is, thus, regarded as a sequence of values S(t), one at each observation time t. Consider now an impulse-like signal I(t) which matches S at an observation time t, but is zero at all other times. The original input signal is the sum of each of its component impulse-like signals (Figure 1). Consequently, the response to the input signal S(t) can be predicted from the response to each of its impulse components.

The scaling property implies that the response to an impulse of height S(t) is exactly S(t) times the response to an impulse of unit height. Furthermore, the stationarity property implies that the response to a unit impulse does not depend on the time at which it is presented. Thus, the response to an impulse of height S(t) presented at time t can be calculated from the response to a unit impulse presented at time zero. This response to a unit impulse is known as the "impulse response", which we will denote by L(t).

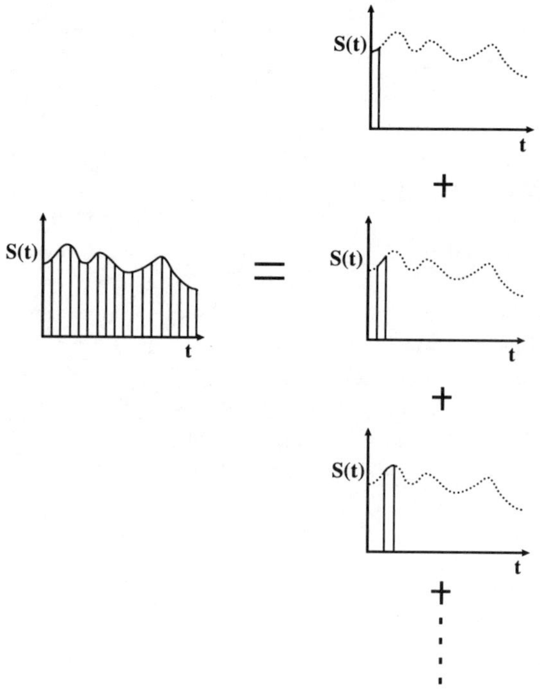

FIGURE 1. Decomposition of an arbitrary stimulus into a sequence of impulses.

The impulse response L(t) may be viewed in two ways. As defined, L(t) is the response of the system to a unit impulse at time zero. Alternatively, our derivation shows that L(t) is the portion of the current response which is due to a unit impulse at time t in the past. That is, the response R(t) is related to the stimulus S(t) by

$$R(t) = \int_0^\infty L(\tau)\, S(t - \tau)\, d\tau \qquad (1)$$

L(t) provides a complete characterization of a linear system. Via Equation 1, it allows a concise prediction of responses to all inputs.

C. THE TRANSFER FUNCTION

Although characterization in terms of an impulse response has intuitive appeal, other considerations motivate an alternative approach. Frequently one wants to express the behavior of a composite system in terms of the behavior of its components. Consider, for example, a composite system composed of two linear systems in series (Figure 2A). In terms of the impulse responses $L_1(t)$ and $L_2(t)$ of the components, the impulse response L(t) of a composite system may be written as follows:

$$L(t) = \int_0^t L_1(t - \tau)\, L_2(\tau)\, d\tau \qquad (2)$$

This convolution integral is distinctly more awkward than the expression we will derive below for frequency-domain representations.

The derivation of the impulse response as a characterization of a linear system began with the observation that any input signal S(t) can be considered to be a superposition of

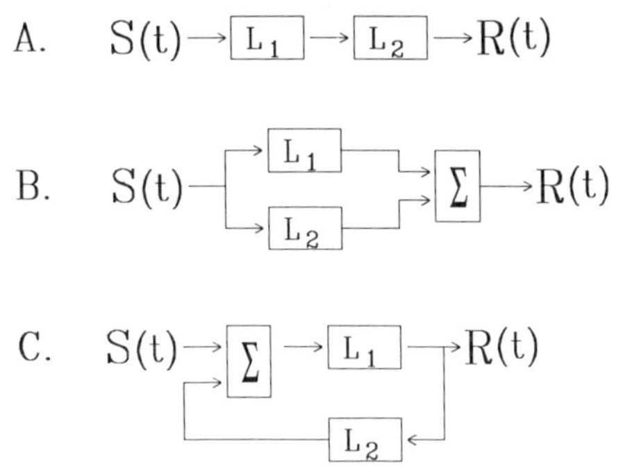

FIGURE 2. Composite systems with components in series (A), parallel (B), and feedback (C).

narrow impulses. In like manner, any input signal can also be considered as a superposition of sine waves, provided that the amplitudes, phases, and frequencies of the sine waves may be freely chosen. This decomposition is known as the Fourier transform (Bracewell 1978). It is closely related to the decomposition of a *periodic* stimulus into a sum of sine waves whose frequencies are integer multiples of the frequency of the stimulus (the Fourier series).

We now use the Fourier transform representation to calculate the response of a linear system to any input signal S(t). The superposition property states that we may sum up the responses to each of the sinusoidal components of S(t). Thus, we need to know the responses to sinusoidal inputs of all amplitudes, phases, and frequencies.

According to the scaling property, two sinusoidal inputs that differ only in amplitude elicit responses that differ only by an overall scale factor. This scale factor is equal to the amplitude ratio of the inputs. Changing the phase of a sinusoidal input is equivalent to a shift in time, and according to the stationarity property, such time shifts lead to corresponding time shifts in the output. Thus, it suffices to consider the responses to sinusoids of unit amplitude and a standard phase.

A not-so-obvious consequence of linearity is that a sinusoidal input at the frequency ω must elicit a sinusoidal output at the same frequency ω. We sketch a way of deducing this nonobvious property directly from the superposition and stationarity properties. The critical observation is that superposition and stationarity imply that if the input satisfies a linear differential equation with constant coefficients and no forcing term, then so must the output. We apply this observation to the differential equation

$$\frac{d^2u}{dt^2} + \omega^2 u = 0 \tag{3}$$

Since the sinusoids of frequency ω constitute all of the solutions to Equation 3, the sinusoid-in, sinusoid-out property follows. The sinusoid-in, sinusoid-out property may also be deduced from the impulse-response characterization (Equation 1).

Since the response to a sinusoidal stimulus is necessarily a sinusoid of the same frequency, the response to a sinusoid of frequency ω of unit amplitude and standard phase is completely described by its amplitude and phase. These amplitudes and phases will depend on the frequency ω. We now develop a convenient representation of these sinusoids, which exploits properties of the complex numbers.

A sinusoidal signal of frequency ω has the general form

$$u(t) = A\cos(\omega t + \phi) \tag{4}$$

where A is the amplitude of the sinusoid and ϕ is its phase. For many purposes, it is convenient to use complex numbers to represent this signal. A complex number, represented as $z = x + iy$, is a vector in the two-dimensional plane (i denotes the imaginary unit, $\sqrt{-1}$). The complex number z can equally well be represented in polar notation $A\ e^{i\phi}$, where A is the length of the vector and ϕ is the angle between the vector and the x-axis. With standard rules for manipulation of complex quantities, Equation 4 can be rewritten

$$u(t) = \mathrm{Re}\{ze^{i\omega t}\} \tag{5}$$

where "Re" denotes "the real part of". Thus, the complex number $z = A\ e^{i\phi}$ contains the amplitude and phase information required to reconstruct the signal u(t).

The representation (Equation 5) of a sinusoid $A \cos(\omega t + \phi)$ by a complex number z has several important properties. The sinusoid of unit amplitude and a phase shift of zero is represented by the complex number 1. Addition of two sinusoids $u_1(t)$ and $u_2(t)$ corresponds to addition of their representations as complex numbers, z_1 and z_2. Multiplication of a sinusoid by a factor b corresponds to multiplication of its complex representation by the real number b. Shifting the phase of a sinusoid by an amount ϕ corresponds to multiplication of its complex representation by the complex number $e^{i\phi}$.

With this understanding, the amplitude and phase of the response of a system to a unit sinusoid of frequency ω may be represented by a complex number, which we will call $\tilde{L}(\omega)$. The key benefit of this maneuver is that the action of the system on a sinusoid of frequency ω corresponds to multiplication of its complex representation by the quantity $\tilde{L}(\omega)$. This is because the amplitude of $\tilde{L}(\omega)$ is the amplitude of the response elicited by a sinusoid of unit amplitude, and the phase of $\tilde{L}(\omega)$ is the phase of the response elicited by a sinusoid of zero phase.

The set of values $\tilde{L}(\omega)$, one for each frequency ω, constitutes the "transfer function" of a linear system. Since a value of the transfer function is the complex representation of the response to a unit sinusoid, it may be calculated from the impulse response. Application of Equation 1 to $S(t) = A \cos(\omega t)$ leads to the relationship between the impulse response L(t) and the transfer function $\tilde{L}(\omega)$:

$$\tilde{L}(\omega) = \int_0^{\infty} L(\tau)e^{-i\omega\tau}\, d\tau \tag{6}$$

That is, the transfer function $\tilde{L}(\omega)$ is the Fourier transform of the impulse response L(t). Consequently, the impulse response L(t) may be recovered from the transfer function by inversion of the Fourier transform:

$$L(\tau) = \frac{1}{2\pi} \int_{-\infty}^{\infty} \tilde{L}(\omega)e^{i\omega\tau}\, d\omega \tag{7}$$

Equations 6 and 7 show that the impulse response and the transfer function are equivalent representations of a linear system.

The properties of the representation of sinusoids by complex numbers lead to simple rules for how the behavior of a composite linear system is related to the behavior of its components. We reconsider the above example of a composite system composed of two linear systems in series (Figure 2A). The transfer function $\tilde{L}(\omega)$ of the composite system is merely the product of the transfer functions $\tilde{L}_1(\omega)$ and $\tilde{L}_2(\omega)$ of the components,

$$\tilde{L}(\omega) = \tilde{L}_1(\omega) \cdot \tilde{L}_2(\omega) \tag{8}$$

This result can be verified in a simple manner. By definition, $\tilde{L}(\omega)$ represents the response of the composite system to the unit sinusoid at frequency ω. After the action of the first linear system, the unit sinusoid has been transformed to a sinusoid represented by $\tilde{L}_1(\omega)$. The action of the second linear system on any sinusoid corresponds to multiplication by its transfer function, $\tilde{L}_2(\omega)$. Thus, the action of the composite system on a unit sinusoid corresponds to serial multiplication, first by $\tilde{L}_1(\omega)$ and then by $\tilde{L}_2(\omega)$.

Similar combination rules hold for parallel and feedback connections. A composite system consisting of a component with transfer function $\tilde{L}_1(\omega)$ in parallel with a component with transfer function $\tilde{L}_2(\omega)$ (Figure 2B) has a transfer function

$$\tilde{L}(\omega) = \tilde{L}_1(\omega) + \tilde{L}_2(\omega) \tag{9}$$

A composite system consisting of a component with transfer function $\tilde{L}_1(\omega)$ whose output is fed back via a component with transfer function $\tilde{L}_2(\omega)$ (Figure 2C) has a transfer function

$$\tilde{L}(\omega) = \frac{\tilde{L}_1(\omega)}{1 - \tilde{L}_1(\omega)\,\tilde{L}_2(\omega)} \tag{10}$$

Thus, the frequency domain representation permits a direct translation of the "wiring diagram" of a system into an algebraic manipulation of its component transfer functions. In the time domain, the composition law for parallel combination is also simple addition of the impulse responses (analogous to Equation 9), but the composition laws for serial combination and feedback involve integral equations.

In choosing between the time domain and the frequency domain representations, there is also an important practical issue. The impulse response has the appeal that it is a single measurement which completely characterizes the system. However, this measurement requires a stimulus of limitingly short duration. In order to generate a response of sufficient size to allow accurate measurement, it is often necessary to choose a test impulse of large magnitude. Such brief large stimuli may drive a system into a nonphysiological range. While analysis of saturations and other phenomena that occur under such extreme conditions may, indeed, provide useful probes of the system, this strategy is inconsistent with the goals of analyzing the linear behavior of a system.

The frequency-domain alternative is to measure the response to sinusoids over a range of frequencies ω. Since the response of a linear system to a sinusoid is anticipated to be another sinusoid of the same frequency, filtering techniques can be used to enhance the signal-to-noise ratio. Compared with a brief large pulse, a sinusoidal input of moderate amplitude is a smooth signal which is less likely to drive the system to saturation. Of course, the frequency domain approach requires measurement of responses to many different sinusoidal stimuli (to cover the frequency range of interest) and, thus, does not have the simplicity of the direct measurement of an impulse response via a single experiment. On the other hand, measurement of the responses to single sinusoids provides data (such as the presence of higher harmonics in the response) which can be used as an initial *test* for linearity. Early examples of the application of sinusoidal stimuli to characterize a neural transduction include the work of Pringle and Wilson (1952) and De Lange (1952).

D. SYSTEMS WITH MORE THAN ONE INPUT

We now consider how the above ideas may be extended to systems with two or more spatially distributed inputs. This extension is critical for study of the visual system. Indeed, study of the visual system traditionally focuses on its analysis of spatial, rather than temporal, patterns.

The main consideration in extending the analysis of linear systems with one input to linear systems with two or more inputs is an extension of the idea of superposition. For single-input systems, a stimulus is considered to be a time series of numbers, and the sum of two stimuli is calculated by adding stimulus values at corresponding points in time. For systems with spatially distributed inputs, a stimulus is a collection of such time series, one for each input. The sum of two stimuli is then calculated by adding stimulus values at each input and at corresponding points in time.

For single-input systems, the superposition property allowed us to describe the system behavior in terms of an impulse response or a transfer function. For multiple-input systems, superposition implies that responses to each input can be considered separately. Thus, the behavior of a multiple-input linear system can be described by a family of impulse responses or transfer functions, one for each input. This approach is useful, for example, in an analysis of the receptive field of a ganglion cell in which the receptive field center is considered to be one input and the receptive field surround is considered to be the second input.

The multiple-input formalism can be used for any number of inputs. For example, the entire receptive field can be divided into a large number of tiny regions, each of which represents a separate input. As the input regions become small, it is more useful considering the stimulus to be continuous in space, as well as in time. The impulse responses that characterize the system consist of responses to infinitesimally narrow pulses in time, confined to infinitesimally small regions of space — a spatiotemporal impulse response.

It is natural to ask how the frequency domain characterization extends to the spatiotemporal impulse response. The key elements in the development of the transfer function were superposition and stationarity. As we have just seen, the superposition property has a natural extension from time to space, but we have not yet extended the stationarity principle. The natural extension of the stationarity principle to space is that a spatial shift in the stimulus leads to a spatial shift in the response — translation invariance. In the typical single-unit experiment, one records from only one neuron, but with the implicit assumption that this neuron (and its relationship to its neighbors) is representative of other nearby neurons of the same class. Translation invariance formalizes this notion.

With translation invariance assumed, the above arguments leading from the impulse response to the transfer function representation readily extend to the spatial domain. A stimulus which is a spatiotemporal sinusoid [e.g., $\cos(\omega t + kx)$] must produce a response which is a spatiotemporal sinusoid of the same frequency, but perhaps of differing amplitude and phase. This leads to the notion of a spatiotemporal transfer function. This is a complex-valued function of temporal and spatial frequency that specifies the amplitude change and phase shift. The composition properties of the temporal transfer function readily extend to spatiotemporal transfer functions.

There is one aspect of the temporal theory that does not extend to the spatiotemporal theory. We have implicitly assumed that systems under study are causal — a stimulus cannot affect responses that have already occurred. Consequently, the impulse response $L(t)$ must be zero for negative values of t. This, in turn, induces relationships between the real and imaginary parts of transfer function $\widetilde{L}(\omega)$ — the Kramers-Kronig relations (De Groot and Mazur 1961, pp. 143–150). But there is no corresponding constraint in the spatial domain: in principle, a stimulus confined to one region in space can affect responses in all directions; that is, a spatiotemporal impulse response may be nonzero for both positive and negative values of its spatial arguments. Consequently, the constraints on real and imaginary parts of the temporal dependence of a transfer function do not extend to its spatial dependence.

III. THE KERNEL APPROACH TO NONLINEAR SYSTEMS ANALYSIS: THEORETICAL BACKGROUND

A nonlinear system is a system which does not necessarily obey superposition. By removing the superposition property, we have eliminated the crucial assumption required to characterize systems by an impulse response or a transfer function. At first, it might seem that this leaves little structure left for building a theory. However, as we will see below, many of the main elements of the linear theory generalize, to a limited but useful extent.

A. FUNCTIONAL EXPANSIONS

We characterized a linear system by discretizing time and decomposing the stimulus S into a sum of impulses of height S(t) at each time t. The superposition property allowed us to consider the contribution of each impulse independently. In characterizing a nonlinear system, we also begin by decomposing the stimulus into a sum of impulses. However, the effects of these impulses on the output cannot be considered independently. Rather, we must consider and classify interactions of these impulses.

B. THE VOLTERRA APPROACH

In the Volterra approach (Volterra 1932), we assume that the system under study becomes more and more nearly linear, as the size of the stimulus becomes small. More precisely, we assume that the potential interaction of impulses at any set of times $t_1, t_2, \ldots, t_n$ is proportional to the product of their heights $S(t_1), S(t_2), \ldots, S(t_n)$. The contribution of this n-fold interaction to the system response at time t is given by the product of their heights and a function $L^{(n)}$ of the time-lags $t-t_1, t-t_2, \ldots, t-t_n$. This function, $L^{(n)}(\tau_1, \tau_2, \ldots, \tau_n)$, is known as the nth order Volterra kernel of the system. More formally, the total contribution of n-fold interactions to the system response is given by

$$V_n[L^{(n)},S](t) = \int_0^\infty \int_0^\infty \cdots \int_0^\infty L^{(n)}(\tau_1,\tau_2, \ldots, \tau_n) \, S(t-\tau_1)S(t-\tau_2)\cdot \cdots \cdot S(t-\tau_n)\, d\tau_1\, d\tau_2 \cdots d\tau_n \tag{11}$$

The objects V_n are called "functionals", to indicate that they are functions which act on functions. The Volterra kernels $L^{(n)}$ represent generalizations of the impulse response. As seen by a comparison of Equation 11 with $n = 1$ to Equation 1, the contribution V_1 to the system response is exactly that of a linear system with impulse response $L^{(1)}(\tau)$.

The Volterra representation has an intrinsic symmetry property. Permutation of the arguments of a Volterra kernel in Equation 11 does not affect the value of the system response. For example, $L^{(2)}(\tau_1,\tau_2)$ leads to the same response as $L^{(2)}(\tau_2,\tau_1)$. Thus, the Volterra kernel $L^{(n)}(\tau_1,\tau_2, \ldots, \tau_n)$ may always be considered to be a symmetric function of its arguments.

If the stimulus S is scaled by a factor b, the size of the n-fold interaction term V_n is scaled by a factor of b^n. Thus, for small stimuli (i.e., scale factors b near zero), the $n = 1$ term dominates the response. That is, the Volterra kernel $L^{(1)}(\tau)$ is the impulse response of the linear system which best approximates the behavior of the system under study for limitingly small stimuli.

The formal Volterra expansion of a nonlinear system μ includes a contribution of one term V_n for each order n. For completeness, a zeroth-order term $V_0[L^{(0)},S](t) = L^{(0)}$ is included, which is the response of the system to the zero stimulus. Thus, the Volterra characterization of a system μ consists of a hierarchy of Volterra kernels, the functions $L^{(0)}$, $L^{(1)}(\tau_1)$, $L^{(2)}(\tau_1,\tau_2)$, ... We write this representation as

$$\mu = \sum_{n=0}^{\infty} L^{(n)} \tag{12}$$

The meaning of this decomposition of μ into a sum of interactions is that its response to a stimulus S(t) is given by

$$R(t) = \sum_{n=0}^{\infty} V_n[L^{(n)},S](t) \tag{13}$$

with V_n defined in terms of the kernel $L^{(n)}$ and the stimulus S according to Equation 11. For a linear system, only $L^{(1)}$ (and hence, only V_1) is nonzero. For a nonlinear system, the remaining terms express deviations from linear behavior that become more and more apparent as signal size grows.

Let us now consider how the above analysis might be used to measure the Volterra kernels in the laboratory. $L^{(0)}$ is the response to the zero input. To measure the first Volterra kernel $L^{(1)}$, one must accurately determine the response of the system to signals consisting of limitingly small impulses. In this regime, contributions of $L^{(2)}$ and higher order terms are negligible in comparison to that of $L^{(1)}$. To measure the second Volterra kernel $L^{(2)}$, one must compare the responses to stimuli consisting of two limitingly small impulses to the superposition of the responses to the impulses presented separately. In this regime, the contribution of $L^{(2)}$ is expected to be proportional to the square of the stimulus size, while contributions of $L^{(3)}$ and higher order terms will be smaller still. The degree of "smallness" required for these experiments is the signal size for which the system is approximately linear. Thus, for systems in which nonlinearities are prominent, there are significant drawbacks to a direct implementation of the Volterra approach. For any system, this approach to measurement of the Volterra kernels requires a comparison of differences between small responses. From the point of view of the experimentalist interested in nonlinear systems, this represents a major drawback.

The Volterra series representation (Equation 13) of an input-output relation is closely analogous to the Taylor series representation of an ordinary function. To make the connection, we consider a system whose response at any given time t depends only on the stimulus value at the same time, and not on prior values. These systems are commonly called static nonlinearities. For static nonlinearities, all of the kernel values $L^{(n)}(\tau_1, \tau_2, \ldots, \tau_n)$ are nonzero only when all time lags τ_i are zero, since prior values of the stimulus have no effect on the current output. Thus, we may write

$$L^{(n)}(\tau_1,\tau_2, \cdots, \tau_n) = a_n\, \delta(\tau_1)\, \delta(\tau_2) \cdot \cdots \cdot \delta(\tau_n) \tag{14}$$

where $\delta(\tau)$ is a delta-function — an idealized impulse of negligible width, infinite height, and unit area. For this particularly simple Volterra kernel, the contribution to the response is

$$V_n[L^{(n)},S](t) = a_n S(t)^n \tag{15}$$

and the Volterra representation (Equation 13) becomes

$$R(t) = \sum_{n=0}^{\infty} a_n S(t)^n \tag{16}$$

This expression is a Taylor series for the function $R(t) = f[S(t)]$, with

$$a_n = \frac{1}{n!}\frac{d^n f}{ds^n}(0) \tag{17}$$

The practical difficulties in estimating the Volterra kernels correspond to the practical difficulties in determining derivatives (Equation 17) from empirical data: highly accurate measurements of small quantities are required.

C. THE WIENER APPROACH

The Wiener (1958) representation of a nonlinear system is an alternative to the Volterra representation. In contrast to the Volterra kernels, the Wiener kernels are defined in terms of the response of the system to large signals, rather than to limitingly small ones. The original Wiener formalism is based on the response of a system to a particular signal, Gaussian white noise. However, the basic idea of the Wiener expansion — an orthogonal series of functionals — makes sense for a wide variety of stimuli (Barrett 1963). We will first describe the Wiener approach in this more general context (Palm and Poggio 1978, Klein and Yasui 1979, Victor and Knight 1979, Yasui 1979) and then consider the implications of a variety of choices of stimuli.

Let us assume that we have observed the responses of a system μ to a particular set Ω of stimuli. We will use these observations to build a sequence of systems μ_0, $\mu_0 + \mu_1$, $\mu_0 + \mu_1 + \mu_2$, . . . which approximate the system μ with increasing accuracy. Each of the systems μ_j is drawn from a stereotyped library of systems. We will describe the first few steps of the procedure in detail and then present the procedure in general.

The initial step is to determine the zeroth-order approximating system, which we will call μ_0. The library of candidates for μ_0 consists of all systems whose Volterra expansion (Equation 13) contains only an $n = 0$ term. Each system μ_0 in this library is thus specified by its zeroth-order Volterra kernel $\mu_0^{(0)}$. The zeroth-order Volterra kernel $\mu_0^{(0)}$ is simply a single value (it is a function with no arguments), which represents the constant output of the system μ_0. More formally, the response of μ_0 to a test signal S(t) is given by its Volterra series,

$$\begin{aligned} R_{\mu_0}(t) &= V_0[\mu_0^{(0)},S](t) \\ &= \mu_0^{(0)} \end{aligned} \tag{18}$$

This is a quite trivial way to approximate a general nonlinear system μ, but it suffices to illustrate how the approximating system is determined. For each test signal in Ω, we consider the difference between the observed response R(t) and the response $R_{\mu_0}(t)$ that would be expected from the approximating system μ_0. We then choose the best approximating system μ_0 as that system which minimizes the mean squared error (MSE) between the observed response R(t) and the approximating response $R_{\mu_0}(t)$. That is, μ_0 is chosen to minimize the mean squared error

$$MSE = \langle |R(t) - R_{approx}(t)|^2 \rangle \tag{19}$$

where the angle-brackets $< >$ indicate an average over the stimulus set Ω, and R_{approx} is the response of an approximating system whose Volterra series consists only of a single term (Equation 18). The zeroth-order Volterra kernel of μ_0, namely $\mu_0^{(0)}$, is the zeroth-order Wiener kernel of the system.

There is an important distinction between the Volterra kernel $\mu_0^{(0)}$ and the initial term $L^{(0)}$ in the Volterra representation of the original system. $L^{(0)}$ is the response of the system to the zero stimulus. $\mu_0^{(0)}$ is the average response of the system to the stimuli drawn from Ω. This distinction will hold for all higher order kernels as well.

The next step is to determine the best approximating system $\mu_0 + \mu_1$. The library of candidate approximating systems consists of all systems whose Volterra expansion (Equation 13) contains only $n = 0$ and $n = 1$ terms. In principle, addition of a first-order term into the approximation may cause the zeroth-order term to be revised. Thus, the "correction" μ_1 to the approximation will typically contain not only a first-order term, but also a zeroth-order term. Therefore, to specify μ_1, we need to consider not only its first-order Volterra kernel $\mu_1^{(1)}(\tau_1)$, but also its zeroth-order Volterra kernel $\mu_1^{(0)}$. The response of μ_1 to a test signal S(t) is given by its Volterra series,

$$R_{\mu_1}(t) = V_0[\mu_1^{(0)},S](t) + V_1[\mu_1^{(1)},S](t)$$

$$= \mu_1^{(0)} + \int_0^{\infty} \mu_1^{(1)}(\tau_1)S(t - \tau_1)\, d\tau_1 \qquad (20)$$

The system μ_1 is obtained by adjusting the values of its zeroth-order Volterra kernel $\mu_1^{(0)}$ and its first-order Volterra kernel $\mu_1^{(1)}(\tau_1)$ so that the mean squared error (Equation 19) is minimized. Here, $R_{approx}(t)$ is given by $R_{\mu_0}(t) + R_{\mu_1}(t)$. The first-order Volterra kernel of μ_1, namely $\mu_1^{(1)}(\tau_1)$, is the first-order Wiener kernel of the system.

We can now state the general procedure for the Wiener expansion. At the kth stage of the procedure, we seek the best approximating system from the library of systems whose Volterra expansions (Equation 13) contain terms up to (and including) order k. This approximating system is represented by $\mu_0 + \mu_1 + \ldots + \mu_k$, so that μ_k is the difference between the best approximation of order $k-1$ and the best approximation of order k. Since the addition of kth-order terms may cause lower-order terms to be revised, μ_k may contain Volterra kernels of orders 0 through k: $\mu_k^{(0)}, \mu_k^{(1)}(\tau_1), \ldots, \mu_k^{(k)}(\tau_1, \ldots, \tau_k)$. The meaning of "best approximating system" is that the MSE (Equation 19) is minimized for this particular choice of Volterra kernels, with $R_{approx}(t)$ given by $R_{\mu_0}(t) + R_{\mu_1}(t) + \ldots + R_{\mu_k}(t)$. The kth order Wiener kernel, by definition, is the kth order Volterra kernel of the best kth order approximation, namely $\mu_k^{(k)}(\tau_1, \ldots, \tau_k)$.

We write the Wiener representation as

$$\mu = \sum_{k=0}^{\infty} \mu_k \qquad (21)$$

The meaning of this representation of μ as a sum of approximating systems is that its response to a stimulus S(t) is a sum of the responses of the approximating systems μ_k:

$$R(t) = \sum_{k=0}^{\infty} G_k[\mu_k,S](t) \qquad (22)$$

Here we have introduced functionals G_k which include the action not only of the highest-order term $\mu_k^{(k)}$, but also of the lower-order correction terms:

$$G_k[\mu_k,S](t) = \sum_{n=0}^{k} V_k[\mu_k^{(n)},S](t) \qquad (23)$$

The Wiener expansion has some important general features and limitations. As in the zeroth-order case, every Wiener kernel depends on the choice of the stimulus set Ω. This is because the Wiener kernels are defined as best approximations, and whether one system is a good approximation to another depends on the choice of stimuli. For linear systems, a stimulus-independent characterization of a system is possible. But this is not the case for nonlinear systems. In the absence of any properties such as superposition or other *a priori* data, measurements of the responses to one set of stimuli cannot provide information on the responses to another set.

There is a rigorous formal sense in which an infinitely long Gaussian white noise stimulus will eventually come arbitrarily close to mimicking any finite stimulus. This is one of the motivations for the use of Gaussian white noise (see below) as a test stimulus. In principle, analysis of the response to this stimulus would result in a set of Wiener kernels that is a complete characterization of the system, at least in a mean-squared sense (Palm and Poggio 1977). However, an empirical Wiener representation of a system built from a finite set of response measurements does not allow for the prediction of the response to arbitrary stimuli.

The difference between the theoretical situation and the empirical one is not merely because of signal-to-noise limitations. Rather, this is a consequence of the fact that the total amount of information available for kernel calculation is finite. As a consequence, minimization of the MSE (Equation 19) fails to determine the Wiener kernels uniquely. The nonuniqueness problem can be completely characterized in a simple way (Victor and Knight 1979). Consider ν, a nonlinear system whose response to all test signals in Ω is zero. No experiment limited to the test signals in Ω can distinguish between the system μ and the composite system of μ and ν in parallel. Conversely, a characterization in terms of kernels built from responses to signals in Ω may be considered as uniquely identifying μ, except for the unavoidable ambiguity of possible parallel combination with such silent systems ν. Naturally, as the stimulus set Ω grows, the possibilities for silent systems ν shrink.

D. COMPARISON OF WIENER AND VOLTERRA REPRESENTATIONS

Though there is no guarantee that the approximating systems μ_0, $\mu_0 + \mu_1$, $\mu_0 + \mu_1 + \mu_2$, . . . uniquely characterize the system μ, the properties of the Wiener representation do guarantee that they form a sequence of increasingly accurate approximations. This is because each successive member of the sequence is chosen from a library of systems which *includes* the library available for the previous approximation. This property allows the Wiener formalism to represent some nonlinear systems that do not have a Volterra representation (see below).

It is worthwhile examining the kth step of the Wiener expansion in more detail. At this step, the approximating system $\mu_0 + \mu_1 + \ldots + \mu_{k-1}$ is adjusted by the addition of the system μ_k. A critical point is that the Volterra expansion of μ_k will, in general, have terms of orders 0, . . . , $k-1$, in addition to a term of order k. That is, addition of a kth order term to the Volterra expansion may require adjustment of the lower-order terms, to retain the "best approximation" property. The basic phenomenon is displayed even by simple nonlinear systems, whose output depends only on the current value of the input. As discussed above, the Volterra representation of these systems reduces to a Taylor series (Equation 16).

We consider a system whose output y is the cube of its input x. In the Taylor series format (Equation 16), the Volterra representation for this system is

$$R(t) = 0 \cdot S(t)^0 + 0 \cdot S(t)^1 + 0 \cdot S(t)^2 + 1 \cdot S(t)^3 \qquad (24)$$

That is, all terms except for the cubic term are zero. Let us now calculate its Wiener expansion

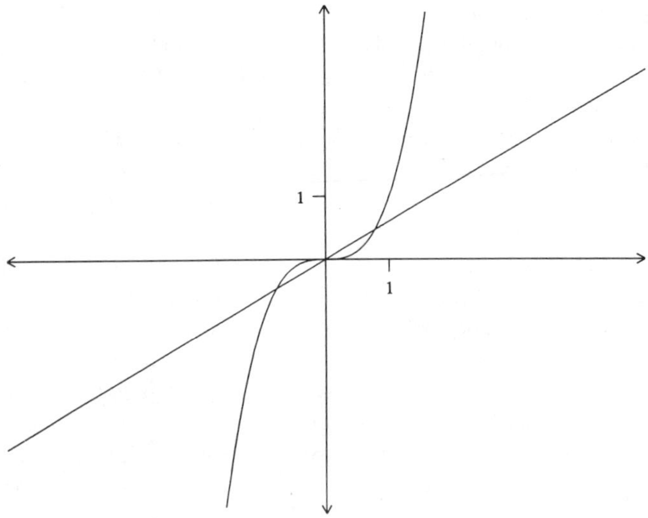

FIGURE 3. Approximation of $y = x^3$ by a linear function $y = 3x/5$.

with respect to a stimulus which consists of all values between -1 and $+1$ with equal probability. The best zeroth-order approximation to this system is the system with a zero output (since negative and positive outputs are equally likely). Therefore, the zeroth-order Wiener kernel $\mu_0(0)$ is zero. The MSE of this approximation, $\langle|x^3|^2\rangle$ over the interval $[-1,1]$, is 1/7 ($\approx$0.14).

At first order, we observe that the input-output law $y = x^3$ may be approximated by a linear relationship $y = \alpha x$. Over the interval $[-1,1]$, the value of α which minimizes the MSE $\langle|x^3 - \alpha x|^2\rangle$ over this interval is 3/5 (Figure 3). Thus, the first-order Wiener kernel $\mu_1^{(1)}$ is $^3/_5\delta(\tau_1)$. The MSE of this approximation is 4/175 ($\approx$0.02). Note that a system whose Volterra representation has no first-order terms has a Wiener representation with a significant first-order term. Indeed, the first-order Wiener representation accounts for 80% of the mean squared error.

At second order, a symmetry argument shows that there is no improvement in the approximation to $y = x^3$ by adding a quadratic term to $y = 3x/5$. Thus, the second-order Wiener kernel of this system is zero.

At third order, further improvement in the approximation does occur, because the system is exactly represented by the third-order Volterra series (Equation 24). The third-order Wiener kernel, which is the third-order Volterra kernel of the newly added term, is therefore $\mu_3^{(3)} = \delta(\tau_1)\delta(\tau_2)\delta(\tau_3)$. But, since the third-order approximation $\mu_0 + \mu_1 + \mu_2 + \mu_3$ is the original system itself, the addition of μ_3 must cancel the first-order component introduced by μ_1. That is, $\mu_3^{(1)} = -\mu_1^{(1)}$. The point here is that the lower-order Volterra terms had to be revised when a third-order approximation was added.

It might seem, then, that the specification of the kth approximating system μ_k in terms of the highest order Volterra kernel $\mu_k^{(k)}$ is incomplete, because this specification omits the lower-order corrections $\mu_k^{(k-1)}, \ldots, \mu_k^{(0)}$. But this is not the case. Rather, equivalent information is contained in the best-approximation constraint. Because the previous approximations $\mu_{k-1}, \ldots, \mu_0$ were optimal for their order, the lower-order corrections in μ_k may be calculated solely from $\mu_k^{(k)}$ and Ω. This procedure is known as "orthogonalization".

The details of the orthogonalization procedure depend on Ω. Since the orthogonalization procedure effectively amounts to a shuffling of contributions among terms, it is not surprising that a Volterra series and an orthogonal series display different kinds of convergence behavior. The general way in which convergence of a Volterra series differs from convergence of an orthogonal is illustrated by considering instantaneous input-output relationships. In this context, the difference in convergence behavior corresponds exactly to the difference in behavior of a Taylor series and an orthogonal series of polynomials. For Gaussian inputs, the appropriate series of orthogonal polynomials is the Hermite polynomials.

We first consider approximations to the relationship $y = e^x$. Figures 4 A to C show a set of polynomial approximations to this function, obtained by truncating an orthogonal series of Hermite polynomials at orders 0, 1, 2, 4, and 8. The weight function for the Hermite polynomials are Gaussians of variance 10.0 (Figure 4A), 1.0 (Figure 4B), and 0.1 (Figure 4C). Figure 4D shows another set of polynomial approximations to this function, obtained by truncating a Taylor series at these same orders. The orthogonal-series approximants for variance 0.1 (Figure 4C) are essentially indistinguishable from the Taylor approximants (Figure 4D). The only evident difference between the approximations for variance 1.0 (Figure 4B) and the Taylor approximants (Figure 4D) is that the first-order approximant, which was tangent to $y = e^x$ at $x = 0$ in the Taylor case, now intersects the curve $y = e^x$ in two places. That is, the first-order approximant is inferior to the Taylor first-order approximant at $x = 0$, but is superior for slightly larger values of x. These differences are more apparent for variance 10.0 (Figure 4A). Here, higher-order approximants are noticeably separate from the desired curve at $x = 0$; in the Taylor case (Figure 4D), all approximants made high-order contacts with $y = e^x$. Essentially, the Taylor approximants are optimal in the infinitesimal neighborhood of $x = 0$, but the orthogonal approximants are optimal (in a least-squares sense) over a range of x-values as determined by the spread of the weight function. For small spreads (Figure 4C), there is little difference between the Taylor and the orthogonal approximants.

By changing from a Volterra (or Taylor) series to a Wiener (or orthogonal) series, we have lost the ''uniqueness'' of the expansion — there is now a dependence of the approximants on the stimulus ensemble (or weight function). But we have gained the ability to represent a wider class of functions. For example, the threshold function $y = \max\{0,x\}$ has no derivatives at zero and, hence, does not have a Taylor representation. But it does have a representation as an orthogonal series, as illustrated in Figure 5. Note that, as in Figure 4, the spread of the weight function controls the region over which the approximation is reasonable.

When a Volterra representation exists, it may be regarded as the limiting case of a family of orthogonal representations as the variance of the test signal becomes small. This limit clearly exists for $y = e^x$, as seen in Figure 4: the expansion for variance 0.1 (Figure 4C) is essentially indistinguishable from that of the Taylor expansion (Figure 4D). In contrast to the case of $y = e^x$, there is no sense in which the approximants approach a limit as the variance approaches zero for $y = \max\{0,x\}$: the expansion for variance 0.1 (Figure 5C) is very different from that for variance 1.0 (Figure 5B) or 10.0 (Figure 5A). This corresponds to the fact that a Taylor series does not exist for a function with a sharp corner.

These idealized examples point out that a Wiener series for a nonlinearity with a singularity such as a sharp corner cannot be converted into a Volterra series. The distinction between Volterra and Wiener approximations *is* of practical importance, even though the typical biological nonlinearity, when regarded on a sufficiently small scale, is likely to be smooth rather than abrupt. This is because for some nonlinearities, a signal size small enough to be sensitive to the ''smoothness'' of the nonlinearity may result in a response which is lost in the noise. Stated another way, a sequence of Wiener approximations built on inputs of successively smaller variances will approach a Volterra limit, provided that the contributions from higher-order components become sufficiently small as the variance decreases.

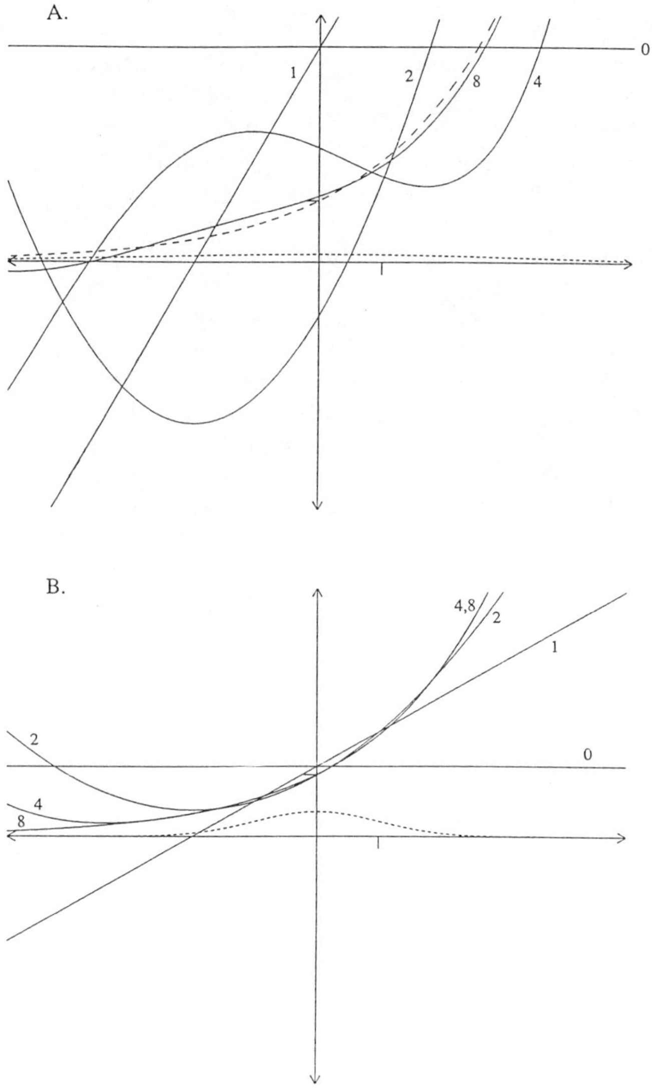

FIGURE 4. Approximation of $y = e^x$ by polynomials of order 0, 1, 2, 4, and 8. A to C show approximations derived from orthogonal polynomials for Gaussians of variance 10.0 (A), 1.0 (B), and 0.1 (C) (short dashed lines); D shows Taylor approximants. The curve $y = e^x$ is shown as a broad dashed line in 4A, and is obscured by the approximants in 4B to D.

This is the behavior which characterizes an "analytic" nonlinearity (such as $y = e^x$). For a "nonanalytic" nonlinearity (such as $y = \max\{0,x\}$), higher-order contributions do not become negligible as the variance decreases. An example of a nonlinearity that appears nonanalytic for signals of limitingly small size is that of the Y cell of the cat retina (Victor and Shapley 1979), whose response consists of second- and fourth-order components down to a contrast on the order of 1%.

IV. METHODS OF KERNEL MEASUREMENT

We now consider several choices of stimulus sets Ω that have proven to be particularly useful in visual neurophysiology and discuss particular features of their corresponding orthogonalization procedures.

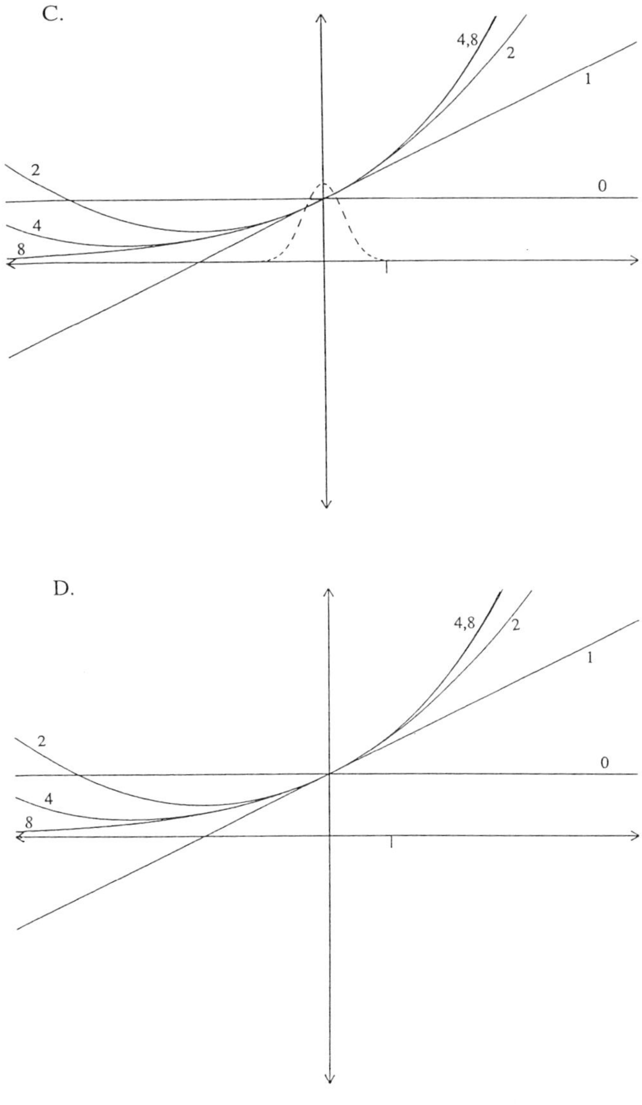

FIGURE 4C AND D.

A. GAUSSIAN WHITE NOISE AS A STIMULUS

A Gaussian white noise signal is an abstract entity whose value at every instant is drawn from a Gaussian distribution. Because each value is independently chosen and each value may in principle have any value (since the probability of a value in either tail of a Gaussian distribution never quite becomes zero), there is a sense in which the Gaussian white noise signal ''contains'' all possible stimuli. However, an ideal Gaussian white noise signal changes arbitrarily rapidly, and in any time period, arbitrarily large values are encountered infinitely often. Such a signal cannot be generated in the laboratory. However, reasonable approximations to such a signal can be generated, provided two steps are taken: (1) the maximum amplitude of the signal is limited to a finite value and (2) the rapidity of the fluctuations of the noise (the ''bandwidth'') is limited to a finite value. The amplitude limits are chosen several standard deviations from the mean, so that these limits would have been exceeded only a small fraction of the time (e.g., 0.25% for three standard deviations). The bandwidth

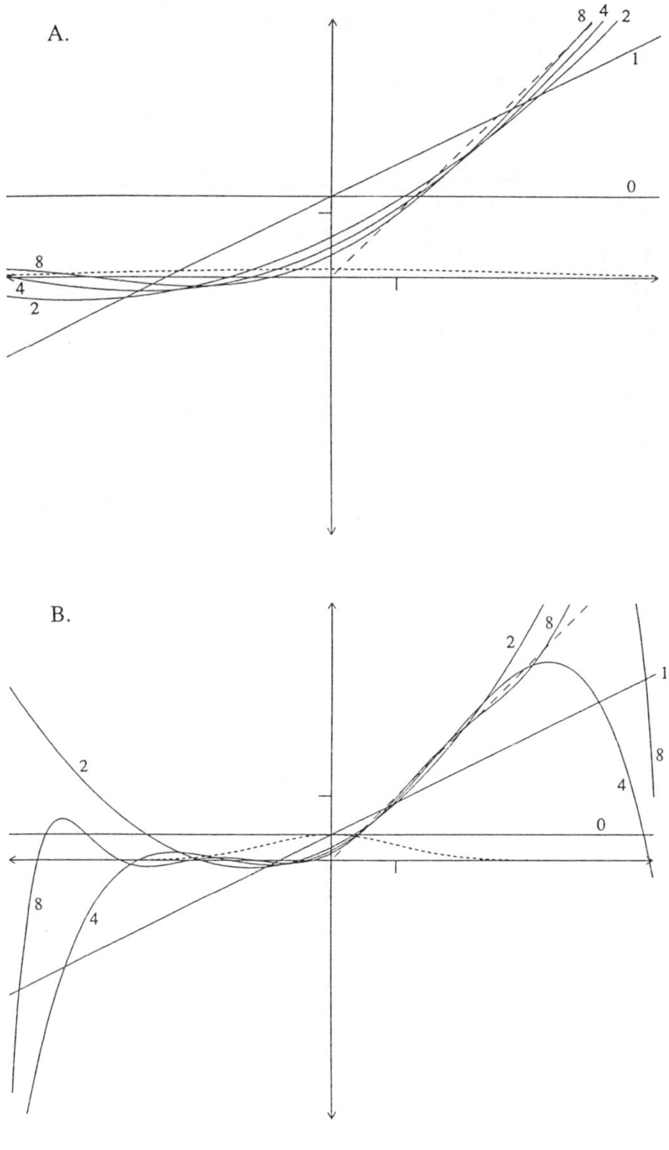

FIGURE 5. Approximation of y = max{0,x} by polynomials of order 0, 1, 2, 4, and 8. A to C show approximations derived from orthogonal polynomials for Gaussians of variance 10.0 (A), 1.0 (B), and 0.1 (C) (short dashed lines); the curve y = max{0,x} is shown as a broad dashed line.

of the signal is typically chosen to be at or above the frequencies to which the system responds. Thus, in some sense, the response of the system is identical to that which would have been produced by a white noise of infinite bandwidth, since the frequencies omitted from the stimulus are the frequencies to which the system does not respond. Here, we will consider a Gaussian white noise as characterized solely by its power P per unit bandwidth, and not focus on corrections related to amplitude and bandwidth limits. The consequences of amplitude and bandwidth limitations are discussed in Marmarelis and Marmarelis (1978).

We consider three aspects of the white noise theory: the orthogonalization, the calculation of the kernels from experimental data, and the kernels of particular nonlinear systems.

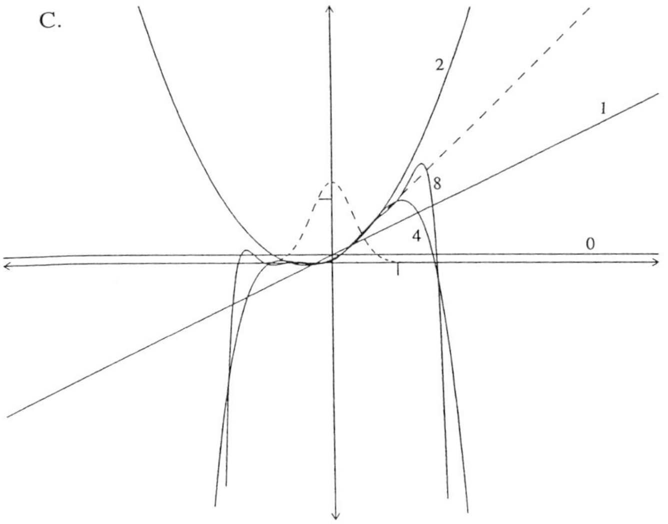

FIGURE 5C.

The orthogonalization process determines the lower-order corrections $\mu_k^{(k-1)}$, . . . , $\mu_k^{(0)}$ that must be added to $\mu_k^{(k)}$ to complete the Volterra expansion of μ_k:

$$\mu_k = \sum_{n=0}^{k} \mu_k^{(n)} \tag{25}$$

For a Gaussian white noise stimulus set Ω with a power P per unit bandwidth, the first few kernels $\mu_k^{(n)}$ may be expressed in terms of $\mu_k^{(k)}$ as follows:

k = 0

$$\mu_0^{(0)} = \mu_0^{(0)} \tag{26}$$

k = 1

$$\mu_1^{(1)}(\tau_1) = \mu_1^{(1)}(\tau_1)$$
$$\mu_1^{(0)} = 0 \tag{27}$$

k = 2

$$\mu_2^{(2)}(\tau_1,\tau_2) = \mu_2^{(2)}(\tau_1,\tau_2)$$

$$\mu_2^{(1)}(\tau_1) = 0$$

$$\mu_2^{(0)} = -P \int_0^\infty \mu_2^{(2)}(\tau,\tau)\, d\tau \tag{28}$$

k = 3

$$\mu_3^{(3)}(\tau_1,\tau_2,\tau_3) = \mu_3^{(3)}(\tau_1,\tau_2,\tau_3)$$

$$\mu_3^{(2)}(\tau_1,\tau_2) = 0$$

$$\mu_3^{(1)}(\tau_1) = -3P \int_0^\infty \mu_3^{(3)}(\tau,\tau,\tau_1)\, d\tau$$

$$\mu_3^{(0)} = 0 \tag{29}$$

k = 4

$$\mu_4^{(4)}(\tau_1,\tau_2,\tau_3,\tau_4) = \mu_4^{(4)}(\tau_1,\tau_2,\tau_3,\tau_4)$$

$$\mu_4^{(3)}(\tau_1,\tau_2,\tau_3) = 0$$

$$\mu_4^{(2)}(\tau_1,\tau_2) = -6P \int_0^\infty \mu_4^{(4)}(\tau,\tau,\tau_1,\tau_2)\, d\tau$$

$$\mu_4^{(1)}(\tau_1) = 0$$

$$\mu_4^{(0)} = 3P^2 \int_0^\infty \int_0^\infty \mu_4^{(4)}(\tau,\tau,\tau',\tau')\, d\tau\, d\tau' \quad (30)$$

The structure of the above expressions becomes clearer when we restrict attention to static nonlinearities, whose response R(t) at time t depends only on the stimulus S(t) at time t, and not on prior values. Any best-approximating system must also be a static nonlinearity, since dependence on prior values cannot improve an estimate. That is, the values of the Wiener kernels are nonzero only when all time lags τ_i are zero. As in the Volterra case (Equation 14), we may write

$$\mu_k^{(k)}(\tau_1,\tau_2, \cdots, \tau_k) = b_k\delta(\tau_1)\delta(\tau_2) \cdot \cdots \cdot \delta(\tau_k) \quad (31)$$

However, the contribution of the kth order Wiener kernel to the response R(t) is not simply $b_k\ S^{(k)}$, since the lower-order correction terms must be added. The expressions in Equations 26 to 30 and their extensions provide a calculation of these correction terms:

$$R(t) = \sum_{k=0}^{\infty} b_k h_k[S(t)] \quad (32)$$

where $h_k(x)$ are Hermite polynomials:

$$h_0(x) = 1$$

$$h_1(x) = x$$

$$h_2(x) = x^2 - P$$

$$h_3(x) = x^3 - 3Px$$

$$h_4(x) = x^4 - 6Px^2 + 3P^2 \quad (33)$$

The Hermite polynomials (Equation 33) are the orthogonal polynomials corresponding to a Gaussian weighting of variance P. That is, truncations of the series (Equation 32) after a finite number of terms provide for an optimal approximation in the mean-squared sense with respect to a Gaussian weighting of variance P. Essentially, the Wiener representation (Equation 21) is a generalization of the orthogonal series approximation (Equation 32) of functions $R(t) = f[S(t)]$ to more general stimulus-response relationships.

Next, we consider how the Wiener kernels may be calculated from experimental data. The definition of Wiener kernels was in terms of an optimal approximation. In principle, optimization of the mean squared error (Equation 19) would require a search over all possible Volterra kernels of a particular order, which would entail a multidimensional least-squares procedure. The dimension of the space to be searched would in principle be very large —

one dimension for each value $\mu_k^{(k)}(\tau_1, \ldots, \tau_k)$ to be determined. However, the orthogonality properties of the kernels allow a much simpler approach. As shown by Lee and Schetzen (1965), the Wiener kernels may be estimated by cross-correlating the response of the system and values of the white noise stimulus at various time lags. This procedure is particularly simple when none of the time lags $\tau_1, \ldots, \tau_k$ are equal. In this case,

$$\mu_k^{(k)}(\tau_1,\tau_2, \ldots, \tau_k) = \frac{1}{P^k k!} \langle R(t) \cdot S(t - \tau_1) \cdot S(t - \tau_2) \cdot \cdots \cdot S(t - \tau_k)\rangle \quad (34)$$

For values of the kernels where two or more of the lags τ_j are equal (the so-called "diagonal" case) this expression must be modified to take into account the interaction of higher-order and lower-order terms. For example,

$$\mu_2^{(2)}(\tau,\tau) = \frac{1}{2P^2} \langle [R(t) - \mu_0^{(0)}] \cdot S(t - \tau) \cdot S(t - \tau)\rangle \quad (35)$$

Early applications of the Wiener theory to a biological system made use of white noise stimulus and the Lee and Schetzen algorithm (Marmarelis and Naka 1972, McCann 1974, Naka et al. 1975).

The third aspect of the white noise theory to be emphasized is that for some composite systems, the kernels have a simple algebraic form. The Wiener kernels of a composite system μ composed of nonlinear systems μ' and μ'' in parallel are the sums of the Wiener kernels of its components:

$$\mu_k = \mu_k' + \mu_k'' \quad (36)$$

Serial composition of nonlinear systems, as well as feedback composition, in general does not lead to a simple algebraic expression for the Wiener kernels, although the Volterra kernels for such systems can be calculated in a variety of ways (Bedrosian and Rice 1971, Korenberg 1973). The one important exception in which the Wiener kernels do have a simple algebraic form is the linear-nonlinear-linear cascade. For a serial cascade of a linear system of impulse response $L'(t)$ a static nonlinearity N, and a second system of impulse response $L''(t)$, the Wiener kernels (Korenberg 1973) are given by

$$\mu_k^{(k)}(\tau_1, \ldots, \tau_k) = c_k \int_0^\infty L'(\tau_1 - \tau)\, L'(\tau_2 - \tau) \cdot \cdots \cdot L'(\tau_k - \tau)\, L''(\tau)\, d\tau \quad (37)$$

Here, the factor c_k depends on both the defining equation of the nonlinearity N and the power P which characterizes the white noise signal.

As seen from Equation 37, Wiener kernels of second and higher orders depend in different fashions on $L'(\tau)$ and $L''(\tau)$. Thus, the second-order Wiener kernel can identify which linear transformations occur before a nonlinearity and which ones occur after a nonlinearity. For this reason, the relationship (Equation 37) and its frequency-domain analogue (see below) have been used to analyze a variety of nonlinear biological transductions, particularly in vision (Spekreijse et al. 1977, Victor and Shapley 1979, Sakuranaga and Naka 1985a–c, Sakai et al. 1988).

B. OTHER ORTHOGONAL EXPANSIONS

We have presented the "white noise method" as an example of a general method of orthogonal functional expansions. As we have seen, the statistical properties of Gaussian white noise allow dramatic simplifications of three aspects of the theory: orthogonalization, kernel estimation, and modeling. However, many other test signals also lead to similar simplifications. This permits the investigator to tailor the stimulus set to the needs of the study.

One general category of test signals is that of signals which are white, but not necessarily Gaussian (Klein and Yasui 1979). As is the case for Gaussian white noise, these test signals have values which are independently chosen at each step in time. However, the values are not chosen from a Gaussian distribution. Most commonly, the values are chosen with equal probability from a small number of possibilities (Marmarelis 1975, 1977). A two-level signal has been used most commonly (Emerson et al. 1987, Gaska et al. 1989, Szulborski and Palmer 1990), but more recently, three-level signals have also been used (Emerson et al. 1990). For these signals, the probability distribution for the signal values is always even-symmetric. That is, a particular stimulus and its negative are equally likely to occur.

Another kind of stimulus in this category is the random impulse train (Krausz 1975), which has particular value in analyzing synaptic transduction properties. A random impulse train may be thought of as consisting of a sequence of binary values, with the nonzero value chosen a small fraction λ of the time. As λ approaches zero and the time discretization become short, this process results in a Poisson train.

For all of these white but non-Gaussian signals, the orthogonalization relations assume a form similar to that of Equations 26 to 30. That is, $\mu_k^{(n)}$, a function of n arguments, is derived from $\mu_k^{(k)}$, a function of k arguments, by performing integrals over the remaining $k - n$ arguments. Provided that the underlying probability distribution is even-symmetric, then the integrals are always over paths in which the $k - n$ arguments are held equal in pairs. For this reason, even-order kernels $\mu_k^{(k)}$ lead only to even-order correction terms $\mu_k^{(n)}$, and odd-order kernels lead only to odd-order correction terms.

This parity principle only applies to white noises based on even-symmetric probability distributions, and not to those based on asymmetric distributions such as the Poisson impulse train. In studying the visual system, it is usually advantageous to use even-symmetric test signals, so that the separation of odd-order and even-order interactions is as complete as possible. This is particularly so when comparisons of on- and off-pathways are of interest. If these pathways are mirror images of each other, then the only nonlinearities present will be of odd order. Conversely, asymmetries between on- and off-pathways, in a formal sense, will always be manifest as even-order nonlinearities.

As in the Gaussian white noise case, the numerical coefficient associated with the correction term $\mu_k^{(n)}$ is related to the coefficient of x^n of a kth-order orthogonal polynomial. For the Gaussian case, the appropriate family of orthogonal polynomials is the Hermite polynomials (Equation 33). More generally, these coefficients are derived from the orthogonal polynomials based on the distribution of values in the stimulus. For example, if the stimulus values were uniformly drawn from the interval $[-1,1]$, then the appropriate family of orthogonal polynomials would be the Legendre polynomials.

An interesting issue arises when the orthogonalization process is applied to stimuli whose values are drawn from a finite number (e.g. N) of alternatives. The sequence of orthogonal polynomials cannot be continued beyond order $N - 1$. This is because polynomials of the form x^N (and higher powers of x) are indistinguishable from linear combinations of lower powers of x, for values of x confined to the particular choices present in the test signal. For example, a cubic can only be distinguished from a lower-order polynomial if it is examined at four or more values of x. If the relationship $y = x^3$ is tested only for $x = -1$, 0, or 1, it could equally well be modeled by $y = x$. Because the orthogonal polynomials cannot be continued beyond degree $N - 1$, certain values of the Wiener kernels are undefined. In this context, "undefined" means that within the confines of an N-level test signal, they are

(1) irrelevant to characterization of the system and (2) not measurable. The undefined kernel values are those of the form $\mu_k^{(k)}(\tau_1, \ldots, \tau_k)$ for which at least N of the τ_j are equal.

In implementing a white noise technique, the experimenter is free to choose how finely time is discretized, i.e., how often a new value of the stimulus is chosen from its defining distribution. Any physical system has a limited range of frequencies over which it operates. One would expect, therefore, that discretization of time at frequencies substantially in excess of this cutoff frequency would provide no further information. Once the switching rate of the stimulus exceeds the resolution of the system, it may be regarded as a smoothly varying signal, whose effective value at any given time is equal to the value of the original discretized stimulus averaged over the minimum period of resolution. Provided that the averaging period includes many samples of the discretized stimulus, the distribution of values in the smoothed stimulus necessarily approaches a Gaussian distribution. This is a consequence of the Central Limit Theorem (Feller 1968). Thus, in the rapid-sampling limit, the white but non-Gaussian stimulus approaches a Gaussian white noise, whose power depends on the frequency range to which the system is sensitive. In a parallel fashion, the kernels derived from the discretized signal may be shown to approach the Wiener kernels that would have been obtained from a Gaussian white noise of appropriate power (Victor 1990). In this rapid-sampling limit, the problem that some kernel values are undefined evaporates. This is because values of the kernels remain defined and measurable at points immediately adjacent to the "diagonal" points, such as $\mu_2^{(2)}(\tau,\tau+\delta\tau)$, where $\delta\tau$ is the sampling interval. As the sampling interval $\delta\tau$ becomes short in comparison to all time constants of the system under study, it becomes reasonable to use the near-diagonal points to estimate an effective value of the on-diagonal points.

For non-Gaussian white noises, the Lee and Schetzen cross-correlation algorithm (Equation 34) may be used to estimate the off-diagonal values of the Wiener kernel. Not surprisingly, the on-diagonal terms require corrections whose details depend on the nature of the test signal. Klein and co-workers (Emerson et al. 1987) have advanced a modification of this procedure for two-level and three-level input signals in which "partial kernels" are calculated by accumulating the cross-correlations (Equation 34) only for certain values of the input. For example, for ternary inputs, the first-order kernel $\mu_1^{(1)}$ is broken down into a partial kernel $\mu_{1,p}^{(1)}$, which only includes terms of Equation 34 in which the input is positive, and a partial kernel $\mu_{1,n}^{(1)}$, which only includes terms of Equation 34 in which the input is negative. The complete first-order kernel $\mu_1^{(1)}$ can be reconstituted by adding the partial kernels $\mu_{1,p}^{(1)}$ and $\mu_{1,n}^{(1)}$. For linear systems, these two partial kernels are identical. For nonlinear systems, the partial kernels may be different. The difference of the partial kernels, $\mu_{1,p}^{(1)} - \mu_{1,n}^{(1)}$, may be expressed in terms of diagonal values of the complete second-order kernels. This difference indicates the extent to which the response to a negative-going signal fails to be the mirror image of the response to the corresponding positive-going signal. Thus, the partial kernels provide a way to interpret certain values of higher-order kernels.

Although an orthogonal functional series can be constructed based on any set of test signals, Gaussian white noise retains a special place for more than historical reasons. As discussed above, there is a rigorous sense in which discretized white noise signals based on other amplitude distributions converge to a Gaussian white noise, and the orthogonal expansions based on such signals converge correspondingly to the orthogonal expansion based on Gaussian white noise. But there are other important special properties of the orthogonal expansions based on Gaussian white noise.

In passing from a Volterra representation to a Wiener expansion, the possibility of a stimulus-independent characterization is abandoned in exchange for a procedure which can be implemented in the laboratory. Though any orthogonal expansion will depend on the statistics of the stimulus, this dependence is particularly simple for Gaussian white noise

(Klein and Yasui 1979, Klein 1987). The dependence of the kth order kernel $\mu_k^{(k)}(\tau_1, \ldots, \tau_k)$ on the mean of the white noise test signal is proportional to an integral $\int \mu_{k+1}^{(k+1)}(\tau_1, \ldots, \tau_k, \tau)d\tau$ over a one-dimensional set of values of the kernel one order higher. The dependence of the kth order kernel $\mu_k^{(k)}(\tau_1, \ldots, \tau_k)$ on the power of the white noise test signal is proportional to an integral $\int \mu_{k+2}^{(k+2)}(\tau_1, \ldots, \tau_k, \tau, \tau)\, d\tau$ over a one-dimensional set of diagonal values of the kernel two orders higher. These relationships provide a way to interpret particular regions of high-order kernels in terms of adaptation and gain-control processes.

C. STOCHASTIC SYSTEMS

Another feature of the Gaussian white noise orthogonal expansion relates to how the approximating systems handle variability in the response of the system. Consider an empirical set of input-output observations which are to be used to estimate the orthogonal expansion of a particular system up to some order k. We recognize that truncation of the complete Wiener representation (Equation 21) at order k will yield an approximating system μ' that will deviate in a systematic way from the actual system μ. That is, the k-term approximating system μ' will not reproduce the input-output observations exactly. The extent of this departure is summarized by the MSE. Two kinds of phenomena may contribute to the MSE: (1) deterministic components omitted from the orthogonal series and (2) intrinsic variability of the measured response of the system. This variability may be due to one or more of several factors, including measurement error, intrinsic stochastic components of the system, and inputs to which the experimenter does not have access.

Until now, we have only considered how a ''noiseless'' system is represented. Here, we consider how to represent the stochastic elements of a system. For this purpose, the observed input-output pairs are best regarded as samples drawn from a probability distribution in the space of all possible inputs and outputs. A deterministic representation of a system with a nonzero MSE may be reinterpreted as a representation of a distribution of input-output observations, by considering the MSE to represent an additive noise of appropriate variance.

How can one separate the component of the MSE that represents deterministic higher-order nonlinearities and the component that represents intrinsic variability? The direct approach is to collect additional replicate runs of responses to the same stimulus. In the absence of such data, one may use concepts from information theory to interpret the MSE. Kernels calculated up to order k are based on measurements of moments up to order k. We wish to formalize the idea that we are ignorant of higher moments. (If we knew higher moments, then we could calculate higher-order kernels.)

We proceed as follows. We assume that we have measured the total variance in the output and that the statistics of the input signal are known at all orders (since we have generated it). The cross-correlation quantities (Equation 34) are moments of the observed input-output distribution, which we assume that we have measured up to order k. To construct a stochastic order k approximation, we assume that higher input-output moments are unknown. We then search all possible input-output distributions for the distribution which (1) has moments which match those determined from the input-output observations and (2) is of maximum entropy among distributions constrained by these moments. The maximum entropy criterion essentially formalizes the assumed ignorance of higher moments (Levine and Tribus, 1979). The best-approximating distribution in the maximum-entropy sense represents not only the average response of the system, but also some elements of its variability. The maximum-entropy input-output distribution may be considered a kth-order representation of the system, including some of its stochastic aspects. Improved representations can be obtained by using additional experimental data to constrain this distribution.

The maximum-entropy distribution of order k may now be compared to the kth-order Wiener or Volterra approximation, reinterpreted as a stochastic system by considering the MSE to represent additive noise. In general, these two representations will be quite different. But for a Gaussian white noise input, these two kinds of approximations must coincide (Victor and Johannesma 1986). That is, a truncated Wiener representation constructed with respect to a Gaussian white noise input is not only a best-fit approximation in the least-squares sense, but also formalizes ignorance of the unused moments.

D. THE M-SEQUENCE APPROACH

In techniques based on a stochastic input, the orthogonalization procedure depends on the assumption that the statistics of the input are equal to that of the ensemble Ω from which it is drawn. The unavoidable deviation of the statistics of the actual input signal from those of the ensemble Ω can be an important source of inaccuracy in kernel measurement (Marmarelis and Marmarelis 1978). One approach to this problem is a modification of the orthogonalization algorithm to account for the actual statistics of the input (Korenberg 1988, Korenberg and Hunter 1990). This provides a more accurate system characterization for the particular input used, at the expense of a small potential for bias in the estimate of kernel values. Here, we focus on another strategy, which relies on input signals whose statistics are known exactly.

A very promising method of this kind has been devised by Sutter (1987). Sutter's method relies on the properties of M sequences, or maximum-length shift register sequences (Golomb 1968). These are sequences whose low-order statistics are matched to those of white noise. A binary M sequence of order n is a cyclic sequence of $2^n - 1$ terms b_i, each of which is 0 or 1. The assignment of a value to the term b_i is done according to a binary recursion rule

$$b_i = \sum_{k=1}^{n} c_k b_{i-k} \pmod 2 \tag{38}$$

with the initial terms chosen by $b_0 = 1, b_1 = 0, \ldots, b_{n-1} = 0$. The sequence generated by a recursion relation such as Equation 38 is guaranteed to be cyclic for any choice of coefficients c_k. For this sequence to constitute an M sequence, these coefficients are chosen so that the repeat period of the sequence b_i is maximal, i.e., of length $N = 2^n - 1$.

Recursion rules such as in Equation 38 are readily implemented in software or in hardware. Indeed, this is the basis for many schemes for computer generation of "random" numbers. For such applications, the order n is chosen to be sufficiently large (perhaps 127) so that the ultimately repetitive nature of the sequence never becomes apparent. But for the present application, we choose a smaller n (typically in the range 8 to 16), so that the cyclic nature of the sequence may be exploited.

The cycle $\{b_0, b_1, \ldots, b_{N-1}\}$ is used to define a cyclic input signal $\{s(0), s(1), \ldots, s(N-1)\}$ of unit variance by choosing $s(i) = 1 - 2b_i$. This cycle of values has several useful properties. Because this sequence is maximal length, it contains all of the 2^n n-tuples of $+1$s and -1s exactly once, except for the n-tuple of all $+1$s (which does not appear). As a consequence, this sequence contains almost exactly the same number of $+1$s and -1s:

$$\langle s(i) \rangle = -\frac{1}{N} \tag{39}$$

More importantly, the M sequence is nearly orthogonal to any nontrivial shift of itself. That is,

ELIHU BURRITT LIBRARY
CENTRAL CONNECTICUT STATE UNIVERSITY
NEW BRITAIN, CONNECTICUT 06050

$$\langle s(i)s(i-j)\rangle = -\frac{1}{N} \tag{40}$$

provided $j \neq 0 \pmod N$).

Equations 39 and 40 may be summarized by stating that the first-order and second-order statistics of the M sequence agree with those of ideal Gaussian white noise to within 1/N. In contrast, the typical deviation of these statistics for a sequence of N binary digits chosen at random would be approximately $1/\sqrt{N}$; thus, the M sequence represents a considerable improvement.

Higher-order statistics do not agree so closely with those of Gaussian white noise, but the situation remains susceptible to analysis. The key observation is

$$s(i-j_1)\cdot \cdots \cdot s(i-j_r) = s[i - F_r(j_1, \cdots, j_r)] \tag{41}$$

where the function F_r depends on the coefficients c_k which defined the M sequence (Equation 38). This property follows from the relationship of M sequences to Galois fields. For any choice of coefficients c_k, F_r must be translation invariant:

$$F_r(j + j_1, \cdots, j + j_r) = j + F_r(j_1, \cdots, j_r) \tag{42}$$

as seen by substituting $i - j$ for i in Equation 41.

Higher-order autocorrelations may be calculated from Equation 41. For third-order correlations, for example, we find

$$\langle s(i)s(i-j_1)s(i-j_2)\rangle = \begin{cases} -\frac{1}{N}, & F_2(j_1,j_2) \neq 0 \\ 1, & F_2(j_1,j_2) = 0 \end{cases} \tag{43}$$

That is, most third-order autocorrelations of the M sequence are also $-1/N$ (again, a good approximation to Gaussian white noise), while selected values are equal to 1.

The above relationships may be used to calculate the nondiagonal values of the Wiener kernels via the Lee-Schetzen algorithm (Equation 34). We first consider the discrete-time version of the cross-correlation algorithm (Equation 34). For convenience, we assume we have chosen time units so that the stimulus changes only at integer values of time t and that the response is measured only at those discrete instants. (In a laboratory application, the time unit is usually chosen to be short in comparison to the dynamical aspects of the system to be studied.) With these understandings, the Wiener kernels $\mu_k^{(k)}(\tau_1,\tau_2, \ldots, \tau_k)$ are only defined for integer values of their arguments and may be computed by the cross-correlation algorithm (Equation 34).

As a consequence of Equation 41, cross-correlations of all orders that appear in Equation 34 reduce to a cross-correlation between a single response value and a single lagged input:

$$\mu_k^{(k)}(\tau_1,\tau_2, \cdots, \tau_k) = \frac{1}{k!}\langle R(t)\cdot s[t - F_k(\tau_1,\tau_2, \cdots, \tau_k)]\rangle \tag{44}$$

The meaning of equation 44 is made more obvious through the use of relative-time coordinates $\Delta\tau_2 = \tau_2 - \tau_1, \ldots, \Delta\tau_k = \tau_k - \tau_1$. In these coordinates, the kernel values $\mu_k^{(k)}(\tau_1,\tau_1 + \Delta\tau_2, \ldots, \tau_1 + \Delta\tau_k)$ express the contribution to the effective impulse-response $L(\tau_1)$

attributable to interactions of a test impulse at time zero with stimulus values at times $\Delta\tau_2, \ldots, \Delta\tau_k$ in the past. For fixed values of the relative-time coordinates, these contributions constitute a slice parallel to (but distinct from) the main diagonal slice $\mu_k^{(k)}(\tau_1,\tau_1, \ldots, \tau_1)$ of the kth-order kernel. These kernel values may be derived from Equation 44 via the relationship of Equation 42, or:

$$\mu_k^{(k)}(\tau_1,\tau_2 + \Delta\tau_2, \cdots, \tau_1 + \Delta\tau_k) = \frac{1}{k!}\langle R(t) \cdot s[t - \tau_1 - F_k(0,\Delta\tau_2, \cdots, \Delta\tau_k)]\rangle \tag{45}$$

This result simultaneously expresses both the power and the limitations of the M-sequence approach. An efficient means for cross-correlating a response with an M sequence has been devised by Sutter (1987), based on the Fast Walsh Transform. The off-diagonal slices of the higher-order kernels may be read from this cross-correlation, with the position of the off-diagonal slice encoded by the function $F_k(0,\Delta\tau_2, \ldots, \Delta\tau_k)$. Thus, the procedure provides for a rapid means of extracting both first-order and higher-order kernels from a single cross-correlation.

The limitations are apparent from some simple theoretical considerations. The cross-correlation of the M-sequence stimulus with the output of the system consists of exactly N ($= 2^n - 1$) values. Since the full Wiener expansion may be thought of as an infinite stack of off-diagonal slices, each of which is potentially infinite in length (i.e., no upper bound on τ_1), the cross-correlation cannot supply an unambiguous characterization of the system. Intuitively, one would expect that the cross-correlation (Equation 45) represented a set of nonzero kernel values for a variety of orders k, time lags τ_1, and relative times $\Delta\tau_i$. However, in principle, an equally valid interpretation is that the entire cross-correlation is one very long first-order kernel $\mu_1^{(1)}(\tau_1)$, for τ_1 ranging from 0 to $N - 1$.

Though there is no *a priori* way to resolve the cross-correlation (Equation 45) unambiguously into its component kernel slices, the ambiguity can be resolved through the use of reasonable *a priori* assumptions. We assume that the output of the system is only influenced by an input signal if at least some of the signal has occurred with a range T of prior times. Then, kernel values $\mu_k^{(k)}(\tau_1,\tau_1 + \Delta\tau_2, \ldots, \tau_1 + \Delta\tau_k)$ can only be nonzero for values of τ_1 that are less than this assumed maximum lag T. Since the cross-correlation (Equation 45) consists of N values, up to N/T kernel slices of length T may be separately resolved, provided that the corresponding values of the encoding function $F_k(0,\Delta\tau_2, \ldots, \Delta\tau_k)$ differ by at least T. In the absence of noise, a system whose kernels consisted of these slices alone would have a cross-correlation function made up of these kernel slices at positions $F_k(0,\Delta\tau_2, \ldots, \Delta\tau_k)$, with intervening values all equal to zero.

There is still, however, room for ambiguity. This is because distinct kernel slices $\mu_k^{(k)}(\tau_1,\tau_1 + \Delta\tau_2, \ldots, \tau_1 + \Delta\tau_k)$ and $\mu_j^{(j)}(\tau_1,\tau_1 + \Delta\tau_2', \ldots \tau_1 + \Delta\tau_j')$ may share the same value of the encoding function. To circumvent this problem, one typically makes the assumption that the only kernels that are significantly different from zero are those of particular low orders, for example, order 1 and order 2. Thus, the number of possible encoded values — $F_1(0)$ and $F_2(0,\Delta\tau)$ — is limited and manageable. The validity of these assumptions can be checked by several means. The simplest is to use an M sequence of sufficient length N so that N/T exceeds the number of slices assumed to be present. If the assumptions concerning the number and length of the nonzero kernel slices are valid, the portions of the cross-correlation (Equation 45) *not* occupied by these kernel slices should be zero. These intervening regions may then be inspected to verify that they are, indeed, near zero. More elegant approaches to checking the validity of these assumptions (Sutter

1987) include (1) use of two different recursion relations (Equation 38), so that the resulting encoding functions F_k are different and (2) use of the original M sequence, but with inversion of polarity. This maneuver will invert response components driven by odd-order interactions, but not those driven by even-order interactions.

It should be emphasized that the need to make assumptions in order to measure kernels is a generic one and is not unique to the M-sequence approach. Measurement of Wiener kernels is intrinsically underdetermined, unless one makes some *a priori* assumptions. The M-sequence approach makes it plain how these assumptions interact with the measurement procedure. Only a finite number of kernel slices can be calculated from a set of N values. The maximum number of kernel slices depends on the assumed memory T of the system. Greater quantities of data are required to obtain internal checks for consistency (such as signal-to-noise, or significance of long-lag or high-order components).

E. FREQUENCY DOMAIN METHODS

A major appeal of the frequency domain approach for linear systems is that the transfer function of a composite system is related in a simple fashion to the transfer functions of its components (Equations 8 to 10). A consequence of these rules is that transfer functions of systems built out of familiar passive components (such as resistors and capacitors) are polynomials and quotients of polynomials. The zeroes of these polynomials then provide clues to the internal composition of a linear system. One motivation for frequency-domain analogues of the Wiener approach is to extend this advantage of the transfer function representation to nonlinear systems. As we will see below, this is partially successful. The frequency-domain approach also permits precise knowledge of the statistical properties of the input signal, with consequent advantages for the accuracy of the orthogonalization.

In this approach, the input signal Ω is a sum of sinusoids of frequencies $\omega_1, \ldots, \omega_Q$. The response of a linear system to this signal, according to the superposition principle, is also a sum of sinusoids at the same set of frequencies. The ratio of the output amplitudes to the input amplitudes, and the phase difference between the output and the input, are determined by the transfer function of the system. Thus, the sum of sinusoids signal provides an efficient way to characterize the linear aspects of a biological transduction (Young and Stark 1963).

The efficiency of the approach also extends to the characterization of nonlinear systems (Victor 1979, Victor and Knight 1979, Victor and Shapley 1980). The response of a nonlinear system μ whose Volterra expansion consists of terms (Equation 11) up to order n is somewhat more complex than that of a linear system. Fourier analysis of this response reveals not only the frequencies ω_j that are present in the input, but also integer combinations of these frequencies, of the form

$$\omega = \sum_{j=1}^{Q} a_k \omega_k \qquad (46)$$

where the a_k are positive or negative integers for which absolute values sum to n or less. Such frequencies ω are called "lattice frequencies" of order n. For example, a system whose Volterra series terminates at order 2 will have responses at frequencies $2\omega_j$, $-2\omega_j$, $\omega_j + \omega_k$, $\omega_j - \omega_k$, and $-\omega_j - \omega_k$, in addition to the input frequencies ω_j, and a zero-frequency (steady) response. We will use the term "primitive lattice frequency" of order n to denote a frequency which is a lattice frequency of order n, but not of lower order. For example, $2\omega_j - \omega_k$ and $\omega_j + \omega_k + \omega_l$ are primitive lattice frequencies of order 3: they may be present in the response of any Volterra system of order 3 or greater, but not in a Volterra system of order 2 or less.

While it is possible to define the Wiener expansion in terms of the general orthogonalization procedure described above, it is far simpler to exploit special properties of this input signal. The situation is simplest in which all of the frequencies ω_j are incommensurate and an infinitely long stretch of data is available. (Modifications in the approach related to the use of commensurate frequencies are detailed in Victor and Knight [1979] and Victor and Shapley [1980]). In the incommensurate case, lattice frequencies corresponding to distinct sets of integer coefficients a_k are always different, and the response at each distinct lattice frequency contributes independently to the MSE. The first-order term in the orthogonal representation of μ is the best-approximating linear system. Because each lattice frequency contributes independently to the MSE, the best approximating linear system is one whose responses match those of μ at primitive first-order lattice frequencies, and whose responses at other lattice frequencies are zero. The second-order term in the Wiener expansion consists of the additional response components that must be added to achieve the best approximation among all Volterra systems terminating at second order. Since lattice frequencies contribute independently to the MSE, this term corresponds to a system whose responses match those of μ at primitive second-order lattice frequencies, and whose responses at other lattice frequencies are zero. Similarly, the nth-order term in the Wiener expansion corresponds to a system whose responses match those of μ at primitive nth-order lattice frequencies, and whose responses at other lattice frequencies are zero.

Thus, the nth-order term in the Wiener expansion is characterized by Fourier components of the responses at all primitive nth-order lattice frequencies. These Fourier coefficients constitute the "frequency kernels" (Victor and Knight 1979). The nth-order frequency kernel $K_n(f_1, \ldots, f_n)$ is (apart from a normalization factor) the Fourier coefficient of the response at a primitive lattice frequency $\omega = \Sigma f_r$. In order for $\omega = \Sigma f_r$ to represent a lattice frequency $\omega = \Sigma a_k\omega_k$ (Equation 46), it is necessary for each argument f_r of the frequency kernel to match one of the input frequencies ω_k or $-\omega_k$. To ensure that $\omega = \Sigma f_r$ represents a primitive lattice frequency, we require that if a particular frequency ω_k appears among the arguments of K_n, then its negative $-\omega_k$ does not. That is, frequency kernel values such as $K_2(\omega_j, -\omega_j)$ and $K_3(\omega_j, -\omega_j, \omega_k)$ are not defined.

The frequency kernels are closely analogous to the time-domain kernels measured with a Gaussian noise. Like the time-domain kernels, they are symmetric functions of their arguments. The need for special orthogonalization procedures for the time-domain kernels when two arguments are equal (e.g., Equation 35) corresponds precisely to the undefined nature of the frequency kernels when two arguments add to zero. The correspondence between the two kinds of kernels can be made rigorous: as more and more sinusoids are added to the sum-of-sinusoids signal, the frequency kernels approach the multi-dimensional Fourier transform of the time-domain Wiener kernels (Victor and Knight 1979). Furthermore, the asymptotic approach of frequency kernels measured with a finite number of sinusoids to frequency kernels measured with an infinite number of sinusoids is paralleled precisely by the asymptotic approach of time-domain kernels based on discrete noise to time-domain kernels based on ideal white noise (Victor 1990).

A frequency kernel essentially represents a sampling of the Fourier transform of the corresponding Wiener kernel at a lattice of points. Since the statistical properties of the sum of sinusoid signals are known exactly, the orthogonalization process is exact. Signal-to-noise is enhanced because the response of the system must be concentrated at the lattice frequencies. The tradeoff is that information at intermediate frequencies is necessarily not available. However, for the typical biological transduction whose dynamics are broadly tuned, characterization at a sparse set of frequencies may suffice (Victor 1979).

Time-domain methods that facilitate kernel measurement typically require input signals whose frequency spectrum is approximately flat. The sum-of-sinusoids signal technique can be implemented for input signals whose spectrum is tailored according to the needs of the study and the response properties of the system (Brodie et al. 1978a,b). One clear drawback of the sum-of-sinusoids signal arises when one attempts to use this approach to study a multiinput system. The frequency domain strategy requires assigning a unique set of frequencies to each input. This becomes impractical in situations in which there are more than two inputs. In such multiinput systems, the broadband nature of time-domain techniques provides a more practical way of simultaneously isolating contributions from each input (Sutter 1985, 1987, 1988, Emerson et al. 1987, Reid et al. 1989, Szulborski and Palmer 1990).

F. RELATED APPROACHES

Perhaps because of the apparently daunting complexity of the alternatives, it is still widespread practice in single-unit physiology to use a sinusoidal stimulus to examine the response of what is known to be a nonlinear system. In some sense, the single sinusoid may be regarded as a special case of the sum-of-sinusoids technique, in which the number of sinusoids is reduced to 1. In this case, the statements concerning the asymptotic approach of the frequency-domain method to the white noise method cannot be considered valid, since they hold only when the number of sinusoids is large. More important from a practical standpoint, the use of a simple signal such as a single sinusoid limits one's ability to distinguish among candidate models, since the range of the behavior of the system under study is less thoroughly explored. In some cases (Zemon and Ratliff 1983), the use of two sinusoids provides sufficient information to suggest nonlinear models, though the responses obtained under these conditions are not guaranteed to approximate the Wiener kernels.

There are at least two other approaches to the analysis of nonlinear systems that may be viewed as intermediates between the single-sinusoid approach and Wiener-like approaches based on rich stimulus ensembles. Regan and Regan (1988a,b) developed an approach based on an input signal composed of two sinusoids, often of relatively similar frequency. They characterized a nonlinear system by its responses to various integer combinations $mf_1 + nf_2$ of the input frequencies as the amplitude of the input frequencies was allowed to vary. The very sparse sampling of the frequency domain limits the ability of this approach to characterize dynamics. But because the anticipated output frequencies are correspondingly sparse, frequency-domain filtering affords a dramatic enhancement of signal to noise, even for high orders of nonlinearity (e.g., $|m| + |n| = 8$). Thus, this approach is well suited for noisy, highly nonlinear systems, such as the visual evoked potential.

Spekreijse and co-workers (Spekreijse and Oosting 1970, Spekreijse et al. 1977) considered the response of a nonlinear system to a single sinusoidal input at the frequency f with and without the addition of Gaussian noise or a second sinusoid. Without the auxiliary noise input added, a typical nonlinear system will generate large responses both at the input frequency f and its higher harmonics 2f, 3f, . . . When the auxiliary noise is added, the relative size of the responses at the higher harmonics is diminished relative to the size of the response at the input frequency f. Quantitative analysis of this phenomenon, known as "linearizing", provides a means to analyze a nonlinearity. In the context of the sum-of-sinusoids approach, the reduction of size in the higher harmonics of the input frequency f corresponds to the redistribution of power from a single frequency $n \cdot f$ to all lattice frequencies of order n.

G. NONLINEAR SYSTEMS WITH MORE THAN ONE INPUT

The Volterra and Wiener approaches to the analysis of nonlinear systems with a single input may be extended to systems with two or more spatially distributed inputs. The extension is analogous to the extension of the linear theory to multiinput systems. For convenience, we assume that the inputs correspond to discrete points in space, or discrete regions of the receptive field.

The presence of multiple inputs changes the formulation of the Volterra expansion. The single-input formulation (Equation 11) was based on the assumption that the potential interaction of impulses at any set of times $t_1, t_2, \ldots, t_n$ is proportional to the product of their heights $S(t_1), S(t_2), \ldots, S(t_n)$. We now must consider more general n-fold interactions among input values at n times in the past, where these n input values are partitioned among one or more of H inputs. These partitioned n-fold interactions can be classified by a vector of integers $\vec{n} = (n_1, n_2, \ldots, n_h, \ldots, n_H)$, where n_h is the order of interaction in input h and $n = \Sigma\, n_h$. We represent the signal presented to input h by $S_h(t)$. With these conventions, the multiinput Volterra kernel of order $\vec{n}$ is a function of n time-lags $L^{(\vec{n})}(\tau_1, \tau_2, \ldots, \tau_n)$. Its total contribution to the system response is given by

$$V_{\vec{n}}[L^{(\vec{n})},S](t) = \int_0^\infty \int_0^\infty \cdots \int_0^\infty L^{(\vec{n})}(\tau_1,\tau_2, \ldots, \tau_n)$$

$$S_{h_1}(t - \tau_1) S_{h_2}(t - \tau_2) \cdot \ldots \cdot S_{h_n}(t - \tau_n)\, d\tau_1\, d\tau_2 \cdots d\tau_n \qquad (47)$$

where the first n_1 values h_j select S_1, the next n_2 values h_j select S_2, etc.

As in the single-input case, reorganization of a Volterra series into a Wiener series is motivated by the fact that true Volterra kernels can only be measured by resolving limitingly small interactions of limitingly small inputs. The reorganization of a Volterra series into a Wiener series proceeds along precisely the same lines as for the single-input case (Equations 26 to 30).

The most straightforward way of extending the white noise procedure for kernel measurement to a multiinput system is to assume that the samples of white noise applied to each input are independent. Such an input signal is typically called ''spatiotemporal white noise''. Kernels may be calculated from the response to spatiotemporal white noise by a straightforward generalization of the Lee and Schetzen algorithm (Equation 34). For off-diagonal terms, the cross-correlation estimate of the Wiener kernel of order $\vec{n}$ is given by

$$\mu_{\vec{n}}^{(\vec{n})}(\tau_1,\tau_2, \ldots, \tau_n) = \frac{1}{P^n \Pi n_h!} \cdot$$

$$\langle R(t) \cdot S_{h_1}(t - \tau_1) \cdot S_{h_2}(t - \tau_2) \cdot \ldots \cdot S_{h_n}(t - \tau_n)\rangle \qquad (48)$$

where the h_n is determined from $\vec{n}$ as in (Equation 47). The special considerations in Equation 35 related to ''diagonal'' terms apply to the multiinput case only when the kernel reflects interactions from identical lags *and* identical inputs.

Though appealingly elegant, the use of spatiotemporal white noise magnifies the statistical problems associated with single-input white noise. The main issue is that in any finite sample of spatiotemporal white noise, the inputs are unlikely to be mutually uncorrelated. Unintended first-order correlations across inputs will cause a response due to one input to appear, in part, as spurious contributions from a second input. Unintended higher-order correlations will cause linear responses to appear, in part, as spurious higher-order kernels intended to represent interactions across inputs (and vice versa). Often, the spurious nature of these contributions is signaled by strong dependence on the precise sample of white noise used, or by some nonphysiologic behavior (such as extremely long time lags) which is counter to one's *a priori* knowledge of the system.

As in the single-input case, the M-sequence technique provides both an efficient practical alternative and some additional insight into this statistical problem. The plan (Sutter 1985,

1987, 1988) is to use a single M sequence to define all of the H inputs to the system, but to separate them sufficiently widely in time so that they appear independent. The single M sequence defines the H inputs through a set of delays $D_1, D_2, \ldots, D_H$. That is, the value of the signal s_h at the time i is $s(i - D_h)$, where D_h is the delay assigned to the input h and s(i) is related to the M sequence by $s(i) = 1 - 2b_i$.

As a consequence of Equation 41 and the fact that all inputs are derived from time shifts of the same M sequence, cross-correlations of all orders that appear in Equation 48 are reduced to a cross-correlation between a single response value and a single lagged input:

$$\mu_{\vec{n}}^{(\vec{n})}(\tau_1,\tau_2, \ldots, \tau_n) = \frac{1}{\Pi n_n!} \cdot$$

$$\langle R(t) \cdot s\,[t - F_n(D_{h_1} + \tau_1, D_{h_2} + \tau_2, \ldots, D_{h_n} + \tau_n)]\rangle \tag{49}$$

This formula extends the result of Equation 44 and provides for extraction of all kernels from all inputs from just a single cross-correlation. It also indicates that the number of kernel slices that can be uniquely determined is limited by N/T, where N is the length of the M sequences and T is the time lag beyond which all kernels are assumed to be zero.

Some approximate calculations will underscore the volume of data required for measurement of high-order kernels of multiinput systems. Consider, for example, a Wiener series description of the response of a neuron in visual cortex to inputs which vary in both space and time. An estimate of the number of spatial inputs H can be obtained by considering the ratio of the receptive-field size to the bar width of the finest resolvable grating. A ratio of 16:1, typical for cortical cells, thus implies that subdivision of the (two-dimensional) receptive field into $H = (16)^2 = 256$ spatially distinct inputs would be required for adequate spatial sampling. This is comparable to the number of inputs employed in physiologic investigations of this sort (Gaska et al. 1989, Reid et al. 1989).

Ideally, the number of time intervals T required to characterize dynamics can similarly be estimated from the ratio of the longest time constant to the shortest time constant (corresponding to the highest temporal frequency resolved) of the neuron under study. For practical purposes, probing of the high-frequency behavior is often limited by the frame rate of a video display (60 Hz). Furthermore, one important aspect of the behavior of visual neurons is the presence of processes whose time constants are in the range of seconds (the cortical contrast gain control [Ohzawa et al. 1985] and minutes (light and dark adaptation [Shapley and Enroth-Cugell 1984]). Thus, a complete characterization would require analysis over a ratio of timescales T approaching 1000. As we will see below, practical considerations force the investigator to limit the analysis to a timescale ratio T in the range 16 to 128. This essentially amounts to ignoring the possible interactions of slow processes, such as gain controls, with the more rapid excitatory and inhibitory processes that determine a visual neuron's response.

As seen in Table 1, most kernels of order n are of the type {1,1, . . . ,1}, which corresponds to interactions that are first-order in n of the H inputs. The number of parameters required to specify completely these kernels is approximately $H^nT^n/n!$ Thus, for $H = 256$ and $T = 16$, the number of parameters required at first order is 4096, and the number of parameters required at second order is approximately 8×10^6. For the more modest spatiotemporal resolution of $H = 16$ and $T = 16$, the number of parameters required at second order is approximately 3×10^4, and the number of parameters required at third order is approximately 3×10^6. Thus, we see that with spatiotemporal resolution which is at best marginally adequate for the analysis of a cortical neuron's receptive field, complete determination of the Wiener kernels beyond order 2 is beyond the reach of experimentation.

TABLE 1
The Number of Parameters Contained in Wiener Kernels of a Multiinput System

Order and type	Number of kernels	Number of points per kernel	Number of parameters
Order 1			
{1}	H	T	HT
Order 2			
{2}	H	$\frac{1}{2}T(T + 1)$	$\frac{1}{2}HT^2 + \ldots$
{1,1}	$\frac{1}{2}H(H - 1)$	T^2	$\frac{1}{2}H^2T^2 + \ldots$
Order 3			
{3}	H	$\frac{1}{6}T(T + 1)(T + 2)$	$\frac{1}{6}HT^3 + \ldots$
{2,1}	$H(H - 1)$	$\frac{1}{2}T^2(T + 1)$	$\frac{1}{2}H^2T^3 + \ldots$
{1,1,1}	$\frac{1}{6}H(H - 1)(H - 2)$	T^3	$\frac{1}{6}H^3T^3 + \ldots$
Order 4			
{4}	H	$\frac{1}{24}T(T + 1)(T + 2)(T + 3)$	$\frac{1}{24}HT^4 + \ldots$
{3,1}	$H(H - 1)$	$\frac{1}{6}T^2(T + 1)(T + 2)$	$\frac{1}{6}H^2T^4 + \ldots$
{2,2}	$\frac{1}{2}H(H - 1)$	$\frac{1}{4}T^2(T + 1)^2$	$\frac{1}{8}H^2T^4 + \ldots$
{2,1,1}	$\frac{1}{2}H(H - 1)(H - 2)$	$\frac{1}{2}T^3(T + 1)$	$\frac{1}{4}H^3T^4 + \ldots$
{1,1,1,1}	$\frac{1}{24}H(H - 1)(H - 2)(H - 3)$	T^4	$\frac{1}{24}H^4T^4 + \ldots$

Note: The system is assumed to have H inputs, and the kernels are assumed to be measured at T time points. The designation of order type by $\{j_1, j_2, \ldots\}$ indicates that the nonzero coordinates $(n_1, \ldots, n_h, \ldots)$ are equal to the values $\{j_1, j_2, \ldots\}$ in some order. For example, order type {4} corresponds to a fourth-order kernel which is fourth order in one input and independent of all other inputs, and order type {2,1,1} corresponds to a fourth-order kernel which is second order in one input, first order in two other inputs, and independent of all others.

This calculation is not intended to imply that the application of kernel methods to nonlinear visual neurons is a hopeless exercise. Rather, its purpose is to emphasize that the Wiener approach does not substitute for the design of experiments specific to the aspects of a model being tested. For example, by restricting the spatial resolution to 16 parallel bars, Emerson and co-workers (1987) were able to draw conclusions concerning the generation of motion signals in cortical complex cells of the cat. Another general strategy which can be effective in dealing with multidimensional kernel data is to test hypotheses for separability in space and time (Gaska et al. 1989). Yet another approach to the problem of the large number of parameters that make up a set of kernels is to design specific stimulus ensembles that sift out particular kinds of interactions, an approach that has been used to study aspects of feature extraction (Diaconis and Freedman 1981, Gagalowitz 1981, Julesz 1981, Victor 1985, Victor and Zemon 1985, Victor 1986, Victor and Purpura 1992).

V. CLOSING REMARKS: THE IMPORTANCE OF NONLINEARITIES

The information carried by the firing pattern of an individual visual neuron is necessarily ambiguous. For example, an increment in the firing of an on-center retinal ganglion cell may represent an increase in luminous flux over the center or a decrease in luminous flux over the surround. Similarly, color and luminance information are confounded in the discharge of a single neuron, even one with chromatically opponent receptive field elements. Feature extraction requires resolution of such ambiguities, presumably by analysis of firing patterns across many afferent units. This is an intrinsically nonlinear process (e.g., Reichardt 1961, Victor 1988).

A second problem that needs to be solved by a sensory system is that of cross-modal unification. Adding up signals cannot provide the requisite complexity for cross-modal unification, so one is led to postulate nonlinear interactions. Firing patterns more sophisticated than simple rate codes (Optican and Richmond 1987, Abeles and Gerstein 1988, Aertsen et al. 1989) might be used to convey information useful either for feature extraction or cross-modal unification (Freeman and Van Dijk 1987, Gray et al. 1989). Creation and manipulation of such temporal codes must involve nonlinear processes. A second general framework for both feature extraction and cross-modal unification is that of connectionist models (Cowan and Sharpe 1987). These models show us that simple nonlinearities embedded in an appropriate network of spatial and temporal filters may give rise to sophisticated processing.

We have discussed the Wiener theory as an organizing principle for the experimental characterization of a nonlinear system. The experimentally determined Wiener kernels should not be considered as a model for a nonlinear system, but rather, as a tool to test models. An attractive feature of the Wiener approach is that qualitative features of the kernels provide immediate tests of the adequacy of simple model nonlinear systems. But for really ''interesting'' nonlinear systems, such as those likely to be involved in the processes discussed above, a similar qualitative approach is as yet lacking. Some advances in this direction have been made, for example, the relationship of Wiener kernels to descriptions of networks in terms of nonlinear difference or differential equations (Pinter 1987). However, the general issue of relating characterizations in terms of Wiener kernels to complex nonlinear models remains an area for future research.

ACKNOWLEDGMENTS

This work was supported in part by NIH grant EY7977. The author thanks M. Conte for assistance in the preparation of this manuscript.

REFERENCES

Abeles, M. and Gerstein, G. L. (1988). Detecting spatiotemporal firing patterns among simultaneously recorded single neurons, *J. Neurophysiol.*, **60**, 909–924.

Aertsen, A. M., Gerstein, G. L., Habib, M. K., and Palm, G. (1989). Dynamics of neuronal firing correlation: modulation of effective connectivity, *J. Neurophysiol.*, **61**, 900–917.

Barrett, J. F. (1963). The use of functionals in the analysis of nonlinear systems, *J. Electron. Control,* **15**, 567–615.

Bedrosian, E. and Rice, S. O. (1971). The output properties of Volterra systems (nonlinear systems with memory) driven by harmonic and Gaussian inputs, *Proc. IEEE,* **59**, 1688–1707.

Billings, S. A. (1980). Identification of nonlinear systems: a survey, *IEE Proc.*, **127**, 272–285.
Bracewell, R. N. (1978). *The Fourier Transform and Its Applications,* McGraw-Hill, New York.
Brodie, S. E., Knight, B. W., and Ratliff, F. (1978a). The response of the *Limulus* retina to moving stimuli: a prediction by Fourier synthesis, *J. Gen. Physiol.*, **72**, 129–166.
Brodie, S. E., Knight, B. W., and Ratliff, F. (1978b). The spatiotemporal transfer function of the *Limulus* lateral eye, *J. Gen. Physiol.*, **72**, 167–202.
Cowan, J. D. and Sharp, D. H. (1987). *Neural Nets,* Reprint LA-UR-87-4098, Los Alamos National Laboratory, Los Alamos, NM.
De Groot, S. R. and Mazur, P. (1961). *Non-Equilibrium Thermodynamics,* North-Holland, Amsterdam.
De Lange, H. (1952). Experiments on flicker and some calculations on an electrical analogue of the foveal systems, *Physica,* **18**, 935–950.
Diaconis, P. and Freedman, D. (1981). On the statistics of vision: the Julesz conjecture, *J. Math. Psychol.*, **24**, 112–138.
Emerson, R. C., Citron, M., Vaughn, W. J., and Klein, S. (1987). Nonlinear directionally selective subunits in complex cells of cat striate cortex, *J. Neurophysiol.*, **58**, 33–65.
Emerson, R. C., Bergen, J., and Adelson, E. H. (1992). Directionally selective complex cells and the computation of motion energy in cat visual cortex, *Vision Res.*, **32**, 203–218.
Feller, W. (1968). *An Introduction to Probability Theory and Statistics,* 3rd ed., John Wiley & Sons, New York.
Freeman, W. J. and Van Dijk, B. (1987). Spatial patterns of visual cortical fast EEG during conditioned reflex in a rhesus monkey, *Brain Res.*, **422**, 267–276.
Gagalowitz, A. (1981). A new method for texture fields synthesis: some applications to the study of human vision, *IEEE Trans. Pattern Anal. Mach. Intelligence,* **3**, 520–523.
Gaska, J. P., Jacobson, L. D., Chen, H.-W., and Pollen, D. A. (1989). Structural modeling of nonlinear networks in V1 of the macaque monkey, *Soc. Neurosci. Abstr.*, **15**, 1056.
Golomb, S. W. (1968). *Shift Register Sequences,* Holden-Day, San Francisco.
Gray, C. M., Konig, P., Engel, A. K., and Singer, W. (1989). Oscillatory responses in cat visual cortex exhibit inter-columnar synchronization which reflects global stimulus properties, *Nature (London),* **338**, 334–337.
Julesz, B. (1981). Textons, the elements of texture perception, and their interactions, *Nature (London),* **290**, 91–97.
Klein, S. (1987). Relationships between kernels measured with different stimuli, in *Advanced Methods of Physiological System Modeling,* Vol. 1, Marmarelis, V. Z., Ed., University of Southern California, Los Angeles.
Klein, S. and Yasui, S. (1979). Nonlinear systems analysis with non-Gaussian white stimuli: general basis functions and kernels, *IEEE Trans. Inf. Theory,* **25**, 495–500.
Korenberg, M. J. (1973). Cross-correlation analysis of neural cascades, in Proc. 10th Annu. Rocky Mountain Bioeng. Symp., 47–51.
Korenberg, M. J. (1988). Exact orthogonal kernel estimation from finite data records: extending Wiener's identification of nonlinear systems, *Ann. Biomed. Eng.*, **16**, 123–142.
Korenberg, M. J. and Hunter, I. W. (1990). The identification of nonlinear biological systems: Wiener kernel approaches, *Ann. Biomed. Eng.*, **18**, 629–654.
Krausz, H. I. (1975). Identification of nonlinear systems using random impulse trains, *Biol. Cybern.*, **19**, 217–230.
Lee, Y. N. and Schetzen, M. (1965). Measurement of the kernels of a nonlinear system by cross-correlation, *Int. J. Control,* **2**, 237–254.
Levine, R. D. and Tribus, M. (1979). *The Maximum Entropy Formalism,* MIT Press, Cambridge.
Marmarelis, P. Z. and Marmarelis, V. Z. (1978). *Analysis of Physiological Systems: The White-Noise Approach,* Plenum Press, New York.
Marmarelis, P. Z. and Naka, K.-I. (1972). White noise analysis of a neuron chain: an application of the Wiener theory, *Science,* **175**, 1276–1278.
Marmarelis, V. Z. (1975). Identification of nonlinear systems through multi-level random signals, in *Proc. 1st Symp. on Testing and Identification of Nonlinear Systems,* Marmarelis, V. Z., Ed., California Institute of Technology, Pasadena.
Marmarelis, V. Z. (1977). A family of quasi-white random signals and its optimum use in biological system identification. I. Theory, *Biol. Cybern.*, **27**, 49–56.
McCann, G. D. (1974). Nonlinear identification theory models for successive stages of visual nervous systems of flies, *J. Neurophysiol.*, **37**, 869–895.
Naka, K.-I., Marmarelis, P. Z., and Chan, R. (1975). Morphological and functional identification of catfish retinal neurons. III. Functional identification, *J. Neurophysiol.*, **38**, 92–131.
Ohzawa, I., Sclar, G., and Freeman, R. D. (1985). Contrast gain control in the cat's visual system, *J. Neurophysiol.*, **54**, 651–667.
Optican, L. M. and Richmond, B. J. (1987). Temporal encoding of two-dimensional patterns by single units in primate inferior temporal cortex. III. Information theoretic analysis, *J. Neurophysiol.*, **57**, 162–178.

Palm, G. and Poggio, T. (1977). The Volterra representation and the Wiener expansion: validity and pitfalls, *SIAM J. Appl. Math.*, **33**, 195–216.

Palm, G. and Poggio, T. (1978). Stochastic identification methods for nonlinear systems: an extension of the Wiener theory, *SIAM J. Appl. Math.*, **34**, 524–534.

Pinter, R. (1987). Kernel synthesis from nonlinear multiplicative lateral inhibition, in *Advanced Methods of Physiological System Modeling,* Vol. 1, Marmarelis, V. Z., Ed., University of Southern California, Los Angeles.

Pringle, J. W. S. and Wilson, V. J. (1952). The response of a sense organ to a harmonic stimulus, *J. Exp. Biol.*, **29**, 220–234.

Regan, D. and Regan, M. P. (1988a). Objective evidence for phase-independent spatial frequency analysis in the human visual pathway, *Vision Res.*, **28**, 187–191.

Regan, M. P. and Regan, D. (1988b). A frequency domain technique for characterizing nonlinearities in biological systems, *J. Theor. Biol.*, **133**, 293–317.

Reichardt, W. (1961). Autocorrelation, a principle for the evaluation of sensory information by the central nervous system, in *Sensory Communication,* Rosenbluth, W. A., Ed., John Wiley & Sons, New York.

Reid, R. C., Shapley, R. M., and Victor, J. D. (1989). A new two-dimensional pseudorandom stimulus for the study of receptive fields in the LGN and striate cortex, *Soc. Neurosci. Abstr.*, **15**, 323.

Rugh, W. J. (1981). *Nonlinear System Theory: The Volterra/Wiener Approach,* Johns Hopkins University Press, Baltimore.

Sakai, H., Naka, K.-I., and Korenberg, M. J. (1988). White noise analysis in visual neuroscience, *Visual Neurosci.*, **1**, 287–296.

Sakuranaga, M. and Naka, K.-I. (1985a). Signal transmission in the catfish retina. I. Transmission in the outer retina, *J. Neurophysiol.*, **53**, 373–389.

Sakuranaga, M. and Naka, K.-I. (1985b). Signal transmission in the catfish retina. II. Transmission to type-N cell, *J. Neurophysiol.*, **53**, 390–410.

Sakuranaga, M. and Naka, K.-I. (1985c). Signal transmission in the catfish retina. III. Transmission to type-C cell, *J. Neurophysiol.*, **53**, 411–428.

Schetzen, M. (1980). The Volterra and Wiener Theory of Nonlinear Systems, *John Wiley & Sons, New York.*

Shapley, R. M. and Enroth-Cugell, C. (1984). Visual adaptation and retinal gain controls, *Prog. Retinal Res.*, 3, 263–346.

Spekreijse, H., Estevez, O., and Reits, D. (1977). Visual evoked potentials and the physiological analysis of visual processes in man, in *Visual Evoked Potentials in Man,* Desmedt, J. E., Ed., Clarendon Press, Oxford, 16–89.

Spekreijse, H. and Oosting, J. (1970). A method for analyzing and synthesizing nonlinear systems, *Kybernetik,* **7**, 23–31.

Sutter, E. (1985). Multi-input VER and ERG analysis for objective perimetry, in IEEE 7th Ann. Conf. Engineering in Medicine and Biology Society, Chicago, 414–420.

Sutter, E. (1987). A practical nonstochastic approach to nonlinear time-domain analysis, in *Advanced Methods of Physiological System Modelling,* Marmarelis, V. Z., Ed., University of Southern California, Los Angeles.

Sutter, E. (1988). Field topography of the visual evoked response, *Invest. Ophthalmol. Vis. Sci.*, **29**, (Suppl.), 433.

Szulborski, R. G. and Palmer, L. A. (1990). The two-dimensional spatial structure of nonlinear subunits in the receptive fields of complex cells, *Vision Res.*, **30**, 249–254.

Victor, J. D. (1979). Nonlinear systems analysis: comparison of white noise and sum of sinusoids in a biological system, *Proc. Natl. Acad. Sci. U.S.A.*, **76**, 996–998.

Victor, J. D. (1985). Complex visual textures as a tool for studying the VEP, *Vision Res.*, **25**, 1811–1827.

Victor, J. D. (1986). Isolation of components due to intracortical processing in the visual evoked potential, *Proc. Natl. Acad. Sci. U.S.A.*, **83**, 7984–7988.

Victor, J. D. (1988). Models for preattentive texture discrimination: Fourier analysis and local feature processing in a unified framework, *Spatial Vision,* **3**, 263–280.

Victor, J. D. (1991). Asymptotic approach of generalized orthogonal functional expansions to Wiener kernels, *Ann. Biomed. Eng.*, **19**, 383–399.

Victor, J. D. and Johanessma, P. I. M. (1986). Maximum-entropy approximations of stochastic nonlinear transductions: an extension of the Wiener theory, *Biol. Cybern.*, **54**, 289–300.

Victor, J. D. and Knight, B. W. (1979). Nonlinear analysis with an arbitrary stimulus ensemble, *Q. Appl. Math.*, **37**, 113–136.

Victor, J. D. and Purpura, K. (1992). Alterations of form vision at isoluminance identified through the use of visual textures, in *Nonlinear Vision: Determination of Neural Receptive Fields, Function, and Networks,* Pinter, R. B. and Nabet, R., Eds., CRC Press, Boca Raton, FL.

Victor, J. D. and Shapley, R. M. (1979). The nonlinear pathway of Y ganglion cells in the cat retina, *J. Gen. Physiol.*, **74**, 671–689.

Victor, J. D. and Shapley, R. M. (1980). A method of nonlinear analysis in the frequency domain, *Biophys. J.*, **29**, 459–484.

Victor, J. D. and Zemon, V. (1985). The human visual evoked potential: analysis of components due to elementary and complex aspects of form, *Vision Res.*, **25**, 1829–1842.
Volterra, V. (1932). *Theory of Functionals and of Integral and Integro-Differential Equations,* Blackie, London.
Wiener, N. (1958). *Nonlinear Problems in Random Theory,* John Wiley & Sons, New York.
Yasui, S. (1979). Stochastic functional Fourier series, Volterra series, and nonlinear systems analysis, *IEEE Trans. Autom. Control,* **24**, 230–242.
Young, L. R. and Stark, L. (1963). Variable feedback experiments testing a sampled data model for eye tracking movements, *IEEE Trans Hum. Factors Electron.*, **4**, 38–51.
Zemon, V. and Ratliff, F. (1983). Intermodulation components of the visual evoked potential: response to lateral and superimposed stimuli, *Biol. Cybern.*, **50**, 401–408.

Chapter 2

NONLINEAR OPERATOR NETWORK MODELS OF PROCESSING IN THE FLY LAMINA

A. C. James

TABLE OF CONTENTS

I. Introduction 40

II. Operator Network Models 40

III. Polynomial Time-Invariant Operators 42

IV. The Insect Visual Lamina 44
- A. Receptor Kernels 46
- B. LMC Kernels 47

V. Structural Modeling of Nonlinear Systems 50
- A. Cascading of Monomial Time-Invariant Operators 51
- B. Linear-Nonlinear-Linear Cascade Models 52
- C. *L* + *LNL* Models 55

VI. Receptor-LMC Transmission 58
- A. The Primary Transmission Pathway 58
- B. Dynamics of Receptor-LMC Transmission 61
- C. Transmission of Receptor Nonlinearity 62
- D. Parallel *LNL* Models 64

VII. Surround Effects 68
- A. Optical Cross-Talk Component 68
- B. Lateral Antagonism 69

VIII. Conclusion 71

References 73

I. INTRODUCTION

The lamina of the insect visual system is the site of the first synaptic transmission of the dynamic visual image sampled by the photoreceptors of the retinal array. At this site the image is recoded before transmission to the next visual ganglion, the medulla.

This chapter is an analysis of the nonlinear aspects of the image/signal processing taking place in the lamina system. The relationship between quantities is modeled by operators, of a type which may be termed polynomial time-invariant operators. These time-invariant operators are analogous to the polynomials used to represent static functions, with the role of the coefficients being played by the *kernels* of the operators.

The quantitative modeling is based on kernels derived from white-noise stimulation. White-noise analysis is now an established technique in visual science, recently reviewed by Sakai et al. (1988). It produces an empirical description of the behavior of the system in a compact form, making a minimum of assumptions. The experiments in this case involved two stimulus channels, matched to the known spatial structure of the receptive fields, and the recorded intracellular responses of the first two cell classes of the visual pathway, the photoreceptors and the large monopolar cells (LMCs) of the lamina.

The empirically derived kernels form the basis for further modeling. Properties of the transmission stage can be inferred, both for linear and for nonlinear components. The introduction of inferred intermediate quantities allows the decomposition of the observed system into components consisting of linear dynamic mappings and static nonlinearities. This structural modeling is best considered with close reference to data from other experimental techniques, anatomical, pharmacological, and electrophysiological, allowing associations to be made between elements of the structural model and the biophysical processes at work in the system.

The extended modeling undertaken is facilitated by the use of a compact notation for polynomial time-invariant operators, while models involving mappings between a number of quantities are represented with clarity by mapping diagrams.

II. OPERATOR NETWORK MODELS

In a static description of the behavior of a system, quantities are represented by single numbers, such as the amplitude of an applied stimulus and the maximum voltage of a cell's response. The relationship between quantities is described by a function mapping one number to the other. For a dynamical description, each quantity is represented as a function of time, and a relationship can be represented by an operator, mapping one function to the other. Thus if x and y are functions describing the time course of two quantities, the dependence of y on x may be represented by an operator F, and we write

$$y = Fx$$

with operator application being denoted by juxtaposition, although the operator may be nonlinear.

The functions are treated as elements of linear spaces, with addition and scalar multiplication defined pointwise on time. A quantity might then be modeled as the sum of a number of components. Addition in the space of output functions, in turn, induces an addition operation for operators, such that

$$(F_1 + F_2)x = F_1x + F_2x$$

and we can speak of an operator as being the sum of a number of components, such as

$$F = F_1 + F_2$$

A quantity may be modeled as being dependent on more than one other quantity, either as a sum of components, such as

$$y = F_1x_1 + F_2x_2$$

or as a nonadditive combination of inputs, represented by an operator taking multiple inputs, such as

$$y = F(x_1,x_2)$$

The output of one operator may be the input to another, forming the composition of the two operators, which we will denote by juxtaposition, thus, GF is defined by:

$$GFx = G[F(x)]$$

When a number of quantities and their interactions are represented, a network of functions and mappings is produced. As the complexity increases, graphical representation of the model is useful. In engineering and much systems analysis literature the block diagram is the traditional graphical representation, in which operators are drawn in boxes with connecting lines indicating the flow of signals. The system $y = Fx$ is thus drawn

$$\xrightarrow{\;x\;} \boxed{F} \xrightarrow{\;y\;}$$

As the number of elements increases, the block diagram representation becomes cumbersome. We will instead represent a model by writing the functions involved at the nodes of a directed graph and drawing arrows to represent the presence of operators. Thus, the elemental relationship $y = Fx$ is drawn as

$$x \xrightarrow{F} y$$

This diagram is dual to the above block diagram, in that the roles of edges and nodes have been exchanged. The diagram is similar to the mapping diagrams used in mathematical literature, although we place a function at each node, rather than a function space. A more significant difference is that the convergence of mappings onto a node will imply the summation of components, or a multiinput relationship, so that

$$\begin{array}{cc} x_1 & x_2 \\ \downarrow & \swarrow \\ y & \end{array}$$

represents a relationship of the form $y = F_1x_1 + F_2x_2$, or $y = F(x_1,x_2)$. When the identifiers for the operators are generated by the convention described below, the operators need not appear explicitly on a mapping diagram. Such a diagram shows in a simple and uncluttered form the causality inherent in a model.

The mapping diagram also gives a different emphasis; in an engineering context the operators generally represent known devices, while the signals are more ephemeral entities. In system identification experimentation the recorded signals are the primary data, while the operators represent hypothetical models inferred from the data.

III. POLYNOMIAL TIME-INVARIANT OPERATORS

The previous section describes in an abstract way an algebraic approach to the description of dynamical system behavior. To produce quantitative descriptions a particular class of operators will be considered, which may be termed polynomial time-invariant operators. Time invariance means that a time shift of the input signals produces an output with the same time shift, hence, the operators commute with the time-shift operator. They may be termed polynomials, as they consist of sums of terms involving powers and products of the input signals just as polynomial functions involve powers and products of the input variables, where the powers and products are here taken in a tensorial sense, defined below. Truncated Wiener and Volterra series are of this form; the class, thus, includes the empirical models fitted by white-noise experimentation. To facilitate dealing with networks of these operators, a compact notation will be defined.

An operator of first degree is the familiar linear time-invariant system, involving convolution of the input with a first-order kernel, $k_1(t)$:

$$y(t) = \int k_1(t_1)x(t - t_1)\, dt_1$$

where integration is over the domain of the kernel k_1, reflecting the memory of the system. All operators considered here will be finite memory, causal operators; thus, the kernels have as domain some finite interval starting at zero. This is also represented by the convolution symbol *:

$$y = k_1 * x$$

In general, the degree two operator maps two inputs to an output. A bilinear time-invariant operator mapping inputs x_1,x_2 to an output y creates the following relationship:

$$y(t) = \int\int k_2(t_1,t_2)x_1(t - t_1)x_2(t - t_2)\, dt_1dt_2$$

with integration taken over the two-dimensional domain of the kernel k_2. Expressions such as this occur frequently, and a more compact yet rigorous notation will be of great use.

The above mapping is termed bilinear because it is linear in each input if the other input is held constant. A fundamental result is that a bilinear function of (x_1,x_2) is a linear function of the tensor product, $x_2 \otimes x_2$. This is the function of two arguments defined by

$$x_1 \otimes x_2(t_1,t_2) = x_1(t_1)x_2(t_2)$$

The integral is equal to the two-dimensional convolution of the kernel with this tensor product, but evaluated along the diagonal only

$$y(t) = k_2 * (x_1 \otimes x_2)(t,t)$$

We will represent the linear operator mapping $x_1 \otimes x_2$ to y by placing the kernel in square brackets, thus, writing the bilinear relationship as

$$y = [k_2]x_1 \otimes x_2$$

Note that the tensor product has the highest precedence, followed by operator application, and lastly function evaluation, so that parsing occurs as in the following:

$$[k_2]x_1 \otimes x_2(t) = ([k_2](x_1 \otimes x_2))(t)$$

When a single input function is repeated, the product becomes a tensor power, which will be denoted as

$$x^{2\otimes} = x \otimes x$$

and similarly for the general tensor power of degree n.

The general order n time-invariant operator with kernel k_n is written as $[k_n]$. Application to an input x is defined by

$$[k_n]x^{n\otimes}(t) = \int \cdots \int k_n(t_1, \ldots, t_n)x(t - t_1) \cdots x(t - t_n)\, dt_1 \cdots dt_n$$

where integration is over the n-dimensional domain of the kernel k_n. Sums of monomials such as this form polynomial time-invariant operators, and the corresponding infinite series is the Volterra series.

In this notation the Volterra series in x is written as

$$y = \sum_{n=0}^{\infty} [k_n]x^{n\otimes}$$

showing clearly the relationship with the power series form of a function:

$$y = \sum_{n=0}^{\infty} \alpha_n x^n$$

When the order of the kernel is clear from context, the tensor exponentiation of the input can be assumed as implicit, thus:

$$[k_n]x = [k_n]x^{n\otimes}$$

and for any kernel k_n, we can refer to the corresponding mapping from input to output simply as the operator $[k_n]$. The use of square brackets is chosen as being reminiscent of the box in which a kernel might be drawn in a block diagram. Once familiarity is gained with this type of expression, the square brackets can be omitted from equations, and juxtaposition of a kernel and the appropriate inputs can be assumed to imply the application of a monomial time-invariant operator. This leads to equations with a particularly ''light'' appearance; however, this chapter will continue with the more explicit notation for clarity.

For a discrete-time model, the integrals are replaced by summations. If the memory considered is over the interval $0 \ldots m - 1$, then

$$[k_n]x(t) = \sum_{t_1,\ldots,t_n=0}^{m-1} k_n(t_1, \ldots, t_n)x(t - t_1) \cdots x(t - t_n)$$

This involves the kernel in terms of unit time steps; the kernel can be scaled to give the appropriate physical units.

This notation can also be translated directly into an algorithmic language for numerical simulations, with functions being defined to calculate the mappings of each degree, taking as inputs the kernel values and the input waveforms and returning the output waveform.

IV. THE INSECT VISUAL LAMINA

The retina of insects consists of an array of ommatidia, generally hexagonally arranged, which forms a dynamic neural image of the visual scene. Proximal to the retina lies the lamina, organized as an array of cartridges, in one-to-one correspondence with the retinal array (see Figure 1, Chapter 14, this volume).

In the fly, the axons of six receptors sampling the same region of space terminate in one lamina cartridge. Each cartridge contains one member of each of 11 other cell classes and also contains amacrine cell processes (reviewed in Shaw, 1984). The principal output neurons are the two LMCs, denoted L1 and L2, which terminate in different layers of the next visual ganglion, the medulla. The LMCs give a graded response to synaptic input from the six terminating receptors.

The transformation of the visual image occurring during receptor-LMC transmission can be characterized by four properties: inversion of the signal, amplification, temporal highpass filtering, and some lateral antagonism. These properties have been observed using impulse and step stimulation, (e.g., Zettler and Järvilehto, 1972; Laughlin, 1974; Laughlin and Osorio, 1989), and by stimulation with moving sinusoidal gratings (Dubs, 1982).

This chapter is based on data obtained by white-noise stimulation. White-noise stimulation on a checkerboard domain produced estimated spatiotemporal first-order kernels for the system, showing precisely the distribution of sensitivity and antagonism on space and time. These results are described elsewhere (James, 1990 and in preparation). To quantify nonlinear properties of the system, a separate two-channel center-surround white-noise stimulus was used (Figure 1), providing the basic data used in this chapter.

Binary white-noise signals modulated the contrast of each region; that is, on each frame, spot and annulus were set independently to be light or dark with equal probability. Stimulus contrast, defined as deviation from the mean divided by the mean, is thus ± 1 at each time step. The stimulus contrast on center and surround are denoted by the time series $a(t)$ and $b(t)$, respectively. The stimulus had a period of 32,768 frames and was repeated eight times per run.

Intracellular recordings were made in separate experiments from the terminals of receptors and from LMCs in the lamina of two species of fly, *Eristalis tenax* and *Calliphora vicina*. The intracellular voltage was sampled in synchrony with the stimulus, to give time series denoted by $r(t)$ for receptor cells, and by $l(t)$ for LMCs.

As a large number of kernels are considered in the following, a systematic way of generating the identifiers is of use. A convenient system is to use subscripts to indicate the output and superscripts to indicate the inputs, with multiple superscripts indicating kernels of order greater than one. Kernel superscripts are thus paired with input functions, reminiscent of the repeated index notation used in tensor calculus. The base symbol h will indicate a kernel labeled by this system. Thus, for example, the identifier h_r^{aa} denotes a second-order kernel which is combined with the repeated input signal a to produce a component of the receptor response r. The corresponding operator mapping $a \rightarrow r$ is denoted by $[h_r^{aa}]$.

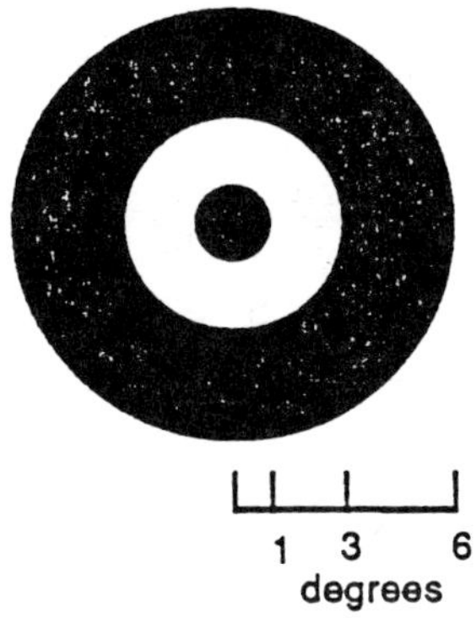

FIGURE 1. Spatial form of the white-noise stimulus. The spot of radius 1° is placed on receptive field center; the annulus extends from radius 3° to 6°. The stimulus was generated on a Joyce oscilloscope at 500 Hz frame rate. The mean intensity of illumination was estimated to produce 10^6 absorbed photons per second per receptor. A lower mean intensity was also used, producing 10^4 absorbed photons per second per receptor.

The operator model was fitted to third degree in a, to second degree in b, and with the second-degree cross-term in a and b. The fitted model predicting receptor response in terms of the two stimulus channels can be written as follows, omitting the constant term:

$$
\begin{aligned}
r &= F(a,b) \\
&= [h_r^a]a + [h_r^b]b \\
&\quad + [h_r^{aa}]a^{2\otimes} + [h_r^{ab}]a \otimes b + [h_r^{bb}]b^{2\otimes} \\
&\quad + [h_r^{aaa}]a^{3\otimes}
\end{aligned}
$$

The equivalent model is fitted for LMC response data. Additional terms can be added to achieve orthogonalization of the series with respect to the white-noise input, as described by Wiener (1958) for the Gaussian white-noise case. We will not consider these terms in the following, as they consist only of constant terms, associated with the kernels h^{aa} and h^{bb} and a linear term associated with h^{aaa} which has a small magnitude relative to the first-order kernel h^a.

The data are sampled on discrete time; hence, the above equation translates into summations over the period of memory of the kernels, which was modeled to 50 ms. As noted above, the model is composed of linear mappings of various tensor products and powers of the input functions. That is, the equation is linear in the kernel elements, although it is nonlinear in the input values. Kernel estimation is thus a linear estimation problem and can be formulated as a linear regression problem, where the response $r(t)$ is regressed on various pointwise products of the stimuli $a(t)$ and $b(t)$. Initial estimates are obtained by the cross-correlation procedure (Lee and Schetzen, 1965). These are then corrected by an iterative procedure to obtain the exact least-squares estimates, giving the closest possible fit to the data for this model (James, 1990 and in preparation).

The inputs are measured in terms of contrast. Although this is dimensionless, it can for convenience be represented by the unit C. Response is in millivolts, mV. For plotting, the order n kernels is divided by Δt^n, where $\Delta t = 2$ ms is the time step, to give kernel units $\mathrm{mV/(Cms)}^n$.

The mean-square values of the fitted components give an indication of the degree of nonlinearity within the system. The statistical properties of the response components are

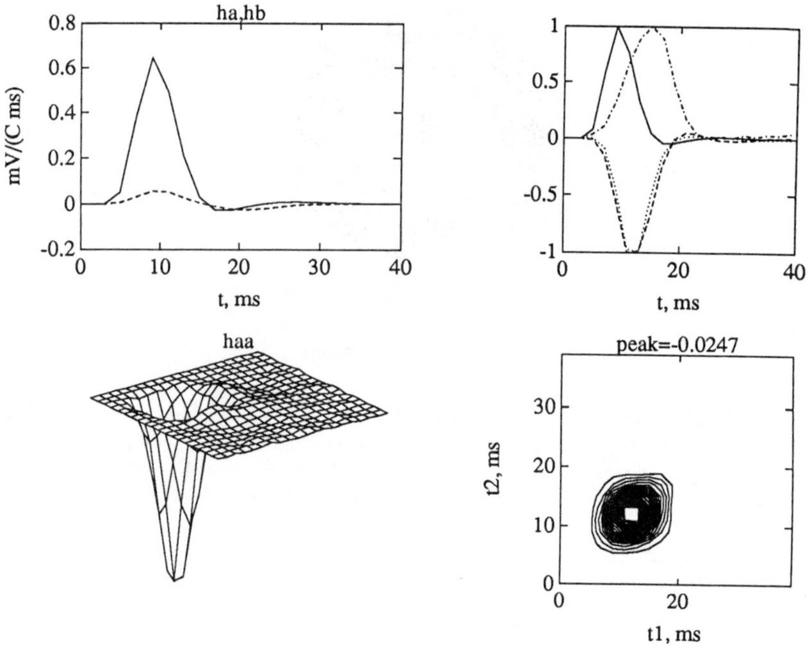

FIGURE 2. First-order and center-only second-order kernels for an *Eristalis* receptor. Upper left, the first-order kernels for center and surround, h_r^a and h_r^b in units mV/(Cms). Upper right, normalized kernels and sections as follows: solid line, center kernel h_r^a; dashed line, section through h_r^{aa} primary peak parallel to an axis; dotted line, diagonal of h_r^{aa}; dot-dash, section through h_r^{aaa} primary peak parallel to an axis, inverted for LMC data to allow comparison with other kernels. Lower, the second-order kernel h_r^{aa}. Left, in perspective. Right, contours; the peak value is indicated in units mV/(Cms)^2. Contour step size is 5% of the peak; zero contour is excluded.

presented in detail elsewhere (James, 1990 and in preparation). The average distribution of power can be summarized as follows: the receptor response was 95% linear, 2.5% quadratic, with 0.3% in cubic and in residual components. The LMC response was 80% linear, 10% quadratic, with 5% in cubic and in residual components. The system is, thus, primarily linear, but with consistent nonlinearity in the LMC response.

A. RECEPTOR KERNELS

The kernels for photoreceptors are relatively simple. Figure 2 shows the first-order and center-only second-order kernels. The first-order kernels are slightly biphasic, with the surround kernel h_r^b having a peak corresponding to the peak of h_r^a, suggesting that the annulus in these experiments did produce a response in the axial photoreceptors through optical cross-talk. The amplitude of h_r^b varied up to a maximum of one fifth the amplitude of h_r^a. When normalized, h_r^b can be seen to have a larger negative peak relative to h_r^a, indicating that the receptor response is not exactly space-time separable.

The second-order kernel h_r^{aa} has a large negative part, with peak around 3 ms after the peak of the first-order kernel. Small positive parts lie axial to this peak. When the first- and second-order kernels are expressed in units of mV/(C ms) and mV/(C ms)^2 respectively, the ratio of the amplitudes of h_r^{aa} over h_r^a varies from 0.01 to 0.04.

It may be worth pausing for a moment to dwell on the meaning of the quadratic time-invariant operator. The kernel h_r^{aa} generates the following component of response in the model:

$$[h_r^{aa}]a^{2\otimes}(t) = \sum_{t_1=0}^{m-1} \sum_{t_2=0}^{m-1} h_r^{aa}(t_1,t_2)a(t - t_1)a(t - t_2)$$

Taking one element of this sum, the output at time t is influenced by the product of the inputs at times $t - t_1$ and $t - t_2$, $a(t - t_1)a(t - t_2)$, with weighting $h_r^{aa}(t_1,t_2)$. Along the diagonal $t_1 = t_2$ this product is nonnegative; hence, the negative values in the main lobe of the kernel h_r^{aa} will always generate a hyperpolarizing component of response, irrespective of input polarity. At these latencies the first-order kernel h_r^a is positive, and so the second-order term reinforces the response to negative contrasts, while opposing the response to positive contrasts. The slower time-course of the second-order kernel produces a model response that is smaller, but faster, given a positive stimulus flash, and larger, but slower, given a negative stimulus flash, as is seen when impulsive stimuli are used. The inclusion of the quadratic operator, thus, introduces polarity asymmetry into the model system. Just as in polynomials, odd-order operators are odd-valued operators in that an inversion of input produces an inverted output, while even-order operators are even-valued mappings. The first-order Wiener kernel provides the best linear model of the system, in the least-squares sense for the white-noise input, and is, thus, a compromise, attempting to describe the response to positive and negative stimuli. The second-order kernel can be used to describe asymmetry observed in the response to incremental and decremental light stimuli.

The kernels h_r^{ab} and h_r^{bb} are variable and smaller in amplitude and are not illustrated. The third-order kernel h_r^{aaa} has as its principal feature a broad positive part, with peak around 5 *ms* after the first-order kernel peak. The amplitude, in $mV/(C\ ms)^3$ is from 0.0005 to 0.001 of the amplitude of h_r^a.

B. LMC KERNELS

The LMC kernels are more complex and show more variation between cells. One complete kernel set is presented in Figures 3 and 4. More extensive presentation of results is given elsewhere (James, 1990 and in preparation). The first- and second-order kernels are plotted aligned in the right-hand column to facilitate comparison of the kernels on time.

In discussing the kernels a distinction will be made between the parts of a kernel and the components of a kernel. The *parts* of a kernel will mean the kernel on nonoverlapping regions of its domain, the parts generally being delineated by the polarity of the kernel. For example, the first-order LMC kernel has two parts, the primary negative part, of large amplitude and short duration, and the positive secondary part, of smaller amplitude and longer duration. Parts of the second-order kernels are defined on two-dimensional regions of the domain, and so on. The parts of a kernel can be identified by direct inspection of the kernel by itself.

The *components* of a kernel, in contrast, are defined with reference to some proposed structural model of the system. The components of the structural model produce component kernels, which when added produce the overall model kernel, attempting to reconstruct the experimentally derived kernel.

A given kernel component may be largely associated with some kernel parts, but, in general, components will be nonzero on overlapping regions of the domain. The components are then confounded, and decomposition into components is ambiguous. One use of the kernels of order greater than one is to distinguish components which are confounded in the first-order kernels, as will be seen below.

Referring to the LMC kernel set of Figure 3, the negative primary part of the first-order kernel h_l^a reflects the inversion taking place at the receptor-LMC synapse, while the presence of an antagonistic secondary part reflects temporal high-pass filtering of the signal. The surround kernel h_l^b has a relatively large positive part, reflecting lateral antagonism, or spatial high-pass filtering of the image.

In the second-order kernel for the center, h_l^{aa}, we can distinguish four parts. The primary part is a compact negative region centered on the diagonal at latency slightly less than the primary peak of the first-order kernel. There is a positive secondary part extending parallel

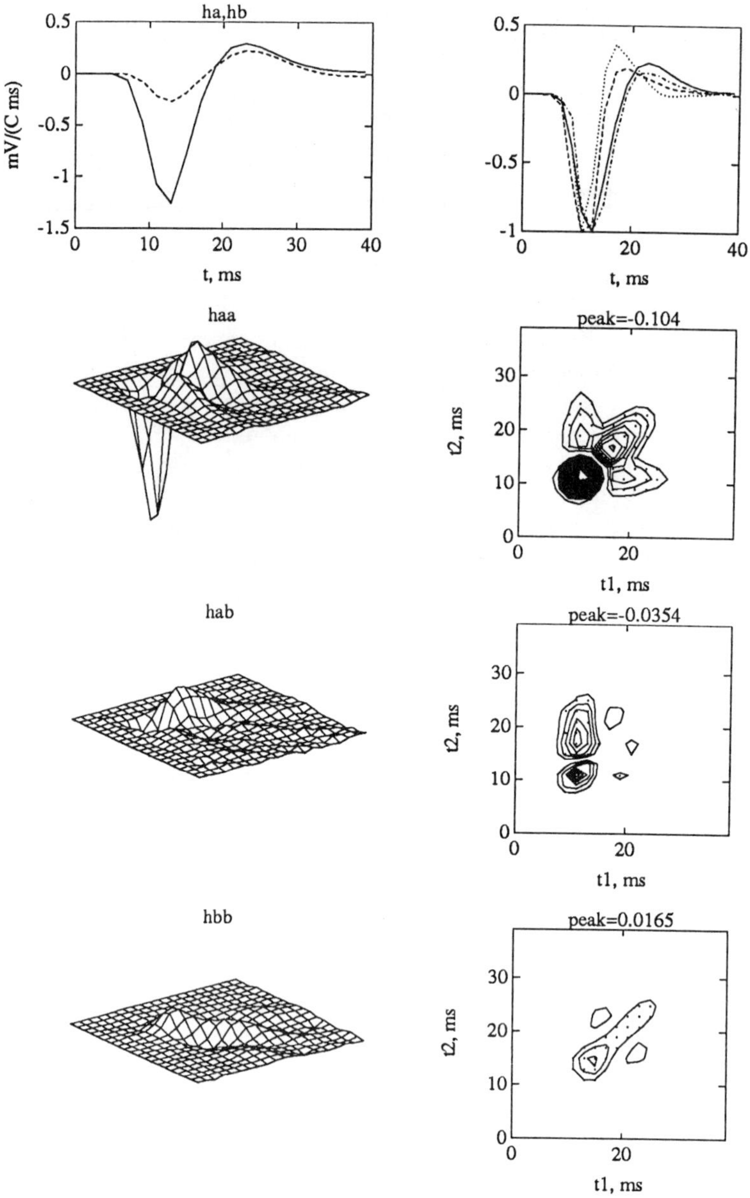

FIGURE 3. First- and second-order kernels for an *Eristalis* LMC. Upper left, the first-order kernels for center and surround, h_l^a and h_l^b in units mV/(Cms). Upper right, normalized kernels and sections as follows: solid line, center kernel h_l^a; dashed line, section through h_l^{aa} primary peak parallel to an axis; dotted line, diagonal of h_l^{aa}; dot-dash, section through h_l^{aaa} primary peak parallel to an axis, inverted for LMC data to allow comparison with other kernels. Lower three, the second-order kernels h_l^{aa}, h_l^{ab}, and h_l^{bb}. Perspective plots are to same scale; peak values are indicated above contour plots in units mV/(Cms)^2. Contour step size is 5% of the peak over all three kernels; zero contour is excluded; dots indicate positive regions. Note the difference in timing between first- and second-order kernels.

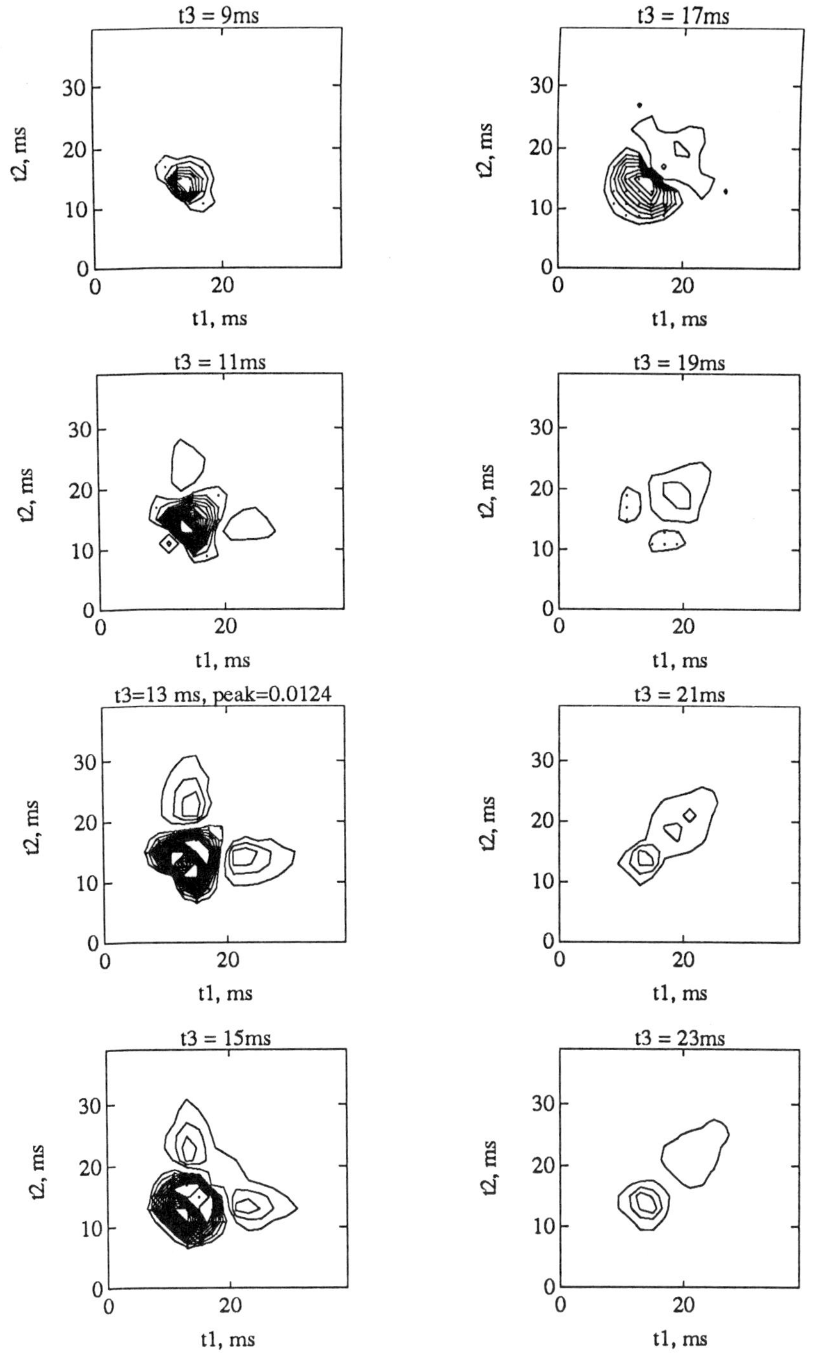

FIGURE 4. The third-order kernel h_l^{aaa} from the *Eristalis* LMC of Figure 3. Plotted are sections of the kernel at the latencies t_3 as indicated. Contour step size is 5% of the overall peak as indicated. Regions of positive values are marked with dots. The main lobe, blackened with contours, is positive. This example also has a small negative peak at $t_1 = t_2 = t_3 = 11$ ms, not generally seen.

to the axes, along a line through the primary peak. This part appears as two mirror images due to the symmetrization. These will be called the axial extensions. A positive diagonal peak appears at a latency ahead of the positive peak of the first-order kernel. Last, we will distinguish a positive diagonal extension, as latencies beyond the positive diagonal peak.

These four parts, with varying amplitudes and timings, can describe the features of most of the observed h_l^{aa} kernels. The other part sometimes seen is a negative tertiary part appearing on each side of the diagonal extension, as in the kernel shown in Figure 13. The greatest variability in the kernels lies in the relative amplitude of the two diagonal peaks. In some cases the positive diagonal peak is entirely absent, while for other cells the positive peak is larger than the negative primary peak.

The second-order cross-kernel h_l^{ab} in some cases has a negative primary part corresponding in latency to the negative primary part of h_l^{aa}; however, this part is not generally the largest in amplitude and is sometimes absent. The most consistent part is a positive part at latency t_1 corresponding to the negative primary peak of h_l^{aa}, and t_2 corresponding to the positive diagonal peak of h_l^{aa}. Thus, in this case the peak would give a depolarizing contribution at time t in response to a pulse of stimulus in the surround at $t - 19$ ms followed by a pulse on center at $t - 11$ ms. Other parts occur in some cross-kernels, the next most common being a negative peak located along the same off-diagonal as the positive peak, at a greater latency.

The second-order kernel for the surround, h_l^{bb}, is very variable between cells and generally small in amplitude relative to the other two second-order kernels. The most consistent feature is a positive peak, and sometimes extension, on the diagonal, corresponding in latency to the positive diagonal parts of h_l^{aa}. This is seen in Figure 3.

The associated third-order kernel, h_l^{aaa}, is displayed as a sequence of sections in Figure 4. These sections can be reassembled mentally to imagine the kernel on its three-dimensional domain. To facilitate visualization, the kernel can be considered as the three-dimensional analog of the typical second-order kernel, h_l^{aa}, but inverted in sign and slower in timing. The primary feature is a positive part on the diagonal coinciding in latency with the primary part of the first-order kernel. Negative axial parts extend from the primary part parallel to each of the three time axes, at latencies corresponding to the secondary part of the first-order kernel. A negative part can be seen extending up the main triple diagonal $t_1 = t_2 = t_3$, in the plots for $t_3 = 17 \ldots 23$ ms.

Figure 3 includes a one-dimensional section of the third-order kernel through the primary peak parallel to the axes. The sections are shown normalized in the upper right plots allowing comparison with other kernels and sections. The close agreement with the first-order kernel time course can be seen, contrasting with the faster time course of the second-order kernel sections. The major lobe and axial extensions of the third-order kernel are, thus, comparable in time course to the first-order kernel h_l^a raised to the third tensor power $(h_l^a)^{3\otimes}$.

V. STRUCTURAL MODELING OF NONLINEAR SYSTEMS

Having studied the empirical form of the kernels derived from experimentation, we now consider what structural models can be constructed, consisting of simpler components which when connected produce overall kernels in agreement with the observed kernels. Structural modeling has as a goal the association of the structural components with biophysical mechanisms and proceeds in concert with the consideration of data from other sources elucidating the mechanisms present. The association of components with mechanisms is, however, a further inferential step, is tentative and, in some cases, incomplete.

The structural modeling of the early visual system studied here is basically to factorize the mapping from stimulus to LMC response into a transduction step, from stimulus to receptor response, and a transmission step or steps, describing the transformation of neural image taking place between receptor and LMC. For linear systems this would amount to a deconvolution; for the nonlinear system found here it is a more involved procedure.

A. CASCADING OF MONOMIAL TIME-INVARIANT OPERATORS

Some further theoretical results are required, concerning the operation of composing, or cascading, monomial time-invariant operators. As is well-known, the cascading of two linear time-invariant operators, $[p]$ and $[q]$ produces an overall mapping which is a linear time-invariant operator, with kernel given by the convolution of the two component kernels:

$$[q][p] = [q * p]$$

The composition of a degree n operator with a degree m operator produces an operator of degree mn, just as squaring followed by cubing produces a sixth power, when monomial functions are considered. Thus, if

$$[h] = [g_n][f_m]$$

then $[h]$ is of degree mn. The form of the kernel is complicated, although it follows easily from the definitions. It can be represented as follows. Let the mn arguments of h be written as an $m \times n$ array t and let $t_{.j}$ be column j of the array. If the n arguments of g_n are written as a vector v with elements v_j, then

$$h(t) = \int g_n(v) \prod_{j=1}^{n} f_m(t_{.j} - v_j)\, dv$$

with integration taken over the n variables v_j, up to the memory length of the operator $[g_n]$.

The following special cases will be of use in this chapter. First, consider the case where a linear operator with kernel p is followed by a degree n operator with kernel g_n:

$$a \xrightarrow{[p]} x \xrightarrow{[g_n]} y$$

The function $a(t)$ represents the input, $x(t)$ represents an intermediate signal, and $y(t)$ the final output. Algebraically, we have the following

$$x(t) = \int p(u)a(t - u)\, du$$

$$y(t) = \int g_n(t)_1, \cdots, t_n)x(t - t_1) \cdots x(t - t_n)\, dt_1 \cdots dt_n$$

where integrals are taken over the relevant memory lengths. Substitution and a change of variables of integration shows that the mapping from a to y is an order n kernel operator, with the following kernel:

$$\int g_n(v_1, \cdots, v_n)p(t_1 - v_1) \cdots p(t_n - v_n)\, dv_1 \cdots dv_n$$

This integral is a convolution, and the result can, thus, be summarized in terms of compact notation as

$$[g_n][p] = [g_n * p^{n\otimes}]$$

where $*$ is an n-dimensional convolution and $p^{n\otimes}$ is the n-fold tensor power of the kernel p.

A special case of this occurs when the nonlinear operator $[g_n]$ is an *instantaneous,* or *static,* nonlinearity, whereby the value of x at each time t is raised to the power n and scaled by a coefficient:

$$y(t) = \gamma_n x(t)^n$$

In this case the overall relationship is

$$y = [\gamma_n p^{n\otimes}]a^{n\otimes}$$

If the static nonlinearity is a polynomial or Taylor series with coefficients γ_n, this can be generalized to

$$y(t) = \sum_{n=0}^{\infty} \gamma_n x(t)^n$$

$$= \sum_{n=0}^{\infty} [\gamma_n p^{n\otimes}]a^{n\otimes}(t)$$

This is a Volterra series, with the kernel of order n being $\gamma_n p^{n\otimes}$. The presence of a tensor product structure in a set of kernels, thus, suggests the possibility of fitting this type of model, which has been termed the Wiener model. The axial parts of the LMC kernels h_l^{aa} and h_l^{aaa} are suggestive of this structure.

The other case we will consider is when the nonlinear operator is followed by a linear operator $[q]$

$$x \xrightarrow{[g_n]} y \xrightarrow{[q]} z$$

producing a final output $z(t)$. Substitution of the definitions and change of variable in this case show that the resulting mapping is a time-invariant monomial operator of order n with kernel k_n, such that

$$k_n(t_1, \ldots, t_n) = \int q(u) g_n(t_1 - u, \ldots, t_n - u)\, du$$

This combination can be thought of as "filtering along each diagonal" of the kernel g_n. The term "diagonal" here means any one-dimensional section of the form $g_n(t_1 - u, \ldots, t_n - u)$ for fixed $t_1, \ldots, t_n$ and varying u. In this sense, each diagonal of k_n is the corresponding diagonal of g_n convolved with the kernel q. This structure can sometimes be suggested by visual inspection of a kernel, where mass appears to be spread diagonally across the domain, for example, along the main diagonal of the kernel h_l^{aa} of Figure 3.

B. LINEAR-NONLINEAR-LINEAR CASCADE MODELS

A subset of the space of polynomial time-invariant operators is formed by that consisting of the composition of linear dynamic operators and static nonlinear mappings. It is known that any nonlinear operator can be approximated by sums of terms comprising linear mappings followed by pointwise multiplication of signals; Wiener's original proposal was to construct a polynomial series with each kernel being approximated as the sum of tensor products of Laguerre functions (Wiener, 1958). The aim of structural modeling of this type is rather to see what is the *simplest* structure which can give a reasonable approximation to the observed kernels.

One of the simplest structures to consider is termed the sandwich system, or LNL system. It consists of a cascade of a linear system and a static nonlinear mapping, as in the Wiener model, then followed by a further linear system. This form of cascade is analyzed in Korenberg and Hunter (1986) and includes a procedure for fitting the model from input and output data.

The model has the following structure:

$$a \xrightarrow{[p]} x \xrightarrow{N} y \xrightarrow{[q]} x$$

Between the linear systems $[p]$ and $[q]$ is a static nonlinearity N. If the function N has a Taylor series with coefficients γ_n, then the mapping from a to y is a Volterra series with kernels $\gamma_n p^{n\otimes}$, as we have seen above. The subsequent linear operator $[q]$ "filters these kernels along each diagonal", to produce the resulting overall Volterra kernel of order n,

$$k_n(t_1, \cdots, t_n) = \gamma_n \int q(v)p(t_1 - v) \cdots p(t_n - v)\, dv$$

as shown by Billings and Fakhouri (1978).

To fit such a model is difficult, being akin to a deconvolution, but involving nonlinear operators. Note that the Volterra kernel k_n is a linear function of the kernel q, but involves the kernel p to degree n. Before proceeding, it is worth testing if the model is likely to provide a fit by seeing if some necessary conditions are satisfied. A particularly simple test is described in Chen et al. (1986). They arrive at the test by means of a complicated argument via the frequency domain; in fact, the result can be derived easily by considering the time domain Volterra kernels of order one and two. From the above, these are

$$k_1(t_1) = \gamma_1 \int q(v)p(t_1 - v)\, dv$$

$$k_2(t_1,t_2) = \gamma_2 \int q(v)p(t_1 - v)p(t_2 - v)\, dv$$

Consider the function obtained by taking a marginal integral of k_2, that is, integrating over one of the latency dimensions. Exchanging the order of integration, and then substituting $w = t_2 - v$, this becomes:

$$\begin{aligned}
\int k_2(t_1,t_2)\, dt_2 &= \gamma_2 \int q(v)p(t_1 - v) \left(\int p(t_2 - v)\, dt_2\right) dv \\
&= \gamma_2 \left(\int p(w)\, dw\right) \int q(v)p(t_1 - v)\, dt_1 \\
&= \gamma_2 \left(\int p(w)\, dw\right) \gamma_1^{-1} k_1(t_1)
\end{aligned}$$

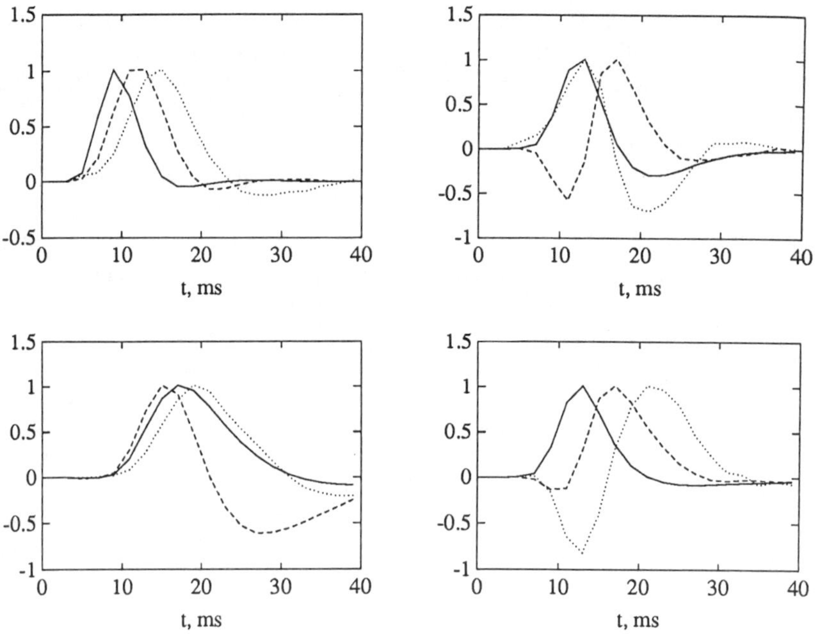

FIGURE 5. Test for *LNL* structure by comparing proportionality of time courses. Solid line, first-order kernel; dashed line marginal integral of second-order kernel; dotted line, marginal integral of third-order kernel. All curves normalized, and possibly inverted, to give unit maximum value. Upper left, *Eristalis* receptor, high intensity. Upper right, *Eristalis* LMC, high intensity. Lower left, the same *Eristalis* LMC at low intensity. Lower right, *Calliphora* LMC, low intensity.

In short, the marginal integral of the second-order kernel is proportional, as a function of latency, to the first-order kernel (Chen et al., 1986). This test will be robust unless the operator $[p]$ is high-pass, making the integral of p small, or the coefficient γ_1 is small.

This result immediately generalizes to degree n, whereby the marginal integral of the kernel k_n, integrated over $n - 1$ dimensions, must be proportional to the first-order kernel k_1.

The test can be applied to Wiener kernels due to the following very useful result: for a system with *LNL* structure, the Wiener kernel of each order is proportional to the corresponding Volterra kernel (Korenberg and Hunter, 1986). Wiener kernels must, thus, also satisfy the proportionality constraints to represent an *LNL* system.

This result does not hold precisely for Wiener kernels corresponding to binary white-noise, where the equations are complicated by additional terms due to the difference in correlation properties relative to Gaussian white-noise. This difference is small, however, and tends to zero as the time discretization becomes finer. Note that the marginal integrals of the higher-order kernels involve all elements of the kernels, so that the interpolated diagonal elements receive little weight in the result. The test is, thus, more robust than considering kernel diagonals alone.

To check for proportionality, Figure 5 shows the first-order kernels and marginal integrals of second- and third-order kernels for a variety of experiments in this study. The curves are normalized, and inverted where necessary, to have unit maximum value. It is clear that proportionality does not hold, with the time courses of the components of different degree being distinctly different.

Chen et al. (1986) suggest illustrating the proportionality of two time courses graphically by plotting their ratio as a function of time. This will give unstable results where the

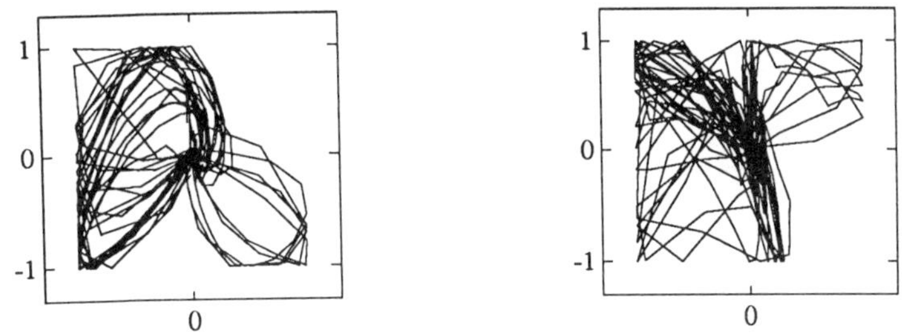

FIGURE 6. Trajectories testing the *LNL* model for the database of higher-order kernels. Left, for each experiment, the marginal integral of the second-order kernel plotted against the first-order kernel, as a trajectory over time. Right, the marginal integrals of the third-order kernels plotted against the first-order kernels. All data are normalized to unit amplitude.

denominator is small in amplitude. As an alternative, the values of the two functions can be plotted against each other, defining a trajectory on the plane. If they are proportional, the curve will be a straight line through the origin. The entire database of higher-order kernels is plotted in this way in Figure 6, showing clearly the lack of proportionality.

It can, thus, be concluded that neither the *LNL* model, nor any submodel, such as the *LN* (Wiener) or *NL* (Hassenstein) models, is likely to give a useful fit to these data. It is necessary to look at more complex structural models, whereby the kernels of different orders may follow different time courses, and possibly where parallel pathways may contribute different components to the same kernel.

C. *L* + *LNL* MODELS

The next-simplest model to consider is to add a linear operator in parallel with the nonlinear cascade. The overall first-order kernel will then be the sum of a component passing through the cascade, $\gamma_1 p * q$ in the notation above, added to the parallel linear operator. We will analyse the second-order kernels for receptors and LMCs for possible *LNL* structure.

Figure 2 shows a typical receptor kernel, h_r^{aa}. Inspection of the kernel h_r^{aa} suggests that it is close to having tensor product form, having cross-sections of constant profile. The kernel is symmetric, but is negative along the main diagonal. We can attempt to fit it with the negative of a tensor square, $-p^{2\otimes}$, for some p a function of time. The function p is then a tensor square root of $-h_r^{aa}$.

This is a nonlinear model, and the Marquardt-Levenburg procedure (Press et al., 1988) was used to produce the least-squares fit. Figure 7 shows the resulting function p along with the kernel h_r^a. The second-order kernel could, thus, be produced by a structural model of the Wiener form

$$a \xrightarrow{[p]} x \xrightarrow{N} r$$

where the nonlinearity N is $r(t) = -x(t)^2$. For some receptor data the presence of a final linear filter $[q]$ is necessary to produce a very good fit, giving a small amount of "smoothing along the diagonal".

We now turn to the LMC kernel h_l^{aa}, to look for possible *LNL* structure. Inspection of a kernel such as that in Figure 3 suggests the type of kernels p and q which would be required to provide a fit. The presence of parts extending parallel to the axes is suggestive of tensor product structure, hence, of significant filtering before the nonlinearity. The section taken along the length of this part is biphasic, implying that the initial first-order kernel p must be biphasic. The spread of mass up the diagonal indicates significant filtering after the

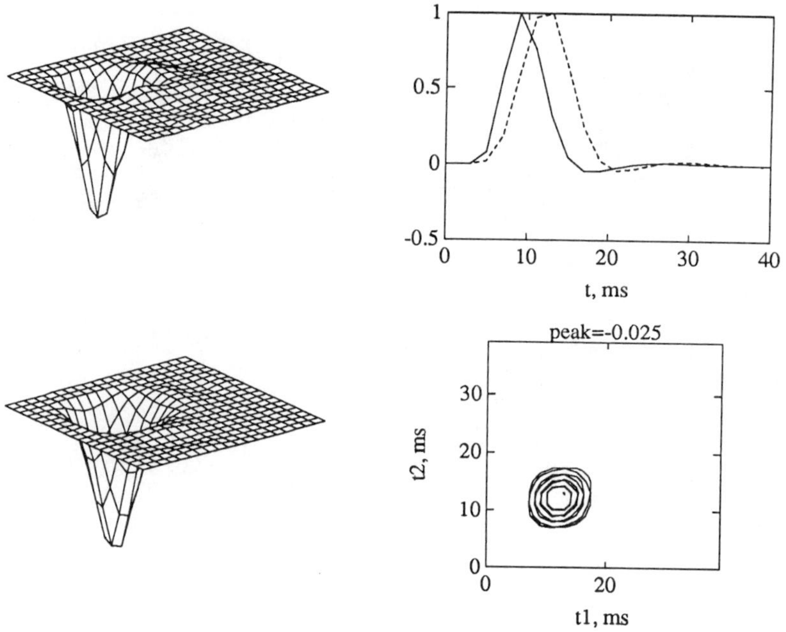

FIGURE 7. Fit of a receptor kernel h_r^{aa} by a tensor square $-p^{2\otimes}$. Upper left, the kernel h_r^{aa} from Figure 2. Upper right, the kernel h_r^{a} (solid line) and the tensor square root p of $-h_r^{aa}$ (dashed line), normalized in amplitude. Lower left, the fitted kernel, $-p^{2\otimes}$. Lower right, observed and fitted kernel, 20% contours superimposed.

nonlinearity, and to generate the large positive peak on the diagonal a biphasic final kernel q will be required.

On discrete time, the kernel $h_l^{aa}(t_1,t_2)$ is to be fitted by a function f of the form

$$f(t_1,t_2) = \gamma_2 \sum_{u=0}^{m_q-1} q(u)p(t_1 - u)p(t_2 - u)$$

where m_q is the memory length of the kernel q, and t_1,t_2 range over the domain of the kernel h_l^{aa}. This model is linear in the values of q, but nonlinear in p. It also contains two forms of redundancy. The amplitudes of p and q and the coefficient γ_2 are confounded; hence, we set γ_2 and the maximum value of the kernel p equal to one. There is also a redundancy in phase; a shift of p on the time axis can be canceled by an equal shift of q in the other direction. We will try to find p and q, giving a good fit and both being plausible physiological responses.

The fit was performed in three ways. First, parametric forms were taken for the kernels p and q. To generate the biphasic waveforms required, a quite general form is needed. Extensive modeling of first-order kernels suggested that a wide range of biphasic waveforms can be fitted by the linear combination of two gamma density functions. The gamma density function with parameters α and λ is given by

$$\frac{\alpha^\lambda}{\Gamma(\lambda)} t^{\lambda-1} e^{-\alpha t}$$

This is more suitable for fitting when reparameterized in terms of the time-to-peak, $t_p = (\lambda - 1)/\alpha$, in place of α, and when normalized to unit height, giving the form

$$(t/t_p)^{\lambda-1} e^{-(\lambda-1)(t/t_p-1)}$$

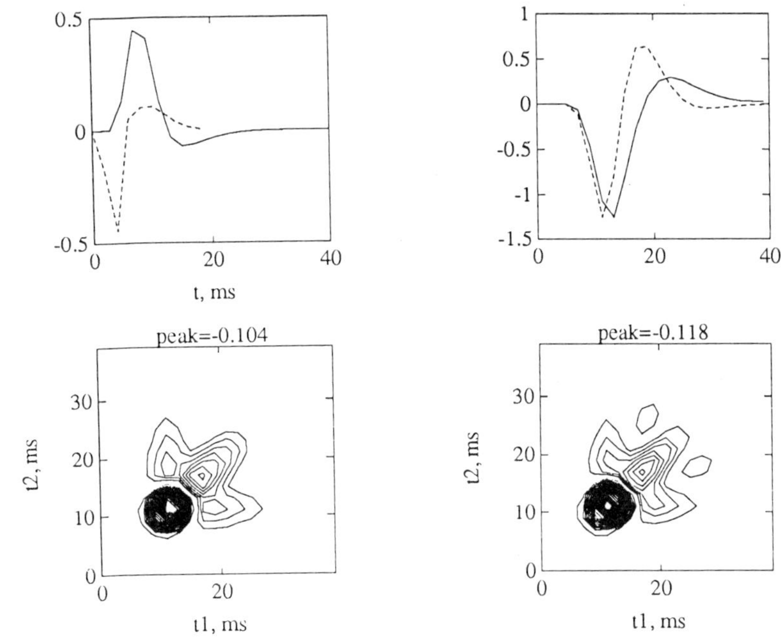

FIGURE 8. Fit of *LNL* model to the LMC kernel h_l^{aa} from Figure 3. The two first-order kernels are generated as linear combinations of two gamma density functions, as described in text. Upper left, the fitted kernels p (solid line) and q (dashed line). The kernel amplitudes have been balanced to be equal. Upper right, the kernel h_l^{a} in mV/(Cms) and the convolution $p * q$ arbitrarily scaled for comparison. The linear pathway in parallel with the *LNL* cascade adjusts for the discrepancy between these two waveforms. Lower left, the measured kernel h_l^{aa}, peak as indicated, contour step size is 5% of peak value, zero contour excluded. Lower right, the fitted second-order kernel, with same contour levels.

Setting p and q each equal to a linear combination of two such functions gives a total of 12 parameters, less one to fix the amplitude of p. Since the aim is to see if the kernel can be fitted by an *LNL* structure, we do not mind using a large number of parameters, as long as the resulting kernels are plausible. First derivatives can be calculated allowing the Marquardt-Levenberg fitting procedure to be used. The resulting fit to an LMC kernel h_l^{aa} is shown in Figure 8.

In subsequent stages of fitting, the kernel q was allowed to take on any values, over a domain from 0 to 20 ms. Last, the kernel p was also allowed to take on any values over its domain. In the case illustrated, this led to very little change in fit relative to the parametric fit.

The *LNL* fit is reasonable in this case; however, not all measured kernels could be satisfactorily fitted. The basic problem is the large variability in the relative size of the negative and positive peaks along the diagonal; in many cases the positive peak is larger in amplitude, which cannot be well fitted. A later section considers a model in which the two peaks are generated by independent mechanisms, which can account for this variation in amplitude.

Although it is of some interest to see if the LMC kernel sets can be reduced to this simple structural form, the approach is limited in that it treats the LMC kernel in isolation. The LMC is postsynaptic to receptors, and it is of more interest to relate the LMC kernels to the recorded receptor kernel sets.

VI. RECEPTOR-LMC TRANSMISSION

The receptor-LMC synapse is the site of the first transformation of the neural image in the fly visual system. The transformation recodes the visual information into a form optimized to the requirements of the system, removing low-frequency information, amplifying the signal, and averaging the signals from the six convergent receptors within each cartridge (reviewed in Laughlin, 1989). The principal problem in characterizing this transformation is the great difficulty of simultaneous recording from a receptor and a postsynaptic LMC; hence, the properties of transmission must be inferred from separate recordings made from the two cell classes. The remaining sections of this chapter show what inferences can be drawn from the empirically derived kernels, by examining the transmission and generation of linear and nonlinear components and developing model systems consisting of networks of polynomial time-invariant operators.

A. THE PRIMARY TRANSMISSION PATHWAY

The principal transmission pathway from receptor to LMC involves the neurotransmitter histamine (Hardie, 1987, 1989), producing an increase in chloride conductance in the LMC membrane (Zettler and Straka, 1987), thus leading to the inversion of the signal. The signal is amplified at this synapse, with a gain of around 6 (Laughlin et al. 1987). A synapse acting at this high gain is likely to introduce considerable nonlinearity of response, due both to nonlinearity between presynaptic voltage and the rate of transmitter release, and to self-shunting of the response in the postsynaptic cell.

The rate of transmitter release is dependent on the presynaptic membrane potential; the difference between intracellular and extracellular voltage. Each lamina cartridge is isolated by surrounding glial cells, suggesting that fluctuations in the extracellular voltage can play a significant role in signal processing in this system. The effect of current flow from neighboring cartridges is a possible mechanism for generating lateral inhibition (Shaw, 1975; Laughlin and Osorio, 1989). Extracellular voltage may also generate the temporal high-pass filtering present in the system, if it follows a time course delayed relative to the intracellular signal. The receptor terminal membrane voltage will be represented by $m(t)$, with the kernel h_m^a relating stimulus to m.

The postsynaptic response of the LMC to the transmitter appears to be rapid. Impulse response data suggests a time constant of 0.5 to 1 ms, with a delay of 0 to 0.5 ms at some point in the transmission (Laughlin et al. 1987). Estimates of the time constant of the LMC membrane are similar, at 0.5 to 1 ms (Laughlin and Osorio, 1989). Given the high bandwidth of the transmission from receptor membrane potential to LMC response, relative to the transduction process, we will simplify the subsequent modeling by ignoring the temporal dynamics of the map $m \to l$ and assuming that it is a static nonlinearity. Transmission through the histamine pathway is thus modeled as a Wiener system

$$r \xrightarrow{[h_m^r]} m \xrightarrow{N} l$$

whereby the linear high-pass filter $[h_m^r]$ is followed by a static nonlinearity N. In some models a pure delay of 1 ms is included to represent the synaptic transmission dynamics.

The synaptic transfer was modeled as a static nonlinear mapping by Laughlin et al. (1987) in order to fit the characteristic curve relating the amplitude of step responses measured intracellularly in receptors and LMCs. The model is similar to the model of Falk and Fatt (1972) for vertebrate photoreceptor synapses.

The postsynaptic conductance, $g(t)$, of the channels activated by the transmitter is assumed to increase exponentially with the receptor membrane voltage m:

$$g = a \exp(bm)$$

The postsynaptic voltage l is related hyperbolically to this conductance, due to the finite reversal potential, l_{max}, for the response, which is negative for chloride. If l is defined to be zero when g is zero, the relationship is of the form

$$l = l_{max} \frac{kg}{kg + 1}$$

Combining these two gives the relation

$$l = l_{max} \frac{ka \exp(bm)}{ka \exp(bm) + 1}$$

as in Laughlin et al. (1987).

If m is defined so that it is zero when $kg = 1$, this simplifies further to

$$l = \frac{l_{max}}{1 + \exp(-bm)}$$

involving only the reversal potential l_{max} and the sensitivity b. An example of the curve can be found in Figure 9 (lower).

If the presynaptic voltage fluctuates about some set level m_0, so that $m = m_0 + \Delta m$, the postsynaptic voltage is then

$$l = \frac{l_{max}}{1 + \exp(-b(m_o + \Delta m))}$$

By differentiation of this expression the coefficients of the Taylor series can be derived, as a function of the set point m_0. These are illustrated in Figure 9 (upper right).

From the Taylor series coefficients and initial and final first-order kernels, the Volterra kernels of the *LNL* cascade can be calculated, as shown above. Truncated Taylor and Volterra series are the appropriate means of approximation in the limit for small fluctuations of signal amplitude. The observed Wiener kernels, however, relate to signals with significant amplitude. If the receptor is assumed to be a linear transducer integrating the stimulus over an interval covering several frames, then the receptor voltage will have an amplitude distribution close to Gaussian, by the central limit theorem. Histograms of recorded receptor voltages are in agreement. The static nonlinear mapping N is then best approximated by a truncated Hermite series, giving the least-squares-error approximation. Combining the coefficients of this Hermite series with the initial first-order kernel produces the overall Wiener kernels for the *LN* cascade model. Thus, if $[h_m^a]$ represents the linear mapping from stimulus a to the receptor membrane potential m, transmission through the nonlinear synapse produces a component of the order n Wiener kernel of the mapping $a \rightarrow l$ given by

$$\gamma_n (h_m^a)^{n\otimes}$$

where γ_n is the Hermite coefficient of degree n.

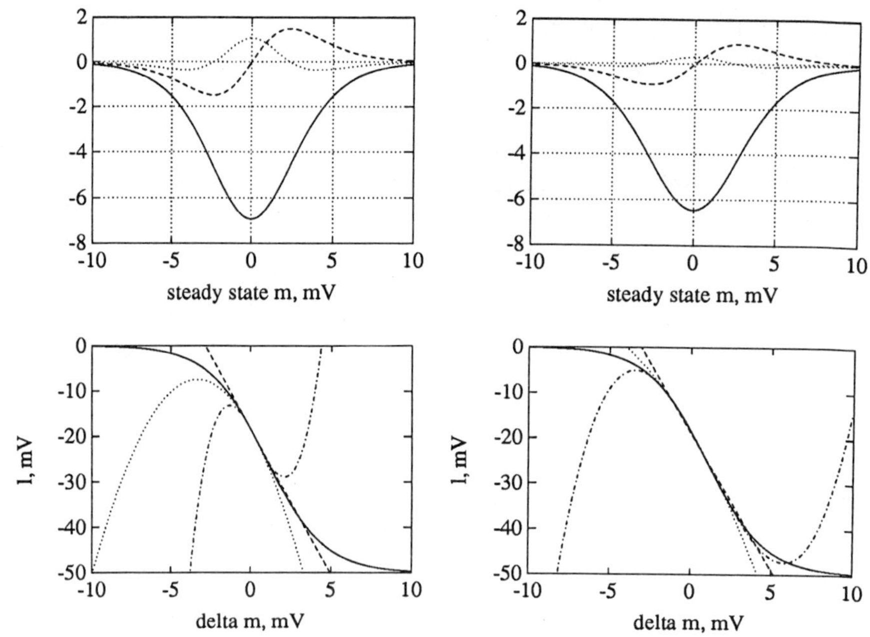

FIGURE 9. Series approximations to a model characteristic curve for the receptor-LMC synapse given in the text. The parameters of the curve are $l_m = -50$ mV, $b = 0.55$ mV^{-1}. Left, Taylor series. Right, Hermite series when receptor membrane voltage fluctuation has standard deviation 1 mV. Upper, the first-, second-, and third-degree coefficients of the series as a function of steady state value m_0, (solid, dashed, and dotted lines, respectively). Lower, model characteristic curve (solid line) and first-, second-, and third-degree approximations for fluctuations about a steady state value $m_0 = -1$ mV (dashed, dotted, and dot-dash, respectively).

The values of the Hermite coefficients depend on the set point, m_0, about which the membrane potential fluctuates. An illustration of this dependence is given in Figure 9 (upper right). The primary part of the observed LMC kernel h_l^a is negative. The kernel h_l^{aa} also has a negative primary part, while the third-order kernel h_l^{aaa} has a positive primary part. This pattern of signs is in correspondence with the Hermite coefficients for values of m_0 in an interval immediately to the left of zero, suggesting that the receptor-LMC synapse is operating around a point slightly less than halfway down its characteristic curve. Note that near $m_0 = 0$ the quadratic coefficient γ_2 is changing rapidly with m_0, and so variability in the size of the quadratic component is to be expected. Figure 9 (lower) shows the fit of the Taylor and Hermite series to the example characteristic curve for a steady-state value in this interval.

We can test the adequacy of this model in the following way. We assume the estimate $\gamma_1 = -6$ (dimensionless). For a given set of LMC kernels, we then have

$$h_m^a = 1/\gamma_1 h_l^a$$

The higher-order kernels are then fitted by linear regression, according to the models

$$h_l^{aa} = \gamma_2 (h_m^a)^{2\otimes}$$

$$h_l^{aaa} = \gamma_3 (h_m^a)^{3\otimes}$$

Figure 10 shows a typical example of the fit obtained.

The primary part and axial extensions of the third-order kernel are well fitted by the model, with a value of γ_3 which is compatible with the values predicted by the Hermite

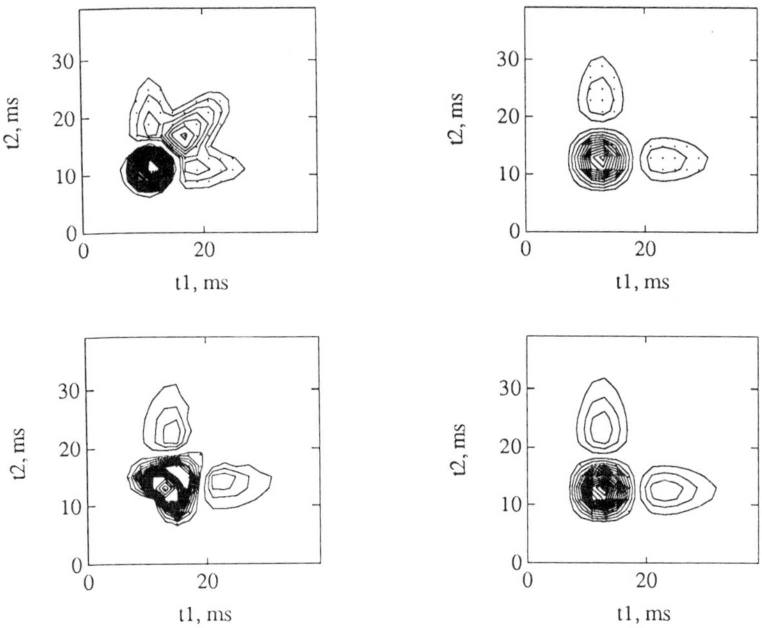

FIGURE 10. Fit of the *LN* cascade model of the primary transmission pathway to the kernels of the *Eristalis* LMC of Figure 3. Upper left, observed kernel h_l^{aa}. Contour step size is 5% of peak as indicated. Upper right, fit of form $\gamma_2(h_m^a)^{2\otimes}$ where h_m^a is estimated from the first-order kernel and $\gamma_2 = -1.5\text{mV}^{-1}$, from regression. Contour levels as in upper left. Lower left, axial cross-section through peak point of the third-order kernel h_l^{aaa}. Contour step size is 5% of peak value as indicated. Lower right, fit of form $\gamma_3(h_m^a)^{3\otimes}$, where $\gamma_3 = 1.1\text{mV}^{-2}$ from regression. Note fit to axial parts of third-order kernel is reasonable while the primary and axial parts of the quadratic are not fitted.

coefficients calculated in Figure 9. As expected, the second-order kernel deviates from the fit both due to the presence of the diagonal parts and due to the faster time course of the primary and axial secondary parts. This is a general result, and in all cases we see that the axial section through the peak of the third-order kernel is in good agreement with the first-order kernel, while the axial section of the second-order kernel is consistently different in waveform.

B. DYNAMICS OF RECEPTOR-LMC TRANSMISSION

This section will present a simple parametric model for the receptor-LMC first-order kernel h_l^r for use in subsequent modeling.

Laughlin et al. (1987) fitted this transmission by considering impulse response data and fitting a transmission impulse response of double exponential form: a rapid hyperpolarizing component with a brief delay and a slower depolarizing component with an additional delay. Of the parameters involved, the delays fitted were close to zero, while the time constant of the hyperpolarizing process was 0.5 ms to 1 ms.

With this in mind the model is simplified in the following way. A considerable variation in the time scale of the receptor kernel has been observed; hence, the time to peak t_p of the receptor kernel model will be fitted, while the synaptic delay will be set to a fixed value of 1 ms, and the time constant of the LMC hyperpolarizing response will be set to zero. These three parameters are largely confounded; thus, the precise optimum values for the fit are not of interest.

No additional delay will be included in the antagonistic component. This component is assumed to be due to fluctuation in the extracellular voltage within the lamina cartridge, denoted by $e(t)$. If e is dependent on r according to first-order linear dynamics, then the kernel h_e^r is a decaying exponential starting at $t = 0$. There seems to be no established notation for this useful function, so we define the function nexp as the negative exponential starting at zero:

$$\text{nexp}(x) = \begin{cases} e^{-x} & \text{if } x > 0 \\ 1/2 & \text{if } x = 0 \\ 0 & \text{if } x < 0 \end{cases}$$

In the form $h(t) = a \, \text{nexp}(t/\tau)$, this is then the required kernel. We note parenthetically that the value $^1/_2$ at $x = 0$ ensures that the kernel $h(t)$ and the function

$$H(\omega) = \frac{a\tau}{1 + i\omega\tau}$$

are related as a Fourier transform pair. This is relevant in the discrete-time context, to make frequency domain calculations agree with time-domain results.

The model kernel for the mapping $r \rightarrow m$ is, thus,

$$h_m^r(t) = \delta(t) - a_e \, \text{nexp}(t/\tau_e)$$

where the dirac δ function produces the identity operator, $[\delta]$, and the kernel for the overall transmission $r \rightarrow l$ is

$$h_l^r(t) = \gamma_1 \left(\delta(t - d_l) - a_e \, \text{nexp}((t - d_l)/\tau_e)\right)$$

where the pure delay is $d_l = 1$ ms.

As it was not possible to record simultaneously from a presynaptic receptor, some model of the receptor kernel is required. It was found that all receptor kernels h_r^a could be fitted by a two-parameter family of functions, with variable amplitude, a_r, and time-to-peak, t_p (James, 1990 and in preparation). This parametric function is, thus, used for h_r^a, and the two parameters are estimated along with the parameters defined above. The resulting estimates are then checked to ensure that they lie within the observed ranges for the empirical receptor kernels.

Convolution with functions involving nexp is efficiently done using recursive digital filtering, thus avoiding the need to transform to the frequency domain for simulation. For each LMC kernel h_l^a the Nelder-Mead algorithm (Nelder and Mead, 1965) varied the two nonlinear parameters t_p and τ_e. The receptor kernel and convolution are calculated with a temporal discretization four times finer than the data, so $\Delta t = .5\ ms$. Decimation by four and linear regression then give estimates of the amplitudes a_r and a_e. The coefficient of synaptic transmission is again taken as $\gamma_1 = -6$. The Nelder-Mead algorithm is convenient to implement for a small number of nonlinear parameters, as it does not require derivatives of the model function. Regression can be used at each step to estimate the linear parameters.

The model fits all LMC kernel data with an RMS error of less than 4% of peak value. Figure 11 shows examples of the fit, with the constituent components, at the high and the low mean intensity.

C. TRANSMISSION OF RECEPTOR NONLINEARITY

In factoring the LMC response into transduction and transmission, we must consider not only the nonlinearities introduced in the transmission, but also the nonlinear components

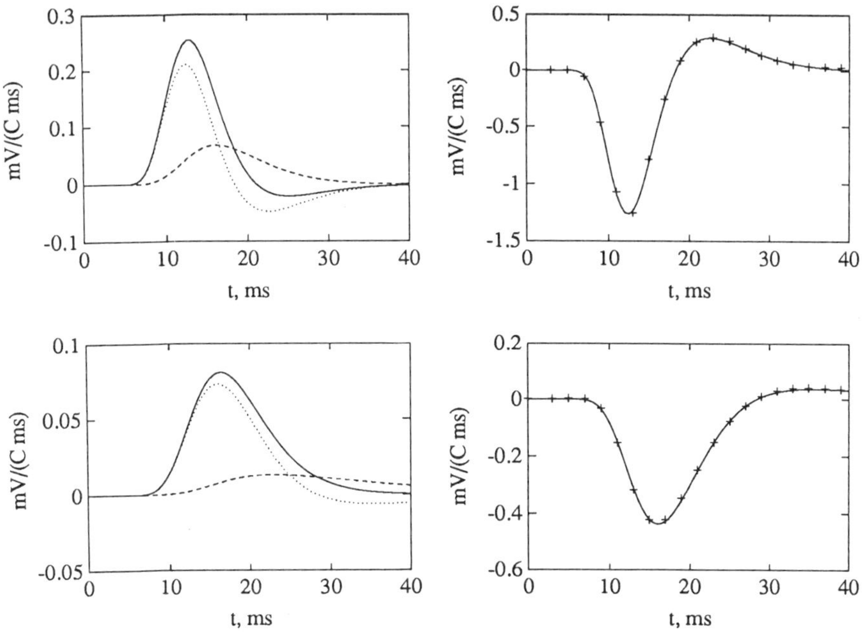

FIGURE 11. Parametric fit for first-order transmission kernel h_l^r. Left, solid line is fitted receptor kernel; dashed line is fitted kernel describing extracellular response, h_e^a; dotted line is fitted kernel describing membrane potential, $h_m^a = h_r^a - h_e^a$. Right, plus signs are observed LMC kernel, h_l^a; solid line is fitted kernel. Upper, fit to the kernels of Figure 3: $t_p = 11.8$ ms, $\tau_e = 6.9$ ms, $a_r = 0.24$ mV/(Cms), $a_e = 0.068$ ms^{-1}. Lower, fit to kernels from the same cell at the low mean intensity of illumination: $t_p = 15.4$ ms, $\tau_e = 15$ ms, $a_r = 0.081$ mV/(Cms), $a_e = 0.25$ ms^{-1}.

in transduction, which are transmitted to the LMC. Although the receptor nonlinearities are smaller than those of the LMC response, the kernel h_r^{aa} is large enough to make a significant contribution to the LMC kernel h_l^{aa}.

The composition of polynomial operators can be expanded in the same way as a composition of polynomial functions. It is probably helpful to consider this simpler case first. Assume that the mappings from stimulus a to receptor voltage r to LMC voltage l were static functions, equal to the following polynomials, omitting terms of zero degree and noting only terms to degree three:

$$r = \alpha_1 a + \alpha_2 a^2 + \alpha_3 a^3 + \cdots$$

$$l = \beta_1 r + \beta_2 r^2 + \beta_3 r^3 + \cdots$$

Substituting and noting terms to degree three, we have

$$\begin{aligned} l = \beta_1\alpha_1 a + \beta_1\alpha_2 a^2 &+ \beta_1\alpha_3 a^3 && + \cdots \\ + \beta_2(\alpha_1 a)^2 &+ 2\beta_2(\alpha_1 a)(\alpha_2 a^2) && + \cdots \\ &+ \beta_3(\alpha_1 a)^3 && + \cdots \end{aligned}$$

The formulation for polynomial time-invariant operators is analogous and can be readily written down using the notational conventions developed. We will consider cascaded Volterra systems for simplicity, omitting the lower-degree terms within each Wiener functional. If stimulus, receptor, and LMC signals have the relationships

$$r = [h_r^a]a + [h_r^{aa}]a^{2\otimes} + [h_r^{aaa}]a^{3\otimes} + \cdots$$

$$l = [h_l^r]r + [h_l^{rr}]r^{2\otimes} + [h_l^{rrr}]r^{3\otimes} + \cdots$$

then the overall mapping from stimulus to LMC response is

$$\begin{aligned} l = {} & [h_l^r][h_r^a]a + [h_l^r][h_r^{aa}]a^{2\otimes} + [h_l^r][h_r^{aaa}]a^{3\otimes} + \cdots \\ & + [h_l^{rr}]([h_r^a]a)^{2\otimes} + 2[h_l^{rr}]([h_r^a]a)\otimes([h_r^{aa}]a^{2\otimes}) + \cdots \\ & + [h_l^{rrr}]([h_r^a]a)^{3\otimes} + \cdots \end{aligned}$$

The leading terms of these lines have already been discussed, as the transmission of the linear receptor component through the nonlinear synaptic mechanisms.

The second term of the first line represents the linear transmission of the receptor quadratic component. For a given LMC kernel set this component can be fitted as follows. The first-order kernel is decomposed into transduction and transmission steps as described in the previous section. From the estimated peak time and amplitude of the fitted transduction kernel h_r^a, an estimate of the receptor kernel h_r^{aa} is made. This is then "filtered along each diagonal" by the linear transmission operator $[h_l^r]$, which can be done efficiently by recursive digital filtering techniques. The observed kernel h_l^{aa} is regressed on the resulting component along with the quadratic transmission component $(h_m^{aa})^{2\otimes}$. The regression coefficients produce estimates of the coefficient γ_2 and an adjustment to the amplitude of h_r^{aa}. Figure 12 shows an example of these fits.

The kernel h_r^{aa} has a large negative part, with peak around 3 ms after the linear peak. When transmitted this produces a positive diagonal part with a smaller negative part at longer latencies. The amplitude of receptor second-order kernels varied from 0.01 to 0.04 times the amplitude of the first-order kernels, in their respective units. The ratio produced by this fitting procedure lies in this range. This component partly accounts for the positive diagonal peak seen in LMC second-order kernels; however, we see that the kernel h_l^{aa} is still not well fitted.

The other two terms, of degree three, representing transmission of receptor nonlinearities were found to be of insignificant amplitude and are excluded from further modeling.

It might be of theoretical interest to note the second term on line two of the expansion, indicating that when the sum of a linear term and a quadratic term undergoes quadratic transmission, a cubic term is produced, among others. This is, of course, obvious when polynomial functions are considered, illustrating the utility of the analogy.

D. PARALLEL *LNL* MODELS

The structural models considered so far have failed adequately to account for the LMC kernel h_l^{aa}. If the system is to be fitted by a structural model involving *LNL* cascades, then at least one additional pathway must be added in parallel to those already considered.

A parallel *LNL*, or parallel sandwich model, has the form

$$[q]N[p] + [q']N'[p']$$

where p,p',q,q' are first-order kernels and N,N' are static nonlinear mappings. The first

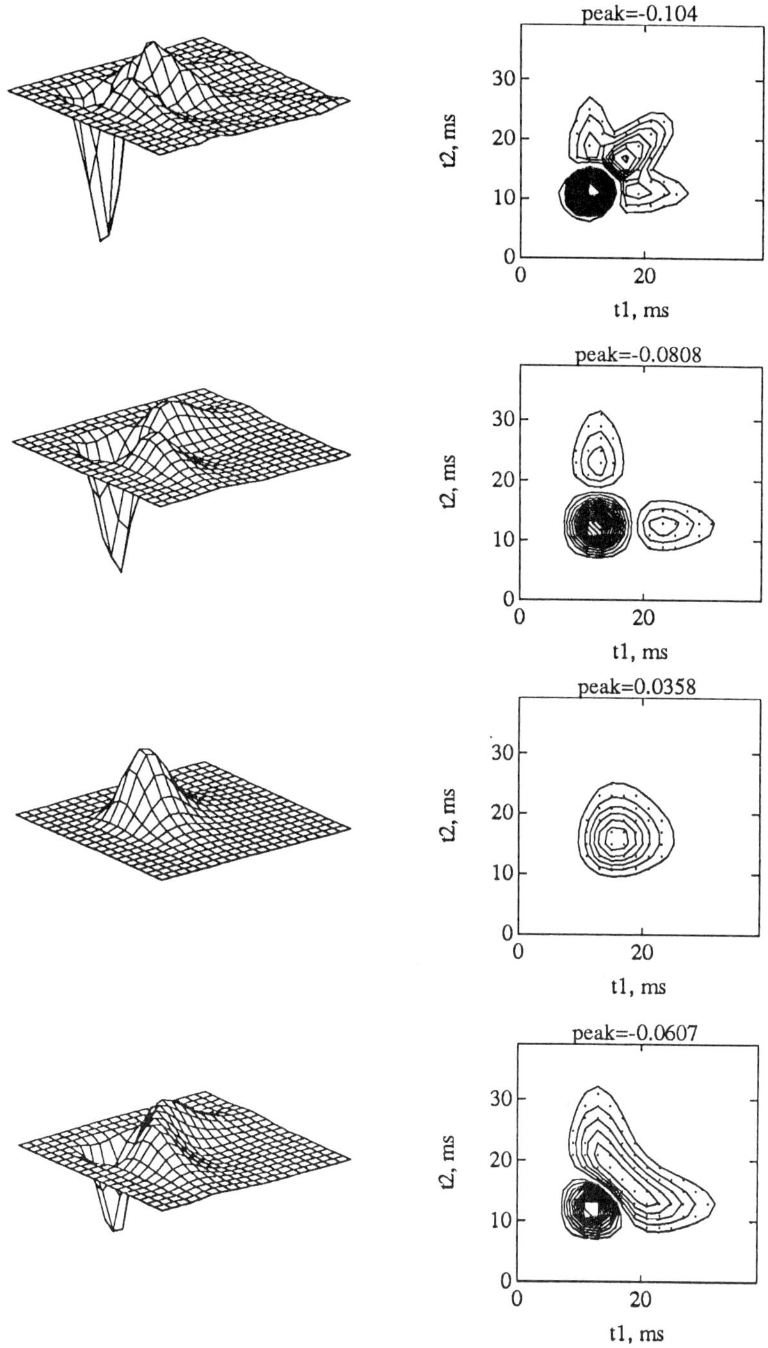

FIGURE 12. The effect of the receptor kernel h_r^{aa} transmitted to the LMC. Upper, the kernel h_l^{aa} from Figure 3. Second row, the estimated component from quadratic transmission, $\gamma_2(h_m^{aa})^{2\otimes}$. The coefficient is estimated as $\gamma_2 = -1.7\ \mathrm{mV}^{-1}$. Third row, the estimated component due to linear transmission of the receptor kernel h_r^{aa}. Regression produces an estimate of -0.025 for the ratio of h_r^{aa} peak to h_r^a peak, which is within the observed range. Lower, the total fitted kernel, which is significantly different from the data, h_l^{aa}.

point to note is that at least two of the pairs $p,p'q,q'$; and N,N' must consist of distinct elements, otherwise the model can be factored into a single *LNL* cascade. For example, if the corresponding first-order kernels are equal, then

$$[q]N[p] + [q]N'[p] = [q](N + N')[p]$$

and a single cascade results.

Consideration of the range of observed kernels h_l^{aa} suggests that there are two components present, with large variation in their relative amplitudes between cells. One component produces the negative primary and positive axial extensions. This is a good example of the kernel produced by an *LN* model, in which a linear operator with a biphasic kernel is followed by a nonlinearity with a negative second-degree coefficient. We have seen such a component in Figure 10, but to fit the observed kernel the biphasic kernel must be faster in time course.

The other component produces the positive diagonal peak and extension. This suggests a significant final linear filter, which "filters each diagonal", thus, spreading the diagonal peak along the main diagonal to produce the extension. A delay is also incurred somewhere along the pathway, but due to commutativity this may be before or after the nonlinearity. The second-degree coefficient for this component is positive. If the first-degree coefficient is negative, then this component reinforces the primary component of the first-order kernel. The delay and smoothing mean that this component adds to the first-order kernel at longer latencies. We, thus, have the possibility that the histamine pathway generates a fast component of response, following the time course of the axial section of the second-order kernel, while the second pathway generates a slightly slower response, adding to the histamine component in the first-order kernel to produce the broader primary part while opposing the histamine component in the second-order kernel, producing the positive diagonal peak and extension.

If the second pathway is driven by the receptor membrane potential m, which we have modeled with a biphasic response to stimulation, then this component will, in addition, generate negative parts in the second-order kernel in regions extended axially from the positive peak. These parts are, in fact, observed in several cases, such as in the case shown in Figure 13.

A mechanism which would produce a component like this is suggested by the results of Laughlin and Osorio (1989) and Weckström et al. (1989). In these studies current-clamp techniques were used to analyze the response to step stimulation. The first paper finds that polarizing conductances with positive reversal potentials contribute to large-amplitude responses to negative contrasts, adding to the reduction in chloride conductance driven by the histamine pathway. As this effect contributes to depolarization when the stimulus is negative, and seems to be less important when the stimulus is positive, the characteristic curve for the transmission is positive when the stimulus is negative and small when the stimulus is positive. The Hermite series for such a curve has a negative linear coefficient and positive quadratic coefficient. Without knowing more about the shape of the curve it is not possible to predict the sign of the cubic coefficient. The depolarizing conductance pathway also appears to operate with slightly longer latency than the histamine pathway. The reported depolarizing pathway, thus, has properties in agreement with the proposed second *LNL* cascade.

Association of the second *LNL* cascade with the reported depolarizing input is hypothetical, and the neural components generating the depolarizing input are at present also uncertain. Let us in this section simply propose a hypothetical quantity z which is driven by the membrane potential m and which effects the LMC voltage l.

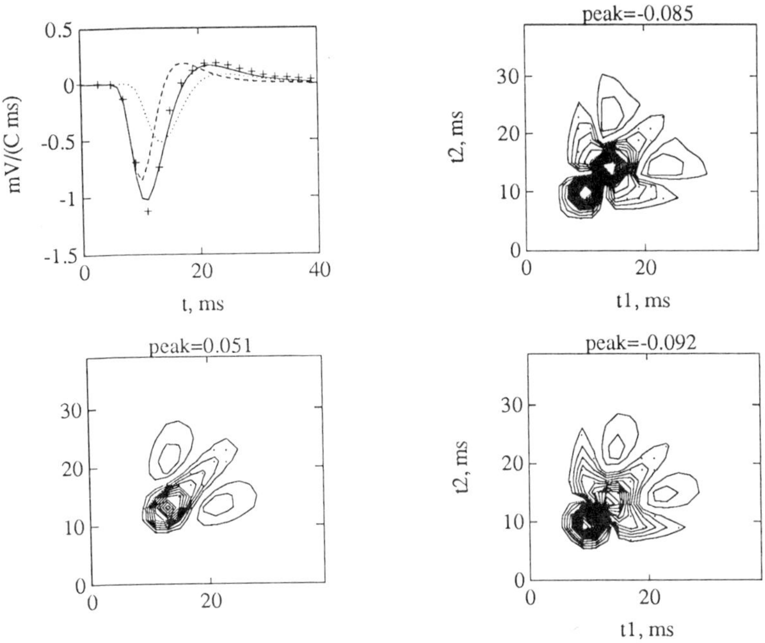

FIGURE 13. Fit of parallel *LNL* model, as described in text. Data are from an *Eristalis* LMC. Contour step size is 5% of peak value of kernel h_l^{aa} (upper right). Upper left, fit to first-order kernel: plus signs, observed kernel *hal*; solid line, total fitted kernel; dashed line, component transmitted through histamine pathway; dot-dash, component transmitted through second pathway, $m \rightarrow z \rightarrow l$. Upper right, observed kernel h_l^{aa}. Lower left, component transmitted through second pathway. The final linear filter $[h_l^z]$ creates the diagonal extension. Lower right, total fit to h_l^{aa}. The nonlinear parameters used are t_p = 8.0 ms, a_e = 0.061 ms^{-1}, τ_e = 7.5 ms, τ_z = 3.6 ms, d_z = 2.6 ms.

The parallel *LNL* model is represented by the following mapping network:

$$
\begin{array}{ccc}
a & & \\
\downarrow & & \\
r & & \\
\downarrow & & \\
m & & \\
\downarrow\downarrow & \searrow\searrow & \\
l & \leftarrow & z
\end{array}
$$

It was fitted to sets of LMC kernels, with the primary transmission pathway being modeled as before: the map a $\rightarrow$ r has as first-order kernel a double log-normal curve with variable parameter t_p and amplitude, and the receptor-LMC primary transmission has a high-pass filter followed by a static nonlinearity with linear coefficient $\gamma_1 = -6$.

The second pathway is also driven by the receptor membrane potential, *m*. It then has a static nonlinear mapping N_z, and a final linear operator $[h_l^z]$.

$$m \xrightarrow{N_z} z \xrightarrow{[h_l^z]} l$$

We let the kernel h_l^z embody a delay followed by an exponentially decaying response:

$$h_l^z(t) = a_z \,\mathrm{nexp}((t - d_z)/\tau_z)$$

A component due to transmission of the receptor quadratic nonlinearity is also included.

First- and second-order kernels were fitted simultaneously using the Nelder-Mead algorithm to vary five nonlinear parameters and linear regression to fit amplitudes. Figure 13 shows the fit that can be obtained.

Consideration of the second-order kernel has, thus, suggested a second pathway of transmission from receptor to LMC, which corresponds in many properties with a pathway suggested by current-clamp results. This raises an interesting parallel; in neural systems there are often multiple mechanisms contributing to an observed behavior, and one aim of neurophysiological investigation is to distinguish components which are normally confounded. Just as the use of current and voltage clamp techniques can separate the roles of contributing mechanisms, consideration of the higher-order kernels can also isolate components.

VII. SURROUND EFFECTS

The effect of stimulation in the surround of the receptive field was tested by an annular region of stimulus extending from 3 to 6°, modulating in contrast according to a binary white-noise signal $b(t)$, independently of the central spot. We now consider what structural models are suggested by the kernels involving the surround channel.

A. OPTICAL CROSS-TALK COMPONENT

The first-order surround kernel for receptors, h_r^b, has a positive primary part which corresponds closely with the positive part of the center kernel h_r^a, but with smaller amplitude. This suggests that light from the annulus is being absorbed in the recorded receptor and is transduced in an equivalent manner to light from the spot. This pathway will be referred to as optical cross-talk, and an optical cross-talk coefficient η can be estimated for each experiment, giving the effectiveness of the annulus on the central receptors relative to the effectiveness of the spot. An example of receptor first-order kernels is shown in Figure 2. The receptor kernels h_r^a and h_r^b are close to being proportional, and so the receptor linear response is modeled as

$$r = [h_r^a]a + [h_r^b]b$$

where the surround kernel is

$$h_r^b = \eta h_r^a$$

for some η. The coefficient η was estimated by considering kernel values up to and including the peak of the primary part, and regressing the surround values on the center values. The receptor data gave values of η from 0.08 to 0.4.

This optical cross-talk component is transmitted to LMCs, influencing the first- and second-order kernels involving surround. Many of the kernels h_l^b have a negative primary part, whose leading edge corresponds closely in latency and waveform with the leading edge of the centre kernel h_l^a. For these experiments the parameter η is estimated as above, yielding estimates from zero to 0.5.

To predict the effect of optical cross-talk on the second-order LMC kernel, we consider the LMC response to second degree in the receptor response; the mappings involved can be summarized by the following diagram:

$$\begin{array}{ccc} a & & b \\ \downarrow & \swarrow & \\ r & & \\ \downarrow\downarrow & & \\ l & & \end{array}$$

Algebraically, we write r in terms of l, and substitute the above expression for r to get

$$\begin{aligned} l &= [h_l^r]r + [h_l^{rr}]r^{2\otimes} \\ &= [h_l^r * h_r^a]a + [h_l^r * h_r^b]b \\ &\quad + [h_l^{rr} * (h_r^a)^{2\otimes}]a^{2\otimes} + [2h_l^{rr} * (h_r^a \otimes h_r^b)]a \otimes b + [h_l^{rr} * (h_r^b)^{2\otimes}]b^{2\otimes} \end{aligned}$$

If the cross-talk gives the relationship $h_r^b = \eta h_r^a$, then the linear component of the surround effect becomes

$$\begin{aligned} h_l^r * h_r^b &= \eta h_l^r h_r^a \\ &= \eta h_r^a \end{aligned}$$

as expected.

The cross-kernel above, operating on $a \otimes b$, becomes equal to

$$\begin{aligned} 2h_l^{rr} * (h_r^a \otimes h_r^b) &= 2\eta h_l^{rr} * (h_r^a)^{2\otimes} \\ &= 2\eta h_l^{aa} \end{aligned}$$

We would, thus, expect a component in the cross-kernel h_l^{ab} with the same form as h_l^{aa} but scaled by 2η. Inspection of the cross-kernels reveals that in every case where the first-order kernels indicate nonzero optical cross-talk, there is a negative part in the kernel h_l^{ab} at latencies corresponding to the negative primary part of h_l^{aa}, and when cross-talk is absent, this part is absent.

The second-order kernel for surround, h_l^{bb}, has a component proportional to η^2, which is not consistently observed.

The space-time inseparability of the LMC response properties means that optical cross-talk only produces a component of the surround kernels as has been seen, and other pathways must now be considered.

B. LATERAL ANTAGONISM

The annulus stimulates a large number of receptors surrounding the central ommatidium. First-order spatiotemporal kernels show that there is a range of forms of lateral antagonism from these receptors. Consideration of the nonlinear components of lateral antagonism is simplified by looking at those cells for which the optical cross-talk appears to be absent. The other cells presumably combine components for optical cross-talk and lateral antagonism.

The network of mappings must be expanded, and we introduce a signal $s(t)$ representing the strength of response in the surround receptors stimulated by the annulus. The mapping from spot to central receptors is presumably similar to the mapping from annulus to surround receptors; thus, $a \rightarrow r$ and $b \rightarrow s$ will be taken as equivalent.

Comparison of center and surround first-order kernels where optical cross-talk is absent suggests that the surround kernel could be obtained from the center kernel by inversion, attenuation, delay, and smoothing. Comparison of the kernels h_l^{aa} and h_l^{ab} in these cases also suggests a relationship where inversion and attenuation have occurred and where delay and smoothing have occurred along the t_2 dimension. The most remarkable case is shown in Figure 14, where the cross-kernel reproduces in inverted and delayed form every detail of the second-order center kernel. The first-order spatiotemporal kernels indicated that this was the *Eristalis* LMC with the strongest lateral antagonism.

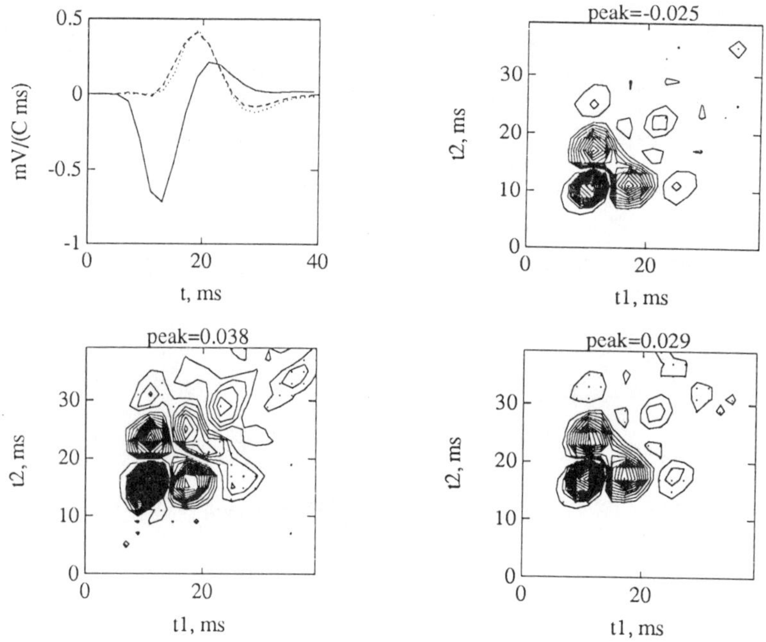

FIGURE 14. Fit of first- and second-order kernels according to lateral antagonism structural model described in text. The kernels are from an *Eristalis* LMC. Contour step size is 5% of peak value of kernel h_l^{ab} (lower left). Upper left, first-order kernels: solid line, center kernel h_l^a, dashed line, surround kernel h_l^b; dotted line, center kernel filtered by [*f*]. Parameters of *f* are $a_s = -0.23$ ms^{-1}, $\tau_s = 1.7$ ms, $d_s = 6.4$ ms. Upper right, observed center kernel h_l^{aa}. Lower left, observed cross-kernel h_l^{ab}. Lower right, kernel h_l^{aa} filtered along t_2 dimension by [2*f*].

The modeling of the primary transmission pathway has suggested that the main source of nonlinearity is the synapse between receptor and LMC. The results noted here can be accounted for by assuming that the lateral antagonism acts presynaptically, before the nonlinearity. This could possibly be by the surround receptor signal s contributing to the extracellular potential in the recorded cartridge, thus adding a negative component to the potential across the receptor membrane *m*.

This leads to the following model network:

$$
\begin{array}{ccc}
a & & b \\
\downarrow & \swarrow & \downarrow \\
r & & s \\
\downarrow & \swarrow & \\
m & & \\
\downarrow\downarrow & & \\
l & &
\end{array}
$$

where the mapping $s \rightarrow m$ gives a component antagonistic to the central pathway. Consider the case when the optical cross-talk is zero, eliminating the upper diagonal mapping $b \rightarrow r$. Rather than giving the full expression for this network, let us consider just the terms of interest.

The linear components of LMC response are

$$[h_l^m * h_m^r * h_r^a]a$$

$$[h_l^m * h_m^s * h_s^b]b$$

for center and surround, respectively. Assuming that $h_r^a = h_s^b$, the only difference is the replacement of h_m^r by h_m^s. The kernel h_m^r reflects the presynaptic high-pass filtering assumed along the primary transmission pathway. The lateral antagonism is represented by the kernel h_m^s, which contributes an inverted, attenuated, delayed, and smoothed surround signal to m.

The center self-quadratic and the cross terms produced are

$$[h_l^{mm} * ([h_m^r * h_r^a] \otimes [h_m^r * h_r^a])]a \otimes a$$

$$[2h_l^{mm} * ([h_m^r * h_r^a] \otimes [h_m^s * h_s^b])]a \otimes b$$

Note that with $h_r^a = h_s^b$ the kernels will be the same except that h_m^r is replaced by $2h_m^s$ in the cross-kernel. The difference between these two is equivalent to applying different filtering along the t_2 dimension to generate the overall second-order kernels, reflecting the inversion, attenuation, delay, and smoothing along the t_2 axis of the cross-kernel relative to the center kernel. The factor 2 in the cross-kernel implies that the cross-kernel can have larger amplitude than the center-only second-order kernel, which was observed in one case.

The model was fitted in the following way to kernels showing no optical cross-talk. We assume that the surround first-order kernel might be derived from the center kernel filtered by an operator [f] with a delayed exponential as kernel. This assumes the relationship

$$h_m^s = h_m^r * f$$

where f is defined by

$$f(t) = a_s \operatorname{nexp}(t - d_s/\tau_s)$$

The kernel f is estimated using the Nelder-Mead algorithm. The center kernel h_l^{aa} is then filtered along the t_2 dimension with the operator [$2f$] to produce a predicted kernel h_l^{ab} due to lateral antagonism. The fit obtained to one kernel set is illustrated in Figure 14.

VIII. CONCLUSION

This chapter has presented a study of the signal processing taking place within a neural system in terms of polynomial time-invariant operators.

The analysis was based on kernels measured by white-noise experimentation. Companion experiments using a spatiotemporal white-noise stimulus defined on a 15 × 15 checkerboard had yielded first-order kernels on a general space-time domain. Nonlinear properties were then analyzed using the lower-dimension two-channel center-surround stimulus, matched to the observed organization of the receptive fields. With modern techniques the spatiotemporal kernel can be calculated in near-realtime and then used to define stimulus regions on which a more specific stimulus can be displayed for analysis of nonlinear components. This two-stage procedure keeps the total number of kernel elements to a tractable limit at each stage and produces higher-order kernels with a good signal-to-noise ratio.

The kernels calculated are then treated as the basic empirical measurements of the study. Visual inspection of the kernels allows inferences to be drawn as to the underlying mechanism within the system. Structural models in terms of linear and static nonlinear mappings can be fitted. The preliminary use of quantitative tests, such as the use of marginal integrals (Chen et al., 1986), is of use in eliminating inadequate structures from consideration.

Consideration of LMC kernels in relation to receptor kernels provides information on the generation and transmission of response components. The principal source of nonlinearity appears to be the receptor-LMC synapse, although the second-order kernel cannot be accounted for by the properties of the primary transmitter histamine alone. A second transmission pathway is suggested, which is in agreement with results from voltage-clamp recording.

Stimulation in the receptive field surround appears to influence LMC response via two pathways. Optical cross-talk generates a component equivalent to that of the central pathway, while lateral antagonism appears to oppose the signal passing through the central pathway at a site before the introduction of the principal nonlinearity.

The models considered are represented in a simple way by mapping diagrams, depicting the assumed pathways of signal transfer. The notation defined for representing polynomial time-invariant operators, which omits the integral signs and variables of integration of the traditional representation, leaves only the essential elements: the kernels and the functions with which they are combined. This has been found to be of great utility in simplifying the expressions involved and, thus, promoting the fluent use of this class of nonlinear relationships.

REFERENCES

Billings, S. A. and Fakhouri, S. Y. (1978). Identification of a class of nonlinear systems using correlation analysis, *Proc. IEE,* **7**, 691–697.

Chen, H. W., Ishii, N., and Suzumura, N. (1986). Structural classification of nonlinear systems by input and output measurements, *Int. J. Systems Sci.*, **17**, 741–774.

Dubs, A. (1982). The spatial integration of signals in the retina and lamina of the fly compound eye under different conditions of luminance, *J. Comp. Physiol. A,* **146**, 321–343.

Falk, G. and Fatt, P. (1972). Physical changes induced by light in the rod outer segments of vertebrates, in *Handbook of Sensory Physiology,* Vol. **7/1**, H. J. A. Dartnell, Ed., Springer-Verlag, Berlin.

Hardie, R. C. (1987). Is histamine a neurotransmitter in insect photoreceptors?, *J. Comp. Physiol.*, **161**, 201–213.

Hardie, R. C. (1989). A histamine-activated chloride channel involved in neurotransmission at a photoreceptor synapse, *Nature (London),* **339**, 704–706.

James, A. C. (1990). White-Noise Studies in the Fly Lamina, Ph.D. thesis, Australian National University.

Korenberg, M. J. and Hunter, I. W. (1986). The identification of nonlinear biological systems: LNL cascade models, *Biol. Cybern.*, **55**, 125–134.

Laughlin, S. B. (1974). Neural integration in the first optic neuropile of dragonflies. III. The transfer of angular information, *J. Comp. Physiol.*, **92**, 377–396.

Laughlin, S. B. (1981). Neural principles in the peripheral visual system of invertebrates, in *Handbook of Sensory Physiology,* Vol. VII/6B, H. Autrum, Ed., Springer-Verlag, Berlin.

Laughlin, S. B. (1989). Coding efficiency and design in visual processing, in *Facets of Vision,* Stavenga, D. G. and Hardie, R. C., Eds., Springer-Verlag, Berlin.

Laughlin, S. B. and Osorio, D. (1989). Mechanisms for neural signal enhancement in the blowfly compound eye, *J. Exp. Biol.*, **144**, 113–146.

Laughlin, S. B., Howard, J., and Blakeslee, B. (1987). Synaptic limitations to contrast coding in the retina of the blowfly *Calliphora, Proc. R. Soc. Lond. B,* **231**, 437–467.

Lee, Y. W. and Schetzen, M. (1965). Measurement of the Wiener kernels of a non-linear system by cross-correlation, *Int. J. Control,* **2**, 237–254.

Nelder, J. A. and Mead, R. (1965). A simplex method for function minimization, *Computer J.*, **7**, 308–313.

Press, W. H., Flannery, B. P., Teukolsky, S. A., and Vettering, W. T. (1988). *Numerical Recipes in C,* Cambridge University Press, New York.

Sakai, H. M., Naka, K.-I., and Korenberg, M. J. (1988). White-noise analysis in visual neuroscience, *Vision Neurosci.*, **1**, 287–296.

Shaw, S. R. (1975). Retinal resistance barriers and electrical lateral inhibition, *Nature (London),* **255**, 480–483.

Shaw, S. R. (1984). Early visual processing in insects, *J. Exp. Biol.*, **112**, 225–251.

Weckström, M., Kouvalainen, E., Djupsund, K., and Järvilehto, M. (1989). More than one type of conductance is activated during responses of blowfly monopolar neurones, *J. Exp. Biol.*, **144**, 147–154.

Wiener, N. (1958). *Nonlinear Problems in Random Theory,* John Wiley & Sons, New York.

Zettler, F. and Järvilehto, M. (1972). Lateral inhibition in an insect eye, *Z. Vergl. Physiol.*, **76**, 233–244.

Zettler, F. and Straka, H. (1987). Synaptic chloride channels generating hyperpolarising responses in monopolar neurones of the blowfly visual system, *J. Exp. Biol.*, **131**, 435–438.

Chapter 3

LINEAR AND NONLINEAR MECHANISMS OF MOTION SELECTIVITY IN SIMPLE CELLS OF THE CAT'S STRIATE CORTEX

Robert C. Emerson and Mark C. Citron

TABLE OF CONTENTS

I. Introduction 76

II. Methods 76
- A. Physiological Preparation 76
- B. Receptive Field Classification: Conventional Stimuli 76
- C. Random Stimuli 77
- D. Response Measurements 77

III. Results 78
- A. Searching for Directionally Selective Neurons 78
- B. 1-Bar Responses in Simple Cells 78
- C. Predictions of Responses to Movement 80
- D. 2-Bar Response Components in Simple Cells 81
- E. Comparison between Simple- and Complex-Cell Directionally Selective Properties 84

IV. Discussion and Conclusions 86
- A. 1-Bar Responses and Receptive Field Classification 86
- B. 1-Bar Responses as a Basis for Directional Selectivity 86
- C. 2-Bar Interactions as a Basis for Directional Selectivity 86
- D. Mechanism of Spatiotemporal Nonlinearity 87

Acknowledgments 87

References 88

I. INTRODUCTION

Our earlier work in *complex* cells (Hubel and Wiesel, 1962) has shown that directional selectivity (DS), and even velocity selectivity, depends strongly on nonlinear interactions between responses to optimally oriented 2-bar combinations that provide samples of a motion sequence (Emerson et al., 1992a; Emerson et al., 1987; Emerson and Coleman, 1981). These neurons typically show no asymmetries in their 1-bar responses that could contribute to their DS, nor that could even suggest the preferred direction for a moving, optimally oriented bar.

In *simple* cells (Hubel and Wiesel, 1962), however, others have shown agreement between the preferred direction of a moving image and the direction of a latency shift in 1-bar responses at different receptive field (RF) positions (Reid et al., 1987, 1991) and even for the preferred velocity of the cell (McLean and Palmer, 1989). Some of our earlier work with flashed bars (Emerson and Gerstein, 1977b; Emerson and Coleman, 1981) suggested that DS simple cells might also show strong dependence on nonlinear interactions. Our goal here was to determine whether the mechanisms were similar to those we have found in complex cells (Emerson et al., 1987). By testing both simple and complex cells with 1- and 2-bar tests, we have found that (1) classification criteria that distinguish simple from complex cells depend mainly on differences in their 1-bar responses; (2) DS simple cells always show 2-bar spatiotemporal nonlinearities; (3) DS simple cells can lack significant 1-bar contributions to DS, depending heavily on nonlinear mechanisms, as complex cells do; and (4) nonlinear 2-bar interactions in DS simple cells are distributed in a temporal and spatial pattern similar to that in DS complex cells. Some of these results have been reported in briefer form (Emerson and Citron, 1989).

II. METHODS

Single-unit extracellular recordings of DS cells in the striate cortex of anesthetized cats provided responses to conventional bar and edge stimuli that were used for classifying cells as simple or complex and for assessing DS. In addition, we studied motion selectivity in simple and complex cells with 1- and 2-bar tests that were part of a 16-bar "white-noise" experiment in which each bar was modulated randomly and independently of the other bars. Details of the preparation and stimuli have been reported elsewhere (Emerson and Gerstein, 1977a; Emerson and Coleman, 1981; Citron and Emerson, 1983; Emerson et al., 1987; Emerson, 1988).

A. PHYSIOLOGICAL PREPARATION

Briefly, adult cats were preanesthetized with a 10-mg/kg intramuscular injection of ketamine, intubated, and surgically anesthetized as necessary with thiopental sodium for exposing area 17 and installing on the skull a fluid-tight cylindrical well. Contact and trial-case lenses focused the eyes on a tangent screen or cathode ray tube (CRT) screen at 115 cm. Before single-cell recording, we switched the animal to a continuous intravenous infusion of 1- to 2-mg/kg/h of thiopental, mixed with 0.125 mg/kg/h of Alloferin (Roche) for muscular paralysis of the eyes. Heart rate was monitored continuously to ensure that anesthetic dosage was sufficient to keep the animal in a relaxed and unresponsive state.

B. RECEPTIVE FIELD CLASSIFICATION: CONVENTIONAL STIMULI

Initially, we hand-plotted on the tangent screen cortical RFs in the dominant eye to determine location, size, presence of DS, and approximate preferred orientation. Then we used a computer-controlled mirror-galvanometer system (Emerson and Gerstein, 1977a;

Emerson and Coleman, 1981) to measure responses to moving edges and moving and flashing light bars to assess the type of RF and the extent of DS. These stimuli are standard for classifying cortical RFs (Schiller et al., 1976; Sherman et al., 1976; Henry, 1977; Citron et al., 1981a; Emerson and Coleman, 1981). Simple cells showed separate regions of response to presentation (ON responses) and removal (OFF responses) of a light stimulus, whereas cells of the complex family gave both ON and OFF responses to either light or dark stimuli (Hubel and Wiesel, 1962; Ikeda and Wright, 1975; Movshon et al., 1978a; Movshon et al., 1978b; Emerson and Coleman, 1981). We then inserted a mirror into the path to the tangent screen to center the RF on a 4° by 5° region of a Tektronix 608 CRT monitor that was driven by a video recorder and a dedicated interface that generated video-sweep signals and supplied synchronized 1-ms clock signals for data acquisition by a (DEC) LSI-11/23. A two-beam laser pointer, also reflecting from the mirror, helped locate the optimal position and orientation for the monitor. A drifting square-wave grating on the CRT screen helped us fine-tune the preferred orientation and direction, when there was a preferred direction.

C. RANDOM STIMULI

Having classified the RF, we then presented a prerecorded video random-bar white-noise stimulus that has been described previously (Citron and Emerson, 1983), except that our newer version could run up to 30 min without interruption. It consisted of 16 contiguous bars that were 0.25° by up to 5°, each of which could change its luminance randomly and independently of the other bars every video frame (about every 16 ms). This random grating included only three intensities: a light level at 360 cd/m^2, a mean luminance at 180 cd/m^2, and a dark level at nearly 0 cd/m^2. Each of the three luminances occurred with equal probability, as determined by a 32-bit linear congruential random number generator (Sect. 3.2.1 and 3.3.3. of Knuth, 1969). Thus, the stimulus was a pseudorandom ternary spatiotemporal grating with a uniform probability distribution. The stimulus was generated offline on an LSI-11/23 and recorded through a Matrox frame buffer on a frame-by-frame video recorder (GYYR) for playback in real time during the experiment. Although the contrast of the stimulus was theoretically 100%, [$C = (L_{max} - L_{min})/(2L_{mean})$], the 16-ms frame time was sufficiently shorter than the integration period of the (human) visual system that the luminance of bars appeared to be modulated continuously, and over a narrower contrast range. As all possible combinations of dark and light bars could occur in a given frame, the grating included narrow and wide bars of both contrast signs, moving up and down in short sequences at random regions of the RF, over a wide range of speeds whose values were limited only by the 16-ms temporal quantization and the 0.25° bar width. These short sequences of motion were sampled by our 2-bar stimulus as the most elemental components of directional motion. We always presented bars at the optimal orientation for the RF, and bars were long enough to fill or extend beyond the RF and wide enough to fill contiguous bar positions (here, 0.25°). Therefore, these experiments were two dimensional, one in space and one in time.

D. RESPONSE MEASUREMENTS

The times of neural impulses were measured with a resolution of 1 ms and stored on disk for offline analysis. We have used Wiener-kernel analysis to calculate 1-bar responses and nonlinear interactions between 2-bar responses. Wiener-kernel analysis is a well-established general method for analyzing nonlinear interactions (Marmarelis and Marmarelis, 1978; Citron et al., 1981b; Marmarelis et al., 1986; Citron and Marmarelis, 1987) and has been used in the cortex by us previously (Citron and Emerson, 1983; Emerson et al., 1987; Mancini et al., 1990; and Emerson et al., 1992b). Kernels were calculated by cross-correlating the neural response with the stimulus at a given position or pair of positions to generate (Wiener) kernels of zeroth order (h_0, the mean response), first order (h_1, the linear estimate of the impulse response), and nonlinear kernels. The nonlinear interaction plots were limited to second order (h_2), which implies the number of stimulus events that contribute to the cross-correlation and for intensive nonlinearities can include two bars that coincide in space and time. In the text, we have used the terms "1-bar" and "2-bar" rather than "linear" and "nonlinear," because we wish to emphasize the spatial and temporal nonlinear interactions elicited by motion. Therefore, except in Figure 1E, we have included the Wiener second-order intensive nonlinearity in our 1-bar responses.

III. RESULTS

In this report we have selected a single example of a DS simple cell that represents a subgroup of five of the ten DS simple cells we have analyzed thoroughly with white noise. We compare this cell with a DS complex cell that we have analyzed previously and which represents the five of seven DS cells of the complex family that we have analyzed thoroughly.

A. SEARCHING FOR DIRECTIONALLY SELECTIVE NEURONS

To find DS neurons we presented an oscillating light bar, oriented optimally for the RF. Figure 1B shows the bimodal responses of a simple cell that clearly responds more strongly to upward motion (left bins) than to downward motion (right bins). Our best approximation to this smooth movement that remains within the spatial resolution of our 16-bar white-noise stimulus is represented in Figure 1A as a bar-sequence, moving in instantaneous jerks (stepwise) up and down across the RF. We have shown previously that such sequences elicit strong directional response asymmetries in both simple and complex cells (Figure 10 of Emerson and Gerstein, 1977b; Figures 4 to 6 of Emerson and Coleman, 1981; and Figure 1B of Emerson et al., 1987). We used these stepwise samples of motion to help divide responses to moving stimuli into simpler spatial and temporal components.

B. 1-BAR RESPONSES IN SIMPLE CELLS

We began (Figure 1D) by measuring responses to impulsive presentations of single light bars, as represented in Figure 1C, that are isolated members of the stepwise sequence represented in Figure 1A. Responses in Figure 1D are shown in time after the 16-ms presentation of a light bar at each RF position (time course shown below). Rather than plotting the time course of responses at each position, we have drawn isoresponse contour lines for this two-dimensional spatiotemporal function, showing values greater than (solid contours) or less than (dashed contours) the mean firing rate driven by the stimulus. A major advantage of the white-noise stimulus is that its high average power provides a background response rate sufficient to see inhibitory events that would normally be below a neuron's firing threshold.

As mentioned in Section II, the 1-bar response is calculated as the Wiener second-order prediction for a single light bar and includes a mild intensive nonlinear component (the second-order "diagonal" term), but its effect was evident only around central position 7 (not shown separately), where threshold nonlinearities at only the strongest negative positions prevented a strong negative response. Essentially, the multiple bars of the white-noise stimulus act to linearize the simple cell, whose operating point normally is carefully positioned on the edge of a major nonlinearity (see Emerson et al., 1988 and Emerson et al., 1989 for a discourse on threshold nonlinearities in simple cells). Simple cells have long been regarded as carrying a relatively linear representation of image luminance distributed in space (Movshon et al., 1978a; Jones and Palmer, 1987; Emerson, 1988; Emerson et al., 1989). However, the strong threshold nonlinearities in all cortical cells at their normal

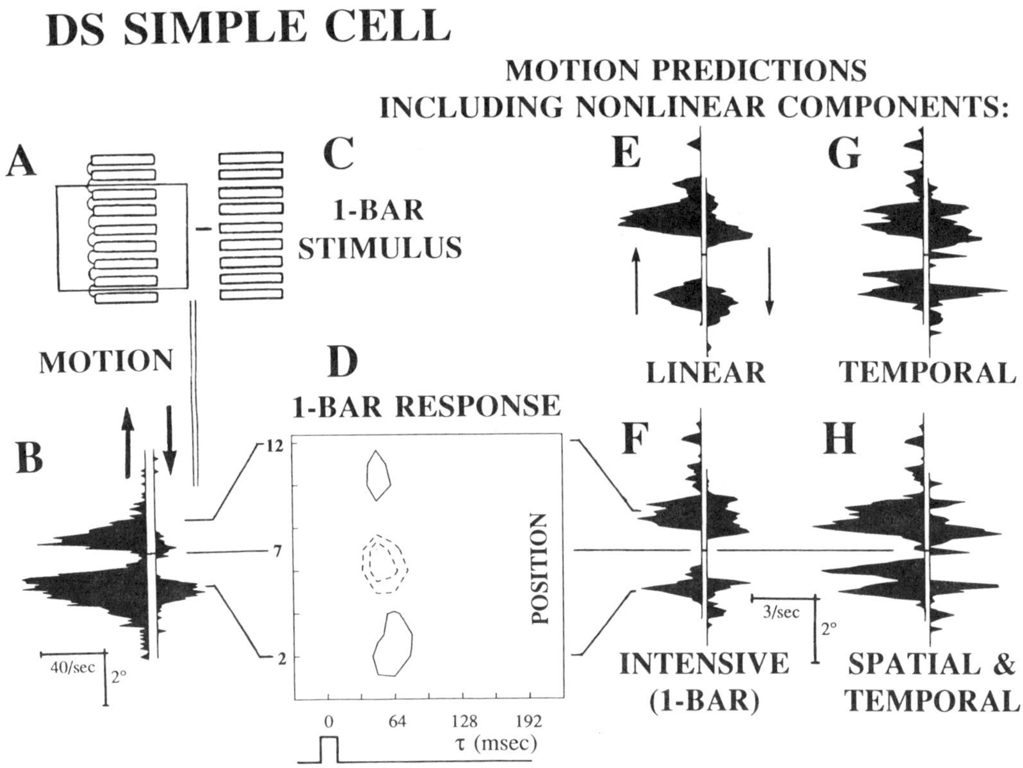

FIGURE 1. A directionally selective (DS) simple cell (cell 8-9). (A) Matrix of sequential bar positions across the 1.5° wide RF (rectangle) approximating the smoothly moving bar that elicited responses in B. (B) DS bimodal response to light bar moving smoothly up (left bins) and down (right bins) 8° across the RF at 3.9°/s. Maximum response was 77.5 spikes/s, averaged over 25 cycles. Bar had luminance of 6.8 cd/m^2 with a contrast of 0.8 log units. (C) Matrix of isolated bar positions assumed by the bar moving stepwise in A and centered on position 7. (D) Contour plot of responses to 16-ms presentations of the bar at isolated positions 1-12 of the RF, a subset of which is depicted in C. Plot shows top view of a surface in which the loci of constant response rates (elevations) are represented by lines, solid for positive (hills) and dashed for negative (valleys). Contour at zero is missing because it is dominated by noise. Origin of temporal axis (τ) is at presentation of stimulus (see stimulus marker, below). Bent lines show spatial correspondence between smooth movement of B and positions 2-12 of D. Range of responses in D: -1.7 spikes/s (three contour levels below mean driven response) to 0.91 spikes/s for a mean luminance of 180 cd/m^2, and 100% pseudorandom modulation (see text). (E) Linear bimodal 1-bar response-prediction for light bar moving as in B, except in 0.25° jerks across the central 4°. Prediction was calculated by convolving stepwise moving stimulus of A with the linear estimate of the 1-bar impulse response (h_1; see text). Negative values were set to zero to mimic zero background rate of conventional stimulus in B. (F) Movement prediction similar to E, except that intensive nonlinearities were included by convolution of stimulus with 1-bar response of D. Bent lines as in D. (G) Movement predictions similar to F, except that temporal nonlinearities from long duration at each position were added by convolution with the temporal kernel, $h_2(\tau, \Delta\tau, s)$. (H) Movement predictions similar to G, except that spatial nonlinearities from one position above and below the reference position were added by convolution with the spatiotemporal kernel, $h_2(\tau, \Delta\tau, s, \Delta s)$. All white-noise measures shown for this cell represent a 28.5-min random-stimulus presentation. Retinal eccentricity of RF was 8.6° in right (ipsilateral) eye. Horizontal dimension of hand-plotted minimum response field in A is shown larger than measured because length summation would have dictated a negative horizontal extent.

operating point implies that this linear representation occurs only intracellularly in the visual cortex.

Note that the locations of the two peaks for movement in Figure 1B agree roughly with the locations of light-excitatory (ON) regions in D, but that the major response in D is light inhibitory, or OFF. (For a discussion of the generation of OFF responses in linear neurons, see Emerson, 1988.) Furthermore, the latencies of all response components in Figure 1D are so consistent across space that the domains show little oblique "orientation" in space-time, which means that these 1-bar responses are relatively separable in two-dimensional space-time. That is, the two-dimensional function can be generated by a row-by-column (outer) product of a vertical spatial function and a horizontal temporal function. Adelson and Bergen (1985) have pointed out that, conversely, the occurrence of oblique orientation would signify that the 1-bar responses carry information about the preferred direction of the unit. Another way of stating this property is that a convolution of the first-order spatial kernels in Figure 1D with the stepwise sequence of Figure 1A, having the same average velocity as smooth movement in Figure 1B, should exhibit no strong DS.

C. PREDICTIONS OF RESPONSES TO MOVEMENT

This expectation is born out by the plot in Figure 1F, in which the components in Figure 1D are shifted and added (convolved) to mimic the stepwise-moving stimulus of Figure 1A. Only the mean firing rate during the white-noise experiment (h_0) has been excluded from this prediction (as well as from Figure 1D) and from other predictions, because the neuron showed no baseline firing when unstimulated. Accordingly, negative prediction values were considered below threshold and set to zero. Although predicted responses in F show small differences in time course of the two directions, the ratio of the number of spikes for upward/downward directions is only 1.41, compared with 6.34 for the movement response in Figure 1B. If calculated as the directional index used by Reid et al. (1987) (i.e., $(R_p - R_o)/(R_p + R_o)$, where R_p and R_o are the numbers of spikes in the preferred and opposite directions, respectively), the indices are 0.170 for the 1-bar prediction and 0.728 for the measured response. This index varies from 0 for nondirectional cells to a maximum of 1. Unexpectedly, a prediction using a *linear* estimate of the cell's impulse response across space (the "linear" kernel, h_1), shown in Figure 1E, exhibits the slightly higher up/down ratio of 1.99, or an index of 0.331. This linear component is computed before convolution and thresholding by subtracting the average dark-bar response at each position from the average light-bar response (Equation A15 from Appendix of Emerson et al., 1987) and has been used also by McLean and Palmer (1989) to study DS. In this case, there were minor inseparable components and spatial asymmetries in the linear kernel that were not visible even in the lowest contour, because their magnitude was only 10% of the deepest valley at position 7, similar to the 1-bar plot of Figure 1D. Their cumulative effects on the prediction became significant under convolution because of a large spatiotemporal area in the kernel. However, linear components rarely account for more than 50% of the DS observed in DS simple cells (Reid et al., 1987, 1991) and here account for 45% of the measured directional index.

To determine whether possible *temporal* nonlinear interactions between 16-ms components of the 64-ms duration at each position contributed to the response (which would correspond more closely to the duration of McLean and Palmer's [1989] stimulus), we added to the intensive prediction of F all the second-order temporal kernel components that could occur in a 64-ms pulse at each position (h_2 components up to a temporal separation ($\Delta\tau$) of 48 ms. Figure 1G shows that this addition did not enhance the DS of the prediction; in fact, it lowered the up/down ratio to 1.23 and the index to 0.102. Finally, to attempt to account for the strong DS of smooth movement in Figure 1B, we included in the prediction shown in Figure 1H all the intensive, temporal, and *spatial* interactions between 2-bar components with a spatial separation up to 1 position above or below the reference position. This addition increased the up/down ratio to 2.64 and the index to 0.450, accounting for 62% of the measured directional index. We have shown previously that complex cells do not show oblique orientation in their 1-bar responses, in agreement with this particular simple cell we have chosen (Emerson et al., 1987). Therefore, in general, 1-bar responses cannot explain

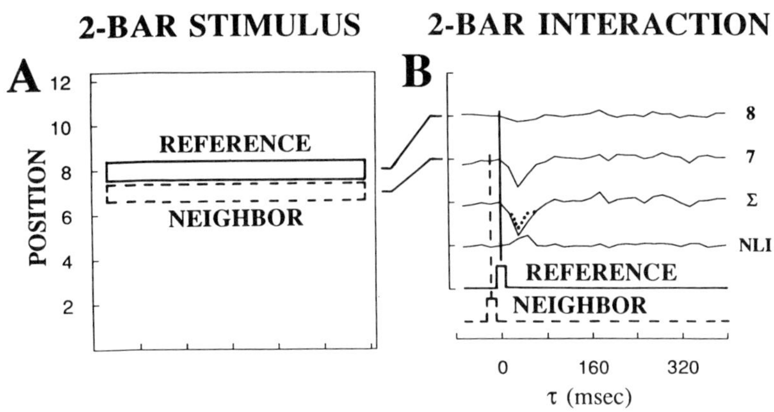

FIGURE 2. Calculating a 2-bar nonlinear interaction. (A) Location of reference bar (solid bar at 8) remains constant, while neighbor bar (dashed bar at 7) in later figures will occupy all active RF positions. (B) 2-bar superposition test showing responses of bars at 7 and 8, their linear sum (solid curve), the time-locked pair response (dotted), and the difference, which represents the nonlinear interaction residual, calculated as the Wiener second-order kernel. The positive interaction agrees with mild facilitation in the preferred direction for the 2-bar sample of upward motion (2-bar time-course shown below). Strongest response in B was −2.02 spikes/s for the linear sum.

DS. The effectiveness of the spatial component shown here in accounting for DS suggests that if we had considered a larger range of spatial separations, we might have accounted for even more of the measured DS (see Section IV).

The generally low magnitude of response predictions derived from white-noise-measured kernels, compared with conventional stimuli, is caused partly by the same high RMS stimulus power that facilitates measuring inhibitory responses. Its high average power increases the average firing rate (h_0), but adapts a neuron such that although the time-course of responses is similar, the incremental gain is lower than with a conventional stimulus (see Citron and Emerson, 1983, and Emerson et al., 1987 for further discussion of cortical adaptation).

The differences among movement predictions in Figures 1 E to H suggest that intensive, temporal, and spatiotemporal nonlinear interactions all play roles in determining the responses of a DS simple cell. In the remaining portion of this chapter, we examine the kernels that express these interactions to obtain a clearer picture of their spatial and temporal limits, and their possible mechanisms.

D. 2-BAR RESPONSE COMPONENTS IN SIMPLE CELLS

The stronger contribution to DS from spatially separated 2-bar response components shown in Figure 1H than from those at a single position indicates that this neuron is using information from at least two RF positions at a time to encode moving images. For an image moving with constant velocity, knowledge about its traverse of any two noncoincident bar positions provides information about direction (order) and speed (interval), just as viewing a moving object through two narrow windows would provide the same perceptual information. We have found previously, using white noise, that in complex cells even a single pair of bars can elicit strong DS (Emerson et al., 1987). Here, we sought more insight into mechanisms of DS in simple cells than we had gained by using isolated conventional bar-pairs (Emerson and Gerstein, 1977b).

Therefore, as shown in Figure 2A, we selected from the white-noise sequence a "reference" position (8) around which to measure interactions with a variable "neighbor" bar. Figure 2B shows responses for the 2-bar sequence in which a 16-ms presentation at neighbor position 7 is followed immediately by a 16-ms presentation at reference position 8, a sample of motion in the preferred, upward direction. The 1-bar responses, shown in the top two plots of Figure 2B are drawn from Figure 1D, but shifted by the 16-ms interval between onset of the two bars. Their linear superposition is shown in the next plot (solid), which in a linear system would be the 2-bar response. As the 16-bar white-noise stimulus presented at the time of the experiment included many such light-bar pairs at the appropriate two positions and the 16-ms temporal separation, we can, in effect, time-lock a response measure to the occurrence of the desired bar-pair in the sequence, similar to a PST histogram averaged over many repetitions of the desired stimulus. The resulting "measured" 2-bar response, shown dotted, is slightly more positive than the linear prediction, which denotes nonlinear facilitation for this sample of movement in the preferred direction. This measurement is calculated as the Wiener second-order estimate of the 2-bar response (see Marmarelis and Marmarelis, 1978, and Appendix of Emerson et al., 1987).

The new information available from the 2-bar test is not the 2-bar response itself, but the difference between the measured response and the expected linear superposition of the separate components. This function, plotted at the bottom of Figure 2B as "NLI", is the Wiener second-order kernel estimate, $h_2(\tau, \Delta\tau = 1, \Delta s = 1)$, for the nonlinear residual from the 2-bar superposition test (see p. 153 of Marmarelis and Marmarelis, 1978). Although this nonlinear facilitation seems too weak to explain the strong DS exhibited in Figure 1B, it was one of the stronger single components included in the prediction for the upward direction of Figure 1H. However, a moving bar elicits possible interactions over many more than one pair of bar positions. To allow plotting of interactions for other neighbor positions, we have summed the interaction over a 64-ms period taken after the later of the two bars to generate a scalar measure of interaction sign and strength. This suppression of the τ variable is possible because the time course of interactions such as in Figure 2B is similar over this spatiotemporal parameter space. That is, the interaction is nearly separable when expressed in τ vs. $\Delta\tau$ variables (Figure 2 of Emerson and Vaughn, 1987 for complex, and Emerson, 1992 for simple cells), where $\Delta\tau$ is the temporal separation between bars.

Figure 3B shows a contour plot of this measure over all 2-bar velocity samples that include the reference bar at position 8. Each interaction value is plotted as a function of neighbor-position along the vertical (as shown dashed in A) vs. the time of presentation of the neighbor bar before or after the reference bar along the horizontal ($\Delta\tau = 0$ when the events are simultaneous). The resulting two-dimensional plot shows beginnings of an obliquely oriented positive ridge passing through neighbor position 8 and $\Delta\tau = 0$, and of a small valley below.

But Figure 3 still does not include all of the possible 2-bar interactions that would be evoked by a moving bar, because such a bar would traverse all possible pairs of bar positions with a temporal order and interval that corresponds to its velocity. Figure 4A shows a plot of all such 2-bar interactions, averaged over a spatial coordinate frame in which all reference-bar positions are superimposed at the vertical center of the plot, which creates a relative spatial axis, Δs. This suppression of the reference-position variable, per se, is possible because the major positive interaction is always centered where the neighbor and reference bars coincide, i.e., at $\Delta s = 0$. In complex cells, the spatial extent of this constancy suggests that DS depends on a family of similar subunits, distributed across the RF (see Figure 3 of Emerson et al., 1987, and Figure 6 of Emerson et al., 1992a). Note that the low spots on the positive ridge of Figure 3B are now completed to produce a continuous straight ridge with valleys on both sides and hints of secondary positive regions still further separated from the central ridge. The large positive interaction values along a horizontal cut through $\Delta s = 0$ denote strong facilitation for temporally separated 16-ms bars, which accounts for the higher overall amplitude in the movement prediction of Figure 1G compared with Figure 1F.

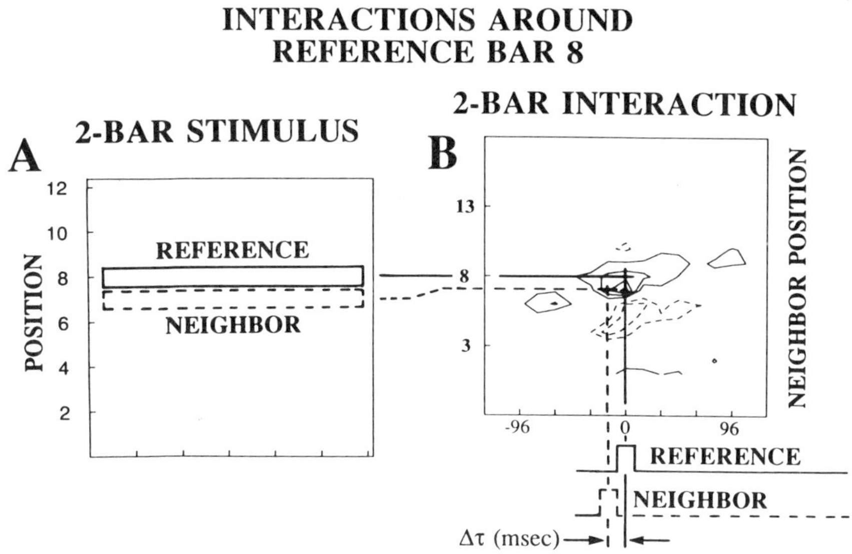

FIGURE 3. Effect of changing the neighbor-bar position. (A) Same bar-pair as in Figure 2A, except that the position occupied by the variable neighbor bar becomes the vertical coordinate for the 2-bar interaction plotted in B. (B) Same Wiener second-order kernel as plotted in Figure 2B, except that interaction values are summed over the 64-ms active period of the 1- and 2-bar responses to allow plotting of interaction strength for all possible bar pairs around reference 8. This period began 16 ms after cessation of the late bar. Horizontal axis is time of occurrence of neighbor bar before or after reference bar at $\Delta\tau = 0$. Range covers 16 neighbor positions of 0.25° each, and 15 temporal separations of 16 ms each. Range of interactions: −1.05 to 1.47/s for the 64-ms summing period.

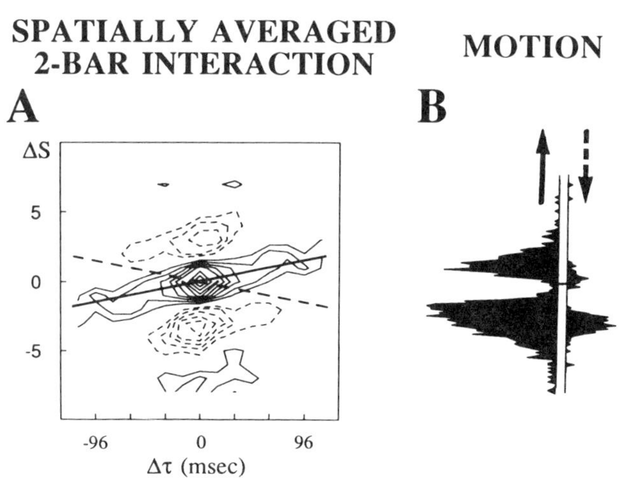

FIGURE 4. 2-bar interaction averaged over all RF space (''motion kernel''). (A) Spatiotemporal dependence of nonlinear interactions as in Figure 3B except that interactions are averaged over all possible 2-bar pairs around reference positions 1-12. (A and B) Solid line shows loci of interactions used by upward velocity in B at the slope of the solid line, and dashed line shows interactions for downward movement in B. This neuron shows preference for upward movement because of strong facilitation at +3.9°/sec (see text). Range of spatially averaged interaction values: −0.72 to 0.77/s for the 64-ms summing period.

We have used this spatially averaged 2-bar interaction to compress the available four-dimensional information about 2-bar motion into a single two-dimensional plot that we have called the motion kernel (Emerson et al., 1987) because of its expression in space vs. time. As suggested previously in the context of complex cells (Emerson et al., 1987), a compelling interpretation of the obliquely oriented ridge-valley structure of the interaction in Figure 4A is that it provides a basis for directional, and perhaps even velocity selectivity. Because the relationship is plotted in space vs. time (for the neighbor bar), constant-velocity bar movement across the RF includes interaction loci occurring only along a straight line whose slope equals the velocity in degrees per second. This line must intersect the central locus $\Delta s = \Delta \tau = 0$, because at some point in the traverse all 2-bar motion samples around a fixed reference position must coincide (in space and time). Velocities whose slope matches that of the oblique positive ridge are facilitated, and others are either less facilitated or suppressed by interactions occurring in the valleys.

Smooth bar motion in this unit provides an example. The solid oblique line in Figure 4A represents interactions occurring for motion in the preferred direction and at the speed corresponding to responses under the upward arrow in Figure 4B, and the dashed line in Figure 4A corresponds to the downward arrow in Figure 4B. Therefore, this unit responds strongly at the upward velocity of Figure 4B because of strong nonlinear facilitation from the positive ridge of Figure 4A, and responds weakly at the downward velocity of Figure 4B because of the lack of facilitation along dashed loci, combined with the predominantly negative (OFF) 1-bar responses shown in Figure 1D. The solid locus of interactions in Figure 4A samples the kernel so completely that kernel contributions to the upward prediction lose their individual temporal identity. The consequence is shown in Figure 1H. Even for only a single horizontal kernel slice at $\Delta s = +1$ and one at $\Delta s = -1$, the contributions merge to produce a continuous pedestal of facilitation for the preferred direction that, before the measured cell's threshold, adds to the response for that direction.

E. COMPARISON BETWEEN SIMPLE- AND COMPLEX-CELL DIRECTIONALLY SELECTIVE PROPERTIES

To determine whether DS simple cells use the same nonlinear spatiotemporal mechanisms as DS complex cells, we have compared in Figure 5 the 1- and 2-bar responses of the simple cell we have analyzed here against responses of a DS complex cell that we have analyzed previously (Citron and Emerson, 1983; Emerson et al., 1987). One-bar responses of Figure 5A vs. C are clearly different, and establish the chief basis for classifying the two cell types. Simple-cell responses of Figure 5A show phase sensitivity (changes in sign as the briefly flashed bar is positioned across space). This accounts for their segregated regions of response to light and dark edges (Sherman et al., 1976), which was tested for this cell, but not shown here.

Complex-cell responses of Figure 5C, on the other hand, show only a single excitatory 1-bar response region. This phase-independent response to single dark bars, along with similar responses to single light bars (Movshon et al., 1978b), usually associated with ON-OFF responses, has prompted us (Emerson et al., 1987) and others (Movshon et al., 1978b) to propose that complex cells accomplish a full-wave rectification of their 1-bar inputs. After further analysis, we have recently proposed that the transformation is a squaring of the input signals (Emerson et al., 1989; Emerson et al., 1992a, 1992b). This difference in linearity of 1-bar responses between simple and complex cells is visible in our white-noise experiment only if one examines the spatial distribution of intensive nonlinearities in both classes, as mentioned in the context of Figure 1D. Simple cells show 1-bar threshold nonlinearities over only a limited region, because the self-conditioning nature of the 16-bar white-noise stimulus increases the background firing rate enough to eliminate threshold discontinuities

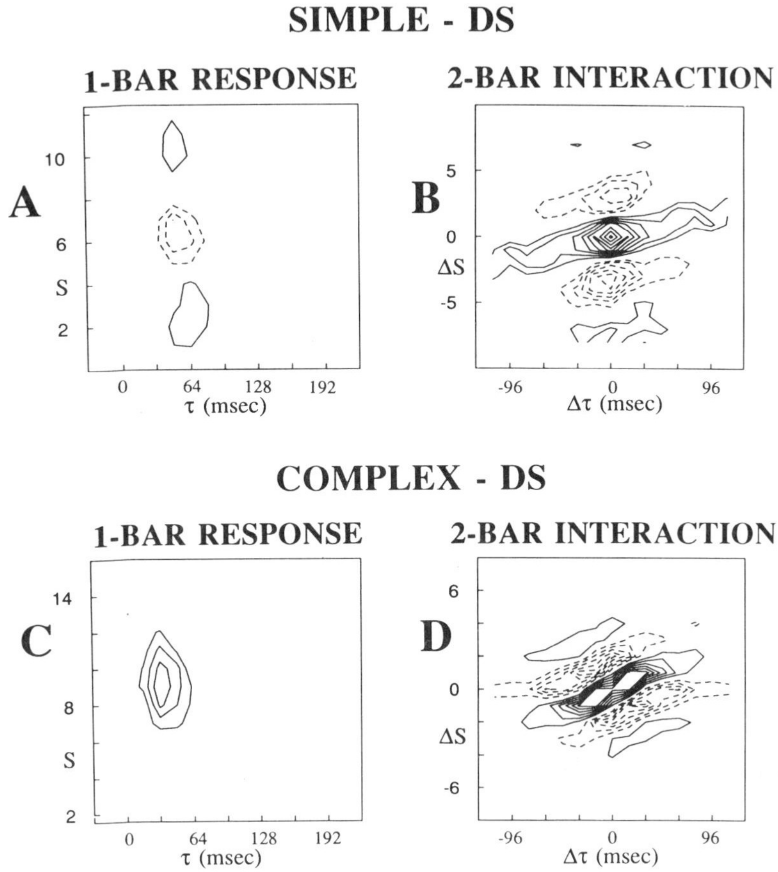

FIGURE 5. Comparison between DS simple (A and B) and complex (C and D) units. (A and C): 1-bar responses (kernel in C is for a dark bar, because it was a more effective stimulus than the light bar of A). Note that the simple cell in A shows separate positive and negative areas, whereas the complex cell in C (cell 3-1) shows only a positive response over space, the main distinction used for classifying these units. Maximum response in C, 2.64/s. (B and D): 2-bar interactions are more similar, which indicates that DS subunits in simple and complex units are similar and that both depend on spatiotemporal nonlinearities. Wiener second-order kernels for the measurement of D have been averaged over reference positions 7-11. Range of spatially averaged interaction values in D: -1.60 to $+1.98$/s as summed over the 80-ms period immediately following cessation of the later bar.

in all but the strongest negative regions (Figure 1C of Emerson et al., 1989). The essential intensive nonlinearity of a complex cell, on the other hand, is distributed uniformly across space and seems to be independent of the background firing rate or of the strength of the stimulus.

Although the 1-bar responses of simple and complex cells differ from each other in obvious ways, the 2-bar interactions of Figure 5B vs. D show striking similarities. These neurons each preferred upward bar motion and exhibited similar obliquely oriented ridge-valley structure, the chief difference being that the suppressive valleys in the simple cell of B were more confined to the vertical axis. Interactions along the vertical axis itself are tested by spatially separated, simultaneous events and reflect the center-surround geometry of retinal and lateral geniculate subunits converging onto the cortical neuron (Movshon et al., 1978b; Citron et al., 1981b; Emerson et al., 1987). The lack of temporal suppression along the horizontal axis ($\Delta s = 0$) and the accompanying lack of oblique extension for suppressive valleys in Figure 5B vs. D probably result from the low-frequency-pass vs. bandpass properties of simple- vs. complex-cell early filters (see discussion below about mechanisms).

IV. DISCUSSION AND CONCLUSIONS

A. 1-BAR RESPONSES AND RECEPTIVE FIELD CLASSIFICATION

We have examined both 1-bar responses and 2-bar nonlinear interactions in white-noise tests of simple and complex cells in the cat's striate cortex. One-bar responses of simple cells differ from those of complex cells by showing separate regions of response to light and dark image elements (flashing bars, and moving bars and edges), a property that is weighted heavily in most RF classification criteria (Emerson, 1988).

B. 1-BAR RESPONSES AS A BASIS FOR DIRECTIONAL SELECTIVITY

Although not shown here, five of the ten DS simple cells we have analyzed thoroughly showed 1-bar responses to white noise that differed from those shown in Figure 1D by having obvious obliquely oriented domains that would confer somewhat larger *modulation* of the response to a bar or grating moving in the preferred direction, as compared with the null direction (see Figures 8F and 13 of Adelson and Bergen, 1985). The direction of latency shift that leads to the orientation seems always to agree with the measured preferred direction. In this respect we agree with Reid et al. (1987, 1991) and with McLean and Palmer (1989) that, as even shown here in Figure 1E, DS simple cells can show inseparable 1-bar responses that contribute to differences in responses to opposite directions of movement. But to explain the difference in *number* of responses for opposite directions in Figure 1B, one must resort to a nonlinear transformation. van Santen and Sperling (p. 305, 1985) clearly justify the need for nonlinearities in any motion detection scheme, as have earlier workers in the insect's visual system (Poggio and Reichardt, 1976). Even a neural threshold qualifies as a nonlinear transformation, although it is usually considered a trivial one. In fact, as mentioned in Section III, even the linear prediction of Figure 1E has a kind of threshold, because we did not show values below 0 spikes per second. Because the response modulation for upward movement was larger than for downward movement, the number of spikes above "threshold" was 99% larger, whereas the raw response was the same (linearity requires that the hidden negative excursion for the upward direction is larger than for the downward direction to preserve a constant total for each direction). Note that the quantitative comparisons between predictions and measurements should be interpreted loosely, because a slight underestimate in the level of tonic inhibition below threshold in the measured response of Figure 1B (none assumed) may have severely decreased the DS values for our models. This phenomenon is often called the iceberg effect. Albrecht and Geisler (1991) and Heeger (1991 and in press) have shown how useful for enhancing selectivity is an accelerating nonlinearity with an exponent and around 2.

We believe that progress in understanding of neural processes has been hampered by a lack of agreement on a clear definition of motion selectivity, and even of DS. A good beginning would be to distinguish between directional asymmetries in number of spikes and in response modulation.

C. 2-BAR INTERACTIONS AS A BASIS FOR DIRECTIONAL SELECTIVITY

To avoid many of these problems of interpretation about strictly intensive transformations, we chose here one of five DS simple units that showed little 1-bar contribution to their DS movement responses. In all five DS units, the pattern of 2-bar spatiotemporal interactions, exemplified here in Figure 4A, predicts the preferred direction elicited by a moving bar. This demonstration eliminates the necessity of having inseparable 1-bar responses. In general, we have found that the slope of any oriented 1-bar domains in simple cells agrees with the (inverse of the) slope of the 2-bar ridge in interactions that always resemble that of Figure 4A. Therefore, to the extent that 1-bar inseparability exists, and the increased modulation can be used to enhance motion selectivity, 1- and 2-bar properties may act synergistically, as Reid et al. (1987, 1991) have asserted for linear and nonlinear mechanisms. McLean and Palmer (1989) have shown recently that the preferred velocity agrees with the slope of 1-bar oblique domains (when they exist), a result that would be expected from the dependence of the preferred velocity on both 1- and 2-bar response components. Our recent measured data do not bear on the issue of preferred speed because we did not measure responses over a sufficiently wide range of bar speeds to compare with slopes of 2-bar interactions.

D. MECHANISM OF SPATIOTEMPORAL NONLINEARITY

The similarity in 2-bar interactions of simple vs. complex cells in Figure 5B vs. D makes it likely that our understanding of mechanisms in DS complex cells can be applied to DS simple cells. To study mechanisms in DS complex cells, elsewhere, we have compared 1- and 2-bar tests of DS complex cells with 1- and 2-bar tests of an "energy" model (Adelson and Bergen, 1985) that seems to mimic well the movement responses in complex cells (Emerson et al., 1992a). The model requires that the outputs of two inseparable spatiotemporal filters (plotted in coordinates of Figure 1D) that are separated in phase by 90° (in quadrature) are squared and summed to remove modulation of the output, but retain DS. (See also similar quadrature models of Hamilton et al., 1989; Pollen et al., 1989; and Heeger, 1991 and in press.) Both 1- and 2-bar responses of complex cells, which are sensitive tests of cortical connectivity, matched those of the energy model. Because the 2-bar interaction averaged across the RF always shows even symmetry and can be shown to equal the autocorrelation function of the spatiotemporal input filters for such a model (see Appendix of Emerson et al., 1992a), we must assume that simple and complex cells have inseparable input filters that are similar.

If simple and complex cells have similar inseparable input filters, then why do simple cells often exhibit those properties in 1-bar responses, while complex cells never do? One possibility is that the squared outputs of the simple-cell quadrature filters are not precisely balanced. This imbalance would allow some of the phase sensitivity of one of the inseparable filters to survive the summation that in a more balanced complex cell would uniformly flatten the ridges and fill the valleys. Compared with strong separable components, the weak inseparable 1-bar components that we have seen in our simple-cell measurements would support this interpretation.

A second possibility is that the subunit filters that contribute to DS in a complex cell occur in larger numbers and are distributed over a larger region in space than in simple cells. Summation of a number of such filters would tend to confuse the phase information more than if the filters exhibited more coherent phase by grouping more compactly. On the other hand, the complete 2-bar interaction-picture for the simple cell shown in Figure 4A requires averaging over many reference and neighbor positions to obtain the fully developed motion kernel. Therefore, mechanisms of DS in this simple cell are apparently distributed to some extent across space.

Further distinctions between mechanisms in simple and complex cells will require modeling of candidate subunit filters, their distribution in space, and the nonlinear transformations that follow the subunits. However, at this point the similarity among the subunits of all DS cortical cells suggests that a primary function of the cat's striate cortex is to generate nonlinear spatiotemporal interactions necessary for processing moving images.

ACKNOWLEDGMENTS

We are indebted to W. J. Vaughn for programming of kernel calculations and to C. M. Cappiello for preparing figures. This research was supported by EY06679 to RCE, EY01319 to Center for Visual Science, and EY00250 to MCC.

REFERENCES

Adelson, E. H. and Bergen, J. R. (1985). Spatiotemporal energy models for the perception of motion, *J. Opt. Soc. Am.* **A2**, 284–299.

Albrecht, D. G. and Geisler, W. S. (1991). Motion selectivity and the contrast-response function of simple cells in the visual cortex, *Visual Neurosci.*, **7**, 531–546.

Citron, M. C. and Emerson, R. C. (1983). White noise analysis of cortical directional selectivity in cat, *Brain Res.* **279**, 271–277.

Citron, M. C., Emerson, R. C., and Ide, L. S. (1981a). Spatial and temporal receptive-field analysis of the cat's geniculocortical pathway, *Vision Res.*, **21**, 385–397.

Citron, M. C., Kroeker, J. P., and McCann, G. D. (1981b). Nonlinear interactions in ganglion cell receptive fields, *J. Neurophysiol.*, **46**, 1161–1176.

Citron, M. C. and Marmarelis, V. Z. (1987). Applications of minimum-order Wiener modeling to retinal ganglion cell spatiotemporal dynamics, *Biol. Cybern.*, **57**, 241–247.

Emerson, R. C. (1988). A linear model for symmetric receptive fields: implications for classification tests with flashed and moving images, *Spatial Vision,* **3**, 159–177.

Emerson, R. C. (1992). Do directionally selective simple cells use motion-energy subunits? *Invest. Ophthalmol. Vis. Sci.*, Suppl. 33, 953.

Emerson, R. C., Bergen, J. R., and Adelson, E. H. (1992a). Directionally selective complex cells and the computation of motion energy in cat visual cortex, *Vision Res.*, **32**, 203–218.

Emerson, R. C. and Citron, M. C. (1989). Linear and nonlinear mechanisms of motion selectivity in single neurons of the cat's visual cortex, in Kleinman, D. L. (Ed.), *Proc. IEEE International Conference on Systems, Man, and Cybernetics;* IEEE Cat. # 89-CH2809-2, (pp. 448–453).

Emerson, R. C., Citron, M. C., Vaughn, W. J., and Klein, S. A. (1987). Nonlinear directionally selective subunits in complex cells of cat striate cortex, *J. Neurophysiol.*, **58**, 33–65.

Emerson, R. C. and Coleman, L. (1981). Does image movement have a special nature for neurons in the cat's striate cortex?, *Invest. Ophthalmol. Visual Sci.*, **20**, 766–783.

Emerson, R. C. and Gerstein, G. L. (1977a). Simple striate neurons in the cat. I. Comparison of responses to moving and stationary stimuli, *J. Neurophysiol.*, **40**, 119–135.

Emerson, R. C. and Gerstein, G. L. (1977b). Simple striate neurons in the cat. II. Mechanisms underlying directional asymmetry and directional selectivity, *J. Neurophysiol.*, **40**, 136–155.

Emerson, R. C., Korenberg, M. J., and Citron, M. C. (1988). Measurement of a simple-cell threshold function in cat's striate cortex, *Soc. Neurosci. Abstr.*, **14**, 899.

Emerson, R. C., Korenberg, M. J., and Citron, M. C. (1989). Identification of intensive nonlinearities in cascade models of visual cortex and its relation to cell classification, in *Advanced Methods of Physiological System Modeling,* Marmarelis, V. Z., Ed., Plenum Press, New York, 97–111.

Emerson, R. C., Korenberg, M. J., and Citron, M. C. (1992b). Identification of complex-cell intensive nonlinearities in a cascade model of cat visual cortex. *Biol. Cybern.*, **66** (4), 291–300.

Emerson, R. C. and Vaughn, W. J. (1987). Nonlinear spatiotemporal analysis of directionally selective cortical neurons, in *Advanced Methods of Physiological System Modeling,* Marmarelis, V. Z., Ed., USC Biomedical Simulations Resource, Los Angeles, 172–204.

Hamilton, D. B., Albrecht, D. G., and Geisler, W. S. (1989). Visual cortical receptive fields in monkey and cat: spatial and temporal phase transfer function, *Vision Res.*, **29**, 1285–1308.

Heeger, D. J. (1991). Nonlinear model of neural responses in cat visual cortex, in *Computational Models of Visual Processing,* Landy, M. and Movshon, J. A., Eds., MIT Press, Cambridge, 119–133.

Heeger, D. J. (In press). Half-squaring in responses of cat striate cells, *Visual Neurosci.*

Henry, G. H. (1977). Receptive field classes of cells in the striate cortex of the cat, *Brain Res.*, **133**, 1–28.

Hubel, D. H. and Wiesel, T. N. (1962). Receptive fields, binocular interaction and functional architecture in the cat's visual cortex, *J. Physiol. (London),* **160**, 106–154.

Ikeda, H. and Wright, M. J. (1975). Spatial and temporal properties of 'sustained' and 'transient' neurones in area 17 of the cat's visual cortex, *Exp. Brain Res.*, **22**, 363–383.

Jones, J. P. and Palmer, L. A. (1987). An evaluation of the two-dimensional Gabor filter model of simple receptive fields in cat striate cortex, *J. Neurophysiol.*, **58**, 1233–1258.

Knuth, D. E. (1969). *The Art of Computer Programming,* Vol. 2, Addison-Wesley, Reading, MA.

Mancini, M., Madden, B. C., and Emerson, R. C. (1990). White noise analysis of temporal properties in simple receptive fields of cat cortex, *Biol. Cybern.*, **63**, 209–219.

Marmarelis, V. Z., Citron, M. C., and Vivo, C. (1986). Minimum-order Wiener modeling of spike-output systems, *Biol. Cybern.*, **54**, 115–123.

Marmarelis, P. Z. and Marmarelis, V. Z. (1978). *Analysis of Physiological Systems: The White-Noise Approach,* Plenum Press, New York.

McLean, J. and Palmer, L. A. (1989). Contribution of linear spatiotemporal receptive field structure to velocity selectivity of simple cells in area 17 of cat, *Vision Res.*, **29**, 675–679.
Movshon, J. A., Thompson, I. D., and Tolhurst, D. J. (1978a). Spatial summation in receptive fields of simple cells in the cat's striate cortex, *J. Physiol. (London)*, **283**, 53–77.
Movshon, J. A., Thompson, I. D., and Tolhurst, D. J. (1978b). Receptive field organization of complex cells in the cat's striate cortex, *J. Physiol. (London)*, **283**, 79–99.
Poggio, T. and Reichardt, W. (1976). Visual control of orientation behavior in the fly. II. Towards the underlying neural interactions, *Q. Rev. Biophys.* **9**, 377–438.
Pollen, D. A., Gaska, J. P., and Jacobson, L. D. (1989). Physiological constraints on models of visual cortical function, in *Models of Brain Function,* Cotterill, R. M. J., Ed., Cambridge University Press, London, 115–135.
Reid, R. C., Soodak, R. E., and Shapley, R. M. (1987). Linear mechanisms of directional selectivity in simple cells of cat striate cortex, *Proc. Natl. Acad. Sci. U.S.A.*, **84**, 8740–8744.
Reid, R. C., Soodak, R. E., and Shapley, R. M. (1991). Directional selectivity and spatiotemporal structure of receptive fields of simple cells in cat striate cortex. *J. Neurophysiol.*, **66**, 505–529.
Schiller, P. H., Finlay, B. L., and Volman, S. F. (1976). Quantitative studies of single-cell properties in monkey striate cortex. I. Spatiotemporal organization of receptive fields, *J. Neurophysiol.*, **39**, 1288–1399.
Sherman, S. M., Watkins, D. W., and Wilson, J. R. (1976). Further differences in receptive field properties of simple and complex cells in cat striate cortex, *Vision Res.*, **16**, 919–927.
van Santen, J. P. H. and Sperling, G. (1985). Elaborated Reichardt detectors, *J. Opt. Soc. Am.* **A2**, 300–321.

Chapter 4

NONLINEAR FUNCTIONAL REPRESENTATIONS FOR MOTION DETECTION AND SPEED ESTIMATION SCHEMES

S. H. Courellis and V. Z. Marmarelis

TABLE OF CONTENTS

I. Introduction 92

II. Local Motion Detection Mechanisms 92

III. Volterra Functionals 93
- A. Volterra Series and Kernels 93
- B. Interpretation of Volterra Kernels 93
- C. The Notion of Cross-Kernel 94

IV. Kernels and Motion Extraction Mechanisms 95
- A. Cross-Kernels and Motion Detection 95
- B. Design Issues of Motion Detecting Cross-Kernels 98
- C. Speed Range Issues of Motion Detecting Cross-Kernels 100
- D. Speed Estimation in Motion Detecting Cross-Kernels 102

V. Kernels of Known Motion Detectors 104
- A. Correlation Model 104
- B. Motion Energy Models 106

VI. Conclusions 107

Acknowledgment 108

References 108

I. INTRODUCTION

Local mechanisms for the extraction of motion from visual information have gained wide acceptance as schemes that biological systems use to accomplish this task. Although feature identification may play a role in extracting motion from visual information (Braddick, 1980), it is generally not necessary. Psychological and behavioral experiments in various species, using periodic or random moving patterns, indicate that no prominent characteristics, as required by feature identification techniques, of the moving pattern are necessary to detect motion. Therefore, local schemes, which in the simplest case operate directly on local distribution of light intensity, may constitute the primary means for the representation and implementation of motion extraction systems.

In this article, we will explore the use of (Volterra) functionals in modeling and implementing local motion detection mechanisms (LMDMs). In Section II, we discuss the operational principles of LMDMs, which serve as the foundation for the employment of functional series as a tool for analysis and design of such mechanisms. In Section III, we summarize the background of the particular functional series we use in this article (Volterra Series). In Section IV, a theoretical approach to functional representation for motion detection and speed estimation schemes is developed. In Section V, we discuss the functional analysis of well-known movement detectors. Summary and conclusions comprise the last section.

II. LOCAL MOTION DETECTION MECHANISMS

LMDMs receive (preprocessed) light intensity signals, elicited by motion in the visual field, and produce information about the direction and/or speed of moving objects. The feature which enables such schemes to provide direction information is called directional selectivity. An elementary LMDM is directionally selective if its output responds to motion in a particular direction (preferred direction) and remains ''silent'' for all other directions. There are several ways to implement directional selectivity, and all of them share the same principles of operation. In order for a mechanism to qualify as a directionally selective LMDM, it has to satisfy certain requirements (Poggio and Reichardt, 1973; Reichardt, 1987; Borst and Egelhaaf 1989).

First, it needs to employ at least two inputs. The direction of motion can be represented as a vector, and, therefore, at least two points are necessary for its specification. A single input would be unable to distinguish between an object moving from left to right, from one moving from right to left or from a stationary object of time varying intensity.

Second, the signals from the two inputs should be asymmetrically processed. Otherwise, the two inputs can be interchanged without any effect on the output, resulting in a nondirectionally selective mechanism.

Third, a nonlinear interaction between the two inputs is required. The reason is that in order to obtain an estimate of direction with smaller fluctuations, the output of an LMDM is often averaged over time. If linear interaction between the two inputs were assumed, the output would be equivalent to the linear interaction of the two averaged inputs. However, averaged input signals may lose their characteristic temporal variation and, hence, behave as if the same signal propagates down both input paths. As a consequence, the two inputs could be interchanged without any difference on the output, and, thus, no directional selectivity could be achieved.

These three principles of operation of an LMDM establish the minimum requirements for its implementation. Therefore, for detection of motion, it suffices to use a nonlinear asymmetric dynamic system with two inputs, or, phrased differently, two inputs from two spatial locations which interact in a nonlinear asymmetric dynamic manner. Although these statements are equivalent, the first one alludes to a functional representation while the second one to a hardwired configuration of the LMDM. The following two sections will provide further insight into such representations.

III. VOLTERRA FUNCTIONALS

In many engineering applications, we study nonlinear transformations by considering Taylor series expansions (power series) and maintaining terms up to such order that a satisfactory approximation is achieved or the problem remains tractable. Similarly, for nonlinear dynamic systems (continuous or discrete time), we can construct an expansion of the system functional to a series of canonical functionals and maintain as high an order as necessary for the series to describe the behavior of the system reasonably well. The series of functionals to be considered here is the Volterra series. There is a conceptual and mathematical connection between Taylor series and Volterra series, where the latter can be thought of as a generalization of the former (Marmarelis and Marmarelis, 1978).

A. VOLTERRA SERIES AND KERNELS

Consider a single-input, time invariant, nonlinear, dynamic system whose output is denoted by $y(t)$, $t \epsilon [0,\infty)$ and input by $x(t)$. The Volterra series expansion of the system is expressed as (Volterra, 1930):

$$y(t) = \sum_{n=0}^{\infty} \int_0^{\infty} \cdots \int_0^{\infty} k_n(\tau_1, \tau_2, \cdots, \tau_n) \prod_{k=1}^{n} x(t - \tau_k)\, d\tau_k \qquad (1)$$

The quantities $k_n(\tau_1, \ldots, \tau_n)$, $n = 0,1,2, \ldots$ constitute the Volterra kernels of the system, and they fully characterize the nonlinear dynamics of the system. A sufficient condition for the convergence of the Volterra series is (Marmarelis, 1987):

$$\int_0^{\infty} \cdots \int_0^{\infty} |k_n(\tau_1, \tau_2, \cdots, \tau_n)|\, d\tau_1 \cdots d\tau_n \leq A_n \qquad A_n \epsilon R^+ \qquad (2)$$

where the scalar value $\{A_n\}$ form a convergent series.

In general, the Volterra series include an infinite number of terms, but for practical purposes only a finite number is considered, normally limited to the zeroth, first-, and second-order kernels. This will prove to be adequate for the purposes of this study, as shown in Section IV.

B. INTERPRETATION OF VOLTERRA KERNELS

Volterra kernels are quantities characteristic of a system, without any dependence on the input. In addition to reproducing the system output for a given input, their structure conveys information about the internal configuration of the system, albeit in a nonunique way (i.e., different configurations may correspond to similar kernels). A general property of Volterra kernels is their symmetry for any permutation of their arguments.

Analytical derivations of Volterra kernels for a number of configurations are available. For a static nonlinearity of the form $y(t) = f[x(t)]$, the Volterra kernels are

$$k_n(\tau_1, \cdots \tau_n) = a_n \delta(\tau_1)\delta(\tau_2) \cdots \delta(\tau_n), \quad n = 0,1, \cdots \infty \qquad (3)$$

where $\{a_n\}$ are the coefficients of a power series expansion of f(·).

The configuration of a linear system with impulse response $g(t)$ followed by static nonlinearity $f(\cdot)$ (LN configuration) yields Volterra kernels of the form:

$$\begin{aligned} k_0 &= a_0 \\ k_n &= a_n g(\tau_1) g(\tau_2) \cdots g(\tau_n), \quad n = 1,2, \cdots, \infty \end{aligned} \tag{4}$$

For LN configurations, the Volterra kernels are separable functions.

More complicated configurations, such as LNL (Korenberg, 1973; Marmarelis and Marmarelis, 1978) or nonlinear feedback (Marmarelis, 1989), have been considered, and kernel expressions are available for them. Finally, it should be noted that the first-order Volterra kernel does not correspond to the impulse response of the system. To demonstrate this property substitute $x(t) = \delta(t)$ in Equation 1. Then

$$y(t) = k_0 + k_1(t) + k_2(t,t) + \cdots + k_n(t,t, \cdots, t) + \cdots \neq k_1(t) \tag{5}$$

The estimation of Volterra kernels is not an easy task, even though this task is simplified in the cases of systems up to the second order. To address the problem, Wiener orthogonalized the Volterra series for Gaussian white-noise input and decoupled the various terms allowing estimation of the Wiener kernels involved in each orthogonal functional (Wiener, 1958). The result was the Wiener series with Wiener functionals orthogonal to each other. Generally, a Wiener series is expressed as:

$$y(t) = \sum_{n=0}^{\infty} G_n[h_n(\tau_1, \cdots, \tau_n); x(t'), t' \leq t] \tag{6}$$

where $h_n(\tau_1, \ldots, \tau_n)$, $n = 0,1, \ldots$ are the Wiener kernels, G_n the Wiener functionals, and the expected value $E\{G_n G_m\} = 0$, $m \neq n$. An explicit relationship between Wiener and Volterra kernels provides the latter when the former have been determined. Note, however, that second-order nonlinear systems have identical Wiener and Volterra kernels. Estimation of Wiener kernels can be performed in more than one way, but this matter is beyond the scope of the present study.

C. THE NOTION OF CROSS-KERNEL

So far only single-input/single-output (SISO) systems have been considered. In general, functional representations and kernels for multiple-input/multiple-output (MIMO) systems can be derived (Marmarelis and Marmarelis, 1978). When the effect of one input on one output of the MIMO system is involved, the corresponding kernels are characterized as *self-kernels*. When the effect of more inputs on one output is concerned, the term *cross-kernels* is used. A cross-kernel describes the dynamic interaction of the inputs involved, as it affects the system output. For instance, if the component of the output attributed to the nth cross-kernel is denoted by $y_{cn}(t)$, then:

$$y_{cn}(t) = \int_0^{\infty} \cdots \int_0^{\infty} k_{cn}(\tau_1, \cdots, \tau_n) x_1(t - \tau_1) x_2(t - \tau_2) \cdots x_n(t - \tau_n)\, d\tau_1 d\tau_2 \cdots d\tau_n \tag{7}$$

where $k_{cn}(\tau_1, \ldots, \tau_n)$ is the nth-order cross-kernel, and $x_1(t), \ldots, x_n(t)$ are the n-interacting inputs associated with the kernel.

The simplest case of a cross-kernel is a second-order one for a system with two inputs and one output. As an example, consider the system in Figure 1. Two inputs $x_1(t)$ and $x_2(t)$

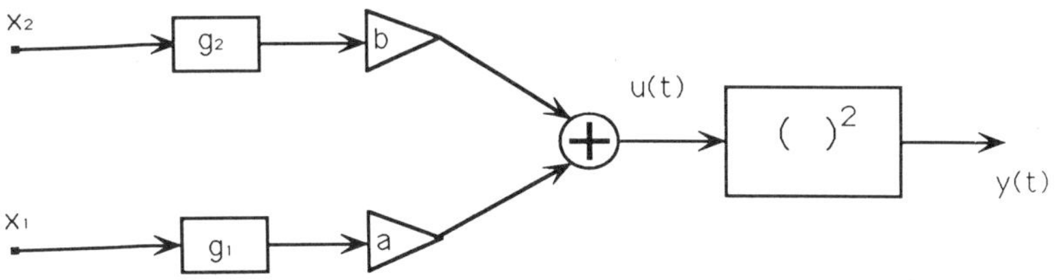

FIGURE 1. A two-input/one-output second-order nonlinear system.

are independently processed by linear filters with impulse responses $g_1(t)$ and $g_2(t)$ and multiplied by a and b, respectively. Then, they are added to yield $u(t)$ which, in turn, is squared via a second-order static nonlinearity to yield the output $y(t)$:

$$y(t) = a^2 \int_0^\infty \int_0^\infty g_1(\tau_1)g_1(\tau_2)x_1(t - \tau_1)x_1(t - \tau_2)\, d\tau_1 d\tau_2 +$$

$$+ 2ab \int_0^\infty \int_0^\infty g_1(\tau_1)g_2(\tau_2)x_1(t - \tau_1)x_2(t - \tau_2)\, d\tau_1 d\tau_2 +$$

$$+ b^2 \int_0^\infty \int_0^\infty g_2(\tau_1)g_2(\tau_2)x_2(t - \tau_1)x_2(t - \tau_2)\, d\tau_1 d\tau_2 \tag{8}$$

The system has two second-order self-kernels, each associated with an input:

$$k_{11}(\tau_1,\tau_2) = a^2 g_1(\tau_1)g_1(\tau_2)$$

$$k_{22}(\tau_1,\tau_2) = b^2 g_2(\tau_1)g_2(\tau_2) \tag{9}$$

and one second-order cross-kernel which depicts the interaction between the two inputs:

$$k_{12}(\tau_1,\tau_2) = 2ab\ g_1(\tau_1)g_2(\tau_2) \tag{10}$$

Note that both self-kernels are symmetric around the diagonal $\tau_1 = \tau_2$, while the cross-kernel is not.

As indicated in Section II, the minimum requirements for an effective motion detection scheme call for a two-input asymmetric nonlinear system formulation. Therefore, a two-input second-order Volterra functional representation is compatible with these requirements.

IV. KERNELS AND MOTION EXTRACTION MECHANISMS

As stated above, the minimum configuration for motion detection requires two interacting inputs and one output, as well as an asymmetric second-order cross-kernel that captures this interaction. In the following, we will explore some cross-kernel candidates for such motion detection schemes.

A. CROSS-KERNELS AND MOTION DETECTION

Consider an elementary motion detection mechanism with two inputs x_1 and x_2 and one output y. The interaction of the two inputs is represented by a cross-kernel $k_{12}(\tau_1,\tau_2)$ as:

$$y(t) = \int_0^\infty \int_0^\infty k_{12}(\tau_1,\tau_2)x_1(t - \tau_1)x_2(t - \tau_2)\, d\tau_1 d\tau_2 \tag{11}$$

In the case of motion from x_1 to x_2, $x_2(t)$ is a delayed version of $x_1(t)$ by time t_0, as long as it takes for a point on the moving pattern to travel from x_1 to x_2. Therefore, Equation 11 becomes

$$y_{12}(t) = \int_0^{\infty}\int_0^{\infty} k_{12}(\tau_1,\tau_2)x_1(t - \tau_1)x_1(t - \tau_2 - t_0)\, d\tau_1 d\tau_2 \tag{12}$$

For motion in the opposite direction, similar reasoning leads to:

$$y_{21}(t) = \int_0^{\infty}\int_0^{\infty} k_{12}(\tau_1,\tau_2)x_1(t - \tau_1)x_1(t - \tau_2 + t_0)\, d\tau_1 d\tau_2 \tag{13}$$

To endow the motion detection mechanism with the property of directional selectivity, an appropriate cross-kernel $k_{12}(\tau_1,\tau_2)$ needs to be chosen. Depending on what sign of the output signal indicates the direction of motion, cross-kernels can be designed to implement oriented motion schemes (i.e., motion in a specific direction elicits activity at the output, while motion in the opposite direction does not) or opponent motion schemes (i.e., motion in a specific direction produces activity of a specific sign at the output, while motion in the opposite direction produces activity of the opposite sign). In order to obtain response of a particular sign, appropriate values should be assigned to the cross-kernel in the regions of the (τ_1,τ_2) domain, over which the product of $x_1(t-\tau_1)x_2(t-\tau_2)$ is nonzero for all t and for motion in either direction.

In the case of motion in a specific direction and for positive $x_1(t-\tau_1)$, $x_2(t-\tau_2)$ signals, selection of nonzero values for the cross-kernel over the region of nonzero $x_1(t-\tau_1)x_2(t-\tau_2)$ product, in the (τ_1,τ_2) domain, yields nonzero output. For motion in the opposite direction, selection of zero values for the cross-kernel over the region of nonzero $x_1(t-\tau_1)x_2(t-\tau_2)$ product, elicits zero output. Such a cross-kernel implements an oriented motion detection scheme. Furthermore, if motion from x_1 to x_2 elicits activity (preferred direction) while motion from x_2 to x_1 does not (null direction), Equations 12 and 13 yield:

$$y_{12}(t) \neq 0,\ y_{21}(t) = 0,\ \forall t \in R_0^+ \tag{14}$$

When positive values for the cross-kernel over the region of nonzero $x_1(t-\tau_1)x_2(t-\tau_2)$ product are considered for motion in a specific direction, positive $x_1(t-\tau_1)$, $x_2(t-\tau_2)$ signals yield positive output. In addition, negative values for the cross-kernel over the region of nonzero $x_1(t-\tau_1)x_2(t-\tau_2)$ product for motion in the opposite direction, elicit negative output. Such a cross-kernel implements an opponent motion detection scheme, and if motion from x_1 to x_2 elicits positive activity, and motion from x_2 to x_1 produces negative activity, Equations 12 and 13 yield:

$$y_{12}(t) > 0, \quad y_{21}(t) < 0,\ \forall\, t \in R_0^+ \tag{15}$$

As a consequence, to design a motion detecting cross-kernel, one needs to determine the above-mentioned regions in the (τ_1,τ_2) domain, where the product $x_1(t-\tau_1)x_2(t-\tau_2)$ assumes nonzero values for motion in both directions. To do so, we will consider motion from x_1 to x_2 first. In this case, as shown in Figure 2a, the product of the two input signals lies at all times diagonally on the left of the line defined by the equation $\tau_2 = \tau_1 + t' - t_0$ (Region I). In this equation, t_0 is the relative delay between signals (as explained in previous section) and t' the effective temporal extent of $x_1(t)$ and $x_2(t)$ (i.e., after t' time, the values of $x_1(t)$ and $x_2(t)$ become negligible). For motion in the opposite direction, as

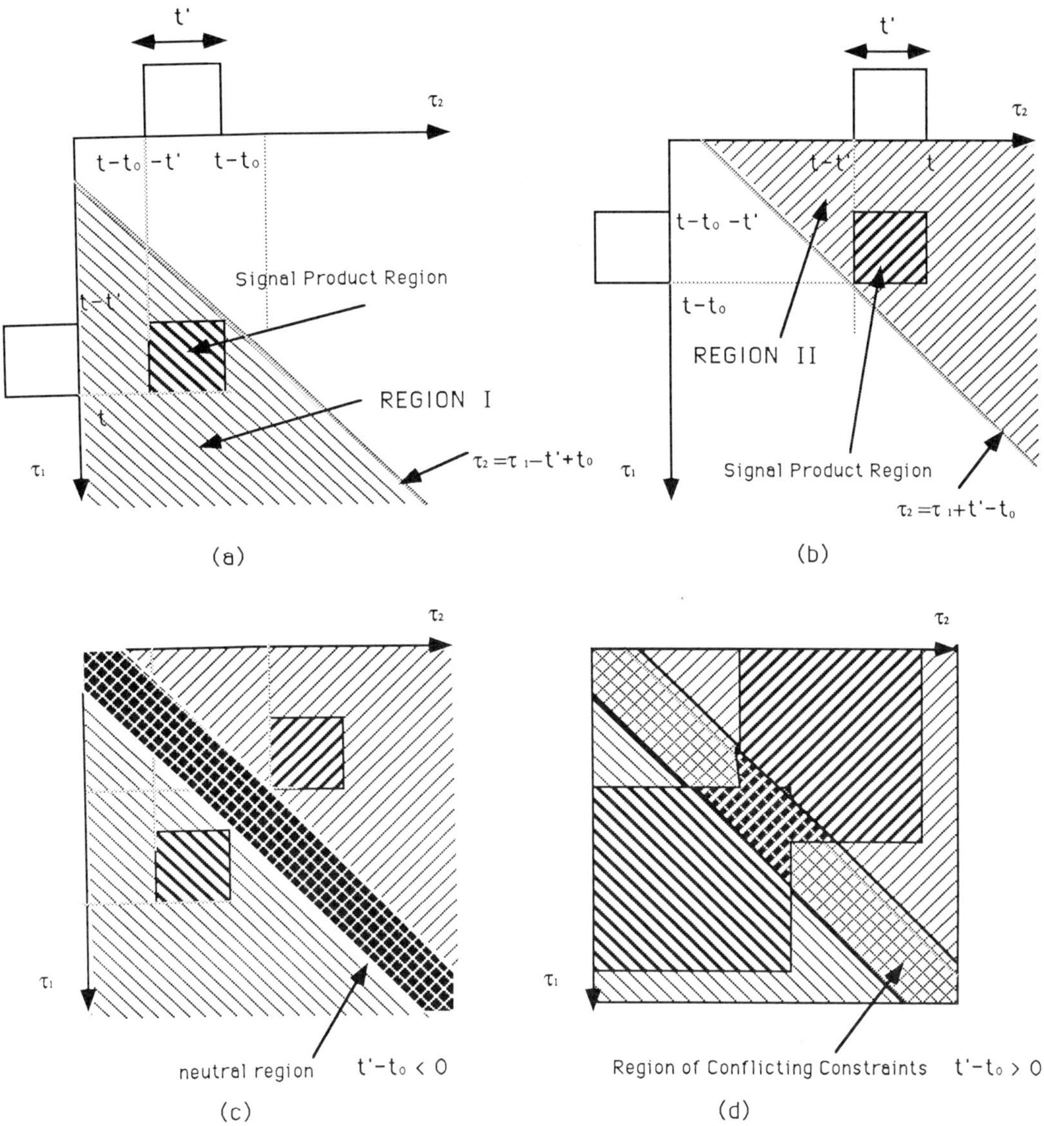

FIGURE 2. Regions that the signal product of the two inputs may lie in the (τ_1, τ_2) domain for motion in one direction (a) and the opposite direction (b). Combining the above two regions a neutral stripe (c) or a stripe of conflicting constraints (d) may result.

depicted in Figure 2b, the nonzero product region lies at all times diagonally to the right of the line defined by the equation $\tau_2 = \tau_1 - t' + t_0$ (Region II). Therefore, a cross-kernel implementing an oriented motion detection scheme, with the preferred direction from x_1 to x_2, will satisfy:

$$k_{12}(\tau_1, \tau_2) \begin{cases} \neq 0 \text{ for } \tau_2 < \tau_1 + t' - t_0 \\ = 0 \text{ for } \tau_2 > \tau_1 - t' + t_0 \end{cases} \tag{16}$$

while a cross-kernel of the opponent motion detection type, with positive response for motion from x_1 to x_2, will satisfy:

$$k_{12}(\tau_1,\tau_2)\begin{cases} > 0 \text{ for } \tau_2 < \tau_1 + t' - t_0 \\ < 0 \text{ for } \tau_2 > \tau_1 - t' + t_0 \end{cases} \tag{17}$$

It is evident from Figure 2a and b that depending on the difference $t' - t_0$, there may be a stripe between Regions I and II (neutral region), in which the cross-kernel is not constrained to assume any specific values (Figure 2c), or there may be overlap between Region I and Region II (Figure 2d) leading to conflict between the conditions imposed for motion detection on second-order cross-kernels (Equations 16 and 17).

Such conflicts can be resolved by relaxing the condition on the cross-kernel for one of the directions of motion, or by reaching a compromise between the two of them. A very obvious compromise consists of using the diagonal $\tau_1 = \tau_2$ as a common boundary between Region I and Region II. To illustrate this compromising solution we cite two examples.

Consider a directionally selective unit (Figure 3a) in a way that motion in the preferred direction evokes a strong output signal, while emotion in the null direction evokes a relatively weak one. A bar moving in the preferred direction is sensed by x_1 first and then by x_2. At a certain time instant, the profile of the product $x_1(t-\tau_1)x_2(t-\tau_2)$ will look in the (τ_1,τ_2) domain as Figure 3b shows. When the bar moves in the opposite direction the corresponding profile appears in Figure 3c. A cross-kernel which has all zero values on one side of the diagonal $(\tau_1 = \tau_2)$ will accomplish this task (Figure 3d). The cross-kernel depicted in Figure 3d, operating on the inputs, will result in a strong output signal for the preferred direction and a weak output signal for the null direction.

Now, suppose that the direction of motion were to be indicated by a change in the sign of the output activity. A different cross-kernel would be employed, and Figure 3e provides one possibility. The cross-kernel is positive on one side of the diagonal and negative on the other. Replicating the rationale developed previously, activity of opposite sign results at the output for opposite directions of motion.

In this section, we provided fundamental principles for the design of minimum-order (second-order) cross-kernels for motion detection. Higher-order kernels are possible for this task, but they are more complicated in structure, tractability, and implementation.

B. DESIGN ISSUES OF MOTION DETECTING CROSS-KERNELS

In the previous section, it was shown that certain specifications on motion detecting cross-kernels may lead to conflicts (Inequalities 16 and 17). Depending on the values of t' and t_0, there may be regions in the (τ_1,τ_2) domain where both conditions in Inequalities 16 or 17 hold simultaneously. To quantify this problem, we need to specify t' and t_0 explicitly. If l_x is the distance between the two inputs x_1 and x_2 of motion detection mechanism, and the object moves with speed c, then $t_0 = l_x/c$. Furthermore, if the spatial extent (size) of the moving pattern in the direction of motion is l_1, then $t' = l_1/c$. Thus, for fixed distance l_x, the difference $t_0 - t' = (l_x - l_1)/c$ will depend on the length of the pattern l_1 and the speed of motion c. This expression defines the region of conflicting conditions.

According to Inequalities 16 or 17, for certain values of $t_0 - t'$, there exists a region:

$$\tau_1 - (t_0 - t') < \tau_2 < \tau_1 + (t_0 - t') \tag{18}$$

such that $k_2(\tau_1,\tau_2)$ can assume any values without any impact on motion detection (neutral

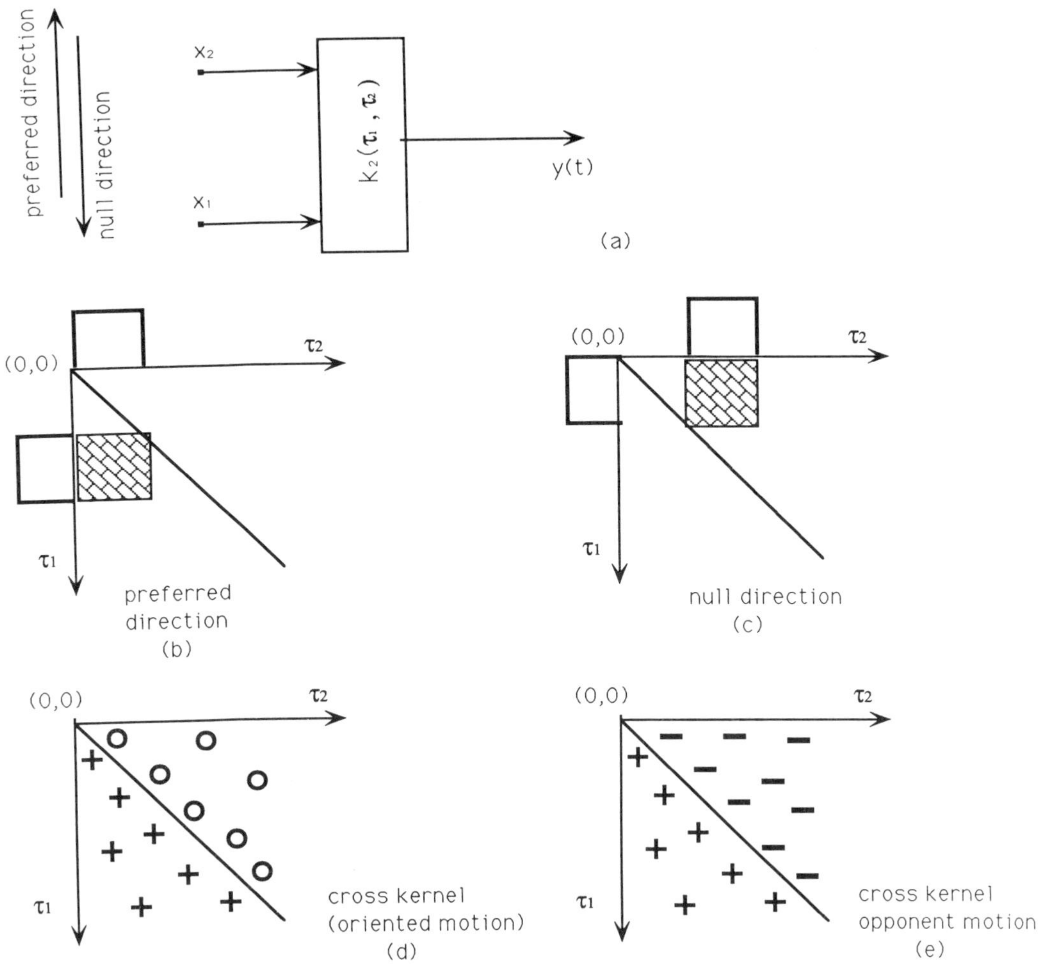

FIGURE 3. Two inputs combined via a second-order cross-kernel (a) provide a signal product region diagonally to the left (b) for motion in the preferred direction and diagonally to the right (c) for motion in the null direction. Oriented motion detection is implemented if the cross-kernel is zero on one side of the diagonal (d) and opponent motion detection when the cross-kernel has opposite signs on the two sides of the diagonal (e).

region; Figure 2c). Observing Equation 18, we conclude that such regions are possible when $t_o - t' > 0$. This inequality implies that:

$$l_x - l_I > 0 \Rightarrow l_x > l_I \tag{19}$$

In the case that $l_x < l_I$, regions of conflicting constraints appear (Figure 2d). Consequently, regions of conflict develop when the size (length) of the pattern in the direction of motion exceeds the distance between the inputs of the motion detecting mechanism. It should be noted that speed is not a limiting factor, although it may provide a regulatory effect. This effect is manifested when the difference $l_x - l_I$ is small and the speed c is large enough to make the ratio $(l_x - l_I)/c$ very small. In such cases, large values of c restrict the region of conflict to a narrow stripe which may introduce negligible error in detecting the correction direction of motion.

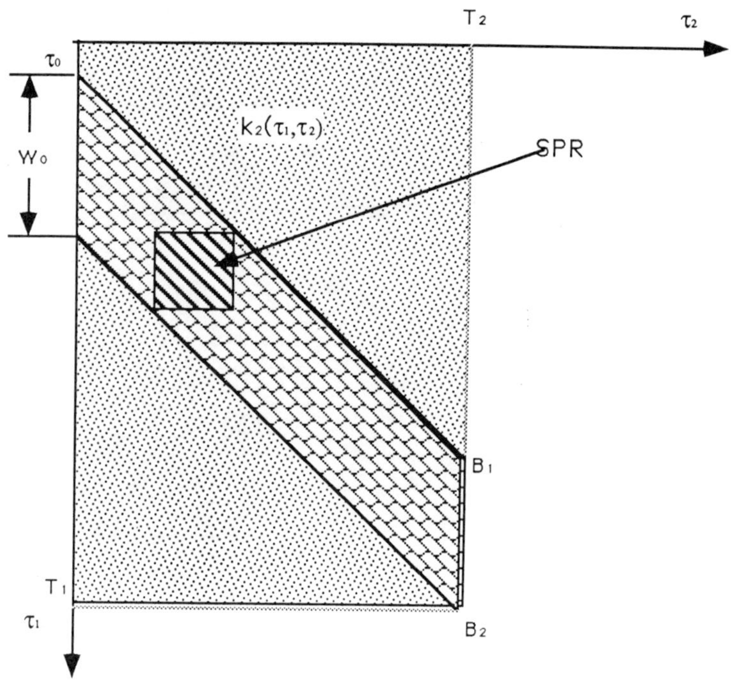

FIGURE 4. The size (w_0) and the location (τ_0) of the stripe traced by the signal product region (SPR) vary with the size and the speed of the object in the direction of motion.

When neither Inequality 19 holds nor the ratio $(l_x - l_l)/c$ is small, a compromising solution, as explained in the previous section, can be sought. It should be further noted that for proper representation of moving patterns, spatial sampling according to the Nyquist criterion (i.e., for proper discrete representation, continuous spatial patterns should be sampled at least twice as fast as the highest frequency component they contain) makes inequality (Equation 19) invalid, and, therefore, regions of conflict are present. Deviation from the Nyquist criterion can eliminate these regions by introducing undersampling of the spatial pattern.

C. SPEED RANGE ISSUES OF MOTION DETECTING CROSS-KERNELS

One of the major concerns in designing motion detecting cross-kernels is the range of speeds they can accommodate. As it will be shown in this section, the range of speeds is associated with the temporal extent of the cross-kernel $k_{12}(\tau_1,\tau_2)$ (i.e., its memory).

To demonstrate the relationship between speed range and kernel memory, let us assume the two input arrangement of Figure 3a, with an input signal profile in the (τ_1,τ_2) domain such as the one in Figure 4, for motion in the preferred direction. In Section IV.B, we determined the boundaries specifying the areas in which the signal activity lies, for motion in the preferred and null directions including the neutral or conflict zone. However, throughout the motion event (with constant speed) and for motion in the preferred direction, the signal product region (SPR) will lie between the above-mentioned boundary (B1; Figure 4) and the boundary traced over time by the diagonally left-most point of the SPR (B2; Figure 4). Mathematically, the area specified by these two boundaries can be expressed as:

$$\tau_1 - (t' + t_0) < \tau_2 < \tau_1 + (t' - t_0) \tag{20}$$

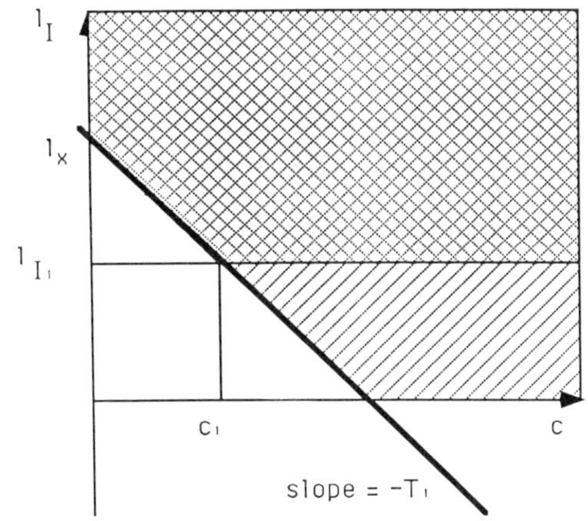

FIGURE 5. The shaded area specifies the region of sizes and speeds of an object moving in a specific direction necessary for a cross-kernel with temporal extent T_1 in the τ_1-direction to successfully detect motion.

or

$$\tau_1 - \frac{l_I + l_z}{c} < \tau_2 < \tau_1 + \frac{l_I - l_x}{c} \tag{21}$$

Consequently, the moving SPR regions will lie inside the stripe specified by Inequality 21. This stripe changes area and location when l_I and c vary. The offset (τ_0) in the τ_I-dimension:

$$\tau_0 = t_0 - t' = \frac{l_x - l_I}{c} \tag{22}$$

and the width w_0 of the stripe along th τ_I-dimension

$$w_0 = 2\,t' = 2\frac{l_I}{c} \tag{23}$$

can be considered as key parameters of change. If $k_{12}(\tau_1,\tau_2)$ extends over T_I in the τ_I-direction,

$$\tau_0 < T_1 \Rightarrow \frac{l_x - l_I}{c} < T_1 \tag{24}$$

for motion to be detected. If T_2 is the temporal extent of $k_{12}(\tau_1,\tau_2)$ in the τ_2-direction, for motion in the opposite direction to be detected with opposite sign, reasoning similar to the above, leads to $(l_x - l_I)/c < T_2$.

Graphically, Inequality 24 represents the area defined by the intersection of the first quadrant $(c > 0,\ l_I > 0)$ in the (c,l_I) domain and the positive half-plane of the line $l_I = -c\,T_I + l_x$ (Figure 5). Figure 5 also provides the minimum speed of motion (c_I) successfully detected for a specific pattern size l_{II} and temporal extent T_I. In the case of spatial under-sampling $(l_x > l_I)$, large values of T_I enable $k_{12}(\tau_1,\tau_2)$ to accommodate lower speed values. However, large T_I values reduce the sensitivity of the motion detection mechanism by disabling it to distinguish between successive motion events happening very close to each other in time.

An upper bound on the speed range is imposed by the desired amplitude (strength) of the output signal. High-speed motion reduces the size of the SPR without altering the values of the signal product, and, therefore, the result of the convolution of the signal product with the cross-kernel yields small values. This last problem can be partially rectified by assigning large values to the cross-kernel near the appropriate boundary in the (τ_1,τ_2) domain.

D. SPEED ESTIMATION IN MOTION DETECTING CROSS-KERNELS

The previous sections discussed how appropriately designed second-order cross-kernels can discriminate the direction of motion. To move one step further, we will explore specifications on motion detecting cross-kernels enabling them to extract speed information. Having the direction of motion encoded in the sign of the output signal, we can further encode speed in the amplitude of the output signal.

In order to accomplish this, appropriate motion detecting cross-kernels need to be selected, which, in addition to the direction of motion, will provide estimates of the speed, irrespective of changes in the magnitude and/or shape of the input signals, resulting from variations in the size, intensity, and/or speed of moving patterns. When the second-order cross-kernel is considered, Equation 11 is used to implement both detection of direction and estimation of speed. Therefore, proper speed estimation will be achieved if we first eliminate the dependence of the output signal $y(t)$ on variations in magnitude and/or shape of the inputs $x_1(t)$ and $x_2(t)$. To eliminate this dependence, we must require that signals entering the motion detection mechanism be constrained in shape and size irrespective of the moving pattern. As a result, signal waveforms characterized by fixed shape and effective duration (t_d) emerge for any moving pattern. A method ensuring such invariability is described in Courellis and Marmarelis (1990). According to this method, moving spots and edges are processed to produce spikes, the standard input waveforms required by the proposed velocity estimation scheme. To extract speed information from such constrained inputs, we focus on the relative temporal displacement t_0 between the input signals. The relative temporal displacement, or relative delay, depends on the speed of the moving pattern. Small speed values correspond to large relative delays and large speed values to small relative delays. It is apparent that a monotonic relationship exists between speed values and relative delays, and, therefore, motion detecting cross-kernels should be equipped with the ability to monotonically map this relative delay to the amplitude of the output signal.

One way to implement this mapping is by using a cross-kernel $k_{12}(\tau_1,\tau_2)$ with magnitude dependent on the difference of its arguments, i.e., $k_{12}(\tau_1 - \tau_2)$. If we let $\tau = \tau_1 - \tau_2$, Equation 11 yields:

$$y(t) = \begin{cases} \int_{\pm t_0 - t_d}^{\pm t_0 + t_d} k_{12}(\tau)\, R_{12}(|\tau|)\, d\tau & \text{for } t > t_0 \\ 0 & \text{for } t < t_0 \end{cases} \tag{25}$$

where $R_{12}(|\tau|)$ is the cross-correlation function between $x_1(t)$ and $x_2(t)$, and the sign in front of t_0 at the limits of the integral depends on the direction of motion. For motion in the preferred direction $x_2(t) = x_1(t - t_0)$, and Equation 12 provides

$$y_{12}(t) = \begin{cases} \int_{t_0 - t_d}^{t_0 + t_d} k_{12}(\tau)\, R_{11}(|t_0 - \tau|)\, d\tau & \text{for } t > t_0 \\ 0 & \text{for } t < t_0 \end{cases} \tag{26}$$

where $R_{11}(|\tau|)$ represents the autocorrelation function of $x_1(t)$. For motion in the opposite direction, similar reasoning leads to

$$y_{21}(t) = \begin{cases} \int_{-t_0-t_d}^{-t_0+t_d} k_{12}(\tau)\, R_{11}(|t_0 + \tau|)\, d\tau & \text{for } t > t_0 \\ 0 & \text{for } t < t_0 \end{cases} \tag{27}$$

In both cases, the autocorrelation function is translated at t_0 and $-t_0$, respectively, weighed by $k_{12}(\tau)$ in a neighborhood of $2t_d$ around the point of translation and integrated over that neighborhood. For positive translations (motion in the preferred direction), positive values of $k_{12}(\tau)$ lead to positive output signal. For negative translations (motion in the opposite direction), zero values for $k_{12}(\tau)$ yield zero output, and negative values yield negative output. These are general guidelines for the design of motion detecting cross-kernels able to estimate speed as well. To illustrate the use of motion detecting cross-kernels in extracting speed, we cite two examples:

(1) Assume the configuration of Figure 3a which includes two inputs x_1 and x_2 combined through a cross-kernel $k_{12}(\tau_1,\tau_2)$ to an output $y(t)$. A spot $\delta(x)$ moving with speed c from x_1 to x_2 is sensed by the two inputs as $x_1(t) = \delta(t)$ and $x_2(t) = \delta(t - l_x/c)$, with l_x the distance between the inputs. Substituting these values for x_1 and x_2 in Equation 11, we obtain

$$y(t) = k_{12}\left(t, t - \frac{l_x}{c}\right) \tag{28}$$

To detect motion from x_1 to x_2 with positive sign and from x_2 to x_1 with negative sign, an antisymmetric cross-kernel $[k_{12}(\tau_1,\tau_2) = -k_{12}(\tau_2,\tau_1)]$ would suffice. As we mentioned before, antisymmetry around the diagonal is required for directionally selective second-order cross-kernels encoding the direction of motion on the sign of the output signal. In addition, a monotonic function with respect to the difference $\tau_1 - \tau_2$ for $|k_{12}|$ should provide an estimate of speed, for instance:

$$k_{12}(\tau_1,\tau_2) = \alpha \frac{1}{\tau_1 - \tau_2} \tag{29}$$

Substitution of Equation 29 into Equation 28 yields

$$y(t) = \frac{\alpha}{l_x} c \tag{30}$$

For $\alpha = l_x$, the output becomes equal to c. It is clear that for motion in the opposite direction, $x_1(t) = \delta(t - l_x/c)$ and $x_2(t) = \delta(t)$. Consequently, following the same steps as above, we get $y(t) = -c$. The same result is obtained by direct substitution of the input signals in Equations 26 and 27.

(2) Consider the same motion conditions as described in (1), but this time the delta functions are substituted by pulses of fixed width t_d and amplitude b. Then, the autocorrelation function will be of the form

$$R_{11}(|\tau|) = \begin{cases} b^2(t_d - |\tau|) & \text{for } |\tau| < t_d \\ 0 & \text{for } |\tau| > t_d \end{cases} \tag{31}$$

Using the motion detecting cross-kernel of the form

$$k_{12}(\tau) = \begin{cases} \alpha e^{-s_1\tau} & \text{for } \tau > 0 \\ 0 & \text{for } \tau < 0 \end{cases} \quad (32)$$

Equations 26 and 27 for motion in the preferred and opposite directions yield

$$y_{12}(t) = 2\alpha b^2 \frac{(s_1 t_d - 1)sinh(s_1 t_d) + cosh(s_1 t_d)}{s_1^2} e^{-s_1 t_0}, \quad y_{21}(t) = 0 \quad (33)$$

Substituting l_x/c for t_0 and selecting α appropriately, we obtain $y_{12}(t) = e - s_1 l_x/c$ in steady state. This last relationship expresses a one-to-one mapping between the amplitude of the output and the values of the speed. As the speed values increase the amplitude of the output decreases monotonically and provides unique encoding of speed in the preferred direction.

These two examples demonstrate how motion detecting cross-kernels can be designed to extract the speed of motion. Although only the category of motion detecting cross-kernels depending on the difference of their arguments was examined in some detail, the general framework for designing velocity estimating cross-kernels was outlined.

V. KERNELS OF KNOWN MOTION DETECTORS

In this section, kernels of widely known motion detectors will be evaluated. The mathematical model representations of these motion detectors will be used to determine analytically the corresponding self- and cross-kernels. Note that many experimental studies have led to kernel estimates of motion detecting cells in the past (e.g., Marmarelis and Naka, 1973; Marmarelis and McCann, 1973; Emerson et al., 1987) with the results corroborating the theoretical postulates of the previous section in terms of cross-kernel configuration.

A. CORRELATION MODEL

The correlation model was introduced by Hassenstein and Reichardt (1953) as the result of behavioral experiments conducted on flies. It consists of two inputs, one of which delays the signal relative to the other. The two signals are multiplied and the result is integrated. The overall operation reflects evaluation of the autocorrelation function of the input signal at different delays. The total delay depends on the direction (relative delay due to asymmetrical processing between the two inputs) and speed (time elapsing for a point to travel from one input to the next) of the moving pattern. The larger the delay the smaller the value of the autocorrelation function. Therefore, motion in the preferred direction corresponds to small total delay, while motion in the null direction to large overall delay. A simple version of the correlation model is shown in Figure 6a (van Santen and Sperling, 1984) and is described by

$$y(t) = \int_0^\infty \int_0^\infty g_1(\tau_1)g_2(\tau_2)x_1(t - \tau_1)x_2(t - \tau_2)\, d\tau_1 d\tau_2 \quad (34)$$

with g_1 responding slower than g_2. Clearly, this model is fully described by only one second-order cross-kernel, $k_{12}(\tau_1,\tau_2) = g_1(\tau_1)\, g_2(\tau_2)$, shown in Figure 6c. Visual inspection shows

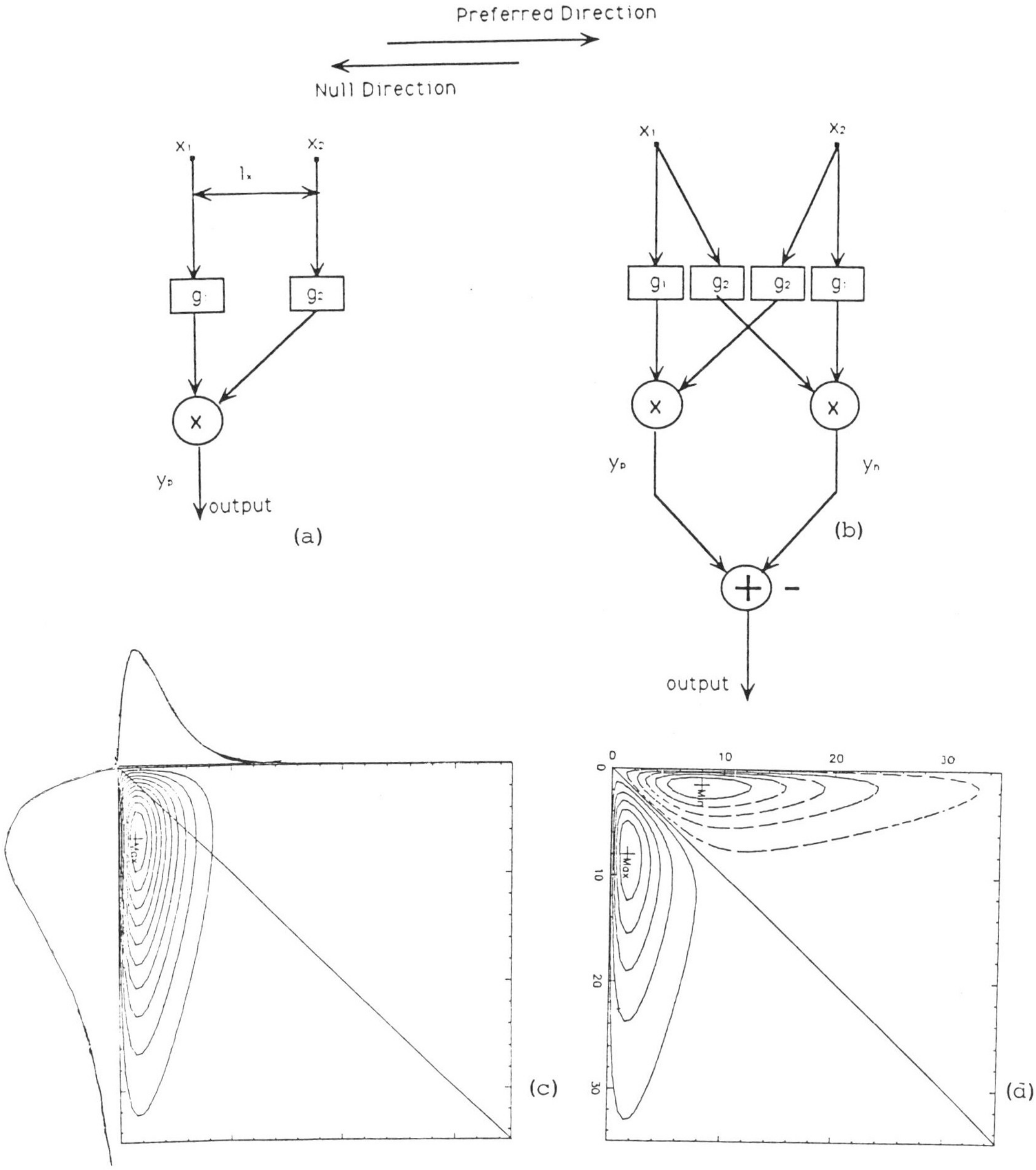

FIGURE 6. One version of the correlation model detecting oriented motion (a) and opponent motion (b) with the corresponding cross-kernels (c and d).

that most of the significant values of the cross-kernel lie on one side of the diagonal. This particular version of the correlation model (oriented motion model) is designed to elicit very low response for motion in the null direction.

An alternate version of the correlation model, providing negative activity for motion in the null direction, is shown in Figure 6b (opponent motion model). This model is described by

$$y(t) = \int_0^\infty \int_0^\infty [g_1(\tau_1)g_2(\tau_2) - g_1(\tau_2)g_2(\tau_1)]x_1(t - \tau_1)x_2(t - \tau_2)\, d\tau_1 d\tau_2 \qquad (35)$$

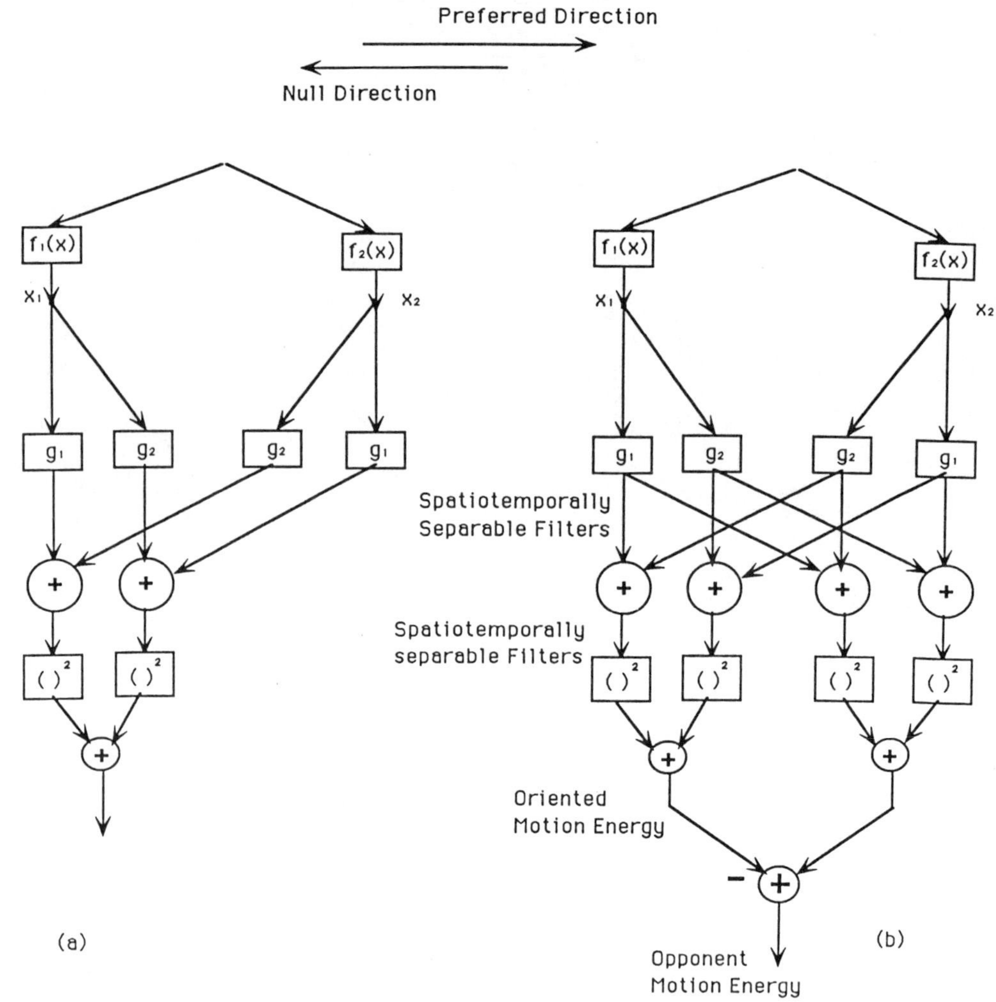

FIGURE 7. One version of the motion energy model employing two inputs and detecting (a) oriented motion energy and (b) opponent motion energy. Note that $f_1(x)$ and $f_2(x)$ imply spatial filtering before temporal processing.

The quantities g_1, g_2 are similar to the ones in the oriented motion model. This configuration is also characterized by only one second-order cross-kernel of the form $k_{12}(\tau_1,\tau_2) = g_1(\tau_1) g_2(\tau_2) - g_1(\tau_2) g_2(\tau_1)$. This cross-kernel is shown in Figure 6d and exhibits antisymmetry about the diagonal, having positive values on one side and negative on the other. Thus, both versions of the correlation model can be implemented by a single second-order cross-kernel, the minimum configuration required for motion detection.

B. MOTION ENERGY MODELS

The motion energy models (Adelson and Bergen, 1985) operate as oriented edge detectors on the trace that moving patterns draw in the spatiotemporal domain. Like the correlational models, they also come in oriented motion and opponent motion configurations. One possible realization of an energy model is shown in Figure 7. The output of the left oriented motion configuration is expressed as

$$y_{lo}(t) = \int_0^\infty \int_0^\infty \{[g_1(\tau_1)g_1(\tau_2) + g_2(\tau_1)g_2(\tau_2)]\, x_1(t - \tau_1)\, x_1(t - \tau_2) +$$

$$+ [g_1(\tau_1)g_1(\tau_2) + g_2(\tau_1)g_2(\tau_2)]\, x_2(t - \tau_1)\, x_2(t - \tau_2) +$$

$$+ [g_1(\tau_2)g_2(\tau_1) - g_1(\tau_1)g_2(\tau_2)]\, x_1(t - \tau_1)\, x_2(t - \tau_2)\}\, d\tau_1 d\tau_2 \qquad (36)$$

Equation 36 clearly indicates that the system is fully described by three kernels. Two second-order self-kernels ($k_{2x_{1y}}$, $k_{2x_{2y}}$), one for each input, and one second-order cross-kernel ($k_{2x_1x_{2y}}$). In particular:

$$k_{2x1y}(\tau_1,\tau_2) = k_{2x1y}(\tau_1,\tau_2) = g_1(\tau_1)g_2(\tau_1) + g_1(\tau_2)g_2(\tau_2) \qquad (37)$$

$$k_{2x1x2y}(\tau_1,\tau_2) = g_1(\tau_2)g_2(\tau_1) - g_1(\tau_1)g_2(\tau_2) \qquad (38)$$

Inspection of the cross-kernel reveals its similarity to the one derived for the correlational model. The opponent version of the energy model (Figure 7) yields no self-kernels but one second-order cross-kernel which comprises a scaled version of the cross-kernel derived for the opponent correlational model. Considering that for these configurations, detection of motion is attributed solely to the cross-kernel, the similarity between the cross-kernels of the correlational and the motion energy models indicates the functional equivalence between the two models.

After determining kernels for certain motion detection models, it is quite clear that the key descriptor qualifying these configurations as directionally selective has been the cross-kernel. Hence, direct realization of the cross-kernels would suffice to implement selection of the direction of motion. In all of the above models, no provisions have been made for estimating speed, and, hence, they do not reflect such a property. However, as it was shown in Section IV.D, the possibility of constructing cross-kernels to implement directional selectivity and estimate speed at the same time exists and offers the means for incorporating speed estimation in motion detection mechanisms.

VI. CONCLUSIONS

Functional representations of motion detection and speed estimation mechanisms provide a general formulation of the problem and allow comparison of their similarities, abilities, and limitations. Using the Volterra functional representation in this article, we explored several aspects of analysis and design of velocity estimation mechanisms.

It was shown that second-order cross-kernels constitute the essential and minimum factors responsible for motion detection, although higher-order cross-kernels can also be employed. They satisfy the operating principles of a motion detection mechanism stated in Section II. A cross-kernel involves at least two inputs (first principle) and is the result of the presence of a nonlinear interaction between the inputs (third principle). Furthermore, for motion detecting cross-kernels to provide directional selectivity, they ought to be asymmetrical about their diagonals, a feature leading to asymmetrical processing of the different input signals (second principle).

It was also shown that second-order cross-kernels (the minimum order cross-kernels) can adequately model and represent motion detection mechanisms. This indicates the sufficiency of second-order interactions between two inputs for the purpose of motion detection. As an application, the functional (not structural) equivalence between correlational models and one type of motion energy models was established. Furthermore, conditions for the design of second-order motion detecting cross-kernels were introduced, and potentialities and limitations of such cross-kernels were examined.

In addition to motion detection, we explored the estimation of speed. We introduced specifications on motion detecting cross-kernels to endow them with the ability to extract speed. A class of velocity estimating cross-kernels was considered in some detail and procedures were established for their design.

Higher-order kernels for motion detection and speed estimation are possible. In particular, when further signal processing features are needed, the use of combinations of higher-order cross-kernels may become necessary.

ACKNOWLEDGMENT

This work was supported by Grant No. RR01861 from the National Center for Research Resources of the National Institutes of Health.

REFERENCES

Adelson, E. and Bergen, J. R. (1985). Spatiotemporal energy models for the perception of motion, *J. Opt. Soc. Am.* **A2(2)**, 284–299.

Borst, A. and Egelhaaf, M. (1989). Principles of visual motion detection, *TINS,* **12(8)**, 297–306.

Braddick, O. J. (1980). Low level and high level processes in apparent motion process in apparent motion, *Philos. Trans. R. Soc. London Ser. B,* **290**, 137–151.

Courellis, S. H. and Marmarelis, V. Z. (1990). An artificial neural network for motion detection and speed estimation, in Proc. Int. Joint Conf. Neural Networks, Vol. 1, 407–421.

Emerson, R. C., Citron, M. C., Felleman, D. J., and Klein, S. A. (1987). Nonlinear directionally selective subunits in complex cells of cat striate cortex, *J. Neurophysiol.,* **58**, 33–65.

Hassenstein, B., and Reichardt, W. (1953). Systemtheoritische Analyse der Zeit- Reinenfolgen- und Vorzeichenauswertung bei der Bewegungspertzeption des Ruselkafers, *Chlorophanus. Z. Naturf.,* **11**, 550–565.

Korenberg, M. J. (1973). Identification of biological cascades of linear and static nonlinear systems, in Proc. 16th Midwest Symp. Circuit Theory, 1–9.

Marmarelis, P. Z. and Marmarelis, V. Z. (1978). *Analysis of Physiological Systems: The White Noise Approach,* Plenum Press, New York.

Marmarelis, P. Z. and McCann, G. D. (1973). Development and application of white-noise modeling techniques for studies of insect visual nervous system, *Kybernetik,* **212**, 74–89.

Marmarelis, P. Z. and Naka, K.-I. (1973). Nonlinear analysis and synthesis of receptive field responses in the catfish retina. III. Two-input white-noise analysis, *J. Neurophysiol.,* **36**, 634–648.

Marmarelis, P. Z. and Naka, K.-I. (1973). Nonlinear analysis and synthesis of receptive-field responses in the catfish retina. II, *J. Neurophysiol.,* **36(4)**, 617–619.

Marmarelis, V. Z. (1987). Nonlinear and nonstationary modeling of physiological systems: an overview, in *Advanced Methods of Physiological System Modelling,* Vol. 1, Marmarelis, V. Z., Ed., Biomed. Simul. Resource, University of Southern California, Los Angeles.

Marmarelis, V. Z. (1989). Volterra-Wiener analysis of a class of nonlinear feedback systems and application to sensory biosystems, in *Advanced Methods of Physiological System Modelling,* Vol. 2, Marmarelis, V. Z., Ed., Plenum Press, New York.

Poggio, T. and Reichardt, W. (1973). Considerations on models of motion detection, *Kybernetik,* **13**, 223–227.

Reichardt, W. (1987). Evaluation of optical motion information by movement detectors, *J. Comp. Physiol.,* **161**, 533–547.

Sakai, H. M. and Naka, K.-I. (1987). Neuron network in catfish retina: 1968–1987, in *Progress in Retinal Research,* Vol. 7, Osborn, N. N. and Chader, G. J., Eds., Pergamon Press, Oxford, 149–208.

van Santen and Sperling (1985). Elaborated Reichardt detectors, *J. Opt. Soc. Am.,* **A2**, 300–321.

Volterra, V. (1930). *Theory of Functions of Integral and Integro-Differential Equations,* Dover Publications.

Wiener, N. (1958). *Nonlinear Problems in Random Theory,* Technology Press of MIT and Wiley, New York.

Chapter 5

OPTIMIZING THE ESTIMATION OF NONLINEAR KERNELS

Stanley A. Klein

TABLE OF CONTENTS

I. Introduction 111

II. Orthogonal Wiener Expansion for Sparse Stimulation 112

III. Estimating the Kernels 115
- A. The "Full History" Expansion 116
- B. The Nonorthogonality Problem Due to Finite Duration 117
- C. The Lee-Schetzen "Solution" 119
- D. Least Squares Estimation of Kernels 119
- E. Sutter's "Full History" Walsh Transform Method 120
- F. Hybrid Method 123
- G. M-Sequences 123
- H. How to Deal with the Anomalies 124
- I. How to Choose the Stimulus 125
  - 1. Autocorrelation Properties 126
  - 2. Stimulus Sampling Time, Δ_S 126
  - 3. Multiple Inputs 128

IV. Units and Mean Square Error 128
- A. Flash Kernels 128
- B. Adaptive Systems and Contrast Units 129
- C. Response Variance 131
- D. "Mean Square Error" for Spiking Systems 132

V. Conditioned Kernels 133
- A. Volterra Expansion 133
- B. General Expansion 134
- C. Effect of Changing the Stimulus 135
- D. An Example: Center-Surround Interactions in a Cat Ganglion Cell 136
  - 1. Conditioned Kernels 136
  - 2. Wiener Kernels of the Ganglion Cell Data 140

VI. Effect of Different Intensity Distributions 144
- A. The Special Role of Diagonal Elements 145
- B. Change of Basis: Exclusive Expansions 145
- C. Inclusive Summations 146
- D. Gaussian Kernels As a Function of Luminance and Power Density 148

E. Advantages and Disadvantages of Different Stimuli 151
1. Symmetric Binary 151
a. Advantages 151
b. Disadvantages 152
2. Asymmetric Binary (Poisson) Stimuli 152
a. Advantages 152
b. Disadvantages 153
3. Ternary Stimuli 153
a. Advantages 154
b. Disadvantages 155
4. Gaussian Stimuli 156
a. Advantages 156
b. Disadvantages 156
5. Summary of Relative Merits of Different Amplitude Distributions 156
F. Radius of Convergence 156

VII. Kernel Variance 157
A. General Formula 157
B. Variance of a General Kernel 159
C. Dark Background Adaptive Systems 160
D. Special Role for Diagonal Elements 161
E. Other Types of Noise 162
F. Sparse-Stimulation 162

VIII. System Identification 164
A. Sandwich Model 164
B. Luminance Gain-Control Model 164

IX. Summary 167

Acknowledgments 168

References 169

I. INTRODUCTION

Wiener kernels and white noise stimuli provide a powerful approach for analyzing nonlinear input-output characteristics of biological systems (Stark, 1968; McCann and Marmarelis, 1975; Hung and Stark, 1977; Marmarelis and Marmarelis, 1978). Although there have been many interesting and important advances using this approach, it hasn't yet received wide acceptance. Some of the reasons for the slow acceptance involve the nonoptimal methods for estimating the kernels that have been used in the past, as will be discussed in this chapter. Another reason is that this approach is mathematically forbidding to most researchers. The complicated mathematics makes it difficult to develop an intuitive feeling for the approach. Without this intuition it is difficult to optimize the stimulus and the analysis to take into account the specific characteristics of the system under investigation. The goal of this chapter is to provide an approach to Wiener kernel analysis that overcomes past difficulties, keeping an eye on its practical application. I hope that as a by-product the kernels will become friendlier.

Section II, following this introduction, develops the general Wiener formalism and shows how a general stationary nonlinear system can be expanded in terms of a set of orthogonal basis functions. A nonstandard notation is used in order to be able to sample the stimulus less frequently than the response, a technique to be called sparse-stimulation.

Section III shows how the expansion coefficients (the kernels) can be calculated from the response of the system to a white noise input. The nonorthogonality problems due to the finite length of the experiment are discussed, and several solutions are proposed. The most promising solution to the orthogonality problem is to use complete M-sequences as the random number generator. Sutter's development of the fast M-transform (Sutter, 1991) enables one to calculate hundreds of kernels (higher-order and/or multiple input) in about the same time that a single first-order kernel could be calculated with previous methods. The near-perfect orthogonality property of M-sequence noise makes it ideal for a wide range of uses. Methods for dealing with the "anomalies" associated with M-sequences are discussed.

The fourth section is about units and about the mean square error, goodness of fit metric. A proper choice of stimulus units can increase the meaningfulness of the kernels and can make the power density a meaningful measure of the strength of the stimulus. Proper use of the goodness of fit metric is informative about the signal-to-noise ratio and nonlinear-to-linear ratio of the system. The mean square error does not work for systems whose output are spikes, so a maximum likelihood goodness of fit metric is discussed.

The fifth section introduces the notion of conditioned kernels. The idea here is that the higher-order kernels produce transient modifications of the lower-order kernels. The formalism for conditioned kernels is developed in full generality. This section provides a quite different way of looking at kernels. Data from an experiment on cat ganglion responses provide a graphic example of the sparse-stimulation method and the use of conditioned expansions. The response to a stimulus transition depends critically on the previous history of the system.

The sixth section examines the relative merits of different amplitude distributions and how the kernels measured with different amplitudes are connected. Under what circumstances should one choose Gaussian or binary or ternary distributions? How does a kernel depend on the stimulus power density? How are kernels obtained with Gaussian noise related to kernels obtained with binary flashes? A formalism for ternary stimuli and ternary kernels will be developed.

The seventh section discusses the kernel variance. This section is of great importance to the experimenter who wants to optimize the stimulus to get the best possible signal-to-noise ratio. Counterintuitive ideas will be found. For example, for a flashing stimulus the stimulus power density can be increased by reducing rather than increasing the flash rate. That is, fewer flashes can reduce the kernel variance. Usable estimates of the kernel variance will be proposed. This section also describes how the use of sparse-stimulation can improve the signal-to-noise ratio of kernels above first order.

Section 8 will compare a system consisting of an LNL sandwich model (a nonlinearity between two linear filters) to a system with a gain control. The sparse-stimulation method will be shown to lead to surprising results for the first-order kernel.

The last section summarizes the highlights of the chapter. The novel points will be reiterated.

The text by Marmarelis and Marmarelis (1978) provides an excellent introduction to the Wiener methodology. The present chapter is meant to supplement that text in a way that loosens up the formalism so that researchers can break away from the confines of Gaussian amplitude distributions and perfectly white stimuli. Different biological systems are best tested with stimuli that are specially tailored to the particular system. Our Section VII on kernel variance supplements the Marmarelis and Marmarelis book which has a large and important section on this topic. However, their formulas are so complicated that their practical usefulness is limited. The goal here is to develop formulas for kernel variance that can guide researchers to improve their stimuli so that the best possible kernel estimates can be obtained. In some cases the views expressed here will differ from those held by Marmarelis and Marmarelis (e.g., my advocacy of M-sequences). These differences of opinion are valuable since they make research interesting.

II. ORTHOGONAL WIENER EXPANSION FOR SPARSE STIMULATION

This chapter examines how a nonlinear system responds to white noise. A white stimulus is one whose second-order correlation function is a delta function; that is,

$$E[x(t')\, x(t' - \Delta\tau)] = \overline{x^2} \text{ for } \Delta\tau = 0$$

and

$$E[x(t')\, x(t' - \Delta\tau)] = \overline{x}^n \text{ for } \Delta\tau \neq 0 \tag{1}$$

The expectation value in Equation 1 is defined by:

$$E\{f[x(t')]\} = \overline{f(t')} = 1/T \sum_{t'=0}^{T} f[x(t')]\, \Delta_s \tag{2}$$

where Δ_s is the time between stimulus samples and T is the total time of the experiment. The present analysis is based on the time indices t' and $\Delta\tau$ being discrete rather than continuous variables. This assumption does not severely limit the analysis since most stimuli and responses are sampled at discrete intervals. The summation in Equation 2 and the values taken by the time delay $\Delta\tau$ in Equation 1 are over the *stimulus* sampling times. This point will become important throughout this chapter where we discuss the advantages of sampling the response faster than the stimulus. Even though different sampling rates are possible, whenever an expectation is taken involving a stimulus, the sampling rate will be that of stimulus. Only the mean square error calculation requires the response sampling rate.

It is also assumed that stimuli at different times are independent, so that the expectation value of a product of stimuli with different time delays equals the product of the expectation values, in the limit of infinite duration stimuli:

$$E[\prod_{i=1}^{n} x(t' - \Delta\tau_i)] = \bar{x}^n \text{ for all the } \Delta\tau_i \text{ being different} \tag{3}$$

Before getting to the mathematical Wiener expansion, it is useful to put it into words. The idea is to express the response y(t) as a polynomial expansion of the stimulus, similar to a Taylor's expansion for a general function. The zeroth-order kernel, h_0, is the average value of the response. The first-order kernel is the linear impulse response. For a white noise stimulus, it can be obtained by time-locking the response to the stimulus. Equation 11b will show that an alternative, but equivalent, definition of the first-order kernel is given by the cross-correlation between the stimulus and the response. The second-order kernel is obtained by time-locking the response to pairs of stimuli after having subtracted off the linear response.

In mathematical language, the usual Wiener orthogonal expansion for a zero-mean stimulus is given by:

$$\begin{aligned} y(t) = h_0 &+ \int_0^\infty h_1(\tau)x(t - \tau)\, d\tau + \int_0^\infty \int_0^\infty h_2(\tau_1,\tau_2) \\ &[x(t - \tau_1)x(t - \tau_2) - \overline{x^2}\delta(\tau_1 - \tau_2)] \\ &d\tau_1 d\tau_2 + \text{third- and higher-order terms} \end{aligned} \tag{4}$$

where $\delta(\tau_1 - \tau_2) = 0$ for $\tau_1 \neq \tau_2$, and its integral is unity. A word of warning is needed about the use of the word "Wiener". In previous writings (Klein and Yasui, 1979; Klein, 1987) Wiener kernels referred specifically to kernels that were obtained using a stimulus with a Gaussian amplitude distribution. In the present chapter, "Wiener kernels" will be used more loosely as the kernels in any orthogonal expansion, even if the amplitude distribution is not Gaussian. When a specific distribution is being discussed the kernels will, for example, be called "Gaussian kernels" or "ternary kernels".

The general Wiener expansion is given by:

$$y(t) = y(t' + t_r) = \sum_{k,n_j,\sigma_j}^{SC} \underline{H}_n(\tau_1^{n_1} \cdots \tau_k^{n_k})\, \Delta_s^n \prod_{i=1}^{k} X_{n_i}(t' - \sigma_i) \tag{5}$$

where $\tau_i = \sigma_i + t_r$. The notation $\tau_i^{n_i}$ means the same time delay, τ_i, is repeated n_i times (an n_ith-order diagonal). The sum of all n_i equals n, the order of that kernel. This expansion differs from that given by Klein and Yasui (1979) and Klein (1987) because of the presence of t_r, the residual time. The residual time is $t_r = t - t'$ where the time t' is the time of the stimulus that is at or just precedes the response time t. Thus t_r is always less than Δ_s. This new variable, t_r, is introduced because of the importance of sampling the response faster than the rate at which the stimulus is changing. The variables t, t_r, and τ_i are sampled at the rapid response sampling interval Δ_r. The variables t', and σ_i are sampled at the stimulus sampling interval Δ_s. The notation SC over the summation sign stands for single counting. The single counting is achieved by time ordering (Klein and Yasui, 1979) so that $\tau_i < \tau_j$ if $i < j$. The first few terms of this expansion are

$$y(t' + t_r) = \underline{H}_0(t_r) + \sum_{\sigma_1} \underline{H}_1(\sigma_1 + t_r)X_1(t' - \sigma_1)\Delta_s +$$
$$\sum_{\sigma_1<\sigma_2} \underline{H}_2(\sigma_1 + t_r,\sigma_2 + t_r)X_1(t' - \sigma_1)X_1(t' - \sigma_2)\Delta_s^2 +$$
$$\sum_{\sigma_1} \underline{H}_2[(\sigma_1 + t_r)^2]X_2(t - \sigma_1)\Delta_s^2 + \tag{6}$$

third- and higher-order terms plus noise

The orthogonal polynomials, X_i, are given by ratios of simple determinants (Klein and Yasui, 1979; Klein, 1987). The zeroth order polynomial is unit: $X_0(t') = 1$. The first two polynomials are

$$X_1(t') = \begin{vmatrix} 1 & \bar{x} \\ 1 & x(t') \end{vmatrix} / |1| \tag{7a}$$

$$X_2(t') = \begin{vmatrix} 1 & \bar{x} & \overline{x^2} \\ \bar{x} & \overline{x^2} & \overline{x^3} \\ 1 & x(t') & x^2(t') \end{vmatrix} / \begin{vmatrix} 1 & \bar{x} \\ \bar{x} & \overline{x^2} \end{vmatrix} \tag{7b}$$

where a matrix between vertical lines means the determinant of the matrix. Thus, $X_1(t') = x(t') - \bar{x}$. The higher-order polynomials are constructed in a manner similar to Equation 7. The proof of orthogonality (in the limit of infinite duration stimuli) is given by Klein and Yasui (1979) and Klein (1987). The underlined kernels in Equation 5 are the *true* system kernels. The next section shows how the kernels can be estimated from experimental data. The *estimated* kernels (without the underlining) do not equal the *true* kernels because of biological noise and because of noise that may be introduced by the kernel estimation procedure (see Section III).

The form of the expansion in Equations 5 and 6 is different in several regards from the usual expansion given in Equation 4.

1. Summations are used rather than integrations since the response is typically sampled at equally spaced intervals. Furthermore, in vision studies the stimulus is also typically generated discretely, as would be the case if a 60-Hz computer monitor produced the stimuli.
2. The diagonal elements are kept separate from the off-diagonal. As will be seen in Section VI, the diagonal kernel elements will turn out to play an especially important role in connecting kernels obtained with different stimuli. For this reason we would like to estimate the diagonal elements of the third- and fourth-order kernels with greater accuracy than the off-diagonal terms. A further reason to treat the diagonals separately is that for non-Gaussian stimuli, the diagonal elements need not be continuous with the off-diagonal elements even if they are continuous for the Volterra kernels (see Section VI).
3. The summations are time-ordered to avoid double-counting in Equation 5. This will simplify many of the forthcoming formulas by removing the combinatoric factors. The Wiener kernels used here are capitalized to distinguish them from the usual (lower case) kernels in Equation 4 that are defined in terms of an expansion without time-ordering. The relationship between the two definitions of kernels is given by:

$$H_n(\cdots) = h_n(\cdots)\, n!/\prod_{i=1}^{k} n_i! \tag{8}$$

where k is the number of different time indices and n_i is the number of repeats of the *i*th index.

4. The summations in Equation 5 have been arranged so that the stimulus sampling interval, Δ_s, can be different from the response sampling interval, Δ_r. As an example, the stimulus might be produced at a 60-Hz video rate (16.7 ms) and the response might be sampled five times as fast, every 3.33 ms. In Equation 5, the variables t and τ are sampled at the faster response rate, and the variables t′ and σ_i are sampled at the stimulus rate. Only minor adjustments were made in Equation 5 to allow different sampling rates for the stimulus and the response. The zeroth-order kernel, for example, has acquired a time argument (Equation 6).

This scheme of sparse-stimulation had been previously called superresolution (Klein, 1979) because the kernels can be measured with greater resolution than the resolution of the white-noise stimulus. Two reasons for introducing sparse-stimulation are to improve the signal-to-noise ratio of the higher-order kernels (see Section VII) and to avoid biases in kernel estimation due to the nonorthogonality of the stimuli (see Section III). The ganglion cell data to be shown in Section V shows the power of the sparse-stimulation technique. Section III discusses how the sparse-stimulation kernels can be estimated from a white-noise experiment. Section VIII shows how sparse-stimulation can be used to identify the type of nonlinearity inside a system.

There may at times appear to be an ambiguity about whether a summation interval should be the response sampling interval or the stimulus sampling interval. Unless otherwise indicated it will always be the *stimulus* interval that is wanted. Only in Section IV in connection with calculating the mean square error will a summation be taken using the response sampling interval.

III. ESTIMATING THE KERNELS

The purpose for introducing the orthogonal polynomials, Equation 7, is to facilitate the estimation of the kernels by cross-correlation (Lee and Schetzen, 1965). The general Wiener kernel is given by:

$$H_n(\tau_1^{n_1} \cdots \tau_k^{n_k}) = E[y(t' + t_r) \prod_{i=1}^{k} X_{n_i}(t' - \sigma_i)] / \prod_{i=1}^{k} P_{n_i} \tag{9}$$

where the factors in the denominator are

$$P_{n_i} = E\{[(X_{n_i}(t')]^2 \Delta_s^{n_i}\} \tag{10}$$

Since $P_0 = 1$, this factor will typically be omitted. For $n_i = 1$, Equation 10 is the power density, and the subscript will usually be omitted: $P_1 = P$. The power density equals the variance of the stimulus, $\overline{x^2} - \bar{x}^2$ times the stimulus sampling time, Δ_s. It is an important quantity, appearing throughout Wiener kernel analysis. Improved units for specifying x(t′) and P will be considered in Section IV.

From Equation 9, the first few kernels are

$$H_0(t_r) = E[y(t' + t_r)] \tag{11a}$$

$$H_1(\sigma_1 + t_r) = E[y(t' + t_r)X_1(t' - \sigma_1)]/P \tag{11b}$$

$$H_2(\sigma_1 + t_r, \sigma_2 + t_r) = E[y(t' + t_r)X_1(t' - \sigma_1)X_1(t' - \sigma_2)]/P^2 \tag{11c}$$

$$H_2[(\sigma_1 + t_r)^2] = E[y(t' + t_r)X_2(t' - \sigma_1)]/P_2 \tag{11d}$$

where the expectation value is defined in Equation 2. In the conventional formalism there would have been extra factors of $n_1! \ldots n_k!/n!$ on the right-hand sides of Equation 11 (a factor of $1/2$ in Equation 11c) because of the normalizations stated in Equation 8. These factors do not appear in the present formalism because the time-ordering discussed above avoids the double-counting that would otherwise have led to these factorial factors. As in Equation 5, the variable t_r only covers a single stimulus sampling interval ($t_r < \Delta_s$). The other variables are defined at the stimulus sampling intervals.

The kernels can be calculated according to Equations 9 and 11 because of an assumed orthogonality of the basis functions:

$$E[X_i(t)X_j(t')] = 0 \text{ if } i \neq j \text{ or if } t \neq t' \tag{12}$$

The derivation of these orthogonality conditions as well as extensions to generalized, nonlinear functions of the stimuli, x(t), are given by Klein and Yasui (1979) and Klein (1987). For a finite duration stimulus the orthogonality of Equation 12 is not exact, and the last half of this section is based on examining the consequences of the lack of orthogonality.

A. THE "FULL HISTORY" EXPANSION

An alternative way of writing the Wiener expansion of Equation 5 and the inversion cross-correlations of Equations 9 and 11 will be useful. Equation 5 can be written as:

$$y(t' + t_r) = \sum_{\{n_i\}} F_{n_{M-1} \cdots n_0}(t_r) \prod_{\sigma=0}^{M-1} X_{n_\sigma}(t' - \sigma)\Delta_s^{n_\sigma} \tag{13}$$

where t' is the time of the stimulus that is at or just preceding the response, and $t_r = t - t'$ is the residual time in response sampling intervals between the time, t, of the response and the time, t', of the preceding stimulus. Thus, if the stimulus is sampled every second and the response is sampled every 0.2 s, then t_r will have one of the five values of 0, 0.2, 0.4, 0.6, and 0.8 s. In the conventional white-noise formalism where the stimulus and the response are sampled at the same rate, then $t_r = 0$, and the F kernels would not depend on the t_r parameter. The subscript n_σ is the number of times the time delay, σ, is repeated.

Consider a system whose memory is three stimulus units. The full history extends back two stimulus time intervals ($M - 1 = 2$) and the "full history" expansion (Equation 13) becomes:

$$y(t' + t_r) = \sum_{\{ni\}} F_{n_2 n_1 n_0}(t_r)X_{n_0}(t')X_{n_1}(t' - 1)X_{n_2}(t' - 2)\Delta_s^n \tag{14}$$

where $n = \Sigma_i n_i$. In Equation 14 we have assumed that the stimuli are sampled every integer value of time ($\sigma = 0, 1, 2 \ldots$).

By comparing Equations 13 and 5 the connection between the F kernels and the H kernels can be established:

$$H_n(t_r^{n_0}, (1 + t_r)^{n_1}, (2 + t_r)^{n_2} \cdots (M - 1 + t_r)^{n_{M-1}}) = F_{n_{M-1} \cdots n_1 n_0}(t_r) \tag{15}$$

If $n_i = 0$, that term is omitted from the argument of H_n. A few examples should clarify the notation. Suppose the stimulus is sampled every integer so that t_r is the fractional part of the time interval:

$$H_1(0.2) = F_{000001}(0.2)$$
$$H_1(1.2) = F_{000010}(0.2)$$
$$H_1(4.2) = F_{010000}(0.2)$$
$$H_2(1.2,4.2) = F_{010010}(0.2)$$
$$H_4(1.2,4.2^3) = F_{030010}(0.2) \qquad (16)$$

The sum of the indices, n_i, is n, the order of the Wiener kernel. Further examples of these connections will be given in Table 1. For a stimulus with q levels, n_i in Equations 13 and 14 must be less than q. Thus, for binary stimuli, n_i can be either 0 or 1. For ternary stimuli, n_i can be 0, 1, or 2. It takes some effort to get used to the F kernel formalism, but the effort is worthwhile.

The expansion in Equation 13 can be inverted by the same procedure as Equations 9 and 11:

$$F_{n_{M-1}\cdots n_0}(t_r) = E[y(t' + t_r) \prod_{\sigma=0}^{M-1} X_{n_\sigma}(t' - \sigma)\Delta_s^{n_\sigma}] / \prod_{\sigma=0}^{M+1} P_{n_\sigma} \qquad (17)$$

For example:

$$F_{10000}(t_r) = E[y(t + t_r)X_1(t' - 4)\Delta_s]/P$$
$$F_{10050}(t_r) = E[y(t' + t_r)X_1(t' - 4)\Delta_s X_5(t' - 1)\Delta_s^5]/P\,P_5 \qquad (18)$$

Later in this section the advantage of this F notation will be described. Notice that the expansion in Equation 14 is time ordered, so the F kernels do not involve the double-counting that is present with the h kernels of the standard notation. Thus, the F kernels have the same normalization (Equation 8) as the H kernels. It is for this reason that combinatoric factorial factors are not present in Equations 14 or 17.

B. THE NONORTHOGONALITY PROBLEM DUE TO FINITE DURATION

The cross-correlation method of Equation 9 assumes that the orthogonality expressed by Equation 12 is valid. This assumption is not valid for a random stimulus of finite duration. Let us consider the calculation of the off-diagonal second-order kernel given by Equation 11c. For simplicity we assume the stimulus has a zero mean ($\bar{x} = 0$), and we assume for the remainder of this section that the response and stimulus are sampled at the same rates. Then equation 11c becomes

$$H_2(\sigma_1', \sigma_2') = E[y(t')x(t' - \sigma_1')x(t' - \sigma_2')]/P^2 \qquad (19)$$

By plugging the expansion for y(t) given by Equation 5 into Equation 19, and remembering $\bar{x} = 0$, one obtains

$$H_2(\sigma_1', \sigma_2') = E\{[\underline{H}_0 + \sum_{\sigma_1} \underline{H}_1(\sigma_1)X_1(t' - \sigma_1)\Delta_s +$$
$$\sum_{\sigma_1<\sigma_2} \underline{H}_2(\sigma_1,\sigma_2)X_1(t' - \sigma_1)X_1(t' - \sigma_2)\Delta_s^2]x(t - \sigma_1')x(t - \sigma_2')\}/P^2 + \cdots \qquad (20)$$

$$= \underline{H}_0 A_2(\sigma_1' - \sigma_2')/P^2 + \sum_{\sigma_1} \underline{H}_1(\sigma_1) A_3(\sigma_1 - \sigma_1', \sigma_1 - \sigma_2')\Delta_s/P^2 +$$

$$\sum_{\sigma_1\sigma_2} \underline{H}_2(\sigma_1,\sigma_2) A_4(\sigma_1 - \sigma_1', \sigma_1 - \sigma_2', \sigma_1 - \sigma_2)\Delta_s^2/P^2 + \cdots \quad (21)$$

$$= \underline{H}_2(\sigma_1',\sigma_2') + \text{terms of order } 1/\sqrt{R} \quad (22)$$

where the functions A_n are the *n*th order stimulus autocorrelation functions, and R is the number of stimulus samples in the record. Equation 22 is obtained because

$$A_4(\sigma_1 - \sigma_1', \sigma_1 - \sigma_2', \sigma_1 - \sigma_2) = \delta(\sigma_1 - \sigma_1')\delta(\sigma_2 - \sigma_2') +$$
$$\text{terms of order } 11\sqrt{R} \quad (23)$$

for $\sigma_2 > \sigma_1$ and $\sigma_2' > \sigma_1'$. The terms of order $1/\sqrt{R}$ are present because for a finite duration experiment the autocorrelation functions are not perfect delta functions. The main purpose of this section is to discuss methods to reduce or eliminate the intrusion of the $1/\sqrt{R}$ terms. If the stimulus is sampled at 10-ms intervals and the run is 100 s, then R = 10,000 and the nonleading terms would be about $10{,}000^{-0.5}$ = 1% of the dominant term. Although this small value may seem negligible, several factors can make the nonorthogonal contributions quite bothersome:

1. Summations over expectation values (see Equations 20 and 21) will increase the magnitude of the unwanted terms.
2. If the first-order kernel elements are much larger than the second-order kernels in Equation 21, then unwanted first-order terms can dominate and the estimated second-order kernel can differ significantly from the true kernel.
3. A popular use of white noise in physiological experiments is to map receptive fields (see Emerson and Citron, 1992 for references) or to do topographic mapping of the electroretinogram (ERG) (Sutter, 1992). The latter experiments use 256 simultaneous inputs, each with an "independent" white-noise flicker. The number of parameters to be estimated is huge. It is reasonable to suppose that each first order ERG kernel has 20 elements (100-ms record with 5-ms samples). Sutter, using a 16.7-ms stimulus interval found that the main nonlinearities are confined to two kernel slices, for a total of 40 more elements to be estimated. Thus, the total number of parameters to be estimated is 60 × 256 = 15,360. If the stimulus sampling interval had been reduced to 5 ms rather than 16.7 ms, more than 2 second-order kernel slices would undoubtedly have been important, forcing a need for yet more kernel elements. In that case, with so many inputs the problems of slight deviations from orthogonality become important. Techniques for avoiding the extra terms of Equation 21 become essential for obtaining high-quality kernels.

The orthogonality problem discussed in the preceding paragraphs is probably the most serious problem facing the white-noise methodology. It is more important than is usually appreciated. To the mathematically minded, the problem is phrased in terms of deviations from the orthogonality conditions in Equation 12. Biologists with common sense will phrase the problem in different language. They will worry that the response to prior and succeeding stimuli add noise to the time-lock average obtained from the cross-correlation. One might try to reassure the biologist that since the prior and succeeding stimuli are uncorrelated with the present stimulus, the responses to the irrelevant stimuli will cancel out. The problem, however, is that the cancellation is complete only if the experiment runs infinitely long. Thus, the problem of nonorthogonality affects not only the estimations of second- and higher-order kernels (Equation 12 c and d), but also Equation 12b, the estimation of the first-order kernel. The common-sense biologist has put his finger on a most serious problem of the white noise approach. Several proposed solutions to the problems of nonorthogonality are now examined.

C. THE LEE-SCHETZEN "SOLUTION"

In an attempt to minimize the contributions of lower-order kernels to the estimates of higher-order kernels, Lee and Schetzen (1965) proposed a method for obtaining the kernels that is slightly different from Equation 9. Instead of cross-correlating stimulus products with the response, y(t), they first subtract off the lower-order contributions to the response. Thus, the second-order kernel in Equation 12c would be

$$H_2(\sigma_1,\sigma_2) = E\{[y(t') - y_1(t')]x(t' - \sigma_1)x(t' - \sigma_2)\}/P^2 \tag{24}$$

where $y_1(t')$ is the first-order response corresponding to the zeroth-order H_0) and first-order (H_1) terms of Equation 5. However, before subtracting the $y_1(t)$ term as in Equation 24, one must be careful that the first-order kernel has been estimated well. If the first-order kernel is not well estimated, then including the $y_1(t)$ term in the cross-correlation could do more damage than good. A poorly estimated first-order kernel could occur if the memory of the kernel is taken to be much longer than it really is, so that most of the kernel is noise. To see the problem, one can substitute the estimate of H_1 into the expansion for $y_1(t)$ and discover that the inclusion of y_1 introduces many new extraneous terms into the estimate of H_2. The Lee-Schetzen algorithm also does not do anything to improve the estimation of the first-order kernel. Thus, Equation 24 does not offer much help in removing nonorthogonality noise. Simulations by Korenberg (1987), and Sutter (1992) demonstrate the impressive improvement in kernel quality that is possible by going beyond the Lee-Schetzen method.

D. LEAST SQUARES ESTIMATION OF KERNELS

Korenberg (1987) and Goussard et al. (1987, 1991) propose dealing with the expansion in Equation 5 as a straightforward least squares estimation problem rather than using the Lee and Schetzen (1965) cross-correlation method. It is a problem of linear regression since the response is linearly related to the kernels being estimated. The nonlinear connection between the response and the stimulus does not affect the simple linear connection between response and kernels. For example, suppose the memory of the system is only two stimulus samples long. Then, instead of Equations 5 or 14 a general expansion for y(t) would be

$$\begin{aligned} y(t) = {} & F_{00} + F_{01}X_1(t) + F_{10}X_1(t - 1) + F_{02}X_2(t) + \\ & F_{20}X_2(t - 1) + F_{11}X_1(t)X_1(t - 1) + F_{03}X_3(t) + \text{higher-order terms} \end{aligned} \tag{25}$$

where the $X_i(t)$ is defined in Equation 7 [$X_1(t) = x(t) - \bar{x}$]. The coefficients F_{ij} use the kernel notation given in Equation 14. Equation 25 assumes the stimulus and response are being sampled at all integer values of t. The task of estimating the coefficients F_{ij} (the kernel elements) is standard multivariate linear regression. When there are a lot of terms in Equation 25, stability of the matrix inversion is improved if the basis functions are close to being orthogonal. That is why the expansion functions are the $X_i(t)$ functions rather than simply products of x(t). The small deviations from orthogonality of the $X_i(t)$ that are associated with finite record length can cause severe problems if there are a large number of terms (recall there were more than 15,000 kernel elements to be estimated in Sutter's ERG experiments). Goussard (1987) and Korenberg (1987) are aware of the problem of large numbers of kernel elements. Goussard proposes an iterative approach to inverting the matrix needed to solve for the kernel elements. However, it is not clear whether the iterations converge when the rank of the matrix is very large. These issues are discussed by Goussard et al. (1991).

An important advantage of this least squares (linear regression) approach is that the expansion functions, $X_i(t)$, do not need to be orthogonal. In fact, the stimulus does not need to be close to white noise. If the number of terms is not excessive, then the linear regression will work just fine, even with large deviations from orthogonality. Except for the presence of biological noise, the least squares approach is able to estimate the kernels *exactly*.

What is clearly needed is for someone to carry out the Goussard or the Korenberg procedures in cases where there are *many thousands* of kernels to be estimated, based on tens of thousands of data points and in the presence of sizable noise. If this demonstration were successful, it would overcome one of the main obstacles facing the present white-noise approach. However, as the number of kernel elements to be estimated increases, the stability of the least squares solution decreases. Sutter's Walsh transform method (discussed next) may have achieved this goal of finding an efficient and stable method for doing the least squares kernel estimation.

E. SUTTER'S "FULL HISTORY" WALSH TRANSFORM METHOD

In a chapter in this volume Sutter (1992) proposes a method for calculating kernels obtained with nonwhite stimuli in a manner that removes the contribution of the nonorthogonality noise. Since this method calculates all possible stimulus configurations that can affect the response (the full history) and since it uses the very rapid, fast Walsh transform, we will refer to this method as the "Full History Walsh method". The *history* of a response is the window that includes all the stimuli affecting that response. A major feature of this method is that it can be used with nonwhite stimuli. A nonwhite stimulus has the property that within the history, not all stimulus sequences are presented an equal number of times. For the sequences that are presented more than once it is easy to compensate by using the *average* response rather than the *summed* response of the Lee-Schetzen algorithm.

Table 1 presents an example that illustrates many of the features of the Full History Walsh method. This table will also be used in Section VII to discuss our sparse-stimulation method as applied to a system with a luminance gain control.

The numbers in Table 1 come from an imaginary experiment, the details of which will be discussed in Section VIII. Column 6 gives the response of the system at time t. Columns 1 to 3 are the eight possible stimuli at the three time intervals before and including time t. While the actual stimuli are flashes of strength 0 and L, the stimulus, X(t), is chosen to have zero mean by taking on values of -1 and $+1$, corresponding to contrast units of -100% and $+100\%$ (see Section IV for further discussion of contrast units). This definition of X(t) has the opposite sign from the convention used by Sutter (1992). See Section VI for a further discussion of the sign convention. X(t) is the orthogonal polynomial defined in Equation 7. The subscript on $X_n(t)$ is not needed since the stimulus is binary and $X_n(t)$ vanishes for $n > 1$. Column 4 shows the total number of times that a particular triplet of stimuli occurred in the run, and column 5 is the time-locked total of all the responses to the particular stimulus configuration. Thus, for example, whenever the flash did not fire at time t (rows 0, 2, 4, and 6), the response was 10 units. There were $N_2 = 790$ occurrences of the 0, L, 0 stimulus configuration (row 2, column 4), so that the summation of all the

TABLE 1

	1	2	3	4	5	6	7	8	9	10
row	X(t − 2)	X(t − 1)	X(t)	N	R	y(t)	H kernels	F	L − S	Sutter
0	−1	−1	−1	780	7,800	10	H0	F_{000}	40.100	40
1	−1	−1	1	801	80,100	100	H1(0)	F_{001}	30.150	30
2	−1	1	−1	790	7,900	10	H1(1)	F_{010}	−9.728	−10
3	−1	1	1	810	48,600	60	H2(01)	F_{011}	−9.769	−10
4	1	−1	−1	795	7,950	10	H1(2)	F_{100}	−5.321	−5
5	1	−1	1	782	62,560	80	H2(02)	F_{101}	−5.393	−5
6	1	1	−1	798	7,980	10	H2(12)	F_{110}	0.149	0
7	1	1	1	802	32,080	40	H3(012)	F_{111}	0.171	0

responses is $R_2 = 7900$ (column 5). Column 7 lists the names of the eight kernels that contribute to the response. Column 8 gives the alternate method for naming the kernels that was discussed in connection with Equations 13 to 18. Notice that in this example there is no kernel argument, t_r because the response is being sampled at the same rate as the stimulus. Figures 1 to 4 of Section V and Tables 3 and 4 of Section VIII will provide examples using sparse-stimulation in which t_r is used. Columns 9 and 10 are the kernels calculated from the responses in rows 5 and 6, respectively. The difference between the results in columns 9 and 10 is due to the difference between the Lee-Schetzen algorithm and Sutter's Full History Walsh method as will now be discussed.

The Lee-Schetzen algorithm is given by Equation 9; the response is convolved with the appropriate stimuli and then divided by $T = \Sigma_i N_i = 6358$, the total number of stimuli in the record. The notation Σ_i means the summation over the 8 rows of Table 1. Thus, the value of H_0 is $H_0 = \Sigma_i N_i y_i / \Sigma_i N_i = (7{,}800 + 80{,}100 + 7{,}900 + 48{,}600 + 7{,}950 + 62{,}560 + 7{,}980 + 32{,}080)/T = 40.1$ The value of y_i is another notation for y(t). It is the average value of the response to the *i*th stimulus configuration (the entry in row i, column 6 of Table 1). The value of $H_2(0, 2)$ would be $H_2(0, 2) = \Sigma_i N_i y_i X(t - 0)X(t - 2)/T = (7{,}800 - 80{,}100 + 7{,}900 - 48{,}600 - 7{,}950 + 62{,}560 - 7{,}980 + 32{,}080)/T = -5.393$. As another example, the value of $H_1(1)$ is $H_1(1) = \Sigma N_i y_i X(t - 1)/T$.

The Full History Walsh method differs from the Lee-Schetzen method in that instead of summating the raw responses as in the preceding paragraph, the number of responses to each stimulus configuration is normalized (Sutter, 1992; see also the final paragraph of the Appendix of Emerson et al., 1987). Thus, instead of using the total response to the 0, L, 0 stimulus configuration (7,900 in row 2, column 5), we would take the *average* response to the configuration (7,900/790 = 10 in column 6). When the average values are used in the cross-correlation, then the kernel values in the tenth column are obtained. The new calculation of the kernel element $H_2(0, 2)$ becomes: $H_2(0, 2) = \Sigma y_i X(t - 0)X(t - 2)/8 = (10 - 100 + 10 - 60 - 10 + 80 - 10 + 40)/8 = -5$. The kernels in column 10 turn out to be the *exact* Wiener kernels. The kernel estimates in column 9 include noise due to the nonorthogonality of the stimulus, caused by the different numbers of the different configurations that are tested.

Klein (1993a) provides some additional comparisons between the least squares method, the Full History Walsh method, and the Lee-Schetzen method. One of the examples in Klein (1993a) is a model system given by $y_m(t) = h_0 + h_1\, x(t) + n(t)$, where h_0 and h_1 are the zeroth and the first order kernels, n(t) is noise and the binary stimulus has values of ± 1. This "toy" system with only two kernels provides many insights into the bias, variance and covariance of the kernel estimates. It also allows us to explore the effects of truncating the Wiener expansion, keeping only the lower order kernels (h_0 in the toy model). The least square estimate of h_0 is $\hat{h}_0 = \bar{y} - \hat{h}_1 \bar{x}$, where $\hat{h}_i$ is the estimated kernel value (including the effect of noise). Notice that the estimate of $\hat{h}_0$ is affected by $\hat{h}_1$. If $\hat{h}_1$ is dominated by noise, then it would have been better to set it equal to zero and to take the Lee-Schetzen value for $\hat{h}_0$ given by $\hat{h}_0 = \bar{y}$. If the stimulus were perfectly white (as would be the case with an M-sequence) then $\bar{x} = 0$ and the lower order kernels ($\hat{h}_0$) would not be contaminated by noisy estimates of the higher order kernels ($\hat{h}_1$). Klein (1993a) discusses how the Full History Walsh method can deal with a truncated Wiener expansion by using an iteration scheme proposed by Sutter (1992).

A comment is in order about the units of the kernels. Columns 9 and 10 were calculated by taking sums and differences of the responses without any factors depending on the stimulus strength, X(t), or the sampling interval, Δ_s. These factors did not appear (see Equation 9) because both the stimulus magnitude and the sampling interval have been normalized to unity in this calculation (P = 1). Section IV will deal with how stimulus units affect the magnitude of the kernels.

Note that the ordering of the kernels in Table 1 (column 7) might at first glance appear to be unusual. It is, in fact, straightforward, as can be seen by the natural order in terms of the F kernel notation in column 8.

Sutter realized the important point that column 10 is the Walsh transform of column 6 (Sutter, 1987, 1991). The Walsh matrix is given by Equation 18 of Sutter (1992) and is equal to the product of X_i factors in the numerator of Equation 17 for the case of binary stimuli. This Walsh transform connection between the responses and the kernels requires the particular ordering of responses and kernels as shown in Table 1. The kernel labeling that was developed in Equation 14 (column 8 of Table 1) was done specifically for the Walsh transform. Further details and examples of how the Walsh matrix transforms the data into Wiener kernels are given in Section V in connection with multiinput ganglion cell data. An important benefit of using this approach is that there is a very fast version of the Walsh transform, similar to the fast Fourier transform. An additional speed advantage of the Walsh transform is that because of the binary nature of the transform, most of the operations are additions and subtractions rather than the multiplications associated with nonbinary stimuli. The speed of this algorithm is important when large numbers of kernels are being estimated and when an iteration procedure is used to compensate for incomplete testing of all possible relevant configurations, as will be discussed next.

In the example shown in Table 1, stimulus configurations of only three time periods were considered. Thus, the nonorthogonality of the stimulus was corrected only for time delays up to two stimulus sampling times. This would be satisfactory if the memory of the system was three time intervals long. A more realistic case would have a system memory of 20 time samples. For this case, instead of Table 1 having $2^3 = 8$ rows it would have $2^{20} = 1{,}048{,}576$ rows. If the run length is 2^{15} intervals (5.5 min for a 10-ms sampling interval), then instead of having about 800 entries (Table 1) for each stimulus configuration there would be 2^{-5} entries per configuration. Thus, only 1/32 of the configurations would have a single sample, and the remaining configurations would not be tested. Sutter (1992) uses an iterative algorithm for estimating the missing responses based on an assumption that the system has kernels only up to a particular order. This iterative method can be used not only to fill in the responses that were never tested, but also to fill in for the sequences whose responses are obliterated by noise artifacts (e.g., eye blinks).

In the absence of noise the two methods (least squares & Sutter's) are identical since they produce the original true kernels without interference from the non-orthogonalities discussed in this section. Klein (1993a) showed that in the presence of noise, the Full History Walsh method is still identical to the least squares method as long as the least squares model has the same number (2^n) of kernels as the Walsh method. For a truncated Wiener expansion (e.g. terms above second order and terms with large latency are dropped) Sutter's iterative method can be used to give the same results as the least squares approximation (Klein, 1993a). If only a relatively small number of kernel elements are desired, then the least squares method would be faster and better conditioned than the Full History Walsh method. The reverse would be true if only a few terms are truncated. The methods by Goussard, Korenberg, and Sutter discussed in this section are all designed to "clean up" the bad autocorrelation functions of random stimuli. However, these methods may not be capable of handling massively multiinput systems such as the 256 input ERG experiments of Sutter and Tran (1991). For these types of experiments one may require M-sequence noise which, as will be discussed soon, has built-in, excellent autocorrelation properties.

F. HYBRID METHOD

For certain systems, a hybrid approach might be best. Consider a single input system with a strong, large memory first-order kernel and with many higher-order kernels that are relatively weak. Because of the large first-order kernel one would want to estimate it accurately. The best technique would probably be a least squares estimation which would involve a matrix inversion whose order is the length of the first-order kernel. Once the first-order kernel is estimated, it can be removed from the response, as proposed by Lee and Schetzen (1965). One must, however, remember the danger of removing the first-order kernel (or any comparable Gram-Schmidt method) if the kernel is noisy. The higher-order kernels could then be estimated by any of the above algorithms. By first removing the dominant response all the above methods that involve iteration should converge faster.

G. M-SEQUENCES

The neatest method for eliminating the autocorrelation noise is to use M-sequences. Sadly, this method has been largely ignored. If the noise is generated using a full-cycle M-sequence, then the bothersome autocorrelation functions in Equation 21 effectively vanish, just as they would for an infinite record length of regular white noise. It is because of the special properties of these M-sequences that Sutter and Tran (1991) have been able to measure simultaneously first- and second-order ERG kernels at 256 retinal locations. The ideal autocorrelation properties of the M-sequences eliminate the self-noise nuisance that has plagued the white-noise approach. In addition, Sutter (1987) and Sutter and Tran (1991) were able to use special properties of the M-sequences to calculate arbitrary kernels from multiinput experiments in shockingly short amounts of time. Although most of the experiments with M-sequences have used binary stimuli, it is straightforward to combine pairs of adjacent binary levels to produce a quaternary (four level) stimulus or stimuli with other numbers of levels (Sutter, 1992; Klein, 1993b). The reader is referred to Sutter (1987, 1992) and Barker (1970) for further details on the properties of complete sequences and how they are generated.

Sutter's new Full History Walsh technique does not require an M-sequence stimulus (or even a random stimulus), so one might ask what the advantage is of using M-sequences. There are several:

1. M-sequences *uniformly* test all possible stimulus sequences. It is advantageous to test as many stimulus sequences as possible. The M-sequence stimulus is efficient in that it tests all stimulus patterns once (up to a length given by the number of bits in the shift register). De Bruijn sequences also have this property, as will be discussed later in this section.
2. Full-cycle M-sequences have near-perfect autocorrelation functions. This property minimizes the problems of nonorthogonality discussed earlier.
3. When the memory of the system is larger than the number of bits of the shift register, it is possible for kernel elements to overlap (the anomaly problem discussed next). As

pointed out by Sutter (1987), this problem must always occur (though it is often disguised), even for stimuli that are not M-sequences. The advantage of the M-sequence is that one can determine at which points and in which kernels the overlap occurs. As will be discussed later in this section it is better to know the details of the overlap than to have overlap and be ignorant of it.

Given their important advantages, one wonders why the M-sequence approach is not more widely used. The answer is *anomalies*. As Sutter (1987) pointed out, it is too bad that the word "anomaly" was applied to one of the most beautiful properties of the binary M-sequences. For a binary sequence in which the stimulus takes on values of -1 and $+1$, the beautiful property is as follows:

$$x(t)x(t - \sigma) = x(t - \sigma') \tag{26}$$

That is, the product of the stimulus and a displaced version of the stimulus is equal to the stimulus displaced by a different amount. Thus, Equation 19 for the second-order kernel becomes

$$H_2(\sigma_1 + \sigma,\sigma_1) = E[y(t')x(t' - \sigma_1)x(t' - \sigma_1 - \sigma)]/P^2$$

$$= E[y(t)x(t' - \sigma_1 - \sigma')]/\Delta_s^2 \tag{27}$$

$$= H_1(\sigma_1 + \sigma')/\Delta_s \tag{28}$$

where σ' is given by Equation 26. The power density is replaced by Δ_s since $P = \overline{x^2}\Delta_s$ and $\overline{x^2} = 1$ for a binary stimulus that is either $+1$ or -1. Equation 28 shows that the second-order kernel is equal to a first-order kernel shifted by the displacement σ'. The value of the displacement σ' can be as large as almost the full record length and is discussed by Sutter (1992). By using the condition in Equation 26 it is apparent that any arbitrary higher-order kernel can be found in a shifted first-order kernel. Sutter uses this principle to calculate any kernel very rapidly.

A problem occurs if the shift, σ', causes two nonnegligible kernels to run into each other, causing an *anomaly*. This possibility of overlap increases if the memory of the system (the length of a kernel) is large. For a single input stimulus the possibility of overlap is remote. A real problem arises in the multiinput case. Suppose 256 stimuli are being independently modulated. In order to obtain the desired "perfect" autocorrelation functions, the 256 stimuli must come from shifted versions of the same stimulus (Sutter, 1992). If the stimuli are generated with a 17 bit shift register, then 256 equally spaced first-order kernels would be separated by 512 time intervals. The 512 interval is large enough that a carefully chosen M-sequence could position a few second-order self-kernels and cross-kernels within the interval without having any overlap of kernels (Sutter and Tran, 1991). However, if the interaction between kernels extended further than between nearest spatial neighbors (creating many cross-kernels), or if the number of self-kernels was large, then the density of the kernels would be so high that overlap would be very likely. In that case the number of multiple inputs would have to be reduced.

H. HOW TO DEAL WITH THE ANOMALIES

M-sequences offer the most promising solution to the orthogonality problem. Sutter (1987, 1992) uses M-sequences to calculate multiinput kernels with much lower noise than would be possible by other techniques. If one had a cure for the kernel overlap (anomalies), one could, in addition, calculate some of the higher-order kernels of his 256-input system.

Sutter has made important progress in this area, and further research on this overlap problem is critically needed to allow the full power of white noise to flourish. Some of the "cures" are listed here:

1. For an M-sequence of order k, there is a window of size k that is totally free of overlaps. By using sparse-stimulation, to be discussed in Section VII, this window can be increased in size. The ganglion cell data to be analyzed in Section VI shows an example of this approach.
2. There are many M-sequences from which to choose. For each different system being studied and for each stimulus used, there would be different optimal M-sequences, since different systems are dominated by different linear and nonlinear kernel elements. Barker et al. (1970, 1972) investigated many M-sequences to determine which sequences gave minimum overlap of kernels of a given order.
3. Sutter (1987) pointed out that there are a variety of tricks that can be used to separate maximally the kernels that might be overlapping. The use of inverse-repeat sequences (where the original sequence is followed by the same sequence with reversed polarity) completely separates the odd- from the even-order kernels. This technique increases the separation between kernels since the even and odd kernels normally alternate in order.
4. To me, the most appealing method for dealing with the anomalies may be the simplest. One should simply repeat the experiment twice (or more), using *different* inverse-repeat M-sequences. One will then calculate *multiple* sets of kernels that should be identical except for different biological noise and different patterns of overlap. This scheme has the advantage that the *pair* of kernels provides an estimate for the standard error of the kernels since the stochastic noise will be different in the multiple runs. The systematic "overlap noise" would also be visible because the different M-sequences would produce different overlaps. When the two kernel estimates are similar, one gains confidence in the validity of the estimate. If, on the other hand, the two kernel estimates have a region where there is a big difference, then the researcher knows that to be a region of ignorance. If that region is a particular slice of the second-order kernel, then it could be filled in by interpolation from adjacent kernels, it could be left unestimated, or it could be estimated in a third run using a third sequence. It is better to know that one is ignorant about a kernel value than what happens with the usual estimation techniques where one does a single run, obtains noisy kernels, and then naively believes in them because a second independent run was not done. If one is dealing with a stationary system for which multiple runs can be done, it would be useful to repeat a run with the *identical* M-sequence as well as with different M-sequences to obtain an estimate of the intrinsic noise, independent of the autocorrelation noise.

I. HOW TO CHOOSE THE STIMULUS

Having just read several methods for estimating the kernels, the reader might wonder which method is best. The optimal method depends on the characteristics of the system, the stimulus characteristics, and the question being asked. This section considers how the choice of the stimulus affects the type of analysis used for estimating kernels. The choice of stimulus will depend on the considerations to be discussed in Section IV on units, Section VI on the effect of different intensity distributions, and Section VII on the role of kernel variance in guiding the choice of stimulus. For present considerations the stimulus will be chosen to be binary. Three further stimulus characteristics must be determined:

1. Autocorrelation Properties

Three choices are be considered here: random sequences, DeBruijn sequences, and M-sequences. The autocorrelation properties are

Random sequences — In these sequences,

$$\begin{aligned} A_2(\tau) &= E[x(t)x(t + \tau)]/E[x^2(t)] \\ &= 1 \text{ for } \tau = 0 \\ &= r(\tau) \text{ for } \tau \neq 0 \end{aligned} \tag{29}$$

where $r(\tau)$ is a random number with a standard deviation of order $1/\sqrt{R}$ where R is the number of stimulus samples. There is very little reason to ever use truly random sequences (equivalent to M-sequences whose cycle time is much larger than the length of the run). Two reasons for using them are (1) the researcher does not want to invest the time needed to learn how to use DeBruijn or M-sequences, and (2) the system being studied has a variable lifetime (such as a delicately held neuron), so it is unlikely to hold it for a full, long cycle.

DeBruijn sequences — These stimuli are intermediate between random sequences and M-sequences. They provide a "complete" test of the system for a given stimulus window size, but they do not have the full orthogonality benefits of M-sequences. A "complete" test means that for a given full history window size (to be called the "generating window") all possible stimulus configurations are presented. For example, for a generating window that is 8 samples long there are 256 possible binary stimuli. A $k = 8$, DeBruijn sequence would have 256 samples comprising all the stimulus configurations. Sutter (1992) has a thorough discussion of how complete sequences can be used for estimating kernels. The autocorrelation function for this class of stimuli is

$$\begin{aligned} A_2(\tau) &= 1 \text{ for } \tau = 0 \\ &= 0 \text{ for } 0 < |\tau| \leq W \\ &= r(\tau) \text{ for } |\tau| > W \end{aligned} \tag{30}$$

where $r(\tau)$ is the same as in Equation 29, and $W = k\Delta_s$ for a DeBruijn sequence of order k. Vaughn (personal communication), who has made a detailed study of the autocorrelation properties of ternary DeBruijn sequences, made me aware of the interesting properties of these sequences. The higher-order correlation functions have characteristics similar to Equation 30.

M-sequences — A maximum length M-sequence has length $R = 2^k - 1$ and its autocorrelation is

$$\begin{aligned} A_2(\tau) &= 1 \text{ for } \tau = 0 \\ &= -1/R \text{ for } \tau \neq 0 \end{aligned} \tag{31}$$

The slight departure from true orthogonality is not a problem. Errors of order $1/\sqrt{R}$ can have significant impact on kernel estimation, but errors of order 1/R are negligible. These stimuli were discussed extensively earlier so further comments are not needed here.

2. Stimulus Sampling Time, Δ_S

The kernel expansion of Equation 5 allows sparse-stimulation, whereby the stimulus sampling interval, Δ_s, can be larger than the response sampling interval, Δ_r. The use of sparse-stimulation allows Δ_r to be small enough to capture the full dynamic range of the response of the system and allows the selection of Δ_s to be based on considerations connected with improving the quality of kernel estimates. The choice of Δ_s must be done carefully since it has dramatic consequences for kernel variance (see Section VII), completeness of system identification (see Section VIII), and for the choice of estimation method, the topic of the present section. The choice of Δ_s directly affects the number of samples, k_m, in the memory window (the full history) and the number of samples in the generating window, k_g, defined as follows:

$$k_m = T_M/\Delta_s \tag{32}$$

$$k_g = \log_2(T_R/\Delta_s) = \log_2(R) \tag{33}$$

where T_M is the memory of the system and R is the number of stimulus intervals in the experimental run. Consider, for example, the ERG where the system memory under photopic conditions is approximately $T_M = 200$ ms. This memory consists of a response duration of about 120 ms following a stimulus, plus an 80-ms interval during which a prior stimulus can adapt or otherwise modify the response of the system. Actually the memory is much longer than 200 ms since if the interval between stimuli becomes very large, then a regime of dark adaptation takes effect and the memory can be more than 1000 s. The system memory, T_M, is really the memory of interest for the present question being considered. A 200-ms memory for the photopic ERG may be sufficient for many purposes even though it is much less than the true "worst case" system memory.

The run length T_R is chosen to be as long as possible without causing the system to lose its stationarity. For the ERG it would be difficult for the human observer to maintain stationarity for run lengths greater than about 1000 s. For such long runs, special techniques are needed to remove eye blink and eye movement artifacts (Sutter and Tran, 1991).

The choice of Δ_s has different effects on k_m and k_g. For $\Delta_s = 10$ ms Equations 32 and 33 give $k_m = 20$ samples and $k_g = 16.6$ samples. For $\Delta_s = 16$ ms the situation is reversed with $k_m = 12.5$ samples and $k_g = 15.9$ samples. When $k_m \leq k_g$, the memory of the system is sufficiently small to be able to use complete sequences. In that case either DeBruijn sequences or M-sequences will produce kernels without the possibility of anomalies or "nonorthogonality noise". Even random stimuli or other nonwhite stimuli can be used as long as compensation is made for repeated sequences or missing sequences as discussed above (Sutter, 1992). A large fraction of the systems that have been studied in the past fall into this category. It is a shame that these techniques for eliminating the nonorthogonality noise were not used years ago since much cleaner kernel estimates could have been obtained and white-noise methodologies would have gained popularity at a faster rate.

When $k_m > k_g$, the memory of the system is too long to be tested with a complete set of stimulus configurations. This situation will be commonly found when multiple stimuli are used as is discussed next. The DeBruijn sequences only guarantee good autocorrelation properties for k_g stimulus intervals. Since there are a vast number of DeBruijn sequences, it is possible to search for particular sequences that have extended good autocorrelation properties (Vaughn, personal communication). When $k_m - k_g$ is not large, then the matrix inversion techniques of Goussard (1987) and Korenberg (1987) would work especially well. The matrix to be inverted to estimate the first- and second-order kernels involves the autocorrelation functions A_2, A_3, and A_4 as seen in Equation 21. The DeBruijn sequences guarantee perfect autocorrelations for k_g time intervals, thereby increasing the stability of the matrix inversion. If M-sequences are used, then there is no need for a matrix inversion because of their good higher-order autocorrelation properties. One must, however, check to see whether there is overlap of kernels as discussed earlier.

3. Multiple Inputs

The third important consideration for choosing a stimulus is to decide how many stimuli can be used simultaneously. In vision there is always a strong temptation to use multiple stimuli. Multiple inputs play a central role in the work of many vision researchers including that of Sutter and Tran (1991) on ERG interactions and ERG topography and in the work of Emerson and co-workers (1987, 1992).

The use of multiple inputs has devastating effects on the size of the system memory. An n-input system increases the effective memory n-fold. Consider, for example, the 2-input ganglion cell data to be discussed in Section V and assume the memory corresponding to each input is 4 intervals in duration. The effective memory is 8 intervals, and the number of kernels to be estimated increases from $2^4 = 16$ to $2^8 = 256$, a dramatic increase for 1 input vs. 2 inputs. For the above ERG example, if the system memory is $k_m = 20$ samples long, and if 256 simultaneous inputs are used, the effective system memory would be $20 \times 256 = 5120$ samples. If random noise were used as the stimulus, and if least squares matrix inversion techniques were used to overcome the nonorthogonality problem, then a 5120×5120 matrix would have to be inverted just for the first-order kernels, a problematic task. DeBruijn sequences are also not possible since a generating window of $k_g = 5120$ steps would take many more than $\approx 2^{5000} \approx 10^{1500}$ lifetimes of the universe for one complete cycle. The only alternatives seem to be either to use M-sequence noise, shifted 256 times for the 256 multiple stimuli, or not to worry about the nonorthogonality noise.

IV. UNITS AND MEAN SQUARE ERROR

The units of the kernels are straightforward to deduce from Equation 5. Consider the ERG where the stimulus is measured in Trolands (Td), a unit of retinal illumination; the response is measured in microvolts (the voltage on a corneal electrode) and time is measured in seconds. From Equation 5 the units of the nth order kernel are $\mu V/(Td\text{-}s)^n$. These unwieldy units make it difficult for the higher-order kernels to have direct meaningfulness. This section explores a special stimulus (flashing rather than continuous) and a special class of systems (adaptive), each with special units that can give greater meaning to the kernels.

A. FLASH KERNELS

Instead of x(t) being a continuous stimulus, measured for example, in Td, it will be useful to consider a flashing stimulus. The flash intensity is no longer specified in Td units, but rather in the stimulus intensity integrated over the flash. Thus, the units would be Td-s. For simplicity of notation the Td-s combination will be replaced with a new unit that we will call a quant (representing quanta of light). One quant will be defined as 0.1 Td-s. This unit is chosen because, with a 0.1-s exposure duration (approximately the human visual system's integration time) and a retinal illumination of 1 Td, a typical cone in the center of the fovea will absorb about 1 photon of light (making the assumption that about 25% of the photons incident on the cornea get absorbed by the photopigment). Thus, if the stimulus is a 5-quant flash of light, we expect that the average cone would absorb 5 photons (quanta of light). It is of interest that before the term Troland was adopted, 1 Td of retinal illumination was called a *photon* of retinal illumination.

In switching from Td to quants it is worth recalling that the units of Td were originally defined as lumens/(s-m^2-steradian) times the pupil size in mm^2. Thus, the Td is a rate unit since it had s in the denominator. Thus, it is quite natural to multiply Td by s to get back to the *amount* of light rather than the *rate of flow* of light. The reason that the units of the *n*th-order kernel involve time to the *n*th power can be traced back to using stimulus units that are rate units (such as Td). If *quantity* units are used (such as quants), then the kernel units simplify.

To illustrate the use of flash units we will examine the first- and second-order terms of the standard Volterra expansion:

$$y(t) = \int_0^\infty g_1(\tau)x(t - \tau)\, d\tau + \int_0^\infty \int_0^\infty g_2(\tau_1,\tau_2)x(t - \tau_1)x(t - \tau_2)\, d\tau_1 d\tau_2 \qquad (34)$$

If instead of a continuous light distribution specified by x(t) the stimulus is a sequence of flashes at times t_i, then the Volterra expansion can be rewritten as:

$$y(t) = \sum_{t_i \leq t} g_1(t - t_i)q(t_i) + \sum_{t_i \leq t} \sum_{t_j \leq t} g_2(t - t_i t - t_j)q(t_i)q(t_j) \qquad (35)$$

where the summation is over all the flashes at times t_i that occur before the response. The nth-order kernel, for our evoked potential example, would have units of $\mu V/\text{quant}^n$. The simplicity of these units suggest that the use of flashes rather than continuous stimuli is quite natural to the white-noise formalism. We wanted to make this point since our ''sparse-stimulation'' formulation makes use of flashes.

B. ADAPTIVE SYSTEMS AND CONTRAST UNITS

Most biological systems are adaptive. For present purposes we will focus on the luminance domain and define adaptation as meaning that as the background luminance increases, the gain of the system decreases such that the system obeys Weber's law: $\Delta L/L$ is a constant, where ΔL is a just-visible increment of luminance. A 10-ms flash of light might be visible if the intensity of the flash were 10 Td (10 Td $\times$ 0.01 s = 1 quant in the above units) on a 100-Td background. If the background were 100 times brighter, then after adapting to the new light level, a 10-ms flash with a 100-fold increase in intensity would be just visible. The reason for this change is that the cones and bipolar cells of the retina operate with a luminance gain control (see Section VIII for the nonlinear kernels of a luminance gain control system). When the light intensity increases 100-fold, the gain decreases by the same factor. At both luminance levels the detection contrast is 10%, where the contrast is defined as $\Delta L/L$. A constant contrast threshold means that thresholds follow Weber's Law as defined above. Electrophysiological signals such as the ERG and the visual evoked potential (VEP) show similar behavior in that the magnitude of the response (but not necessarily the timing) is a function of the *contrast* change rather than the *luminance* change; that is, a 10% luminance increment leads to approximately the same response level, independent of whether the background luminance was 10 Td or 1000 Td.

It is for this reason that for adaptive systems the stimulus is typically specified in contrast units rather than in luminance units. Using contrast rather than luminance for the stimulus is useful for the Wiener expansions. The reason is that contrast does not carry units whereas luminance does (such as Trolands). The *n*th-order kernel, therefore, has the same units as the response, divided by time raised to the *n*th power.

Another reason for using contrast units is that it allows one to make a quick estimate of whether the system is operating in its nonlinear range. The visual system is often characterized as having a logarithmic nonlinearity of the form: $\ln(L + \Delta L) = \ln(L) + \ln(1 + c)$. The Taylor's expansion of a logarithm

$$\ln(1 + c) = c + c^2/2 + c^3/3 + \cdots \qquad (36)$$

shows that the nonlinear terms begin to dominate when the contrast gets close to unity. For smaller contrasts the linear term strongly dominates.

Our goal in this section has been to replace the $\mu V/(Td\text{-}s)^n$ unwieldy units of the *n*th-order kernel. So far we have succeeded in replacing the Td with a unitless quantity (contrast). We now ask what are good units with which to measure time. The kernels must carry units of inverse time because the kernels are density functions. As seen in Equation 5, the response is given by a summation over kernel elements. If the units in which time is measured are changed from seconds to minutes, then the value of Δ_s in Equation 5 would be divided by 60. For example, if $\Delta_s = 1/100$ s, then the new value would be 1/6000 min. Since a change in units cannot change the response, the magnitude of the kernels must change to compensate for the change in the value of Δ_s. If, instead of changing the units, the stimulus sampling time were changed, say from 1/100 s to 1/200 s, then there would be twice as many terms in the summation and the value of the kernels would not change.

We want the unit of time to be a time interval that is natural to the system, such as the integration time of the system. One method for measuring the integration time would be to compare the response of the system to a brief flash with its response to a step change. Suppose we are measuring the ERG while using a near-threshold stimulus consisting of a 0.3-ms flash of 200% contrast (the luminance of the flash is three times its background level, so $c = \Delta L/L = (3L - L)/L = 200\%$). The flash strength is $0.3 \times 200\% = 60\%$-ms. Suppose further that this flash produces the same magnitude response as a sudden 2% step increase in luminance. We define the integration time to be the duration that would be needed for the 2% increment to have the same total "energy" as the 60%-ms flash. The integration time is the flash strength divided by the step threshold. In our example the integration time is $60\%\text{-ms}/2\% = 30$ ms. We will refer to this unit of time as a "bloch", in memory of Bloch's Law of temporal integration. The duration of a bloch will be system dependent.

Once the unit of time has been chosen, whether it is a second or a bloch, one can ask how the stimulus sampling time affects the size of the kernels. Consider the example given in Table 1. As was discussed, for this example the sampling time, Δ_s, was chosen to be 1 bloch. Several kernels from Table 1 were $H_0 = 40$; $H_1(\tau) = 30, -10, -5$; $H_2(\tau,\tau + 1) = -10, 0$ for $\tau = 0$, 1, and 2 blochs. Suppose, instead, that the same values in Table 1 were obtained if the stimulus sampling interval had been 0.5 blochs. We assume that the stimulus strength stayed the same at ± 1. In that case the kernel estimates from Equation 9 would include the factor of Δ_s^{-n} for the *n*th-order kernel. The kernel estimates in column 10 of Table 1 would now become $H_0 = 40$; $H_1(\tau) = 60, -20, -10$; $H_2(\tau,\tau + 0.5) = -40, 0$ for $\tau = 0$, 0.5, and 1 blochs. This dependence of the kernels on the reciprocal of the sampling interval will be important for the estimates of kernel variance in Section VII.

One of the main reasons for going to "natural" units is to be able to have meaningful units for the power density, P. The following example shows how the natural units work. Suppose the stimulus is Gaussian white noise whose standard deviation is half of the mean luminance. The signal will be clipped at the 2 standard deviation point since the luminance cannot go below zero. In contrast units the standard deviation is 0.5 so the variance $(\overline{x^2} - \bar{x}^2)$ is 0.25. It is not possible to have Gaussian stimuli with larger variance without severely distorting the Gaussian distribution by clipping the stimulus. Suppose further, that the stimulus sampling time is 5 ms which is one sixth of the 30-ms integration time that is being used as our time unit, so $\Delta_s = 0.167$ blochs. The power density is given by

$$P = (\overline{x^2} - \bar{x}^2)\Delta_s \tag{37}$$

In natural units the power density is $P = 0.25 \times 0.167 = 0.042$ blochs. In these natural units the strength of the power density is meaningful. We would expect that a power density of 0.042 blochs indicates a weak stimulus since it is much less than unity. For a weak stimulus we would expect that the linear contribution to the response dominates the nonlinear terms. We should point out again that these considerations are expected to apply to systems whose main nonlinearity is due to adaptation. For systems with an intrinsic nonlinearity such as a full-wave rectifier or a squaring element, the nonlinear terms can dominate even with low power stimuli. We expect, however, that there are a wide class of biological systems that are adaptive. This analysis also works for rectifying systems if the stimulus has been carefully designed to stimulate only one polarity of the rectifier.

The units of power density, P, are also meaningful in the Fourier domain. The stimulus x(t) has a relatively flat amplitude spectrum up to a frequency corresponding to the sampling interval. The power density has units of amplitude squared per Hz (or per cycle/bloch in our natural units). Since the bloch units appear in the denominator of the denominator, the power density has units of blochs, as before.

C. RESPONSE VARIANCE

As an application of our units it is useful to examine the contributions of different terms to the response variance (VR) of the Wiener expansion (Equation 5). Using the orthogonality of the X_n expansion functions, the VR can be separated into contributions from each of the kernel elements:

$$VR = \frac{1}{T}\sum_{t=0}^{T} y(t)^2 \Delta_r \tag{38}$$

$$= \sum_{k,n_j,\sigma_j}^{SC} \left(\prod_{i=1}^{k} P_{n_i}\right) \underline{H}_n^2(\tau_1^{n_1} \cdot \cdot \cdot t_k^{n_k}) \Delta_s^n \Delta_r + \overline{n^2} \tag{39}$$

where $\overline{n^2}$ is the variance of the noise. As in Equation 5, $\tau_i = \sigma_i + t_r$, and the summation over t_r is only over the response samples within a single stimulus interval. P_m is the power density relevant to the *m*th-order diagonals as defined in Equation 10. As before, the subscript will be omitted for m = 1, corresponding to the usual power density P. P_m is the quantity in the denominator of the cross-correlation Equations 9 and 11. For a Gaussian stimulus $P_2 = 2P^2$, while for a binary stimulus $P_2 = 0$. In general, for a stimulus with n levels, $P_m = 0$ for $m \geq n$. With a binary stimulus none of the diagonal kernel terms will contribute to the response, so no information about the diagonal is obtained and the variance of these kernel estimates will be infinite (see Section VII).

A value of P = 0.042 blochs, as in our example above, means that the stimulus was gentle. The power density is low when the sampling time is much smaller than the integration time because rapid fluctuations in the stimulus will get washed out by temporal integration, and the effective contrast of the stimulus will, therefore, be reduced. With a power density of 0.042 blochs, one might expect that most of the response is coming from the first-order kernel. For power densities above 1 bloch, the higher-order kernels will have a bigger role and may dominate the response.

The units of the kernel summations are unusual. If the response y(t) is measured in μV, the first-order contribution

$$VR_1 = \sum_{\tau>0} H_1(\tau)^2 \Delta_r \tag{40}$$

will be measured in μV^2/bloch, which has units of the rate of flow of variance. The summations involving the *m*th-order kernels will have units of $\mu V^2/\text{bloch}^m$. When this term is multiplied by the power density (which has units of blochs), the result is μV^2 in agreement with the left side of Equation 38.

What does all this imply for the practical estimation of kernels? Consider the first few terms of the response variance in Equation 38:

$$VR_{total} = \overline{n^2} + P\ VR_1 + P^2\ VR_2 + \cdots \tag{41}$$

where VR_2 is the contribution to the variance that comes from the off-diagonal elements of the second-order kernel (the second summation in Equation 33). Suppose the system in question has the following characteristics:

$$\begin{aligned} &\text{Intrinsic noise variance: } \overline{n^2} &&= 1\mu V^2 \\ &\text{First order variance:} \quad VR_1 &&= 10\mu V^2/\text{bloch} \\ &\text{Second order variance:} \quad VR_2 &&= 5\mu V^2/\text{bloch}^2 \end{aligned} \tag{42}$$

In order to have the first-order kernel make as large a contribution to the response as the noise contribution would require a stimulus power density of:

$$P = \overline{n^2}/VR_1 = 0.1 \text{ blochs} \tag{43}$$

In order to have the second-order kernel make as large a contribution as the first-order kernel would require a power density of:

$$P = VR_1/VR_2 = 2 \text{ blochs} \tag{44}$$

The goal of this analysis is to show how consideration of the units of the stimulus and the response can provide insight into the appropriate power density levels at which to run the experiment.

D. "MEAN SQUARE ERROR" FOR SPIKING SYSTEMS

One runs into trouble with the concept of response variance when dealing with a neural system in which the response is a spike train. The output of the Wiener expansion is a smooth function that predicts the *probability* of a spike. Suppose the response sampling interval is so small that the highest predicted probability is only 0.1 spike per bin. In that case even if the Wiener expansion provides a good fit to the time-locked average data, the MSE of the predicted response to the actual raw spikes will be poor. It is hard to compare a continuous curve (the predicted probability) to the sparse spikes of the neuron.

Brillinger (1988) points out that for spiking systems the appropriate goodness of fit metric is not the square error summed over each response interval, but rather the maximum likelihood:

$$\text{Log likelihood} = \Sigma_i \log(p_i) \tag{45}$$

where the summation is over each spike in the response and p_i is the predicted probability of a spike from the Wiener expansion. For the case of a spiking system it seems appropriate to estimate the kernel elements by maximizing the log likelihood function given by Equation 45, rather than by minimizing the summed square error as in the usual approach. The two approaches should produce very similar kernel estimates, but further research should be done to compare the two methods.

A plot showing the number of spikes (ordinate) as a function of the predicted probability (abscissa) is quite informative, both about the goodness of fit and also about the sharpness of the spike generator (Brillinger, 1988). Further research is needed to determine how best to use this plot to determine the goodness of fit of the model prediction.

The goodness of fit metric in Equation 45 was based on time-locking to the responses rather than the stimulus. This method, commonly called *reverse correlation,* has now become quite popular when dealing with spiking systems. Possibly the first use of this method was Sutter's (1975) measurement of the three-dimensional first-order kernels of a simple cell in cat visual cortex. His elegant technique did not even use a computer.

V. CONDITIONED KERNELS

One of the drawbacks of the Wiener kernel approach to the analysis of nonlinear systems is that the "meaning" of the higher-order kernels is often not immediately obvious. One could, indeed, develop a catalog of second-order kernels and two-dimensional cuts of higher-order kernels similar to a catalog of fingerprints (Hung and Stark, 1977; Hung et al., 1976). The "fingerprint" catalog could be used to identify the structure of the underlying system. It is often the case, however, that for a higher-order kernel to be meaningful, it must be examined in the context of lower-order kernels. Consider, for example, an adaptive system such as the ERG response to a flash (to be called the primary flash). The magnitude and timing of the ERG response depends on the adaptive state of the system at the time of the flash. If the primary flash was preceded by another flash (the conditioning flash) within 30 ms, then the system at the time of the primary flash would be adapted to a higher luminance background level, and the response to the primary flash would be smaller and faster. The second- and higher-order kernels encode the information about how the conditioning stimuli alter the response to the primary flash. These higher-order kernels are based on the difference in the responses with and without the conditioning stimuli. However, it is much easier to interpret what is going on by looking directly at the response to the conditioned primary rather than looking at differences in response. It is the purpose of this section to develop a formalism that allows one to directly display the responses to a primary flash that is altered in different ways by conditioning flashes.

A. VOLTERRA EXPANSION

Klein (1979) showed that the Volterra expansion

$$y(t) = g_0 + \sum_{n} \sum_{\tau_j} \Delta^n g_n(\tau_1 \cdots \tau_n) \prod_{i=1}^{n} x(t - \tau_i) \tag{46}$$

can be rewritten as

$$y(t) = g_0 + \sum_{\tau_1} \Delta \hat{g}_1(t,\tau_1) x(t - \tau_1) \tag{47}$$

This equation looks simple since it has the form of a linear system. The new quantity $\hat{g}_1(t, \tau_1)$ will be called the first-order conditioned kernel. From Equation 46, it is given by

$$\hat{g}_1(t,\tau_1) = g_1(\tau_1) + \sum_{\tau_2} \Delta \hat{g}_2(t,\tau_1,\tau_2) x(t - \tau_2) \tag{48}$$

Note that τ_2 is summed over all positive time intervals that are both larger than and smaller than τ_1. The second-order conditioned kernel is given in turn by

$$\hat{g}_2(t,\tau_1,\tau_2) = g_2(\tau_1,\tau_2) + \sum_{\tau_3} \Delta \hat{g}_3(t,\tau_1,\tau_2,\tau_3)x(t - \tau_3) \quad (49)$$

The higher-order conditioned kernels are defined similarly.

The leading term of the first-order conditioned kernel is the first-order Volterra kernel. The convolution of the second-order conditioned kernel with the stimulus modifies the first-order kernel. The best way to think about the role of the second-order kernel is that due to the prior history of the stimulus the shape of the first-order kernel fluctuates. The amount of fluctuation is controlled by the second- and higher-order kernels.

The Volterra expansion has only limited practical usefulness since it is only valid for very small values of contrast, x(t). For weak stimuli, the response is dominated by the intrinsic noise of the system, so it is not possible to obtain an accurate estimate of the Volterra kernels. For this reason we now examine conditioned expansions for general stimuli with power densities that are large enough to allow for an improved signal-to-noise ratio.

B. GENERAL EXPANSION

Conditioned kernels can be defined based on the time-ordered expansion of Equation 5, which can be rearranged as (for simplicity the stimulus and response intervals are taken to be equal):

$$y(t) - \underline{H}_0 = \sum_{\sigma_1} \hat{H}_1(t,\sigma_1)X_1(t - \sigma_1)\Delta_s + \sum_{\sigma_1} \hat{H}_2(t,\sigma_1^2)X_2(t - \sigma_1)\Delta_s^2 +$$

$$\sum_{\sigma_1} \hat{H}_3(t,\sigma_1^3)X_3(t - \sigma_1)\Delta_s^3 + \text{fourth and higher-order terms} \quad (50)$$

Note that all the summations are over the single time index σ_1 so that each term in Equation 50 is like a first-order kernel. The difference is that the terms $\hat{H}_n$ for $n > 1$ contain higher powers of x(t). The lack of multiple time indices is a great simplification of Equation 50 over Equation 5. The conditioned kernel $\hat{H}_n$ is given by:

$$\hat{H}_1(t,\sigma_1) = \underline{H}_1(\sigma_1) + \sum_{\sigma_2>\sigma_1} \hat{H}_2(t,\sigma_1,\sigma_2)X_1(t - \sigma_2)\Delta_s +$$

$$\sum_{\sigma_2>\sigma_1} \hat{H}_3(t,\sigma_1,\sigma_2^2)X_2(t - \sigma_2)\Delta_s^2 +$$

$$\text{fourth and higher-order terms} \quad (51)$$

$$\hat{H}_2(t,\sigma_1,\sigma_2) = \underline{H}_2(\sigma_1,\sigma_2) + \sum_{\sigma_3>\sigma_2} \hat{H}_3(t,\sigma_1,\sigma_2,\sigma_3)X_1(t - \sigma_3)\Delta_s +$$

$$\text{higher-order terms} \quad (52)$$

Similar equations can be obtained from Equation 5 for the other conditioned kernels. It is seen that the general equation (Equation 50) is similar to Equations 48 and 49 except that higher-order diagonals contribute.

Ignoring fourth and higher-order terms, the first-order conditioned kernel can be expanded in terms of the true orthogonal kernels as:

$$\hat{H}_1(t,\sigma_1) = \underline{H}_1(\sigma_1) + \sum_{\sigma_2>\sigma_1} \underline{H}_2(\sigma_1,\sigma_2)X_1(t - \sigma_2)\Delta_s +$$

$$\sum_{\sigma_2>\sigma_1} \underline{H}_3(\sigma_1,\sigma_1^2)X_2(t, - \sigma_2)\Delta_s^2 +$$

$$\sum_{\sigma_3>\sigma_2>\sigma_1} \underline{H}_3(\sigma_1,\sigma_2,\sigma_3)X_1(t - \sigma_1)X_1(t - \sigma_2)X_1(t - \sigma_3)\Delta_s^2 \quad (53)$$

The notation $\hat{H}_1(t, \sigma)$ may be unfortunate. The purpose of inventing the conditioned kernels was not to define a set of kernels that depended on a random stimulus x(t). Rather, the goal was to define a set of kernels that depended on a *specific* prior history. Thus, rather than the time index, t, on the left of Equation 53, it would be better to specify a particular history. Consider, for example, a system with only first- and second-order kernels, and suppose we would like to know how the system is affected by a previous stimulus whose value is X_+ at a time $\Delta\sigma$ before the primary flash. The first-order conditioned kernel would be $\hat{H}_1(\text{flash at } \Delta\sigma,\sigma) = \underline{H}_1(\sigma) + \underline{H}_2(\sigma,\sigma + \Delta\sigma)X_+\Delta_s$. Later in this section, in connection with the ganglion cell data, an improved notation for the conditioned kernels will be introduced. The ganglion cell data will strongly reinforce the main idea of this section: that higher-order kernels might be hard to interpret in isolation, but make sense when added together.

C. EFFECT OF CHANGING THE STIMULUS

Equation 53 allows us to determine what happens to the first-order kernel if the previous history is changed. Suppose the original kernels were measured with a stimulus whose mean was $\overline{x}$ and whose power density was P. Suppose that in the present experiment the mean has increased to $\overline{x'}$ and the power density has increased to P′. According to Equation 53 the average value of the first-order kernel that has been conditioned by the new prior history can be determined by taking the time average of Equation 53: The new kernel, to be called H_1', is given by:

$$H'_1(\sigma_1) = \underline{H}_1(\sigma_1) + (\overline{x'} - \overline{x}) \sum_{\sigma_2>\sigma_1} \hat{H}_2(\sigma_1,\sigma_2)\Delta_s + (P' - P)$$

$$\sum_{\sigma_2>\sigma_1} \underline{H}_3(\sigma_1,\sigma_2^2)\Delta_s + (\overline{x'} - \overline{x})^2 \sum_{\sigma_3>\sigma_2>\sigma_1} \underline{H}_3(\sigma_1,\sigma_2,\sigma_3)\Delta_s^2 \cdots \quad (54)$$

since

$$E\{P_2\Delta_s\} = E\{(x'^2 - \overline{x^2})\Delta_s\} = P' - P \quad (55)$$

The terms in Equation 54 with the factors of $(\overline{x'} - \overline{x})$ represent the change in the first-order kernel due to a change in the mean luminance of the stimulus. The terms with the factor $(P' - P)$ represent the change due to the change in the power density.

The expansion given by Equation 54 is different from a similar expansion given by Klein (1987) in that the summation over σ_2 does not include $\sigma_2 = \sigma_1$ nor $\sigma_2 < \sigma_1$. The difference is because of the condition $\sigma_2 > \sigma_1$ in Equation 51. In Klein (1987) and in Section VI the stimulus properties remain changed after the primary stimulus up to the response time. Thus, one must be careful about using Equation 54 since only changes in the power density and luminance preceding the primary flash are included. The full expansion to connect kernels obtained with different stimuli is given in Section VI. It is possible to define a different type of conditioned kernel without the time ordering of Equation 5 that will do a better job in connecting the conditioned kernels to the expansion of Section VI. In fact, the Volterra kernels in Equations 48 and 49 have this improved definition since they avoid the time-ordered summations.

D. AN EXAMPLE: CENTER-SURROUND INTERACTIONS IN A CAT GANGLION CELL

1. Conditioned Kernels

In order to demonstrate the usefulness of a modified conditioned kernel expansion I will present data on ganglion cell responses. The data are also ideal for showing the power of sparse-stimulation. In addition, they provide important insights into the nonlinear gain control properties of ganglion cells. The data were collected in collaboration with Drs. Bill Levick, John Kroeker, and Mark Citron when we were all at the California Institute of Technology. The figures to be presented here are based on my analysis of one 7-minute run on one cat ganglion cell with one particular stimulus that I designed to test the interactions between the center and surround. The time of each spike was recorded, so that the kernels could be analyzed with an arbitrarily small response sampling interval. These data are highly relevant to the present question of conditioned kernels since it will be seen that the previous 100 ms of fluctuating stimulus history has a strong conditioning effect on the cell's response. The dynamic noise method showed that the presence of a conditioning stimulus almost 100 ms before the primary stimulus can reliably modify the time of occurrence of the cell's spike, to millisecond accuracy.

In this experiment two binary white-noise signals independently modulated the center and the surround of the receptive field. Every 30 ms there was a 50% chance that the center would change (from on to off or off to on) and a 50% chance that the center would remain unchanged. The surround simultaneously and independently had a similar possible change. Figure 1 shows the poststimulus histograms that are obtained by time-locking the spiking response onto the 4 possible center transitions and the 4 possible surround transitions, making a total of 16 histograms. Each column in the figures has the same center signal, while each row has the same surround signal. The center stimulus is indicated in the row above the responses, and the surround stimulus is indicated in the column to the right of the responses. The 16 average histograms can be labeled using a notation similar to that developed in Equations 13 to 18 for Wiener kernels. Using the notation $Y_{s_0s_1c_0c_1}(t_r)$, (Y stands for the response histogram), the bottom row of stimuli are $Y_{0000}(t_r)$, $Y_{0001}(t_r)$, $Y_{0010}(t_r)$, $Y_{0011}(t_r)$, where c_1 is 0 or 1 depending on whether the center stimulus at time $t - 1$ is in the low or the high state. Similarly, c_0 refers to the center stimulus at time t, and s_1 and s_0 refer to the surround luminance. Notice that the order of subscripts is reversed from the convention in Table 1, because of the order in which the data in Figures 1 to 4 were plotted. The stimuli in the next row up are $Y_{0100}(t_r)$, $Y_{0101}(t_r)$, $Y_{0110}(t_r)$, $Y_{0111}(t_r)$, and so forth. The argument t_r refers to the responses with sampling intervals of 2.5 ms that lie between the 30-ms stimulus intervals.

As an example, the histograms in the third column, $Y_{s_0s_110}(t_r)$, all correspond to the cases where the luminance in the center of the receptive field makes a transition from dark to bright (notice the *off-on* stimulus at the top of column 3). The histograms in the second row from the bottom, $Y_{01c_0c_1}(t_r)$ all correspond to the cases where the luminance in the surround makes a transition from bright to dark (notice the *on-off* stimulus at the right of row 2). The circled histogram, $Y_{0110}(t_r)$, corresponding to the third column and second from bottom row, has the strongest responses, indicating that this cell has an excitatory center and an inhibitory surround. That histogram was generated by searching throughout the 7-min experiment for all occurrences of the center switching from off to on and the surround switching from on to off. In the 7-min experiment there were about 13,000 stimulus events (one every 30 ms). Since only about 1/4 of these events have the *off-on* center and only 1/4 have the *on-off* surround, there are about 13,000/16 ≈ 800 stimulus repetitions contributing to each histogram. The tickmarks underneath each histogram represent a 10-ms interval. The histograms are plotted with a bin width of 2.5 ms. Each histogram has 14 data points plotted. There are 12 data points within the 30-ms stimulus period and 2 data points for the 5 ms

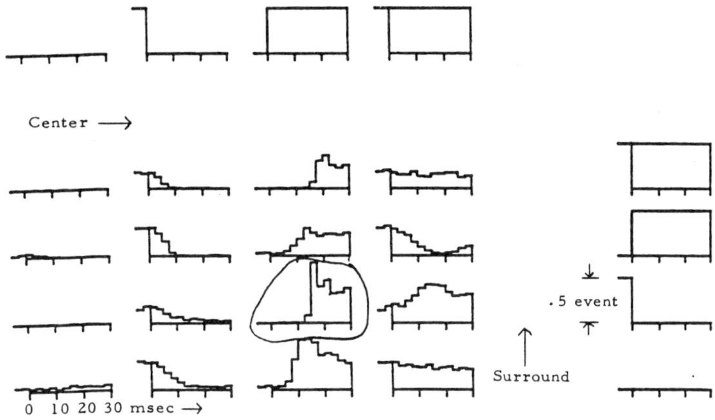

FIGURE 1. Response histograms of a cat ganglion cell. The stimulus was a pair of independently modulated binary stimuli, in the center and the surround of the receptive field. The stimulus randomly changed every 30 ms. The 16 responses are given by: $Y_{s_0s_1c_0c_1}(t_r)$. Each histogram bin is 2.5 ms. The row of the figure (counting from the bottom) is specified by the first two subscripts, and the column is specified by the last two. Thus, for example, $s_0s_1 = 00$ and 01 specifies the lowest and next-to-lowest rows, respectively. The stimulus configurations are shown in the rightmost and topmost rows. The height of the *on* stimulus was chosen to act as a scale for measuring the response; the height corresponds to a 50% chance of a spike falling into one of the 2.5-ms bins. The circled histogram, $Y_{0110}(t_r)$, is of special interest because when examined in more detail in Figures 2 and 3, sharp spiking is revealed.

prior to the stimulus period being plotted. The vertical axis is scaled such that the height of the stimulus (see topmost row or rightmost column) represents 1/2 spike per 2.5-ms bin. Thus, the top peak of the circled figure indicates the cell had a 70% chance of firing within that time interval. Since there were 800 stimulus repetitions, there are about $800 \times 0.7 \approx 550$ spikes in that 2.5-ms bin.

Many interesting and subtle features can be found in Figure 1. Consider, for example, the question of how the surround affects the response to a luminance increment in the center. The third column, $Y_{s_0s_1 10}(t_r)$, contains the relevant information. The bottom histogram, $Y_{0010}(t_r)$, shows how rapidly the cell can fire when the surround remains off. In the 7.5-ms bin (bins will be referred to by the initial time, so the 7.5-ms bin includes spikes occurring 7.5 to 10 ms after the stimulus transition) there is a 30% chance of a spike, and in the 10-ms bin there is a 50% chance of a spike. This is a rapidly responding cell. However, as seen in the second from bottom row, $Y_{0110}(t_r)$, and in the top row, $Y_{1110}(t_r)$, if the surround had been on before the center's transition, then the cell would not have started firing until about the 15-ms bin. Even though the turning off of the surround is excitatory, its turning off does not speed up the response of the center. It is not that the surround is slow, as can be seen in column 4, rows 2 and 3.

Greater insight can be gained by looking at a longer stimulus history, the topic of conditioned kernels, which is the central point of this section. Figure 2 is similar to Figure 1 except that now there are 8 rows and 8 columns. In addition to the primary transition at $t = 0$ ms, a conditioning stimulus at $t = -30$ ms is also shown. The particular stimulus pattern for the center and surround is shown at the very top row and the very right column. Note that to conserve space the stimulus transition is shown at -2.5 ms rather than at -30 ms. In the notation used above, the responses are arranged in the order: $Y_{s_0s_1s_2c_0c_1c_2}(t_r)$. The kernels in the bottom row are $Y_{000000}(t_r)$, $Y_{000001}(t_r)$, $Y_{000010}(t_r)$, $Y_{000011}(t_r)$, $Y_{000100}(t_r)$, $Y_{000101}(t_r)$, $Y_{000110}(t_r)$, $Y_{000111}(t_r)$. This particular ordering of the kernels will be important when in Figure 4 we take the Walsh transform of the average histograms (Y) to obtain the Wiener kernels, F, of the system.

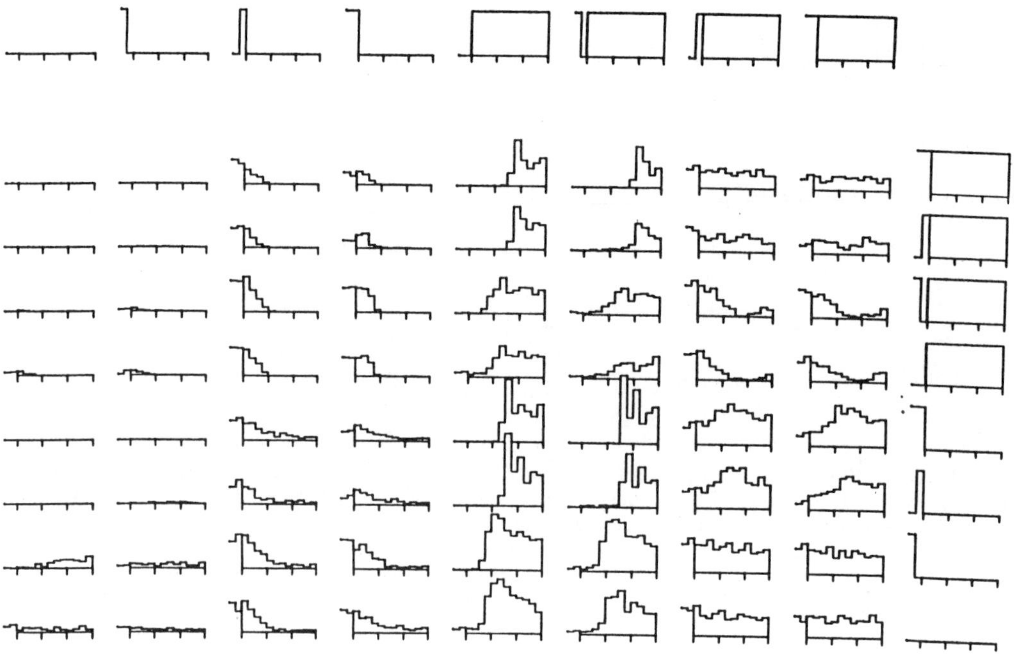

FIGURE 2. Similar to Figure 2 except that the stimulus history being considered is 3 time samples instead of 2 time samples which were used in Figure 1. The 64 responses shown are labeled, $Y_{s_0s_1s_2c_0c_1c_2}(t_r)$. The corresponding stimuli are shown in the top and right rows. The stimulus transition at -30 ms is positioned at -2.5 ms to simplify the presentation. One of the interesting features of the data is the slight shift in the time of the spike between histograms $Y_{010101}(t_r)$ and $Y_{011101}(t_r)$ (6th column and 3rd and 4th rows from bottom). The only difference between the two responses is whether the surround luminance was high or low between 30 and 60 ms before the primary transition.

One of the most interesting features in Figure 2 involves comparing the responses in the 6th column, and 3rd and 4th rows from the bottom, $Y_{01s_2101}(t_r)$. This is the column for which the center has a conditioning *on-off* transition at -30 ms and then a primary *off-on* transition at 0 ms. The surround has an *on-off* primary transition, and we are comparing the two possible surround transitions at -30 ms. If the conditioning luminance before -30 msec was *on*, then there will most likely be a spike in the 15-ms bin. If, on the other hand, the conditioning luminance was *off*, then the spike will tend to be delayed until the 17.5-ms bin.

The effect of the stimulus transition at -60 ms can also be examined as shown in Figure 3. The histograms are arranged in the order: $Y_{s_0s_1s_2s_3c_0c_1c_2c_3}(t_r)$. Shown here are expanded history views of the lower right four boxes of Figure 1 (center on, surround off during the 30-ms response period). Thus, the 64 displayed responses are $Y_{0s_1s_2s_31c_1c_2c_3}(t_r)$. There are a number of histograms where the transition at -60 ms causes small shifts in the timing of the spikes [compare, for example, row 5 from bottom, columns 3 and 4, $Y_{0100101c_3}(t_r)$]. It should be pointed out that the 2.5-ms binning was too coarse for the exquisite time-locking of the spikes. For example, the upper left histogram of Figure 3, $Y_{00011000}(t_r)$, does not show as strong ringing as does the histogram to its right, $Y_{00011001}(t_r)$. However, I am confident that the ringing is just as strong, but the ringing is slightly displaced so that the bin boundaries erase the ringing. The spikes occur approximately every 5 ms, so a 2.5-ms sampling interval is not sufficiently fine.

The top 4 rows of the first column of Fig. 3, $Y_{01s_2s_31000}(t_r)$ tell an interesting story. The probability of firing in the 12.5-ms bin is directly related to the history of activity in the

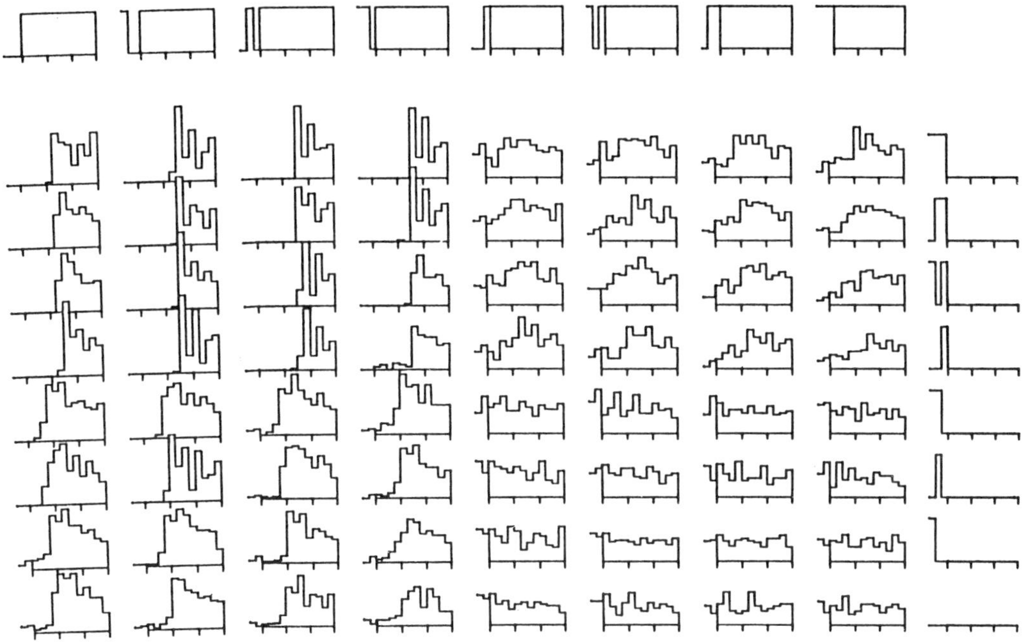

FIGURE 3. Similar to Figure 3 except that the stimulus history being considered is 4 time samples instead of 3 time samples which were used in Figure 1. Instead of showing all 16 × 16 = 256 responses only 64 are shown specified by: $Y_{0s_1s_2s_31c_1c_2c_3}(t_r)$. The center stimulus is *on* and the surround stimulus is *off* during the same 30 ms as the response being shown. The sharpness of the spiking indicates that the response is deterministic when the full history is taken into account.

surround in the two intervals from −90 to −30 ms. If the number of *on* intervals between −90 and −30 ms was 2, 1, or 0, then the probability of spiking at 12.5 ms is about 50%, 30%, and 5%, respectively. This result would indicate that more light in the surround before the *off-on* center transition speeds up the responsiveness of the cell. However, the bottom 4 rows of the first column, $Y_{00s_2s_31000}(t_r)$ remind us that an *off* interval in the surround just preceding the *off-on* center transition speeds up the spiking even more, as discussed earlier. The rebound from the earlier *on* period (before −30 ms) strengthens the effectiveness of the off period from −30 to 0 ms before the primary transition. The responses in the top 4 rows of the first column do not show much ringing. One possible reason is that the 2.5-ms sampling interval was too coarse, and the phase of the ringing happened to fall orthogonal to the bin boundaries. A second reason is that the 4 histograms being considered all have 90 ms of *off* in the center before the *off-on* transition. These histograms might be exceptionally sensitive to whether the period from −120 to −90 ms was *on* or *off*, and we are averaging over these two possibilities. The memory of the system might be quite long if the *off-on* transition is preceded by a long *off* period.

The stimulus in the "distant" past is able to influence the timing of the ganglion cell's spike. This is not a simple nonlinearity such as a saturation or threshold effect. It is a complex luminance and contrast gain control. The word "gain control" is being used in the general sense, meaning that the sensitivity, gain, latency, delays, or other characteristics of the cell are modified by the prior stimulus history of the cell. A simplified contrast gain control model will be discussed in Section VIII.

It is easy to calculate confidence limits on the spike rates. There are 13,000/256 ≈ 50 stimulus occurrences contributing to each histogram in Figure 3. In the histograms with a lot of ringing, the peak corresponds to about a 80% chance of a spike in the 2.5-ms bin.

This implies about 40 spikes at the peaks. Since spikes are a Poisson process, the standard deviation of the spike rate is $1/\sqrt{40} \approx 15\%$, which is much smaller than the amplitude of the oscillations that are seen in the left 4 columns of Figure 3. Thus, the ringing is highly significant. I suspect that if the bin size had been reduced to 1 ms, the ringing would have been even more dramatic and would have been present in more of the histograms.

2. Wiener Kernels of the Ganglion Cell Data

The histograms shown in Figures 1 to 3 are really raw responses rather than conditioned kernels, since the conditioned kernels defined in Equations 51 and 52 involve differences. For example, the conditioned first-order kernel in Equation 51 or 53 involves the difference in the response to a positive primary flash minus the response to a negative primary flash. However, in many cases the raw responses are equal to the conditioned kernels because of a quirk of the ganglion cell data. Consider, for example, the four cells with the striking spiking $Y_{01s_210c_2}(t_r)$. One possible conditioned kernel would be the first-order center response, taking the difference between center-on and center-off at $t = 0$. This could be written as

$$\hat{H}_{c0}(c_1,c_2,s_0,s_1,s_2;t_r) = Y_{s_0s_1s_21c_1c_2}(t_r) - Y_{s_0s_1s_20c_1c_2}(t_r) \tag{56}$$

For the four cells with the spiking the conditioned kernels would be

$$\hat{H}_{c0}(0,c_2,0,1,s_2;t_r) = Y_{01s_210c_2}(t_r) - Y_{01s_200c_2}(t_r) \tag{57}$$

Inspection of Figure 2 shows that the four response histograms $Y_{01s_200c_2}(t_r)$ all vanish. In general, the response is negligible when the center stays off at $t = 0$ and $t = -1$. Thus, for this particular example the response histograms equal a specific set of conditioned kernels. We now examine the more general case in which the Y histograms of Figures 1 to 3 are related to the Wiener kernels defined in Equation 5.

The Wiener kernels and the conditioned kernels are equal to sums and differences of the response histograms. The Wiener kernels to be discussed in this section are based on sampling the stimulus only the instant after the beginning of each stimulus transition interval. We will come back to this point at the end of this section. The connection between the Wiener kernels, F or H, and the responses, Y, is particularly elegant since it is exactly the Walsh transform discussed in Section III and illustrated in Table 1. Figure 4B shows the data of Figure 1 transformed to Wiener kernels using a Walsh transform. Figure 4A is the same as Figure 1 except the data are shown as lines rather than histograms. Lines have the advantage that the responses are smoother. The vertical dotted line goes through the peak of the Y_{0110} responses at $t_r = 15$ ms. This is the location of the first spike that is seen more clearly in Figures 2 and 3. As discussed earlier, the peak of the Y_{0110} distribution is at a probability of 70% chance of firing in the 2.5-ms bin. Figure 4B shows the Full History Wiener kernels, $F_{s_0s_1c_0c_1}(t_r)$. Since the data in Figure 4 involve time-locking onto a single transition, Figures 4A and B are called *single transition* responses and kernels. The kernels in Figure 4B are arranged in the same order as the responses in Figure 4A. Thus, from left to right, the bottom row is $F_{0000}(t_r)$, $F_{0001}(t_r)$, $F_{0010}(t_r)$, $F_{0011}(t_r)$. The next row up is $F_{0100}(t_r)$, $F_{0101}(t_r)$, $F_{0110}(t_r)$, $F_{0111}(t_r)$.

The connection between the F kernels and the Y responses is elegantly given by the Walsh transform and its inverse:

$$Y_{s'_0s'_1c'_0c'_1}(t_r) = \Sigma\, F_{s_0s_1c_0c_1}(t_r)X_{s_0}(s'_0)X_{s_1}(s'_1)X_{c_0}(c'_0)X_{c_1}(c'_1) \tag{58}$$

$$F_{s_0s_1c_0c_1}(t_r) = \Sigma\, Y_{s'_0s'_1c'_0c'_1}(t_r)X_{s_0}(s'_0)X_{s_1}(s'_1)X_{c_0}(c'_0)X_{c_1}(c'_1)/16 \tag{59}$$

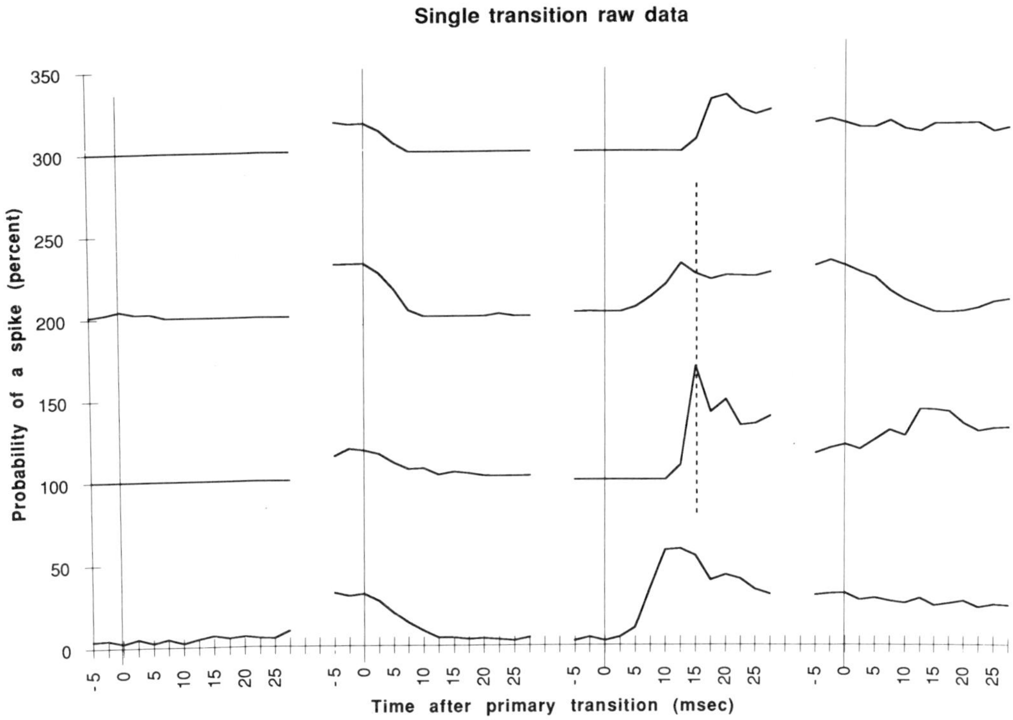

A

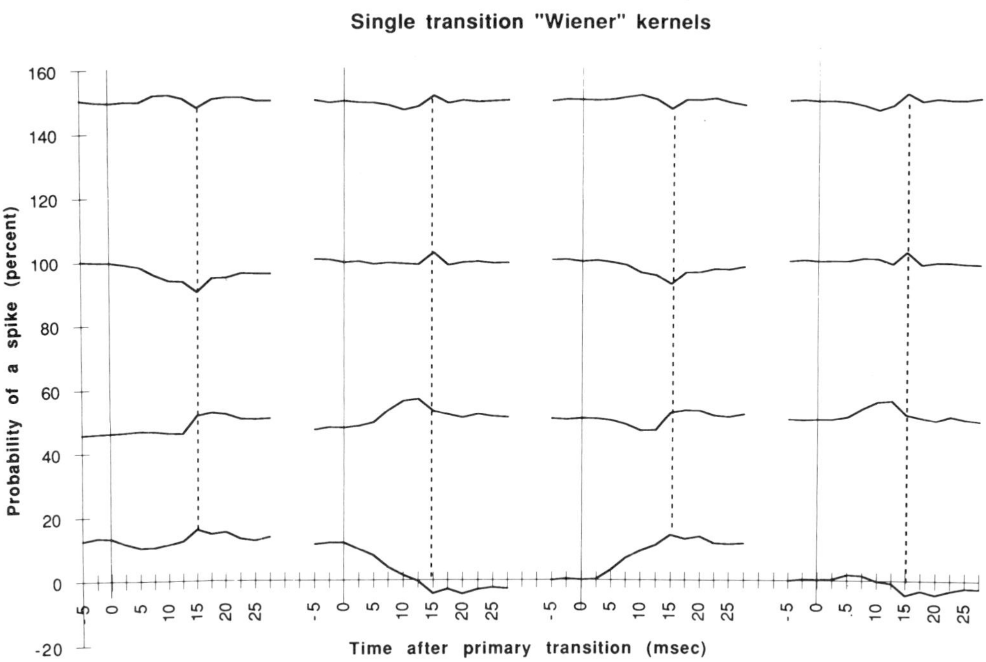

B

FIGURE 4. (A) A replot of Figure 1. Instead of plotting the data as histograms, they are plotted as line drawings to better exhibit the continuity of the data. The vertical dotted line in the third column $[Y_{s0s110}(t_r)]$ is at t_r = 15 ms corresponding to the first spike. (B) The 16 Wiener kernels corresponding to the responses in A. A Walsh transform connects A and B. The vertical dotted line is at t_r = 15 ms, the same as in A. It is seen that the spike in one of the responses contributes to all the kernels, either positively or negatively. This is an argument against using Wiener kernels to display the data when the system is highly nonlinear.

These two equations come directly from Equations 13 and 17. The orthogonal polynomial, $X_i(j)$, is -1 or $+1$ according to:

$$X_i(j) = (2j - 1)^i = (-1)^{i(1-j)} \tag{60}$$

where i and j take on values of 0 and 1. Thus, the zeroth order polynomial ($i = 0$) is always $+1$, and the first-order polynomial ($i = 1$) is -1 or $+1$, depending on whether j is 0 or 1. The value of $X_1(j) = 1$ is reversed in sign from Sutter's (1992) definition. He makes the correspondence $0 \rightarrow 1$ and $1 \rightarrow -1$, whereas our convention is $1 \rightarrow 1$ and $0 \rightarrow -1$. Our definition has the advantage that X corresponds to the contrast of the stimulus, a definition compatible with our discussion in Section IV. Sutter's definition has the advantage that $X_i(j) = (-1)^{ij}$, which is simpler than our formalism in Equation 60. For Sutter, the Walsh matrix element that multiplies the response, Y_a to get the Wiener kernels, F_b (and also the inverse matrix from F to Y), is simply the bitwise parity of a AND b. For example, if $a = 5$ and $b = 9$, then 5 AND 9 = '0101' AND '1001' = $0 \times 1 + 1 \times 0 + 0 \times 0 + 1 \times 1 = 1$, which is odd, so the Walsh matrix connecting the 5th element of Y to the 9th element of F would be -1. For our definition of $X_i(j)$ given in Equation 60, the polarity of all the Y bits must be switched before the AND operation. This sign difference in the definition of X_1 has the consequence that all of Sutter's odd-order kernels are reversed in sign from our odd-order kernels. The even-order kernels have the same sign. The connection between the F kernels and the Y responses is the same as the connection between columns 6 and 10 in Table 1.

The first few Wiener kernels shown in the bottom row of Fig. 4b are

$$H_0(t_r) \quad = F_{0000}(t_r) = \Sigma\, Y_{s'_0s'_1c'_0c'_1}(t_r)/16 \tag{61a}$$

$$H_{1c}(t_r) \quad = F_{0010}(t_r) = \Sigma\, Y_{s'_0s'_1c'_0c'_1}(t_r)X_1(c'_0)/16$$

$$= -\Sigma\, Y_{s'_0s'_1c'_0c'_1}(t_r)(-1)^{c'_0}/16 \tag{61b}$$

$$H_{1c}(1 + t_r) = F_{0001}(t_r) = \Sigma\, Y_{s'_0s'_1c'_0c'_1}(t_r)X_1(c'_1)/16$$

$$= -\Sigma\, Y_{s'_0s'_1c'_0c'_1}(t_r)(-1)^{c'_1}/16 \tag{61c}$$

$$H_{2c}(1 + t_r, t_r) = F_{0011}(t_r) = \Sigma\, Y_{s'_0s'_1c'_0c'_1}(t_r)X_1(c'_1)X_1(c'_0)/16 \tag{61d}$$

This last equation means:

$$\begin{aligned} F_{0011}(t_r) = [&Y_{0000}(t_r) + Y_{0100}(t_r) + Y_{1000}(t_r) + Y_{1100}(t_r) - \\ &Y_{0001}(t_r) - Y_{0101}(t_r) - Y_{1001}(t_r) - Y_{1101}(t_r) - \\ &Y_{0010}(t_r) - Y_{0110}(t_r) - Y_{1010}(t_r) - Y_{1110}(t_r) + Y_{0011}(t_r) + \\ &Y_{0111}(t_r) + Y_{1011}(t_r) + Y_{1111}(t_r)]/16 \end{aligned} \tag{62}$$

The inverse equations giving the responses are similar. The circled histogram of Figure 1 that includes the sharp spiking is given by:

$$Y_{0110}(t_r) = \Sigma F_{s_0s_1c_0c_1}(t_r)(-1)^{s_0}(-1)^{c_1} \tag{63}$$

$$= F_{0000}(t_r) + F_{0100}(t_r) - F_{1000}(t_r) - F_{1100}(t_r) - F_{0001}(t_r) -$$

$$F_{0101}(t_r) + F_{1001}(t_r) + F_{1101}(t_r) + F_{0010}(t_r) + F_{0110}(t_r) -$$

$$F_{1010}(t_r) - F_{1110}(t_r) - F_{0011}(t_r) - F_{0111}(t_r) + F_{1011}(t_r) + F_{1111}(t_r) \tag{64}$$

The dotted line in Figure 4A shows that only response Y_{0110} has the sharp spike. Thus, when the Walsh transform is performed on the responses, all 16 Wiener kernels must participate in the spike. If the Walsh transform had been done on the 3-time period data with the sharper spiking shown in Figure 2, then the spikes would have been distributed among almost all 64 kernel elements. This is a powerful argument pointing out that for certain highly nonlinear systems it is much easier to interpret the raw responses rather than the Wiener kernels.

One advantage of considering the conditioned kernels rather than the raw responses is that the Wiener expansions are simpler. Thus, the conditioned kernel corresponding to the circled histogram in Figure 1 is (as discussed above):

$$\hat{H}_{c0}(0,0,1;t_r) = [Y_{0110}(t_r) - Y_{0100}(t_r)]/2 = \Sigma F_{s_0s_1c_0c_1}(t_r)(-1)^{s_0}(-1)^{c_1}[1-(-1)^{c_0}]/2$$

$$= \Sigma F_{s_0s_1c_0c_1}(t_r)(-1)^{s_0+c_1}c_0 \tag{65}$$

$$= + F_{0010}(t_r) + F_{0110}(t_r) - F_{1010}(t_r) - F_{1110}(t_r) - F_{0011}(t_r) -$$

$$F_{0111}(t_r) + F_{1011}(t_r) + F_{1111}(t_r) \tag{66}$$

Thus, instead of 16 Wiener kernels that contribute to the raw response, there are only 8 terms that contribute to the conditioned linear response. The zeroth-order kernel is missing in Equation 66, which is reasonable since, in general, lower-order kernels do not contribute to higher-order conditioned kernels.

In addition to providing insight into the operation of ganglion cells there are a number of important methodological points that are being told by these data:

1. This section began with a critique of the practice of using the higher-order kernels as "fingerprints" to identify important characteristics of the system. It was pointed out that the higher-order kernels were often difficult to "understand" when isolated from the lower-order kernels. Conditioned kernels as used here offer an alternative way to present the higher-order kernels and can be critical for understanding the nonlinearities of the system. In the comparison between Figure 4 A and B, it was seen that a feature that is localized to a single response histogram was spread uniformly throughout all 16 single-transition kernels. If, instead of presenting the ganglion cell data as conditioned kernels, I would have shown them as the regular Wiener kernels, then the interesting features would have been lost. For the data in Figure 3, we are looking at features that are specific to 4 center and 4 surround transitions, so many kernels through eighth order carry the specific information. The spike timing would be encoded in many kernels through eighth-order kernel in a manner that would make the spikes barely visible. Only by showing the effect in the context of all the lower-order kernels, as in Figure 3, are the effects clear and dramatic. I suspect that situations similar to the one just described are common in biological systems.

2. Sparse-stimulation is essential to drive the system strongly and not lose information about the dynamics of the system. In the present case, $\Delta_s = 30$ ms and $\Delta_r = 2.5$ ms. The small response sampling time was essential to reveal the most interesting properties of the cell. As discussed above, the 2.5-ms interval should have been shortened to 1 ms. The long stimulus interval of 30 ms was essential to drive the cell hard. It is useful to express power density in terms of units in which contrast is used for stimulus strength so that the stimulus variance is unity. The power density of the center or surround stimulus becomes P = 30 ms. This is a high power density since 30 ms is probably somewhat larger than the characteristic adaptation time of the neuron. Because of this large power density it is not surprising that we were able to learn about highly nonlinear aspects of the cell involving kernels up to eighth-order. Further research with a more rapid stimulation would be needed to pin down the timing of adaptation more accurately.
3. A word or caution is needed about our nomenclature. I have been calling the kernels shown in Figure 4B, Wiener kernels. That is not entirely correct. The actual stimulus was constant for 30 ms either at luminance L or 0. Thus, the stimulus autocorrelation function would be a triangle with a 60-ms base. A Lee-Schetzen cross-correlation would totally smear out all the features of the kernels. Instead of cross-correlating the response with the actual stimulus, we cross-correlated the response with a sampled stimulus. No information was lost. As mentioned earlier, what the stimulus does between samples must be explicitly described so that the experiment can be replicated. All that is needed for our sparse-stimulation methodology is to know that the stimulus between samples is perfectly predictable from the stimulus at the sample. Another approach to this situation would have been to time-lock onto *transitions* rather than onto the luminance in a given bin. This has the advantage that the transition is sharp, whereas the luminance step is smeared over a 30-ms interval. The histograms would relate to the cross-correlation of the response with the *derivative* of the stimulus. The problem with the derivative of the stimulus (the *transition* stimulus) is that it is not white, even on the 30-ms time scale. The transition stimulus is ternary (*on-off, off-on,* and *no change*), but transitions are not independent. A zero transition is always possible, but an *on-off* transition is only possible if the preceding nonzero transition was *off-on.* This complexity would make the relationship between the responses and the orthogonal expansion a bit tricky, but a one-to-one connection between the transition kernels and the kernels shown in Figure 4B can be established. There are many occasions in vision research where transition kernels can be useful. Even though the transition stimuli depart seriously from independence and whiteness, our analysis has shown that these stimuli should not be avoided.

VI. EFFECT OF DIFFERENT INTENSITY DISTRIBUTIONS

This section examines how the kernels defined in Sections II and III depend on the amplitude distribution of the stimuli. How, for example, do the kernels obtained with binary stimuli differ from the kernels obtained with Gaussian stimuli? This section is an abbreviated version of material presented elsewhere (Klein, 1987; Klein and Yasui, 1979).

There are two aspects in which the intensity distribution affects the kernels.

1. The number of intensity levels the stimulus can assume places a limit on how many diagonal elements can be calculated. In general, if the stimulus has n levels, then kernels with n or more repeated time indices cannot be estimated. If a binary stimulus is used, then none of the diagonal elements can be calculated. This is because kernel

elements with two or more equal time delays do not contribute to the response. For a ternary stimulus, kernels with 3 or more repeated indices cannot be estimated. This section will examine the special and important role played by the diagonal elements. The advantages and disadvantages of limiting the number of intensity levels will be discussed.

2. The value of the kernels does depend on the amplitude distribution of the stimulus. The kernels obtained using different stimuli can be related to each other by an expansion similar to a Taylor's expansion, as will be spelled out in this section.

A. THE SPECIAL ROLE OF DIAGONAL ELEMENTS

Life would be simple if we could neglect the diagonal kernel elements and stick with binary stimuli. Advantages of binary stimuli were discussed in Section III (kernel estimation), and further advantages will be offered in this and in the following section (Section VII on kernel variance). There are, however, two problems with using binary stimuli:

1. It is not always possible to avoid the diagonals. This section shows how the diagonal elements play the critical role in connecting kernels obtained using different stimuli. For example, later in this Section I will show that the connection between the first-order Wiener kernels, obtained using a stimulus with high-power density and the Volterra kernels corresponding to low-power density, involves diagonal elements of the third-order kernel. If the kernels are smooth, one might be able to calculate the diagonal elements by interpolation. Interpolation may be difficult when using sparse-stimulation since the kernel slices could be far apart. On the other hand, the use of Poisson binary stimuli (not sparse) facilitates interpolation, but has variance problems, as will be discussed.
2. It will become clear that different stimuli are appropriate for different purposes. There are occasions where Gaussian or ternary stimuli are preferred. The formalism to be developed in this section shows how the kernels obtained with different stimuli are related.

B. CHANGE OF BASIS: EXCLUSIVE EXPANSIONS

In order to clarify how kernels obtained with different stimuli can be related to each other, consider a simple system with zeroth- through second-order kernels. Equation 5 gives the general expansion of the response of the system to the stimulus x(t) and the methods of Section III are used to estimate the kernels (the cross-correlation method of Equation 9 is fine since we are not worrying about nonorthogonality noise in this section). If we now stimulate the system with a different stimulus, w(t), the general expansion given by Equations 5 and 6 can be written as (assuming for simplicity the stimulus and response intervals are equal):

$$y(t) = \underline{H}_0 + \sum_{\tau_1} \underline{H}_1(\tau_1)X_1(t - \tau_1)\Delta_s + \sum_{\tau_1<\tau_2} \underline{H}_2(\tau_1,\tau_2)X_1(t - \tau_1)X_1(t - \tau_2)\Delta_s^2 + \sum_{\tau_1} \underline{H}_2(\tau_1^2)X_2(t - \tau_1)\Delta_s^2 + \text{plus third- and higher-order items} \qquad (67)$$

where

$$X_1(t) = w(t) - \bar{x} \qquad (68)$$

and

$$X_2(t) = w^2(t) - w(t)\frac{\overline{x^3} - \overline{x^2}\,\overline{x}}{\overline{x^2} - \overline{x}^2} + \frac{\overline{x^3}\,\overline{x} - \overline{x^2}^2}{\overline{x^2} - \overline{x}^2} \tag{69}$$

The estimated first-order kernel in which the expansion had been based on stimulus w(t), H_1^w is given by Equation 11b:

$$H_1^w(\tau_1) = E\{y(t)[w(t - \tau_1) - \overline{w}]\}/(\overline{w^2} - \overline{w}^2)\Delta_s \tag{70}$$

$$= \underline{H}_1^x(\tau_1) + (\overline{w} - \overline{x}) \sum_{\tau_1<\tau_2} \underline{H}_2^x(\tau_1,\tau_2)\Delta_s + \underline{H}_2^x(\tau_1^2)\,\Delta_s\left(\frac{\overline{w^3} - \overline{w^2}\,\overline{w}}{\overline{w^2} - \overline{w}^2} - \frac{\overline{x^3} - \overline{x^2}\,\overline{x}}{\overline{x^2} - \overline{x}^2}\right) \tag{71}$$

The first term shows that to leading order, the kernels obtained from different stimuli are, indeed, the same. The second term shows that there is a contribution to the new kernel that is proportional to the change in mean luminance, $(\overline{w} - \overline{x})$, due to the shift from x(t) to w(t). The coefficient of the luminance change is a sum over the second-order kernels. The last term is the contribution from the diagonal element of the second-order kernel times a combination of moments of the old and the new stimulus.

Notice that the summation in the second term excluded the diagonal. For this reason the expansion given by Equation 71 is called an exclusive expansion (Klein and Yasui, 1979; Klein, 1987). The general exclusive expansion is given by:

$$H_n^w(\tau_1^{n_1} \cdots \tau_k^{n_k}) = \sum_{p\geq k,m_j} \prod_{i=1}^{p} C_{n_i,m_j}^{x,w} \sum_{\tau_j}^{SC} \Delta_s^{p-k} H_m^x(\tau_1^{m_1} \cdots \tau_p^{m_p}) \tag{72}$$

The notation used here is the same as Klein and Yasui (1979) where τ^n indicates that the τ index is repeated n times (an *n*th order diagonal). On the left and right sides of Equation 72 are the kernels obtained using stimulus w(t) and x(t), respectively. The order of diagonals on the right is always greater than or equal to the order on the left ($m_i \geq n_i$). The kernels on the right side have $p - k$ extra time delays that are summed over. The notation SC on the top of the summation sign indicates "single counting" whereby the time delays, τ_i, are time ordered, as discussed in Section II, to avoid factorial factors. For every time delay on the right side there is a coefficient factor, C, given by:

$$C_{n,m}^{x,w} = \overline{X_m W_n}/\overline{W_n W_n}\,\Delta_s^{m-n} \tag{73}$$

For $n = 0$ the exponent of Δ_s is $m - 1$ rather than m, since an extra factor of Δ_s is already included in Equation 72. The basis functions, X_m, are similar to Equations 68 and 69 where w(t) has replaced x(t) in Equation 7. For example, $X_1(t) = w(t) - \overline{x}$. For the basis functions W_n, both x(t) and the expectation values of x(t) in Equation 7 are replaced by w(t). For example, $W_1(t) = w(t) - \overline{w}$. The expansion functions given by Equation 73 are, thus, relatively easy to compute based on the determinant formalism of Equation 7. A mathematically more elegant expansion can be obtained by using *inclusive* summations in which summations over τ are allowed to include diagonal elements as is discussed next.

C. INCLUSIVE SUMMATIONS

As was discussed following Equation 7, the usual Wiener formalism differs in several regards from the expansion we are using in Equation 5. One major difference is that our expansion avoided the combinatoric factors by using a time-ordered expansion. To clarify this difference we used H_n rather than h_n as the expansion kernels. These two kernels were equal except for a normalization factor given by Equation 8. For the second-order example of Equation 67 the normalizations relating the two-kernel definitions are

$$\begin{aligned} \underline{H}_1(\tau) &= \underline{h}_1(\tau) \\ \underline{H}_2(\tau,\tau) &= \underline{h}_2(\tau,\tau) = \underline{h}_2(\tau^2) \\ \underline{H}_2(\tau,\tau') &= 2\ \underline{h}_2(\tau,\tau') \end{aligned} \tag{74}$$

When Equation 74 is substituted into Equation 71 and the summation over τ' is not time ordered and includes $\tau' \le \tau$, then the expansion of the w-kernels (h^w) in terms of the x-kernels (h) becomes:

$$h_1^w(\tau) = \underline{h}_1(\tau) + (\overline{w} - \overline{x})\, 2\sum_{\tau'} \underline{h}_2(\tau,\tau')\Delta_s + [Q_{1,2}(w) - Q_{1,2}(x)]\underline{h}_2(\tau^2) \tag{75}$$

where the factor of 2 in Equation 75 comes from the normalization in Equation 8. This normalization occurs because there are two possible orderings of τ and τ' in the general expansion, but in Equation 75 only one of the orderings is chosen. The factor $Q_{1,2}(x)$ is related to the third-order cumulant of the x distribution:

$$Q_{1,2}(x) = \Delta_s\left(\frac{\overline{x^3} - \overline{x^2}\,\overline{x}}{\overline{x^2} - \overline{x}^2} - 2\overline{x}\right) = \Delta_s(\overline{x^3} - 3\overline{x^2}\,\overline{x} + 2\overline{x}^3)/(\overline{x^2} - \overline{x}^2) \tag{76}$$

$$= \Delta_s\overline{(x(t) - \overline{x})^3}/\overline{(x(t) - \overline{x})^2} = \text{skewness/variance} \tag{77}$$

The first and second subscripts on the Q coefficient are 1 and 2 because the number of repeated τ indices on the left and right sides of the expansion are 1 and 2, respectively. Equation 77 is the third-order cumulant of the x(t) distribution (the skewness) divided by the second-order cumulant (the variance). The connection of the expansion coefficients with the distribution cumulants was discussed in greater detail by Klein and Yasui (1979).

The same analysis that was done in Equation 75 can be extended to the general situation. The result is

$$H_n^w(\tau_1^{n_1} \cdots \tau_k^{n_k}) = \sum_{p>k,m_j} \prod_{i=1}^{p} [Q_{n_i,m_j}(w) - Q_{n_i,m_j}(x)] \sum_{\tau_j}^{INC} \Delta_s^{p-k} H_m^x(\tau_1^{m_1} \cdots \tau_p^{m_p}) \tag{78}$$

This expansion is similar to Equation 72, except that the summations over τ include the diagonal, and the expansion coefficient $C_{n_i,m_j}^{x,w}$ is replaced by $Q_{n_i,m_j}(w) - Q_{n_i,m_j}(x)$.

Table 2 of Klein and Yasui lists some of the expansion coefficients for inclusive summation for the case in which x(t) and w(t) have zero mean:

$$Q_{i,i}(x) = 1 \text{ for all } i \tag{79a}$$

$$Q_{0,1}(x) = \overline{x} = 0 \tag{79b}$$

$$Q_{0,2}(x) = \overline{x^2}\Delta_s = P \text{ the power density} \tag{79c}$$

$$Q_{0,3}(x) = \overline{x^3}\Delta_s^2 = \text{the skewness} \tag{79d}$$

$$Q_{0,4}(x) = (\overline{x^4} - 3\overline{x^2}^2)\Delta_s^3 = \text{the kurtosis} \tag{79e}$$

$$Q_{0,5}(x) = (\overline{x^5} - 10\overline{x^2}\,\overline{x^3})\Delta_s^4 \tag{79f}$$

$$Q_{0,6}(x) = (\overline{x^6} - 15\overline{x^2}\,\overline{x^4} - 10\overline{x^3}^2 + 30\overline{x^2}^3)\Delta_s^5 \tag{79g}$$

$$Q_{0,7}(x) = (\overline{x^7} - 21\overline{x^2}\,\overline{x^5} - 35\overline{x^3}\,\overline{x^4} + 140\overline{x^3}\,\overline{x^2}^2)\Delta_s^6 \tag{79h}$$

$$Q_{1,n}(x) = Q_{0,n+1}(x)/Q_{0,2}(x) \tag{79i}$$

$$Q_{2,3}(x) = (\overline{x^5} - \overline{x^3}\,\overline{x^2})/(\overline{x^4} - \overline{x^2}\,\overline{x^2})\Delta_s \tag{79j}$$

In general, a *new* time index introduced in the summation of Equations 72 or 75 (a τ on the right without a matching value on the left) is associated with the cumulants $Q_{0,i}$. A single time index or off-diagonal index on the left of Equation 72 or 75 is associated with $Q_{1,n}$, given by the ratio of cumulants as shown in Equation 79i. For diagonal indices (a pair of repeated τ values on the left), the expansion coefficients, $Q_{2,n}$, are not cumulants, and further research is needed to determine these higher-order coefficients. For binary stimuli both the numerator and denominator of Equation 79j vanish, but that is okay since diagonal kernels do not contribute to the expansion for binary stimuli.

D. GAUSSIAN KERNELS AS A FUNCTION OF LUMINANCE AND POWER DENSITY

One of the clearest demonstrations of the usefulness of this formalism is seen with Gaussian stimuli. A Gaussian stimulus is totally specified by its mean luminance, L, and its power density, P. Two sets of kernels obtained with different Gaussian stimuli are easily related because all the Gaussian cumulants, Q_{0n}, except $Q_{01} = L$ and $Q_{02} = P$ vanish. Thus, the general expansion for Gaussian kernels, denoted by G^{PL}, is given by:

$$G_n^{PL}(\tau_1 \cdots \tau_k) = \sum_{r,s} \frac{(P - P_0)^r}{r!} \frac{(L - L_0)^s}{s!} \sum_{\tau_i, i>k}$$

$$\Delta_s^{r+s} G_m^{P_0 L_0}(\tau_1 \cdots \tau_k \, \tau_{k+1}{}^2 \cdots \tau_{k+r}{}^2 \tau_{k+r+1} \cdots \tau_{k+r+s}) \tag{80}$$

This expansion looks exactly like a Taylor expansion, with the second summation in Equation 80 being the *r*th and *s*th derivative of G_n^{PL} with respect to P and L, respectively. It has a simpler form than what has been presented elsewhere (Marmarelis and Marmarelis, 1978) because of the kernel normalization being used and because of a convention adopted for diagonal elements. Equation 80 uses the notation G_n^{PL} for the kernels rather than h_n or H_n for two reasons. First, the letter G stands for Gaussian noise, rather than the letter H which we have been using for general noise. Second, the normalization is slightly different from what was specified by Equation 8. For the "off-diagonal" terms where none of the values of τ on the right-hand side of Equation 75 are equal, then the normalization is given by Equation 8. By this device extra factorial factors such as the factor of 2 in Equation 75 are absorbed in the definition of G. For the diagonal elements, however, the normalization of Equation 8 is not used. Rather, the diagonal elements are normalized as if they were off-diagonal. This point will be clarified in the next paragraph using a first-order kernel as an example. The summations over τ_i are not time ordered (similar to Equation 75), and the factorials in the denominator are needed to compensate for the double counting.

By setting $P = 0$ one obtains the special case of the expansion of Volterra kernels, G^{0L}, in terms of Wiener kernels, G^{PL}. For the inverse relationship of the expansion of Wiener kernels in terms of Volterra kernels, the value of P_0 rather than P, is set equal to zero. The first-order Wiener kernel has the following expansion in terms of Volterra kernels:

$$G_1^{PL}(\tau) = G_1^{0L}(\tau) + P \sum_{\tau_1} \Delta_s G_3^{0L}(\tau,\tau_1^2) + P^2/2! \sum_{\tau_1,\tau_2} \Delta_s^2 G_5^{0L}(\tau,\tau_1^2,\tau_2^2) \tag{81}$$

Hybrid kernel normalization and summation conventions are used in Equation 81. The normalization is given by Equation 8:

$$G_n(\tau,\tau_1^2 \cdots \tau_p^2) = g_n(\tau,\tau_1^2 \cdots \tau_p^2)\, n!/2^p \tag{82}$$

Contrary to what might be expected from Equation 8, the higher-order diagonal elements in Equation 81 continue to use the normalization of Equation 82. That is, the third-order diagonal in Equation 81 is given by $G_3^{0L}(\tau,\tau^2)$ rather than by $G_3^{0L}(\tau^3)$ which would have a different normalization according to Equation 8.

In order to clarify our normalization convention, it is useful to apply the definition of $G_1^{PL}(\tau)$ as given by Equation 11b to the Volterra expansion as given by Equation 46. Equation 11b becomes

$$\begin{aligned} G_1^{PL}(\tau) &= H_1(\tau) \\ &= E\{[(y(t) - g_0)][x(t-\tau) - \bar{x}]\}/P \end{aligned} \tag{83}$$

$$= E\{\sum_t \sum_n \sum_{\tau_j} \Delta^n g_n(\tau_1 \cdots \tau_n)[x(t-\tau) - \bar{x}] \prod_{i=1} x(t-\tau_i)\}/P \tag{84}$$

For the present section, x(t) is a symmetric Gaussian random variable so that the expectation value (the average over t) of all odd moments vanishes. The expectation value of a product of x(t) is given by the summation of products of delta functions summed over all possible pairs of τ values. For the $n = 3$ term in Equation 84 one gets:

$$E[\Delta^3 \sum_t x(t-\tau) \prod_{i=1}^{3} x(t-\tau_i)] =$$

$$P^2[\delta(\tau-\tau_1)\delta(\tau_2-\tau_3) + \delta(\tau-\tau_2)\delta(\tau_1-\tau_3) + \delta(\tau-\tau_3)\delta(\tau_1-\tau_2)] \tag{85}$$

The contribution of the first few terms to Equation 84 becomes

$$G_1^{PL}(\tau) = g_1^{0L}(\tau) + 3P \sum_{\tau_1} \Delta_s g_3^{0L}(\tau,\tau_1^2) + \text{fifth and higher order terms} \tag{86}$$

The factor of 3 comes from the three terms in Equation 85. The factor of 3 is present for the terms with $\tau \neq \tau_1$ because only one of the three orderings of the arguments of g_3 is kept in Equation 86. The diagonal term ($\tau = \tau_1$) which is included in the summation of Equation 86 also has a factor of 3 because for Gaussian stimuli $\overline{x^4} = 3\overline{x^2}^2$. In going from Equation 86 to Equation 81, the normalization factor of Equation 82 (same as Equation 8) has been used to make the expansion have the simple form of a Taylor's expansion.

It is useful to consider a specific example of a third-order system specified by the single third-order Volterra kernel:

$$y(t) = a\, x(t)x(t-1)x(t-2) \tag{87}$$

Its orthogonal Wiener expansion is

$$y(t) = a[X(t) + \bar{x}][X(t - 1) + \bar{x}][X(t - 2) + \bar{x}] \quad (88)$$

$$= a\{\bar{x}^3 + \bar{x}^2[X(t) + X(t - 1) + X(t - 2)] + \bar{x}[X(t)X(t - 1) + X(t)X(t - 2) + X(t - 1)X(t - 2)] + X(t)X(t - 1)X(t - 2)\} \quad (89)$$

Notice that since there are no repeated time indices in Equation 87, only the first moment of the stimulus appears in the general Wiener expansion. For this reason, the general kernels depend only on the mean luminance of the stimulus and not on the power density, skewness, or other higher moments. Since diagonal elements are not present, the Gaussian expansion in Equation 80 is valid for *general* stimuli. Thus, the general kernels can be calculated either from Equation 80 or directly from Equation 89. It is instructive to calculate a few of the kernels by both methods. Equation 80 gives:

$$H_0^{PL} = (L - L_0)^3/3!\ \Delta_s^3[H_3^{00}(0,1,2) + H_3^{00}(0,2,1) + H_3^{00}(1,0,2) + H_3^{00}(1,2,0) + H_3^{00}(2,0,1) + H_3^{00}(2,1,0)]$$

$$= L^3\ \Delta_s^3 H_3^{00}(0,1,2) \quad (90a)$$

$$H_1^{PL}(0) = (L - L_0)^2/2!\ \Delta_s^2[H_3^{00}(0,1,2) + H_3^{00}(0,2,1)]$$

$$= L^2\ \Delta_s^2 H_3^{00}(0,1,2) \quad (90b)$$

where H_3^{00} is the Volterra kernel, since $L_0 = P_0 = 0$ for the Volterra expansion. Equation 90 a and b is in agreement with Equation 89 which gives $H_0^{PL} = a\,\bar{x}^3$ and $H_1^{PL}(0) = a\,\bar{x}^2$ since $L = \bar{x}$ and $H_3^{00}(0,1,2) = a/\Delta_s^3$. The other kernels can also be calculated from either Equation 80 or Equation 89 with both methods giving the same result.

For the Volterra expansion of Equation 87 the general kernels did not depend on the power density of the stimulus. This is because diagonal Volterra kernels were not present. It is useful to consider a system that illustrates the special role that diagonal elements play in the dependence of kernels on the power density. Suppose the Volterra expansion is

$$y(t) = a\ x(t)x^2(t - 1) \quad (91)$$

This Volterra expansion can be reexpressed in terms of the orthogonal polynomials X_1 and X_2 by first inverting Equation 7a and b:

$$x(t) = X_1(t) + \bar{x} \quad (92a)$$

$$x^2(t) = X_2(t) + X_1(t)\ [Q_{12}(x) + 2\bar{x}] + \overline{x^2} \quad (92b)$$

where $Q_{12}(x)$ is skewness/variance as given by Equations 76 and 77, assuming $\Delta_s = 1$. Using Equation 92 a and b, Equation 91 becomes:

$$y(t) = a(X_1(t) + \bar{x})[X_2(t - 1) + X_1(t - 1)(Q_{12}(x) + 2\bar{x}) + \overline{x^2} - \bar{x}^2 + \bar{x}^2] \quad (93)$$

The general kernels can be read directly from Equation 93 by comparing the terms with Equation 5.

$$H_0^{PL} = (\bar{x}^3\Delta_s^3 + \bar{x}P\Delta_s^2)H_3^{oo}(0,1,1) \quad (94a)$$

$$H_1^{PL}(0) = (\bar{x}^2\Delta_s^2 + P\Delta_s)H_3^{oo}(0,1,1) \quad (94b)$$

$$H_1^{PL}(1) = [2\bar{x}^2 + \bar{x}Q_{12}(x)]\Delta_s^2H_3^{oo}(0,1,1) \quad (94c)$$

$$H_2^{PL}(0,1) = [2\bar{x} + Q_{12}(x)]\Delta_sH_3^{oo}(0,1,1) \quad (94d)$$

$$H_2^{PL}(1,1) = \bar{x}\Delta_sH_3^{oo}(0,1,1) \quad (94e)$$

$$H_3^{PL}(0,1,1) = H_3^{oo}(0,1,1) \quad (94f)$$

where $H_3^{oo}(0,1,1) = a/\Delta_s^3$. It is seen that the general kernels H_n^{PL} depend on the power density and skewness as well as on the luminance of the stimulus. We will come back to the examples in Equations 87 and 92 when we consider kernel variance.

E. ADVANTAGES AND DISADVANTAGES OF DIFFERENT STIMULI

This section has shown that kernels obtained using stimuli with different amplitude distributions can be related to each other. We now ask what the relative merits are of the different stimuli.

1. Symmetric Binary

It should be pointed out that a symmetric binary stimulus does not necessarily appear symmetric. Consider a temporal square-wave presentation of a tiny spot on a computer monitor with one frame on and one frame off. Because of the raster display, the *on* frame is actually much briefer than the 16.7-ms frame rate. Thus, the actual display appears as about 1 ms *on* and 32.3 ms *off*. Even though the stimulus as seen by a photometer is not at all symmetric, it is symmetric for our purpose which looks at a single sample of the stimulus with the stimulus sampling interval of 16.7 ms. The brief pulses that are produced by raster displays are ideal for the sparse-stimulation methodology discussed in this chapter, in which the stimulus is sampled at a slower rate than the response. The brief pulses have the advantage that they produce sharp responses.

a. Advantages

1. High power density. By alternating between the highest and lowest luminances that the stimulus generator can produce one obtains maximal power densities. This is desirable since the kernel variance decreases as the stimulus power density increases. If for some reason one wants reduced power density, it is preferable to reduce the stimulus sampling interval rather than reducing the luminance range since a smaller sampling interval improves the temporal resolution of the kernel estimates.
2. Improved orthogonality for kernel estimation by using complete binary M-sequence white noise as the stimulus. This point has been discussed in detail earlier.
3. Very rapid kernel estimation techniques using the Walsh transform are available as discussed earlier.

b. Disadvantages

1. The main limitations of binary stimuli are associated with the missing diagonal elements. It is often claimed that binary stimuli are missing more than the diagonal elements. Since binary stimuli only test the system at two amplitude levels, they seem to provide an incomplete test of the system. This can, indeed, be a serious shortcoming for systems in which the first stage is highly nonlinear. However, the incompleteness of the test is fully attributable to the missing diagonals. If the stimulus sampling interval is small enough that an initial linear filter smears successive stimuli into each other, then it is likely that the binary test is adequate to test the system fully since diagonal elements can be estimated by interpolation.
2. For systems with a strong *on* vs. *off* asymmetry such as would be found in neurons with sharp thresholds, it would be useful to use a multilevel stimulus, such as ternary, to test the two polarities separately. See the following discussion about ternary stimuli.

2. Asymmetric Binary (Poisson) Stimuli

An asymmetric binary stimulus is one in which the probability, *p,* of the stimulus being at the higher luminance differs from 0.5. The difference between an asymmetric binary and a Poisson stimulus is that for a finite sampling interval the Poisson stimulus will have a nonvanishing probability to have more than 1 flash per interval, whereas the asymmetric binary stimulus is limited to a single flash per interval. In the limit of an arbitrarily small sampling interval and a fixed number of flashes per second, the asymmetric binary stimulus becomes indistinguishable from a Poisson stimulus.

In general, the asymmetric binary stimulus is specified by three numbers: x_L, the lower luminance; x_H, the higher luminance; and p, the probability of occurrence of the higher luminance. In terms of these three variables, the mean luminance, power density, and skewness are

$$L = \bar{x} = p\, x_H + (1 - p)x_L = x_L + p(x_H - x_L) \tag{95a}$$

$$P = (\overline{x^2} - \bar{x}^2)\Delta_s = p(1 - p)(x_H - x_L)^2\Delta_s \tag{95b}$$

$$Q_{12} = (\overline{x^3} - 3\overline{x^2}\bar{x} + 2\bar{x}^3)/(\overline{x^2} - \bar{x}^2)\Delta_s = (1 - 2p)(x_H - x_L)\Delta_s \tag{95c}$$

Notice the simple form taken by Q_{12} for the asymmetric binary stimulus. As expected, when $p = 0.5$, the skewness, given by Equation 93, vanishes. The $p = 0$ (Poisson) case is discussed in more detail by Klein and Yasui (1979).

a. Advantages

One of the main advantages of flashes on a dark background with $p \ll 0.5$ is their high-power density when expressed in contrast units. Suppose the low and high stimulus values are chosen so that the mean is zero:

$$\bar{x} = px_H + (1 - p)x_L = 0 \tag{96}$$

This implies $x_H = -(1 - p)x_L/p$. The largest contrast that can be produced is to have the low stimulus be totally black. If x is measured in contrast units (see Section IV), then $x = \Delta L/L$ where L is the mean luminance. The most negative contrast will be $x_L = -L/L = -1$ or -100% contrast. The corresponding high contrast value is $x_H = (1 - p)/p$. For $p = 0.5$ the low and high values are -100 and $+100\%$ as expected. For $p = 0.25$ the low and high values are -100 and $+300\%$.

For an asymmetric binary stimulus the power density as given by Equation 95b is

$$P = \Delta_s(1 - p)/p \tag{97a}$$

The power density can be rewritten as:

$$P = \Delta_A(1 - p) \tag{97b}$$

where $\Delta_A = \Delta_s/p$ is the average time between flashes. If p is very small (a Poisson stimulus), then the power density is equal to the average interflash interval. The meaningfulness of p being in the denominator of the power density (Equation 97a) will be discussed in Section VII.

In comparing the asymmetric binary stimulus to a symmetric binary stimulus it is useful to consider two regimes: (1) If both stimuli have the same stimulus sampling interval, Δ_s, then the asymmetric (Poisson) stimulus ($p < 0.5$) has a greater power density than that of a symmetric binary stimulus ($p = 0.5$). The larger power density improves the kernel variance of the first- and higher-order kernels. (2) Suppose, on the other hand, both stimuli have the same power density by having approximately the same average flash rate, Δ_A. In this case the symmetric stimulus would have a larger sampling interval ($\Delta_s = p\Delta_A$). Since the variance of the first-order kernel depends on the power density and not on Δ_s, the variance of the first-order kernels will be the same for the two stimuli. The asymmetric stimulus has the advantage that the time slices of the higher-order kernels will have higher resolution (but with poorer variance as discussed next and in Section VII).

b. Disadvantages

The kernel variance of second- and higher-order kernels is higher for Poisson stimuli than for symmetric binary stimulus with the same average interstimulus interval (Section VII). A symmetric binary stimulus with a $p = 0.5$ chance of flashing every 50 ms will have the same *average* interstimulus interval ($\Delta_A = 100$ ms) as an asymmetric binary stimulus with $p = 0.1$ every $\Delta_s = 10$ ms. The effects of p on the kernel variance will be discussed in Section VII. The trouble with the smaller Δ_s is that there are fewer pairs of stimuli with a given flash separation (say 100 ms for our example). The kernel variance of the 50-ms kernel slice would be smaller by a factor of 5, for the case with $p = 0.5$ vs. $p = 0.1$.

3. Ternary Stimuli

Stimuli with more than two levels can be very useful. As has been discussed earlier, the diagonals of second- and higher-order kernels do not contribute to the response when using binary stimuli. A primary advantage of ternary stimuli is that the second-order diagonals can be estimated. A formalism for ternary stimuli was developed by Emerson et al., (1987) in which the second-order diagonal was treated at the same level as the first-order kernel. The present discussion presents a summary and clarification of the Emerson et al. results.

For ternary stimuli the expansion functions $X_n(t)$ vanish for $n > 2$ so only the terms with $X_1(t)$ and $X_2(t)$ remain. Making use of the identity:

$$ab + cd = 0.5[(a + c)(b + d) + (a - c)(b - d)] \tag{98}$$

Equation 5 can be rewritten as:

$$y(t) = \underline{H}_0 + \sum_{\tau>0} [\underline{H}_+(\tau)X_+(t - \tau) + \underline{H}_-(\tau)X_-(t - \tau)]\Delta_s +$$

$$\sum_{\tau>0,\tau'>0} [\underline{H}_{++}(\tau,\tau')X_+(t - \tau)X_+(t - \tau') +$$

$$\underline{H}_{--}(\tau,\tau')X_-(t - \tau)X_-(t - \tau') +$$

$$\underline{H}_{+-}(\tau,\tau')X_+(t - \tau)X_-(t - \tau') + \underline{H}_{-+}(\tau,\tau')X_-(t - \tau)X_+(t - \tau']\Delta_s^2 \quad (99)$$

where

$$X_q(t) = 1/2\ [X_1(t) + qX_2(t)] \quad (100)$$

and

$$\underline{H}_q(\tau) = \underline{H}_1(\tau) + q\underline{H}_2(\tau^2)\Delta_s \quad (101a)$$

$$\underline{H}_{qr}(\tau,\tau') = \underline{H}_2(\tau,\tau') + q\underline{H}_3(\tau^2,\tau')\Delta_s + r\underline{H}_3(\tau,\tau'^2)\Delta_s + qr\underline{H}_4(\tau^2,\tau'^2)\Delta_s^2 \quad (101b)$$

where q and r take on the values ± 1. For a symmetric ternary stimulus in which the three levels are $x = -1$, 0, and $+1$ with equal probabilities, then $X_1(t) = x(t)$ also takes on the values -1, 0, $+1$ and $X_2(t) = x^2(t) - \overline{x^2}$ takes on the values 1/3, $-2/3$, 1/3 since $\overline{x^2} = 2/3$ (see Emerson et al., 1987 for details). The combination $X_+(t) = 1/2\ [X_1(t) + X_2(t)]$ in Equation 100, takes on the values $-2/3$, $-2/3$, $+4/3$ and $X_-(t)$ has values of $+4/3$, $-2/3$, $-2/3$. Thus, $X_+(t)$ and $X_-(t)$ single out the + and the − state of the ternary stimulus, respectively.

The best way to clarify the ternary formalism is to provide an example. Table 2 gives the responses of a system that has the following Wiener expansion:

$$y(t) = [X_1(t) + X_2(t) + 2/3]\{7.5[X_1(t - 1) + X_2(t - 1)] + 35\} + 10 \quad (102)$$

Based on Equation 100 the ternary expansion is

$$y(t) = [2X_+(t) + 2/3][15X_+(t - 1) + 35] + 10 \quad (103)$$

Columns 1, 2, and 3 are similar to columns 2, 3, and 6 of Table 1 in that column 3 gives the "full history" time-locked averages. The stimuli in columns 1 and 2 take on the three possible ternary values. Column 7 contains the Wiener kernels that can be calculated using Equation 9. An easier method to calculate the Wiener kernels is to read them directly off Equation 102, by comparing Equation 102 to Equation 5.

The ternary kernels in column 5 can be obtained by comparing Equation 103 to the expansion in Equation 99 that defines the ternary kernels; or the connection with Wiener kernels given in Equation 101 can be used. For the nonlinearities that characterize the system presented in Table 2, it is clear that the ternary kernels provide a tidier description of the system than is given by the Wiener kernels. Only one ternary kernel above first order is needed whereas six Wiener kernels are needed including two third-order and one fourth-order kernel. A simple examination of column 5 tells the story that the system responds only to the high-level stimulus $[X(t) = +1]$ and that there is a nonlinearity caused by a prior high level stimulus. This example shows the benefits of the ternary expansion.

a. Advantages

An important advantage of the ternary expansion is that important elements of the higher-order Wiener kernels are included in the lower-order ternary expansion terms. This becomes

TABLE 2

	1	2	3	4	5	6	7
Row	X(t − 1)	X(t)	y(t)	$H_{ternary}$		H_{Wiener}	
0	−1	−1	10	H_0	33.33	H_0	33.33
1	−1	0	10	$H_+(0)$	70	$H_1(0)$	35
2	−1	1	70	$H_+(1)$	10	$H_1(1)$	5
3	0	−1	10	$H_-(0)$	0	$H_2(00)$	35
4	0	0	10	$H_-(1)$	0	$H_2(11)$	5
5	0	1	70	$H_{++}(0,1)$	30	$H_2(01)$	7.5
6	1	−1	10	$H_{+-}(0,1)$	0	$H_3(011)$	7.5
7	1	0	10	$H_{-+}(0,1)$	0	$H_3(001)$	7.5
8	1	1	100	$H_{--}(0,1)$	0	$H_4(0011)$	7.5

especially useful for systems that have a strong polarity asymmetry, such as neurons with a half-wave rectification. The system in Table 2 responds at time t only to a +1 stimulus at time t. Its response is enhanced by a + stimulus in the preceding time interval. This system has a complicated Wiener expansion but a simple ternary expansion. It is clearly advantageous to use the ternary expansion.

The second-order diagonal has a special role when using ternary stimuli. Equation 101a shows that the diagonal is the difference between the two "first-order" responses to the +1 and −1 stimulus levels. The diagonal, thus, acts as a first-order kernel. The special role of the second-order diagonal also is relevant to calculations of kernel variance. The variance of the second-order kernel diagonal can be the same as the variance of the first-order Wiener kernel when using a ternary stimulus (see Equations 129 and 130 for more details). This is a useful property if one would like to know the diagonal to greater accuracy than the off-diagonal kernels.

b. Disadvantages

As discussed by Emerson et al. (1987), the ternary basis functions, X_+ and X_-, are not orthogonal to each other, so the simple Lee-Schetzen cross-correlation cannot be used to calculate the kernels. This is not a major disadvantage since the calculation of the ternary kernels is straightforward, as was shown in connection with Table 2. If one does desire to calculate the kernels by cross-correlation, then the combinations ($X_+ + 0.5\,X_-$) and ($X_- + 0.5\,X_+$) can be used since they are orthogonal to X_- and X_+, respectively.

An important disadvantage of ternary stimuli is that some of the beautiful properties of binary M-sequences are not available for ternary stimuli. There do exist ternary "M-sequences" (Barker and Obidegue, 1973) with near-perfect autocorrelation properties. However, the techniques developed by Sutter (1992) using the fast Walsh transform are not directly applicable to ternary stimuli. As discussed by Sutter (1992) it is easy to use a binary stimulus generator to produce 4-level (quaternary) stimuli which would have all the advantages of the ternary stimuli that have been discussed so far. Klein (1993b) has derived a method for relating the quaternary kernels (including diagonal elements) for arbitrary stimulus luminance levels to binary self-kernels and cross-kernels.

Ternary stimuli have lower power densities, P, than binary stimuli. As will be discussed in the next section, the variance of the *n*th-order kernel is proportional to P^{-n}. The power density for a binary stimulus with levels 0 and L is $\overline{x^2}\Delta_s = L^2\Delta_s/4$, where Δ_s is the stimulus sampling interval. For a ternary stimulus with equiprobable levels 0, L/2, and L the power density is $L^2\Delta_s/6$.

4. Gaussian Stimuli

The textbook amplitude distribution for white noise is Gaussian. While Gaussian noise is commonly found in nature and in statistics courses, it may not be the best stimulus to use for testing nonlinear systems. Here we look at its advantages and disadvantages.

a. Advantages

The expansion coefficients given by Equations 78 and 79 are simpler for Gaussians, since all cumulants above the second vanish. It is for this reason that all Gaussian kernels at different luminances and power densities can be related to each other in a fairly simple manner, as shown in Equation 80. However, even though I am a great fan of Equation 80, it has limited practical use since it requires knowledge of many higher-order kernels. Thus, this particular advantage of Gaussians is more aesthetic than practical.

As will be discussed in Section VIII, if one is dealing with an LNL sandwich model system, then Gaussian kernels are especially useful since the Volterra kernels are proportional to the Wiener kernels for Gaussian stimuli (Korenberg, 1973). The problem with this particular point is that when dealing with complex systems one never knows whether the system is of the LNL form. As will be discussed in Section VIII, it is easy to be misled about the nature of the system if there are not many independent points in the second-order kernel.

Another possible advantage is that Gaussian stimuli can be used to estimate diagonal kernel elements of arbitrarily high order. However, since it is rare that one would ever go beyond the third-order diagonals, Gaussian stimuli are no better than quaternary stimuli.

b. Disadvantages

In order to have a realizable stimulus, one must truncate the distribution because it is not possible to have negative amounts of light. If the truncation point is at two standard deviations, then one standard deviation is L/4, since the mean would be at L/2, where L is the maximum luminance. In that case the power density is $P = (L/4)^2\Delta_s$. This value is 4 times smaller than the binary case. Since the variance of the second-order kernel is proportional to P^{-2}, the variance for Gaussian stimuli would be 16 times worse than for binary stimuli with the same sampling interval, Δ_s.

Another serious disadvantage is that M-sequences do not exist for Gaussian or any other stimuli with a continuous amplitude distribution. Therefore, it will be difficult to correct for the nonorthogonality noise, especially when using multiinput stimuli. This limitation also implies that the calculation of the kernels for Gaussians will be much slower than for binary or quaternary stimuli.

5. Summary of Relative Merits of Different Amplitude Distributions

The point of view I am taking is that the different kernels estimated using stimuli with different amplitude distributions are equivalent. Kernels obtained with Gaussian stimuli are not "more correct" than kernels obtained with binary stimuli. It is true that with binary stimuli the diagonal kernels cannot be estimated, but in many situations that is a small price to pay for the reduced variance. It is also true that the expansions based on Gaussian stimuli have a mathematical beauty that is not present for other stimuli (except Poisson kernels have a different type of elegance and binary kernels have the simplicity of totally avoiding the diagonals). There might be a temptation to search for the "correct" kernels of a system. There is no correct kernel. Gaussian kernels are the correct kernels of the system in response to Gaussian noise, and binary kernels are the correct kernels in response to binary noise.

F. RADIUS OF CONVERGENCE

An important consideration that has been avoided so far is the question of the radius of convergence of the expansions that relate different kernels. Consider, for example, Equation

80 that relates Gaussian kernels with different power densities and different mean luminances. For the expansion to be valid the kernels on the left side of Equations 72, 78, or 80 must be within the radius of convergence of the expansion on the right side. Notice that Equation 80 looks exactly like a Taylor's expansion, and the notion of radius of convergence is of central importance whenever dealing with such expansions.

A simple example will clarify the problem. Consider the expansion given in Equation 81 in which a Wiener kernel for a Gaussian stimulus of power density, P, is expressed in terms of Volterra kernels associated with power density $P_0 = 0$. Suppose the system being studied has a small threshold at the input to the system such that the output of this first stage is zero whenever $-0.01\sigma < x < 0.01\sigma$ where σ is the standard deviation of the noise for the power density P. Otherwise the output of this first stage is equal to the input. This threshold will have a negligible effect on the Wiener kernels of power density P since the threshold has a very limited range. However, the threshold will cause all the Volterra kernels to vanish! The Volterra kernels have a very small radius of convergence. Conversely, the Wiener kernels, would require a great many terms to exhibit the threshold reliably.

It is important for researchers to assess the radius of convergence of the system before investing much time studying the system. A simple test would be to test the system at two luminances separated by a factor of two and to see whether the second-order kernel accounts for the change in first-order kernel according to Equation 71. The same test could be done with a pair of power densities to check whether the dependence of the first-order kernel on power density can be predicted from the summation over the diagonal of the third-order kernel.

VII. KERNEL VARIANCE

A. GENERAL FORMULA

There is a big difference in kernel variance depending on whether M-sequences or random sequences are used as the stimulus. If the latter are used together with the Lee-Schetzen cross-correlation method of kernel estimation, then a major contribution to the kernel variance comes from the kernels themselves. Marmarelis and Marmarelis devote a large chapter of their book (1978) to examining contributions to the kernel variance that come from the nonorthogonal terms of the Lee-Schetzen method. The formulas for kernel variance are complicated. If, however, M-sequences are used, or if the kernels are estimated using a least-squares method or some other means to avoid the nonorthogonality noise, then the kernel variance simplifies greatly, as will be seen in this section.

We begin by recasting Equation 9 in a manner that will simplify the calculation of kernel variance. As an example of our approach, consider the third-order kernel that was defined in Equations 91 and 93. It would be estimated as a time average according to Equation 9.

$$H_3(0 + t_r, 1 + t_r, 1 + t_r) = E[y(t' + t_r)X_1(t')X_2(t' - 1)]/P_1P_2 \qquad (104)$$

where P_m was defined in Equation 10, with $P_1 = P$, the power density, and $P_2 = E\{X_2^2(t')\Delta_s^2\}$. We have reintroduced the sparse-stimulation variable, t_r, just to show that sampling the response faster than the stimulus does not cause any problems. Rather than averaging over time, it is easier to collect the responses to all possible stimulus configurations that are within the history of the system.

$$H_3(0 + t_r, 1 + t_r, 1 + t_r) = R^{-1}\Sigma(Rp_{i_0}p_{i_1} \cdots P_{i_{m-1}})$$
$$Y_{i_{m-1}, \cdots, i_1, i_0}(t_r)X_1(i_0)X_2(i_1)/P_1P_2 \qquad (105)$$

The notation $Y_{i_{m-1}, \cdots i_1, i_0}(t_r)$ which was used in Section V refers to the average response at time t_r after the full history of preceding stimuli, where i_0 is the stimulus level at $t = 0$; the subscript i_{m-1} corresponds to the stimulus level at $t = -m + 1$ where m is the memory of the system; p_{i_k} is the probability of finding the stimulus at $t = -k$ at the level i_k. R is the number of stimulus samples. The product $(Rp_{i_0}p_{i_1} \cdots p_{i_{m-1}})$ is the number of times that particular stimulus configuration appears in the record. The summation is over all the values of all the i_k. For a ternary stimulus with a total memory that is m samples in duration, the number of terms in the summation is 3^m.

The variance of the kernel estimate is given by:

$$\text{var}[(H_3(0,1,1))] = (RP_1P_2)^{-2}\Sigma(Rp_{i_0}p_{i_1} \cdots p_{i_{m-1}})$$
$$\text{var}(Y_{i_{m-1}, \cdots i_1, i_0})[X_1(i_0)X_2(i_1)]^2 \qquad (106)$$

The sparse-stimulation time-delay t_r has been omitted from Equation 106 and the following equations for simplicity. It does not introduce any essential changes in the equations. The factor $(RP_1P_2)^{-2}$ and the $X_j(i_k)$ factors are squared because when taking the variance of a quantity any constant factors multiplying the random variable become squared. The factor $Rp_{i_0}p_{i_1} \cdots p_{i_{m-1}})$ is not squared because it gives the number of occurrences of a particular stimulus pattern. This number is unchanged whether one is calculating the kernel as in Equation 105 or calculating its variance as in Equation 106.

The beauty of carrying out the sum over the complete history of stimuli is that there is no randomness in the response due to prior stimuli that has not been included in the prediction. Thus, the only contribution to the variance of $Y_{i_{m-1}, \cdots i_1, i_0}$ comes from biological noise in the system. This is the step at which our analysis differs from previous estimates of kernel variance (Marmarelis and Marmarelis, 1978). In previous approaches, the stimulus x(t) would have been considered to be a random variable in addition to the response being a random variable; the variance of the response would not only get contributions from the noise, n(t), but also from the response to previous stimuli.

We now make the assumption that the zero-mean noise is added at the output of the system, so that it is independent of the stimulus:

$$\text{var}(Y_{i_{m-1}, \cdots i_1, i_0}) = \overline{n^2} \qquad (107)$$

In previous approaches where the stimulus was a random variable, the right side of Equation 107 would be the total VR as given, for example, by Equation 39. If the Lee-Schetzen (1965) algorithm is used in which lower-order responses are subtracted before the higher-order kernels are calculated, then the variance is reduced somewhat, but further errors are introduced (see discussion following Equation 24). Our approach of using complete test sequences (such as M-sequences) avoids this problem and produces a simpler and smaller kernel variance. Since the variance in Equation 107 is a constant, Equation 106 becomes:

$$\text{var}[H_3(0,1,1)] = \overline{n^2}R^{-1}(P_1P_2)^{-2}\Sigma p_{i_0}p_{i_1} \cdots P_{i_{m-1}}[X_1(i_0)X_2(i_1)]^2 \qquad (108)$$

The summation in Equation 108 is easy since

$$\Sigma p_{i_k} = 1 \text{ for } k > 1 \qquad (109)$$

$$\Sigma p_{i_0}[X_1(i_0)]^2 = P_1/\Delta_s \qquad (110a)$$

$$\Sigma p_{i_1}[X_2(i_1)]^2 = P_2/\Delta_s^2 \tag{110b}$$

The last two equations come from the definition in Equation 10. The kernel variance simplifies to:

$$\text{var}[H_3(0,1,1)] = \text{var}[H_3(0,1^2)] = \overline{n^2}/(RP_1\Delta_s P_2\Delta_s^2) \tag{111}$$

B. VARIANCE OF A GENERAL KERNEL

The variance of a general kernel is calculated by the same method to be:

$$\text{var}[H_n(\tau_1^{n_0}, \cdots, \tau_k^{n_k})] = \overline{n^2}/R\Delta_s^n P_{n_0} \cdots P_{n_k}) \tag{112}$$

Further insight into the derivation of kernel variance is given by Klein (1993a) where the two-kernel toy model leads to Equation 112. Since one often wants to compare the variance for two runs of equal duration, but with different stimulus conditions, one must express the number of stimulus samples, R, in terms of the run duration T: $T = R\Delta_s$. The general kernel variance becomes:

$$\text{var}[H_n(\tau_1^{n_0}, \cdots, \tau_k^{n_k})] = \overline{n^2}/(T\Delta_s^{n-1} P_{n_0} \cdots P_{n_k}) \tag{113}$$

Although we will use Equation 112 for the following analysis, one must remember that Equation 113 may have greater relevance when choosing stimulus properties.

For an asymmetric binary stimulus only the off-diagonal kernel elements contribute to the response, so only the simple power density, $P_1 = P$, appears. From Equation 95b the power density is $P = p(1 - p)L^2\Delta_s$ where $L = x_H - x_L$, the difference in luminance between the two levels. From Equation 112 the general variance becomes:

$$\text{var}[H_n(\tau_1, \cdots, \tau_n)] = \overline{n^2}/[R(P\Delta_s)^n] \tag{114}$$

$$= \overline{n^2}/\{R[p(1 - p)]^n(L\Delta_s)^{2n}\} \tag{115}$$

The standard error of the kernel is given by the square root of this equation:

$$\text{SE}[H_n(L\Delta_s)^n] = \{\overline{n^2}/R[p(1 - p)]^n\}^{0.5} \tag{116}$$

The $L\Delta_s$ factors were brought to the other side of the equation so that the left side of the equation is the contribution of a kernel element to the response as seen in Equation 5. In the limit of $p \ll 1$, Equation 116 becomes

$$\text{SE}[H_n(L\Delta_s)^n] = (\overline{n^2}/N)^{0.5} \tag{117}$$

where $N = p^nR$ is the number of times that n flashes occur with a particular configuration of time delays. The right-hand side is simply the statement that the standard error is equal to the standard deviation divided by the square root of the number of samples being averaged.

In order to get a better feeling for the magnitude of the kernel variance it is useful to compare the variance to the magnitude of an average kernel element. The total contribution of the *n*th order off-diagonal kernels to the response is

$$\overline{y_n^2} = P^n \Sigma H_n^2(\cdots)\Delta_s^n \tag{118}$$

The mean square kernel element is, thus

$$\overline{H_n^2} = \overline{y_n^2}/(MP\Delta_s)^n \tag{119}$$

where M is the memory length of kernel in samples

$$\overline{H_n^2} = n!^{-1}M^{-n}\Sigma H_n^2(\cdot \cdot \cdot) \tag{120}$$

The n! factor is present because of our single counting convention in Equation 5. The ratio of Equation 114 to 119 gives the ratio of the variance to the mean square kernel:

$$\text{var}(H_n)/\overline{H_n^2} = n!(\overline{n^2}/\overline{y_n^2})M^n/R \tag{121}$$

Suppose the *n*th order kernels contribute 10% of the total MSE of the response and that the intrinsic noise contributes 50%. Suppose further that the length of the run is $R = 10^4$ samples and that the system memory is $M = 20$ samples. The fractional error of a kernel estimate is given by the square root of Equation 121:

$$\begin{aligned} SE(H_n)/H_n &= [n!(50\%/10\%)\ 20^n/10^4]^{0.5} \\ &= 0.1 \text{ for } n = 1 \\ &= 0.63 \text{ for } n = 2 \\ &= 4.9 \text{ for } n = 3 \end{aligned} \tag{122}$$

This example shows that the signal-to-noise ratio of the second- and third-order kernels is inadequate to estimate the kernels properly. In order to reduce the kernel variance it would be advantageous to increase the stimulus sampling time, Δ_s, and thereby reduce M, the number of samples in the system memory. The introduction of sparse-stimulation allows this by sparsely sampling the stimulus but keeping arbitrarily fine sampling of the response.

C. DARK BACKGROUND ADAPTIVE SYSTEMS

For systems that adapt (see Section IV), the system responds to contrast rather than to absolute stimulus units, so the stimulus x should be measured in % contrast. If, between flashes, the luminance is black, then simple formulas for power density result. From Equations 97a and b the power density is $P = \Delta_s(1 - p)/p = \Delta_A(1 - p) = \Delta_A - \Delta_s$, where $\Delta_A = \Delta_s/p$ is the average time between flashes. Equation 114 becomes:

$$\text{var}[H_n(\tau_1, \cdot \cdot \cdot, \tau_n)] = \overline{n^2}/\{R[(1 - p)\Delta_s^2/p]^n\} \tag{123}$$

$$= \overline{n^2}/\{R[(1 - p)\Delta_A\Delta_s]^n\} \tag{124}$$

Further understanding of the variance comes by going to the limit of $p \ll 1$ and taking the square root. Equation 123 becomes:

$$SE[H_n(\tau_1, \cdot \cdot \cdot \tau_n)]\Delta_s^n = (\overline{n^2}p^n/R)^{0.5} \tag{125}$$

The inclusion of the time factors, Δ_s^n, makes the left side of Equation 125 have the same units as the contribution of a kernel element to the response in Equation 5.

The right-hand side of Equation 125 shows extra powers of p for higher-order kernels.

At first sight this is surprising since it indicates that for small p, the standard error of the higher-order kernels is smaller than the lower kernels. It is also disturbing that the kernel variance vanishes in the limit of p going to zero. This behavior occurs because when the stimulus is in contrast units the power density is very high for small p, thereby creating a very high signal-to-noise ratio. This small p behavior prompts us to issue a caveat about adapting systems. The formula for power density given by Equation 97a was based on defining the mean luminance as being the time averaged luminance. When p is very small, so that the flashes are farther apart than the characteristic adaptation time of the system, then the assumption about the effective mean luminance is not reliable. The system probably has its own "dark noise" or "resting" state of adaptation even in total darkness. The assumption that the power density is proportional to 1/p (Equation 97a), or to the average interflash interval (Equation 97b), must be modified (reduced) when the average interflash interval becomes larger than the typical system adaptation time.

D. SPECIAL ROLE FOR DIAGONAL ELEMENTS

The variance of diagonal elements can be quite different from the nearby off-diagonal elements. From Equation 112, the variance of the second-order diagonal is

$$\mathrm{var}[H_2(\tau^2)] = \overline{n^2}/(R\Delta_s^2 P_2) \tag{126}$$

where from Equation 10

$$P_2 = E[(x^2 - \overline{x^2})(x^2 - \overline{x^2})\Delta_s^2] \tag{127}$$

$$= (\overline{x^4} - \overline{x^2}^2)\Delta_s^2 \tag{128}$$

If the stimulus is Gaussian, then $P_2 = 2P^2$. If it is a symmetric ternary stimulus with a probability, p, of being in the $x = -L$ or $+L$ state and $1 - 2p$ of being in the middle state, $x = 0$, then

$$P = 2pL^2\Delta_s \tag{129}$$

$$P_2 = [2pL^4 - (2pL^2)^2]\Delta_s^2$$

$$= (1 - 2p)2pL^4\Delta_s^2$$

$$= [1/(2p) - 1]\, P^2 \tag{130}$$

If $p = 0.5$, the stimulus becomes a binary rather than a ternary stimulus, and P_2 vanishes, which means that the diagonal kernel elements have infinite variance. This is to be expected since the diagonal elements do not contribute to the response if the stimulus is binary.

If $p = 1/6$ then Equation 130 become $P_2 = 2P^2$, the same as for a Gaussian stimulus. Equation 128 implies $P_2 - 2P^2 = (\overline{x^4} - 3\overline{x^2}^2)\Delta_s^2$. Thus, $p = 1/6$ is the condition for the kurtosis of a symmetric ternary stimulus to vanish. This is a constraint that has been advocated by Victor (1991). It has the advantage that the ternary kernels will be closer to the Gaussian kernels, but it has the disadvantage that the power density will be lower than that of kernels with larger values of p. A large power density may be needed for estimating kernels of noisy systems. On the other hand, for systems with a strong *contrast* gain control the response of the system might be relatively independent of the power density since the system would turn its gain down when the power density is high. For these systems a lower power density may be advantageous, in which case values of p less than 1/6 may be desired.

If p is much less than 1/6, then P_2 will be much larger than what it would be for a Gaussian. For example, suppose $p = 0.05$. Then $P_2 = 9P^2$, which is 4.5 times larger than for a Gaussian of the same power density. The variance of the second-order diagonal would be 4.5 times smaller than for the Gaussian. The Gaussian comparison is especially useful because for Gaussian stimuli the diagonal variance is the same as the off-diagonal variance (taking into account our normalization in Equation 8). Thus, if we were especially interested in knowing the diagonals as opposed to off-diagonals, as was discussed in connection with Table 2 of Section VI, then sparse ternary stimuli have added benefits. This would especially be true for systems with a strong *contrast* gain control.

E. OTHER TYPES OF NOISE

The discussion up to this point has assumed that the noise was added to the output of the system. This is the simplest case since the noise is independent of the activity of the system. For this case the noise variance, $\overline{n^2}$, is easy to estimate by simply turning off the stimulus so that the response comes wholly from the noise.

The visually evoked scalp potential provides an interesting example where noise is not simply added at the output. If it were, then the noise should be the same with or without a stimulus. However, in the absence of a stimulus the brain can easily produce giant potentials (alpha waves) that are extinguished when the white noise stimulus is turned on. Another example would be a gain control system in which the total output of the system is kept relatively constant. Thus, when the system is unstimulated the output noise will be large, but when the system is responding to a large stimulus the output noise will be suppressed. For these types of systems the noise cannot be estimated by turning off the stimulus. Rather, the noise estimate must be made either from (1) the residual VR after the contributions from the estimated kernels are taken into account or (2) repeating the experiment two or more times using either same or different M-sequences. Using different M-sequences has the added advantage that contamination of the kernels due to kernel overlap (suppose the system memory was larger than expected) would be included in the estimate of kernel variance.

In the real world the noise need not be additive as assumed above. There might well be multiplicative noise at the output of the system. In this type of multiplicative noise the kernel variance would simply be proportional to the square of the kernel. More complicated sources of noise could enter in the middle of the system and could make different contributions to different kernels.

Further research is clearly needed for how to estimate kernel variance from realistic systems. The above analysis was presented to show how output noise is easier to deal with than is commonly assumed.

F. SPARSE-STIMULATION

Sparse-stimulation is the name given to the situation when the stimulus is sampled less often than the response. The same concept had been previously called superresolution (Klein, 1979) since the response is sampled more frequently than the white-noise stimulus. The use of sparse-stimulation overcomes the main disadvantage of white-noise stimuli, namely, the poor signal-to-noise ratio of higher kernels. We believe that the second- and third-order kernels are of critical importance for understanding the dynamics of nonlinear systems, and their estimation must be improved. This belief is the driving force behind our development of sparse-stimulation.

The details of the sparse-stimulation expansion have already been presented in Section II. The formalism developed in that section was careful to allow a direct extension to the case where the response was sampled more often than the stimulus. Similarly, the details of how to estimate the sparse-stimulation kernels have already been presented in Section

III. Section V and Figures 1 to 3 presented data on ganglion cell dynamics for which sparse-stimulation was essential. An explicit example of a system being analyzed using sparse-stimulation together with the resulting kernels will be presented in the next section in Table 4. The present section shows how sparse-stimulation can play a major role in decreasing kernel variance.

The relevance of sparse-stimulation to reducing kernel variance can be seen from Equations 114 and 121. The variance of an off-diagonal kernel is given by Equation 114 to be

$$\mathrm{var}[H_n(\tau_1, \cdots, \tau_n)] = \overline{n^2}/\{R[(\overline{x^2} - \overline{x}^2)\, \Delta_s^2]^n\} \tag{131}$$

For $n = 2$ (second-order kernel) the variance is proportional to Δ_s^{-4}. Thus, by increasing the stimulus sampling interval from $\Delta_s = 10$ to 30 ms one achieves a $3^4 = 81$-fold improvement in kernel variance. For the third-order kernel it would be a $3^6 = 729$-fold reduction in variance. As discussed in connection with Equation 113, for a fixed run duration, T, the variance of the second- and third-order kernels is proportional to Δ_s^{-3} and Δ_s^{-5}, so there would be 27- and 243-fold reductions, respectively, in variance.

One's first (and proper) reaction to a slower stimulus sampling is a fear that information about the high temporal frequency aspects of the response of the system is being lost. As discussed in Section II, it is possible to sample the response at an arbitrarily high rate, while the stimulus generator is sampled at a much lower rate. Suppose, for example, that the stimulus is sampled at intervals of 1 bloch (a bloch is a temporal unit characteristic of the system that was discussed in Section IV), whereas the response is sampled at 0.1-bloch intervals. The second-order kernel, $H_2(\sigma_1 + t_r, \sigma_2 + t_r)$, would have t_r sampled in increments of 0.1 and σ_1 and σ_2 sampled in increments of 1.

By having the stimulus generator sparsely sampled, the power density is being focused in fewer kernel elements than would be the case if the stimulus generator were sampled more frequently. As an example, suppose the system memory is 3 blochs long and the stimulus generator is sampled every 0.1 bloch. In that case there would be $30 \times 31/2 = 465$ independent elements of the second-order kernel. If, on the other hand, the stimulus is sampled at 1-bloch intervals while the response is still sampled at 0.1-bloch intervals, then the diagonal would have 30 elements, and the off-diagonal at $\sigma_2 - \sigma_1 = 1$ and 2 blochs would have 20 and 10 elements for a total of 60. The reduction from 465 to 60 corresponds to the reduction in kernel variance that is achieved by sparse-stimulation. Since fewer kernel elements are being used to generate the response, the kernel elements that are being used receive a greater proportion of the power density than would otherwise be the case. The kernel variance of the tested kernel elements is reduced, while the kernel variance of the untested element is increased (to infinity).

An alternative method for improving the signal-to-noise ratio is to stimulate at the faster rate and then to pool data from neighboring bins. This method is somewhat similar to sparse-stimulation except that it degrades the temporal resolution of the kernels. This method is effectively what is achieved by low-pass filtering the white-noise stimulus and response (Marmarelis and Marmarelis, 1978). The sparse-stimulation method does not degrade the temporal resolution of the kernels, but rather it sparsely samples the kernels.

In the discussion of sparse-stimulation I have sometimes been sloppy and referred to an increased stimulus sampling interval. A more careful statement would have been to refer to an increased interval of the stimulus generator; that is, between the intervals Δ_s, the stimulus could be constant. What precisely the stimulus is doing between the stimulus sampling intervals must be clearly stated. For a binary stimulus such as would be generated on a video monitor, the stimulus sampling interval might be 16.7 ms with darkness between stimulus flashes. Or as with the ganglion cell data, the stimulus can remain at the level of the sample until the next stimulus. There is a problem if the full vertical extent of the video monitor is being used, in that the raster is slowly (on the millisecond time scale) moving from top to the bottom of the screen. Sutter and Tran (1991) discuss how they compensated for the slow raster in their ERG experiments. For a ternary stimulus the level between the stimulus samples might be at the middle level or it might be at the dark level. This aspect of the stimulus would need to be specified in order to replicate the experiment, but it would not be needed in computing the kernels. The specification of the intermediate luminance would also be needed if one wanted to use contrast units for the stimulus since contrast units require knowledge of the mean luminance level. The response of the system would, of course, depend on what the stimulus did between samples.

VIII. SYSTEM IDENTIFICATION

A. SANDWICH MODEL

When a zero-memory nonlinearity is sandwiched between two linear filters, the resulting kernels have special properties. A great deal of effort has been devoted to estimating the parameters of this sandwich model (Korenberg and Hunter, 1986; Victor and Shapley, 1979; Spekreijse et al., 1977). As was discussed in connection with the advantages of Gaussian noise, it turns out that if Gaussian noise is used for a sandwich model system, then the Wiener kernels are proportional to the Volterra kernels (Korenberg, 1973). The sandwich model often does a fairly good job of fitting the second-order kernel (Korenberg and Hunter, 1986). This success can be misleading, however. The second-order kernel is often quite smooth. Let us suppose it has 5 $\times$ 5 somewhat independent, low-pass filtered kernel elements. Because of the symmetry of the kernel, that would be 15 kernel elements (counting the diagonal). The sandwich model would have 9 independent parameters (5 for each linear filter minus 1 scaling parameter). Thus, the 9 parameters of the sandwich model should do a good job in fitting many smooth, second-order kernels. The widespread use of the sandwich model reminds one of the gentleman who is searching for his lost keys under the streetlight even though he dropped them a distance away in the dark alley. A simple test of the sandwich model will be discussed next in connection with the gain-control model.

B. LUMINANCE GAIN-CONTROL MODEL

Table 1 showed the responses to a symmetric binary stimulus with a history that is three stimulus intervals in length. It is readily verified that the responses in column 9 of Table 1 must come from the following system:

$$y(t) = 10 + [X_1(t) + 1][30 - 10\,X_1(t - 1) - 5\,X_1(t - 2)] \qquad (132)$$

Further insight into this model is obtained if sparse-stimulation is used, whereby the response is sampled faster than the stimulus. As in Table 1 we will continue to have the stimulus either flash or not flash every stimulus sampling time, which is taken to be 1 bloch. The response is now sampled ever 0.2 blochs. Table 3 gives the response of our hypothetical system.

There are several differences between Tables 1 and 3. All entries in column 4 are equal to 512 except for rows 0 and 7 in which the three preceding stimuli were all 1 or all -1. These numbers are the result of using an 11-bit inverse-repeat M-sequence to generate the stimuli. In a full cycle M-sequence all possible configurations of 11 bits occur once except for the sequence with 11 consecutive values of 1. The inverse-repeat sequence means that the first M-sequence is repeated with all bits reversed in polarity, so now the sequence with 11 consecutive values of -1 is missing. The purpose for using the inverse-repeat technique

TABLE 3

	1	2	3	4	5	6	7	8	9	10	11	12	13
Row	X_{-2}	X_{-1}	X_0	N_i	y_0	$y_{-0.2}$	$y_{-0.4}$	$y_{-0.6}$	kernels	H_0	$H_{-0.2}$	$H_{-0.4}$	$H_{-0.6}$
0	−1	−1	−1	511	10	10	10	10	H0	40	30	20	10
1	−1	−1	1	512	100	70	40	10	H1(0)	30	20	10	0
2	−1	1	−1	512	10	10	10	10	H1(1)	−10	−6.7	−3.3	0
3	−1	1	1	512	60	43	27	10	H2(01)	−10	−6.7	−3.3	0
4	1	−1	−1	512	10	10	10	10	H1(2)	−5	−3.3	−1.7	0
5	1	−1	1	512	80	57	33	10	H2(02)	−5	−3.3	−1.7	0
6	1	1	−1	512	10	10	10	10	H2(12)	0	0	0	0
7	1	1	1	511	40	30	20	10	H3(012)	0	0	0	0

is to reduce efficiently the overlap of kernels as discussed in Section III. By using the M-sequence the distribution of number of occurrences of each stimulus pattern is much more uniform than what was shown in Table 1 where a random stimulus had been used. Columns 5 to 8 show *average* responses time-locked onto the various stimulus configurations. The kernels in columns 10 to 13 are the Walsh transforms of columns 5 to 8, as discussed in Section III. The responses in Table 3 could have been produced by the following two-stage system:

$$w(t') = [X(t') + 1][1 + \Sigma_{t''} f_1(t' - t'')X(t'')] \tag{133}$$

$$y(t) = 10 + \Sigma f_2(t - t')\, w(t') \tag{134}$$

The linear filters are

$$f_1(t') = 0, -4/9, \; -2/9, 0, 0, 0 \cdots$$

$$\text{at } t' = 0, 1, 2, 3, 4, 5, \cdots \text{ blochs}$$

$$f_2(t) = 45, 30, 15, 0, 0, 0, 0, \cdots$$

$$\text{at } t = 0, 0.2, 0.4, 0.6, 0.8, 1.0, 1.2, \cdots \text{ blochs}$$

The block diagram for this system is shown in Figure 5. In order to clarify the system and sparse-stimulation, Table 4 presents an example of X(t), w(t), y(t) and also the resultant first- and second-order kernels. Four stimulus samples are shown in the row labeled X(t′) in Table 4. The stimuli are $X(t') = 1, -1, 1$, and 1 at $t' = 0, 1, 2, 3$ blochs. Since the stimulus interval is $\Delta_s = 1$ bloch, Δ_s does not appear in Equations 132 to 134. It is assumed that the stimulus was negative at earlier times: $X(-1) = X(-2) = -1$. That is, for $t' < 0$, there were no flashes. A convention has been used such that the time variables t′ and t″ in Equations 133 and 134 only take on integral values (at the stimulus sampling times). The variable, t, is defined at the response sampling times (0.2 blochs in our example). The filter $f_1(t)$ may well be a continuous function, but because of the factor $[X(t) + 1]$ in Equation 101, the only occasions that w(t′) is defined are at the sparse-stimulus sampling times.

The row for $H_0(\tau)$ clarifies the claim made in Equation 5 that the zeroth-order kernel has time dependence. Notice that H_0 only has values for the first five response samples, which lie within a single stimulus-sampling interval. One of the most interesting features of Table 4 is the ringing in the first-order kernel. The initial response of 30, 20, 10, 0, 0 is directly related to the second filter, $f_2(\tau)$. Thus, subsequent responses, time-locked to the

TABLE 4

	0	0.2	0.4	0.6	0.8	1.0	1.2	1.4	1.6	1.8	2.0	2.2	2.4	2.6	2.8	3.0	3.2
$X(t')$	1					−1					1					1	
$1 + X(t)$	2					0					2					2	
$w(t')$	2					0					14/9					10/9	
$y(t)$	100	70	40	10	10	10	10	10	10	10	80	57	33	10	10	60	43
$H_0(\tau)$	40	30	20	10	10												
$H_1(\tau)$	30	20	10	0	0	−10	−6.7	−3.3	0	0	−5	−3.3	−1.7	0	0	0	0
$H_2(\tau,1)$	−10	−6.7	−3.3	0	0	0	0	0	0	0	0	0	0	0	0	0	0
$H_2(\tau,2)$	−5	−3.3	−1.7	0	0	0	0	0	0	0	0	0	0	0	0	0	0

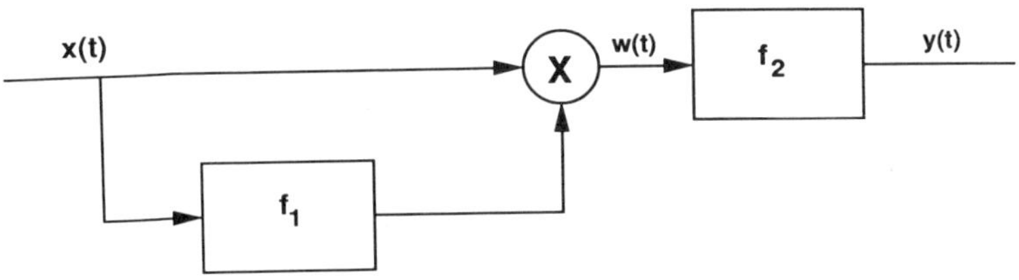

FIGURE 5. A system with a luminance gain control. The low-pass filter f_1 attenuates the incoming stimulus, x(t), to produce w(t), the input to the filter f_2.

stimulus-sampling intervals are caused by the changing state of adaptation that is dependent on the prior history seen by the system. It is very easy to determine whether ringing in H_1 is caused by a rapid ringing in the system or caused by a slow gain-control change. One simply changes the stimulus-sampling rate. If the ringing interval changes correspondingly, then one is assured that it is caused by a gain control mechanism. It should be noted that sparse-stimulation was needed to produce the ringing. Emerson and Vaughn (1991) have reported just this sort of ringing in the response of cortical neurons to a ternary white-noise stimulus. I interpret this as evidence for a gain control mechanism of the sort shown in Figure 5. Further research in this area is needed to show how the characteristics of the first- and higher-order kernels can pin down the type of gain control that is operating. Equation 80, for example, shows how the higher-order kernels modify the lower-order kernels when the luminance changes. Any change in kernel shape or magnitude is another consequence of a luminance gain control. It is quite likely that gain-control stages operate at several levels of the visual pathway, and it would be useful to probe their characteristics with white-noise stimuli.

IX. SUMMARY

My goal in this chapter has been to present a variety of techniques for optimizing kernel estimation. This section will summarize the central five topics of this chapter: sparse-stimulation, elimination of the nonorthogonality noise in kernel estimation, use of full history or conditioned kernels for increasing the meaningfulness of kernels, methods to relate kernels obtained using different stimuli, and, finally, a simplified formula for kernel variance.

Sparse-stimulation — The orthogonal expansion in Equation 5 took care to keep the response-sampling rate separate from the stimulus-sampling rate. In the usual white-noise methodology the stimulus and the response are sampled at the same rate. This chapter repeatedly emphasized the advantages of having the stimulus change at a slower rate than the response-sampling interval. The ganglion cell results shown in Figures 1 to 4 presented an excellent example of sparse-stimulation with the stimulus generator making its decision every 30 ms, and the response being sampled every 2.5 ms. Many interesting properties of the center-surround interactions were revealed by this technique. Sparse-stimulation produces dramatic improvements in the signal-to-noise ratio of the higher-order kernels. Sparse-stimulation also facilitates the elimination of nonorthogonality noise as discussed next.

Elimination of nonorthogonality noise — Almost all past researchers using white-noise techniques have used the Lee-Schetzen method for estimating the underlying kernels. Because of the finite length of the recording time, the autocorrelation function of the stimulus is not a perfect delta function. This deficiency introduces "nonorthogonal" noise into the kernel estimates. I believe that this type of noise was *dominant* in many past studies. Section III discussed a variety of techniques that are able to remove this contribution to the noise. If these techniques had been present 20 years ago, I have no doubt that the white-noise methodology would be much more popular than it is at present. The most promising method for removing the nonorthogonality noise is to use full cycle M-sequences. When these stimuli are combined with Sutter's Fast Walsh transform and his method for removing artifacts, the white-noise methodology becomes a powerful tool for learning about nonlinear systems and multiinput systems.

Conditioned and full-history kernels — The ganglion cell data made the point that there are systems for which the Wiener expansion is not appropriate. The histograms in Figures 1 to 3 revealed many interesting features of how the ganglion cell depended on the past stimulus history. When the Walsh transform of the responses was taken to produce the Wiener kernels, many of the interesting features were washed out. The white-noise stimulus was still important for rapidly intermixing all the different conditions. So this is not a critique of the white noise methodology; rather it is a critique of displaying the data as Wiener kernels. Very often the second- and higher-order kernels should be added to the lower-order kernels for them to be meaningful. One of the important features of Section V was the development of a formalism for handling the connection between raw data, Wiener kernels, and intermediate (conditioned) kernels.

Relationship of kernels obtained using different stimuli — One wants to ask different questions of different systems. The optimal stimulus for each question and each system will be different. Section VI was directed to developing a formalism for relating kernels based on stimuli with different amplitude distributions. The important role of the diagonal elements was discussed. Cumulants of the stimulus distribution are the expansion coefficients relating the different kernels to each other. The relative merits of different stimuli were examined. In order to get the best signal-to-noise ratio binary stimuli are best, but binary stimuli are unable to estimate the second-order diagonals. A good compromise stimulus is ternary or quaternary white noise. The separate *on* and *off* channels in vision provide further rationale for using three-level stimuli. Ternary stimuli have an added advantage that the variance of the diagonal elements of the second-order kernel can be made smaller than the variance of the off-diagonal elements.

Kernel variance — One of the most important sections of this chapter is Section VII, dealing with simplified formulas for kernel variance. An "intuitive" understanding of kernel variance is of great importance for the practical estimation of good-quality kernels. The most critical step in simplifying the variance equations is the use of M-sequence noise, or some other technique to avoid the contamination of the nonorthogonality noise discussed earlier.

The rapid gain-control dynamics of biological systems has resisted detailed examination and understanding. The recent development of many new techniques in the Wiener kernel technology gives one an exciting feeling that new methods are now available for learning about these systems and subsystems. I expect that during the next 10 years these new methods will be applied to a wide variety of complex systems.

ACKNOWLEDGMENTS

Supported by grant AFOSR-89-0238 from the Air Force Office of Scientific Research. I also thank Derek Fender for his support during the 1970s when most of this work (except for Section III and sparse stimulation) was done. Bill Levick, John Kroeker, and Mark Citron were responsible for collecting the ganglion cell data. I thank Erich Sutter for many conversations clarifying his findings that were reported in Section III and Jonathan Victor for a delightful series of e-mailings related to the range of validity of the expansions in Section VI. Brent Beutter provided helpful comments on the manuscript.

REFERENCES

Barker, H. A., Obidegue, S. N., and Pradisthayon, T. (1972). Performance of antisymmetric pseudorandom signals in the measurement of second-order Volterra kernels by crosscorrelation, *Proc. IEE,* **119**, 353–362.

Barker, H. A. and Obidegue, S. N. (1973). Combined crosscorrelation method for the measurement of 2nd-order Volterra kernels, *Proc. IEE,* **120**, 114–118.

Brillinger, D. R. (1988). The maximum likelihood approach to the identification of neuronal firing systems, *Ann. Biomed. Eng.,* **16**, 3–16.

Emerson, R. C., Citron, M. C., Vaughn, W. J., and Klein, S. A. (1987). Nonlinear directionally selective subunits in complex cells of cat striate cortex, *J. Neurophysiol.,* **58**, 33–65.

Emerson, R. C. and Citron, M. C. (1992). Linear and nonlinear mechanisms of motion selectivity in simple cells of the cat's striate cortex, in *Nonlinear Vision,* Pinter, R. B. and Nabet, E. B., Eds.

Goussard, Y. (1987). Wiener kernel estimation: a comparison of cross-correlation and stochastic approximation methods, in *Advanced Methods of Physiological System Modeling,* 1, Marmarelis, V. Z., Ed., Biomedical Simulations Resource, University of Southern California, 289–302.

Goussard, Y., Krenz, W. C., Stark, L., and Demoment, G. (1991). Practical identification of functional expansions of nonlinear systems submitted to non-Gaussian inputs, *Ann. Biomed. Eng.,* in press.

Hung, G., Stark, L., and Eykhoff, (1976). On the interpretation of kernels. I. Computer simulation of responses to impulse pairs, *Ann. Biomed. Eng.,* **5**, 130–143.

Hung, G. and Stark, L. (1977). The kernel identification method (1910–1977) — Review of theory, calculation, application, and interpretation, *Math. Biosci.,* **37**, 135–190.

Klein, S. A. (1979). Interpreting nonlinear systems: the third order kernel of the eye movement control system, in *10th Annu. Pittsburgh Conf.* on Modeling and Simulation, 215–221.

Klein, S. A. (1987). Relationship between kernels measured with different stimuli, in *Advanced Methods of Physiological System Modeling,* 1, Marmarelis, V. Z., Ed., Biomedical Stimulations Resource, University of Southern California.

Klein, S. A. and Yasui, S. (1979). Nonlinear systems analysis with non-Gaussian white stimuli: general basis functionals and kernels, *IEEE Trans. Inf. Theor.,* **25**, 495–500.

Korenberg, M. J. (1973). Crosscorrelation analysis of neural cascades, *Proc. 10th Annu. Rocky Mountain Bioeng. Symp.,* 47–52.

Korenberg, M. J. (1987). Functional expansions, parallel cascades and nonlinear difference equations, in *Advanced Methods of Physiological System Modeling,* 1, Marmarelis, V. Z., Ed., Biomedical Simulations Resource, University of Southern California, 221–240.

Korenberg, M. J. (1988). Identifying nonlinear difference equations and functional expansion representations: the fast orthogonal algorithm, *Ann. Biomed. Eng.,* **16**, 210–214.

Korenberg, M. J. and Hunter, I. W. (1986). The identification of non-linear biological systems: LNL cascade models, *Biol. Cybern.,* **55**, 125–134.

Lee, Y. W. and Schetzen, M. (1965). Measurement of the Wiener kernels of a nonlinear system by cross-correlation, *Int. J. Contr.,* **2**, 237–254.

Marmarelis, P. Z. and Marmarelis, V. Z. (1978). *Analysis of Physiological Systems,* Plenum Press, New York.

Marmarelis, P. Z. and Naka, K. I. (1972). White noise analysis of a neuron chain: an application of the Wiener theory, *Science,* **175**, 1276–1278.

McCann, G. D. and Marmarelis, P. Z., Eds. (1975). *Proc. 1st Symp.* on Testing and Identification of Nonlinear Systems, California Institute of Technology.

Ream, N. (1970). Nonlinear identification using inverse-repeat *m*-sequences, *Proc. IEE,* **117**, 213–218.

Spekreijse, H., Estevez, O., and Reits, D. (1977). Visual evoked potentials and the physiological analysis of visual processes in man, in *Visual Evoked Potentials in Man,* Desmedt, J., Ed., Clarendon Press, Oxford, 16–89.

Stark, L. (1968). *Neurological Control Systems, Studies in Bioengineering,* Plenum Press, New York.

Sutter, E. E. (1975). A revised conception of visual receptive fields based on pseudorandom spatiotemporal pattern stimuli, in *Proc. 1st Symp.* on Testing and Identification of Nonlinear Systems, McCann, G. D. and Marmarelis, P. Z., Eds., California Institute of Technology, 353–365

Sutter, E. E. (1987). A practical nonstochastic approach to nonlinear time-domain analysis, in *Advanced Methods of Physiological System Modeling,* 1, Biomedical Simulations Resource, Univ. So. Calif., Marmarelis, V. Z., Ed., 303–315.

Sutter, E. E. (1991). The fast M transform: a fast computation of cross-correlations with binary M-sequences, *SIAM J. Comput.,* **20**, 686–694.

Sutter, E. E. (1992). Deterministic analysis of nonlinear systems, in *Nonlinear Vision,* Pinter, R. B. and Nabet, E. B., Eds.

Sutter, E. E. and Tran, (1991). The field topography of ERG components in man. I. The photopic luminance response, *Vision Res.*, **32**, 433–446.

Victor, J. D. and Shapley, R. M. (1979). The nonlinear pathway of Y ganglion cells in the cat retina, *J. Gen. Physiol.*, **74**, 671–689.

Victor, J. D. (1991). Asymptotic approach of generalized orthogonal functionals expansions to Wiener kernels, *Ann. Biomed. Eng.*, **19**, 383–399.

Chapter 6

A DETERMINISTIC APPROACH TO NONLINEAR SYSTEMS ANALYSIS

Erich E. Sutter

TABLE OF CONTENTS

I. Introduction 172

II. General Considerations Concerning Systems Analysis 173

III. Linear Systems 175
- A. Testing Strategies 175
- B. Computation of the Two Kernels 176

IV. Deterministic Binary Analysis of Nonlinear Systems 177

V. Comparison with the Binary Lee and Schetzen Technique 182

VI. Completion of Incomplete Data Sets 184

VII. Methods to Overcome the Limitations of Binary Testing 188
- A. Multilevel Analysis Using Binary Components 189
- B. Quaternary Stimuli 189

VIII. Signal-to-Noise Considerations 190

IX. Testing Efficiency 191
- A. Complete Sequences 191
- B. Binary M-Sequences 192
  1. Generation and Structure 192
  2. Fast Computation of the Cross-Correlation Cycle 194
  3. M-Sequence Stimulation and Incomplete Data Sets 197
  4. Orthogonalization Procedures 198
  5. The Anomaly Problem 200

X. Multiple Input Systems 202
- A. General Considerations 203
- B. Partial Orthogonalization through Use of the Same Sequence for All Inputs 204
- C. The M-Sequence Approach to Multiinput Systems 205
  1. The Basic Technique 205
  2. Orthogonalization Procedures 206
  3. Generation of Multiinput M-Sequence Stimuli 207
  4. Quaternary Stimulation with Binary M-Sequences 208
  5. Sparse-Stimulation 209
  6. Artifact Removal 210

7. Examples of Multiinput Analysis with Binary M-Sequences....211

XI. Summary and Conclusions....213

Acknowledgments....215

References....215

Appendix A....217

Appendix B....218

I. INTRODUCTION

The electrophysiological investigation of the visual system poses two major problems for systems analysis and identification. First, for the most part the visual system is highly nonlinear, and second, it deals with a two-dimensional array of inputs that interact with each other along the visual pathway. The analysis of this complexity requires elaborate testing strategies and long recording times. The methods required are those of single and multiinput nonlinear systems analysis. While these methods have been under development of several decades their application to the complex problems of vision is still in the beginning stages. This chapter has two main objectives. One goal is to illuminate the old subject of white-noise systems analysis from a different angle in the hope of generating some new insights. The second goal is to communicate methods that have proven extremely useful in visual electrophysiology.

The distinction made here between stochastic and deterministic techniques is based on the way the response properties are extracted. In a stochastic approach the system is stimulated by means of a random process, and the extraction of the response characteristics is based on the stochastic properties of the stimulus. In a deterministic analysis, on the other hand, one uses the exact input signal.

Perhaps the most prominent nonstochastic approach in general use today is the sum of sinusoids technique (Victor and Knight, 1979). There are interesting analogies between this frequency domain method and the binary M-sequence technique discussed in the later parts of this chapter. In both cases the test input is designed to achieve exact orthogonality of the kernels and kernel elements. Here the focus will be exclusively on the time domain.

In the early days of nonlinear systems analysis the opinion prevailed that random stimulation provides the best coverage of the large input ensembles. In most cases kernel extraction proceeded by means of stochastic analysis of the response using the Lee and Schetzen cross-correlation techique (Lee and Schetzen, 1961, 1965). While the approach works reasonably well with relatively simple systems, application to nonlinear multiinput systems can become unwieldy. The enormous size of the input ensembles associated with such systems often permits only a relatively sparse coverage in the available recording times. This results in poor quality of the kernel estimates. In addition, the computational problems in the implementation of the higher order cross-correlations are in stark contrast to conceptual simplicity of the technique.

What is clearly needed is the development of better testing strategies and faster extraction of the response characteristics from the raw data. Both of these goals prompted reconsideration of various aspects of nonlinear systems analysis using deterministic rather than stochastic concepts. This has led to testing strategies that can be tailored to specific problems and an ultrafast computational technique for the extraction of kernels. This chapter contains an outline of these methods.

With a somewhat unconventional point of view, this presentation of methods hopes to contribute to a clearer and more intuitive understanding of some important aspects of system testing and kernel derivation. In particular, this treatment may lead to a better grasp of the power as well as the limitations of the methodology and facilitate the design of experiments best suited for specific systems.

The approach taken here is strictly binary in nature. It should be stressed, however, that this in no way implies a restriction to binary test inputs that result in a limited resolution of kernels. It only means that inputs are considered in terms of binary components and that the kernel computation exploits the enormous computational advantages a binary representation of the test input can bring (see Section VII).

II. GENERAL CONSIDERATIONS CONCERNING SYSTEMS ANALYSIS

As is customary, we restrict our considerations to systems with the following properties:

1. Stationarity
2. Limited memory m
3. Limited input cutoff frequency f_i
4. Limited output cutoff frequency f_r

Property 3 implies that input frequencies above f_i have no effect on the response of the system and property 4 means that no frequencies above f_r are observed in the response.

Experimental records of analog data are obtained by periodic sampling of the signals. The following treatment uses discrete time only, and all integrals are replaced by sums. This avoids a layer of mathematical complexity that is of little concern to the experimental biologist. The input and output of the system are, thus, sampled at finite intervals t_i and t_r, respectively. Except in Section XIII, t_i and t_r are considered equal: $t_i = t_r = t$.

Owing to property 2, the response amplitude of the system at a given time t_o is determined by the input within a time window w $< m$ preceding t_o.

```
--+----------+------+---->
  t_m        t_d    t_0
```

The distinction between system memory and input window is not commonly made. However, it is important when designing the appropriate strategy for testing and analysis of highly nonlinear systems. The system memory m is clearly not the critical factor in determining the size of the input ensemble. Consider, for example, a system consisting of a pure delay line generating a delay of d time bins. While its memory is $m = d + 1$, its input window is only one bin wide. In vision research the distinction is particularly important because of the observed delays of signal onset. Knowing that the onset latency of visual evoked cortical responses is about 40 to 50 ms significantly reduces the size of the complete test.

Note also that the impulse response may provide a good estimate for the delay d but not for the window size k. A pulse stimulus may have a significant effect on the response to a second pulse presented after the impulse response has died down.

The size of the input window measured in time bins t *is* $k = (m - d)/t$. The response at time t is a function of k variables:

$$R_{t_0} = R(x_{t_d}, x_{t_{d+1}}, \cdots, x_{t_m}) \tag{1}$$

Under certain conditions, the general function R can be described as a linear combination of a complete set of familiar functions. The most common of such representations is the power series expansion.

The best-known power series representation of a function is the Taylor series, or its functional analog the Volterra series, which, in the discrete case considered here, is simply the power series expansion of the multivariable function $R(x_d, \ldots, x_m)$. When truncated, it provides the best representation in the neighborhood of a selected point. The direct derivation of the terms of the Volterra series would require testing of the system with very small stimuli. Clearly, this cannot lead to satisfactory results since the responses could not be detected in the noise.

The fact that the Volterra series is generally not attainable does not constitute a serious problem, since it is usually more desirable to derive a truncated power series that fits best within a given range of inputs, i.e., within a given input ensemble.

Let us focus now on the derivation of the kernels in the power series expansion

$$\begin{aligned} R_j = K^0 &+ \sum_{i=d}^{m} K^{(1)}_{i_1} x_{j-i_1} \\ &+ \sum_{i_1=d}^{m} \sum_{i_2=d}^{m} K^{(2)}_{i_1,i_2} x_{j-i_1} x_{j-i_2} \\ &+ \sum_{i_1=d}^{m} \sum_{i_2=d}^{m} \sum_{i_3=d}^{m} K^{(3)}_{i_1,i_2,i_3} x_{j-i_1} x_{j-i_2} x_{j-i_3} \\ &+ \cdots \end{aligned} \tag{2}$$

Note that the kernels are symmetrical with respect to permutation of their indices.

The kernel elements $\{K^{(0)}, K^{(1)}{}_i, \ldots, K^{(r)}{}_{ij,\ldots,l}\}$ can be determined experimentally by testing the system with different values of the input variables x_i. For each configuration of these k variables a response amplitude $R(X_d, \ldots, X_m)$ is measured. By inserting an input configuration together with the corresponding response into the above expansion one obtains an equation for the determination of the kernels. Clearly, the number of configurations that need to be included in the test depends on the number of kernel elements in the expansion that are different from zero. In general, the tests lead to a system of equations that can be represented as

$$[X] \cdot \vec{K} = \vec{R} \tag{3}$$

Where K is the set of kernel elements of all orders, X is a matrix whose rows are obtained by inserting the test input configurations into expansion of Equation 2, and R is the set of corresponding response amplitudes.

III. LINEAR SYSTEMS

A. TESTING STRATEGIES

Consider a first-order system. Its response properties are completely described by the first two terms in the expansion:

$$R_j = K^{(0)} + \sum_{i_1=n}^{m} K_{i_1}^{(1)} x_{j-i_1} \tag{4}$$

In this case there are a total of k + 1 kernel elements to be determined; i.e., k + 1 independent data points need to be collected to set up the necessary euqations. This can be done in an infinite number of different ways that all lead to the same answer. In the following we only consider binary configurations. The input variables can, thus, take on only two values, e.g., 0 and 1. It is particularly easy to perform separate tests for each time bin in the input window.

Because of the continuity of physical time, the efficiency in implementing such a series of measurements depends strongly on the sequence in which the individual tests are performed. We cannot instantaneously place a configuration into the input window. It is always shifted in from the future and leaves the window into the past. In the process of changing from one input to another, the contents may assume many unnecessary intermediate configurations. This problem is seen, for example, in frequency domain testing where it is clearly impossible to expose the system to only the necessary Fourier components.

The above series of individual tests for each time bin can be implemented most efficiently by presenting the test configurations in the sequence specified by the rows of Equation 5.

$$\begin{array}{ccccc} x_d & x_{d+1} & x_{d+2} & \cdots & x_m \\ 0 & 0 & 0 & \cdots & 0 \\ 1 & 0 & 0 & \cdots & 0 \\ 0 & 1 & 0 & \cdots & 0 \\ 0 & 0 & 1 & \cdots & 0 \\ \cdot & \cdot & \cdot & \cdots & \cdot \\ 0 & 0 & 0 & \cdots & 1 \end{array} \tag{5}$$

This sequence of tests represents a single pulse traveling through the input window. It can be implemented most efficiently, since after the generation of a single pulse at the input it will automatically assume all the positions in the input window. With each advance in time by one time bin another one of the required configurations is generated, and the test is completed in the shortest possible time.

Testing efficiency through proper sequencing of the test configurations may not be important in the case of this simple linear system where the required input ensemble is small. However, in studies of the highly nonlinear multiinput visual system the input ensemble frequently becomes enormous in size. Often only partial coverage is possible in a single experiment, and the sequencing as well as the selection of the subset used in a particular experiment become important. In this context deterministic approaches become increasingly important, since random stimulation cannot provide sufficient control over these factors. Selection and sequencing of test-input configurations will be important considerations throughout this chapter.

When inserted in the expansion Equation 4, the test of Equation 5 leads to the following equations:

$$
\begin{array}{llllllll}
K^{(0)} & & & & & & = & R_0 \\
K^{(0)} & + K_1^{(1)} & & & & & = & R_1 \\
K^{(0)} & & + K_2^{(1)} & & & & = & R_2 \\
K^{(0)} & & & + K_3^{(1)} & & & = & R_3 \\
\cdots & \cdots & \cdots & \cdots & \cdots & \cdots & & \\
K^{(0)} & & & & & + K_k^{(1)} & = & R_k
\end{array}
\qquad (6)
$$

This system is trivial to solve.

When signal-to-noise considerations are unimportant, this direct measurement of the impulse response is undoubtedly the simplest and most elegant way to derive the two time domain kernels, $K^{(0)}$ and $K^{(1)}$, of a linear system. However, in practice, with noisy biological systems this simple and efficient test is usually not the best approach. In such cases it is desirable to increase the stimulus power by using test input configurations with more than one pulse in the test window. In principle any set k of linearly independent input configurations can be used for the kernel determination. However, while some sets such as single pulses lead to an easily solvable system of equations, others do not.

In general, the computational problem boils down to inversion of the matrix X in Equation 3. There is a class of matrices that are particularly simple to invert, namely, the orthogonal matrices. They are matrices whose rows and columns are mutually orthogonal (uncorrelated). Their inverse is proportional to their transposition (exchange of rows and columns).

$$[O][O]^T = [O]^T[O] = n \cdot [I] \qquad (7)$$

In this case the computation of kernels is reduced to the simple multiplication of the transposed matrix X^T with the response vector. When dealing with linear systems, the rows of the matrix X are the input configurations themselves. It is, thus, desirable to use inputs that are orthogonal functions over a time interval equal or larger than the input window. While all of these orthogonal sets share the computational advantage, most of them lead to inefficient testing. A good example is the frequency domain test mentioned above. For efficiency, a new member of the orthogonal set should appear in the input window with each advance in time by one bin. The discrete input function is then orthogonal to all its shifts, and its autocorrelation function is given by:

$$\sum_{i=1}^{k} x_i x_{i+j} = c \cdot \delta_{i,j}$$

$$\delta_{i,j} = \begin{cases} 1 \; for \; i = j \\ 0 \; for \; i \neq j \end{cases} \qquad (8)$$

Most of the following discussion will focus on binary stimulation. When dealing with linear systems, binary test inputs offer some important advantages. They can provide the most stimulus power whenever the amplitude range of the signal is limited and, thus, permit shorter recording times than test inputs with multilevel or Gaussian amplitude distributions. Amplitude limits are often imposed to avoid saturation of the transducers used for stimulation or saturation of the biological system itself. An additional benefit of binary stimulation is that the extraction of the impulse response from the record is significantly simpler and faster.

B. COMPUTATION OF THE TWO KERNELS

Complete sets of orthogonal binary functions with values of $+1$ and -1 are called Hadamard sets, and the linear transforms which they mediate are Hadamard transforms. A well-known example is the Walsh transform which is, in a sense, the binary analog of the Fourier transform. It is easily seen that Hadamard functions that satisfy Equation 8, i.e., that are all related to one another through shifts, are not possible. Complete orthogonal sets of functions with values +1, −1 always include the "DC" component represented by the function which assumes only one of the two values. Clearly, this function cannot be obtained from the others through a shift. However, one easily finds Hadamard matrices whose rows and columns satisfy Equation 8 if one eliminates this single function from the set. Consider the following example:

$$\begin{matrix} + & + & + & + & + & + & + & + \\ + & - & + & + & - & + & - & - \\ + & + & + & - & + & - & - & - \\ + & + & - & + & - & - & - & + \\ + & - & + & - & - & - & + & + \\ + & + & - & - & - & + & + & - \\ + & - & - & - & + & + & - & + \\ + & - & - & + & + & - & + & - \end{matrix} \qquad (9)$$

When the matrix (9) is stripped of its first row and column, the remainder of its rows (columns) are, indeed, related to one another by cyclical shifts. The cycle $x_i (i = l, \ldots k)$ that defines such a matrix is called a binary M-sequence. Other important properties of these sequences are discussed in Section X.B.

Exact orthogonality of the M-sequence matrix requires the completion to the 8 × 8 matrix (9). In terms of systems analysis the first row represents a separate test needed to determine the DC component. Since biological signals are usually somewhat high-pass filtered (AC coupled), this test is often omitted as its outcome is known. If the length of the input window of the system is 7 or less rather than 8, then the first column has no effect on the amplitude measurements of the subsequent tests. The tests can, thus, be performed by stimulating the system with abbreviated rows. These tests can be efficiently combined as outlined above through stimulation with a complete binary M-sequence.

Because the matrix (9) is orthogonal and satisfies Equation 7, no matrix inversion but only matrix multiplication is needed to solve Equation 5 for the derivation of the two kernels. In the next section it is shown how this matrix multiplication can be executed most efficiently by means of a single fast Walsh transform (FWT).

IV. DETERMINISTIC BINARY ANALYSIS OF NONLINEAR SYSTEMS

Consider now the testing of nonlinear systems. Here we can no longer assume that the higher-order kernel elements in the expansion (Equation 2) are zero. The ensemble of test inputs has to be enlarged significantly to obtain the necessary independent equations for their determination.

In the general case, solving this system of equations (which may have tens of thousands of unknowns) by inverting the matrix X can be difficult, and the computation times may be prohibitive. The computation is, therefore, usually limited to kernels of order 3 or less. However, when the test is restricted to binary inputs, the inversion of the matrix X is almost trivial and extremely fast. In this case the matrix is nonsingular and contains only matrix elements with the values *+1* and *−1*. It is of the same type as the matrix (9). The computation is based on the fact that any system of *k* linearly independent equations with *k* unknowns in which the unknowns only appear with coefficients of *+1* or *−1* can be solved by means of a Walsh-Hadamard transform. This transform can be executed with the well-known FWT algorithm.

It is important to note that the restriction to binary inputs limits the Volterra kernel elements that can be resolved. This limitation and the resulting relationship between the binary kernels and the Volterra kernels will be discussed below and in Section VII. In Section VII we will see how the limitations associated with binary stimulation can be overcome and how the binary analysis technique outlined below can be generalized to multilevel stimulation.

The number of different binary configurations in an input window of k time bins is 2^k. On the other hand, the total number of binary kernel elements for this input window is also 2^k. By means of a complete binary test it should, therefore, be possible to determine the entire binary kernel series. Such a test provides exactly the same number of equations as there are unknowns.

To illustrate the binary testing strategy and the kernel computation let us consider a third-order system with an input window of size 3. In order to simplify the formalism we assume $d = 0$. In this case Equation 2 becomes:

$$\begin{aligned} R_j = K^{(0)} &+ \sum_{i=1}^{3} K_i^{(1)} x_{j-i} \\ &+ \sum_{i_1=1}^{3} \sum_{i_2=1}^{3} K_{i_1,i_2}^{(2)} x_{j-i_1} x_{j-i_2} \\ &+ \sum_{i_1=1}^{3} \sum_{i_2=1}^{3} \sum_{i_3=1}^{3} K_{i_1,i_2,i_3}^{(3)} x_{j-i_1} x_{j-i_2} x_{j-i_3} \end{aligned} \tag{10}$$

We insert the $2^3 = 8$ different binary configurations into the Equation 10. Choosing of $+c$ or $-c$ for the values x_i in the triplet (x_1, x_2, x_3) we obtain one equation. For example, the choice $(-c, c, -c)$ yields the equation

$$\begin{aligned} R(-c,c,-c) = &\{ K^{(0)} + c^2K_{1,1}^{(2)} + c^2K_{2,2}^{(2)} + c^2K_{3,3}^{(2)} \} \\ &- \{ cK_1^{(1)} + c^3K_{1,1,1}^{(3)} + c^3K_{1,2,2}^{(3)} + c^3K_{1,3,3}^{(3)} \} \\ &+ \{ cK_2^{(1)} + c^3K_{2,1,1}^{(3)} + c^3K_{2,2,2}^{(3)} + c^3K_{2,3,3}^{(3)} \} \\ &- \{ cK_3^{(1)} + c^3K_{3,1,1}^{(3)} + c^3K_{3,2,2}^{(3)} + c^3K_{3,3,3}^{(3)} \} \\ &- \{ c^2K_{1,2}^{(2)} \} - \{ c^2K_{2,3}^{(2)} \} + \{ c^2K_{1,3}^{(2)} \} + \{ c^3K_{1,2,3}^{(3)} \} \end{aligned} \tag{11}$$

The terms in brackets always have the same sign irresepctive of the choice of the binary variables $\{x^1, x^2, x^3\}$. The system of equations resulting from a complete binary test can be solved for the values of the brackets but not for the terms contained in them, i.e., the binary test is, thus, unable to distinguish the Volterra kernel elements within each bracket. However, the signs in front of the brackets depend on the input configuration. In fact, each input configuration leads to a different set of signs such that all the 2^3 possible combination of the signs occur when the system is tested with the 2^3 possible binary input configurations. As we will see below, the test will permit easy determination of the value of each bracket. The brackets will be called binary kernel elements denoted by $B^k_{ij,...,l}$ where k is the order and $i,j,\ldots,l$ are the indices of the lowest-order Volterra kernel element appearing in the bracket. In terms of the binary kernels $B^k_{ij,...,l}$ Equation 11 takes the form:

$$R(-c,c,-c) = B^{(0)} - B_1^{(1)} + B_2^{(1)} - B_{1,2}^{(2)} - B_3^{(1)} - B_{2,3}^{(2)} + B_{1,3}^{(2)} + B_{1,2,3}^{(3)} \tag{12}$$

Note that each even (or odd) binary kernel B^k contains only even (or odd) Volterra kernel elements. If the binary kernels are known, one obtains from their definition in Equation 12 two independent sets of equations for the determination of the Volterra kernels.
For the even-order kernels:

$$\begin{aligned}
&K^{(0)} + c^2K^{(2)}_{1,1} + c^2K^{(2)}_{2,2} + c^2K^{(1)}_{3,3} = B^{(0)} \\
&K^{(2)}_{1,2} = B^{(2)}_{1,2} \\
&K^{(2)}_{2,3} = B^{(2)}_{2,3} \\
&K^{(2)}_{3,1} = B^{(2)}_{3,1}
\end{aligned} \tag{13}$$

For the odd-order kernels:

$$\begin{aligned}
&cK^{(1)}_1 + c^3K^{(3)}_{1,1,1} + c^3K^{(3)}_{1,2,2} + c^3K^{(3)}_{1,3,3} = B^{(1)}_1 \\
&cK^{(1)}_2 + c^3K^{(3)}_{2,1,1} + c^3K^{(3)}_{2,2,2} + c^3K^{(3)}_{2,3,3} = B^{(1)}_2 \\
&cK^{(1)}_3 + c^3K^{(3)}_{3,1,1} + c^3K^{(3)}_{3,2,2} + c^3K^{(3)}_{3,3,3} = B^{(1)}_3 \\
&c^3K^{(3)}_{1,2,3} = B^{(3)}_{1,2,3}
\end{aligned} \tag{14}$$

Note that each Volterra kernel element appears in only a single equation. This is because the binary kernel series is obtained from the Volterra series by dropping all pairs of equal indices and substituting for them a factor c^2. Each Volterra kernel element with pairs of equal indices (diagonal kernel elements), thus, becomes an additive contribution to a single binary kernel element lower by two orders for every pair of identical indices.

$$K^n_{i,\,\dots\,,j,j,\,\dots\,,k} \rightarrow c^2 b^{n-2}_{i,\,\dots\,k} \tag{15}$$

where $b^{n-2}_{i,\dots,k}$ is an additive contribution to the binary kernel element $B^{n-2}_{i,\dots,k}$.
For any binary input x_i of amplitude c the response can be exactly represented by the series

$$\begin{aligned}
R_j = B^{(0)} &+ \sum_{i_1=n}^{m} B^{(1)}_{i_1} x_{j-i_1} \\
&+ \sum_{i_1=n}^{m}\sum_{i_2=n}^{m} B^{(2)}_{i_1,i_2} x_{j-i_1} x_{j-i_2} \\
&+ \sum_{i_1=n}^{m}\sum_{i_2=n}^{m}\sum_{i_3=n}^{m} B^{(3)}_{i_1,i_2,i_3} x_{j-i_1} x_{j-i_2} x_{j-i_3} \\
&+ \cdots
\end{aligned} \tag{16}$$

Even a complete binary test is, thus, unable to resolve all the Volterra kernel elements. If the Volterra series converges fast and all kernels of an order higher than n are zero, then the binary kernel elements of the highest even and odd orders (n and $n - 1$) are identical to the Volterra kernel elements. Depending on the system at hand, the missing diagonal elements of these kernels (elements with pairs of identical indices) may be approximated by interpolation procedures. The interpolated elements can then be used to determine the lower-order Volterra kernels. If interpolation cannot be used, a test with multilevel signals becomes necessary (see Section VII).

Let us now focus on the derivation of the binary kernels from measurements obtained in a complete binary test. Each test configuration contributes one equation for the determination of the binary kernels B^k. Returning to our example of a sytem with an input window of length 3, we obtain all eight equations of type in Equation 12 corresponding to the eight binary input configurations. The system of equations is written out in matrix form

$$[X]\vec{B} = \vec{R} \tag{17}$$

The order of the rows and columns in the matrix $[X]$ is quite arbitrary. Each row represents a specific test configuration, and each column contains the coefficients of the binary kernel elements. For reasons that will become clear below, we chose to order the rows and columns as follows: each 1 and -1 in a test configuration is translated into a binary number by means of the substitution $-1 \rightarrow 0$ and $1 \rightarrow -1$. This number specifies the column of the matrix $[X]$ that contains the coefficients of the binary kernel elements for this test configuration as well as the component of the vector R which contains the measured amplitude for this configuration. Note that, according to this convention, the numbering always starts at 0. In our example the coefficients corresponding to the test configuration $(-1,\ 1\ -1)$ are placed in the 5th column of X and the corresponding amplitude in the 5th component of R. Together these entries determine the 5th equation in the system.

The order in which the coefficients of the binary kernel elements are arranged in the columns is determined in the following way: each binary kernel element is uniquely identified by the pattern of interacting time bins in the input window, i.e., the values of the lower indices. The kernel K^0 represents the output offset (DC component) and corresponds to the test configuration with zero amplitude in all bins of the input window. The first-order kernel K^1 depends only on the content of one single time bin whereas higher-order kernels depend on the interaction of two or more bins. This bin pattern is translated into a binary number which specifies the column number for the coefficients of each kernel element. For instance:

Kernel element B^0	$\rightarrow$	Binary pattern (000)	$\rightarrow$	Column 0
Kernel element $B^2_{2,3}$	$\rightarrow$	Binary pattern (110)	$\rightarrow$	Column 6

This was, in fact, the ordering used in Equation 12. Using the expansion (Equation 10) and the definition of the binary kernels through Equations 11 and 12, one easily verifies that the matrix X defined by the complete binary test is given by

	Row	0	1	2	3	4	5	6	7
	Test	(000)	(001)	(010)	(011)	(100)	(101)	(110)	(111)
Kernel	*Delays*								
$B^{(0)}$	(000)	+	+	+	+	+	+	+	+
$B^{(1)}_1$	(001)	+	−	+	−	+	−	+	−
$B^{(1)}_2$	(010)	+	+	−	−	+	+	−	−
$B^{(2)}_{1,2}$	(011)	+	−	−	+	+	−	−	+
$B^{(1)}_3$	(100)	+	+	+	+	−	−	−	−
$B^{(2)}_{1,3}$	(101)	+	−	+	−	−	+	−	+
$B^{(2)}_{2,3}$	(110)	+	+	−	−	−	−	+	+
$B^{(3)}_{1,2,3}$	(111)	+	−	−	+	−	+	+	−

(18)

Given the response amplitudes R^i derived from all test configurations, the kernel elements can be computed by inversion of this matrix [X] according to:

$$\vec{B} = [X]^{-1} \cdot \vec{R} = [X]^{T} \cdot \vec{R} \tag{19}$$

Note that the bit positions of the binary row and column numbers represent the time bins of the input window. The 1s in the binary row number, thus, represent the time bins whose amplitudes are multiplied together for the corresponding binary kernel element, while a 1 in the binary column number represents a negative test amplitude for this bin. The matrix elements of X are, thus, equal to the parity of the logical AND of the binary row and column numbers. This is also the definition of the Hadamard-Walsh matrix in its natural ordering (Fino and Algazi, 1976; Henderson, 1970). It defines the Walsh transform which is proportional to its inverse:

$$[X]^2 = n \cdot [I] \qquad \begin{array}{ll} n = 2^k & \textit{number of rows} \\ k & \textit{size of the input window} \\ [I] & \textit{identity} \end{array} \tag{20}$$

Clearly, these considerations are not dependent on the size k = 3 of the input window selected for this illustration. It, thus, follows: the binary kernels can be derived from any complete binary test by means of a single Walsh transform. The Walsh transform can be computed by means of the FWT algorithm (see Appendix A). Even for very large data sets this highly efficient algorithm permits derivation all of the binary kernels at once in seconds or minutes.

Computing the binary kernel elements over an input window of size k (solving the matrix Equation 17, thus, boils down to the following simple procedure:

Deterministic Binary Procedure

1. Insert all the measured response values into an array of size 2^k. Each value is added into the array address formed from the k elements in the binary stimulus sequence which immediately precede the response value:

Stimulus sequence $\cdots$ 0 1 0 0 0 1 1 0 1 1 1 0 1 0 1 0 0 1 $\cdots$

$|\leftarrow \quad k \quad \rightarrow|$

t

Response amplitude $R_t \rightarrow$ *Array address* (011011101) = 221 (21)

If, in this process, the same binary address occurs several times, the best estimate for the entry at this address is the average of the corresponding response amplitudes. It is, therefore, generally necessary to keep track of the number of times each address is serviced and divide the content in the end by this number.

2. A fast Walsh-Hadamard transform (natural Walsh basis) is applied to the array.
3. The binary kernel elements are found in the transformed array at the addresses formed by the delay patterns by which they are defined. The number of 1s in the address is equal to the kernel order. The spacing between the 1s is equal to the delays. Addresses of kernel elements are most easily generated for kernel elements on lines parallel to the main kernel diagonal. Such a string of kernel elements will be called a kernel slice. It is read from consecutive binary addresses obtained by shifting the appropriate bit pattern through a register of size k. For instance:

1st order kernel addresses:	(00000001), (00000010), (00000100),...
2nd order kernel, 1st slice:	(00000011), (00000110), (00001100),...
2nd order kernel, 2nd slice:	(00000101), (00001010), (00010100),...
etc. (similar for higher-order kernels).	

Note that a kernel slice ends when the first *1* in the bit pattern drops off the right end of the address register. Naturally the kernel slices become shorter for larger separations between the first and last *1* in the address. This is merely a consequence of the fact that events outside this window (represented here by the binary address register) cannot affect the response amplitude.

This simple and ultrafast procedure delivers all binary kernels associated with an input window of length k. In the absence of contaminating noise the kernel estimates are exact and not just approximations, but only if the input window of the system is of size smaller than or equal to k and if the binary test is indeed complete. Other kernel estimation procedures are stochastic in nature and yield approximate kernels whose accuracy depends on the exact procedure used.

As pointed out earlier, kernel computation requires solving a very large system of equations. In the special case of binary inputs the system is linear and can be solved exactly. Usually matrix inversion for the solution of such systems requires iterative procedures that are highly computationally intensive and converge toward the exact solution. However, the special nature of the matrix generated by the complete binary test permits its exact solution by means of the FWT (see also Sutter, 1991).

V. COMPARISON WITH THE BINARY LEE AND SCHETZEN TECHNIQUE

The deterministic approach outlined above can be directly compared with conventional techniques. Consider the popular cross-correlation technique introduced by Lee and Schetzen. It is meant to be used with random white-test stimuli and relies on the stochastic properties of the test sequence, i.e., on properties of infinitely long random sequences which are only approximate when the test is limited in length. An infinite random binary sequence is white, and so are all its auto-products (element-wise products of the sequence with shifted replica of itself) as long as the factors are not pair-wise identical. This follows from the orthogonality properties of the product sequences $X^{(r)}_{(\ldots)i}$:

If

$$X^{(r)}_{(k_1,k_2,\ldots,\ k_r),\ i} = x_{i-k_1}\, x_{i-k_2} \cdots x_{i-k_r} \tag{22}$$

then

$$\lim\,(n\rightarrow\infty)\,\frac{1}{n}\sum_{i=1}^{n} X^{(r)}_{(k_1k_2,\ \cdots\ ,k_1),i} = \begin{cases}1 & \textit{if } k_i \textit{ pairwise equal}\\ 0 & \textit{otherwise}\end{cases} \tag{23}$$

Any two product sequences X^r_i and Y^s_i of order r and s whose factors are not pairwise equal, thus, have the following auto- and cross-correlation properties:

Auto-correlation

$$\lim\,(n\rightarrow\infty)\,\frac{1}{n}\sum_{i=1}^{n} X^{(r)}_i\, X^{(r)}_{i-j} = \begin{cases}1 & \textit{for } j = 0\\ 0 & \textit{for } j \neq 0\end{cases} \tag{24}$$

Cross-correlation

$$\lim (n \rightarrow \infty) \frac{1}{n} \sum_{i=1}^{n} X_i^{(r)} Y_{i-j}^{(s)} = 0 \tag{25}$$

Because of these properties, a slice of a binary kernel of any order can be estimated as the cross-correlation function with the corresponding product sequence (Lee and Schetzen procedure):

$$B^{(r)}_{(k_1,k_2,\ \cdots\ ,k_r);k} = \sum_{i=1}^{k} X_{i-k}^{(r)} R_i \ \{k = 0, \cdots, m\} \tag{26}$$

Note that only a limited range of the cross-correlation function between response and product sequences needs to be computed, since the kernel slices become zero as soon as the largest lag between the response and any of the factors in the product becomes larger than the system memory m.

With the deterministic binary approach the length of the kernel slices is automatically limited by the size of the Walsh array. But within this range its computation is almost exactly equivalent to the Lee and Schetzen estimation. The relationship between the two can be derived from Matrix 18. Assuming that each configuration of length m occurs exactly once in the stimulus sequence, the second, third, and fifth rows in Matrix 18, which represent the elements of the first-order kernel, are identical to the first three elements of the cross-correlation function between the response and the binary input. A similar correspondence is found for higher-order kernel slices. The difference between the two procedurs relates only to multiply occurring input configurations. In the cross-correlation technique, amplitudes measured at different occurrences are summed, while the deterministic binary approach correctly uses the best estimate, namely, the average. The Lee and Schetzen technique relies on the stochastic properties of the input sequence, i.e., on the fact that if an infinitely long white input sequence were used, all configurations would be equally likely. However, in the relatively short sequences used in laboratory experiments, the number of actual occurrences varies.

The connection between the Walsh kernels derived by summing rather than averaging and the zero, first, and higher-order cross-correlation functions always holds, independent of the size of the Walsh array. To illustrate this we shall consider the use of the smallest nontrivial Walsh array which is of dimension 2. In this case the input configurations consist of only one time bin. According to the deterministic binary procedure a measured response amplitude is placed into location 0 of the Walsh array if the input amplitude in the time bin exactly d time units prior was a 1, and in location 1 otherwise. The number of occurrences of both input configurations is very large and approximately equal. The Walsh transform is trivial, and the two kernel elements it produces are the zero-order kernel, i.e., the average of all the data points and the weighted average using the input at time bin d preceding as a weighting function. The second kernel element is, thus, a single point of the first-order cross-correlation function that corresponds to the correlational shift d. In order to obtain other points of the cross-correlation function, the same procedure has to be applied for different shifts d. Clearly, better estimates for the kernel elements could be obtained if one compensated for the unequal occurrence of the two configurations.

If the deterministic binary procedure is applied using the next larger Walsh array size of 4, then the Walsh transform produces the two elements of the first-order cross-correlation function, $B_d^{(1)}$ and $B_{d+1}^{(1)}$ at once and, in addition, the single second-order kernel element $B_{d,d,+1}^{(2)}$. Again, by choosing different values of d it is possible to derive the necessary stretches of the cross-correlation associated with these elements.

For larger Walsh arrays this method yields longer stretches of the cross-correlation functions at once as well as cross-correlation functions for higher-order kernels and kernel slices further removed from the main diagonals of the kernels. If one does not compensate for the number of occurrences, the kernel slices are exactly the same, regardless of the size of the Walsh array used to obtain them. This is generally not the case if the data sets are normalized (division by occurrences). It is important to remember that exact binary kernels are obtained only with a complete, normalized data set and only if the size of the Walsh array is 2^k where k is equal or larger than the length of the input window. Note also that the delay d has to be chosen such that the k time bins cover the entire window.

Because the deterministic approach makes it possible to keep track of the number of occurrences of each configuration, it can also be used with nonwhite inputs. This considerably widens the range of possible applications. For instance, the technique can be applied to systems where endogenous signals are considered as inputs, i.e., where the transfer characteristics between two biological signals are to be determined. The signal that plays the role of the input is converted into a binary series or decomposed into two or more binary components. These binary signals are then converted into Walsh array addresses for storage of the output amplitudes and the kernels are derived by means of the FWT. One can also use the same signal as input as well as output in order to investigate higher-order autocorrelations in an endogenous signal itself.

In this chapter the input is always described in terms of amplitude levels within k equally sized time bins of the input window. The cross-correlation technique is dependent on this specific choice of the input parameters. However, with the deterministic binary approach, in principle, any binary parametrization of the input configurations can be used. This opens up a wealth of largely unexplored approaches to nonlinear analysis. The usefulness of this flexibility is demonstrated with the following simple example.

It is often observed that the higher frequencies are only contained in the early response components and that for all the kernels the regions some distance from the origin only contain low-frequency components. This is true, e.g., for the visual evoked cortical response. In such cases the sampling grid for the kernels needs to be much finer near the kernel origin than in the more remote regions. With the cross-correlation technique this differentiation cannot be made, since the input is treated as a continuous time series sampled at periodic intervals. The deterministic binary technique, on the other hand, permits sampling of the input window at arbitrary intervals. Sampling intervals can easily be stretched out toward the end of the window as long as proper local antialiasing filtering is used. According to the deterministic binary procedure the sampled input is then converted to a binary address for storage of the corresponding response amplitude. Note that the coordinates on which the kernel elements are ultimately plotted should have the exact scaling used for the sampling of the input window. With this technique it is possible to perform complete binary tests on systems with relatively large input windows as long as they satisfy the above condition.

Other alternate binary parametrizations might involve decomposition into orthogonal function sets over the input window.

VI. COMPLETION OF INCOMPLETE DATA SETS

The exact binary kernels of all orders can only be computed if the system is tested with all possible configurations of length k. If the test is incomplete, one ends up with fewer equations than unknown binary kernel elements, and the system clearly has no unique solution. The data array on which the FWT algorithm operates has holes, and the transform into the kernel domain can be executed only if values are placed into these vacant array locations.

Incompleteness of the test and the resulting missing information is, indeed, one of the major problems facing nonlinear systems analysis. It is easy to see that for biological systems a truly complete test is rarely feasible, even with restriction to binary stimulation. For most such systems the size of the input window is considerably larger than 24 time bins. With sampling intervals of 1 ms a complete binary test would, thus, require recording times larger than 2^{24} ms = ~2 h. Often the initial assumption of time invariance of the system no longer holds for time intervals of this size. In addition, processing of such enormous, nonredundant data sets is not always feasible on commonly available laboratory computers, not so much because of the computation times but because of the memory requirements. Execution of the FWT requires allocation of the entire array in real memory. In practice the array size is usually limited to about 2^{20} data points, but instability of the biological preparation and other considerations may place more stringent limitations on the size of the record.

Consider, now, the case where the size k of the input window is sufficiently small such that the array of 2^k points fits into memory, but the number of recorded data points is less than 2^k. Insertion of the recorded data into the array may leave it relatively poorly populated. To fill the remaining array locations requires additional information not contained in the test results. This information could be supplied in the form of assumptions concerning convergence of the kernel series or in the form of a more specific model. It should be pointed out, however, that the use of such information changes the flavor of the analysis somewhat. The white-noise approach loses some of its general and unbiased nature that is often touted as its most important asset.

The easiest way to complete a partially populated array is to leave all the array locations that have not been accessed at zero. One should be aware that this choice introduces as much of an experimental bias as others. The zeroed locations are determined by the test sequence, and the resulting kernels, therefore, depend, at least to some extent, on the particular choice of the sequence. In general, the effect is a noisy appearance of the kernels, with conspicuous streaks in the direction of the main diagonal. The same type of streaking is caused if one sums rather than averages the response amplitudes to multiply occurring input configurations. In both cases the effect is due to the fact that input configurations always shift through the input window and, thus, affect several consecutive elements of a given kernal slice.

Note that filling the holes in the data set with zeros and summing rather than averaging the amplitudes measured in response to multiply occurring input configurations leads to exactly the same kernels as the Lee and Schetzen cross-correlation method. The streaky appearance of the kernel estimates found, e.g., in Marmarelis and Marmarelis (1978, pp. 285 and 286) is, thus, easily explained.

It would be preferable if one had additional information that could be used to fill holes in the test rather than filling them with zeros. Fast convergence of the kernel series is the most commonly used assumption in kernel analysis. Rarely are kernels of orders higher than two or three computed when more conventional techniques are used. If the assumption is not correct, this early truncation of the series is clearly not recommended.

Consider a more general case of a system that is known to fit a certain model. This model may imply certain relations between kernels, kernel slices, or kernel elements. Because of the incomplete test these relations are not satisfied in the computed kernels. The goal is to modify the unfilled locations in the raw data array until the kernels satisfy the relations imposed by the model. When this is achieved, the kernel estimates have been improved by the supplemental information supplied in form of the model. If the data set together with the supplemental information are equivalent to a complete data set, the kernel elements approach the exact values.

Within the framework of the Lee and Schetzen cross-correlation technique the condition that higher-order kernels are zero cannot easily be exploited to improve kernel estimates.

However, in the context of the deterministic binary technique, the supplemental information can be incorporated into the analysis. This is possible because the technique delivers kernels of all orders on which the model can be imposed. There are two properties of the transformation X between raw data and the full set of corresponding binary kernels that are important in this context:

1. The transformation is its own inverse; i.e., two consecutive applications of the FWT will reproduce the original raw data set.

$$[X] \times [X] = \frac{1}{2^k} [I] \tag{27}$$

   Note that the transform can be executed in fixed-point arithmetic. It is, thus, not subject to rounding errors that would propagate through repeated applications.
2. The transform is extremely fast.

These properties make it possible to move back and forth rapidly between the representation of the test results as raw data and as kernels in order to implement the following iterative procedure:

1. The computed kernel series is modified and brought into compliance with the model.
2. The modified kernel series is transformed back to the raw data domain by means of the FWT. The untested array locations are now filled with values that bring the raw data into agreement with the model. However, the new signal estimate is based on the original, incomplete data set and, thus, is not exact. For this reason the measured data points also have been altered.
3. The measured data points are restored.
4. The resulting data array is converted into improved kernel estimates by means of the FWT.

If the imposed model is correct, iteration of this procedure will lead to more accurate kernels. Various modifications of the iteration procedure are possible which accelerate convergence. Note that the originally measured data are never modified in this procedure, but only supplemented by data estimated on the basis of additional information.

To demonstrate the effectiveness of the procedure we apply it to a specific second-order Wiener model. The exact second-order kernel is shown in Figure 1a. We assume that the experimenter knows that the size of the system's input window is 16 time bins and that all kernels of order higher than second are negligible. The system is stimulated with a random binary sequence, and the record length is limited to 2048 data points.

The second-order kernel in Figure 1b has been computed by imbedding the data array in a Walsh array of size 2^{16} according to Step 1 of the deterministic binary procedure. The missing diagonal was interpolated with a cubic spline. This kernel is slightly more accurate, but barely indistinguishable from the kernel computed by means of cross-correlation (not shown). In this example the difference is very small since the data set is sparse; even double occurrences are rare.

In this model the measured data points fill only about 1.5% of the Walsh array. After the Walsh transform the effect of the missing data is expected to be distributed over the entire kernel array as apparent noise. Since, in this case, the first- and second-order kernels only account for less than 0.5% of the kernel elements, most of this noise power is in higher-order elements.

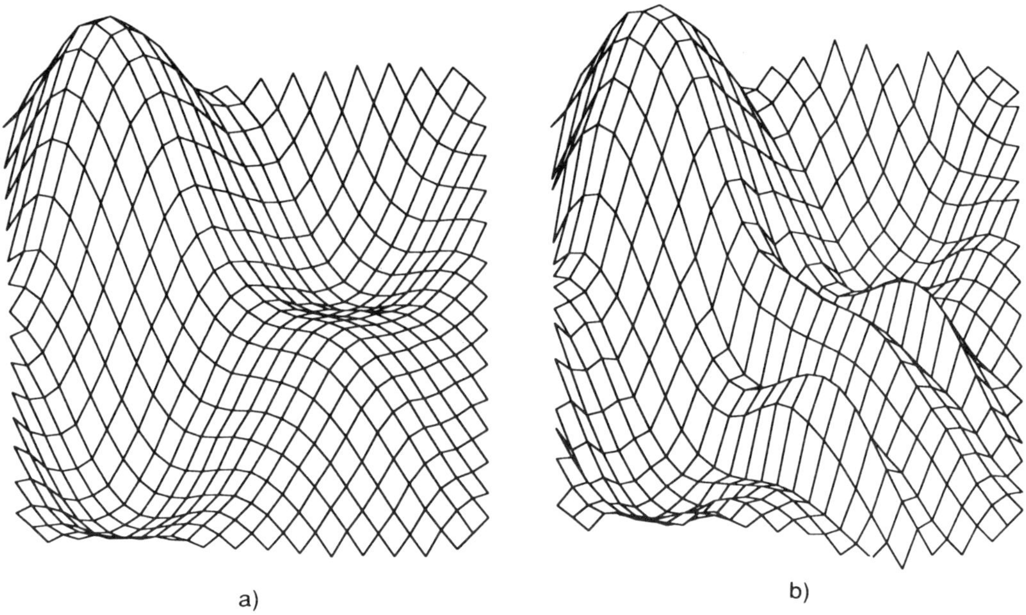

FIGURE 1. (a) Exact Volterra kernel of a second-order Wiener model. (b) Second-order binary kernel obtained from a data set of 2048 data points. The stimulation was random binary. The 16 × 16 point kernel computed by means of the deterministic binary technique. The missing diagonal was interpolated with a cubic spline. This kernel is slightly more accurate, but barely indistinguishable from the kernel computed by means of cross-correlation (not shown).

If the Walsh array were completely populated, the first- and second-order kernels would be the only ones significantly different from zero. They would, however, be larger by approximately a factor K/D where $K = 2^k$ is the size of the Walsh array (in this example 2^{16}) and D is the number of locations filled with recorded data. The best estimate for the missing data is, thus, obtained by multiplying the first- and second-order kernels by K/D, zeroing all the higher-order kernel elements and transforming the resulting kernel array back to the raw data domain. The back transform deviates from the original data, since it is based on kernels estimated from an incomplete data set. The measured data points are more accurate than the back transform and are now restored such that the back transform is only used to fill in the missing data. A subsequent forward transform results in a greatly improved second approximation to the exact first- and second-order kernels as well as in a significant reduction in the power of all higher-order kernels.

This process can be iterated until all the higher-order kernel elements approach zero. Note, however, that in subsequent iterations the factor K/d is equal to 1, since the Walsh array is now completely populated. The original data are never changed in this procedure, but only supplemented by additional hypothetical data selected to minimize the power of all the higher-order kernels. The result of 32 iterations applied to the data of Figure 1 is shown in Figure 2. The resulting kernels are now very close to the exact kernels shown in Figure 1a.

If the equation set up by the data (measured response amplitudes to 1024 randomly chosen input configurations) is sufficient to determine the 257 elements of the 0-, 1st-, and 2nd-order kernels, then the kernels obtained by the procedure must converge toward the exact kernels.

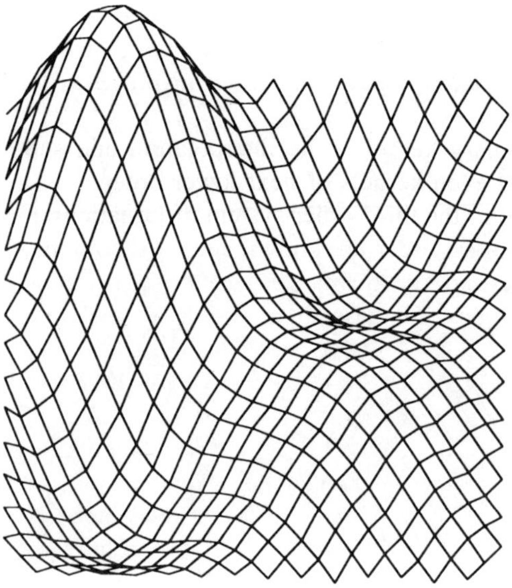

FIGURE 2. The kernel of Figure 1b improved by means of an iteration procedure. The procedure makes use of the knowledge that the system is of second order (see text). The kernel is now almost indistinguishable from the exact Volterra kernel of Figure 1a.

The iteration procedure used in this example is similar to the least squares approximation to low-order Volterra kernels (Korenberg, 1988; Klein, 1991). In both cases it is assumed that kernels of higher order than first or second are zero. Note, however, that in the frame work of the determinisitc binary analysis fast convergence of the Volterra series is not the only condition that can be imposed. Any known relationship between kernels can be exploited since binary kernels of all orders are immediately available.

Variations and refinements of the above basic iteration procedure that accelerate convergence will be discussed elsewhere.

It is important to keep in mind that the improvement from the kernels of Figure 1b to those of Figure 2 was possible only thanks to the knowledge that the system was of second order. If no such assumption is made and the possibility of non-zero kernels up to highest order is left open, then the kernel of Figure 1b cannot be improved. It contains ambiguities due to insufficient data, and its appearance depends not only on the system at hand, but also on the particular choice of the sparse data set, i.e., on the test stimulus.

VII. METHODS TO OVERCOME THE LIMITATIONS OF BINARY TESTING

We have seen in Section IV that each binary kernel element is a linear combination of Volterra kernel elements of equal and higher order and that each Volterra kernel element appears in exactly one binary kernel. Even if the binary test is complete, it is unable to differentiate between the Volterra kernel elements which appear in the same binary kernel element (compare Equations 13 and 14). The information derived from a binary test is, thus, incomplete. The seriousness of this problem depends on the system at hand. In many cases, the missing information can be substituted from prior knowledge or a working model of the system.

Often interpolation procedures can be helpful. We have seen earlier that the nonzero off-diagonal binary kernel elements of highest odd and even order are always equal to the

Volterra kernels. If the missing diagonal elements of these kernels can be estimated by interpolation, then the entire Volterra kernel is known. Such interpolation is permissible, for example, when significant low-pass linear filtering precedes the nonlinearities. The lower-order Volterra kernels can be estimated by subtraction of interpolated diagonal elements from the lower-order binary kernels. The interpolation and subtraction procedure is then repeated for the odd and even kernels of next lower order.

Consider, for example, the third-order system of Equations 13 and 14. The off-diagonal elements of second and third order are known and, if the input window were larger, could be used to derive the diagonal elements by interpolation. Equations 13 and 14 could then be used to estimate the first- and zero-order Volterra kernels.

A. MULTILEVEL ANALYSIS USING BINARY COMPONENTS

Sometimes it may be necessary to discriminate the Volterra kernel elements within the binary kernels experimentally. This requires the use of multilevel test inputs. Fortunately multilevel analysis can be implemented without abandoning the elegance, speed, and accuracy of the deterministic binary technique. Multilevel sequences can be decomposed into binary components, and the kernels can be computed in terms of these components.

It is important to keep in mind that the transition to a multilevel test input always implies much larger ensembles of test input configurations and, thus, longer tests. The difficulties of achieving good coverage of the ensemble is, therefore, much more severe than in the binary case discussed above. In the following we will discuss the transition to a test input with four equally spaced amplitude levels.

B. QUATERNARY STIMULI

A quaternary sequence has four equally spaced levels. They can be decomposed into the sum of two binary sequences with an amplitude ratio of 1:2. They are easily generated by means of a 2-bit digital-to-analog converter whereby each digital input is modulated independently with a binary sequence.

For a complete quaternary test of an input window of length k, all possible configurations of the four level amplitudes within the k time bins have to occur. This is clearly equivalent to a complete binary test of a window of size $2k$. With a test of a given length there is, thus, a trade-off between window size and amplitude resolution. While a complete binary test for a window of size k requires 2^k input configurations, a complete quaternary test encompasses an ensemble of $4^k = 2^{2k}$. In return, the test will provide a system of 4^k independent equations for the determination of 4^k kernel elements. The gain in the number of kernel elements results in the discrimination of Volterra kernel slices on or near the main diagonal of higher-order kernels. Clearly, completeness of the quaternary test is only possible for a relatively small window size k.

In principle there are many different kernel representations that can be derived from the data set. If they are complete, i.e., if they contain a total of 4^k independent kernel elements, then they must be equivalent and can be transformed into each other. The easiest kernels to compute are those associated with the complete set of $2k$ binary variables, since they can be derived by means of the FWT. For a third-order system the kernel series associated with the binary components are sufficient to compute the exact, complete Volterra series. The representation of the Volterra kernels in terms of the component kernels is derived in Appendix B.

It is important to keep in mind that the goal of nonlinear analysis is not the derivation of the Volterra series, but an understanding of the system in terms of a realistic model. This interpretation of the data in physiological terms definitely poses the most difficult problem. It can be greatly facilitated by the specific format in which the data are presented. While the Volterra series is undoubtedly the best-known representation, it is by no means the only one to choose from. In some cases the self- and mutual kernels associated with the binary components may actually generate more insight into the properties of the system than the Volterra kernels.

VIII. SIGNAL-TO-NOISE CONSIDERATIONS

The only contamination of the kernels which has been considered so far is that due to incompleteness of the test. If we assume that the stimulation is random, the missing data points are randomly distributed over the array of raw data, and, after the FWT, all the binary kernels are randomly affected. The signal contamination with additive extraneous noise has very similar effects on all the kernels. This type of contamination has very interesting consequences in regard to the kernel representation and signal detection. We shall consider this problem in the light of the binary analysis technique.

When a Walsh array is filled with recorded data, it usually contains both of the above types of noise: addresses that have not been accessed contain zeros, and the measured data points contain experimental errors due to the additive noise. In general, the response as well as the noise is fairly uniformly distributed over the entire array. The content of the array is the sum of signal and noise vectors. The length of these vectors is defined as the RMS of the array. Note that when random binary testing is used this RMS value is not exactly equal to the RMS of the response, since the response amplitudes to multiply occurring input configurations have been averaged rather than summed. We now consider what happens to the signal and noise vectors when the data are transformed into the kernel domain.

The Walsh transform used to convert the raw data array into an array of kernel elements has two properties that are important here: first, it is a linear transform, and second, it preserves the RMS measure. It, thus, acts on the signal and noise components of the data vector independently and preserves their respective RMS or power.

$$\begin{aligned} [WT]\,(\vec{S} + \vec{N}) &= \vec{K}_s + \vec{K}_n \\ |\vec{S}| &= |\vec{K}_s| \\ |\vec{N}| &= |\vec{K}_n| \end{aligned} \qquad \text{where} \begin{cases} \vec{S},\ \vec{N} & \textit{signal and noise in raw data} \\ \vec{K}_s,\ \vec{K}_n & \textit{signal and noise in kernel array} \\ [WT] & \textit{Walsh transform} \end{cases} \tag{28}$$

Because of the random nature of the noise, its power becomes randomly distributed over the entire kernel array, while the signal power is usually contained in just a few kernels. The signal-to-noise ratio of these kernels depends, thus, on the total number of kernel elements over which the response is distributed. For example, if there are a total of K nonzero kernel elements computed from a data set of A data points, then the gain in signal-to-noise ratio achieved in the transition from raw data to kernels is roughly A/K. Clearly, the higher the degree of nonlinearity of a system, i.e., the larger the total number of kernel elements over which the response power is distributed, the larger the data set necessary to detect the kernels in the noise. This is particularly important in vision, where the signal-to-noise ratio in the raw signal is generally poor.

This signal detection problem can best be illustrated using the extreme case of a highly selective system that responds to only a single one of the stimulus configurations presented in the test. The response might be clearly visible in the raw data with excellent signal-to-noise ratio within the time interval R where the response occurred. However, since selectivity of this type of system requires nonzero kernels of all orders, the kernel computation distributes this response power over the entire kernel array. The signal-to-noise ratio of the kernels is, thus, lowered in the process by roughly a factor of A/R. In this fictitious case the kernel extraction is clearly counterproductive, as the signal is diluted with noise recorded during ineffective stimulation. The kernels would probably disappear in the noise, and the system would be judged nonresponsive. That this type of system behavior can be found in the visual system is exemplified by the ganglion cell data presented by Klein (Figure 4, Chapter 5 of this volume). Klein finds conspicuous responses in the raw data while the kernels are relatively small.

The white-noise approach to systems analysis is often touted as being highly unbiased, since it tests the system with a very large number of randomly chosen input configurations. While this may be true for the test itself, it is clear from the above considerations that the representation of the result in the form of a kernel series introduces a significant bias in favor of fast-converging models. It is important to keep this in mind when applying such techniques to the investigation of higher visual processing. Neurons in some visual areas have been reported to be highly selective. In such cases it may be necessary to enrich the test input ensemble in those configurations that elicit responses. In the framework of the deterministic binary approach, tailoring of the test input is possible. Examples of such techniques are given in Section X.

Highly selective response characteristics are not the only mechanisms that distribute the response power over a large number of kernel elements. The nonzero kernels observed in recordings of visual evoked responses may extend up to the sixth order or higher and span a significant percentage of the total kernel array derived from an average-sized record. As the signal-to-noise ratio in the raw data is generally poor, the record length required to discern the kernel structures in the noise can be quite large.

In order to facilitate the detection as well as interpretation of the responses one might well consider abandoning the traditional kernel representation. One might seek a transformation into a basis of functionals in which the data set takes a particularly simple form; i.e., a basis where the power is distributred over a small number of components. A much higher signal-to-noise ratio could then be achieved. The nature of the transformation required to achieve this simplification would provide much of the information needed for data interpretation.

IX. TESTING EFFICIENCY

A. COMPLETE SEQUENCES

When test input configurations of length k are chosen at random, completeness can only be achieved by significant overtesting. It can be shown that if 2^k data points are recorded with random binary stimulation, only about 63% of the locations in the Walsh array of size 2^k are expected to be filled. In order to fill 95% of the array, the test would have to be about three times as long. In this case there is much redundancy in the test; i.e., most of the possible binary configurations in the input window are tested more than once. In the presence of noise contamination this redundancy is not completely wasted, since multiple testing helps to improve the kernel estimates through averaging. However, this improvement is bought at the expense of longer records. When dealing with biological systems, the assumption of time invariance is usually only valid for a limited time, and it is necessary to keep the recording times as brief as possible.

The following question arises: Is it possible to design a nonrandom binary input that tests a given input window of length k completely and without redundancy? This would require construction of a binary cycle of length 2^k with the property that the content of a window of k consecutive elements assumes a new configuration for each of the 2k possible shifts of the window along the cycle. Such complete sequences do exist. They are called De Bruijn sequences. For a given length k there are often many different complete sequences available. As long as k is equal to or larger than the size of the input window it does not matter which one is selected. Each one represents the same complete test of the system, and they differ only in the sequence in which the test configurations are presented. Note that most of these sequences are not white. However, we known that for correlational shifts of l to k the auto-correlation functions must be zero. The same holds for higher-order auto-correlation functions. They are zero as long as none of the correlational shifts is larger than k. This follows immediately from correspondence between the deterministic analysis and the Lee and Schetzen technique (Section V) and from the fact that the complete test yields exact kernels.

Note also that if a system with memory smaller than k time bins is stimulated with a complete binary sequence of length 2^k the cross-correlation method and the deterministic binary approach lead to the exact same kernels. The only difference is in the enormous saving in computation time achieved with the FWT.

B. BINARY M-SEQUENCES

1. Generation and Structure

While all complete tests are equivalent in the accuracy of the kernel estimates, the selection of tested input configurations can be very important if the test is incomplete. We have seen in Section VI how missing information can be substituted using models or other constraints such as fast convergence of the kernel series. It is desirable in such cases that the information supplied by the test and the model is complementary and not redundant and that together the information is as nearly equivalent to a complete test as possible. The best choice for the test and the quality of the resulting kernels depends, therefore, on the system at hand and the class of models that may apply, i.e., on prior knowledge about the system. A single test sequence that is best in all cases clearly does not exist. A good testing strategy for a complex system where a complete test is not possible should always be tailored to the problem at hand. There are, however, test inputs that are suitable for a whole class of systems such as single input systems with a fast-converging kernel series or fast-converging multiple input systems. A powerful tool for the construction of such test inputs is found in binary M-sequences (Golomb, 1982; Zierler, 1959 and 1968).

Pseudorandom binary sequences can be generated by means of digital shift registers with a linear feedback. Each of the k stages of the register can assume the states 0 and 1. If the modulo 2 sum (parity) of an even number of specially selected taps in the register is fed back into the serial input, subsequent clocking of the register will carry it through all the 2^k possible configurations except for the zero state which is a cycle in itself (see Figure 3). The output of the register is, thus, a pseudorandom binary sequence of length $2^k - 1$. The configurations of feedback taps that lead to maximal length or M-sequences have been computed and tabulated for a large number of register sizes k (see, e.g., Golomb, 1982).

A binary M-sequence of length $2^k - 1$, thus, contains all possible configurations of k consecutive elements exactly once, except the all-zero configuration. The all-zero configuration can never be part of the sequence, since the modulo 2 sum of any number of zeros is zero. These sequences are commonly called maximum length linear feedback shift register sequences or simply binary M-sequences. Note that each binary M-sequence contains a run of $k - 1$ zeros. If a zero is added to this run, the M-sequence becomes a complete sequence.

Many properties of M-sequences have been carefully studied (Golomb, 1982; Zierler, 1959). We will focus here on those of particular importance to nonlinear systems analysis.

In the following we will consider the sequences obtained through the substitution

$$\begin{array}{ccc} 0 & \rightarrow & +1 \\ 1 & \rightarrow & -1 \\ + & \rightarrow & \times \end{array} \tag{29}$$

Modulo 2 addition now becomes multiplication.

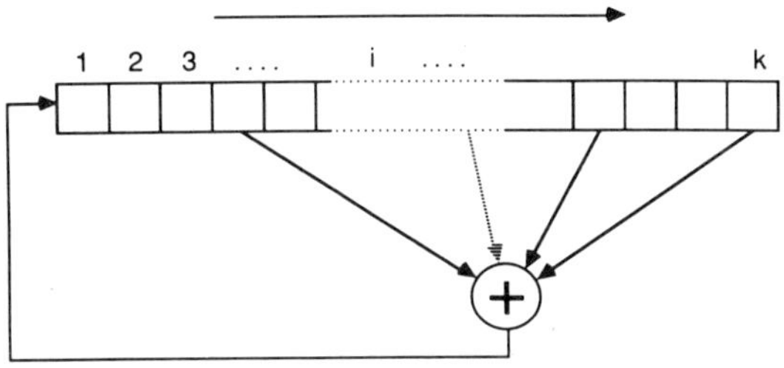

FIGURE 3. Recursive generation of binary M-sequences by means of a linear feedback shift register (generating register). If the modulo 2 sum of an even number of selected stages is fed back into the serial input, the register will assume all possible configurations except for the all-zero state. The sequence derived from any stage of the register is a binary M-sequence.

M-sequences with more than two levels have also been used in systems analysis. Particularly popular are ternary M-sequences (Citron and Emerson, 1983; Ream, 1967; Moller, 1983). They are generated by means of a shift register whose stages can assume the levels 0, 1, 2, whereby the feedback is generated with modulo 3 addition. Note, however, that no direct correspondence can be established between modulo addition and multiplication except for modulus 2. The importance of this correspondence to nonlinear analysis has already been seen in the general deterministic binary approach (Section IV). It will continue to play an essential role in the remainder of this section.

The complete set of binary configurations (1, -1) of length k, as generated, e.g., by means of a De Bruijn sequence, is closed under multiplication. The element-wise product of any two configurations is again a binary configuration of length k and, therefore, contained in the set. With this operation the complete set of configurations, thus, becomes an Abelian (commutative) group. In general, this does not apply to the configurations longer than k obtained from the same sequence. There are, for instance 2^{k+1} possible configurations of length $k + 1$ while only 2^k of these occur in the sequence, and there is no obvious reason why the incomplete set should be closed under the group operation. However, for binary M-sequences the additional element in each configuration is related to the others. It is the product of a specified number of them (see Figure 3). It follows that for every two configurations of $k + 1$ elements that occur in the sequence, their product will also be found. The only exception is the product of a configuration with itself which is a run of $k + 1$ 1s.

This closure under multiplication applies not only to $k + 1$ sequential elements of the sequence but, recursively, to any number of consecutive elements, i.e., to the sequence as a whole. An M-sequence is, thus, identical to all its auto-products with nonzero relative shift of the factors. To achieve closure under element-wise multiplication we have to include a sequence of $2^k - 1$ 1s. We, thus, conclude: The set of $2^k - 1$ sequences obtained from an M-sequence through cyclical shifts becomes an Abelian (commutative) group under element-wise multiplication when completed with a sequence of $2^k - 1$ 1s (unit element of the group). From this property follow immediately the special auto-correlation properties of M-sequences. The first-order auto-correlation function is

$$\sum_{i=0}^{2k-1} x_i x_{i-j} = \begin{cases} 2^k & -1 \; for \; j = 0 \\ & -1 \; for \; j > 0 \end{cases} \tag{30}$$

For nonzero correlational shifts all the product sequences under the summation sign are equal to the shifted M-sequence itself. Since the number of -1s in an M-sequence always exceeds the number of 1s by one, the sum is -1. For zero correlational shift the product under the summation sign is the unit sequence, and its sum is, thus, $2^k - 1$. Note that the excess -1 found in all M-sequences is a consequence of the fact that the sequence contains all configurations of the length k except for the all 1-configuration.

The auto-correlation function of M-sequences is not quite perfect as required for truly white sequences. For this reason they are often considered slightly flawed in the context of systems analysis. However, this small imperfection disappears if we consider the required data set in its entirety. The M-sequence alone does not represent a complete test, as it only contains $2^k - 1$ configurations. A complete test consists of two cycles, the M-sequence cycle and the all-zero configuration which is a cycle in itself. The recursive generation (Figure 3) defines *both* cycles. The zero cycle is needed for the completion of Matrix 9, or 18 with a zeroth column. With this addition the cross-correlation matrix becomes an exact orthogonal Hadamard matrix. It follows that if we include the all-zero configuration in the M-sequence test, the computation of the binary kernels is, indeed, exact. In nonlinear systems analysis where long M-sequences are used, it is of no practical significance whether or not this single data point is included.

Following the Lee and Schetzen technique, all binary kernel slices are computed as cross-correlation functions between the response and the stimulus sequence x_i and all its auto-products X_i. Since the stimulus sequence is an M-sequence, all auto-products are identical to the shifted stimulus cycle. It follows that all kernel slices lie on the same cross-correlation cycle of the response and the M-sequence itself. At the beginning of the cycle (near correlational shift 0), we find the first-order kernel slice. The higher-order slices are found at the locations specified by the shift of the corresponding product function. It follows that when binary M-sequences are used for stimulation, auto-products need not be computed. Only one cross-correlation between the input and response is necessary. However, in order to retrieve all binary kernels, the cross-correlation cycle has to be computed in its entirety. Note that all binary kernel slices for an epoch of length k appear on the cross-correlation function, except for the zero-order kernel. The inverse is also true: every point on the cross-correlation cycle belongs to a kernel slice. This is a consequence of the recursive generation of the sequence (Figure 4) which defines all elements of the sequence as unique products of factors with relative shifts smaller or equal to k.

An example of a cross-correlation cycle between the M-sequence stimulus and the response is shown in Figure 4a. The system is a second-order Wiener model. Since there is no contamination through extraneous noise, the first-order binary kernel at the beginning of the cycle and second-order kernel slices are exact, and all kernel slices of order higher than two are zero. The two kernels are, thus, the exact Volterra kernels. The second-order kernel shown in Figure 4b has been assembled from the slices of Figure 4a. The computation of the cross-correlation cycle and extraction of specific kernel slices is the subject of the next section.

2. Fast Computation of the Cross-Correlation Cycle

We have seen earlier how the same full set of binary kernel slices that makes up the cross-correlation cycle between the binary M-sequence and the response can also be obtained by means of the FWT. It should, thus, be possible to compute the cross-correlation cycle by means of the FWT. One only need find how to string together the kernel slices in proper sequence. This should not be difficult, since their sequencing on the cross-correlation cycle is determined by the correspondence of auto-products and shifts of M-sequence. We will now show how the proper sequence of the kernel slices can be derived using the recursive generation of the sequence as defined by Figure 3.

Consider the generating register for an M-sequence shown in Figure 5, where the sequence is derived from the last stage (t_k). Shifted sequences up to a shift of $k - 1$ steps can be obtained simply by moving the output tap to the left along the register.

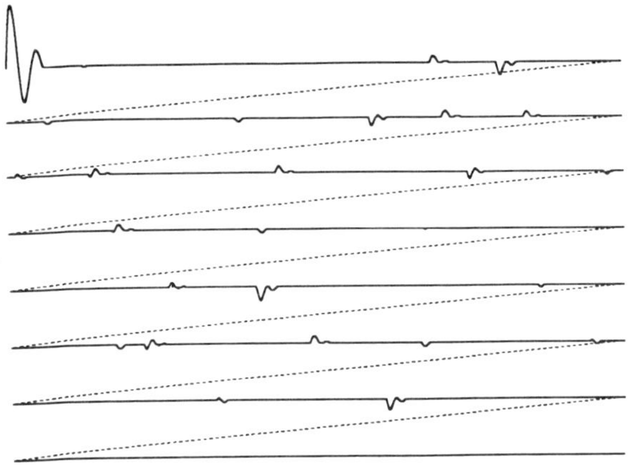

a)

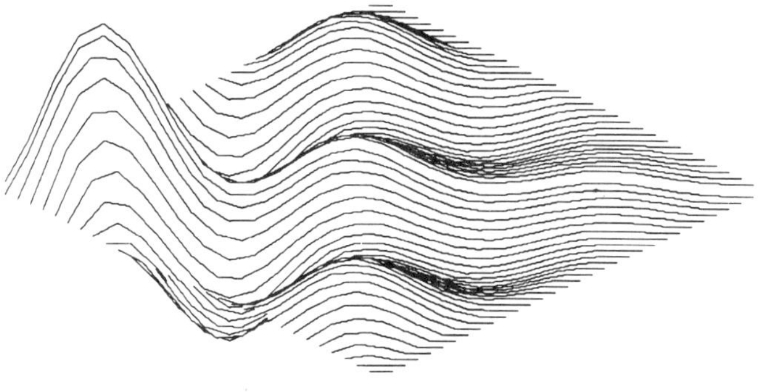

b)

FIGURE 4. (a) These traces represent the entire cross-correlation cycle of 4096 points for a second-order Wiener model computed by means of the fast M-transform algorithm. Computation time was 1.08 s. The cycle contains the first-order kernel at the beginning followed by the second-order kernel slices distributed over the cycle. Note that between the kernel slices the M-transform is exactly zero. (b) The second-order kernel assembled from the slices of the cycle in a. Since the cross-correlation cycle is long enough to accommodate all the second-order kernel slices without overlap, the kernel is exact.

For the *k*th shift the tap drops off the register. However, we know which M-sequence element will enter the register on the next clock cycle. It is the modulo 2 sum of the feedback taps, or, in terms of the $(1, -1)$ representation, the auto-product of the shifted sequences defined by the feedback taps. To generate larger shifts, the entire new tap configuration now has to be moved to the left. Whenever a tap falls off the beginning of the register, it again has to be replaced by the original feedback configuration of the generating register. If in this process a new tap falls on an already existing tap, this tap position can be dropped, since the product of the same sequence with itself always contributes a factor of 1. One easily verifies that this operation between new taps and already existing taps is equivalent to modulo 2 addition or the logic operation of exclusive OR. This generation of tap configurations corresponding to consecutive M-sequence shifts can be implemented in a separate shift register of length k which will be called a tap register (Figure 6).

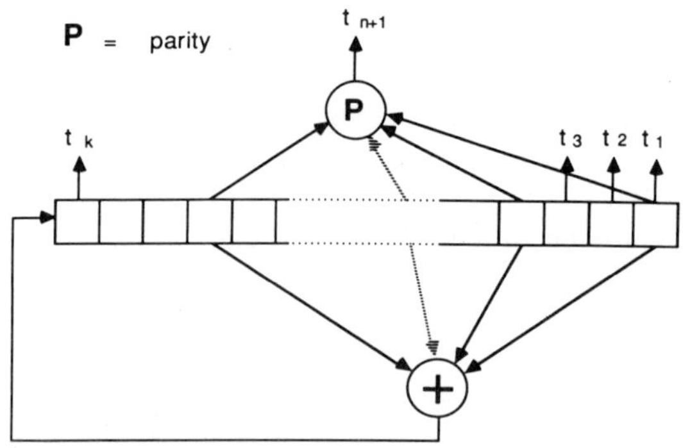

FIGURE 5. The generation of M-sequence lags by means of taps on the M-sequence generating register. Moving the output tap to the left leads to an advance in the sequence. An advance by *n* steps is generated by the feedback taps. Further advances are generated by left-ward displacement of all the existing taps and an exclusive OR operation with the feedback taps whenever a tap drops off the register.

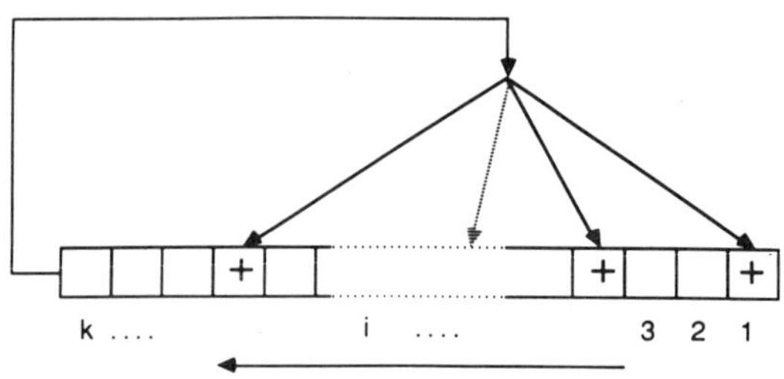

FIGURE 6. This shift register is called the tap register. When initialized with a single right-justified 1 and clocked by *n* steps, the configuration of 1s in the register specifies the stages that have to be summed modulo 2 to produce the M-sequence advanced by *n* steps. For details see text.

The tap register is shifted left one step for each shift in the M-sequence, and the serial output of the register is added modulo 2 to the register stages at the positions of the feedback taps that define the M-sequence. If the tap register is initialized with a right justified *1* and clocked by n steps, then the content of the register specifies the unique auto-product of M-sequence shifts smaller than *k* that corresponds to the shift of n steps. The number of 1s in the register indicates the number of factors in the auto-product, and the positions of the 1s indicate the shift of the factor.

With the correspondence between auto-products and shifts of the M-sequence established by the tap register, we now know exactly where in the Walsh array we can find consecutive points of the cross-correlation cycle. When initialized with a left justified 1 and clocked through $2^k - 1$ steps, the tap register produces the addresses for consecutive points for the entire cycle. In this procedure kernel slices are sequentially generated in the right order. For each bit pattern that travels through the register, a kernel slice is produced. When the first 1 in the pattern drops off the register, a new bit pattern appears in the register, marking the beginning of a new kernel slice on the cross-correlation cycle.

It follows from the above that cross-correlations between a binary M-sequence of length $2^k - 1$ and any real sequence of the same length can be obtained by means of a single FWT using the following simple procedure.

Binary M-Sequence Procedure

1. The temporal modulation of the stimulus is derived from the first stage of the generating register. The recorded data point is stored in an array at the binary address specified by the content of the generating register (compare: Deterministic Binary Procedure, Step 1, Section IV).
2. A Walsh transform is performed on the full data array.
3. The tap register is initialized with a single 1 in its input stage. The cross-correlation cycle between the recorded analog signal and the M-sequence used for stimulation is read from binary Walsh array addresses specified by consecutive tap register configurations.

In Figure 7 the procedure is illustrated using the example of an array size of 2^3. Note that the operations of loading and reading the Walsh matrix are equivalent simple permutations.

When implemented by means of the FWT algorithm this procedure has been called the fast M-transform. For a more detailed mathematical treatment see Sutter (1991).

3. M-Sequence Stimulation and Incomplete Data Sets

When the stimulus is not a binary M-sequence, the kernel slices computed by means of the Walsh transform (Deterministic Binary Procedure) are independent of each other, and their arrangement into a continuous loop would make no sense. However, with M-sequence stimulation they have a specific relationship defined by their position on the cross-correlation cycle. This becomes important when the memory of the system is larger than k time bins. In this case a specific kernel slice on the cross-correlation cycle does not end when the next slice begins. If both slices are nonzero, then they contribute additively in the region where they overlap. If kernel slices are to be resolved on the basis of a single M-sequence data set, such overlap has to be avoided.

Since the sequencing of the kernel slices is determined by multiplicative properties of the M-sequence, the length of the sequence as well as the particular choice of the feedback taps become important. Fortunately, for each length k of the generating register, there are usually many different feedback tap configurations that produce M-sequences. There are many relatively short M-sequences that place the nonzero kernel slices of a particular class of systems sufficiently far apart on the cross-correlation cycle to prevent overlap. Such a sequence may be found through a simple search. Using the tap register (Figure 6 and Appendix A code section: Read Array) M-sequences can be scanned in seconds for the proper spacing of certain kernel slices. For instance, if one knows that kernels of order higher than three are negligible in size, one simply keeps track of the number of 1s in the register and counts the number of steps between the beginning of configurations where the count is less than four.

An M-sequence test of a given length can, thus, yield the exact binary kernels for either a low-order system with a large input window or a higher-order system with a large number of nonzero kernel slices and a short input window. In the simple case of a linear system, the input window can be as long as the M-sequence itself. When the window is equal or shorter than k, there are no restrictions on the degree of nonlinearities, since the test is complete for this window size and delivers all the binary kernels.

The usefulness of the M-sequence technique in deriving exact kernels of specific types of systems from small records is shown in Figure 8. Here an appropriately selected sequence of length $2^{11} - 1$ was applied to a second-order Wiener model. The model was similar to that of Figures 1 and 2, except for a reduction in the damping factor and a corresponding increase memory size of 25 bins. Despite an increase in the number of first- and second-order kernel elements from 136 to 325, perfect orthogonality was achieved. Figure 8a shows

row	t_i	col.
0 →	000	→ 0
1 →	001	→ 1
2 →	010	→ 2
3 →	100	→ 4
4 →	011	→ 3
5 →	110	→ 6
6 →	111	→ 7
7 →	101	→ 5

zero cycle (row 0); m-sequence cycle (rows 1–7)

col.	g_i	row
0 →	000	→ 0
1 →	001	→ 1
2 →	100	→ 4
3 →	010	→ 2
4 →	101	→ 5
5 →	110	→ 6
6 →	111	→ 7
7 →	011	→ 3

$$\underbrace{\begin{bmatrix} 1&0&0&0&0&0&0&0\\ 0&1&0&0&0&0&0&0\\ 0&0&1&0&0&0&0&0\\ 0&0&0&0&1&0&0&0\\ 0&0&0&1&0&0&0&0\\ 0&0&0&0&0&0&1&0\\ 0&0&0&0&0&0&0&1\\ 0&0&0&0&0&1&0&0 \end{bmatrix}}_{[P_2]^T} * \underbrace{\begin{bmatrix} +&+&+&+&+&+&+&+\\ +&-&+&-&+&-&+&-\\ +&+&-&-&+&+&-&-\\ +&-&-&+&+&-&-&+\\ +&+&+&+&-&-&-&-\\ +&-&+&-&-&+&-&+\\ +&+&-&-&-&-&+&+\\ +&-&-&+&-&+&+&- \end{bmatrix}}_{[WT]} * \underbrace{\begin{bmatrix} 1&0&0&0&0&0&0&0\\ 0&1&0&0&0&0&0&0\\ 0&0&0&1&0&0&0&0\\ 0&0&0&0&0&0&0&1\\ 0&0&1&0&0&0&0&0\\ 0&0&0&0&1&0&0&0\\ 0&0&0&0&0&1&0&0\\ 0&0&0&0&0&0&1&0 \end{bmatrix}}_{[P_1]}$$

$$= \underbrace{\begin{bmatrix} +&+&+&+&+&+&+&+\\ +&-&+&+&-&+&-&-\\ +&+&+&-&+&-&-&-\\ +&+&-&+&-&-&-&+\\ +&-&+&-&-&-&+&+\\ +&+&-&-&-&+&+&-\\ +&-&-&-&+&+&-&+\\ +&-&-&+&+&-&+&- \end{bmatrix}}_{[MT]}$$

FIGURE 7. The binary M-transform illustrated with the example of an M-sequence of length $2^3 - 1$. The generating register defines the permutation of the data points (permutation matrix P_1) that can be performed during loading of the array. The tap register defines the permutation of the data points (P_2) performed on reading the result from the array.

the entire cross-correlation cycle with the distribution of first- and second-order kernel slices. Because all higher-order kernels are zero, the second-order kernel shown in Figure 8b is exact, except on the diagonal, where it has been interpolated.

4. Orthogonalization Procedures

When it is not possible to avoid overlap problems with the proper choice of the M-sequence, there are various techinques that permit orthogonalization of certain kernel slices.

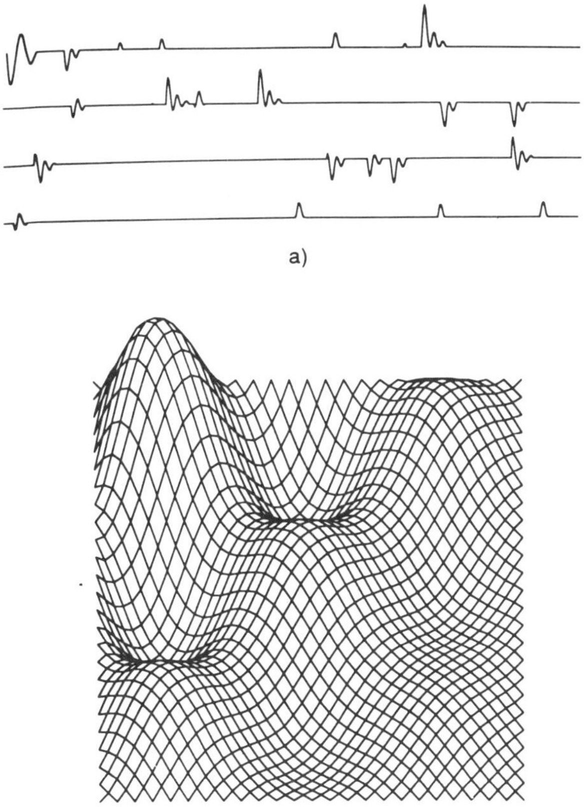

a)

b)

FIGURE 8. (a) These traces represent the complete cross-correlation cycle obtained from the Wiener model with a memory of 25 bins. Even with a short M-sequence (2047) perfect orthogonality of all nonzero kernel slices was achieved. The missing diagonal elements have been filled through interpolation.

The most common one is the inverse repeat techique that has frequently been used with the Lee and Schetzen approach (Ream, 1970). One simply repeats the experiment with the inverse signal polarity. Because of the symmetry properties of the kernels, the cross-correlation cycle derived from the difference of the two data sets only contains odd-order kernel slices. Similarly, the even-order kernel slices are isolated on the cross-correlation cycle derived from the sum of the data sets.

This technique is particularly helpful because even- and odd-order kernel slices alternate along the cross-correlation cycle. The inverse repeat method will, thus, always vacate some space after each nonzero slice. The interleaving of even and odd slices is a consequence of the fact that an even number of shift register feedback taps is needed to generate an M-sequence. One easily verifies that the number of 1s in the tap register (Figure 6) always changes by an odd number at the beginning of a new slice.

Time-reverse repeat is a method for orthogonalization that permits separation even when the two overlapping slices are both of even or both of odd order. The experiment is simply repeated with the M-sequence generated in the reverse direction. Such time reversal also reverses the sequencing of the kernel slices on the cross-correlation cycle. If, in the first experiment, the end of kernel slice A ran into the beginning of kernel slice B, then kernel slice B now runs into kernel slice A. The algebraic separation of the two is trivial as long as the overlap is less than 50% of the longer slice.

Note that the time reversed M-sequence can be generated by means of the tap register of Figure 6 simply by reversing its operation. It is now shifted right, and whenever a *1* reaches the right-most stage (stage 1), a *1* is added modulo 2 to the stages at the feedback positions as well as to stage $k + 1$.

The goal of all these techniques is to achieve exact orthogonality of specific kernel slices in relatively short tests that are incomplete for the input window of the system.

5. The Anomaly Problem

Binary M-sequences have over the years acquired a bad reputation in nonlienar systems analysis. The opinion now widely held is that they are afflicted with serious inperfections and that they do not have the randomness properties that would make them suitable for the task. It follows from the considerations presented in the previous sections that quite the opposite is true. The intricate nonrandom structure of these sequences makes them particularly useful. In an effort to vindicate the much maligned sequences, this section is devoted to reconsideration of the so-called flaws which have led to much controversy in the past.

Binary M-sequences were originally introduced into linear systems analysis because of their whiteness properties (Briggs et al., 1964) in an effort to reduce the record length. Excellent impulse responses were obtained using input-output cross-correlation. Gyftopoulos and Hooper (1964) demonstrated that good estimates of second-order kernels could be obtained from much shorter records than those needed by Widnall (1962) with random stimulation. However, strange bumps appeared in the cross-correlation functions that could not be part of the kernels themselves. These apparent contaminations which depended on the presence of nonlinearities were traced to so-called anomalies in the auto-correlation functions (Ream, 1970; Simpson, 1966). Barker et al. (1970, 1973) studied the auto-correlation properties of test inputs derived from M-sequences and computed the location of anomalies relevant for the second-order kernel by means of binomial division.

In the light of the deterministic considerations presented in the previous sections (see also Sutter, 1987), the origin of the anomaly problem in the case of binary M-sequences is quite obvious. Since the binary M-sequence is identical up to a shift with all its auto-products (assuming nonzero shift of factors), none of their mutual cross-correlation functions are zero. In fact, they also are all identicial up to a shift. All the kernel slices are, thus, segments of the same cross-correlation cycle and can be orthogonal only up to their allotted length. We have also seen that when the length of the M-sequence is $2^k - 1$ where k is the size of the input window, then none of the kernel slices can exceed its allotted length, and perfect orthogonality is guaranteed.

It had been noted that the undesirable "artifacts" appearing in cross-correlation functions of kernel slices were dependent on the presence of nonlinearities (Swerup, 1978), but researchers were apparently not aware that these artifacts were actual overlapping kernel slices. On the other hand, it was known from the recursive generation of M-sequences that there is a correspondence between product sequences and shifts, but the connection between the two facts was apparently not seen. The dependence between the specific choice of the M-sequence and the location of the "anomalies" had been observed and efforts were made to identify the best choices feedback taps (Barker and Pradisthayon, 1970), but the computations by means of polynomial division were cumbersome. Note that the simple relationship between shifts and product sequences defined by the tap register Figure 6 permits computation of the location of all these "anomalies" in seconds.

Much of this oversight can be explained by the mind-set of the time, which was mostly influenced by the Lee and Schetzen approach and the underlying stochastic considerations. M-sequences were investigated primarily as a replacement for truly random test sequences which are much more difficult to generate. It was generally assumed that the best pseudo-random sequences for nonlinear systems analysis were those whose auto-correlation properties most closely resembled those of random sequences.

It was known that the orthogonality of the kernels and their elements breaks down when the test is finite in length, and the effects of this breakdown on the kernel quality were estimated through stochastic analysis (see, e.g., Klein, 1991; Marmarelis and Marmarelis, 1978). Various methods were developed that permit more accurate kernel estimation approaching a true least mean squares approximation. However, all these techniques work well only if the input ensemble is adequately covered by the random stimulus. In vision research the ensemble can be very large, indeed, and good coverage of the relevant subset often requires specially designed deterministic tests. This is where M-sequences can play an important role.

The following simple Gedanken experiment will demonstrate that good randomness properties of the test input can lead to insufficient resolution in the Lee and Schetzen kernel estimates when dealing with highly complex systems. Consider a finite test sequence generated by a random white process. If the cross-correlation method is to resolve accurately all the kernel elements within each kernel slice and without restriction on the length of the slices, the test sequence and all its product sequences have to be white. For the sake of argument we assume that we were very lucky and that this is, indeed, the case. A difficulty now arises from the fact that any two different white sequences of finite length cannot be mutually orthogonal or uncorrelated. Each white sequence in all its cyclical shifts represents a complete orthogonal set of functions or basis in the function space over the test interval. The cross-correlation cycle of the response of the system with any one of the normalized product sequences represents the expansion of the response in an orthonormal basis. Because of the completeness of each basis the cross-correlation functions must contain the entire response power, i.e., the contributions from all kernel slices. While the contribution from the kernel slice associated with a given product sequence is found near the beginning of its cross-correlation cycle, the power of all the other nonzero kernel slices is expected to be distributed randomly over the entire cycle. Some of it will appear superimposed on the kernel slice and, thus, contaminate it. How serious this problem is can be roughly estimated if we know the power P_{ks} and the length delta T_{ks} of the kernel slice, the total response power P_{tot}, and the duration of the test T_{test}. The signal-to-noise ratio of the kernel slice is approximately

$$\frac{S}{N} = \frac{P_{ks}}{(P_{tot} - P_{ks})} * \frac{T_{ks}}{T_{test}} \tag{31}$$

This Gedanken experiment is slightly unrealistic, since random stimulation results in finite sequences and product sequences that deviate from exact whiteness. The result is some degree of mutual contamination of the kernel elements within each slice when computed by means of simple cross-correlation.

It is clear from these considerations how, with random stimulation, the quality of the Lee and Schetzen kernel estimates is limited by the recording time and the number of nonzero kernel elements. This problem is particularly serious in the visual system where the study of interactions between multiple inputs may require computation of a very large number of such elements.

Is it possible to improve the quality of kernels using nonrandom stimulation? In addition, is it possible to achieve exact orthogonality between kernel slices; i.e., can the contamination of a particular kernel slice by all the others be removed? Perhaps test sequences can be designed for which the contaminiating power from all the other kernel slices is not randomly distributed over the cross-correlation cycle but comes to lie outside the region where the kernel slice is located. It would be particularly desirable if the contaminations took the shape of the contaminating kernel slices themselves. In this case kernel slices of all orders would be lined up, one after the other, along the same cross-correlation cycle. One would have only to make sure that the cycle is long enough to accommodate them all.

As shown in the previous section, this is exactly what the intricate structure of binary M-sequences achieves. As long as the kernel slices do not overlap, they are perfectly orthogonal and the resulting binary kernels are exact. If a test of length $2^k - 1$ is possible where k is the size of the input window, then nonoverlap is guaranteed. If not, then it can usually be avoided through careful selection of the M-sequence or through suitable orthogonalization procedures.

The contamination due to the so-called anomalies is clearly related to the fact that the number of kernel elements that can be derived from a data set can never exceed the number of data points measured. Creative application of M-sequences may permit us to get close to the theoretical limit.

The anomalies are really an orthogonality problem that is in no way unique to M-sequence stimulation. With random stimulation the mutual contamination of kernel slices and kernel elements within slices exhibits itself as noise and is difficult to distinguish from contamination by extraneous noise. The experimenter may, thus, never appreciate what could be achieved with a better testing strategy. It is worth noting that the cross-contamination between kernels always has the temporal frequency spectrum of the contaminating kernel slices. It becomes, thus, a perfect mask for kernel slices with similar spectra and aggravates the problem of signal detection.

X. MULTIPLE INPUT SYSTEMS

The visual system is a multiinput system. In principle, every receptor on the retina represents an input channel. The receptors themselves are coupled with each other, and the signals they produce interact in many ways as they find their way through the retina and up the visual pathway. These interactions are essential in understanding the processes that lead to integration of the visual scene and to visual recognition. The interactions can be studied on the level of single receptors, receptor types, groups of receptors, or small retinal patches. In all cases the biological response depends on the spatial as well as temporal characteristics of the stimulus. Multiinput nonlinear analysis is, thus, a very important tool.

Early two-input experiments were carried out by the Caltech group in the early 1970s. Marmarelis and McCann (1973) studied motion detection in the insect visual system using Gaussian white-noise stimulation of two spot stimuli. Similar experiments were performed on bipolar cells in the catfish retina using independent spot and annulus stimulation (Marmarelis and Naka, 1974). But the small number of inputs used in such experiments can provide only very limited information on the spatiotemporal properties of receptive fields. The increasing interest in the complex receptive field properties of single cortical neurons in the mammalian visual system prompted the development of techniques that permit greater spatial resolution. The first full spatiotemporal analysis of cortical receptive fields was performed using optical processing techniques (Sutter, 1974 and 1975). It was capable of good spatial and temporal resolution, but could only be carried out up to first-order kernels.

In recent years more extensive first-order spatiotemporal receptive field studies have been reported by various authors (Citron and Emerson, 1983; Cummings et al., 1990; Freeman and Ohzawa, 1990; Reid et al., 1989). In most cases pseudorandom spatiotemporal noise was used for stimulation, in an effort to emulate true random white noise.

Awareness of the enormous extent of spatiotemporal input ensembles and the extremely sparse coverage that can be achieved has prompted the search for more efficient testing strategies and utimately led to the deterministic methodology presented in this chapter. The deterministic multiinput techniques outlined in this section have been extensively tested in studies of retinal and cortical visual evoked responses in man.

A. GENERAL CONSIDERATIONS

The testing of a system with multiple inputs is, in principle, no different from testing a system with a very large input window that is the union of the input windows of all inputs. Unless we are able to make some special assumptions about the system, we have to consider the possibility that nonlinear interactions exist not only between the time bins belonging to the same input, but also between bins of different inputs. Nonlinearities involving different inputs will be called lateral interactions.

Consider, for example, a two-input system where the two inputs have window size r and s, respectively. The Volterra expansion of the system is given by

$$\begin{aligned} R_t = K^{(0)} &+ \sum_{i=1}^{r} K_i^{(10)} x_i + \sum_{i=1}^{s} K^{(01)} y_i \\ &+ \sum_{i=1}^{r}\sum_{j=1}^{r} K_{i,j}^{(20)} x_i x_j + \sum_{i=1}^{r}\sum_{j=1}^{s} K^{(11)} x_i y_j + \sum_{i=1}^{s}\sum_{j=1}^{s} K^{(02)} y_i y_j \\ &+ \sum_{i=1}^{r}\sum_{j=1}^{r}\sum_{k=1}^{r} K_{i,j,k}^{(30)} x_i x_j x_k + \sum_{i=1}^{r}\sum_{j=1}^{r}\sum_{k=1}^{s} K^{(21)} x_i x_j y_k + \cdots \end{aligned} \tag{32}$$

The upper indices specify the order of the kernels in each input, i.e., the number of lower indices belonging to the input. Kernels containing only indices for one of the inputs (only one upper index is nonzero) are called self-kernels, while the others are called mutual or cross-kernels.

This expansion is formally identical to the expansion (Equation 10) for a system with an input window of size $r + s$, where

$$x_{r+1} = y_1,\ x_{r+2} = y_2,\ \cdots,\ x_{r+s} = y_s \tag{33}$$

In terms of the representation (Equation 10), the fields in the upper index represent the range of one of the lower indices.

$$K^{(O)} = K^{(O_1O_2)}\ O = O_1 + O_2 \tag{34}$$

The kernels of the multiinput expansion (Equation 32) are simply segments of the kernels of expansion (Equation 11). Note that the kernels of Equation 11 are symmetrical with respect to permutation of all their lower indices, while those of Equation 32 are symmetrical only with respect to permutation of indices belonging to the same input.

It is clear from this relationship between the two series how the correspondence between binary kernels and Volterra kernels (Equations 13 and 14) can be extended to multiinput systems. The mutual kernels of order 1 in each input correspond to off-diagonal elements in the expansion of the representation of Equation 10 and do not contribute to lower-order binary kernels.

It is often still believed that independent random stimulation of all the inputs would provide the best coverage of the input ensemble. However, with an increasing number of inputs and the possibility of lateral interactions the number of nonzero kernel elements quickly approaches the maximum number of data points that can be recorded in the available recording time. Relying on chance to cover the input configurations necessary to discriminate all the nonzero elements seems less promising, particularly since the configurations are selected from a set of astronomical size. With only a dozen inputs the input window may easily comprise two or three hundred time bins, and even with restriction to binary stimulation the duration of a complete test would dwarf the age of the universe. To extract the relevant information by means of such a sparse test may require careful design of the input sequence.

To see the problem more clearly, we apply the Gedanken experiment of Section IX.B.5 to the multiinput case. We assume that each input is, in fact, stimulated with an independent random binary sequence and that the sequences for all the inputs happen to be exactly white over the duration of the test. We assume further that all the auto-products and mutual products of the input sequences are also exactly white. Since each one of these normalized, white sequences in all its shifts represents an orthogonal basis in the function space over the recording interval, the power in its cross-correlation cycle with the response is equal to the total response power (see also Section IX.B.5). While the kernel slice associated with a particular white sequence is located in a small region at the begining of the cycle, all the rest of the power is randomly distributed. With many inputs and the presence of nonlinearities each slice may contribute only a very small amount to the total response power. Unless the stimulus sequence is extremely long, the signal-to-noise ratio of a slice estimated according to Equations 27 and 31 must be small. Just as in the single input case, this cross-contamination between kernel slices has the appearance of additive noise with the temporal spectrum of the contaminating kernel slices (see Section IX.B.5). With multiple inputs the problem is now much more severe. Not only is the contamination now multiplied by the number of inputs, but there may also be a large number of nonzero mutual kernels between different inputs.

B. PARTIAL ORTHOGONALIZATION THROUGH USE OF THE SAME SEQUENCE FOR ALL INPUTS

When the number of inputs is large and the number of nonzero kernel slices for each input and between inputs is relatively small, better orthogonality of the kernel slices (less cross-talk between kernels) is achieved when the same white sequence is used for the stimulation of all inputs. In order to render the response contributions from the different inputs uncorrelated it is necessary to introduce a sufficient relative lag between the stimulation of different inputs. These lags have to be equal or larger than the memory associated with the inputs.

To exhibit this more clearly let us consider the simple example of a linear system with 100 different inputs, each having an input window of 64 time bins. Computing the 100 first-order kernel slices that comprise a total of 6400 elements requires collection of a minimum of 6400 data points. Assuming further that we have found a stimulus cycle that is exactly white over its length of 10,000 points, we now stimulate each input with this sequence, with the starting points for consecutive inputs on the cycle delayed by 100 steps. One easily sees that the first-order kernels (in this case impulse responses) of all 100 inputs are found on the cross-correlation cycle of the response with the white simulus sequence. They are all lined up, one after the other, at invervals of 100 points. They are exact, since they do not overlap or contaminate each other in any way.

In order to achieve accuracy it is important that a full cycle of valid data is recorded and that the cross-correlations are performed with cyclical shifts (wrap-around). No windowing of any kind should be used.

Consider now the slightly more complex but also more realistic example of a nonlinear 100-input system with only three nonzero self-kernel slices for each input, namely, the first-order kernel and three second-order kernel slices. The total number of nonzero kernel elements is now approximately 25,600. Clearly, a test comprising 10,000 data points will no longer be sufficient to resolve all these elements, as it will, at best, result in a system of 10,000 equations with almost 25,600 unknowns. If the test is not enlarged, the missing information will result in very severe mixing of kernel slices, which may be hard to distinguish from noise contamination. In fact, when the first-order kernels are computed, the total power of all the second-order kernel slices is now superimposed on the cross-correlation cycle, and almost two thirds of it falls on the intervals where the kernels are located.

This cross-contamination between kernels of first and second order can readily be dealt with. One simply orthogonalizes even- and odd-order kernels by means of the inverse-repeat technique discussed earlier (Section IV.B.4). The same test is repeated with the opposite signal polarity. In this second experiment the contributions from odd-order kernels are reversed while those from even-order kernels remain the same. Even- and odd-order kernels are then separated by addition and subtraction of the two data sets.

In this example the inverse repeat technique eliminates all contamination of the first-order kernel by the two second-order kernel slices. In the absence of extraneous noise they should, thus, be exact since the first-order kernel is the only nonzero kernel of odd order. However, the three second-order kernel slices are still not discriminable. By doubling the data set through inverse repeat only 10,000 of the 20,000 equations contain the 19,200 nonzero even-order kernel elements. We are, thus, clearly dealing with an underdetermined system of equations. Further enlargement of the data set is necessary for computation of exact second-order kernels. Using a longer white sequence certainly helps but is in itself not sufficient to produce orthogonality between the second-order kernel slices. Orthogonalization of the three second-order kernel slices using methods similar to the inverse repeat technique is possible but will not be discussed here. Instead we will turn to M-sequence stimulation of multiinput systems.

It has become clear through the above Gedanken experiment that significantly better performance can be achieved when the same white sequence is used for all inputs. One thereby avoids all cross-contamination between the same kernel slices of different inputs. Note that the improvement over independent random stimulation is larger when the number of nonzero kernel slices is small.

For simplicity, effects of nonwhiteness of finite random sequences and their auto-products have been neglected in the above analysis. However, this has little bearing on the overall validity of the arguments.

C. THE M-SEQUENCE APPROACH TO MULTIINPUT SYSTEMS

1. The Basic Technique

With M-sequence stimulation the same fundamental criterium applies as with any other type of stimulus: the minimal number of data points required for computation of the kernels has to be equal to or larger than the total number of nonzero kernel elements. The main advantage of M-sequences is that it is generally possible to achieve better kernel separation with a limited data set. The control that these sequences can provide over the orthogonality of specific kernel slices (see Section IX.B) makes them particularly useful for the investigation of multiinput systems. The great computational advantages M-sequences offer in signal generation as well as in kernel derivation will also be discussed in this section.

Binary M-sequences are realizations of the hypothetical, perfectly white sequences with white auto-products considered in the above example. Their special multiplicative properties make it possible to achieve exact orthogonality between kernel slices belonging to the same input. Because of the relationship between products and lags of M-sequences, single input kernel slices of all orders are located on the same cross-correlation cycle of the M-sequence and the response (Sections IX.B.1, X.C.3). The relative location of kernel slices is given by the specific M-sequence used and can be controlled, at least to a certain extent, by the choice of the sequence.

To achieve the best possible equivalence in the stimulation of multiple input channels it is desirable to distribute the channel lags (see Section X.B) equally over the entire stimulus cycle. The sets of self-kernel slices belonging to different inputs are then found on the cross-

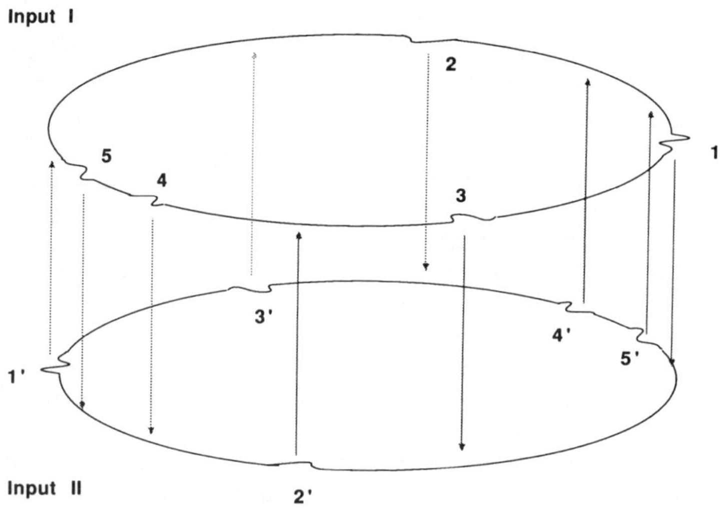

FIGURE 9. The problem of kernel slice overlap in the multiinput case illustrated with the example of a two-input system. The two inputs are stimulated with the same M-sequence with l/cycle relative lag. It is assumed that each input produces 5 nonzero kernel sliced shown separately on the upper and lower cycle. The cross-correlation cycle obtained from a two-input experiment is the sum of both cycles. With a lag of the $^1/_2$ cycle none of the slices from the two inputs overlap in this example.

correlation cycle shifted relative to one another by the channel lag. By carefully selecting the M-sequence one attempts to avoid overlaps between kernel slices belonging to different inputs. This overlap probelm is illustrated in Figure 9 for the simple case of a two input system with five non-zero self-kernel slices for each input.

The only method known at this time for selection of an M-sequence for which there is no overlap between the nonzero kernel slices of a specific mutiinput system is random search. However, because of the fast computation of the entire cross-correlation cycle by means of the M-transform (see Section IX.B.2; and Sutter, 1991) a good choice can often be found in minutes. For each candidate sequence one simply fills the cross-correlation cycle, i.e., the entire Walsh array, with a given constant. Sequentially one then extracts the sets of kernel slices that are expected to carry significant power in a real experiment. Each time only one slice is selected for each input or for each input pair when mutual kernel slices are considered. In this process one always replaces the accessed kernel slices by zeros. If any of the kernel slices tested show stretches of zeros, then there is overlap. It is easy to write a short computer program that tests a number of candidate sequences automatically.

When a sequence has been found for which there is no overlap between the nonzero kernel slices, their orthogonality is guaranteed, and the kernal computation is exact.

2. Orthogonalization Procedures

When overlap of kernel slices cannot be completely avoided through careful choice of the M-sequence, there are many different orthogonalization procedures that permit separation of the slices. Most such techniques take advantage of some information known about the type of system at hand; others such as the inverse repeat and time reverse repeat methods discussed in Section X.B have more general applicability.

With multiple input systems there exists a generalized version of the inverse repeat technique that is particularly useful when the number of inputs is small. The signals to each input can be inverted separately. With n inputs this leads to a total of 2^n different data sets. By adding and subtracting the sets. i.e., by forming all 2^n linear combinations with coefficients of $+1$ and -1, it is possible to orthogonalize a large number of kernel slices. This generalized inverse repeat technique can be very useful in the presence of strong lateral interactions.

In the visual systems the set of multiple inputs used commonly corresponds to retinal locations. Certain sensible assumptions can often be made based on their topological organization. If, for example, lateral interactions have a limited range, then the number of mutual kernels and, thus, the total number of possible nonzero kernel slices is greatly reduced. There may be no need to use the generalized inverse repeat technique to its full extent, and one may choose to orthogonalize only between nearest neighbors. If the stimuli are elements in a square matrix, one might choose all possible stimulus polarities of even and odd rows and even and odd columns.

3. Generation of Multiinput M-Sequence Stimuli

When random stimuli are used for multiinput systems analysis, the stimulus sequence has to be generated ahead of time and stored in a large array for use in the generation of the relatively delayed stimulation of the individual inputs. Multiinput M-sequence stimuli, on the other hand, have the great advantage that they can be generated in real-time, even for a large number of inputs. This is due to the easy generation of evenly spaced channel lags and the fast and elegant recursive derivation of the M-sequence itself. It can be shown that, if the number of inputs is a power of 2, simple decimation of the sequence over the inputs will produce the desired evenly spaced delays. Consider, for example, the sequence of length 15:

$$\begin{matrix} m_1 & m_2 & m_3 & m_4 & m_5 & m_6 & m_7 & m_8 & m_9 & m_{10} & m_{11} & m_{12} & m_{13} & m_{14} & m_{15} \\ 0 & 0 & 0 & 1 & 0 & 0 & 1 & 1 & 0 & 1 & 0 & 1 & 1 & 1 & 1 \end{matrix} \qquad (35)$$

The distribution of consecutive elements m_i over four columns as indicated on the left in Expression 36 yields:

$$\begin{matrix} m_1\, m_2\, m_3\, m_4 & & 0 & 0 & 0 & 1 \\ m_5\, m_6\, m_7\, m_8 & & 0 & 0 & 1 & 1 \\ m_9 \cdots & & 0 & 1 & 0 & 1 \\ \cdots & & 1 & 1 & 1 & 0 \\ & & 0 & 0 & 1 & 0 \\ & & 0 & 1 & 1 & 0 \\ & & 1 & 0 & 1 & 1 \\ & & 1 & 1 & 0 & 0 \\ & & 0 & 1 & 0 & 0 \\ & & 1 & 1 & 0 & 1 \\ & & 0 & 1 & 1 & 1 \\ & & 1 & 0 & 0 & 0 \\ & & 1 & 0 & 0 & 1 \\ & & 1 & 0 & 1 & 0 \\ & & 1 & 1 & 1 & 1 \end{matrix} \qquad (36)$$

Each column contains the M-sequence (Expression 35) with an advance of four steps between consecutive columns. An exception is the advance between the last and first column, which is always one step short, as the M-sequence only contains $2^n - 1$ elements.

4. Quaternary Stimulation with Binary M-Sequences

The multilevel analysis technique using binary components discussed in Section VII can be considered a special application of multiinput analysis whereby each binary component of the stimulus is treated as a separate input. We will now briefly consider the special case where the two inputs are derived from a single binary M-sequence.

According to Sections IX.B.1 and X.C.3, a maximum lag between the two binary components is obtained if they are derived from two consecutive elements of the sequence. Thus, we always generate two consecutive elements of the binary M-sequence at a time. In accordance with Equation 29 we use the substitutions

$$\begin{array}{ll} 0 \rightarrow \ \ c & \qquad 0 \rightarrow \ \ 2c \\ 1 \rightarrow -c & \qquad 1 \rightarrow -2c \end{array} \tag{37}$$

for the first and second elements, respectively, and sum the two values. As an example, the quaternary sequence derived in this manner from the sequence in Expression 35 is shown in Expression 38. Note that a complete stimulation cycle would take us through the M-sequence twice.

$$\begin{array}{cccccccccc} \cdots & 0\,0 & 0\,1 & 0\,0 & 1\,1 & 0\,1 & 0\,1 & 1\,1 & 1\,0 & \cdots \\ | & | & | & | & | & | & | & | & | & \\ & 3c & 1c & 3c & -3c & 1c & 1c & -3c & -1c & \end{array} \tag{38}$$

To achieve complete orthogonality between the self- and mutual kernel slices of the two components, application of the generalized inverse repeat techique can be helpful. In this case four different data sets would have to be recorded using the four possible combinations of signal polarities for the binary components. For the extraction of a specific kernel all four response cycles are summed with a factor of $+1$ or -1. The sign for a specific set is determined by the order of the kernel in each component and the component polarities used in this record. If, for example, the desired kernel is of even order in the first component and of odd order in the second, then it will only change sign with a polarity change of the second component but not the first. One easily verifies that the following simple procedure yields the proper coefficient for each data set:

- A two-bit binary number b_i is assigned to each set. Each bit represents one of the components. A bit is set if the corresponding stimulus component was reversed for this record.
- The order of the desired kernel is similarly translated into a two-bit binary number k_j. Its bits are set if the order of the kernel in the corresponding component is odd.
- The sign of a specific set is equal to the parity of the logic AND of the b_i and k_j.

Note that this procedure is equivalent to application of the 4×4 Walsh transform matrix on a vector whose components are the four data sets ordered according to the binary number b_i (compare with the 8×8 case shown in Matrix 18). The matrix maps this vector into a vector whose components are the data sets corresponding to even/odd kernel order k_j. If the generalized inverse-repeat method were used with a large number of inputs one might consider evaluation of the transform using the FWT algorithm.

After application of this procedure, kernel slices with the different kernel order number k are located on the different cross-correlation cycles. For example, all second-order self-kernels are located on cycle 0. Cycles 1 and 2 contain the first- and third-order self-kernels of component one and two, respectively, and the second-order mutual kernels are found on cycle 3. This distribution greatly improves the chances for achieving orthogonality between all nonzero kernel slices, particularly in the presence of lateral interations.

5. Sparse-Stimulation

As one increases the number of stimulated inputs to the visual system, perhaps to record the response topography of the visual field or to study lateral interactions, one is confronted with a tremendous proliferation of nonzero kernel slices. One soon reaches the point where they can no longer all be resolved from data sets of manageable size. An additional problem arises even under perfect conditions where through judicious application of M-sequence stimulation and orthogonalization procedures all kernel slices are perfectly resolved: the distribution of the response power over such a large number of kernel elements leads to a poor signal-to-noise ratio in the kernels (see Sections VIII and IX.B.5). A solution to both these problems is found in methods that help reduce the number of kernel slices. One such method is sparse-stimulation.

Sparse-stimulation means that the rate of M-sequence stimulation is below the input cutoff frequency of the system. It reduces the number of kernel slices, since only pulses within the input window of the system can interact with each other. At half the stimulation rate this number also is cut in half, and the number of possible kernel slices is reduced to its square root. Since the sampling rate of the system response is not changed, the spacing of data points on each slice remains the same; however, the kernel slices are now found on a coarser grid perpendicular to the diagonal. The effect on a second-order binary kernel is shown in Equation 39. The numbers indicate the kernel slices and the symbols indicate the gaps. In this example the stimulation rate is 1/4 of the response sampling rate.

Clearly, the sparse stimulation with an M-sequence does no longer constitute a complete binary test. The reduction of kernel slices also means a reduction in information. It is, in general, no longer possible to derive the Volterra kernels from sparse kernel sets. However, certain relationships remain intact. It can be shown that for an nth-order system the sparse kernels of order n and n − 1 are points on the corresponding Volterra kernels. Under certain conditions it may be possible to approximate the Volterra kernels through interpolation. However, this may not be necessary since the sparse kernels can be directly interpreted. For a more detailed study of sparse kernels the reader is referred to the work of Klein (1992, Chapter 5, this volume).

$$
\begin{array}{cccccccccccccccc}
- & - & - & - & 1 & - & - & - & 1 & - & - & - & 1 & - & - & - \\
- & - & - & - & - & 4 & - & - & - & 4 & - & - & - & 4 & - & - \\
- & - & - & - & - & - & 3 & - & - & - & 3 & - & - & - & 3 & - \\
- & - & - & - & - & - & - & 2 & - & - & - & 2 & - & - & - & 2 \\
1 & - & - & - & - & - & - & - & 1 & - & - & - & 1 & - & - & - \\
- & 4 & - & - & - & - & - & - & - & 4 & - & - & - & 4 & - & - \\
- & - & 3 & - & - & - & - & - & - & - & 3 & - & - & - & 3 & - \\
- & - & - & 2 & - & - & - & - & - & - & - & 2 & - & - & - & 2 \\
1 & - & - & - & 1 & - & - & - & - & - & - & - & 1 & - & - & - \\
- & 4 & - & - & - & 4 & - & - & - & - & - & - & - & 4 & - & - \\
- & - & 3 & - & - & - & 3 & - & - & - & - & - & - & - & 3 & - \\
- & - & - & 2 & - & - & - & 2 & - & - & - & - & - & - & - & 2 \\
1 & - & - & - & 1 & - & - & - & 1 & - & - & - & - & - & - & - \\
- & 4 & - & - & - & 4 & - & - & - & 4 & - & - & - & - & - & - \\
- & - & 3 & - & - & - & 3 & - & - & - & 3 & - & - & - & - & - \\
- & - & - & 2 & - & - & - & 2 & - & - & - & 2 & - & - & - & -
\end{array}
\qquad (39)
$$

Many aspects of sparse stimulation with binary M-sequences are best considered in the framework of the deterministic binary approach. The contents of the input window of the system now no longer assumes all binary configurations of a given length, but only those of a very special subset. Consecutive elements of the M-sequence are now spaced n time bins apart. During the time one binary configuration travels through the input window n consecutive data points are collected, one for each of its shift positions in the window. There are, thus, n independent M-sequence data sets generated that can be placed in separate Walsh arrays. Each array produces its own binary kernel elements which are subsets of the sparse kernels. In the second-order kernel (Expression 39) the sparse elements are labeled with the number of the array from which they were taken.

The sparse analysis procedure can be summarized as follows:

1. The n consecutive points recorded in each stimulation interval are collected in n different Walsh arrays. They are all placed at the same array addresses specified by the previous k elements of the binary stimulus sequence (compare: Deterministic Binary Procedure Step 1, Section IV and procedure Load Array in Appendix A).
2. Each Walsh array is transformed into the kernel domain.
3. Walsh array addresses for a specific kernel slice are sequentially generated by means of the tap register (Figure 6). For each address n kernel elements are read consecutively from the n arrays (see example Expression 39).

6. Artifact Removal

The same technique introduced in Section VI for completion of partial data sets can be used very effectively for the elimination of certain types of signal contamination. Already known system properties such as fast convergence of the kernel series are used to detect and remove artifacts from the record. The technique described here works best if the contamination is localized in brief time intervals distributed over the record and if the total contamination extends only over a relatively small fraction (perhaps 10 to 15%) of the record length. It is ineffective if the contamination is uniformly distributed over the record. The technique is ideally suited, for example, to the removal of blink and eye movement artifacts from ERG records. No assumptions have to be made about the shape of the artifacts.

In traditional transient ERG recordings, artifacts are commonly removed by eliminating the contaminated trials, or, in steady-state recordings, by cutting out the contaminated segment. This approach is not advisable with the M-sequence technique. Here a complete record is needed for accurate kernel computation. Elimination of portions of the record would lead to an incomplete data set and call for the method described in Section VI. In order to include the signal superimposed on the artifact in the analysis we must start off with the entire contaminated record. Of course, this signal is only available if care is taken during recording that the artifacts do not saturate the amplifier or the analog-to-digital converter.

Consider, for example, a system with a fast-converging kernel series with only a few nonzero kernel slices. The information available for detection and elimination of the artifacts is, thus, that the artifacts are localized in the raw data set and that the response is localized in the kernel domain. On the other hand, the signal is fairly uniformly distributed over the raw data set while the artifacts appear randomly distributed over the entire cross-correlation cycle. By zeroing the entire cross-correlation cycle except for the segments where the nonzero kernel slices are located, most of the noise is eliminated. After a back transform into the raw data domain the artifacts must be greatly reduced while the signal is largely unaffected. The contaminated segments can, thus, easily be detected by comparison with the original data. Note that the back transform is based on the contaminated kernel slices. In the uncontaminated regions of the raw data domain the original signal is more accurate and is now restored. A subsequent transformation into the kernel domain yields greatly improved kernel estimates. This procedure can be iterated.

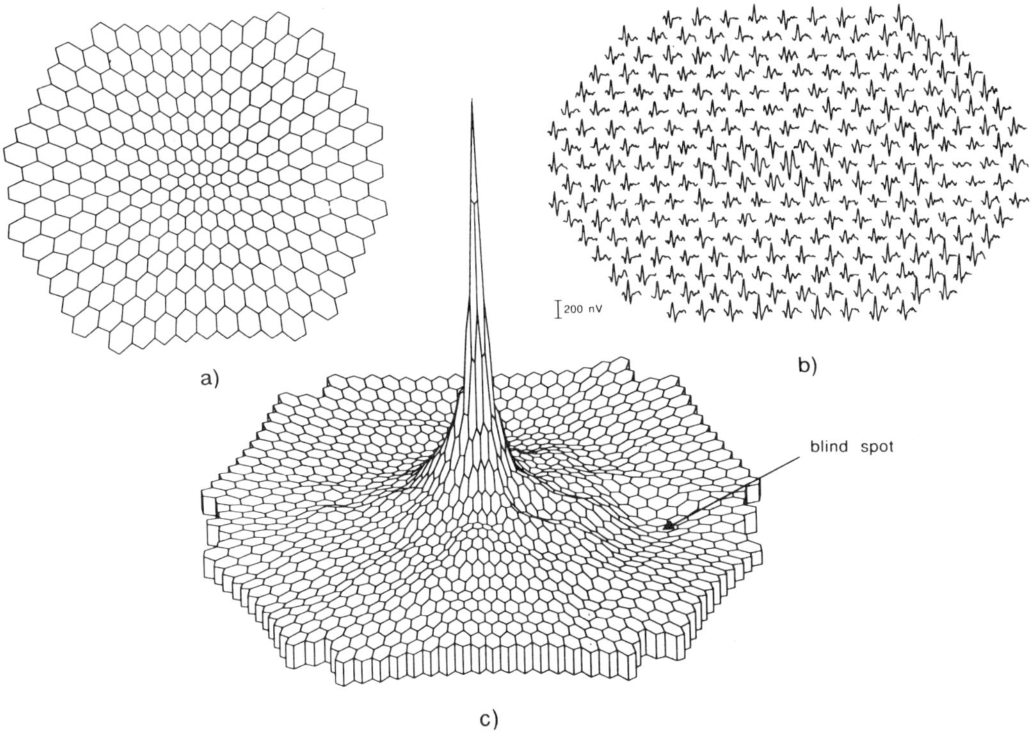

FIGURE 10. The response topography of the electroretinogram derived by means if the multiinput M-sequence technique. The length of M-sequence used was $2^{16} - 1$. The stimulation rate was 67/s and the sampling rate 436/s (sparse-stimulation). The stimulus was an array of 241 flickering hexagonal elements. (a) Stimulus array; (b) array of first-order kernel slices; (c) three dimensional representation of the response density. For details see text.

Practical application of this very general method of artifact elimination is made possible by the fast computation of the cross-correlation cycle by means of the fast M-transform (Sutter, 1991). Even for large data sets this algorithm permits moving back and forth between the raw data and kernel domains in seconds (see Appendix A). The effectiveness of this artifact elimination technique is shown in an example at the end of the next section.

7. Examples of Multiinput Analysis with Binary M-Sequences

Sparse-stimulation is commonly used in visual electrophysiology, often not by choice but because of technical limitations. The versatile CRT-based stimulation devices are capable of frame rates up to about 200/s but usually less. These rates are far below the input Nyquist rate for the system at hand. However, it follows from the considerations of Section X.C.5 that this may be a blessing in disguise. In the case of the multiinput ERG experiments of Figures 10 and 11 sparse-stimulation sufficiently reduced the number of kernel slices to permit high-resolution topographic representation of the responses across the visual field. In both of these experiments the same binary M-sequence was used to stimulate all 241 hexagons in the stimulus array shown in Figure 10a. All hexagons were modulated with the same binary M-sequence between the luminance levels 200 and 3 cd/m^2. The sequence length was $2^{16} - 1$. A lag of 1/256 of an M-sequence cycle between consecutive stimulus elements was obtained by means of the decimation technique (see Example 36). The stimulus was generated in real time and displayed on a CRT screen by means of a modified MacIntosh video board. The stimulation rate (frame rate) was 67/s. The net recording time was, thus, approximately 16 min. The record was obtained in slightly overlapping segments of tolerable

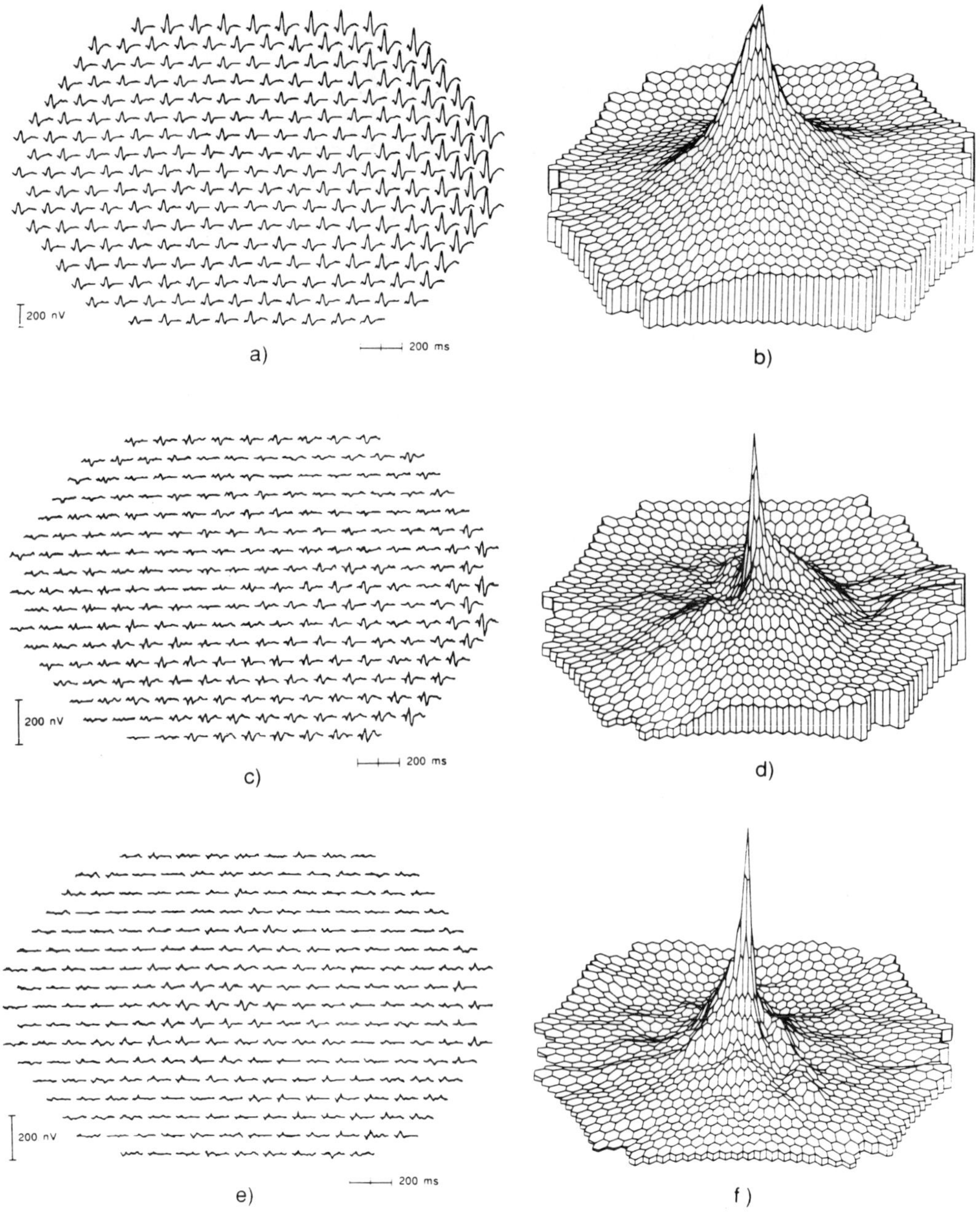

FIGURE 11. Nonlinear analysis of ERG response topography. With dilated pupil, three-kernel slices are detected in a record of the same length as that of Figure 10. (a and b) First-order kernel; (c and d) first slice of second-order kernel; (e and f) second slice of second-order kernel. For details see text.

length (approximately 30 s) and the segments were joined smoothly with a graded splice. The ERG response was derived from the cornea of the eye by means of a Burian-Allen contact lens electrode. The data were sampled at eight times the frame rate while the subject was fixating the center of the display. The processing time required to transform the data array of 2^{19} data points into the kernel array was 17.6 s using a Macintosh IIx.

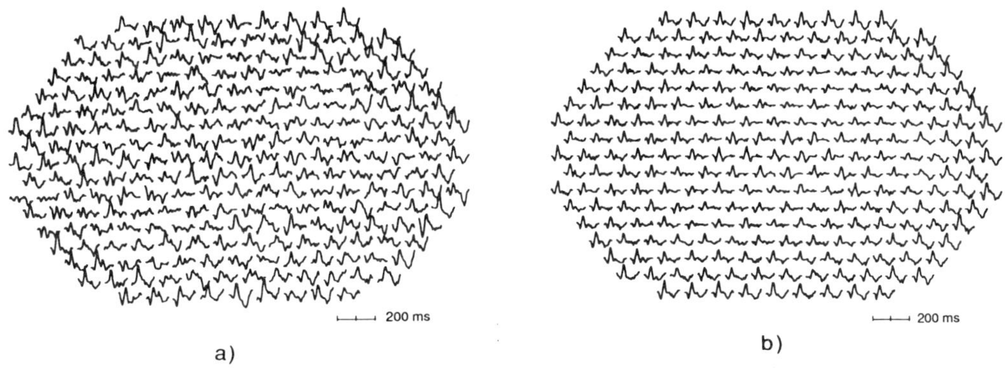

FIGURE 12. Illustration of the effectiveness of the artifact elimination procedure. (a) Response array derived from a record heavily contaminated with artifacts from eye movements and blinks; (b) the same array after removal of the artifacts. For details see text.

The record of Figure 10 was obtained with a natural pupil (approximately 3 mm). Because of the sparse-stimulation only the first-order kernel was significantly above the noise level under these conditions. The array of 241 first-order kernels is shown in Figure 10b. The time difference in the scanning of the hexagons from the top to the bottom of the screen has been corrected. The response amplitudes were measured as the scalar product of the waveform with a response template extracted from the entire data set (for details see Sutter and Tran 1992). Division of each amplitude by the area of the corresponding stimulus hexagon yielded the response density plotted in Figure 10c.

The data of Figure 11 was recorded under identical conditions but with dilated pupil. With the corresponding tenfold increase of retinal illuminance the first two sparse kernel slices of the second-order kernel were also slightly above noise level. The response topography of the first-order kernel is shown in Figure 11 a and b. The corresponding data for the first and second slice of the second-order kernel are found in Figure 11 c, d and d, e, respectively. For the M-sequence used in this experiment there was no overlap between the three nonzero kernel slices, and their discrimination did not require any special orthogonalization procedures such as inverse-repeat or generalized inverse-repeat. All the data shown in Figure 11 are obtained from the same 16-min record. Note the different density topographies of the three kernel slices.

The effectiveness of the artifact elimination procedure described in Section X.C.6 is exemplified by Figure 12. Here the technique was applied to a highly contaminated record. Stimulation and signal derivation were identical to that used in the experiment of Figure 10. With an M-sequence length of $2^{16} - 1$ and eight samples collected per display frame, the 16-min record comprised 2^{19} data points. In this example it was known that only the first-order self kernels carried significant power. Since they account for less than 2% of the cross-correlation cycle, the conditions for successful application of the technique are clearly met. Figure 12 a and b show the array of 241 first-order kernels before and after application of the procedure.

XI. SUMMARY AND CONCLUSIONS

In the early days of nonlinear systems analysis the view was held and promoted by many experts in the field that nonlinear systems are best analyzed with white Gaussian random noise. There was much talk about the good ''ergodic'' properties of such inputs, the even coverage of the input ensemble it produces, and the unbiased nature of the kernel analysis. While a major problem was seen in the tremendous computational effort of kernel derivation, it did not seem insurmountable. Special-purpose hardware for the computation of kernels was developed, and the advent of faster processors promised to cut the computation times down to acceptable size, at least for two or three low-order kernels.

The problems of nonlinear processing in the visual system have led to the realization that this early optimism was premature. The deterministic analysis presented in this chapter brings to light some of the difficulties and limitations of the white Gaussian noise approach, but it also points the way toward ultrafast methods of kernel computation and an arsenal of versatile tools for the design of more efficient testing strategies.

The major conclusions to be drawn from the deterministic considerations presented in this chapter can be summarized as follows:

1. With many complex single and multiinput systems encountered in vision research the input ensemble is much too large for a complete unbiased test. The ensemble coverage possible in a laboratory test is often extremely sparse. When the response power is distributed over a large number of kernel elements, kernel estimates depend not only on the system at hand, but also on the particular choice of the test input. The best results can be achieved by tailoring the test to the system, making use of known system properties. In such cases adaptable deterministic testing strategies can be extremely helpful.
2. Experimental bias is introduced not only by the choice of the test input but also by the representation of the test results. The use of a particular kernel series introduces a strong bias in favor of fast converging models. Even if higher-order kernels are computed, the distribution of the response power over a large number of kernel elements leads to very poor kernel estimates. It may be important to seek other representations of the system characteristics which lead to faster convergence. The deterministic binary approach may facilitate this process, since it yields all the binary kernels at once. Inspection of the series may suggest subsequent transformations of the data set. The Walsh transform may then be viewed as a first step in the analysis that separates the system properties from the properties of the test input. These possibilities are still largely unexplored.
3. When the response of a nonlinear system shows selectivity to a relatively small subset of input configurations, kernel computation can be counterproductive. Since each kernel element is computed as a weighted average of the entire data set, the response is watered down and can easily be lost in the noise. In such cases other data analysis techniques should be considered such as time-locking onto specific events, as suggested by Klein (1992). Once a preliminary knowledge of the selectivity of the system has been gained, deterministic tests can be designed that focus on the relevant regions of the input ensemble.
4. Most importantly, there is no universal technique in nonlinear systems analysis that can be successfully applied to complex systems. There is, thus, no replacement for the experimenter's skill and creativity in designing techniques appropriate for the variety of problems he encounters. It is useful, however, to have available an arsenal of different techniques that can be judiciously applied.

A major goal of this chapter is to increase this arsenal through the application of new concepts and the introduction of fast computational techniques. The approach is based on the following observations: (1) in the binary domain computations can be extremely fast and yield kernels of all orders at once; (2) in any given case numerous binary parametrization of the input ensemble are possible. The result is a set of highly adaptable techniques. On one hand it permits the design of highly efficient tests by means of binary M-sequences whose auto-correlation properties are known in every detail. On the other hand, arbitrary inputs can be used and even endogenous activity can be suitably parametrized for application of the binary algorithms. In both cases it is possible to incorporate prior knowledge about the system in the analysis. In the first case the test input can be tailored to cover those regions of the input ensemble which are of particular importance. In the second case the parametrization can be adapted to the known system properties. For instance, kernels can be sampled with a finer grid in regions where rapid changes are expected. Through methods of sparse-stimulation the techniques permit the trade-off of temporal resolution against spatial resolution (number of stimulated inputs). Similarly temporal resolution can be traded off against amplitude resolution (number amplitude levels in the stimulus) for the derivation of the diagonal elements of higher-order kernels. While such adaptability may be of little concern in the study of relatively simple single input systems, its importance in the investigation of the highly nonlinear multiinput visual system has been demonstrated.

ACKNOWLEDGMENTS

I thank Stanley Klein and Richard Miller for numerous helpful discussions during the preparation of the manuscript. Thanks are also due to Duong Tran for her software support and Lani Hardage for her assistance in editing. This work was supported in part by NIH grant EY06861 and the Smith-Kettlewell Eye Research Institute.

REFERENCES

Barker, H. A., and Pradisthayon, T. (1970). High-order autocorrelation functions of pseudorandom signals based on M-sequences, *Proc. IEEE,* **117**, 1857–1863.

Barker, H. A., Eng, C., and Obidegwu, S. N. (1973). Combined crosscorrelation method for the measurement of the 2nd-order Volterra kernels, *Proc. IEE,* **120**(1), 114–118.

Briggs, P. A. N., Hammond, P. H., Hughes, M. T. G., and Plumb, G. O. (1964-1965). Correlation analysis of process dynamics using pseudo-random binary test perturbations, *Proc. Inst. Mech. Eng.*, **179**, 37–50.

Citron, M. C. and Emerson, R. C. (1983). White noise anlaysis of cortical directional selectivity in cat, *Brain Res.*, **279**, 271–277.

Cummings, J. M., Palmer, L. A., and Wollman, D. E. (1990). First order spatiotemporal receptive field structures in cat striate cortex form a continuum, *Invest. Ophthalmol. Vis. Sci.*, Suppl. 31, 429.

Fino, B. J. and Algazi, V. R. (1976). Unified matrix treatment of the fast Walsh-Hadamard transform, *IEEE Trans. Comp.*, C-25, 1142–1146.

Freeman, R. D. and Ohzawa, I. (1990). On the neurophysiological organization of binocular vision, *Vision Res.*, **30**, 1661–1676.

Golomb, S. W. (1982). *Shift Register Sequences,* Aegean Park Press, Laguna Hills, CA.

Gyftopoulos, E. P. and Hooper, R. J. (1964). Signals for transfer-function measurements in nonlinear systems, in Noise and Nuclear Systems, USAEC/ Symp. Ser. *4, TID-7679,* 335–345.

Henderson, K. W. (1970). Comment on computation of the fast Walsh-Fourier transform, *IEEE Trans. Comp.*, C-19, 950–951.

Klein, S. (1979), Interpreting nonlinear systems: the third order kernel of the eye movement control system, 10th Annu. Pittsburgh Conf. on Modelling and Simulation, 215–221.

Korenberg, M. J. (1988). Identifying nonlinear difference equations and functional expansion representations: the fast orthogonal algorithm, *Ann. Biomed. Eng.*, **16**, 210–214.

Lee, Y. W. and Schetzen, M. (1965). Measurement of the Wiener kernels of a non-linear system by cross-correlation, *Int. J. Control,* **2**, 237–254.

Lee, Y. W. et al. (1961). *Statistical Communication Theory,* MIT, *NTIS,* 60.

Marmarelis, P. Z., and McCann, G. D. (1973). Development and application of white-noise modeling techniques for studies of insect visual nervous systems, *Kybernetik,* **12**, 74–89.

Marmarelis, P. Z. and Naka, K.-I. (1974). Multi-input statistical testing on physiological systems, in Proc. 27th Annu. Conf. Engineering in Medicine and Biology, Philadelphia.

Marmarelis, P. Z. and Marmarelis, V. Z. (1978). *Analysis of Physiological Systems,* Plenum Press, New York.

Moller, A. R. (1983). Use of pseudorandom noise in studies of frequency selectivity: the periphery of the auditory system, *Biol. Cybern.*, **47**, 95.

Ream, N. (1970). Nonlinear identification using inverse-repeat m sequences, *Proc. IEE,* **117**(1), 213–218.

Ream, N. (1967). Testing a 2-input linear system with periodic binary sequences, *Proc. IEE,* **114**(2), 305–308.

Reid, R. C., Shapley, R. M., and Victor, J. D. (1989). A new two-dimensional pseudorandom stimulus for the study of receptive fields in the LGN and striate cortex, *Soc. Neurosci. Abstr.*, **15**, 323.

Schetzen, M. (1974). A theory of non-linear system identification, *Int. J. Control,* **20**(4), 577–592.

Simpson, H. R. (1966). Statistical properties of a class of pseudorandom sequences, *Proc. IEE,* **103**, 2075–2080.

Sutter, E. E. (1991). The fast m-transform: a fast computation of cross-correlations with binary m-sequences, *SIAM J. Comput.*, **20**(4).

Sutter, E. E. (1974). Electrophysiological investigation of visual units in the cat: a revised conception of receptive fields, in Society for Neuroscience Conf. on Testing and Identification of Nonlinear Systems, St. Louis, MO, California Institute of Technology/NSF, 444.

Sutter, E. E. (1975). A revised conception of visual receptive fields based upon pseudorandom spatio-temporal pattern stimuli, in *Proc. 1st Symp. on Testing and Identification of Nonlinear Systems,* Pasadena, CA, California Institute of Technology, 353–365.

Sutter, E. E. (1987). A practical non-stochastic approach to nonlinear time-domain analysis in *Advanced Methods of Physiological Systems Modelling,* Vol. 1, University of Southern California, Los Angeles.

Sutter, E. E. and Tran, D. (1992). The field topography of ERG components in man: 1. The photopic luminance response, *Vis. Res.*, **32**(3), 433–446.

Swerup, C. (1978). On the choice of noise for the analysis of the peripheral auditory system, *Biol. Cybern.*, **29**, 97–104.

Victor, J. and Shapley, R. (1980). A method of nonlinear analysis in the frequency domain, *Biophys. J.*, **29**, 459–483.

Victor, J. D. and Knight, B. W. (1979). Nonlinear analysis with an arbitrary stimulus ensemble, *Q. Appl. Math.*, **37**(2), 113–136.

Widnall, W. S. (1962). Measurement of a second order Wiener kernel in a nonlinear system by crosscorrelation, MIT, Research Laboratory of Electronics, *NTIS,* **65**.

Zierler, N. (1968). On primitive trinomials (Mod 2), *Inf. Control,* **13**, 541–554.

Zierler, N. (1959). Linear recurring sequences, *J. Soc. Indust. Appl. Math.*, **7**, 31–49.

APPENDIX A

This appendix contains the C-code for the fast M-transform, an algorithm for fast computation of cross-correlations with binary M-sequences (Sutter, 1991). It consists of a fast Walsh transform proceeded and followed by special permutations. The first permutation is executed by the procedure Load Array that loads the data into the proper locations of the Walsh array. The second permutation, executed by the procedure Read Array that reads back consecutive points of the M-transform from the array.

```
/* Fast m-Transform */
void fmt(tapw, geni)
int tapw;               /* tap word; defines characteristic        */
                        /* polynomial of m-sequence                */
int geni;               /* initalization if generating register;   */
                        /* defines beginning of m-sequence         */
int input();            /* returns sequential data points of       */
                        /* m-transform input                       */
int output();           /* outputs sequential data points of       */
                        /* m-transform                             */

{
     int mt[];                   /* array must fit into real memory!!   */
     int c1, c2;
     int ln;                        /* order of m-sequence              */
     int nm;                        /* length of m-sequence             */
     int nw;                        /* dimension of Walsh Transform     */

     register intr1, r2, r3, r4, r5, r6;

     for (ln = -1, r1 = tapw; r1; r1 >>= 1)
         ln++;
     nw = 1 << ln;
     nm = nw - 1;

/* --------------- Initialize Tap Register r3 to Match geni --------------- */

     r2 = tapw;                          /* tap word                */
     r3 = 0;                             /* tap register T          */
     r4 = geni;                          /* generating register G   */

     for (r1 = nw >> 1; r1; r1 >>= 1, r2 <<= 1)
         if (r4 & r1)
             r3 ^= r2;
         r3 &= nm;

/* ------------------------------ Load Array ------------------------------ */

     mt[0] = input();

     for (r1 = nw, r2 = tapw, r5 = 0; r5 < nm; r5++) {
         r3 <<= 1;
         mt[r4] = input();
         if (r3 & r1) {
             r3 ^ = r2;
             r4 ^ = r1;
         }
         r4 >>= 1;
     }
```

```
/* ------------------------------------------------- Fast Walsh Transform ------------------------------------------------- */

        for (c2 = 0; c2 < ln; c2++) {
            r4 = 1 << c2;
            r5 = r4 << 1;
            for (c1 = 0; c1 < r4; c1++) {
                for (r1 = c1, r6 = nw; r1 < r6; r1 += r5) {
                r2 = r1 + r4;
                r3 = mt[r2];
                mt[r2] = mt[r1] - r3;
                mt[r1] = mt[r1] + r3;
                }
            }
        }

/* -------------------------------------------------------Read Array ------------------------------------------------------- */

        r1 = 1 << ln;
        r2 = tapw;
        output (mt[0]);
        for (r3 = 1, r5 = 0; r5 < nm; r5++, r3 <<= 1){
            if (r3 & r1) {
                r3 ^= r2;
            }
                output (mt[r3]);
        }
}
```

APPENDIX B

For a third-order system the complete, exact Volterra kernels can be obtained from the response to a complete quaternary input. They are usually derived from the quaternary kernels computed by means of the cross-correlation technique. However, it is computationally much easier to derive the Volterra kernels from the kernels associated with the binary components of the test signal. Those are much more easily computed by means of the FWT algorithm. This derivation of the Volterra kernels is always possible, since the set of 4^k self- and mutual kernel elements associated with an epoch of size k are exactly equivalent to the 4^k quaternary kernel elements.

Here we consider first the more general case of a 4-level test that can be decomposed into the sum of two binary components of different amplitudes c and C. Below, the expressions for the Volterra kernels in terms of the kernels of two binary components are derived for a third-order system. At the end we focus on the usual quaternary case where C = 2c.

Consider the Volterra series of a third-order system for an input that is the sum of the two binary components x_i and X_i.

$$\begin{aligned} R_j = K^{(0)} &+ \sum_i K^{(1)}_{i_1} (x_{i_1} + X_{i_1}) \\ &+ \sum_{i_1}\sum_{i_2} K^{(2)}_{i_1,i_2} (x_{i_1} + X_{i_1})(x_{i_2} + X_{i_2}) \\ &+ \sum_{i_1}\sum_{i_2}\sum_{i_3} K^{(3)}_{i_1,i_2,i_3} (x_{i_1} + X_{i_1})(x_{i_2} + X_{i_2})(x_{i_3} + X_{i_3}) \end{aligned} \tag{40}$$

Where R_j is the response amplitude at time j to the stimulus $x_i + X_i$ contained within the input window. The variables x_i and X_i take the values $(+c, -c)$ and $(+C, -C)$, respectively. Inserting the binary variables into Equation 40 yields:

$$\begin{aligned}
R = \quad & K^{(0)} \\
+ \sum_{i} \quad & K_i^{(1)} (x_i + X_i) \\
+ \sum_{i} \quad & K_{i,i}^{(2)} (c^2 + 2x_i X_i + c^2) \\
+ \sum_{i_1 \neq i_2} \quad & K_{i_1,i_2}^{(2)} (x_{i_1} x_{i_2} + 2x_{i_1} X_{i_2} + X_{i_1} X_{i_2}) \\
+ \sum_{i} \quad & K_{i,i,i}^{(3)} (c + 3C^2)x_i + (3c^2 + C)X_i \\
+ \sum_{i_1 \neq i_2} \quad & K_{i_1,i_1,i_2}^{(3)} (c^2 + C^2)x_{i_2} + 2x_{i_1} x_{i_2} X_{i_2} + 2x_{i_1} X_{i_2} X_{i_2} + (c^2 + C^2)X_{i_1} \\
+ \sum_{i_1 \neq i_2 \neq i_3} \quad & K_{i_1,i_2,i_3}^{(3)} (x_{i_1} x_{i_2} x_{i_3} + 3x_{i_1} x_{i_2} X_{i_3} + 3x_{i_1} X_{i_2} X_{i_3} + X_{i_1} X_{i_2} X_{i_3})
\end{aligned} \quad (41)$$

Terms of equal power in the two variables x and X belong to the same self- and mutual binary kernels. They are collected and labeled as $B_{i,j}^{(rs)}$ where r and s specify the power in x and X, respectively, and the lower indices specify the kernel element:

$$\begin{aligned}
B^{(0)} &= K^{(0)} + (c^2 + C^2) \sum_{i} K_{i,i}^{(2)} \\
B_i^{(10)} &= cK_i^{(1)} + c(c^2 + 3C^2) K_{i,i,i}^{(3)} + c(c^2 + C^2) \sum_{j} K_{i,j,j}^{(3)} \\
B_i^{(01)} &= CK_i^{(1)} + C(3c^2 + C^2) K_{i,i,i}^{(3)} + C(c^2 + C^2) \sum_{j} K_{i,j,j}^{(3)} \\
B_{i,j}^{(20)} &= c^2 K_{i,j}^{(2)} && \{i \neq j\} \\
B_{i,j}^{(02)} &= C^2 K_{i,j}^{(2)} && \{i \neq j\} \\
B_{i,j}^{(11)} &= cCK_{i,j}^{(2)} && \{\forall i, j\} \\
B_{i,j,k}^{(30)} &= c^3 K_{i,j,k}^{(3)} && \{i \neq j \neq k\} \\
B_{i,j,k}^{(03)} &= C^3 K_{i,j,k}^{(3)} && \{i \neq j \neq k\} \\
B_{i,j,k}^{(21)} &= 3c^2 C K_{i,j,k}^{(3)} && \{i \neq j \neq k\} \\
B_{i,i,j}^{(21)} &= 2c^2 C K_{i,i,j}^{(3)} && \{i \neq j\} \\
B_{i,j,k}^{(12)} &= 3cC^2 K_{i,j,k}^{(3)} && \{i \neq j \neq k\} \\
B_{i,i,j}^{(12)} &= 2cC^2 K_{i,i,j}^{(3)} && \{i \neq j\}
\end{aligned} \quad (42)$$

This system of equations can easily be solved for the Volterra kernels $K^{(r)}$:

$$K^{(0)} = B^{(0)} - \frac{1}{cC}\sum_i (c^2 + C^2)B^{(11)}_{i,i}$$

$$K^{(1)}_i = \frac{(3c^2 + C^2)}{2c(C^2 - c^2)}B^{(10)}_i + \frac{(c^2 + 3C^2)}{2C(C^2 - c^2)}B^{(01)}_i + \frac{c^2 + C^2}{4cC}\sum_{j\neq i}(\frac{1}{C}B^{(21)}_{i,j,j} + \frac{1}{C}B^{(12)}_{i,j,j})$$

$$K^{(2)}_{i,j} = \frac{1}{c^2}B^{(20)}_{i,j} = \frac{1}{C^2}B^{(02)}_{i,j} = \frac{1}{cC}B^{(11)}_{i,j} \qquad \{i \neq j\}$$

$$K^{(2)}_{i,i} = \frac{1}{cC}B^{(11)}_{i,i}$$

$$K^{(3)}_{i,j,k} = \frac{1}{c^3}B^{(30)}_{i,j,k} = \frac{1}{C^3}B^{(03)}_{i,j,k} = \frac{1}{c^2C}B^{(21)}_{i,j,k} = \frac{1}{cC^2}B^{(12)}_{i,j,k} \qquad \{i \neq j \neq k\}$$

$$K^{(3)}_{i,i,j} = \frac{1}{c^2C}B^{(21)}_{i,i,j} = \frac{1}{cC^2}B^{(12)}_{i,i,j} \qquad \{i \neq j\}$$

$$K^{(3)}_{i,i,i} = \frac{1}{2cC(c^2 - C^2)}(CB^{(10)}_i - cB^{(01)}_i) \tag{43}$$

Consider now the special case C = 2c which corresponds to stimulation with a 2-bit A/D converter:

$$K^{(0)} = B^{(0)} - \frac{5}{2}\sum_i B^{(11)}_{i,i}$$

$$K^{(1)}_i = \frac{1}{c}\left[\frac{7}{6}B^{(10)}_i + \frac{13}{12}B^{(01)}_i + \frac{5}{8}\sum_{j\neq i}(B^{(21)}_{i,j,j} + \frac{1}{2}B^{(12)}_{i,j,j})\right]$$

$$K^{(2)}_{i,j} = \frac{1}{c^2}(B^{(20)}_{i,j} = \frac{1}{4}B^{(02)}_{i,j} = \frac{1}{2}B^{(11)}_{i,j}) \qquad \{i \neq j\}$$

$$K^{(2)}_{i,i} = \frac{1}{c^2}(\frac{1}{2}B^{(11)}_{i,i})$$

$$K^{(3)}_{i,j,k} = \frac{1}{c^3}(B^{(30)}_{i,j,k} = \frac{1}{8}B^{(03)}_{i,j,k} = \frac{1}{2}B^{(21)}_{i,j,k} = \frac{1}{4}B^{(21)}_{i,j,k}) \qquad \{i \neq j \neq k\}$$

$$K^{(3)}_{i,i,j} = \frac{1}{c^3}(\frac{1}{2}B^{(21)}_{i,i,j} = \frac{1}{4}B^{(12)}_{i,i,j}) \qquad \{i \neq j\}$$

$$K^{(3)}_{i,i,i} = \frac{1}{c^3}\left[\frac{1}{12}(\mathrm{B}^{(01)}_i - 2B^{(01)}_i)\right] \tag{44}$$

Chapter 7

APPLICATIONS TO THE STUDY OF VISUAL SYSTEMS OF THE FAST ORTHOGONAL SEARCH AND PARALLEL CASCADE METHODS

Michael Korenberg

TABLE OF CONTENTS

I. Introduction 222

II. Sinusoidal Series Analysis 223
  A. Preliminaries 223
  B. Searching for Significant Frequencies 223
  C. Fast Orthogonal Search 224
  D. Discussion of Some Search Techniques 227
  E. Example 1 227
  F. Example 2 228

III. Nonlinear System Identification 230
  A. Preliminaries 230
  B. Parallel Cascade Identification 231
  C. Example 3 233
  D. Analyzing a Sequence of Time-Series Data 235
  E. Example 4: A Preliminary Analysis of Some Visual-Tracking Errors 236

IV. Concluding Remarks 240

Acknowledgment 240

References 240

I. INTRODUCTION

A major problem in the analysis of biological signals is the extraction of useful information, such as significant frequency content, from brief noisy sequences which may be unequally spaced or missing some data. Here we describe and illustrate two recent methods, fast orthogonal search (Korenberg, 1987a, 1989a) for spectral analysis, and parallel cascade identification (Korenberg, 1982, 1987b, 1990, 1991) for modeling dynamic nonlinear systems and individual sequences of time-series data. Both methods are robust and offer increased resolution, precision, and versatility over many of the common procedures such as conventional Fourier series analysis and a standard technique for kernel estimation.

Fast orthogonal search, proposed by Korenberg (1987a, 1989a), provides an efficient high-resolution method for obtaining economical but accurate sinusoidal series representations or models of time-series data. The method finds the most significant model frequencies first. Moreover, the model frequencies are not required to be commensurate, nor integral multiples of a fundamental frequency. This frequently enables more concise sinusoidal models to be obtained than a conventional Fourier series wherein all frequencies need be harmonics (integer multiples of the fundamental frequency corresponding to the record length). The frequency resolution of the Fourier series approach is, thus, limited by the magnitude of the fundamental frequency, as is well known. Fast orthogonal search is capable of much finer frequency resolution, coupled with accurate estimation of the corresponding amplitudes and phases. In simulations below, we show that fast orthogonal search is effective with brief, noisy signals and with unequally spaced data, and can achieve eight times the resolution of a Fourier series analysis.

Parallel cascade identification (Korenberg, 1982, 1987b, 1990, 1991) is an efficient accurate method for modeling systems with high-order nonlinearities and/or lengthy memory. It uses a structure suggested by Palm (1979), namely, a sum of cascades of alternating dynamic linear and static nonlinear elements. Palm proved that the parallel cascade structure can be used to approximate a very wide class of nonlinear systems, but did not provide an identification method for obtaining this structure.

In more detail, consider a nonlinear, discrete-time, time-invariant, causal, finite-memory system which is a continuous mapping of its input, in that "small" changes in the system input result in "small" changes in the system output. Palm (1979) showed that any such system can be uniformly approximated over a uniformly bounded set of input signals, to an arbitrary degree of accuracy, by a sum of a sufficient number of LNL cascades. Each LNL cascade comprises a dynamic linear element, followed by a static nonlinearity, which is, in turn, followed by a dynamic linear element.

The parallel cascade structure is related to a model of Wiener (1958) and Bose (1956) for expanding Wiener kernels using a complete set of basis functions (e.g., the Laguerre functions). However, a fixed set of basis functions may not result in rapid convergence to approximate accurately a wide variety of nonlinear systems. In the parallel cascade identification method proposed by Korenberg (1982, 1987b, 1990, 1991), the first element in each cascade, a dynamic linear system, can be defined using a slice of a cross-correlation, so that, in effect, the basis functions change with the system to be identified. The parallel cascade method can also be used to estimate the kernels of a nonlinear system and has been shown (Korenberg, 1990, 1991) to be capable of greater accuracy than the well-known cross-correlation method (Lee and Schetzen, 1965) for estimating Wiener kernels.

In succeeding sections, we briefly describe sinusoidal analysis via fast orthogonal search, and system and signal modeling via parallel cascades. In addition, we illustrate both methods on synthetic data and parallel cascade on data obtained from the human visual tracking system.

II. SINUSOIDAL SERIES ANALYSIS

A. PRELIMINARIES

Let y(n), n = 0, .., T represent time-series data, and let

$$y(n) = \sum_{m=0}^{M} a_m p_m(n) + e(n) \tag{1}$$

represent a sinusoidal series model which is to be developed for the time series. In Equation 1,

$$p_0(n) = 1, \qquad n = 0, \cdot \cdot \cdot , T \tag{2}$$

the remaining p_m (n) are the sine and cosine terms to be selected, and e(n) is the model error. More particularly, for i = 1, 2, ...

$$p_{2i-1}(n) = \cos(\omega_i n) \tag{3}$$

$$p_{2i}(n) = \sin(\omega_i n) \tag{4}$$

where n = 0, ..., T.

B. SEARCHING FOR SIGNIFICANT FREQUENCIES

The frequencies ω_i are found by ''orthogonally searching'' (Korenberg, 1987a, 1989a) through a set of candidate frquencies ω_A, ω_B, ... These candidate frequencies need not be commensurate nor integral multiples of the fundamental frequency corresponding to the record length.

Suppose that Gram-Schmidt (or preferably modified Gram-Schmidt) orthogonalization was used to rearrange Equation 1 into the series

$$y(n) = \sum_{m=0}^{M} g_m w_m(n) + e(n) \tag{5}$$

In Equation 5, the $w_m(n)$ were constructed, from the corresponding $p_m(n)$, to be orthogonal over n = 0, ..., T, and the g_m are the orthogonal expansion coefficients (achieving a least-squares fit). Suppose that

$$T_i = a_{2i-1}\, p_{2i-1}(n) + a_{2i}\, p_{2i}(n) \tag{6}$$

was the last cosine/sine term pair added to the right side of Equation 1. Then, the addition of this term pair reduced the mean-square error (MSE) by the amount

$$Q(i) = g_{2i-1}^2\, \overline{w_{2i-1}^2(n)} + g_{2i}^2 \overline{w_{2i}^2(n)} \tag{7}$$

where the overbar denotes the time average from n = 0 to n = T. Equation 7 follows from the mutual orthogonality of the $w_m(n)$.

To build up the sinusoidal series model of Equation 1, successively set i = 1, 2, ..., and M = 2i, and search for the value of ω_i in Equations 3 and 4 as follows. For each candidate frequency as yet unselected, evaluate the quantity Q(i) in Equation 7. Choose the candidate for which Q is greatest (optionally subject to exceeding a specified threshold level) since this choice will result in the greatest reduction in MSE. Continuing in this way, we can select the model frequencies for the sinusoidal series of Equation 1, and then the coefficients a_m can readily be calculated from the g_m.

This procedure is analogous to an orthogonal search method developed (Korenberg, 1985; McIlroy, 1986) for constructing difference equation models for dynamic nonlinear systems with unknown structure. Previously, Desrochers (1981) had used an analogous approach to build static models of nondynamic (i.e., zero-memory) nonlinear systems. McIlroy (1986) accelerated the searching through use of thresholds; thus, candidate difference equation terms with a potential reduction of MSE below threshold were permanently dropped from consideration, thereby reducing the number of candidates to be subsequently searched.

However, the explicit creation of the orthogonal functions $w_m(n)$ is unnecessary and wastes computing time and memory storage. This is avoided, and other computational savings are realized, in the fast orthogonal search, which is discussed next.

C. FAST ORTHOGONAL SEARCH

Suppose, first, that the $p_m(n)$ terms in Equation 1 were already known. Then, the orthogonal expansion coefficients in Equation 5 are given, for m = 0, ..., M, by

$$g_m = \frac{C(m)}{D(m,m)} \tag{8}$$

where

$$D(0,0) = 1 \tag{9}$$

and for m = 1, ..., M

$$D(m,0) = \overline{p_m(n)} \tag{10}$$

$$D(m,r) = \overline{p_m(n)\,p_r(n)} - \sum_{i=0}^{r-1} \alpha_{ri} D(m,i) \qquad r = 1, \cdots, m \tag{11}$$

$$\alpha_{mr} = \frac{D(m,r)}{D(r,r)} \qquad r = 0, \cdots, m-1 \tag{12}$$

$$C(0) = \overline{y(n)} \tag{13}$$

$$C(m) = \overline{y(n)p_m(n)} - \sum_{r=0}^{m-1} \alpha_{mr} C(r) \tag{14}$$

This is a Cholesky factorization and can be carried out using the following pseudocode (for similar substitute code, e.g., the Cholesky outer product version, see Golub and Van Loan [1989]):

```
D(0, 0) = 1
FOR m = 1 to M
Calculate D(m, 0) from Equation 10
NEXT m
FOR m = 1 to M
FOR r = 0 TO m − 1
Calculate α_mr from Equation 12
Calculate D(m, r + 1) using Equation 11
NEXT r
NEXT m
```

This yields the α_{mr} and D(m, m) so that the g_m can be obtained from Equations 8, 13, and 14. The coefficients a_m in Equation 1 can then be calculated by the following formula (Korenberg, et al., 1988):

$$a_m = \sum_{i=m}^{M} g_i v_i \tag{15}$$

where

$$v_m = 1 \tag{16}$$

$$v_i = -\sum_{r=m}^{i-1} \alpha_{ir} v_r \qquad i = m + 1, \cdots, M \tag{17}$$

To begin constructing the sinusoidal series of Equation 1, use Equation 2 to introduce a constant, and obtain g_0 from Equations 8, 9, and 13. However, in practice the $p_m(n)$ in Equation 1 for $m \geq 1$ are not known *a priori* because the frequencies ω_i, i = 1, 2, ..., in Equations 3 and 4 need to be selected, and this is considered next.

It follows readily (Korenberg, 1987a, 1989a) that if T_i in Equation 6 were the last term pair added to the right side of Equation 1, then the MSE would be reduced by the amount

$$Q(i) = g_{2i-1}^2 D(2i - 1, 2i - 1) + g_{2i}^2 D(2i, 2i) \tag{18}$$

which is analogous to Equation 7.

Therefore, to find ω_i, i = 1, 2, ... in Equations 3 and 4, set M = 2i, and for each candidate frequency for ω_i, use the pseudocode and Equations 8 and 14 to obtain g_m and D(m,m), where m = 2i − 1, 2i. Hence, determine the value of Q(i) in Equation 18 corresponding to the candidate frequency. Chooose the candidate with the largest Q value (optionally, subject to exceeding a threshold level).

In testing the candidates for ω_i, we can avoid repeating calculations already performed at an earlier stage of searching. This is done by carrying out the above pseudocode, and using Equations 8 and 14 to obtain C(m) and g_m, only for m = M − 1, M (where M = 2i) and not for earlier values of m. After the choice for ω_i has been made, repeat the abbreviated pseudocode, and recalculate C(m) and g_m with m = M − 1, M from Equations 8 and 14 for the chosen candidate. This is done to set properly the values of the α_{mr}, D(m,m), C(m), and g_m, m = M − 1, M (r = 0, ..., m − 1), prior to either searching for ω_{i+1} or terminating the search.

For convenience, we have assumed above that the time-series data are equally spaced. When this is true, the time averages on the right sides of Equations 10 and 11 may readily be calculated efficiently (Korenberg, 1987a) using well-known formulas (Dwight, 1960) for summing a sine or cosine time function at equally spaced instants. A slight modification will be discussed shortly for coping with time series having unequally spaced or missing data.

The MSE for the sinusoidal series model of Equation 1 is

$$\begin{aligned} e &= \overline{[y(n) - \sum_{m=0}^{M} a_m p_m(n)]^2} \\ &= \overline{y^2(n)} - \sum_{m=0}^{M} g_m^2 D(m,m) \end{aligned} \tag{19}$$

and this can be used in deciding when to stop the searching. Model development may also be terminated when a preset number of frequencies have been selected, or when no remaining candidate frequency can cause a reduction in MSE exceeding a threshold level.

For example, before choosing a candidate for ω_i, one may optionally require that (Korenberg, 1989b)

$$g_{M-1}^2 D(M-1, M-1) + g_m^2 D(M,M) \geq \frac{10.9}{T+1}\left[\overline{y^2(n)} - \sum_{m=0}^{M-2} g_m^2 D(m,m)\right] \quad (20)$$

where M = 2i. This criterion (which follows immediately from using a standard correlation test with 99.9% confidence limits) helps to avoid choosing frequencies which are merely fitting noise. Thus, suppose that all correct model frequencies had already been selected. If the searching then attempted (erroneously) to fit zero-mean, independent Gaussian noise, the probability that Inequality 20 would be satisfied is only about 0.001. Of course, 10.9/(T + 1) on the right side of Inequality 20 can be replaced by other factors, corresponding to different confidence limits (e.g., 4/(T + 1) for approximately 95% limits). If Inequality 20 is to be used as a requirement for continuing the searching, then one should demand as a further requirement that the MSE exceed a threshold level, e.g., a small percentage of the time-series variance. This is because, in the case of clean time-series data, the right side of Inequality 20 will be negligible once the correct frequencies have been found. Then Inequality 20 may be satisfied due to roundoff errors; therefore, to prevent further (unnecessary) model development, a small MSE is used as a stopping criterion.

Once the searching is terminated, Equations 15 to 17 can be employed to calculate the coefficients a_m in the sinusoidal series of Equation 1. The obtained sinusoidal series can then be used as an approximation of the original time series. Since the sinusoidal series was constructed using only the most significant frequencies, it tends to be significantly less noisy than the given time-series data. (See example below.)

Clearly, once the cosine/sine term pair in Equation 6 has been determined, then both the amplitude and the phase corresponding to the selected frequency ω_i can be calculated.

Related methods of the present author have been applied to real data in projects supervised by him in the Department of Electrical Engineering at Queen's University. In one such study (Ho et al., 1987), to cope with unequally spaced data, the variable n on the right sides of Equations 3 and 4 was replaced with the function t(n) corresponding to the actual timing of the nth data sample. The same strategy can be used in the fast orthogonal search described above to cope with unequally spaced or missing data.

Fast orthogonal search can also be used to build difference equation models of dynamic nonlinear systems (Korenberg, 1989a). Indeed, fast orthogonal search is closely related to the earlier-mentioned orthogonal search method developed (Korenberg, 1985; McIlroy, 1986) for constructing nonlinear difference equation models. Typically, the models have the general form of Equation 1, where y is now the system output, and Equation 2 still holds. For $m \geq 1$ the difference equation terms $p_m(n)$ comprise delayed values of the output y, present and delayed values of the system input x, and cross-products of any number of such x and y terms, including powers (Haber and Keviczky, 1976). The candidate difference equation terms will, therefore, have this structure, and fast orthogonal search can be used to select a concise subset of terms to build an accurate final model. To allow the modeling of nonstationarities in the system, the set of candidate difference equation terms can be enlarged to include cosine and sine terms (as in Equations 3 and 4), delayed y terms, present and delayed x terms, and cross-products of any number of these three types of terms (including powers). Fast orthogonal search can be used to determine the frequencies of the sinusoidal factors in the difference equation terms selected. In place of or in addition to the sinusoids, other sets of basis functions, such as decaying exponentials and exponentially decaying sinusoids, can be used in forming some of the candidate difference equation terms.

D. DISCUSSION OF SOME SEARCH TECHNIQUES

As noted, fast orthogonal search avoids the explicit creation of orthogonal functions and uses a Cholesky factorization in carrying out the search. A similar approach was used by Desrochers (1981) in some valuable work on fitting static models to zero-memory nonlinear systems. However, fast orthogonal search is generally a faster technique (for a comparison see Korenberg, 1989c, Korenberg and Paarmann, 1991). In addition, Desrochers' work did not consider the sinusoidal series modeling of time-series data discussed above.

Abraira and Ibarz (1986) considered the sinusoidal modeling of time-series data, where the required frequencies are chosen by successively scanning a set of candidates. Frequency selection is made according to a best-fitness criterion. However, these authors did not use an orthogonal approach, and it is possible for the same frequency to be chosen more than once, as they point out. The final amplitude for a selected frequency is the sum of the amplitudes for the times that the frequency is selected (taking into account the phase) and is only an approximation. The searching can continue even if all correct frequencies have been chosen once (this may be necessary to improve the amplitude estimates), so that there is the possibility of additional (incorrect) frequencies being selected. Since each sinusoidal component estimated is subtracted from the time series, this may introduce incorrect frequencies and undermine subsequent searching. However, their method offers higher resolution than a conventional Fourier series analysis, does not require the selected frequencies to be commensurate, and does not demand equally spaced data.

Note that fast orthogonal search need select each correct frequency only once, and choosing an incorrect frequency during the searching will not introduce error provided the correct frequencies are eventually chosen. The algorithm will then yield a negligible value for the amplitude corresponding to the incorrect frequency.

The Prony and Pisarenko methods can be used to estimate a sum of sinusoids in the presence of noise with high resolution; however, model order must be selected, as required also by the autoregressive (AR) and maximum entropy methods, and a polynomial equation must be solved (Box and Jenkins, 1976; Kay, 1988; Kay and Marple, 1981; Mohanty, 1986). In fast orthogonal search, the order of the sinusoidal series developed is automatically determined, and no polynomial equation need be solved (Korenberg, 1987a, 1989a, b).

E. EXAMPLE 1

The test time-series data (201 points defined over $n = 0, \ldots, 200$) comprised the sum of five sinusoids, with radian frequencies 0.05, 0.3, 0.32, 0.5, 0.8. Each sinusoidal amplitude was $\sqrt{2}$, and phase was 45°. The first 100 points of the original data are shown in Figure 1. Fast orthogonal search was then examined in tests using noise-free data, unequally spaced data, and noisy data. Similar tests were carried out previously when the method was briefly reviewed by Korenberg and Paarmann (1989).

In the first test, fast orthogonal search was applied to the 201 point (noise-free) time series. The correct model frequencies (and no others) were chosen in the following order: 0.5, 0.8, 0.05, 0.32, 0.3. All sinusoidal amplitudes and phases were also determined precisely. This surpasses the resolution capability of a conventional Fourier series analysis, whose nominal resolution width is $2\pi/201 = 0.031$ and, thus, could not resolve the frequencies 0.3, 0.32. Note that fast orthogonal search does not assume that the phases (or the amplitudes) at the different model frequencies are the same, as happened here.

In the second test, 50% of the original time-series data were randomly deleted. Fast orthogonal search, applied to the 100 remaining (unequally spaced) data points, selected

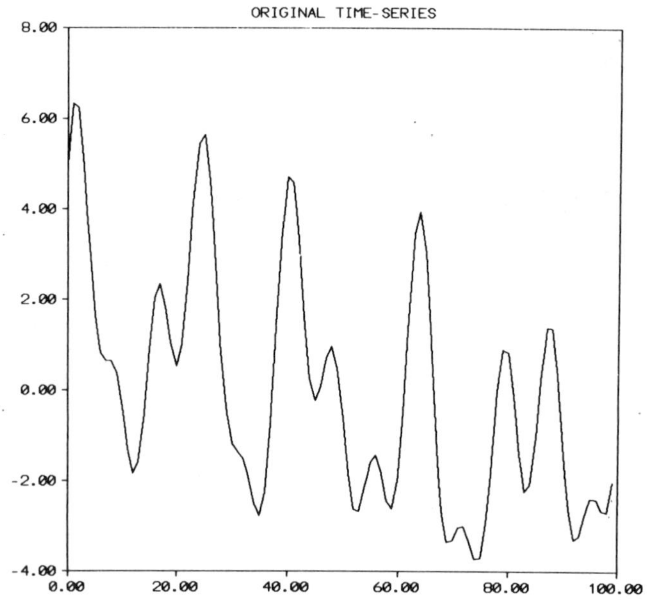

FIGURE 1. The first 100 points of the test time series (Example 1): the sum of 5 sinusoids.

only the correct frequencies, in order: 0.3, 0.05, 0.5, 0.8, 0.32. The amplitudes and phases were again determined precisely.

In the third test, the time series was contaminated by independent zero-mean, uniformly distributed noise. The variance of the noise equaled that of the original time series, for a signal-to-noise ratio of unity (0 dB). The first 100 points of the noisy time series are shown in Figure 2. Fast orthogonal search applied to the 201 point noisy time series detected only 5 significant (and correct) frequencies, in order: 0.32, 0.8, 0.5, 0.3, 0.05. The MSE of the model fit to the noisy time series was 48.48% (expressed as a percentage of the noisy time-series variance). As shown in Figure 3, the model gives a very good approximation to the original (noise-free) time series. The model has an MSE of 3.26% of the original time-series variance, for a signal-to-noise ratio of 30.7 (14.87 dB). This is a dramatic improvement over the signal-to-noise ratio at the start, namely, unity, corresponding to the noisy time series. Thus, fast orthogonal search can be applied directly to a noise-corrupted time series to recover a close estimate of the original time series prior to the noise corruption.

In the above tests, a maximum of 20 frequencies was allowed in the final model, and 100 candidate frequencies equally spaced between 0.01 and 1 (inclusive) were searched. (A larger range of frequencies could obviously be searched, with the largest normalized candidate frequency being π rad/s.) Inequality 20 was used as a requirement for continuing the searching. In addition, if the MSE became less than 0.01% of the time-series variance, model development was terminated automatically.

F. EXAMPLE 2

Inequality 20 and the same 0.01% test of the MSE used in Example 1 were again employed in another study of the resolution of fast orthogonal search. Here a 2-ms sampling interval was assumed, for a sampling rate of 500 Hz. The 125-point-test time series comprised the sum of five sinusoids with frequencies (in Hz): 5, 15, 30, 30.5, 44. The corresponding phases were 60°, 30°, 45°, 45°, 45°; each sinusoidal amplitude was $\sqrt{2}$. Fast orthogonal search applied to the time series selected the five correct frequencies only, in order: 30.5, 5, 44, 15, 30. All amplitudes and phases were correctly determined. The 0.5-Hz resolution

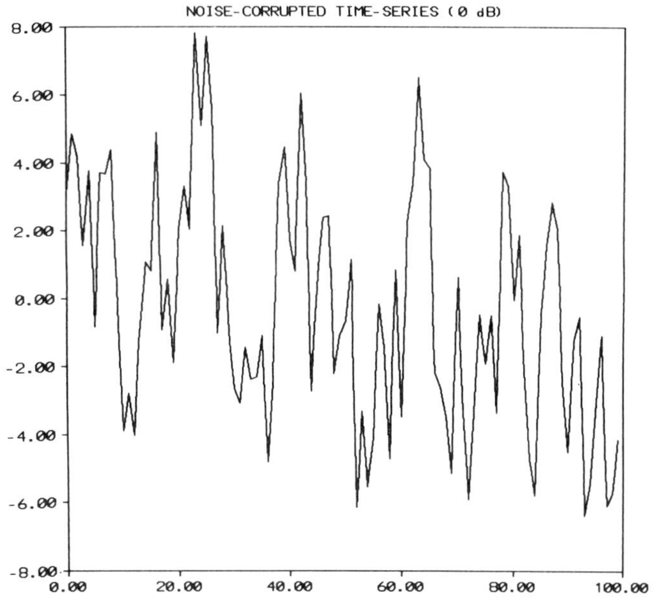

FIGURE 2. The first 100 points of the time series corrupted with additive noise (Example 1): signal-to-noise ratio of unity (0 dB).

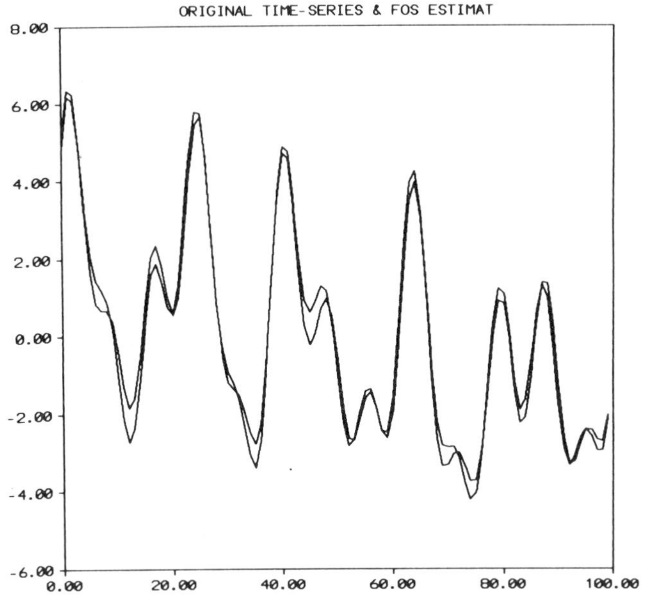

FIGURE 3. The first 100 points of the original time series together with the estimate obtained via fast orthogonal search, using only the noise-corrupted time series (Example 1).

displayed by fast orthogonal search in this example (by resolving the frequencies 30, 30.5) is eight times finer than that achievable by a Fourier series analysis, whose resolution width here would be 500/125 = 4 Hz.

In the above study, 100 candidate frequencies equally spaced between 0.5 and 50 Hz (inclusive) were searched, and a maximum of 20 frequencies were allowed in the model.

Fast orthogonal search is not always as precise as in this example, but in extensive testing has continually surpassed the Fourier series result. Similar tests to the above were previously carried out by Korenberg (1989d) and in the review article by Korenberg and Paarmann (1991). In addition, fast orthogonal search showed superior performance when tested (Korenberg, 1989a) on data previously used by Kay and Marple (1981) in a study of several spectral estimation techniques, which included the Prony method and its spectral line variant. In further comparative tests (Adeney and Korenberg, 1992), fast orthogonal search clearly outperformed the root-MUSIC method (which is known to outperform the MUSIC method).

III. NONLINEAR SYSTEM IDENTIFICATION

A. PRELIMINARIES

Consider a nonlinear, discrete-time, time-invariant, causal, finite-memory system which is a continuous mapping of its input, in that small changes in the system input x result in small changes in the system output y. As noted, Palm (1979) proved that such a nonlinear system can be uniformly approximated over a uniformly bounded set of input signals, to an arbitrary degree of accuracy, by a sum of a sufficient number of LNL (i.e., dynamic linear, static nonlinear, dynamic linear) cascades. Korenberg (1982) showed that, in fact, a sum of LN cascades sufficed to approximate the nonlinear system and demonstrated how this approximation could be built up. [The LN cascade is sometimes referred to as a Wiener model (not to be confused with a Wiener series, discussed below) and comprises a dynamic linear element followed by a static nonlinearity (Hunter and Korenberg, 1986).]

In view of the Stone-Weierstrass theorem [see Dieudonné (1976); Palm (1979)] and the work of Sandberg (1983), the above nonlinear discrete-time system can also be uniformly approximated over a uniformly bounded set of input signals by a finite-order Volterra series of the form

$$y_s(n) = k_0 + \sum_{m=1}^{P} V_m \qquad n = 0, 1, \cdots \tag{21}$$

where for m = 1, ..., P

$$V_m = \sum_{i_1=0}^{R} \cdots \sum_{i_m=0}^{R} k_m(i_1, \cdots, i_m)\, x(n - i_1) \cdots x(n - i_m) \tag{22}$$

Here P is the order (or degree) of the series, V_m and k_m are, respectively, the *m*th order Volterra functional and symmetric kernel, and the constant k_0 is the zero-order kernel. Finally, (R + 1) is the memory length, since the series output $y_s(n)$ depends on input delays from 0 to R lags.

Using the Gram-Schmidt procedure, Wiener (1958) created a functional expansion whose terms were mutually orthogonal for a given white Gaussian input process. The Wiener kernels in his orthogonal series, in general, differed from the Volterra kernels of corresponding order, but a straightforward relation exists (Marmarelis and Marmarelis, 1978) between the two sets of kernels.

The parallel cascade representation has some important advantages over the Volterra and Wiener series descriptions, particularly in convenience of approximating systems with higher order nonlinearities. This is because the nonlinearities in the parallel cascade always occur as static functions. Hence, their estimation is far simpler than that for the multidimensional kernels in the Volterra or Wiener series. For example, if a Wiener series is used to represent a dynamic system with nonlinearities up to some high order, then estimating the Wiener kernels via the Lee-Schetzen (1965) method requires computing cross-correlations up to the same high order. This is more time consuming than estimating static nonlinearities (e.g., represented by polynomials of sufficient order) in the parallel cascade.

Shi and Sun (1990) have remarked that any discrete-time, finite-memory Volterra functional can be exactly represented by a finite sum of LNL cascades. Korenberg (1991) showed that a finite sum of LN cascades sufficed for exact representation.

We will briefly describe a method (Korenberg, 1982, 1987b, 1990, 1991) for obtaining a parallel cascade approximation for any discrete-time finite-memory causal nonlinear system having a Wiener functional expansion. This expansion displays convergence in the mean-square sense, a weaker approximation criterion than the uniform approximation referred to above. We suppose that the system input x(n) and output y(n) are available for the portion of the experimental record extending from n = 0 to n = T. (The experiment may have commenced well before, and continued long after, the available record segment.)

The method is illustrated below by successively approximating the system output, and thereafter the unidentified residue, using a simple cascade model. The final representation is a parallel array of these cascade models, which are identified one at a time. Note that other basic models, such as simple nonlinear difference equations (Korenberg, 1982) can also be individually identified and used in the parallel array in place of, or in addition to, the cascade models considered below.

B. PARALLEL CASCADE IDENTIFICATION

The essence of the parallel cascade identification method (Korenberg, 1982) is to approximate the dynamic nonlinear system by a first cascade of dynamic linear and static nonlinear elements. The residue, i.e., the difference between the nonlinear system and the cascade outputs, is treated as the output of a new nonlinear system, a second cascade is found to approximate the new system, and so on. Under broad conditions, the nonlinear system can be approximated, to an arbitrary degree of accuracy in the mean-square sense, by a sum of the individually obtained cascades. Proof of convergence of a detailed algorithm is presented by Korenberg (1991). It is important to note that the input is not required to be Gaussian or white, nor to have special probability density or autocorrelation properties.

For simplicity, in the description below, each cascade comprises a dynamic linear element followed by a static nonlinearity, but it will be appreciated that any number of alternating dynamic linear and static nonlinear elements can be used in creating a cascade path (Korenberg, 1982, 1987b). Moreover, many variations of the algorithm are possible (see Korenberg, 1987b, 1990, 1991). We assume that the nonlinear system to be identified has finite memory lasting up to R lags (as in Equation 22).

Let $y_i(n)$ denote the residue remaining after fitting the *i*th cascade, so that $y_0(n) = y(n)$. Consider fitting the residue y_{i-1} with the *i*th cascade, where $i \geq 1$.

Let $h_i(j)$ denote the delta (i.e., discrete impulse) response of the first component, a dynamic linear element, in the *i*th cascade. Then $h_i(j)$ can be defined using a first-order input-residue cross-correlation, or a slice of a second- or higher-order cross-correlation to which delta functions are added or subtracted at diagonal values. For example, let $\phi_{xy_{i-1}}(j)$ and $\phi_{xxy_{i-1}}(j_1, j_2)$ be, respectively, the first- and second-order cross-correlations of the input with the residue, computed over n = R, ..., T:

$$\phi_{xy_{i-1}}(j) = \frac{1}{T - R + 1} \sum_{n=R}^{T} y_{i-1}(n)\, x(n - j) \tag{23}$$

$$\phi_{xxy_{i-1}}(j_1, j_2) = \frac{1}{T - R + 1} \sum_{n=R}^{T} y_{i-1}(n)\, x(n - j_1)\, x(n - j_2) \tag{24}$$

Then $h_i(j)$ can be set equal to one of

$$q_1(j) = \phi_{xy_{i-1}}(j) \tag{25}$$

$$q_2(j) = \phi_{xxy_{i-1}}(j,A) \pm C\delta(j - A) \tag{26}$$

for $j = 0, \ldots, R$. Set $h_i(j) = 0$ for $j > R$ since the nonlinear system to be identified is assumed to have memory lasting up to R lags. In Equation 26, the discrete delta function $\delta(j) = 0$ for $j \neq 0$, and $\delta(0) = 1$. Whether Equation 25 or 26 is used to define $h_i(j)$ for $j = 0, \ldots, R$ is decided at random. If Equation 26 is to be used, then the integer A is randomly selected from $0, \ldots, R$, the sign of the δ-term is chosen at random, and C is adjusted to tend to zero as the mean square of the residue approaches zero.

For example, set

$$C = \frac{\overline{y_{i-1}^2(n)}}{\overline{y^2(n)}} \tag{27}$$

where here (and throughout this section) the overbar denotes the time-average from $n = R$ to $n = T$, as in Equations 23 and 24. Note that $h_i(j)$ may also be defined using slices of higher-order cross-correlations (Korenberg, 1982, 1991) with discrete impulse functions added or subtracted at diagonal values, analogously to Equation 26.

Once $h_i(j)$ has been selected, calculate the output of the first element in the *i*th cascade:

$$u_i(n) = \sum_{j=0}^{R} h_i(j)\, x(n - j) \tag{28}$$

The static nonlinearity in the cascade can be represented as a polynomial [having input $u_i(n)$] which is best fit to the residue $y_{i-1}(n)$, using the portion of the record from $n = R$ to $n = T$. The cascade output is then

$$z_i(n) = \sum_{j=0}^{P} b_{ij}\, u_i^j(n) \tag{29}$$

where the b_{ij} are the polynomial coefficients.

The new residue is then

$$y_i(n) = y_{i-1}(n) - z_i(n) \tag{30}$$

and the parallel cascade development can be continued analogously. Since the static nonlinearity was determined by minimizing the mean-square error over $n = R, \ldots, T$, it follows immediately that

$$\overline{y_i^2(n)} = \overline{y_{i-1}^2(n)} - \overline{z_i^2(n)} \tag{31}$$

Hence, adding the *i*th cascade reduces the mean square of the residue by amount $\overline{z_i^2(n)}$, the mean-square of the cascade output.

Consider continuing the process by fitting the residue $y_i(n)$ with the $(i + 1)$th cascade, having output $z_{i+1}(n)$, $i \geq 1$. Before accepting a candidate for the $(i + 1)$th cascade, one may *optionally* require that

$$\overline{z_{i+1}^2(n)} > \frac{4}{T - R + 1}\,\overline{y_i^2(n)} \tag{32}$$

which is a similar criterion to that of Inequality 20. Inequality 32 (a standard correlation test using 95% confidence limits) helps to avoid choosing unnecessary cascades which are merely fitting noise.

Suppose the test is used and a candidate fails to satisfy Inequality 32. Then a new candidate for the $(i + 1)$th cascade can be found by reselecting $h_{i+1}(j)$ and redetermining the static nonlinearity. Again $h_{i+1}(j)$ can be defined using a first-order input-residue cross-correlation, or a slice of a second- or higher-order cross-correlation with delta functions added or subtracted at diagonal values, analogous to Equation 26. Once the static nonlinearity has been best-fit, the new candidate for the $(i + 1)$th cascade may be tested for compliance with Inequality 32.

Parallel cascade development may be terminated when a preset number of cascades have been added or tested, when the MSE is sufficiently small, or when no remaining candidate cascade can reduce the MSE more than a threshold amount.

In summary, the first (linear) element in a cascade path can be defined using a slice of an input-residue cross-correlation (first- or higher-order). However, many other ways exist (Korenberg, 1990, 1991) for defining the delta response of the linear system, such as via basis functions and sums and differences thereof. The static nonlinearity can then be found by best-fitting the residue. As noted (Korenberg, 1982), the cascade development can readily be continued by adding a second dynamic linear element following the static nonlinearity. Mo and Elkasabgy (1984) noted that this significantly increased the speed of convergence in simulated examples. In carrying out the parallel cascade identification (Korenberg, 1982), Mo and Elkasabgy (1984) best-fit a differential equation (to the residue) for the second linear system. Many variations in developing a cascade path are available (Korenberg, 1987b, 1990, 1991), for example, one can continue to add alternating dynamic linear and static nonlinear elements, and the above presentation is merely an introduction.

As noted above, one major advantage of the parallel cascade identification is the ease and rapidity with which dynamic systems having high-order nonlinearities can be modeled. Indeed, the speed of "copying" the dynamic nonlinear behavior of a system is generally much greater using parallel cascades than a functional expansion (Wiener, 1958) approach. Once the parallel cascade representation has been identified, it can be rearranged into a functional expansion and, thus, yield estimates of the kernels in the expansion. Such estimates obtained via parallel cascade identification tend to be significantly more accurate (Korenberg, 1990, 1991) than those found by cross-correlation. This is illustrated in the following example.

C. EXAMPLE 3

The system to be identified was a second-order Volterra series ($P = 2$ in Equation 21) with memory length $(R + 1)$ of 100. Thus, one zero-order, 100 first- and 5050 distinct second-order kernel values were to be estimated using both the cross-correlation (Lee and Schetzen, 1965) and parallel cascade (Korenberg, 1982, 1987b, 1990, 1991) methods. Note that, since the test system is second-order, the corresponding Volterra and Wiener kernels of first and second order are equal. A zero-mean white Gaussian input was applied to the system to generate 10,000 input/output data pairs.

As shown in Figure 4, there is significant error in the cross-correlation estimate of the first-order kernel. Similarly, the actual second-order kernel (Figure 5) was not estimated very accurately via the cross-correlation method (Figure 6). Aside from the jitter in the cross-correlation estimates, the relative sizes of the various kernel peaks are not well preserved.

The identified parallel cascade model had an MSE less than 0.025% of the output variance. As shown in Figures 7 and 8, parallel cascade estimates of the first- and second-order kernels were very close to the actual kernels. These estimates can be made even more precise by adding further cascades to the parallel cascade model.

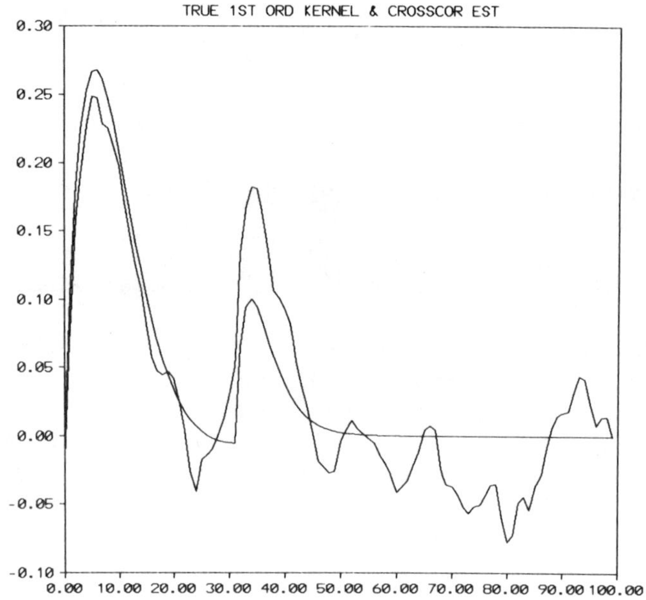

FIGURE 4. Actual first-order kernel and the cross-correlation estimate, obtained using 10,000 points of white Gaussian input (Example 3).

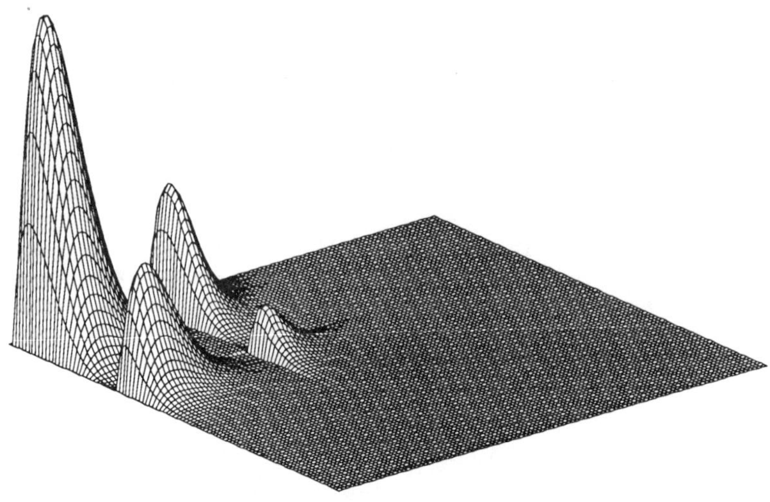

FIGURE 5. Actual second-order kernel (Example 3).

As noted above, the parallel cascade method does not require use of a Gaussian or white input sequence, nor one having special probability density or autocorrelation properties. The method yields values for the kernels which are nearly least-square estimates (Korenberg, 1991) and will, in general, depend on the input used. A white Gaussian input was employed in this example to enable comparison with the cross-correlation method (Lee and Schetzen, 1965) of estimating Wiener kernels. Note that attempting to obtain directly least-square estimates of the kernel values is, in the present example, numerically formidable. It would entail inversion of a symmetric 5151 $\times$ 5151 matrix (which is not Toeplitz nor near-to-Toeplitz so as to enable rapid inversion).

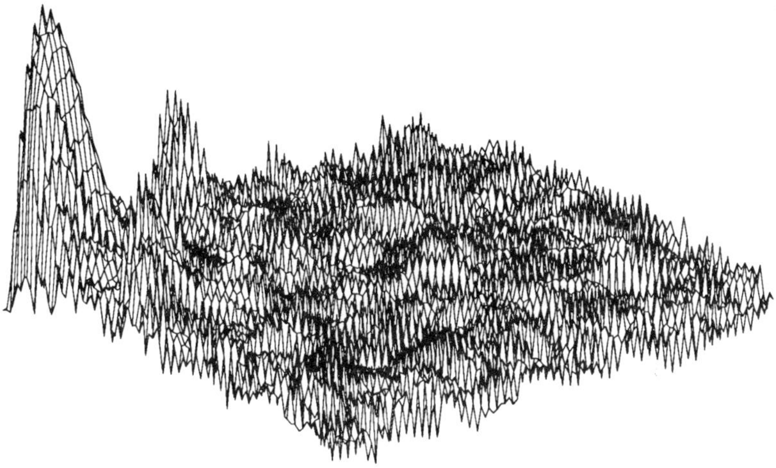

FIGURE 6. Cross-correlation estimate of the second-order kernel, obtained using 10,000 points of white Gaussian input (Example 3).

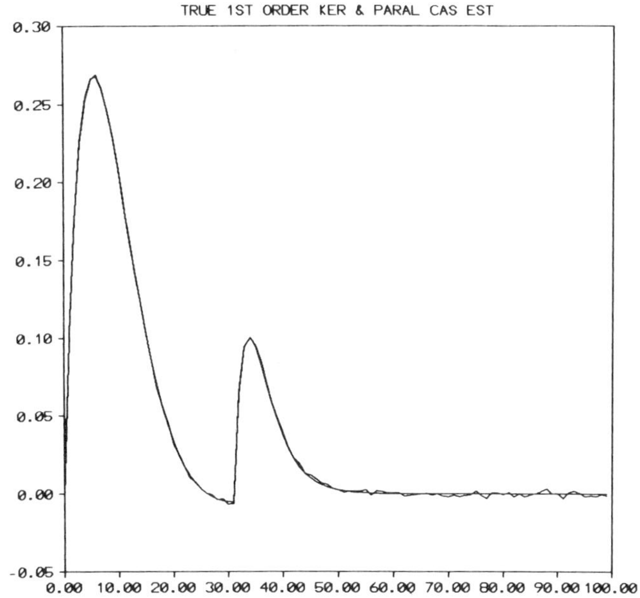

FIGURE 7. Actual first-order kernel and the parallel cascade estimate, obtained using 10,000 points of white Gaussian input (Example 3).

Parallel cascade identification is robust and shows good performance in kernel estimation when noise contaminates the output (Korenberg, 1990), and can conveniently be used to estimate third- and higher-order kernels.

Cross-correlation estimates do approach the actual kernel values with increasing record length. However, in other comparative tests, cross-correlation kernel estimates did not attain the accuracy of parallel cascade estimates, even with a record 100 times longer than that used by parallel cascade.

D. ANALYZING A SEQUENCE OF TIMES-SERIES DATA

A great deal of attention (Glass and Mackey, 1988; Gleick, 1987; Liebovitch, 1989) has been paid to the problem of distinguishing between chaos due to a deterministic nonlinear

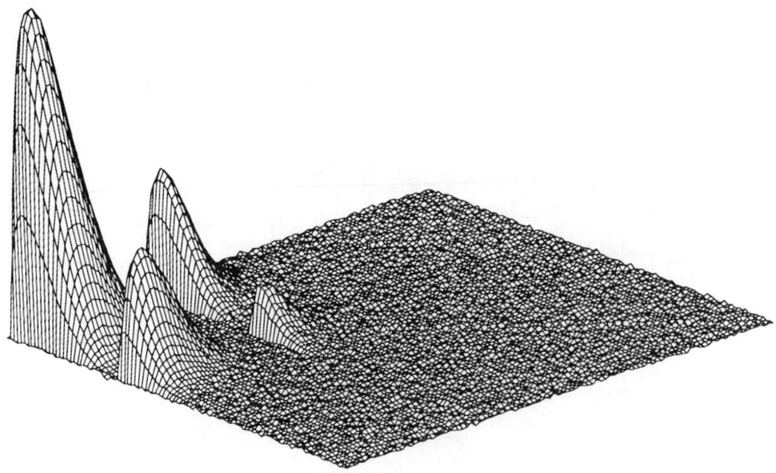

FIGURE 8. Parallel cascade estimate of the second-order kernel, obtained using 10,000 points of white Gaussian input (Example 3).

system and random noise. In the next example, we illustrate how the parallel cascade method can be used to analyze a single sequence of time-series data. The approach involves treating a delayed version of the time series as the system input, and the original (undelayed) time series as the system output; then use the input/output data to identify a nonlinear system. This can be a useful means of distinguishing deterministic chaos from random noise. For example, suppose that the above-created input/output data is fit by a parallel cascade model, and that Inequality 32 is used as a requirement for adding a cascade. Then, if the time series reflects deterministic behavior, a significantly larger number of cascades will generally be chosen, and the model MSE will be smaller than if the time series were independent random noise (Korenberg, 1991). Moreover, the kernels obtained from the parallel cascade will characterize the deterministic chaos, whereas for independent random noise, the kernels will be very noisy and generally have negligible values. The kernel estimates obtained to characterize the time series will, in general, depend on the amount of delay used to create the input signal (Korenberg, 1987c, 1991).

E. EXAMPLE 4: A PRELIMINARY ANALYSIS OF SOME VISUAL-TRACKING ERRORS

Figure 9 shows a portion of the visual tracking of a point target jumping (one dimensionally) between positions to the left and right of center screen, roughly according to a square-wave pattern. The visual tracking experiments were conducted by Peter Beck, Douglas Hofmann, and David Poirier at Queen's University using an infrared eye-movement monitoring device which they had built.

The error signal, i.e., the difference between the visual tracking and the target signal, was the sequence of time-series data to be analyzed. This time series was delayed by one sample and, with the original undelayed time series, formed the input/output data (559 pairs) fit by a parallel cascade model with second-degree polynomial nonlinearities and a memory length of 25. Inequality 32 was used as a requirement for adding a cascade. Then kernels up to second order were computed from the identified model.

The first-order kernel (Figure 10) had a (positive) peak at lag 0 and a (negative) dip of slightly larger magnitude at lag 19. This illustrates that the present value of the error signal is predictable, at least in part, using the error signal one sample earlier, corresponding to lag 0, and 20 samples earlier, corresponding to lag 19. Since the target motion was near

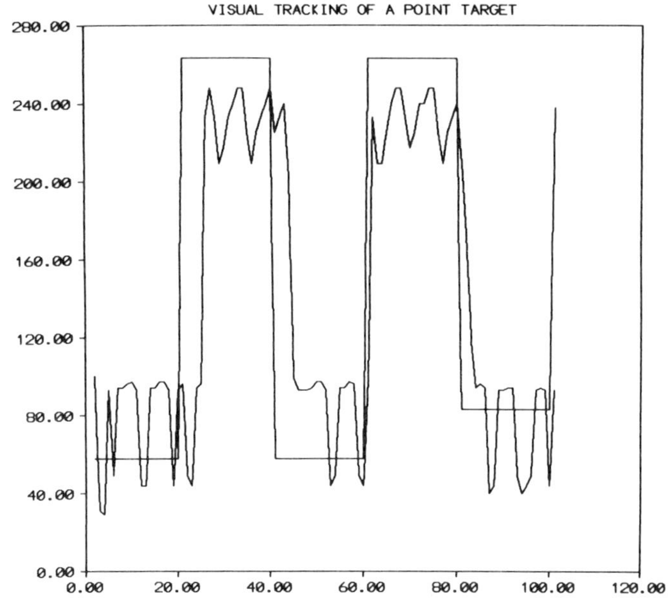

FIGURE 9. Portion of a record showing the visual tracking of a point target jumping one-dimensionally between positions to the left and right of the center screen (Example 4). Although this eye movement signal is corrupted by measurement noise, it was obtained only to give the reader an idea of the signal analysis method used.

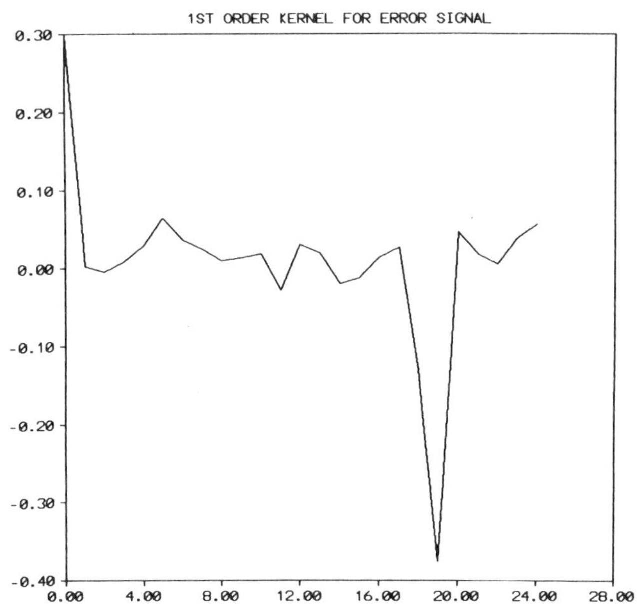

FIGURE 10. The first-order kernel obtained via parallel cascade using the visual tracking error from the same experiment illustrated in Figure 9 (Example 4).

cyclic with a half-period of 20, this accounts for the dip at lag 19. When the analysis was repeated using the error signal from a like (but less noisy) visual tracking experiment, a very similar first-order kernel was obtained. The latter kernel also had a peak at lag 0 and a larger dip at lag 19, but was about 1.5 times the scale of the kernel in Figure 10. Both first-order kernels are shown in Figure 11; their structure is indicative of the symmetry of

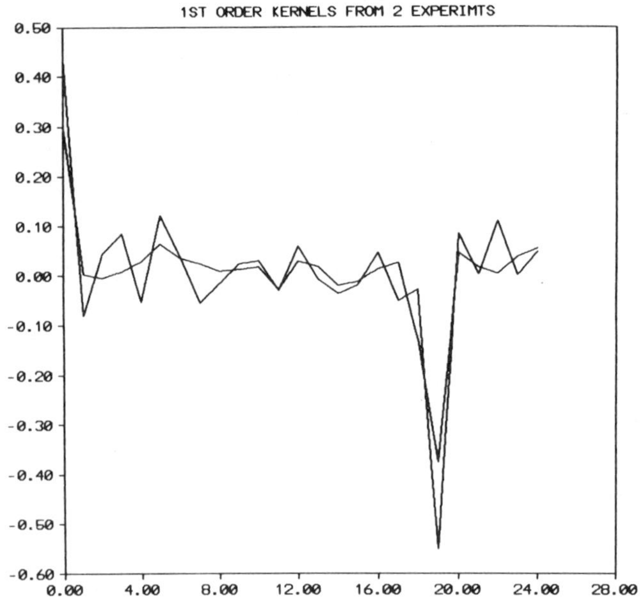

FIGURE 11. The first-order kernel of Figure 10 together with the first-order kernel obtained via parallel cascade from a second visual tracking experiment (Example 4).

the measured visual tracking response to target jumps to the left or to the right of center screen. (In an additional analysis of the error signal from both experiments, the initial large errors occurring immediately after a jump in target position were suppressed. Each of the resulting first-order kernels still had a peak at lag 0 and a dip at lag 19, but the dip was much less deep than previously. Thus, the peak, and probably also the dip, in the first-order kernels of Figure 11 are not due merely to the large error values immediately following target jump.)

Figure 12 shows the diagonal of the second-order kernel, multiplied by 1000, found from the same experiment used to obtain Figures 9 and 10 (no error values suppressed). Note that the diagonal shows a dip at lag 19 (but not also the peak at lag zero present in the first-order kernel). This shape of the kernel diagonal was not clearly reproduced using the error signal from the second visual tracking experiment. There the diagonal of the second-order kernel (Figure 13) showed a dip at lag 0 and a deeper dip at lag 20.

Recall that to analyze the error signal, a parallel cascade with second-degree polynomial nonlinearities was constructed to predict present error values from past values. The presence of nonlinearities in the model proved important in increasing the accuracy of prediction. For the first visual tracking experiment, the parallel cascade with second-degree polynomial nonlinearities had an MSE of 33.4% of the tracking error variance. When the polynomials in the cascade paths were only first degree, the MSE rose to 42.1%. For the second visual tracking experiment, parallel cascades having polynomials of first- and second-degree, respectively, had MSEs of 32.8 and 21.1% of the tracking error variance, respectively.

The above analysis of visual tracking errors is certainly preliminary and used eye movement signals highly corrupted with noise. However, it does appear to suggest that at least some kernel shapes of fair reproducibility are recoverable from the parallel cascade model, that these parallel cascade models show reasonable accuracy, and that nonlinearities are important in attaining this accuracy.

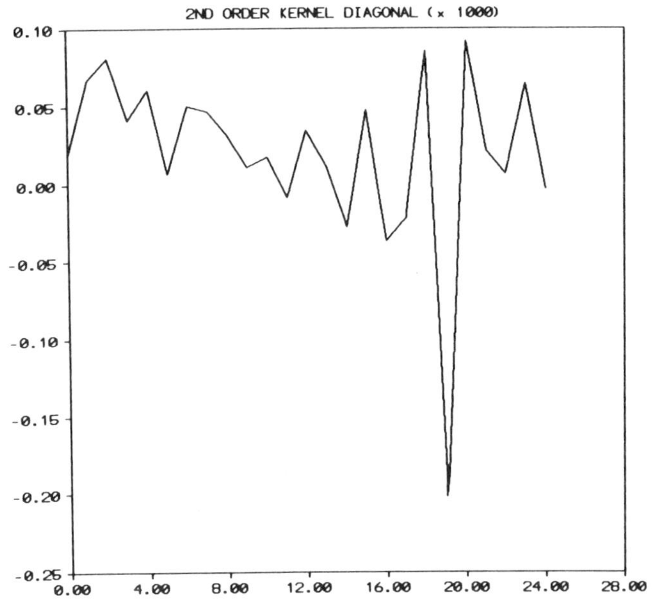

FIGURE 12. The second-order kernel diagonal (× 1000) obtained via parallel cascade using the visual tracking error from the experiment illustrated in Figure 9 (Example 4).

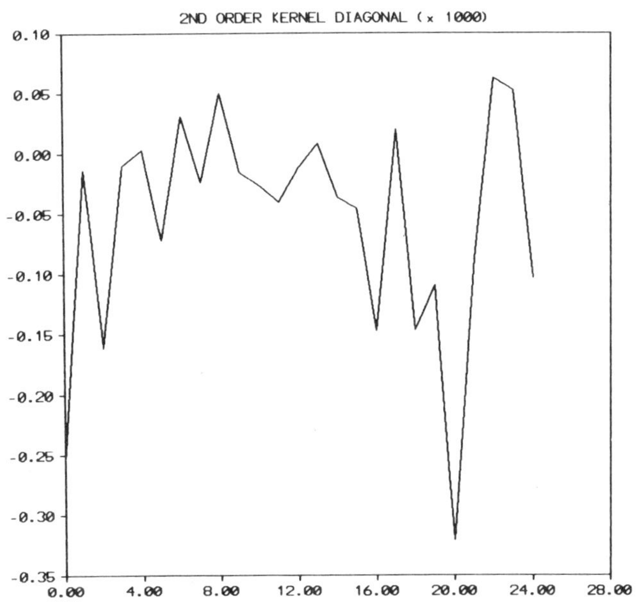

FIGURE 13. The second-order kernel diagonal (× 1000) obtained via parallel cascade from the second visual tracking experiment (Example 4).

IV. CONCLUDING REMARKS

Both fast orthogonal search and parallel cascade identification can usefully be applied to the study of visual and many other biological systems. A very preliminary application of parallel cascade identification to examine visual tracking errors was set out above. Fast orthogonal search has also been used to study a visual system: the present author has collaborated with Drs. Ken-Ichi Naka and Hiroko M. Sakai of the Department of Ophthalmology, New York University Medical Center, to investigate the transition in kernel shape in penetrating from outer to inner catfish retina. A full account is yet to be published, but it has already been noted (Korenberg et al., 1989) that fast orthogonal search detected an evolution in measured kernel shape which is consistent with an overall cascade model structure.

ACKNOWLEDGMENT

This research was supported in part by the Natural Sciences and Engineering Research Council of Canada.

REFERENCES

Abraira, V. and Ibarz, J. M. (1986). Spectral estimation of temporal series at unequal intervals, *Comput. Biomed. Res.*, **19**, 203–212.

Adeney, K. M. and Korenberg, M. J. (1992). Orthogonal search techniques for bearing estimation and spectrum estimation, with comparisons to MUSIC and root-MUSIC, 16th Biennial Symp. Communications, Department of Electrical Engineering, Queen's University, Kingston, Ontario, Canada, 204–209.

Bose, A. G. (1956). A Theory of Nonlinear Systems, Tech. Rep. No. 309, Res. Lab. Elect., MIT, Cambridge, MA.

Box, G. E. P. and Jenkins, G. M. (1976). *Time Series Analysis: Forecasting and Control,* Holden-Day, San Francisco.

Desrochers, A. A. (1981). On an improved model reduction technique for nonlinear systems, *Automatica,* **17**, 407–409.

Dieudonné, J. (1976). *Foundations of Modern Analysis,* Springer-Verlag, Berlin.

Dwight, H. B. (1960). *Tables of Integrals and Other Mathematical Data,* Macmillan, New York.

Glass, L. and Mackey, M. C. (1988). *From Clocks to Chaos,* Princeton University, Princeton, NJ.

Gleick, J. (1987). *Chaos, Making a New Science,* Viking, New York.

Golub, G. H. and Van Loan, C. F. (1989). *Matrix Computations,* 2nd ed., Johns Hopkins University Press, Baltimore.

Haber, R. and Keviczky, L. (1976). Identification of nonlinear dynamic systems, *IFAC Symp. Ident. Syst. Param. Est.,* **1**, 79–126.

Ho, T., Kwok, J., Law, J., and Leung, L. (1987). Nonlinear System Identification, ELEC-490 Project Rep. Department of Electrical Engineering, Queen's University, Kingston, Ontario, Canada.

Hunter, I. W. and Korenberg, M. J. (1986). The identification of nonlinear biological systems: Wiener and Hammerstein cascade models, *Biol. Cybern.,* **55**, 135–144.

Kay, S. M. (1988). *Modern Spectral Estimation: Theory and Application,* Prentice-Hall, Englewood Cliffs, NJ.

Kay, S. M. and Marple, S. L. (1981). Spectrum analysis — a modern perspective, *Proc. IEEE,* **69**, 1380–1419.

Korenberg, M. J. (1982). Statistical identification of parallel cascades of linear and nonlinear systems, *IFAC Symp. Ident. Syst. Param. Est.* (preprints) **1**, 580–585.

Korenberg, M. J. (1985). Orthogonal identification of nonlinear difference equation models, *Proc. Midwest Symp. Circuit Syst.,* **1**, 90–95.

Korenberg, M. J. (1987a). Fast orthogonal identification of nonlinear difference equation and functional expansion models, *Proc. Midwest Symp. Circuit Syst.,* **1**, 270–276.

Korenberg, M. J. (1987b). Functional expansions, parallel cascades, and nonlinear difference equations, in *Advanced Methods of Physiological System Modeling,* Marmarelis, V. Z., Ed., Vol 1, USC Biomedical Simulations Resource, Los Angeles, 221–240.

Korenberg, M. J. (1987c). Exact orthogonal estimation of kernels with biological applications, *IEEE Montech Conf. Biomed. Technol.*, 27–32.

Korenberg, M. J. (1989a). A robust orthogonal algorithm for system identification and time-series analysis, *Biol. Cybern.*, **60**, 267–276.

Korenberg, M. J. (1989b). Fast orthogonal algorithms for nonlinear system identification and time-series analysis, in *Advanced Methods of Physiological System Modeling,* Marmarelis, V. Z., Ed., Vol. 2, Plenum Press, New York, 165–177.

Korenberg, M. J. (1989c). A fast orthogonal search method for biological time-series analysis and system identification, in IEEE Int. Conf. Syst. Man Cybern., Cambridge, MA, 459–465.

Korenberg, M. J. (1989d). Biological time-series analysis via a fast orthogonal search, in Proc. 15th Annu. Northeast. Bioeng. Conf., 149–150.

Korenberg, M. J. (1990). A rapid and accurate method for estimating the kernels of a nonlinear system with lengthy memory, 15th Biennial Symp. Communications, Department of Electrical Engineering, Queen's University, Kingston, Ontario, Canada, 57–60.

Korenberg, M. J. (1991). Parallel cascade identification and kernel estimation for nonlinear systems, *Ann. Biomed. Eng.*, **19**, 429–455.

Korenberg, M. J., Bruder, S. B., and McIlroy, P. J. (1988). Exact orthogonal kernel estimation from finite data records: extending Wiener's identification of nonlinear systems, *Ann. Biomed. Eng.*, **16**, 201–214.

Korenberg, M. J. and Paarmann, L. D. (1989). Applications of fast orthogonal search: time-series analysis and resolution of signals in noise, *Ann. Biomed. Eng.*, **17**, 219–231.

Korenberg, M. J., Sakai, H. M., and Naka, K. -I. (1989). Dissection of the neuron network in the catfish inner retina. III. Interpretation of spike kernels, *J. Neurophysiol.*, **61**, 1110–1120.

Korenberg, M. J. and Paarmann, L. D. (1991). Orthogonal approaches to time-series analysis and system identification, *IEEE Signal Process. Mag.*, **8**, 28–43 July.

Lee, Y. W. and Schetzen, M. (1965). Measurement of the Wiener kernels of a nonlinear system by cross-correlation, *Int. J. Control,* **2**, 237–254.

Liebovitch, L. S. (1989). Introduction to the properties and analysis of fractal objects, processes, and data, in *Advanced Methods of Physiological System Modeling,* Marmarelis, V. Z., Ed., Vol. 2, Plenum Press, New York, 225–239.

Marmarelis, P. Z. and Marmarelis, V. Z. (1978). *Analysis of Physiological Systems. The White Noise Approach,* Plenum Press, New York.

McIlroy, P. J. H. (1986). Applications of Nonlinear Systems Identification, M.Sc. thesis, Queen's University, Kingston, Ontario, Canada.

Mo, L. and Elkasabgy, N. (1984). ELEC-841 Rep., Department of Electrical Engineering, Queen's University, Kingston, Ontario, Canada.

Mohanty, N. C. (1986). *Random Signals Estimation and Identification. Analysis and Applications,* Van Nostrand, New York.

Palm, G. (1979). On the representation and approximation of nonlinear systems. II. Discrete time, *Biol. Cybern.*, **34**, 49–52.

Sandberg, I. W. (1983). Expansions for discrete-time nonlinear systems, *Circuits Syst. Signal Process,* **2**, 179–192.

Shi, J. and Sun, H. H. (1990). Nonlinear system identification via parallel cascaded structure, *Proc. 12th Annu. IEEE Int. Conf. Eng. Med. Biol. Soc.*, **12(4)**, 1897–1898.

Wiener, N. (1958). *Nonlinear Problems in Random Theory,* John Wiley & Sons, New York.

Chapter 8

ALTERATIONS OF FORM VISION AT ISOLUMINANCE IDENTIFIED THROUGH THE USE OF VISUAL TEXTURES

Jonathan D. Victor and Keith Purpura

TABLE OF CONTENTS

I. Introduction 244

II. Background: Isodipole Textures as Visual Stimuli 245
  A. Stochastic Visual Textures and Intracortical Analysis of Form 245
  B. Constraints and a Working Model 247

III. Methods 249
  A. Visual Stimuli 249
  B. Incorporation of Color into Visual Stimuli 250
  C. Measurement of Visual Evoked Potentials 251

IV. Results 251

V. Discussion 255
  A. The Isodipole Response 260
  B. The Appearance/Disappearance Response 261
  C. Significance of Isodipole Stimuli 261

Acknowledgments 262

References 262

I. INTRODUCTION

One emerging theme of central nervous system function is that sensory information is processed in parallel (though not necessarily independent) channels. In vision, physiological (Zeki, 1973; Zeki, 1977; Van Essen and Maunsell, 1983; Hubel and Livingstone, 1987), anatomical (Shipp and Zeki, 1985; Van Essen, 1986; Livingstone and Hubel, 1987a), psychophysical (Lu and Fender, 1972; Treisman, 1982; Cavanagh et al., 1984; Cavanagh et al., 1985; Livingstone and Hubel, 1987b; Schiller and Logothetis, 1990; Schiller et al., 1990), and clinical (Meadows, 1974; Damasio et al., 1980; Zihl et al., 1983; Victor et al., 1989; Farah, 1990) evidence indicate that information is divided among submodalities, including form, color, motion, and depth.

Parallel visual processing begins in the retina (Enroth-Cugell and Robson, 1966; Kaplan and Shapley, 1986; Shapley and Perry, 1986; Shapley, 1990). In the primate cortex (Felleman and Van Essen, 1987), the processing of visual information occurs in two sets of anatomically distinct cortical regions. While there is considerable cross-talk between these two sets of regions, each is dominated by a distinct pattern of connectivity radiating from the striate cortex. One cluster of cortical areas, known as the parietal pathway, is believed to be involved in analyzing spatial location and movement. The other cortical cluster, the inferotemporal pathway, has been implicated in the analysis of color and form. It has been proposed that the parietal pathway is the central continuation of the M cell/magnocellular retinogeniculate pathway and the inferotemporal pathway is the central continuation of the P cell/parvocellular pathway (Felleman and Van Essen, 1987). Recent immunocytochemical studies (De Yoe et al., 1990) have demonstrated that neurons in the M cell/magnocellular retinogeniculate pathway have antigenic determinants which are similar to those of neurons in the cortical parietal pathway. This result supports the concept that the visual system is organized into parallel streams of information processing.

In this parallel pathway framework, understanding the analysis of visual form is particularly critical. For scenes composed of regions of differing hues but identical brightness, visual sensitivity to form is reduced (Livingstone and Hubel, 1987a). This has been interpreted as evidence that form vision is driven by the magnocellular pathway since the responsiveness of the pathway is markedly reduced under such isoluminant conditions. An alternative interpretation of these findings rests on the fact that the activity of all parvocellular neurons is somewhat decreased under isoluminant conditions because of the low cone contrasts inherent in the isoluminant stimuli that can be produced in the laboratory on cathode ray tubes (CRTs). Conversely, only some magnocellular neurons are completely silent for any particular isoluminant stimulus, while the rest (due to their high intrinsic sensitivity and cone input balance) remain active (reviewed in Kaplan et al., 1990). Thus, it is possible that the loss of form vision at isoluminance is a consequence of a loss of parvocellular input to form processing, rather than a loss of magnocellular inputs.

A third view is that changes in form vision at isoluminance have no immediate implications for the relevance of magnocellular or parvocellular inputs: the apparent loss of form vision may simply be a result of the change in the spatial filtering properties of the visual system under isoluminant conditions (Mullen, 1985; Purpura and Victor, 1990). To begin to sort out these possibilities, we explore this third view by examining visual responses to isodipole textures. These stimuli are specifically designed to dissociate spatial filtering from local feature detection.

We will begin with a brief review of the use of isodipole textures in the study of visual processing. We will then present our results concerning isoluminant stimuli. These results demonstrate a selective loss in local form processing at isoluminance which cannot be explained by a selective change in the spatial filtering properties of the visual system: the loss persists while other aspects of pattern processing remain. Finally, we will discuss these findings in the context of the above controversy.

II. BACKGROUND: ISODIPOLE TEXTURES AS VISUAL STIMULI

A. STOCHASTIC VISUAL TEXTURES AND INTRACORTICAL ANALYSIS OF FORM

As recorded on the scalp, the visual evoked potential (VEP) is the sum of a large number of electrical signals generated by a variety of neural processes time-locked to a visual stimulus. As such, it contains contributions both from the postsynaptic response of cortical cells to activity in the geniculate afferents and from intrinsic cortical processing. Our motivation for the use of isodipole textures as visual stimuli was to identify components in the VEP which are due to intracortical interactions and to separate them from VEP components which simply reflected the initial cortical response to the geniculate output.

The rationale for this approach is that the receptive fields of cortical cells are qualitatively different from the receptive fields of cells in the subcortical visual pathways. As a consequence, an appropriate visual stimulus might enable separation of their responses on a physiological basis. Receptive fields of geniculate cells, like their retinal inputs, are reasonably well approximated by linear spatiotemporal filters, or linear spatiotemporal filters combined with rectification (Kaplan and Shapley, 1982 and 1986). The receptive fields of cortical neurons are much more heterogeneous and complex, with selective responses to bars and edges a prominent feature (Hubel and Wiesel, 1959 and 1968). Since the scalp VEP reflects neural activity over a wide region of cortex, stimuli which evoke identical responses from linear filters arrayed over space would elicit equal contributions to the VEP from geniculate afferents, or from their immediate postsynaptic targets. Conversely, any difference between VEPs elicited by such stimuli must reflect intracortical processing of the pattern of afferent activity, and not simply its overall size.

To implement this approach, we needed to construct visually discriminable texture pairs with identical spatial power spectra. The challenge of creating such texture pairs (''isodipole'' textures) is formally identical to that of creating exceptions to Julesz's conjecture (Julesz, 1962) that textures are discriminable only if their (second-order) autocorrelations are distinct. Exceptions (Caelli and Julesz, 1978; Caelli et al., 1978; Julesz et al., 1978) to the Julesz conjecture constitute discriminable textures with the same spatial power spectrum. Furthermore, since exceptions to the Julesz conjecture are discriminated preattentively (Caelli and Julesz, 1978; Caelli et al., 1978; Julesz, 1981), it was likely that interchange of such textures would, indeed, generate distinct VEPs. The difference between such VEPs would then constitute a signal reflecting intracortical processing of form.

Initial studies were based on interchanges between the ''even'' and ''odd'' isodipole textures (Julesz et al., 1978), which share not only second-order, but also third-order autocorrelations, yet are highly discriminable. These textures are illustrated in Figure 1A and formally described below. The raw VEP elicited by interchange between examples of two texture classes may be decomposed into two components (Figure 1B): a ''symmetric component'', consisting of the sum of all even harmonics in the VEP, and an ''antisymmetric component'', consisting of the sum of all odd harmonics. The symmetric component contains contributions from neural mechanisms sensitive to local luminance and contrast changes and is, therefore, similar to the traditional P-100 (Regan, 1989). Because these mechanisms behave like linear spatial and temporal filters and linear filters followed by simple nonlinearities such as rectifiers, their responses to appearance of examples of the two texture classes will be, on the average, the same.

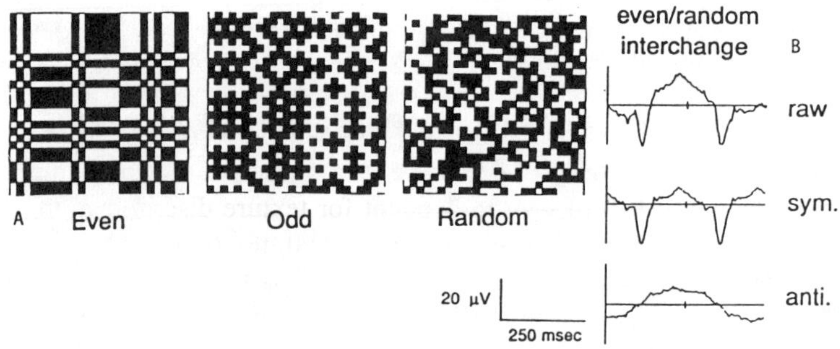

FIGURE 1. (A) Examples of isodipole textures; (B) human VEP responses to interchange of even and random textures, with a check size of 4 min and a contrast of 0.4. Positivity at 0_z (relative to C_z) is plotted downward.

The antisymmetric component, however, contains *differences* between the VEPs elicited by the appearance of examples of one texture class and appearance of examples of the other texture class. Therefore, the antisymmetric component reflects intrinsic cortical processes sensitive to stimulus features (elements of form) other than luminance and spatial contrast. Neural mechanisms that contribute to the antisymmetric component must be more complex than linear filters followed by simple nonlinearities. In summary, the antisymmetric response component reflects the activity of neural elements sensitive to what is different between the two texture classes (elements of form), while the symmetric response component reflects the activity of neural elements sensitive to what is the same between the two texture classes (luminance, contrast, and spatial frequency content).

Interchange between examples of isodipole textures generated a robust antisymmetric VEP component (Victor and Zemon, 1983). The antisymmetric VEP component was more narrowly tuned (both spatially and temporally) than the symmetric VEP component (Victor and Zemon, 1985). One methodological concern (Victor, 1985a) was that the key statistical property of identical power spectra was only true for the ensemble of textures from which particular stimulus examples were drawn, and that the spatial averaging implicit in a scalp VEP was not sufficient to remove local inhomogeneities in the texture stimuli. This was ruled out by the observation that antisymmetric VEP components elicited by different examples of isodipole texture pairs differed by only 10% or so and were similar in phase. With improved apparatus for visual stimulation, we were able to essentially eliminate this residual variability. Rather than average scalp signals across multiple repeats of interchanging a single example of each texture ensemble, we averaged across interchanges between multiple examples (typically 256) of each ensemble.

The distinction between ensemble properties and properties of particular texture examples is subtle but critical. Failure to recognize this difference has led to apparent paradoxes (Yellott and Iverson, 1990). Viewed as an ensemble, the second-order autocorrelation of the even, odd, and random textures are identical. However, an autocorrelation measured from a finite sample of the even texture will certainly differ from an autocorrelation measured from a finite sample of the random texture. (Indeed, autocorrelations measured from two examples of a single texture type will also be different.) Similarly, third-order correlations are identical in an ensemble sense, though not for any finite sample of each texture. Experimentally, the size and character of the antisymmetric response component is relatively independent of whether a single texture example or multiple texture examples are used. The psychophysical task of discrimination between texture is effortless and requires no training on the classes per se; one example (see Figure 1A) suffices. Thus, we conclude that VEP and psychophysical measures of differences between the textures do not depend on the discrepancies between the statistics of a texture example and those of the ensemble from which it is drawn. Rather, texture discrimination must rely on features which consistently distinguish all typical samples of one texture from all typical samples of another. This conclusion is reinforced by the fact that other isodipole texture classes (Victor and Conte, 1991) whose statistics differ from that of the fully random texture first at order 4 cannot readily be distinguished from the random texture — an observation which implies that order-by-order statistics is not a fruitful way to account for texture discrimination.

A second concern was that the statistics of the textures only guaranteed that response of linear neurons would be, on the average, equal across textures; no such guarantee was *a priori* available for cells in which rectification was prominent (Y cells in the cat [Victor and Shapley, 1979] or some M cells in the monkey [Kaplan and Shapley, 1982]). Indeed, since rectification can be approximated by an orthogonal series which includes fourth-order terms, one would anticipate that rectification might provide a basis for discrimination of the even, odd, and random textures. However, a more detailed analysis (Victor, 1985a; Victor and Conte, 1989) demonstrates that the antisymmetric responses generated by a single stage of rectification would be small, and that their qualitative dependence on texture type and check size is inconsistent with experimental data.

The unique dynamics of the antisymmetric VEP component elicited by isodipole interchange provided a clue (Victor, 1985b; Victor and Conte, 1987) that it isolated a stage of visual processing later than simple spatial filtering. The antisymmetric VEP elicited by isodipole interchange was decidedly more low-pass (cutoff at approximately 8 Hz) and had an effective latency approximately 40 ms longer than that of the symmetric VEP component or of appearance-disappearance and contrast-reversal VEPs. This is seen qualitatively in Figure 1B: the symmetric component is characterized by a sharp occiput-positive peak, while the antisymmetric component appears much more low-pass.

Finally, the intrinsic cortical origin of the antisymmetric VEP was directly confirmed in the anesthetized cat: an antisymmetric VEP component could be recorded from the skull while simultaneous records of single- and multiunit activity in the lateral geniculate revealed no difference in responses to "even" and "odd" isodipole textures (Victor, 1986). In contrast, appearance-disappearance checkerboard stimuli elicited differential responses from both cortex and subcortical structures. This was anticipated on the basis of the fact that the population activity of spatial filters would be greatly different for uniform illumination and pattern.

B. CONSTRAINTS AND A WORKING MODEL

We exploited the statistical properties of isodipole stimuli which elicit an antisymmetric response component to deduce constraints for a minimally adequate computational model. As pointed out above, linear filtering alone or rectification along with circularly symmetric spatial filtering could not account for the VEP data (Victor, 1985a). From analysis of responses to even/odd, even/random, and odd/random interchanges and their dependence on check size, we excluded models which consisted of more general linear filters along with a single nonlinearity of essentially arbitrary characteristic (Victor and Conte, 1989).

To further clarify the nature of the spatial interactions critical for eliciting an antisymmetric response component, we created a richer stimulus set by introducing decorrelation into the isodipole textures (Victor and Conte, 1989). Decorrelation was introduced either in a pointwise ("sporadic") fashion, or in a propagated fashion. Sporadic decorrelation is introduced by randomly altering the states of individual checks at a rate ϵ_{spor}. Propagated decorrelation corresponds to values of ϵ_{prop} between 0 and 0.5 in Equation 1. These processes allow for independent manipulation of long-range and short-range correlation structure while retaining the isodipole constraint. The differential dependence of responses on long-range and short-range high-order correlations led us to postulate a model with two nonlinear stages:

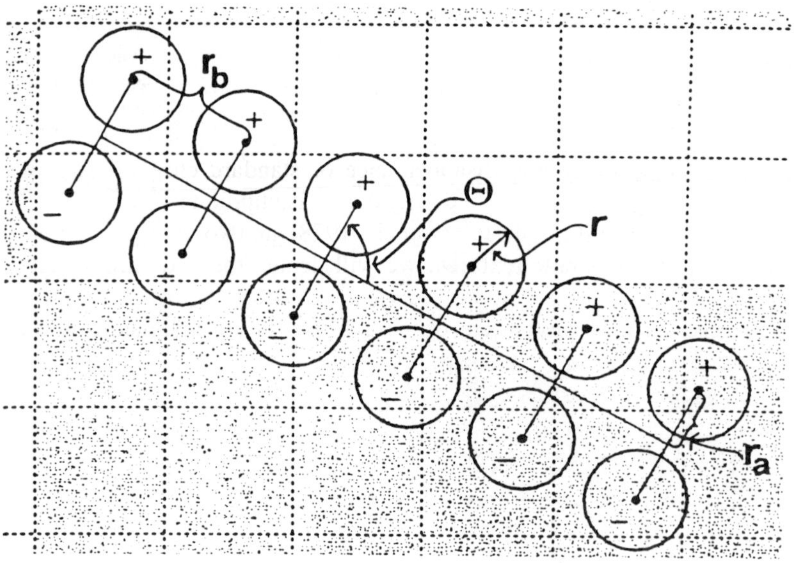

FIGURE 2. A working model for the generation of the antisymmetric response. Subunits consist of rudimentary edge-detectors (circular Gaussian profiles whose outputs are combined antagonistically and then rectified). The sum of the output of N such subunits (here N = 6) is then subjected to a threshold.

a short-range stage, such as rectification of local linear filters, and a second, long-range stage, where local signals were combined in a nonlinear fashion. This model provided the best fits to experimental data if the summation of local signals occurred along a one-dimensional path (Victor and Conte, 1989).

Most recently, we attempted to define the spatial structure of this model by examining responses to other classes of isodipole textures (Victor and Conte, 1991), whose underlying recursion relationship differed from that of Equation 1. Although all of these textures had highly structured fourth-order correlation, only some of them had visually salient structure. Only the textures with visually salient structure generated a significant antisymmetric VEP component. Clearly, the spatial pattern of the correlations, and not just their existence, was crucial. We were able to account for which textures generated antisymmetric responses by a model in which local edge detectors were combined in a cooperative fashion (Figure 2). Crucial features of the model were (1) the edge detectors themselves had to be nonlinear; (2) the combination of edge detector signals had to be nonlinear, and several (typically 6) such signals had to be combined; and (3) the edge detectors had to be aligned in a manner parallel to their preferred orientation. Models based on information-theoretic considerations or which ignored the spatial pattern of the correlation structure failed to account for which textures were visually salient and which were not.

The working model is consistent with the major features of cortical connectivity in visual cortex (Ts'o et al., 1986) and relies on features (long-range interactions, nonlinearities, gain control) thought to be important for cortical function in general. It thus serves as a biologically reasonable framework for the interpretation of our results. But it is important to note that the model (Victor and Conte, 1991) is incomplete in several respects, including the nature of spatial filtering within the subunit, the nature of the cooperative interaction across subunits, and the need for a contrast normalization mechanism to ensure that selectivity is maintained over a wide contrast range. Definitive identification of these aspects of the model are likely to be beyond the reach of VEP and psychophysical studies, and require analysis at the level of single cells.

III. METHODS

A. VISUAL STIMULI

In this study, we recorded steady-state VEPs elicited by patterns whose contrast was modulated along various directions in color/luminance space. We used two kinds of visual stimuli: (1) abrupt appearance and disappearance of standard checkerboard patterns (AP/DIS) and (2) interchange of "even" and "random" isodipole textures (E/R-iso). The isodipole textures (described initially in Julesz et al., 1978) are two-valued colorings of a square lattice. To describe the isodipole textures, we will denote the two chromaticity-luminance values by $+1$ and -1, and the value assigned to the check in row i and column j by $a_{i,j}$. First, we will describe how the pattern of $+1$s and -1s are assigned to the checks, and later we will detail how these states are rendered into chromaticity and luminance values.

The states of checks in the first column, $a_{i,0}$, and the states of checks in the first row, $a_{0,j}$, are chosen in a random, uncorrelated fashion to be -1 or $+1$. Then, states of checks in the interior of the texture are assigned by a recursion rule

$$\text{prob}\left\{a_{i,j} \cdot a_{i-1,j} \cdot a_{i,j-1} \cdot a_{i-1,j-1} = 1\right\} = 1 - \epsilon_{\text{prop}} \qquad (1)$$

In the experiments presented here, we will restrict attention to the "even" texture ($\epsilon_{\text{prop}} = 0$) and the "random" texture ($\epsilon_{\text{prop}} = 0.5$); Figure 1A also shows an example of the "odd" texture ($\epsilon_{\text{prop}} = 1.0$).

All of these textures are uncorrelated at orders 1, 2, and 3. This is the basic property that allows isolation of complex intracortical interactions (Victor 1985a, 1986). To be uncorrelated at order 1 means that each check is independently assigned a luminance about which all modulations are made. In the present context, it implies that each texture class has the same mean luminance. To be uncorrelated at order 2 means that there are no correlations between the luminance values assigned to any pair of checks. In the present context, it implies that each texture class has the same two-dimensional autocorrelation and the same spatial power spectrum. To be uncorrelated at order 3 means that there is no special statistical relationship between the luminance values assigned to any three checks. This does not have a simple intuitive interpretation, but it places constraints on models.

Higher-order correlations have analogous meanings. For the isodipole textures we used, fourth-order correlations are not zero. Indeed, the defining relationship of the "even" texture is that an even number of checks in any 2×2-block of checks are assigned a state of $+1$. This is an explicitly nonzero fourth-order correlation. More generally, fourth-order correlations for these textures are readily described in terms of the correlation parameter c_{prop}, which is related to the error rate ϵ_{prop} by

$$c_{\text{prop}} = 1 - 2\epsilon_{\text{prop}} \qquad (2)$$

At order 4, the nonzero correlations are those of the form

$$c_{I,J} = < a_{i,j} \cdot a_{i-I,j} \cdot a_{i,j-J} \cdot a_{i-I,j-J} > \qquad (3)$$

where $< >$ denotes an ensemble average over the texture. The value of this fourth-order correlation is (Victor and Conte, 1989):

$$c_{I,J} = c_{\text{prop}}^{|IJ|} \; (I, J \neq 0) \qquad (4)$$

B. INCORPORATION OF COLOR INTO VISUAL STIMULI

We incorporated combinations of chromaticity and luminance values into the visual stimli in the following manner. First, we chose a standard background luminance for the three CRT guns (R, G, and B), so that the mixture of the guns' outputs at these luminances appeared white. These luminances will be denoted I_γ, where γ = R, G, or B. The total background luminance was 63 cd/m^2.

The direction and depth of modulation in color-luminance space was defined by a triple of signed contrasts c_γ. The contribution of gun γ to one set of checks ($a_{i,j} = +1$) is $I_{\gamma,j+} = I_\gamma (1 + c_\gamma)$, and its contribution to the other set of checks ($a_{i,j} = -1$) is $I_{\gamma,-} = I_\gamma (1 - c_\gamma)$. Note that contrast for any gun is related to its luminance contributions by $c_\gamma = (I_{\gamma,+} - I_{\gamma,-})/(I_{\gamma,+} + I_{\gamma,-})$. Also, for any set of contrasts (c_R, c_G, c_B), the mean stimulus luminance is always that of the white point (I_R, I_G, I_B).

For example, a stimulus defined by (c_R, c_G, c_B) = (0.2, 0.2, 0.2) has all guns modulated in synchrony and is composed of achromatic checks of luminance 20% above and below background. A stimulus defined by (c_R, c_G, c_B) = (1, 0, 0) is composed of checks with identical unmodulated G-gun and B-gun inputs, but maximally-different R gun inputs: bright pink (+R) checks and dim cyan (−R) checks. A stimulus defined by (c_R, c_G, c_B) = (1, −1, 0) has the R and G guns modulated maximally and in opposite phases. Since the overall luminance contribution of the green gun I_G exceeds that of the red gun I_R, checks with (+R, −G) modulation were dimmer than checks with (−R, +G) modulation. For an observer with the standard CIE luminosity function V_λ, the value of counterphase G contrast which precisely neutralizes the luminance increment provided by $c_R = 1$ is $c_G = -I_R/I_G$, or approximately −0.29. Similarly, a stimulus defined by maximal B contrast and counterphase G contrast is isoluminant for the standard observer when $c_G = -I_B/I_G$, or approximately −0.13.

These stimuli were produced on a Conrac 7351 RGB monitor at a frame rate of 135 Hz. The display subtended a 14° square at a distance of 114 cm. Signals to drive the CRT were generated by specialized hardware modified from that of Milkman et al. (1980). This device provided signals which were independently compensated (via look-up tables) for the nonlinear voltage-intensity characteristics of the R, G, and B guns of the CRT. The luminance and spectral distribution of the phosphor emissions were determined via a Pritchard 703 spectrophotometer; photometric and CIE chromaticity data are provided in Table 1. The background luminances I_γ were chosen so that the maximum gun output required ($2I_\gamma$) was 90% of the available output. Thus, periodic recalibration of the look-up tables allowed us to compensate for intensity loss over time.

The spectral distribution of the phosphor emissions were convolved against the cone absorptions for a standard observer (Wyszecki and Stiles, 1967; Smith and Pokorny, 1975; Schnapf et al., 1987). This provided a relationship between gun contrasts and cone-isolating directions for the standard observer, as indicated in Table 1. As a check of the standard-observer chromatic calibrations, predicted protanopic and deuteranopic color confusions were verified in confirmed dichromatic subjects.

To ensure that the "isoluminant" cardinal directions (Krauskopf et al., 1982; Derrington et al., 1984) were indeed isoluminant for each subject, cone absorptions were corrected for macular and lens absorption to match each observer's empirically determined R/G and B/G brightness match. These brightness matches were determined by flicker photometry of a 2° field at 16.8 Hz. Stimuli in directions related to the cardinal directions of isoluminant space will be described by their cone contrasts. The cone contrast for the β cone (β = L, M, or S) is defined as $c_\beta = (A_{\beta,+} - A_{\beta,-})/(A_{\beta,+} + A_{\beta,-})$, where $A_{\beta,+}$ and $A_{\beta,-}$ represent the maximum and minimum absorptions of the β cone, respectively, as calculated from the adjusted Smith-Pokorny fundamentals.

TABLE 1
Luminance and Chromaticity Data for the Conrac 7351 RGB Display Used

	R gun	G gun	B gun
Mean luminance I_Y (cd/m^2)	13.1	44.6	5.9
CIE chromaticity values			
X	0.6177	0.2773	0.1535
Y	0.3485	0.6004	0.0684
Dominant wavelength (nm)	611	520	466
Purity (%)	76.77	45.32	93.34
Cone excitations produced by unit gun modulations I_Y			
L cone	0.034713	0.088119	0.011023
M cone	0.008122	0.056632	0.009565
S cone	0.000031	0.000223	0.001611
Relative gun modulations for cone-isolating stimuli			
L cone	45.306	−6.504	0.038
M cone	−70.935	28.265	−2.567
S cone	111.187	−123.334	635.833

C. MEASUREMENT OF VISUAL EVOKED POTENTIALS

Electroencephalographic signals were recorded differentially at $C_z - 0_z$ with standard gold cup electrodes applied to the scalp with electrolyte paste, amplified 10,000-fold, and bandpass filtered (0.03 to 100 Hz) prior to digitization. Artifact rejection was performed by the computer prior to averaging, and VEPs were saved for offline analysis. Stimuli were presented in continuous runs of approximately 1 min, and two or three interleaved runs of each condition were obtained for each stimulus condition.

The subject pool for most experiments consisted of three normal volunteers, two male and one female, with normal color vision as tested by Ishihara Plates and the Farnsworth-Munsell 100-hue test (Farnsworth, 1957). Subjects ranged in age from 23 to 35 years and were corrected to normal visual acuity if necessary.

IV. RESULTS

The basic question we address here is whether the processing of visual form is selectively impaired at isoluminance. We therefore examined the VEP elicited by interchange of even and random isodipole textures whose component checks were defined by a range of chromatic and luminance values. Initially, we considered a "sweep" through color-luminance space in which R gun contrast c_R was kept at a maximal value of 1.0, and counterphase G gun contrast $-c_G$ was systematically varied from 0.0 to 1.0.

Representative VEP waveforms for subject KP are shown in Figure 3. The transition to an "even" texture occurs at the beginning of each tracing; the transition to a "random" texture occurs halfway through each tracing, as indicated by the tickmark. For all values of c_G, the E/R-iso response has prominent occiput-positive peaks which occur approximately 140 ms after each texture transition (positive plotted downward). The size and shape of these peaks appear relatively independent of c_G. Other components of the averaged waveform depend on the value of c_G. With no counterphase G modulation ($c_G = 0$), the negativity (upward overshoot) following the transition to the "even" stimulus is much more prominent than the negativity following the transition to the "random" stimulus.

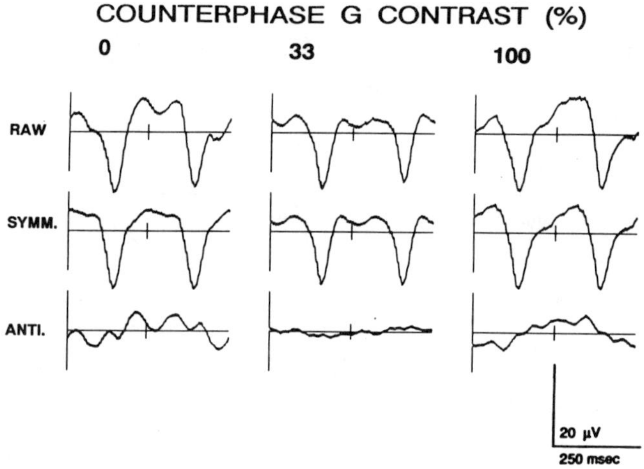

FIGURE 3. VEPs elicited by interchange of even and random textures (E/R-iso). Check size: 13 min. R gun contrast c_R was fixed at 1.0; counterphase G gun contrast $-c_G$ was varied as indicated. The symmetric and antisymmetric components are the sums of the even and odd Fourier components of the raw response. For this subject (KP), psychophysical isoluminance occurred at $c_G = -0.35$. A downward deflection indicates positivity at 0_z relative to C_z.

As discussed above, neural mechanisms sensitive to local luminance and contrast will generate identical responses to the even and random textures, while asymmetries between the responses to even and random textures can only be generated by neural mechanisms sensitive to elements of form. Thus, we decomposed the raw E/R-iso responses shown in the first row of Figure 3 into symmetric and antisymmetric components, by separating the even and odd Fourier components. The symmetric component extracted by this method consisted almost entirely of the occiput-positive transients and was relatively independent of c_G. The antisymmetric component differed dramatically in two respects: (1) the antisymmetric component depended strongly on c_G; the antisymmetric response was similar for all stimuli in which the pink checks had greater luminance ($c_G = 0$) and lesser luminance ($c_G = -1$), but was essentially absent near this subject's isoluminant point for $+R/-G$ modulation; (2) the dynamics of the antisymmetric response differed markedly from the brief transient that characterized the symmetric response.

We carried out a similar analysis for responses for checkerboard AP/DIS stimuli. We used a check size (26 min) which was twice that used for the E/R-iso experiment (13 min). This is because the average horizontal or vertical distance between contours in a random or even texture is twice the check size. Results similar to those described below were also obtained with AP/DIS checks of 13 min.

As seen in Figure 4, the antisymmetric component elicited by AP/DIS was present at all values of c_G, including those near isoluminance. Another feature of the AP/DIS responses is the marked dependence of the symmetric and antisymmetric response waveforms on c_G. This suggests that the VEPs to AP/DIS contain contributions from mechanisms sensitive to both luminance and color, and that these components have different dynamics.

We quantified the symmetric and antisymmetric response components by the sum of their first two harmonics (F2 + F4 for the symmetric response, F1 + F3 for the antisymmetric response). Amplitudes of above 1 μV typically represent significant responses, as judged by trial-to-trial reproducibility. The dependence of these measures on c_G for E/R-iso is summarized for three subjects in Figure 5. For all three subjects, the antisymmetric component becomes undetectable near isoluminance. All subjects showed a decrease in the

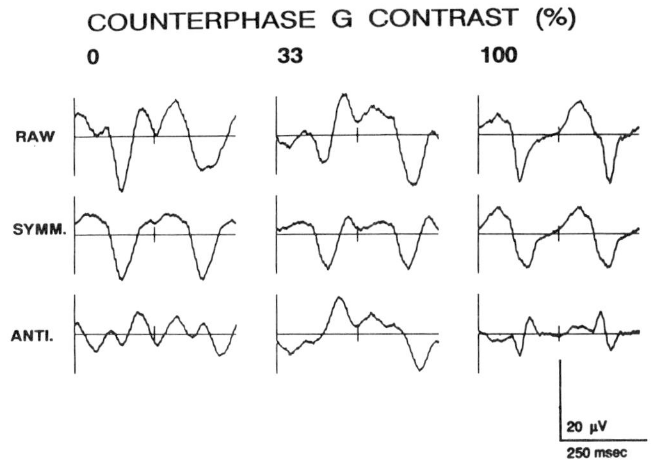

FIGURE 4. VEPS elicited by checkerboard AP/DIS. Check size: 26 min. R gun contrast c_R was fixed at 1.0; counterphase G gun contrast $-c_G$ was varied as indicated. Subject: KP. Symmetric and antisymmetric components were determined as in Figure 3.

size of the symmetric response at isoluminance. The amount of the decrease varied between subjects (approximately 20% for KP, 50% for JV and MC). In all cases, the symmetric response component was readily detectable even at isoluminance, and the change in the size of the antisymmetric response component was out of proportion to the change in size of the symmetric response component. This internal control shows that the loss of the antisymmetric response near isoluminance is not a nonspecific effect of a decrease in the strength of the subcortical input.

AP/DIS data for these subjects are shown in Figure 6. Symmetric and antisymmetric responses show a very different dependence on c_G. In all subjects, the overall size of the symmetric response component decreased near isoluminance. The antisymmetric response component always had a maximum near isoluminance. A comparison of the antisymmetric responses in Figures 5 and 6 indicates that the absence of an antisymmetric response to E/R-iso at isoluminance cannot be accounted for on the basis of an overall loss in sensitivity to contrast. Indeed, it appears that chromatic stimulation is at least as effective as luminance stimulation in driving the AP/DIS antisymmetric response component.

In the data presented so far, isoluminance was approached by adding counterphase modulation of the G gun of the CRT, to maximal modulation of the R gun. We found analogous results in other directions of color-luminance space. Figure 7 shows a summary of symmetric and antisymmetric response components elicited with maximal B gun modulation (c_B = 1) and varying amounts of counterphase G gun modulation. The isoluminant point has shifted to a lower value of counterphase modulation ($-c_G$ = 0.17 for this subject). Again, for E/R-iso, the antisymmetric component becomes almost undetectable near isoluminance, while for AP/DIS, this component is maximal.

Finally, we examined responses to E/R-iso and AP/DIS within the isoluminant plane. A key isoluminant stimulus (Derrington et al., 1984) is the "L-M" cardinal direction, in which L and M cones are driven in antiphase, and the S cones are not modulated. Modulation in any direction in the isoluminant plane produces stimuli which vary in chromatic contrast, not luminance contrast. The amount of chromatic contrast can be quantified by the amount of modulation produced in a particular cone type. The E/R-iso responses are summarized in Figure 8A. The symmetric response component grew steadily with chromatic contrast, while the antisymmetric response component was essentially undetectable at all chromatic

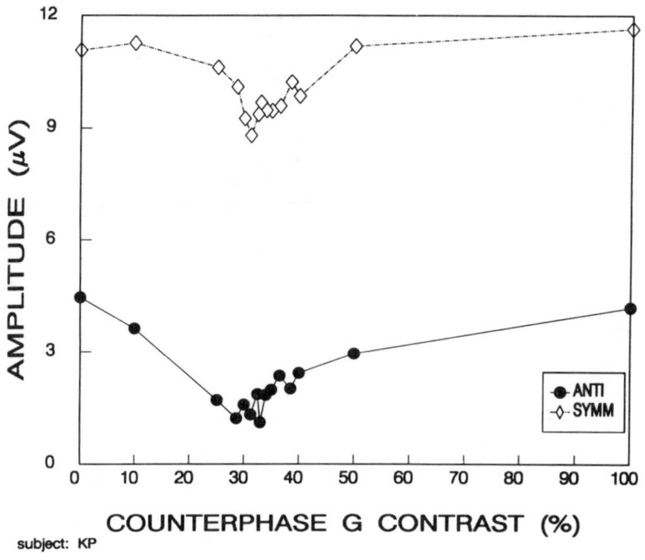

A

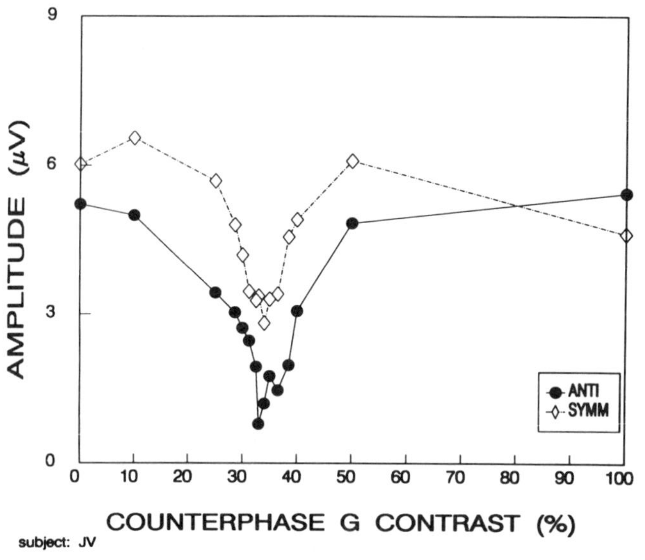

B

FIGURE 5. Symmetric and antisymmetric components of the response to E/R-iso as quantified by their Fourier coefficients (F2 + F4 for the symmetric response, F1 + F3 for the antisymmetric response). R gun contrast c_R was fixed at 1.0; counterphase G gun contrast $-c_G$ was varied as indicated. (A) Subject KP (data from Figure 3); (B) subject JV, isoluminance at $c_G = -0.30$. (C) subject MC, isoluminance at $c_G = -0.33$.

contrasts. For AP/DIS stimuli (Figure 8B), both symmetric and antisymmetric responses were present even at low chromatic contrast. A striking finding was that the antisymmetric response component was larger than the symmetric response component at all contrasts. Similar results were obtained for the other cardinal isoluminant direction (isolated S-cone modulation), as well as oblique directions within the isoluminant plane.

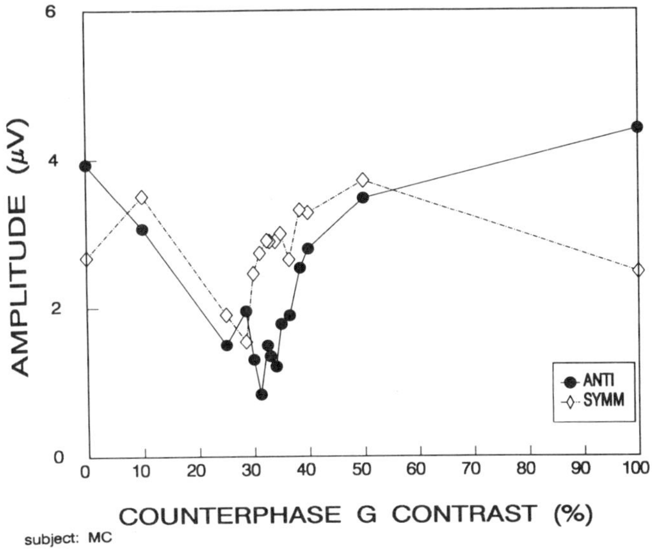

FIGURE 5C.

In all of the AP/DIS data presented so far, the antisymmetric response component exceeded the symmetric response component at or near isoluminance. For all isoluminant stimuli, only relatively small L cone contrasts are accessible. This is because the M cone must be modulated in antiphase, but the L and M cone absorption spectra overlap substantially, and the CRT phosphor emissions are broadband. Thus, it was possible that the larger size of the antisymmetric response was somehow a consequence of low cone contrasts. As a test for this, we repeated the measurement of Figure 8B, but with L and M cone modulation in phase (Figure 8C). Thus, at corresponding modulation depths, the stimuli of Figure 8 B and C are identical in cone contrasts. With in-phase stimulation, the antisymmetric component is always smaller than the symmetric component. This is true both at low cone contrasts (for which corresponding isoluminant stimuli exist) and at higher cone contrasts (which are only accessible in the nonisoluminant direction). Thus, the relative enhancement of the antisymmetric response component to AP/DIS near isoluminance is not a consequence of the low cone contrasts, but rather, of the color-opponent nature of the stimulus.

V. DISCUSSION

Our main finding is that symmetric and antisymmetric components of the VEP elicited by E/R-iso and AP/DIS have dramatically different dependences on chromatic and luminance modulation. For isodipole interchange, the symmetric response component is somewhat reduced at isoluminance, while the antisymmetric response component is essentially eliminated. For AP/DIS, the symmetric response component is also reduced at isoluminance, but the antisymmetric response component is enhanced. These characteristics were found in all three subjects and in multiple directions of modulation in color-luminance space.

To interpret these findings, it is useful to consider what kinds of neural mechanisms contribute to each VEP component. The VEP is a sum of minute time- and space-varying electric fields set up in brain tissue as a consequence of neural activity synchronized to visual stimulation. It contains contributions from the synaptic potentials generated by retinogeniculate afferents, as well as the synaptic potentials generated by corticocortical interactions. Cortical neuron action potentials also produce minute electric fields which contribute to the VEP (Mitzdorf and Singer, 1978, 1979; Kraut et al., 1985; Schroeder et al., 1991).

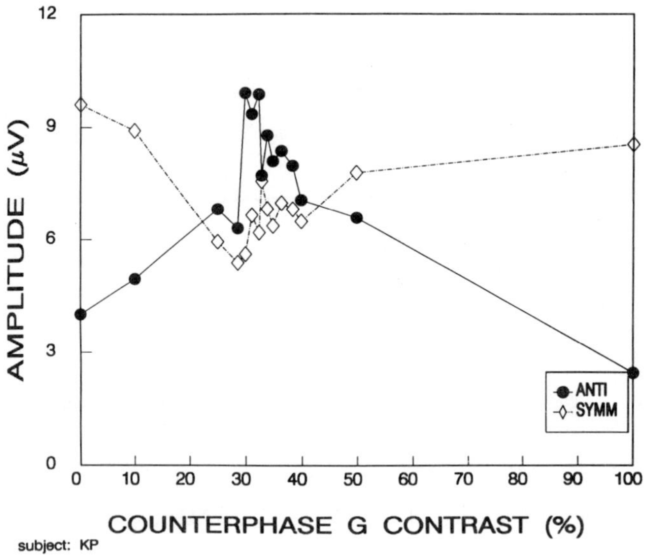

A

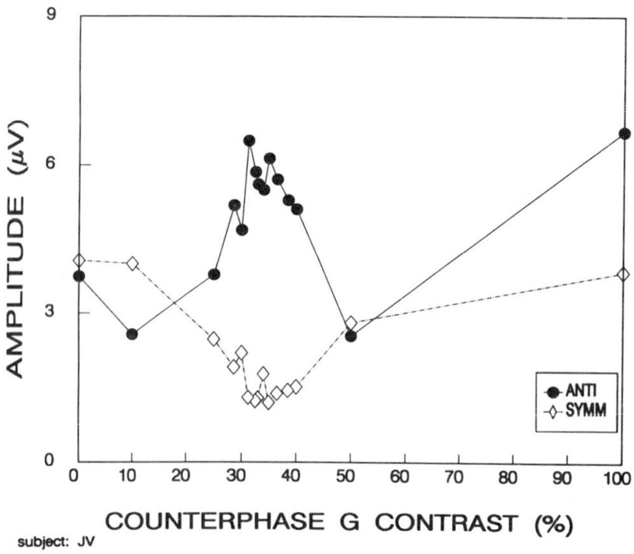

B

FIGURE 6. Symmetric and antisymmetric components of the response to checkerboard AP/DIS as quantified by their Fourier coefficients, as in Figure 5. R gun contrast c_R was fixed at 1.0; counterphase G gun contrast $-c_G$ was varied as indicated. Subjects and isoluminance points as in Figure 5.

Though visual processing occurs in multiple cortical areas, the bulk of the VEP appears to be generated in primary visual cortex, as judged from animal studies (Dagnelie, 1986; Previc 1986; Dagnelie et al., 1989; Schroeder et al. 1991) and from studies examining scalp topography of human VEPs (Lehmann et al., 1982; Srebro, 1985; Maier et al., 1987).

We consider three kinds of mechanisms: mechanisms sensitive to purely local changes in luminance and chromaticity, mechanisms sensitive to spatial contrast, and mechanisms

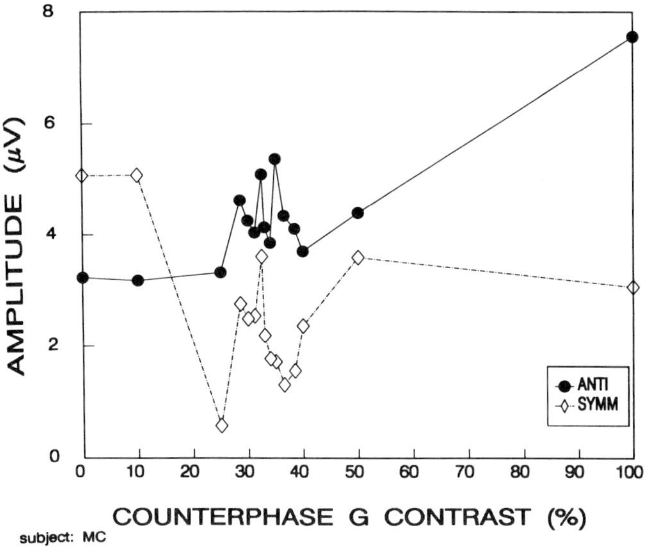

FIGURE 6C.

sensitive to elements of visual form. For simplicity, we discuss achromatic stimuli initially; similar reasoning will apply to stimuli with combined luminance and chromatic modulation.

First, we consider a contrast-reversing checkerboard. At each phase of the reversal, some regions of the stimulus will increase in luminance, while others will decrease. Because of nonlinearities such as a nonlinear intensity/response relationship or the impossibility of a neural response below zero impulses per second, responses of mechanisms sensitive to luminance increase will not be precisely canceled by responses of these mechanisms to neighboring luminance decrease. In addition, mechanisms driven by spatial contrast (e.g., center-surround antagonistic units or units whose receptive field consists of elongated excitatory and inhibitory regions) will be excited by the contrast-reversal of a checkerboard. Similarly, mechanisms sensitive to form elements will be excited at each phase of contrast reversal.

The translational symmetry of the checkerboard implies that both phases of contrast reversal will elicit the same response. Thus, the response to a full cycle of checkerboard contrast-reversal (two reversals) has a symmetric component which contains contributions from mechanisms sensitive to luminance changes, pattern, and contrast, and an antisymmetric component which is zero.

Spekreijse et al. (1973) used AP/DIS stimulation to separate luminance-driven mechanisms from the other mechanisms outlined above. From the point of view of a neuron whose receptive field falls entirely within a single check, both AP/DIS and contrast-reversal checkerboards consist of alternate increases and decreases of local luminance. Furthermore, the sign and magnitude of the luminance change are identical at the time of onset of the pattern and at the time of its disappearance. Thus, a mechanism sensitive only to local luminance would respond in a symmetric fashion to the AP/DIS stimulus and not contribute to the antisymmetric response component. Conversely, mechanisms that are able to detect spatial contrast would be expected to respond differentially to appearance and disappearance of edges. Thus, they would contribute to the antisymmetric component of the response. Such mechanisms might contribute to the symmetric component of the response as well, because their response to appearance and disappearance of edges need not be equal and opposite. For example, a neuron which responds to pattern appearance with a burst and to pattern disappearance with silence will produce symmetric and antisymmetric response components

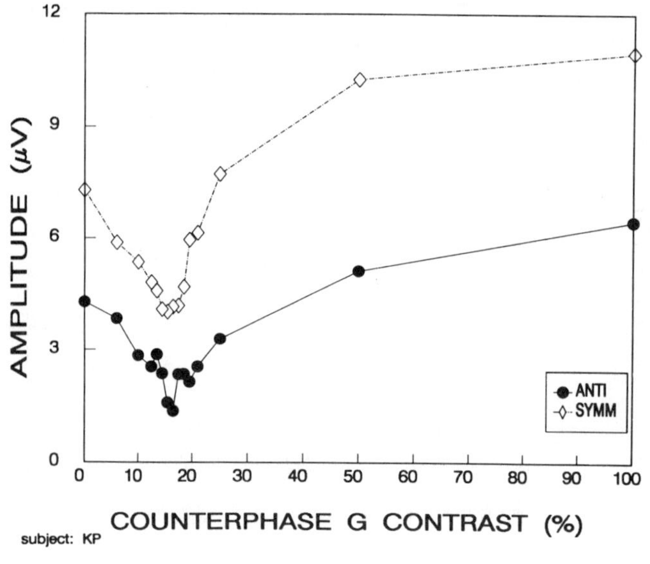

A

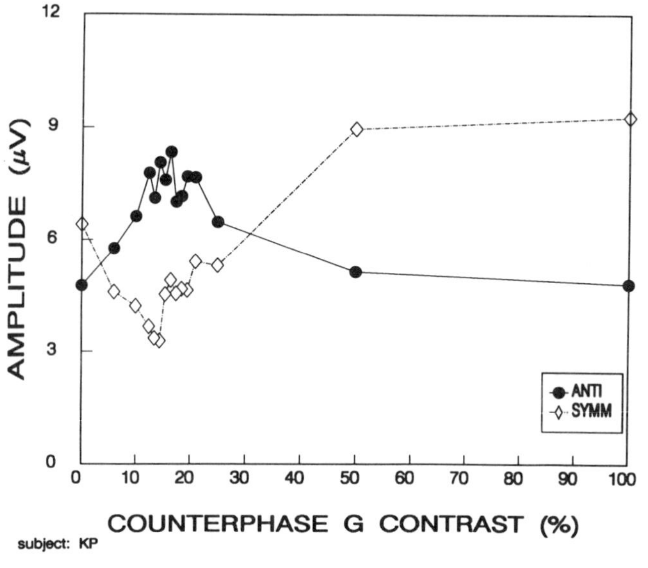

B

FIGURE 7. Symmetric and antisymmetric components of the response to E/R-iso (A) and AP/DIS (B), with B gun contrast c_B fixed at 1.0 and counterphase G gun contrast $-c_G$ varied as indicated. Symmetric and antisymmetric response components were quantified by their Fourier coefficients, as in Figure 5. Subject KP, isoluminance at $c_G = -0.17$.

of similar size. For the same reasons, mechanisms sensitive to form elements would be expected to contribute to both symmetric and antisymmetric response components.

In an analogous manner, the E/R-iso stimulus separates mechanisms sensitive to form from mechanisms sensitive to contrast. Because of their isodipole nature, the even and random stimuli contain the same number of edges and the same total edge length. Thus, mechanisms merely sensitive to local contrast cannot respond in a differential manner to

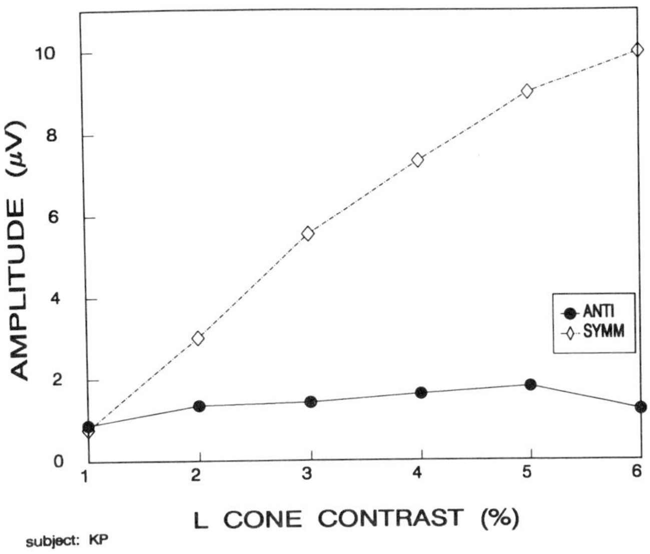

A

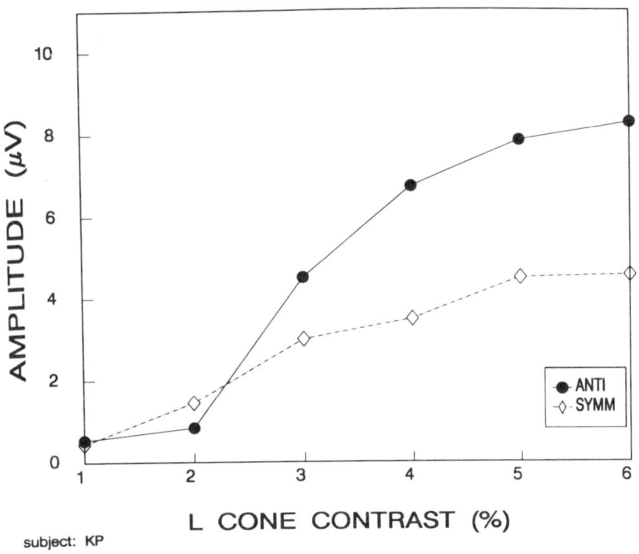

B

FIGURE 8. Symmetric and antisymmetric components of the response to E/R-iso (A) and AP/DIS (B) for stimuli along the L-M cardinal direction of the isoluminant plane. C shows responses to AP/DIS with L and M cones modulated in the same proportion as in B, but in phase. Symmetric and antisymmetric response components were quantified by their Fourier coefficients, as in Figure 5. Subject KP.

these textures and cannot contribute to the antisymmetric response component. The differences between representatives of these two texture classes is, thus, one of form and not local contrast differences. Thus, appropriate mechanisms sensitive to form may contribute to the antisymmetric response component (see Section II).

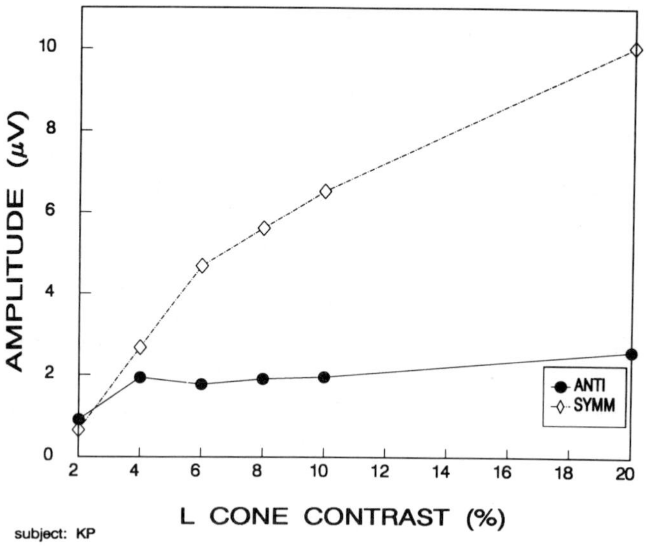

FIGURE 8C.

A. THE ISODIPOLE RESPONSE

With these ideas in mind, the absence of the antisymmetric E/R-iso response at isoluminance suggests an immediate interpretation: mechanisms sensitive to form require luminance differences and not merely chromatic differences. There are two kinds of alternative explanations, however, that require exclusion.

One possibility is that the antisymmetric response is absent simply because of the low cone contrasts required for isoluminant stimulation. This can be ruled out because other response components (the symmetric component of the E/R-iso response and the antisymmetric component of the AP/DIS response) are preserved at isoluminance. Thus, we can be sure that the absence of the antisymmetric response component is not merely a consequence of the low cone contrast of the visual stimuli. A second possibility is that the antisymmetric response disappears at isoluminance because of altered spatial filtering properties of the visual system under these conditions (Mullen, 1985). This is presumably because chromatic contrast sensitivity is determined only by the parvocellular pathway (Shapley and Perry, 1986), while both parvocellular and magnocellular pathways contribute to luminance contrast sensitivity. This explanation can also be eliminated by the fact that the symmetric component of the E/R-iso response was preserved at isoluminance. If the elimination of the antisymmetric response component were due to a change in spatial filtering properties, then the symmetric response component would have been eliminated as well.

A logical consequence of the perceptual difference between textures of similar spatial power spectrum is that spatial phase information is utilized in form perception. Thus, our findings are closely related to those of Troscianko and Harris (1988), who showed loss of sensitivity to relative spatial phase of compound gratings under isoluminant conditions.

The reduction in form vision at isoluminance implies limited utilization of chromatic information by *cortical* form mechanisms. A possible neuroanatomical substrate of this phenomenon is that the cytochrome-oxidase rich blobs of primary visual cortex are relatively nonselective for orientation, but highly sensitive to color, while interblob cells show complementary properties (Livingstone and Hubel, 1987b; Ts'o and Gilbert, 1988).

However, it is likely to be an oversimplification to conclude that form vision is driven only by achromatic information conveyed by the M-cell pathway. One problem with this view is that alterations in chromatic components of the stimulus change the antisymmetric response component (compare first and third antisymmetric response components of Figure 3A). This provides evidence that cortical form mechanisms utilize both luminance and chromatic information, provided that luminance inputs exceed a threshold. Furthermore, the overall size of the antisymmetric response component (Figure 5) is small for a range of R/G balances near isoluminance and then rises abruptly. This is inconsistent with the properties of the magnocellular pathway, which does not have a threshold (Shapley and Perry, 1986).

Furthermore, luminance information is not signaled exclusively by the magnocellular pathway, as demonstrated by preservation of brightness discrimination in macaques with selective magnocellular lesions (Schiller et al., 1990). These lesion studies also demonstrated preservation of pattern and shape discrimination following selective lesions of either magnocellular or parvocellular pathways in most circumstances, supporting the notion that both pathways contribute to form vision. Since the shape discrimination tasks in these studies might have been made on the basis of spatial frequency differences rather than form differences per se, the finding that parvocellular inputs appeared crucial for some discriminations (square vs. circle) can be interpreted on the basis of either cortical or subcortical mechanisms.

B. THE APPEARANCE/DISAPPEARANCE RESPONSE

In view of previous studies (Regan, 1973; Paulus et al., 1986; Previc, 1986; Murray et al., 1987), we anticipated that the AP/DIS would not disappear at isoluminance. Furthermore, since the symmetric component of that response is likely to be driven in large part by luminance mechanisms, a reduction in the symmetric component at isoluminance was also expected (Regan, 1970, 1973).

Our finding that the antisymmetric component of the AP/DIS response was consistently larger at isoluminance than when luminance differences were present was surprising. To a first approximation, the luminance and chromatic sensitivity of neurons in primary visual cortex (Lennie et al., 1990) may be described by their preferred axis in color-luminance space. However, there is no evidence that most neurons prefer isoluminant modulations; on the contrary, most neurons respond optimally to stimuli which contain both luminance and chromatic differences (Lennie et al., 1990). The notion that there are specialized areas which respond to color (Zeki, 1973, 1977) does not immediately account for these results either; one would have to postulate that such chromatic responses would be suppressed by luminance differences, for which there is little evidence.

One possible explanation for this finding (which must be regarded as speculative at present) is that responses to chromatic differences are in some sense more ''nonlinear'' than responses to luminance differences. If this is indeed the case, counterphase modulations in isoluminant directions would have less of a tendency to cancel each other out than counterphase modulations of luminance. Superimposed luminance variations might then reduce the VEP by a phenomenon akin to ''linearization'' of a nonlinear system by an auxiliary input (Spekreijse and Oosting, 1970).

C. SIGNIFICANCE OF ISODIPOLE STIMULI

We find that form processing (as distinguished from spatial filtering) only occurs in the presence of luminance differences. We caution that this finding does *not* imply selective magnocellular participation in form analysis, since (1) chromatic information is likely to be used in form analysis, provided that luminance information exceeds a threshold, and (2) luminance information is carried on both magnocellular and parvocellular pathways.

This study illustrates the usefulness of stimuli with carefully designed statistics for probing the visual system. The requirement for luminance information would not have been apparent had we not been careful to remove spatial frequency cues from the stimulus. Indeed, the AP/DIS response can be driven by exclusively chromatic cues. In a similar vein, we used isodipole stimuli to show that spatial frequency cues serve as tokens for motion correspondence, but form cues devoid of spatial frequency differences do not (Victor and Conte, 1990).

Grating stimuli have served vision research well, not because they are natural or ethologically significant stimuli, but because they are useful tools for analyzing spatial filtering processes (Campbell and Green, 1965; De Valois and De Valois, 1988). In a similar manner, isodipole stimuli are useful, not because they are "natural" stimuli, but because they are analytical tools that can separate spatial-filtering mechanisms from highly nonlinear mechanisms sensitive to elements of form.

ACKNOWLEDGMENTS

This work was supported in part by NIH grant EY7977 and the Revson Foundation. The authors thank A. Canel and M. Conte for their assistance in this work. A portion of this work was presented at the 1990 meeting of the Association for Research in Vision and Ophthalmology (Purpura and Victor, 1990).

REFERENCES

Caelli, T. and Julesz, B. (1978). On perceptual analyzers underlying visual texture discrimination. I, *Biol. Cybern.*, **28**, 167–175.

Caelli, T., Julesz, B., and Gilbert, E. N. (1978). On perceptual analyzers underlying visual texture discrimination. II, *Biol. Cybern.*, **29**, 201–214.

Campbell, F. W. and Green, D. C. (1965). Optical and retinal factors affecting visual resolution, *J. Physiol. (London)*, **181**, 576–593.

Cavanagh, P., Boeglin, J., and Favreau, O. E. (1985). Perception of motion in equiluminous kinematograms, *Perception*, **14**, 151–162.

Cavanagh, P., Tyler, C. W., and Favreau, O. E. (1984). Perceived velocity of moving chromatic gratings, *J. Opt. Soc. Am.*, **A1**, 893–899.

Dagnelie, G. (1986). Pattern and Motion Processing in Primate Visual Cortex, Thesis, University of Amsterdam.

Dagnelie, G., Spekreijse, H., and van Dujk, B. (1989). Topography and homogeneity of monkey VI studies through subdurally recorded pattern-evoked potentials, *Visual Neurosci.*, **3**, 509–525.

Damasio, A., Yamada, Y., Damasio, H., Corbett, J., and McKee, J. (1980). Central achromatopsia: behavioral, anatomic, and physiological aspects, *Neurology*, **30**, 1064–1071.

De Valois, R. and De Valois, K. K. (1988). *Spatial Vision*, Oxford University Press, New York.

DeYoe, E. A., Hockfield, S., Garren, H., and Van Essen, D. C. (1990). Antibody labelling of functional subdivisions in visual cortex: Cat-301 immunoreactivity in striate and extrastriate cortex of the macaque monkey, *Visual Neurosci.*, **5**, 67–81.

Derrington, A. M., Krauskopf, J., and Lennie, P. (1984). Chromatic mechanisms in lateral geniculate nucleus of macaque, *J. Physiol. (London)*, **357**, 241–265.

Enroth-Cugell, C. and Robson, J. G. (1966). The contrast sensitivity of retinal ganglion cells of the cat, *J. Physiol. (London)*, **187**, 517–552.

Farah, M. (1990). Visual Agnosia, MIT Press, Cambridge.

Farnsworth, D. (1957). *The Farnsworth-Munsell 100-Hue Test for the Examination of Color Discrimination: Manual*, Munsell Color Company, Baltimore.

Felleman, D. J. and Van Essen, D. C. (1987). Receptive field properties of neurons in area V3 of macaque monkey extrastriate cortex, *J. Neurophysiol.*, **57**, 889–920.

Hubel, D. H. and Livingstone, M. S. (1987). Segregation of form, color, and stereopsis in primate area 18, *J. Neurosci.*, **7**, 3378–3415.

Hubel, D. H. and Wiesel, T. N. (1959). Receptive fields of single neurons in the cat's striate cortex, *J. Physiol. (London)*, **148**, 574–591.

Hubel, D. H. and Wiesel, T. N. (1968). Receptive fields and functional architecture of monkey striate cortex, *J. Physiol. (London)*, **195**, 215–243.

Julesz, B. (1962). Visual pattern discrimination, *IRE Trans. Inf. Theory,* **IT-8**, 84–92.

Julesz, B. (1981). A theory of preattentive texture discrimination based on first-order statistics of texture, *Biol. Cybern.*, **41**, 131–138.

Julesz, B., Gilbert, E. N., and Victor, J. D. (1978). Visual discrimination of textures with identical third-order statistics, *Biol. Cybern.*, **31**, 137–149.

Kaplan, E., Lee, B. B., and Shapley, R. M. (1990). New views of primate retinal function, in *Progress in Retinal Research,* Vol. **9**, 273–336. Osborne, N and Chader, J., Eds., Pergamon Press, Oxford.

Kaplan, E. and Shapley, R. M. (1982). X and Y cells in the lateral geniculate nucleus of macaque monkeys *J. Physiol. (London),* **330**, 125–143.

Kaplan, E. and Shapley, R. M. (1986). The primate retina contains two types of ganglion cells, with high and low contrast sensitivity, *Proc. Natl. Acad. Sci. U.S.A.*, **83**, 2755–2757.

Krauskopf, J., Williams, D. R., and Heeley, D. W. (1982). Cardinal directions of color space, *Vision Res.*, **22**, 1123–1131.

Kraut, M. A., Arezzo, J. C., and Vaughan, H. G. (1985). Intracortical generators of the flash VEP in monkeys, *Electroencephalogr. Clin. Neurophysiol.*, **62**, 300–312.

Lehmann, D., Darcey, T. M., and Skrandies, W. (1982). Intracerebral and scalp fields evoked by hemiretinal checkerboard reversal, and modeling of their dipole generators, *Adv. Neurol.*, **32**, 41–48.

Lennie, P., Krauskopf, J., and Sclar, G. (1990). Chromatic mechanisms in striate cortex of macaque, *J. Neurosci.*, **10**, 649–669.

Livingstone, M. S. and Hubel, D. H. (1987a). Connections between layer 4B of area 17 and thick cytochrome oxidase stripes of area 18 in the squirrel monkey, *J. Neurosci.*, **7**, 3416–3468.

Livingstone, M. S. and Hubel, D. H. (1987b). Psychophysical evidence for separate channels for the perception of form, color, movement, and depth, *J. Neurosci.*, **7**, 3416–3468.

Lu, C. and Fender, D. H. (1972). The interaction of color and luminance in stereoscopic vision, *Invest. Ophthalmol. Vision Sci.*, **11**, 482–490.

Maier, J., Dagnelie, G., Spekraijse, H., and Van Dijk, B. W. (1987). Principal components analysis for source localization of VEP's in man, *Vision Res.*, **27**, 165–177.

Meadows, J. C. (1974). Disturbed perception of colours associated with localized cerebral lesions, *Brain,* **97** 615–632.

Milkman, N., Schick, G., Rossetto, M., Ratliff, F., Shapley, R., and Victor, J. (1980). A two-dimensional computer-controlled visual stimulator, *Behav. Res. Methods Instrum.*, **12**, 283–292.

Mitzdorf, U. and Singer, W. (1978). Prominent excitatory pathways in the cat visual cortex (A17 and A18): a current source density analysis of electrically evoked potentials, *Exp. Brain Res.*, **33**, 371–394.

Mitzdorf, U. and Singer, W. (1979). Excitatory synaptic ensemble properties in the visual cortex of the macaque monkey: a current source density analysis of electrically evoked potentials, *J. Comp. Neurol.*, **187**, 71–84.

Mullen, K. T. (1985). The contrast sensitivity of human color vision to red-green and blue-yellow chromatic gratings, *J. Physiol. (London),* **359**, 381–400.

Murray, I. J., Parry, N. R. A., Carden, D., and Kulikowski, J. J. (1987). Human visual evoked potentials to chromatic and achromatic gratings, *Clin. Vision Sci.*, **1**, 231–244.

Paulus, W. M., Hömburg, V., Cunningham, K., and Halliday, A. M. (1986). Colour and brightness coding in the central nervous system: theoretical aspects and visual evoked potentials to homogeneous red and green stimuli, *Proc. R. Soc. London Ser.* **B 227**, 53–66.

Previc, F. H. (1986). Visual evoked potentials to luminance and chromatic contrast in rhesus monkeys, *Vision Res.*, **26**, 1897–1907.

Purpura, K. and Victor, J. D. (1990). Dissociation of cortical form mechanisms and spatial filtering at isoluminance, *Invest. Ophthalmol. Vision Sci.*, **31** (Suppl.), 265.

Regan, D., (1970). Objective method of measuring the relative spectral-luminosity curve in man, *J. Opt. Soc. Am.*, **60**, 856–859.

Regan, D. (1973). Evoked potentials specific to spatial patterns of luminance and color, *Vision Res.*, **13**, 2381–2402.

Regan, D. (1989). *Human Brain Electrophysiology,* Elsevier, New York.

Schiller, P. H. and Logothetis, N. K. (1990). The color-opponent and broad-band channels of the primate visual system, *Trends Neurosci.*, **13**, 392–398.

Schiller, P. H., Logothetis, N. K., and Charles, E. R. (1990). Role of the color-opponent and broad-band channels in vision, *Visual Neurosci.*, **5**, 321–346.

Schnapf, J. C., Kraft, T. W., and Baylor, D. A. (1987). Spectral sensitivity of human cone photoreceptors, *Nature (London),* **325**, 439–441.

Schroeder, C. E., Tenke, C. E., Givre, S. J., Arezzo, J. C., and Vaughan, H. G., Jr. (1991). Striate cortical contribution to the surface-recorded pattern-reversal VEP in the alert monkey, *Vision Res.*, **31**, 1143–1157.

Shapley, R. (1990). Visual sensitivity and parallel retinocortical channels, *Annu. Rev. Psychol.*, **41**, 635–658.

Shapley, R. and Perry, V. H. (1986). Cat and monkey retinal ganglion cells and their functional roles, *Trends Neurosci.*, **9**, 229–235.

Shipp, S. D. and Zeki, S. (1985). Segregation of pathways leading from area V2 to areas V4 and V5 of macaque monkey visual cortex, *Nature (London)*, **315**, 322–324.

Smith, V. C. and Pokorny, J. (1975). Spectral sensitivity of the foveal cone photopigments between 400 and 500 nm, *Vision Res.*, **15**, 161–171.

Spekreijse, H. and Oosting, J. (1970). A method for analyzing and synthesizing nonlinear systems, *Kybernetik*, **7**, 23–31.

Spekreijse, H., Van der Tweel, L. H., and Zuidema, T. (1973). Contrast evoked responses in man, *Vision Res.*, **13**, 1577–1601.

Srebro, R. (1985). Localization of visually evoked cortical activity in humans, *J. Physiol. (London)*, **360**, 233–246.

Treisman, A. (1982). Perceptual grouping and attention in visual search for features and for objects, *J. Exp. Psychol. Hum. Percept. Performance*, **8**, 194–214.

Troscianko, T. and Harris, J. (1988). Phase discrimination in chromatic compound gratings, *Vision Res.*, **28**, 1041–1049.

Ts'o, D. Y., Gilbert, C. D., and Wiesel, T. N. (1986). Relationships between horizontal interactions and functional architecture in cat striate cortex as revealed by cross-correlation analysis, *J. Neurosci.*, **6**, 1160–1170.

Ts'o, D. Y. and Gilbert, C. D. (1988). The organization of chromatic and spatial interactions in the primate striate cortex, *J. Neurosci.*, **8**, 1712–1727.

Van Essen, D. C. (1984). Functional organization of primate visual cortex, in *Cerebral Cortex*, Vol. **3**, Peters, A. and Jones, E. G., Eds., Plenum Press, New York.

Van Essen, D. C. and Maunsell, J. H. R., (1983). Hierarchical organization and functional streams in the visual cortex, *Trends Neurosci.*, **6**, 370–375.

Van Essen, D. C., Newsome, W. T., Maunsell, J. H. R., and Bixby, J. L. (1986). The projections from striate cortex (V1) to visual areas V2 and V3 in the macaque monkey: asymmetries, areal boundaries, and patchy connections, *J. Comp. Neurol.*, **244**, 451–480.

Victor, J. D. (1985a). Complex visual textures as a tool for studying the VEP, *Vision Res.*, **25**, 1811–1827.

Victor, J. D. (1985b). The effect of local complexity on the dynamics of the VER, *Proc. 7th Annu.Conf. IEEE Eng. Med. Biol. Soc.*, 420–425.

Victor, J. D. (1986). Isolation of components due to intracortical processing in the visual evoked potential, *Proc. Natl. Acad. Sci. U.S.A.*, **83**, 7984–7988.

Victor, J. D. and Conte, M. M. (1987). Visual evoked potentials elicited by simple and complex textures: distinct components with similar scalp topographies, in *Evoked Potentials III: The Third International Evoked Potentials Symposium*, Barber, C. and Blum, T., Eds., Butterworths, Boston, 183–189.

Victor, J. D. and Conte, M. M. (1989). Cortical interactions in texture processing: scale and dynamics, *Visual Neurosci.*, **2**, 297–313.

Victor, J. D. and Conte, M. M. (1990). Motion mechanisms have only limited access to form information, *Vision Res.*, **30**, 289–301.

Victor, J. D. and Conte, M. M. (1991). Spatial organization of nonlinear interactions in form perception, *Vision Res.*, **31**, 1457–1488.

Victor, J. D., Maiese, K., Shapley, R., Sidtis, J. and Gazzaniga, M. S. (1989). Acquired central dyschromatopsia: analysis of a case with preservation of color discrimination, *Clin. Vision Sci.*, **4**, 183–196.

Victor, J. D. and Shapley, R. M. (1979). The nonlinear pathway of Y ganglion cells in the cat retina, *J. Gen. Physiol.*, **74**, 671–689.

Victor, J. D. and Zemon, V. (1983). Visual evoked responses to discriminable isodipole textures, *Invest. Ophthalmol. Vision Sci.*, **24** (Suppl.), 188.

Victor, J. D. and Zemon, V. (1985). The human visual evoked potential: analysis of components due to elementary and complex aspects of form, *Vision Res.*, **25**, 1829–1842.

Wyszecki, G. and Stiles, W. S. (1967). *Color Science: Concepts and Methods, Quantitative Data and Formulas*, John Wiley & Sons, New York.

Yellott, J. and Iverson, G. (1990). Triple correlation and texture discrimination, *Invest. Ophthalmol. Vision Sci.*, **31** (Suppl.), 561.

Zeki, S. M. (1973). Colour coding in rhesus monkey prestriate cortex, *Brain Res.*, **53**, 422–427.

Zeki, S. M. (1977). Colour coding in the superior temporal sulcus of rhesus monkey visual cortex, *Proc. R. Soc. London Ser. B*, **197**, 195–223.

Zihl, J., von Cramon, D., and Mai, N. (1983). Selective disturbance of movement vision after bilateral brain damage, *Brain*, **106**, 313–140.

Chapter 9

A FREQUENCY DOMAIN METHOD FOR ISOLATING SPECIFIC KINDS OF NONLINEAR NEURAL PROCESSING AND FOR TESTING NONLINEAR MULTINEURON MODELS AGAINST DATA

M. P. Regan and D. Regan

TABLE OF CONTENTS

I. Introduction 266
- A. Frequency-Domain Methods 267

II. Mathematical Basis of the Technique for Sum-of-Two-Sinusoids Input 268
- A. Half-Wave Linear Rectifier: Response to One Sinusoid 268
- B. Half-Wave Linear Rectifier: Response to the Sum of Two Sinusoids 269
- C. Half-Wave Nth Power Rectifier: Response to the Sum of Two Sinusoids 271
- D. Half-Wave Rectifier Combining Accelerating and Compressive Segments 271
- E. Two or More Half-Wave Rectifiers in Series 273
- F. Two Parallel Rectifiers Whose Summed Outputs Pass Through a Third Rectifier: the Dichoptic Case 274
- G. Frequency-Dependent Nonlinearities 278
- H. Testing the Predictions of the Model Against Data: the Importance of High-Order Terms 278

III. Zoom-FFT and Ultrahigh-Resolution Spectra 280
- A. The Limitations of the FFT 280
- B. Zoom-FFT 280

IV. Experimental Results 283
- A. Measurement of Orientation Tuning, Spatial Frequency Tuning, and Temporal Tuning of Cortical Cells 283
  1. Spatial Frequency Bandwidth 283
  2. Orientation Bandwidth 285
  3. Temporal Bandwidth 286
- B. Investigation of Binocular Neurons 287

V. Discussion 288

Acknowledgments 289

References 289

I. INTRODUCTION

Response asymmetry is a type of nonlinearity that is a common property of single neurons at both peripheral and central levels in mammalian sensory pathways. For example, asymmetrical responses to increase vs. decrease of light intensity (Schiller et al., 1986; Hubel and Wiesel, 1962) and spatial contrast (Movshon et al., 1978; Foster et al., 1985), to leftward vs. rightward motion (Barlow and Levick, 1965), and to away/toward motion in depth (Cynader and Regan, 1978; Poggio and Talbot, 1981) are well-documented properties of visual pathway neurons of cat and nonhuman primates.

Because single-neuron electrophysiology is rarely accepted as an ethical procedure in humans, little is known about human single-neuron electrophysiology. There is, however, ample psychophysical and evoked potential evidence that response asymmetries are common in human sensory pathways. For example, the human visual pathway contains psychophysical channels for increasing vs. decreasing luminance (Clynes et al., 1964), increasing vs. decreasing contrast (Spekreijse et al., 1973), leftward vs. rightward motion (Sekuler et al., 1978), and the direction of motion in depth (Beverley and Regan, 1973).

The general class of asymmetrically responding neurons encompasses a variety of nonlinear response characteristics. Asymmetric, compressive characteristics (Figure 1A) have been described in single-cell responses to luminance, contrast (De Valois et al., 1982; Ohzawa et al., 1982), and velocity (Orban, 1985) in the visual pathway, and as a function of tone intensity in the auditory pathway (Brugge and Reale, 1985). Asymmetric, accelerating characteristics (Figure 1B) have been postulated to explain behavioral data in motion detection (Reichardt et al., 1983) and on human line interval discrimination (Morgan and Regan, 1987). An asymmetric characteristic that combines compression and acceleration with a threshold (Figure 1C) has been proposed to describe contrast transduction in human vision (Wilson, 1980; Legge and Foley, 1980).

This response asymmetry can be compared to half-wave recitification, as illustrated in Figure 2A and B. The response of half-wave rectifier to a sinewave input always has the same polarity, positive in the case of Figure 2A and B. Two half-wave rectifiers can be combined to give a symmetric (Figure 2C) or asymmetric (Figure 2D) full-wave rectifier. The output of an asymmetric rectifier can be either entirely of one polarity (Figure 2D) or of opposite polarities (Figure 2E).

A physiological approximation to constant-polarity full-wave rectification (Figure 2C or D) is offered by the ON-OFF organization of the visual pathway in primates. Both increase (ON) and decrease (OFF) of luminance are signaled by excitation along parallel pathways that differ both in connectivity and in terms of neurochemistry (Schiller et al., 1986). According to Schiller et al., the physiological advantage of this arrangement is that it provides a combination of signaling speed and metabolic efficiency. The reference level of luminance is signaled by inactivity of both ON and OFF channels, but if light-increase and light-decrease were represented by opposite-polarity signals within a single channel, then the reference point would be signaled by some resting level of activity and that would be metabolically wasteful. A further advantage of signaling the reference point by zero activity is to reduce "DC drift" of the reference point. Clynes et al. (1964), who first described full-wave rectification in human vision and eye pupil dynamics, coined the term "rein control" to describe its properties.

Rather than activating a single asymmetrically-responding neuron, it is usually the case that a sensory stimulus activates a cascaded series of asymmetrically-responding neurons along a sensory pathway. This chapter includes a theoretical discussion of the cases that these cascaded rectifiers are DC- or AC-coupled, and that successive rectifiers have the same or different characteristics. The case of parallel-cascaded rectifiers is also discussed.

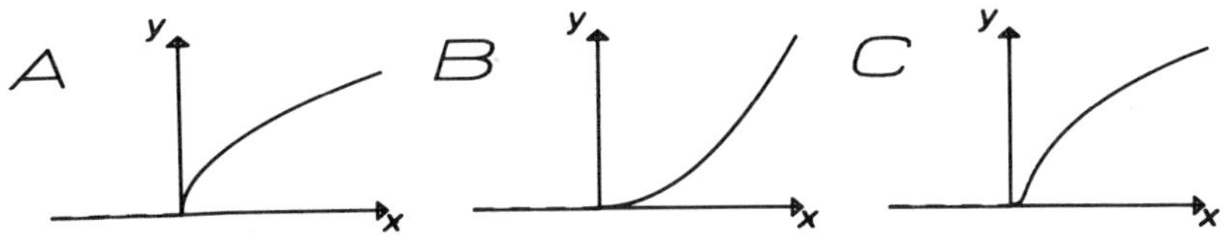

FIGURE 1. (A) Compressive characteristic (square root); (B) accelerating characteristic (square law); (C) combination of accelerating and compressive with a threshold, similar to the contrast transducer function in human vision. The characteristic is a function of x.

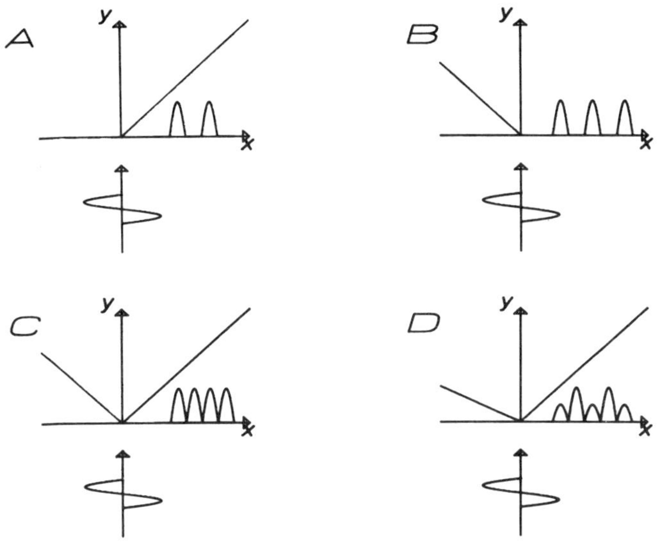

FIGURE 2. (A, B) Linear half-wave rectifiers; (C) symmetric full-wave rectifier; (D) asymmetric full-wave rectifiers. The characteristic of the rectifier is a funciton of x; the input and output signals are functions of time.

In physiological terms, this case corresponds to dichoptic stimulation, where stimulus frequency F_1 is presented to one eye while stimulus frequency F_2 is simultaneously presented to the other eye. If there are nonlinearities both before and after the signals from the two eyes unite, then the situation is clearly different from the monocular or binocular cascaded cases. The approach described here is convenient for analyzing a physiological system from the standpoint of parallel processing, and also in terms of "dissecting" the hierarchical sequence of processing.

A. FREQUENCY-DOMAIN METHODS

Frequency-domain mathematical methods for analyzing different kinds of asymmetric nonlinear characteristics were discussed in the late 1920s and early 1930s. The following analysis is a development of one of the three approaches (specifically, a double Fourier series expansion) described by Bennett (1933). Even when the rectifier limbs are linear, as in Figure 2, one consequence of an asymmetric response characteristic is that the response of a single rectifier to a sinusoidal input of frequency F contains harmonic terms rather than being an F Hz sinewave, as is the case for a linear system. A further consequence is that the response of a single rectifier to an input consisting of the sum of two sinusoids of frequency F_1 and F_2 generally contains multiple discrete terms of frequency $(nF_1 \pm mF_2)$, where n and m are zero or integers (Bennett, 1933).

The mathematical procedure described below shows that if one of the two sinusoidal inputs is held at a constant amplitude while the amplitude of the other input is varied, a family of curves is obtained, one curve for each of the nonlinear cross-modulation ($nF_1 \pm mF_2$) components. The main finding is that, for all the cases investigated so far, this family of curves is characteristic of the responsible nonlinearity. When many high-order* terms are taken into account, the specificity is so high that it may not be fanciful to regard the family of curves generated by a particular nonlinearity as a kind of fingerprint of that nonlinearity.

Several authors have presented mathematical analyses of a single rectifier. One of the other two approaches outlined by Bennett was further discussed by Rice (1944) and developed by Spekreijse (1966) and by Spekreijse and Oosting (1970), who published computation of a single linear and a single square law rectifying element. In a further development, Spekreijse and Reits (1982) obtained the second-order cross-modulation terms (i.e., $\{F_1 + F_2\}$, $\{F_1 - F^2\}$) for a single rectifying element, but did not describe the properties of cross-modulation terms higher than second order. Sternberg et al. (1954) published a set of tables for values of the output of a half-wave linear rectifier for the first-, second-, and fourth-order terms (all odd-order terms above the second are zero); and Sternberg et al. (1955) published tables for values of the ouptut of a half-wave square law rectifier for first-, second-, third-, and fifth-order terms (all even-order terms above the third are zero). In both cases the value of k (the ratio of the amplitudes of the two input frequencies) was less than or equal to 1. In this chapter, the value of k is taken from 0 to 4, and previous work is extended up to the limitation that the characteristic of the single rectifier has a positive power. Further, an analytical treatment of multiple-rectifier models which are cascaded or parallel-cascaded models is presented. These cases are of particular physiological interest because, in principle, they provide more realistic models than does a single rectifier to the known multistage organization of physiological sensory pathways in mammalian brain.

II. MATHEMATICAL BASIS OF THE TECHNIQUE FOR SUM-OF-TWO-SINUSOIDS INPUT**

A. HALF-WAVE LINEAR RECTIFIER: RESPONSE TO ONE SINUSOID

If the input voltage to a half-wave rectifier is given by $e(t) = P\cos(pt + \theta_p)$, then this can be expressed as*** $F(x) = P\cos x$. If the characteristic equation of the half-wave linear rectifier is given by $s = cr,\ r > 0;\ s = 0,\ r \leq 0$, then the output will be

$$f(x) = \begin{cases} F(x), & F(x) > 0 \\ 0, & F(x) \leq 0 \end{cases}$$

$$= \begin{cases} P\cos x, & \cos x > 0 \\ 0, & \cos x \leq 0 \end{cases}$$

This function can be expressed by the Fourier series

$$f(x) = \frac{P}{\pi} + \frac{P}{2}\cos x + \frac{2P}{3\pi}\cos 2x - \frac{2P}{15\pi}\cos 4x + \cdots$$

* The order of a term of frequency $mx \pm ny$ is given by $m + n$, where m and n are integers. For example, $3x \pm 2y$ is a fifth-order term.

** A complete discussion is given in Regan and Regan (1988).

***Throughout this section, the following definitions are used: P and Q are the amplitudes of the two input sinusoids and taken as positive; $k = Q/P$, $p = 2\pi$ (frequency of the input of amplitude P), θ_p = phase, $q = 2\pi$ (frequency of the input of amplitude Q), θ_q = phase, $x = (pt + \theta_p)$, and $y = (qt + \theta_q)$. The constant of proportionality of the rectifier characteristic $c = 1$.

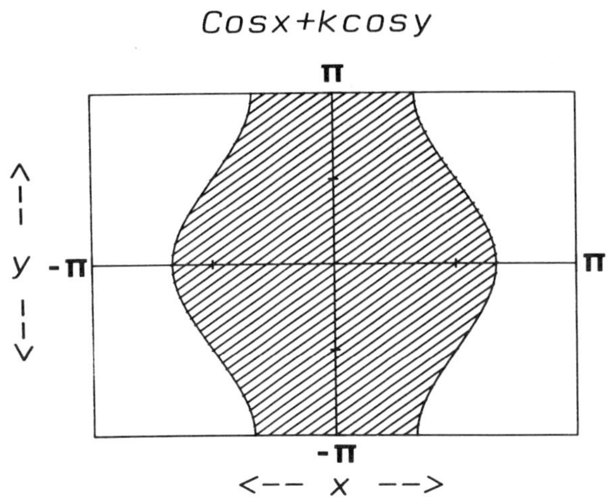

FIGURE 3. Region for which $\cos x + k \cos y \geq 0$.

This shows that, in the output of the rectifier, the DC level is P/π, the amplitude of the first harmonic is $P/2$, the amplitude of the second harmonic is $3P/2\pi$, etc.

B. HALF-WAVE LINEAR RECTIFIER: RESPONSE TO THE SUM OF TWO SINUSOIDS

If the input voltage to the rectifier, whose characteristic is shown in the previous section, is given by

$$\begin{aligned} e(t) &= P \cos (pt + \theta_p) + Q \cos (qt + \theta_q) \\ &= P[\cos (pt + \theta_p) + k \cos (qt + \theta_q)] \end{aligned}$$

where $k = Q/P$, then set

$$f(x,y) = \begin{cases} P(\cos x + k \cos y), & (\cos x + k \cos y) > 0 \\ 0, & (\cos x + k \cos y) \leq 0 \end{cases}$$

where* $x = (pt + \theta_p)$, and $y = (qt + \theta_q)$.

$f(x,y)$ is a surface in and above the x,y plane, bounded by $\cos x + k \cos y = 0$ in the x,y plane. Clearly adding 2π to x or y leaves $f(x,y)$ unaltered. Thus, $f(x,y)$ is a periodic function in both x and y. If $f(x,y)$ is known in the rectangle $(-\pi,\pi) * (-\pi,\pi)$, all its values will be known.

Since $f(x,y)$ is bounded in the rectangle $(-\pi,\pi) * (-\pi,\pi)$ and its first derivatives are bounded, the double Fourier series in x,y of $f(x,y)$ is a valid expansion in this rectangle (Hobson, 1926). If the Fourier series of $f(x,y)$ is valid in the x,y plane, then it is valid on the line $py - qx = p\theta_q - q\theta_p$, found by eliminating t from $x = (pt + \theta_p)$, $y = (qt + \theta_q)$.

The boundaries of $f(x,y)$ are the curves given by $\cos x + k \cos y = 0$, as shown by Figure 3. In the shaded area, $\cos x + k \cos y \geq 0$, elsewhere $\cos x + k \cos y < 0$, giving $f(x,y) = 0$. Since $f(x,y)$ is an even function, its double Fourier expansion will be a cosine series given by

* See footnote *** on p. 268.

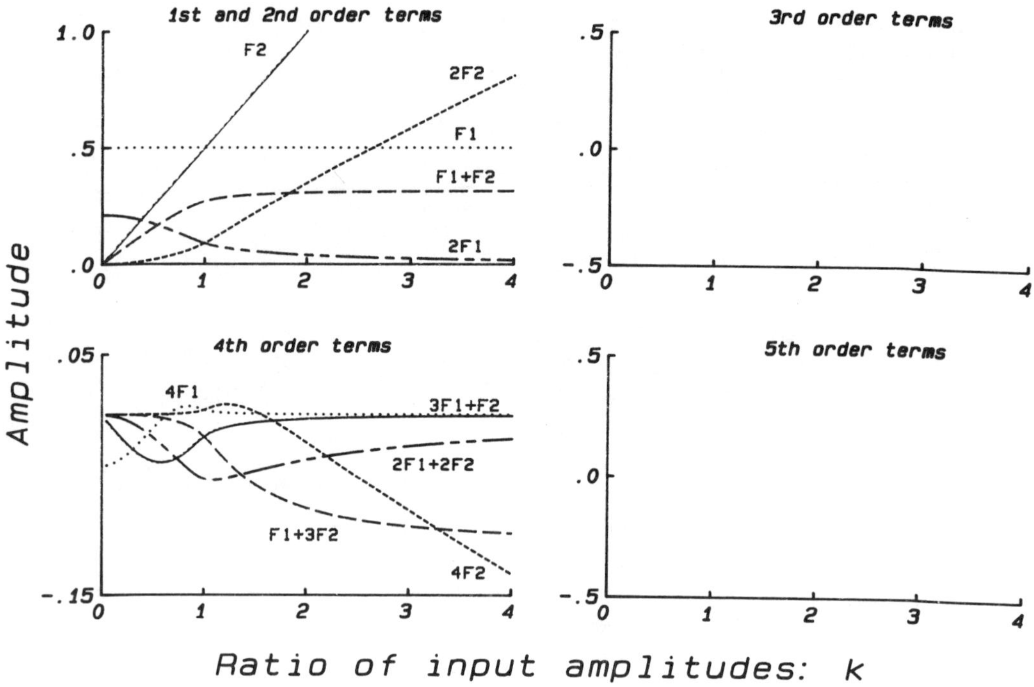

FIGURE 4. The family of curves for a half-wave linear rectifier. This is plotted for the amplitude of each frequency vs. k, where k is the ratio of the input amplitudes of the two frequencies F_1 Hz and F_2 Hz.

$$f(x,y) = \frac{1}{2}A_{00} + \sum_{m=1}^{\infty}\sum_{n=1}^{\infty} A_{\pm mn} \cos(mx \pm ny) + \sum_{m=1}^{\infty} A_{m0} \cos mx + \sum_{n=1}^{\infty} A_{0n} \cos ny$$

where

$$A_{\pm mn} = \frac{1}{2\pi^2}\int_{-\pi}^{\pi}\int_{-\pi}^{\pi} f(x,y) \cos(mx \pm ny)\, dx\, dy$$

$$= \frac{1}{2\pi^2}\int_{-\pi}^{\pi}\int_{-\pi}^{\pi} f(x,y)\, [\cos mx \cos ny \mp \sin mx \sin ny]\, dx\, dy$$

Since the region is symmetrical in both x and y, $A_{\pm mn}$ can be found by using one quarter of the plane. Hence,

$$A_{\pm mn} = \frac{2P}{\pi^2}\int_0^{\pi} \cos ny \int_0^{\arccos(-k\cos y)} (\cos x + k \cos y) \cos mx\, dx\, dy$$

since $f(x,y) = 0$ when $x > \arccos(-k \cos y)$.

The graphs of A_{mn} [the amplitude of the frequency $(nF_1 \pm mF_2)$ in the rectifier output] for $0 \leq m, n \leq 5$ are shown in Figure 4.

C. HALF-WAVE NTH POWER RECTIFIER: RESPONSE TO THE SUM OF TWO SINUSOIDS

In a similar manner, the linear half-wave rectifier case can be extended to the general case of a rectifier of the form $s = cr^N$, $r > 0$ and $s = 0$, $r \leq 0$, where $N > 0$. First set

$$F(x,y) = P(\cos x + k \cos y)$$

as in the case of a linear half-wave rectifier. Then, the output of the rectifier can be considered as the function $f(x,y)$ where

$$f(x,y) = \begin{cases} [F(x,y)]^N, & F(x,y) > 0 \\ 0, & F(x,y) \leq 0 \end{cases}$$

$$= \begin{cases} P^N(\cos x + k \cos y)^N, & \cos x + k \cos y > 0 \\ 0, & \cos x + k \cos y \leq 0 \end{cases}$$

Following the methods in the previous section, $f(x,y)$ is an even function, and its double Fourier expansion will be a cosine series given by

$$f(x,y) = \frac{1}{2} A_{00} + \sum_{m=1}^{\infty} \sum_{n=1}^{\infty} A_{\pm mn} \cos(mx \pm ny) +$$

$$\sum_{m=1}^{\infty} A_{m0} \cos mx + \sum_{n=1}^{\infty} A_{0n} \cos ny$$

where

$$A_{\pm mn} = \frac{1}{2\pi^2} \int_{-\pi}^{\pi} \int_{-\pi}^{\pi} f(x,y) \cos(mx \pm ny)\, dx\, dy$$

$$= \frac{1}{2\pi^2} \int_{-\pi}^{\pi} \int_{-\pi}^{\pi} f(x,y)[\cos mx \cos ny \mp \sin mx \sin ny]\, dx\, dy$$

Since the region is symmetrical in both x and y, $A_{\pm mn}$ can be found by using one quarter of the plane. Hence,

$$A_{\pm mn} = \frac{2P^N}{\pi^2} \int_0^{\pi} \cos ny \int_0^{\arccos(-k\cos y)} (\cos x + k \cos y)^N \cos mx\, dx\, dy$$

since $f(x,y) = 0$ when $x > \arccos(-k \cos y)$. These are shown in Figure 5 for the case $N = 2$ and Figure 6 for the case $N = 0.5$.

D. HALF-WAVE RECTIFIER COMBINING ACCELERATING AND COMPRESSIVE SEGMENTS

For the rectifier illustrated in Figure 1C, which represents the contrast transducer function of the human visual system (Wilson, 1980; Legge and Foley, 1980), the curve equation can be described by

$$s = \begin{cases} 0, & r \leq c \\ d(r - c)^4, & c < r \leq 5c \\ (r - c)^{1/16} - g, & 5c < r \end{cases}$$

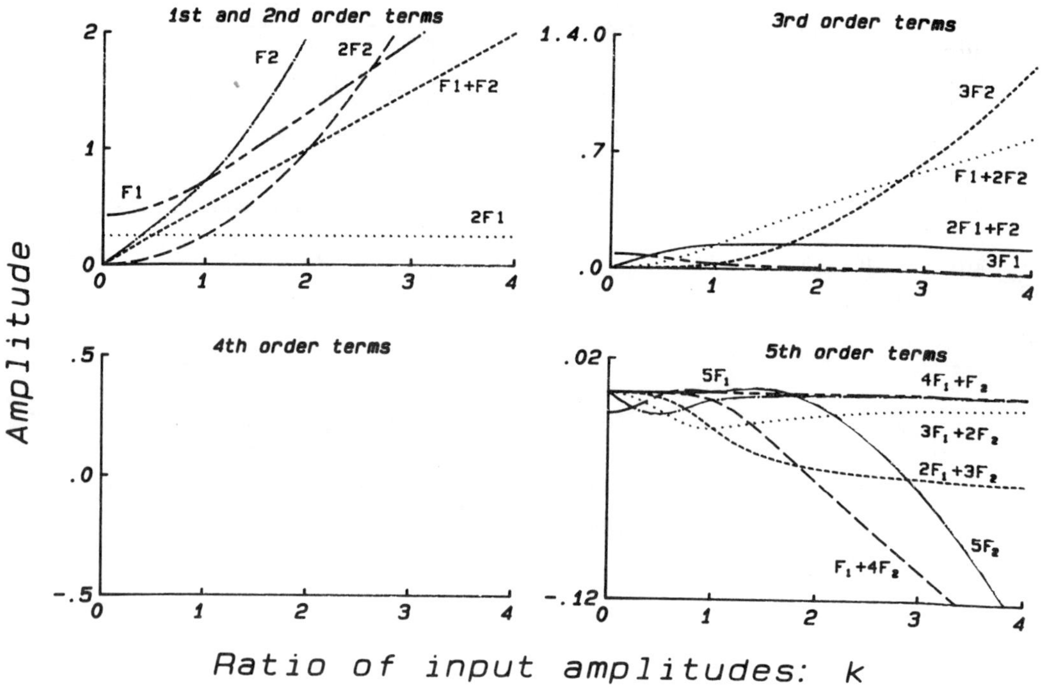

FIGURE 5. The family of curves for a half-wave square law rectifier. This is plotted for the amplitude of each frequency vs. k, where k is the ratio of the input amplitudes of the two frequencies F_1 Hz and F_2 Hz.

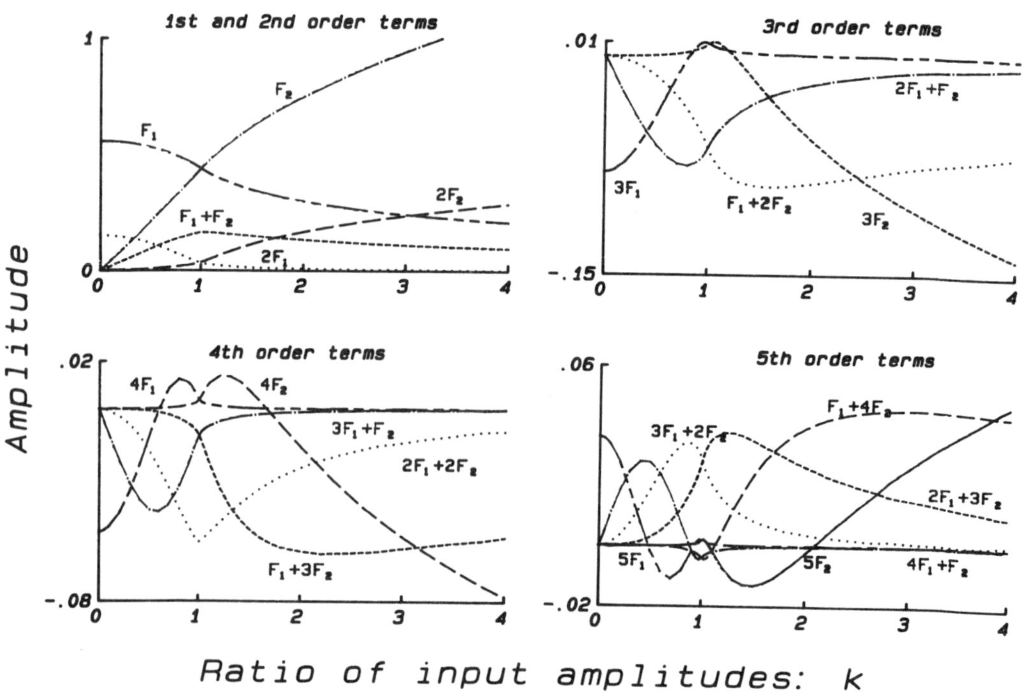

FIGURE 6. The family of curves for a half-wave square root rectifier. This is plotted for the amplitude of each frequency vs. k, where k is the ratio of the input amplitudes of the two frequencies F_1 Hz and F_2 Hz.

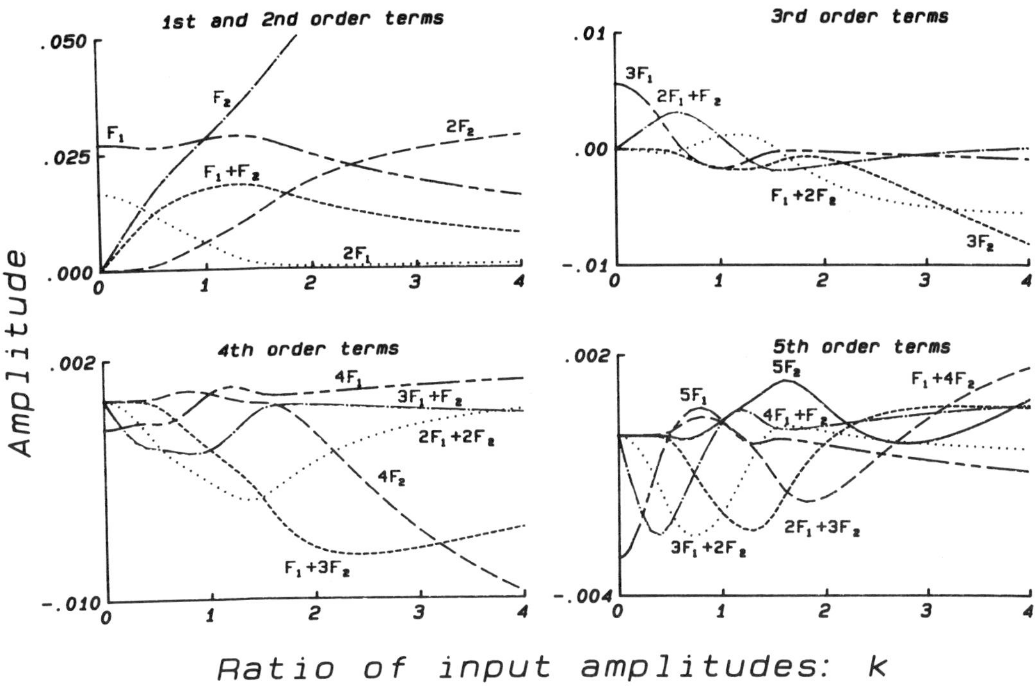

FIGURE 7. The family of curves for a half-wave rectifier whose characteristic combines accelerating and compressive segments with a threshold. This is plotted for the amplitude of each frequency vs. k, where k is the ratio of the input amplitudes of the two frequencies F_1 Hz and F_2 Hz.

where $d = 1/64(4c)^{\frac{63}{16}}$ and $g = \frac{63}{64}(4c)^{\frac{1}{16}}$ and c is chosen suitably. Consequently,

$$f(x,y) = \begin{cases} 0, & \cos x + k \cos y \le c \\ P^4 d(\cos x + k \cos y - \frac{c}{P})^4, & c < \cos x + k \cos y \le 5c \\ P^{\frac{1}{16}}(\cos x + k \cos y - \frac{c}{P})^{\frac{1}{16}} - g, & 5c < \cos x + k \cos y \end{cases}$$

where $x = (pt + \theta_p)$, and $y = (qt + \theta_q)$. So,

$$A_{\pm mn} = \frac{2}{\pi^2} \int_0^{\pi} \cos ny \int_0^{\pi} f(x,y) \cos mx \, dx \, dy$$

This is shown in Figure 7 with $c = \frac{2\pi}{15}$.

E. TWO OR MORE HALF-WAVE RECTIFIERS IN SERIES

If two linear half-wave rectifiers are DC coupled, the output will be the same as a single linear half-wave rectifier. Indeed, if two half-wave rectifiers are DC coupled and the first of the series is a linear rectifier, the final output will be the same as that of the second rectifier alone.

After the two sinusoids pass through the first rectifier ($s = cr^N$, $r > 0$, $s = 0$, $r \leq 0$), the function of the ouptut is given by

$$f(x,y) = \begin{cases} P^N(\cos x + k \cos ny)^N, & (\cos x + k \cos y) > 0 \\ 0, & (\cos x + k \cos y) \leq 0 \end{cases}$$

This has a DC level given by $A_{00}/2$, which is a functional (Volterra, 1959) with parameter k, in the double Fourier series expansion of $f(x,y)$. If these two successive rectifiers are linked by AC coupling, the DC level must be removed, and so the function entering the second rectifier is given by

$$F(x,y) = f(x,y) - A_{00}/2$$

where

$$A_{00} = \frac{2P^N}{\pi^2}\int_0^{\pi}\int_0^{\arccos(-k\cos y)} (\cos x + k \cos y)^N \, dx \, dy$$

After passing through the second rectifier ($s = cr^M$, $r > 0$, $s = 0$, $r \leq 0$), the output is given by

$$\phi(x,y) = \begin{cases} [F(x,y)]^M, & F(x,y) > 0 \\ 0, & F(x,y) \leq 0 \end{cases}$$

This can be represented by a double Fourier series where the coefficients $A_{\pm mn}$ are given by

$$A_{\pm mn} = \frac{2}{\pi^2}\int_0^{\pi} \cos ny \int_0^{\pi} \phi(x,y) \cos mx \, dx \, dy$$

This is represented graphically in Figure 8 for two linear half-wave rectifiers and Figure 9 for $N = 2$, $M = 1/3$.

Clearly this process can be easily extended to obtain the family of curves for three or more cascaded AC-coupled half-wave rectifiers. See Figure 10 for an illustration of three cascaded third-root half-wave rectifiers.

F. TWO PARALLEL RECTIFIERS WHOSE SUMMED OUTPUTS PASS THROUGH A THIRD RECTIFIER: THE DICHOPTIC CASE

In the dichoptic case, the inputs to the two eyes are distinct, i.e., for a two-sinewave input, one eye will see one modulating sinewave of frequency F_1 Hz only and the other eye, the second modulating sinewave of frequency F_2 Hz. In this case the input to rectifier 1 ($s = cr^N$, $r > 0$, $s = 0$, $r \leq 0$) is a single sinewave of frequency F_1, and the input to rectifier 2 ($s = cr^M$, $r > 0$, $s = 0$, $r \leq 0$) in parallel with the first rectifier is a single sinewave of frequency F_2. Then, the ouptut from both rectifiers combine to form the input to the third rectifier ($s = cr^L$, $r > 0$, $s = 0$, $r \leq 0$).

The output of the first rectifier is $f(x)$ where

$$f(x) = \begin{cases} P^N \cos^N x, & \cos x > 0 \\ 0, & \cos x \leq 0 \end{cases}$$

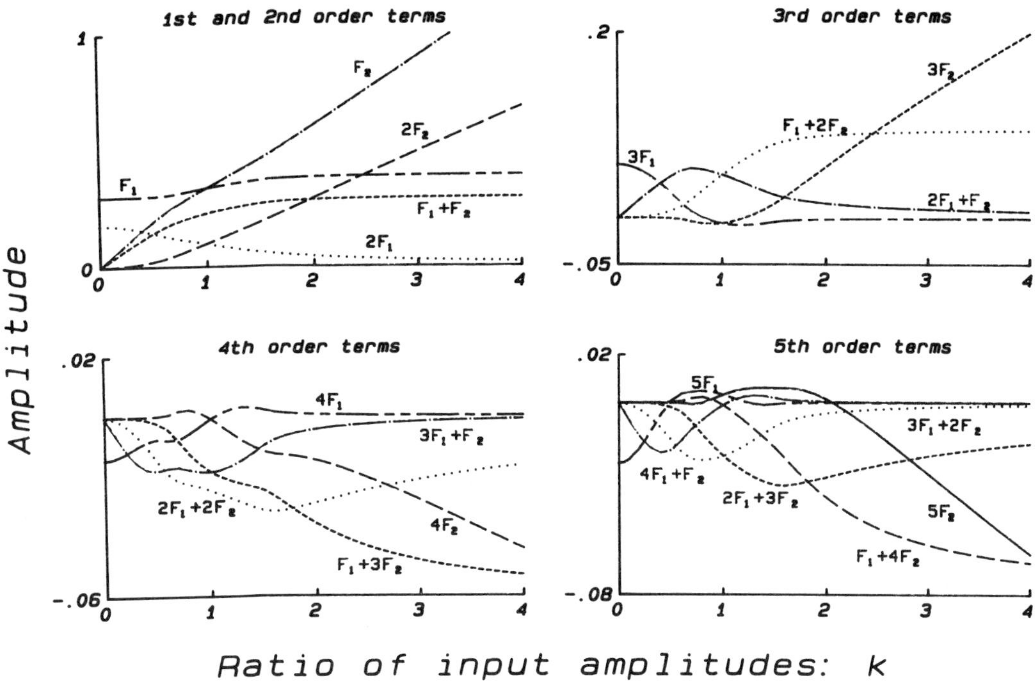

FIGURE 8. The family of curves for two half-wave linear rectifiers in series (AC coupled). This is plotted for the amplitude of each frequency vs. k, where k is the ratio of the input amplitudes of the two frequencies F_1 Hz and F_2 Hz.

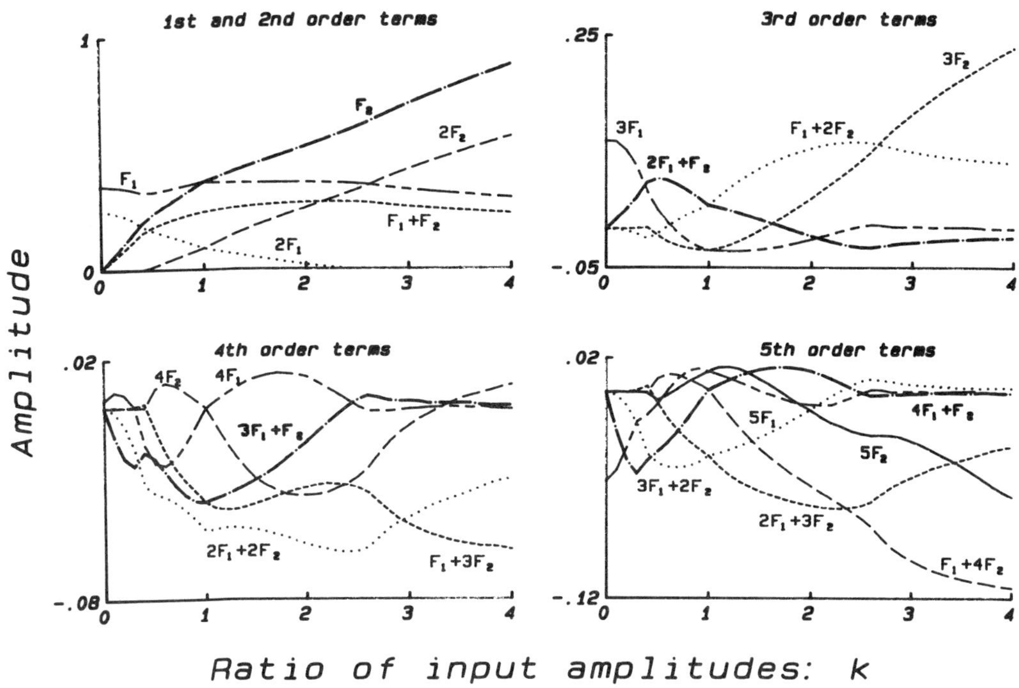

FIGURE 9. The family of curves for a half-wave square law rectifier in series (AC coupled) with a half-wave third root rectifier. This is plotted for the amplitude of each frequency vs. k, where k is the ratio of the input amplitudes of the two frequencies F_1 Hz and F_2 Hz.

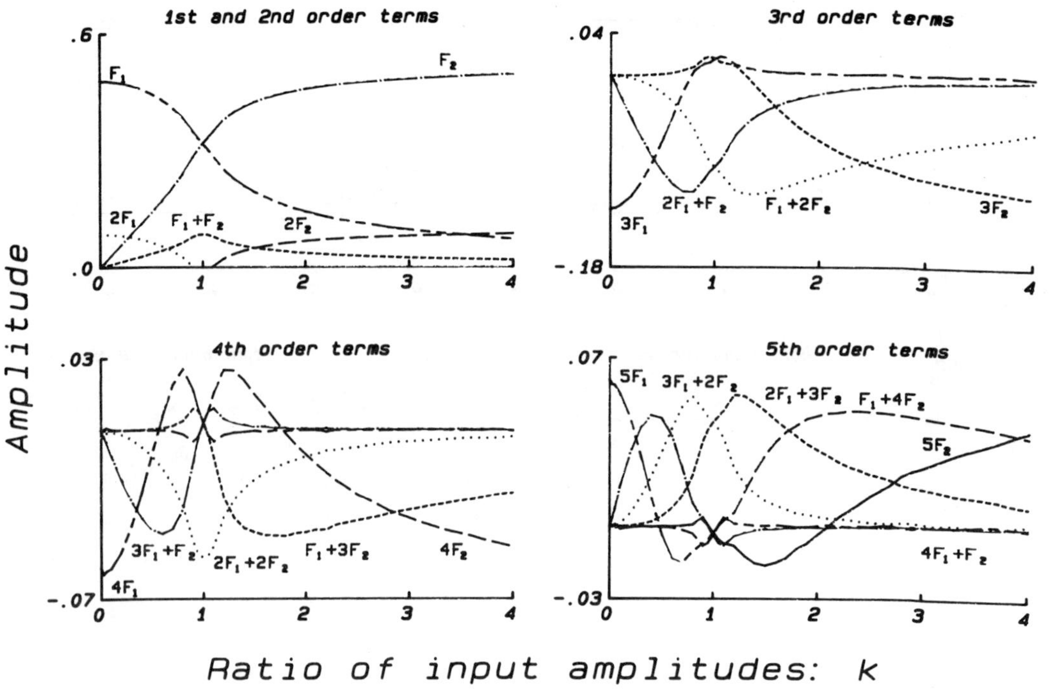

FIGURE 10. The family of curves for three cascaded half-wave third root rectifiers in series (AC coupled). This is plotted for the amplitude of each frequency vs. k, where k is the ratio of the input amplitudes of the two frequencies F_1 Hz and F_2 Hz.

with a DC level of P^N/π. The output of the second rectifier is $g(y)$ where

$$g(y) = \begin{cases} (Pk)^M \cos^M y, & \cos y > 0 \\ 0, & \cos y \leq 0 \end{cases}$$

whose DC level is $(Pk)^M/\pi$. To adjust for the DC level, the input to the third rectifier will be the function

$$h(x,y) = f(x) - P^N/\pi + g(y) - (Pk)^M/\pi$$

The output from the third rectifier is given by

$$H(x,y) = \begin{cases} [h(x,y)]^L, & h(x,y) > 0 \\ 0, & h(x,y) \leq 0 \end{cases}$$

Hence, the coefficients of the double Fourier series can be found, since

$$A_{\pm mn} = \frac{2}{\pi^2} \int_0^\pi \cos ny \int_0^\pi H(x,y) \cos mx \, dx \, dy$$

The dichoptic case modeled as three linear half-wave rectifiers is shown in Figure 11, and Figure 12 shows the case of three half-wave square law rectifiers.

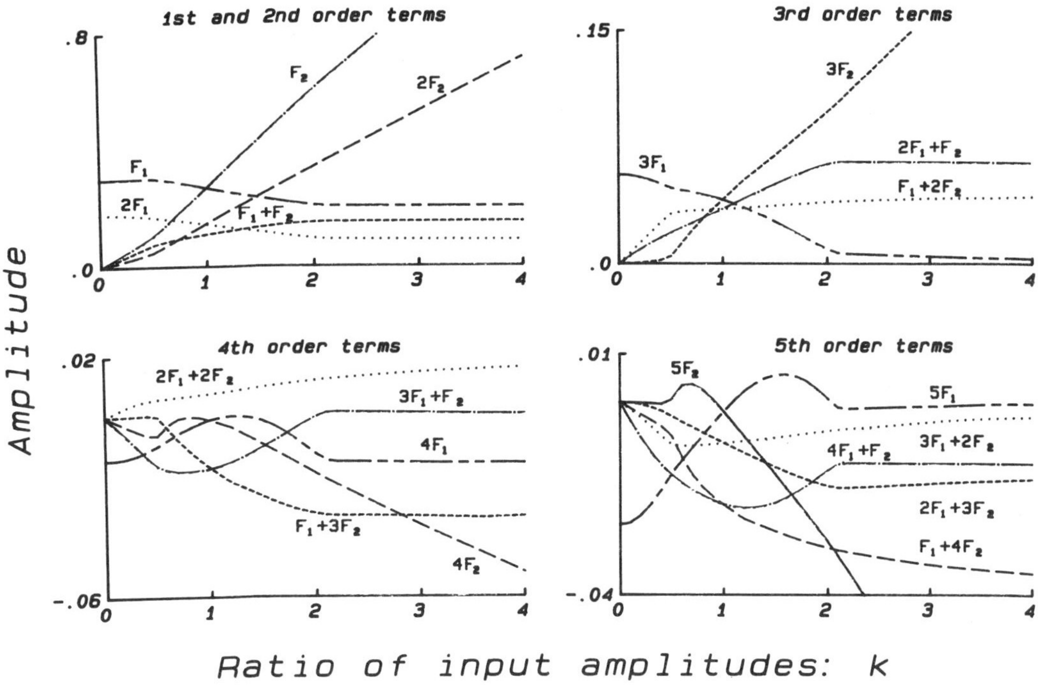

FIGURE 11. The family of curves for three half-wave linear rectifiers joined dichoptically (AC coupled), see text. This is plotted for the amplitude of each frequency vs. k, where k is the ratio of the input amplitudes of the two frequencies F_1 Hz and F_2 Hz.

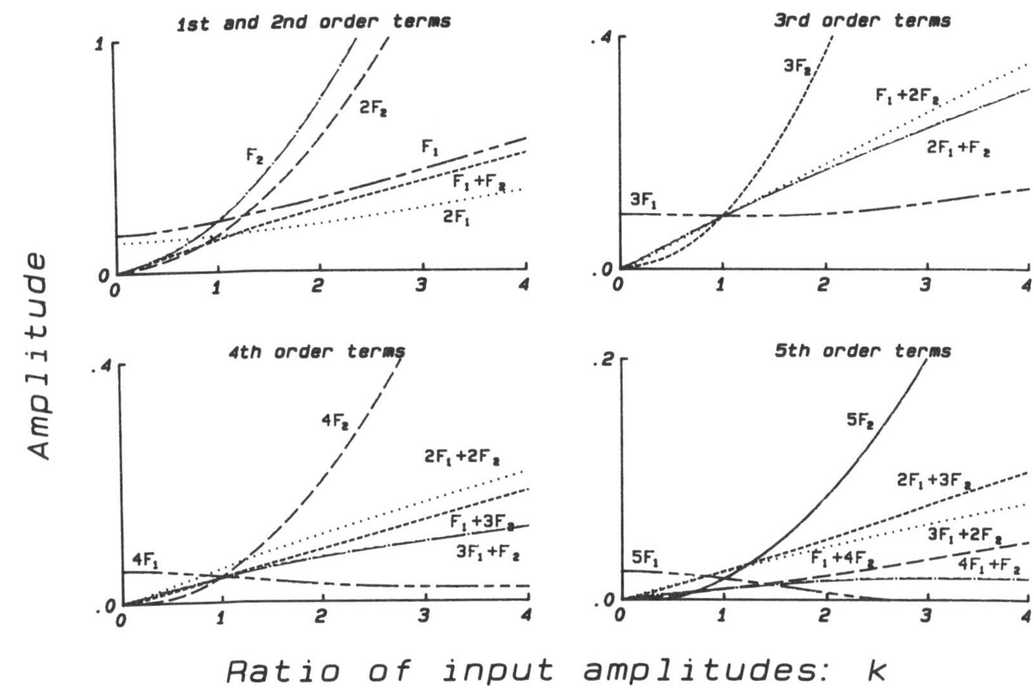

FIGURE 12. The family of curves for three half-wave square law rectifiers joined dichoptically (AC coupled), see text. This is plotted for the amplitude of each frequency vs. k, where k is the ratio of the input amplitudes of the two frequencies F_1 Hz and F_2 Hz.

G. FREQUENCY-DEPENDENT NONLINEARITIES

So far the discussion has been restricted to nonlinearities whose properties do not depend on the temporal frequencies of the input sinusoids. The discussion can be extended to include single frequency-dependent rectifier if the assumption is that such a nonlinearity can be regarded as a frequency-independent nonlinearity that is preceded by a linear frequency-dependent stage (i.e., a frequency filter) and followed by a second linear frequency-dependent stage. This kind of model (called a sandwich model) has been discussed by several authors (Marmarelis and Marmarelis, 1978; Spekreijse, 1966; Spekreijse and Oosting, 1970; Spekreijse and Reits, 1982; Victor and Shapley, 1980), but has been restricted to the case of a single rectifier. The treatment of single rectifiers above can be extended to the frequency-dependent case by applying appropriate linear weighting, frequency-dependent functions, first to the F_1 Hz and F_2 Hz inputs and then to the outputs of the rectifier. Again, if the case of cascaded frequency-dependent rectifiers is reduced to an equivalent cascade of rectifiers whose individual properties are independent of frequency, each of which is preceded and followed by a linear, frequency-dependent stage, then the previous discussion can be extended by applying appropriate linear frequency-dependent weighting functions to the signals entering and leaving each successive rectifier.

H. TESTING THE PREDICTIONS OF THE MODEL AGAINST DATA: THE IMPORTANCE OF HIGH-ORDER TERMS

In order to compare neural models with EP data, it is necessary to measure many simultaneously occurring frequency terms in the electroenceaphalogram (EEG), some of which have very low amplitudes. As will be shown, an ultrahigh-resolution method can be used for recording the steady-state EP. This method (zoom-FFT) allows precise measurement of components whose power is several thousand times less than the overall EEG power.

Figure 13 brings out a point that may not be intuitively obvious, but in practical recording it is of considerable importance. Some response components are only appreciable at specific values of k (the ratio of the two input amplitudes); see, for example, the $(2F_1 + 2F_2)$ and $3F_2$ terms in the case of the linear rectifier followed by a third-root rectifier. Thus, if the stimulus were an F_1 Hz flickering light of 20% modulation depth superimposed on an F_2 Hz flickering light of variable modulation depth, the $(2F_1 + 2F_2)$ EP term would be seen when the F_2 modulation depth was near 8 or 30%, but would be weak or absent when the F_2 modulation depth was below about 4%, between 15 and 18%, or increased above about 40%.

This figure also illustrates the point that different rectifier models can produce families of curves that look rather similar when only low-order terms are considered, but become more and more distinctive as higher- and higher-order terms are taken into consideration. The four rectifier models considered are

1. A linear rectifier followed by a compressive (third-root) rectifier, AC coupled
2. Two compressive (square root) rectifiers in series, AC coupled
3. A rectifier whose shape resembles the proposed contrast transducer function of the human eye (Wilson, 1980; Legge and Foley, 1980), namely, a threshold followed by an initial accelerating segment followed by a compressive segment
4. An accelerating (square law) rectifier followed by a compressive (third-root) rectifier, AC coupled

Figure 13A shows that the second-order curves are fairly similar, but when the third-order terms are taken into account (Figure 13B), the families of curves are easily distinguished.

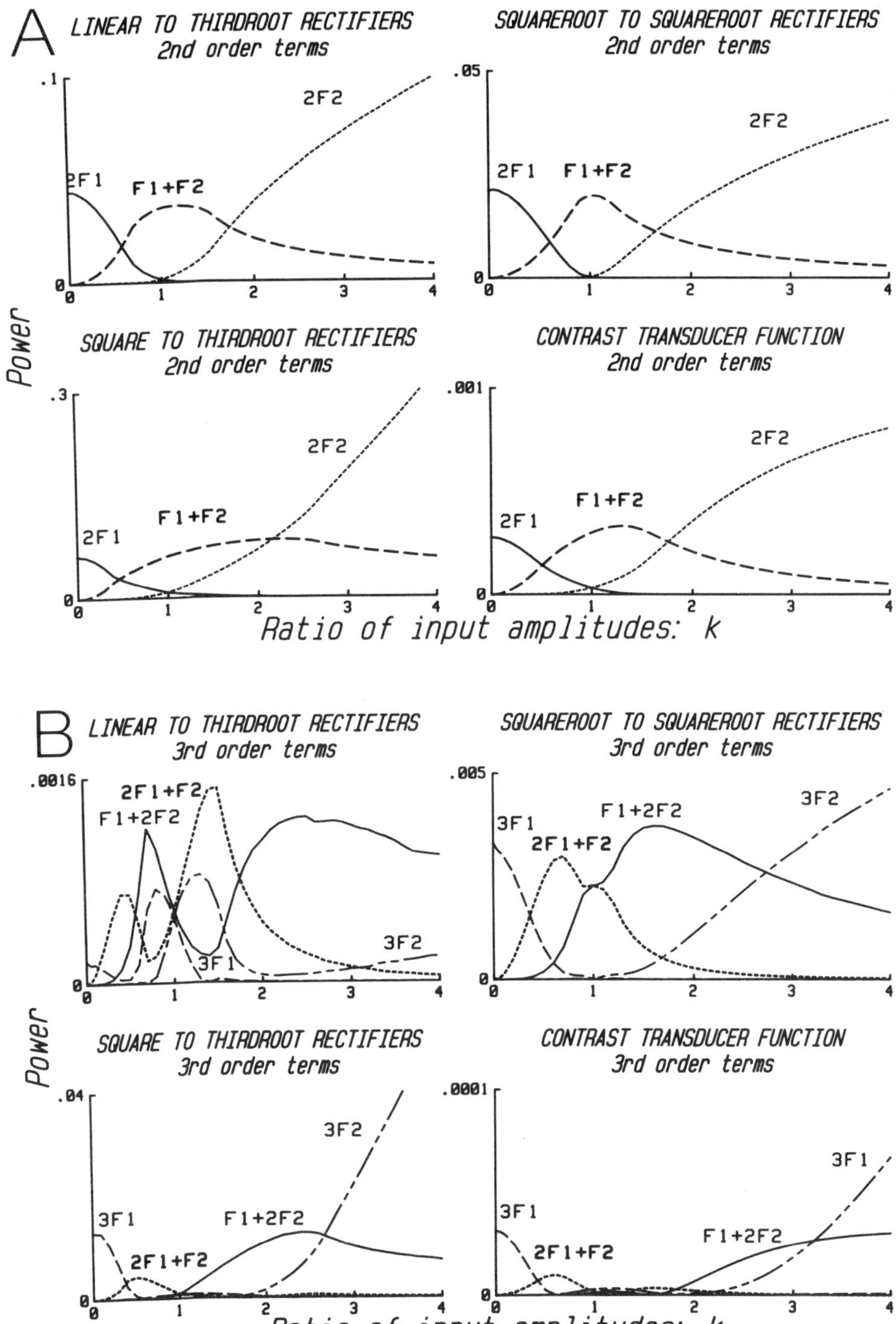

FIGURE 13. (A) Second-order terms for four different rectifier models. The curves are not strikingly different. (B) Third-order terms for the same four rectifier models. The curves are quite different.

Finally, the effects of linear frequency filters are briefly considered. A linear frequency filter placed before the rectifying stage would alter the effective value of k, depending on the absolute values of F_1 and F_2 relative to the tuning curve of the filter, thus, shifting the curves bodily along the abscissa. This shift can be rendered indefinitely small by minimizing the difference between F_1 and F_2. Again, a linear filter placed after the rectifying stage would bodily shift the curves along the ordinate without affecting their shapes, the amount depending on the absolute frequency of the component in relation to the tuning curve of the filter (Spekreijse, 1966; Spekreijse and Reits, 1982). In the experiments described below $(F_1 - F_2)$ was small, and emphasis is put on the shapes of the curves rather than their absolute amplitudes.

III. ZOOM-FFT AND ULTRAHIGH-RESOLUTION SPECTRA

This section discusses the limitations of the FFT which have led to the zoom-FFT method that allows the maximum possible resolution to be achieved in the frequency spectrum. The final section discusses the advantages of frequency-domain averaging over time-domain averaging.

A. THE LIMITATIONS OF THE FFT

The best possible frequency resolution of the power spectrum of the EEG is set by the duration of the EEG time signal: the longer the duration, the better the resolution. In particular, the best possible frequency resolution (ΔF) for an EEG sample of duration T seconds is given by $\Delta F = 1/T$ Hz (Gabor, 1946). (This equation exists in several slightly different versions, depending on how "frequency resolution" and "duration" are defined). Thus, for a DC-100 Hz bandwidth recording of 250-s duration, this theoretical limit is about 0.004 Hz, giving about 25,000 lines over DC-100 Hz bandwidth. However, there is often a practical problem in achieving this best possible resolution or even approaching it. Suppose that the EEG recording bandwidth is chosen to be DC-100 Hz. A digital analyzer samples the EEG waveform at some preset rate (consistent with the Nyquist frequency, which is twice the baseband frequency), and this sampling rate must be set at rather more than twice the highest frequency in the EEG signal. Suppose that the sampling frequency is 250 Hz and the recording duration is 256 s, then the total number of samples will be $250 \times 256 = 64{,}000$. An FFT carried out on these 64,000 samples as a whole would give a spectrum with a resolution of $(256)^{-1}$ s, or approximately 0.004 s, i.e., the theoretical resolution for a 256-s recording duration. Using present technology, however, a computer with sufficient memory to store such a large number of samples, each of which is measured with sufficiently high precision, would be expensive, and the computations required to obtain the spectrum might be unduly time-consuming. Zoom-FFT is a procedure for transcending this technical limitation so as to generate a spectrum with the best possible resolution of $\Delta F = 1/T$ on the basis of a comparatively small number of time samples (Thrane, 1980). It follows that only a correspondingly small part of the original frequency range can be covered in any one analysis.

B. ZOOM-FFT

Fast, special-purpose FFT analyzers are available, but they operate with a rather low number of time samples. Rather than the $N = 64{,}000$ time samples we have just discussed, commercial FFT analyzers are commonly restricted to $N = 1024$ or $N = 2048$ samples. The FFT transforms the N equally spaced time samples of the EEG into $N/2$ equally spaced lines in the frequency domain, so that, at best, the resolution over a DC f_{max} bandwidth is

$2f_{max}/N$ independently of the recording duration. (For technical reasons, however, the situation is even worse: fewer than $N/2$ lines are usually available.) The following paragraph explains how zoom-FFT can be achieved over a limited range of frequencies by frequency shift (heterodyning) followed by low-pass filtering (Regan, 1989; Regan and Regan, 1989).

Suppose we wish to concentrate our entire number of available lines (e.g., 800) within the frequency range $[f_k - (f_k + 3.125)]$ Hz rather than over the entire DC-100 Hz bandwidth. First, the EEG signal is low-pass filtered by the antialiasing filter before being sampled by the analog to digital converter (ADC). This ensures that the highest frequency in the signal is less than half the sampling rate f_8. For our DC-100 Hz EEG, let us take $f_8 = 250$ Hz. The samples from the ADC are then multiplied by a complex exponential function $e^{-i2\pi f_k t}$, i.e., by a unit vector rotating at a rate $-f_k$. Because the FFT gives a spectrum with a lowest frequency of zero (i.e., DC) rather than f_k, this multiplication produces a necessary frequency shift, that moves the frequency span of interest from $[f_k - (f_k + 3.125)]$ Hz to (DC − 3.125) Hz. After this frequency shift, the time samples are fed through a low-pass filter that passes only the frequency span of interest (i.e., DC 3.125 Hz) and resampled at a rate (f_s/z), where z is the zoom ratio. (In this case $z = 100/3.125 = 32$). Finally, the time samples are subjected to FFT.

High zoom ratios are needed to exploit the high signal-to-noise ratios consequent on the very narrow bandwidths of the discrete frequency components of the steady state EP. For example, with an 800-line machine and a record bandwidth of DC-100 Hz, we require a zoom ratio of about 32 to give the 0.0039-Hz bandwidth of Figure 14. Unfortunately, in commercially available equipment that offers zoom ratios greater than 10, the original time function is destroyed by the zoom operation, so that only a limited section of the entire frequency spectrum can be analyzed. Consequently, since data are required at about 30 or more different discrete frequency components that are, in general, not harmonically related and can be located virtually anywhere within our recording bandwidth of typically DC-100 or DC-200 Hz, the zoom-FFT process had to be extended.

The time factor cannot be ignored either. For example, in practice, to analyze any given 3.125-Hz segment of the DC-100 bandwidth into 800 discrete frequency bins at a resolution of 0.0039 Hz calls for a recording duration of at least 256 s. But, if the analysis of each and every 3.125-Hz segment of the DC-100 Hz bandwidth took 256 s, the zoom-FFT analysis of one 256-s recording would take over 2 h.

Software has been developed to work in conjunction with the analyzer that overcomes these problems. The experimental data are stored for easy retrieval and entered into the analyzer at high speed. After the process of zoom-FFT, the required data about the relevant frequencies are then easily read from the analyzer to computer and stored for future use. This gives the amplitudes of the frequency peaks and information about the signal-to-noise ratios.

Figure 14 illustrates part of a DC-100 Hz EEG signal analyzed by this procedure at a resolution of 0.0039 Hz. There are about 12,500 frequency bins between 0.5 and 49.5 Hz in Figure 14, i.e., 12,500 lines. Note that the ordinate in the 34.5- to 48.5-Hz segment is 50 times smaller than in the 0.5- to 18.5-Hz segment, and that the $5F_2$ component is quite clear, even though it is about 300 times smaller than the $2F_2$ component, and its power is thousands of times less than the total EEG power.

The extreme narrowness that a steady-state evoked potential component can attain is illustrated in Figure 15. The subject viewed a counterphase-modulated grating for 640 s, thus, giving the analyzer a resolution of 0.00195 Hz. Figure 15 shows that the experimentally measured EP power was mostly contained in one 0.00195-Hz bin. Thus, the bandwidth of the EP component was no more than 0.00195 Hz, and perhaps even less. (Even if the signal had an indefinitely narrow bandwidth, some power would have spread into adjacent lines

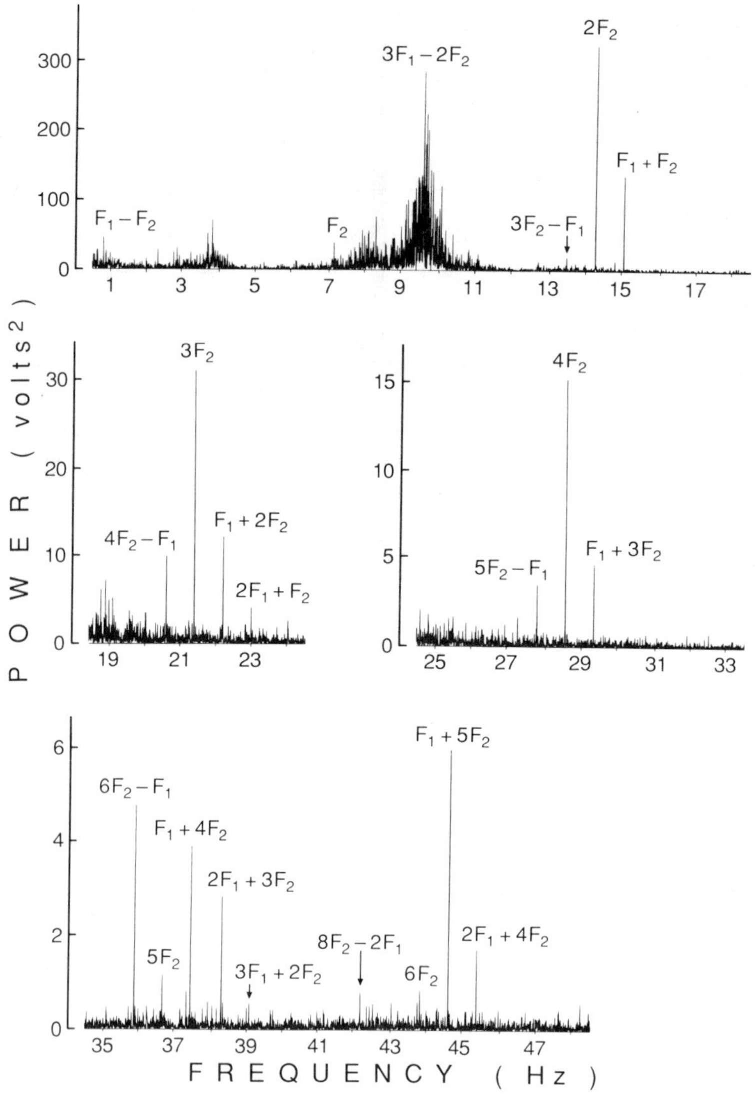

FIGURE 14. A homogeneous circular white patch of light of diameter 10 degrees and mean luminance 250 cdm^{-2} was multiplied by the sum of two sinusoids, one of frequency 8 Hz, giving a modulation depth of 20%, and one of frequency 7 Hz, giving a modulation depth of 60%. The EEG was analyzed by nondestructive zoom-FFT at a resolution of 0.0039 Hz. The zoom ratio was 32. Recording duration was 320 s. This section contains 12,300 lines over 0.5 to 49.5 Hz bandwidth. The steady-state EP consists of discrete frequency components whose bandwidths were less than 0.0039 Hz. Cross-modulation terms are due to nonlinear interactions between neural responses to the two flickers. (From Regan and Regan, 1989, © 1989 by *Can. J. Neurological Sciences.)*

because of overlap between the sensitivity curves of adjacent bins. In particular, the total power in the two adjacent lines cannot be less than about 50% of the power in the central line, even for a signal of indefinitely narrow bandwidth, and this is approximately what we see in Figure 15.) This narrowness in the frequency domain implies that the amplitude and phase variability of the signal was small (Regan, 1989; Regan and Regan, 1989).

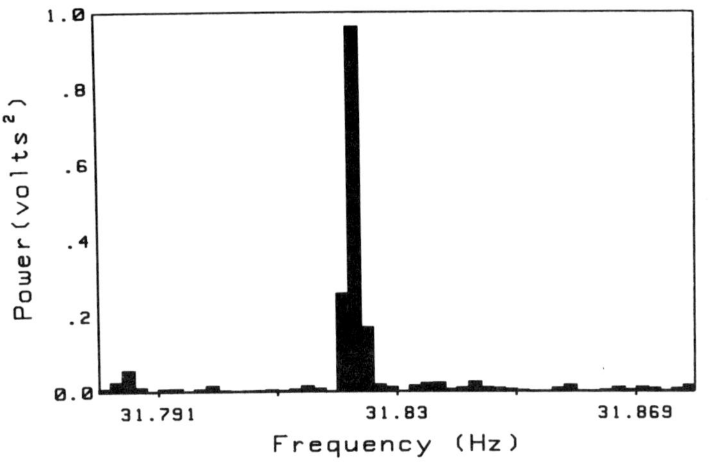

FIGURE 15. A single vertical grating of contrast 40% was counterphase modulated at F ≈ 8 Hz, and viewed binocularly. The EEG was recorded for 640 s, allowing the analyzer to attain a frequency resolution of 0.00195 Hz. The zoom ratio was 64. The plot shows the relative power per bin at and around the 4F Hz evoked potential component, demonstrating the remarkable narrowness of the response component. (From Regan and Regan, 1989, © 1989 by *Can. J. Neurological Sciences.)*

IV. EXPERIMENTAL RESULTS

A. MEASUREMENT OF ORIENTATION TUNING, SPATIAL FREQUENCY TUNING, AND TEMPORAL TUNING OF CORTICAL CELLS

The usual strategy in VEP (visual evoked potential) studies of spatial frequency selectivity is to stimulate the eye with a single sinewave grating. This approach gives rather broad tuning bandwidths. For example, the half-height full bandwidth of the $N_1 - N_2$ onset response is about 2.2 octaves (Russell et al., 1987). Contrast-reversal tuning bandwidths are usually broad also, though narrow spatial frequency tuning can be seen at some temporal frequencies as, for example, in Tyler et al. (1979).

Considerably narrower selectivity is revealed by a different experimental approach in which the eye is stimulated by two superimposed gratings rather than by a single grating (Regan, 1983; Fiorentini et al., 1983). The rationale of this alternative approach is as follows: a two-grating stimulus can produce nonlinear grating-grating interaction terms in the VEP *only if the neural mechanism that generates the VEP "sees" both gratings simultaneously* (Regan, 1983). In other words, nonlinear interaction terms cannot be generated if only one grating is within the spatial frequency and orientation bandpass of the generator mechanism. Given this rationale, the tuning bandwidths of a given grating can be estimated in the following ways (Regan, 1983).

1. Spatial Frequency Bandwidth

Spatial frequency bandwidth can be estimated by superimposing on the reference grating a grating of closely similar orientation and closely similar counterphase-modulation frequency. Then, the spatial frequency of this second grating is varied. Figure 16 illustrates this possibility (Regan, 1983). The following procedure was used. The subject viewed a vertical sinewave grating that was counterphase-modulated at F_1 Hz. This stimulus produced a strong $2F_1$ Hz component in the EEG. The spatial frequency tuning of this $2F_1$ Hz response was very broad (Figure 16A). Then, with this reference grating set to 5.2 c per degree, a second vertical grating of spatial frequency 5.2 c per degree was superimposed on the first grating. This second grating (the variable grating) was counterphase-modulated at a different

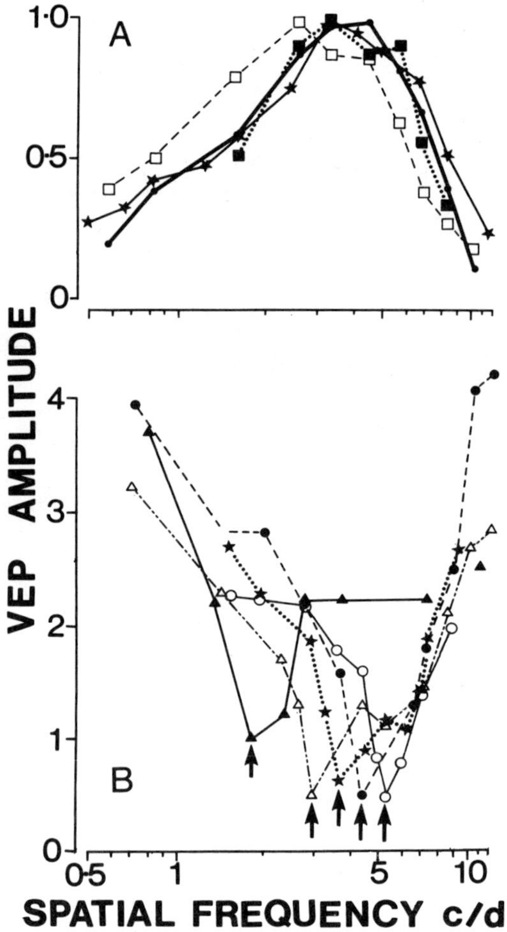

FIGURE 16. Spatially selective subunits. (A) Normalized evoked potential amplitudes for a single grating of fixed temporal frequency (20 reversals per second) plotted vs. the spatial frequency of the grating for contrasts of 15% (large solid squares, dotted line), 30% (open squares, broken line), 60% (solid circles, heavy continuous line), and 100% (stars, fine continuous line). (B) Evoked potential amplitude for a reference grating (No. 1) in the presence of a superimposed variable grating (No. 2). The temporal frequencies of both gratings were fixed, No. 1 at 20 and No. 2 at 18 reversals per second. The spatial frequency of grating 1 was constant, while the spatial frequency of reference grating 2 was varied (abscissa in cycles per degree). The five plots show data for five experiments in which, as indicated by vertical arrows, the spatial frequency of reference 1 was fixed at 1.6, 3.0, 3.7, 4.4, and 5.2 cycles per degree, respectively, in different experiments. Ordinates in microvolts. Amplitudes shown are for the VEP frequency component at the reversal rate of grating 1. (Reprinted from *Vision Res.*, **23**, 1401–1407, Regan, D., © 1983, Pergamon Press plc.)

rate, F_2 Hz. A nonlinear interaction between responses to the two gratings was observed that almost abolished the strong $2F_1$ Hz response, and produced new response terms, including a term of frequency $(2F_1 + 2F_2)$. In a control experiment, the F_1 Hz grating was superimposed, not on a second grating, but on an unpatterned homogeneous field that was flickered at F_2 Hz. The suppression of the $2F_1$ Hz response was not observed, nor was a $(2F_1 + 2F_2)$ Hz component generated, indicating that these effects reflected pattern-specific nonlinear interactions rather than nonlinear interactions between F_1 Hz and F_2 Hz local flicker.

Then, with the reference grating held at a fixed spatial frequency of 5.2 c per degree, the spatial frequency of the second grating was varied over a wide range. Figure 16B shows

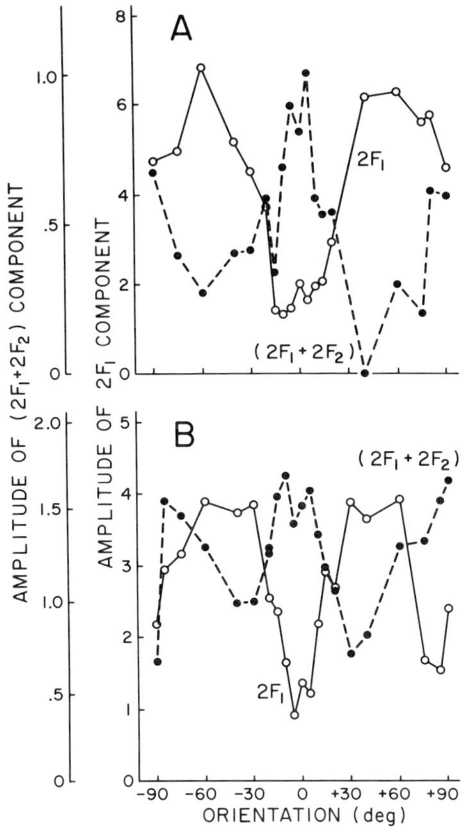

FIGURE 17. Nonlinear interactions between responses to two gratings as a function of orientation difference. A vertical grating was counterphase-modulated at F_1 Hz, and a superimposed variable-orientation grating was modulated at F_2 Hz. Solid symbols plot the amplitude of the nonlinear crossmodulation ($2F_1 + 2F_2$) Hz term in the evoked potential vs. the orientation of the variable grating. Open symbols plot the frequency-doubled $2F_1$ Hz term. Results are shown for two subjects. (Reprinted from *Vision Res.*, **27**, 2181–2183, Regan, D. and Regan, M. P., ©1987, Pergamon Press plc).

that a nonlinear interaction between the two gratings (in particular, the suppression of the $2F_1$ Hz response) was greatest when the variable grating had a spatial frequency of 5.2 c per degree, exactly the same as the reference grating. The experiment was repeated with the reference grating set to five different spatial frequencies, and each time the nonlinear interaction was maximal when the variable grating had exactly the same spatial frequency as the reference grating. These findings can be understood if (1) the visual pathway contains a plurality of neural mechanisms whose tuning curves are much narrower than that shown in Figure 16A and (2) there are sufficiently many overlapping narrowly tuned mechanisms so that there is, effectively, one for each spatial frequency (Regan, 1983).

2. Orientation Bandwidth

Orientation bandwidths can be estimated by superimposing on the reference grating a second grating of closely similar spatial frequency and closely similar counterphase-modulation frequency. Then the orientation of this second grating is varied. Figure 17 illustrates this possibility (Regan and Regan, 1987). One vertical grating of spatial frequency 5 c per degree was modulated at F_1 Hz. A second grating of spatial frequency 5.5 c per degree, superimposed optically on the first grating, was modulated at F_2 Hz and its orientation varied

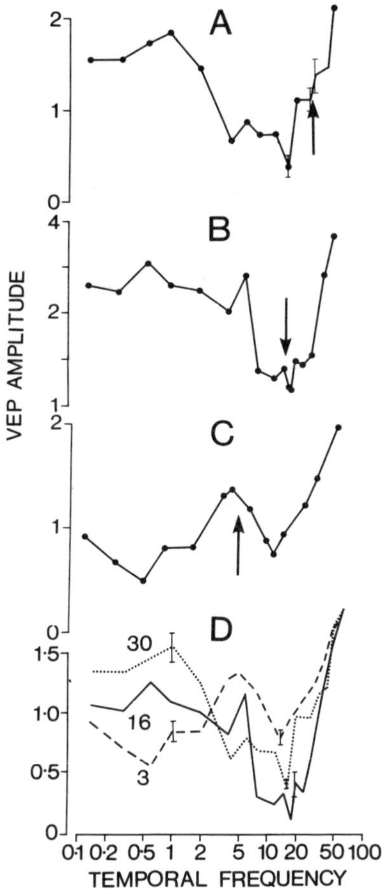

FIGURE 18. Temporal frequency "channels" in grating VEPs. A grating of variable counterphase-modulation frequency was superimposed on a second grating of fixed (*F* Hz) counterphase-modulation frequency (15 Hz in A, 8 Hz in B, and 1.5 Hz in C). Ordinates plot the amplitudes of the 2*F* Hz VEP produced by the fixed grating (30 Hz in A, 16 Hz in B, and 3 Hz in C). Abscissas plot the frequency of the variable frequency grating. The arrows in A to C mark the reversal frequencies of the fixed grating. Panel D is a superimposition of panels A to C. The two gratings were parallel and had the same spatial frequency of 4.1 cycles per degree. (Reprinted from *Vision Res.*, **23**, 1401–1407, Regan, D., ©1983, Pergamon Press plc.)

from vertical to horizontal. The $(2F_1 + 2F_2)$ Hz nonlinear interaction frequency component was large, and the suppression of the $2F_1$ Hz response was strong when the gratings were parallel (both vertical or both horizontal), but the nonlinear interaction fell to a minimum when the orientations differed by about 30 degrees. This gave a half-height bandwidth tuning curve for orientation of about 12 degrees, which can be understood if the $(2F_1 + 2F_2)$ term is generated by cortical neurons tuned to a narrow range of orientations as described by De Valois et al. (1982). The strong nonlinear interaction observed when one grating was horizontal and the other vertical can be understood if vertically tuned neurons interact nonlinearly with horizontally tuned neurons.

3. Temporal Bandwidth

Temporal bandwidth can be estimated by superimposing on the reference grating a second grating of closely similar spatial frequency and orientation. Then, the counterphase-modulation frequency of this second grating is varied. Figure 18 illustrates this possibiilty (Regan,

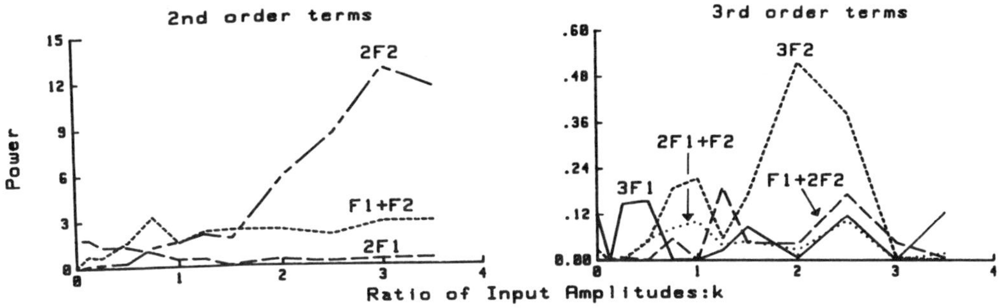

FIGURE 19. Evoked potential data. Ordinates plot the amplitudes of some of the various frequency components in the evoked potential recorded while the subject viewed a homogeneous field flickering at 8 Hz with 20% modulation depth in the left eye, while a second homogeneous field flickering at 7 Hz viewed by the right eye was varied in modulation depth from 0 to 70%. (Regan, M. P. and Regan, D., 1989, ©1989, *Can. J. Neurological Sciences*).

1983). One grating of spatial frequency 4.1 c per degree was modulated at F_1 Hz. A second grating with the same spatial frequency was superimposed on the first grating and was modulated at a different temporal frequency F_2 Hz. The amplitude of the VEP frequency of the first grating (F_1 Hz) was measured. The measurement was then repeated for several different values of F_2 Hz. The strongest interactions between the grating responses *did not necessarily occur when the temporal frequencies were equal.* This finding is quite different from the analogous finding for spatial frequency tuning shown in Figure 16B. We conclude that it is not the case that there is a continuum of neural mechanism along the temporal frequency axis. A proposed explanation for these temporal tuning data is that the neural subunits that respond best to a spatial frequency of 4.1 c per degree differ along a low-pass-bandpass continuum, and that those toward the bandpass (i.e., phasic) extreme all respond best in the range 5- to 10-Hz modulation frequencies.

B. INVESTIGATION OF BINOCULAR NEURONS

The stimulus consisted of two homogeneous circular white fields subtending 10 degrees, of equal luminances, viewed at 100 cm. The luminance of one field was modulated sinusoidally to a depth of 20%, at $F_1 = 8$ Hz, the other was modulated at $F_2 = 7$ Hz, and the modulation depth varied from 0 to 70%. The two fields were presented dichoptically. Great care was taken to ensure tht F_1 flicker was routed entirely to one eye and F_2 flicker entirely to the other (Regan and Regan, 1986a, 1989). The rationale of this experiment has been described before in a psychophysical context (Regan and Beverly, 1973): any cross-modulation frequencies in the response must necessarily be generated after convergence of signals from left and right eyes. Thus, this technique allows binocular processing to be isolated from monocular processing (assuming that we can ignore modulation of lateral geniculate body processing by signals descending from the cortex).

Figure 19 plots powers of EP response components vs. the modulation depth of the F_2 Hz flicker stimulus (from 0 to 70%). The procedure for modeling the responsible nonlinearities is to compare families of curves, such as those shown in Figure 19, with corresponding families of curves that have been theoretically calculated for candidate models.

The models considered for the Figure 19 dichoptic data comprised:

- Model 1: two linear monocular channels converging onto a binocularly driven rectifier
- Model 2: the linear sum of two identical monocular channels, each containing a rectifier
- Model 3: two identical monocular channels, each containing a rectifier, converging onto a binocularly driven rectifier

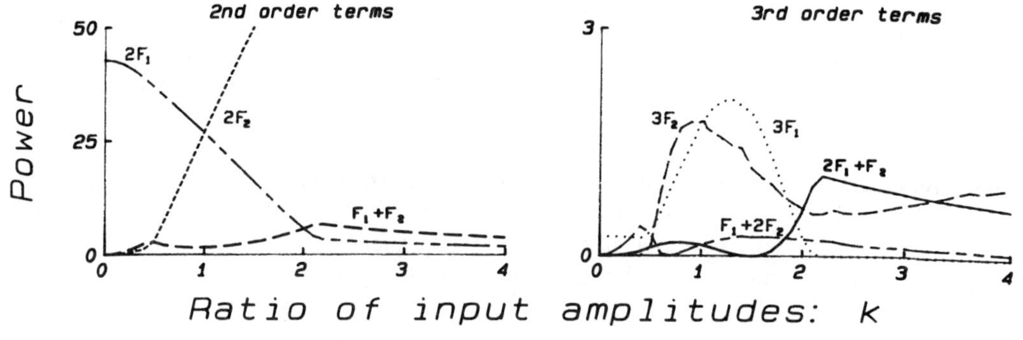

FIGURE 20. The family of curves for two half-wave linear rectifiers joined dichoptically (AC coupled) in a half-wave squareroot rectifier for the second- and third-order terms only. This is plotted for the power of each frequency vs. k, where k is the ratio of the input amplitudes of the two frequencies F_1 Hz and F_2 Hz.

The families of curves for Models 1, 2, and 3 were calculated for the cases of linear, compressive, and accelerating half-wave rectifiers. Model 1 could be rejected when the $3F_2$ term was taken into account as well as second-order terms; see Figures 4 to 7 (note, these are for amplitude and not power). Model 2 could be rejected because it does not provide any of the cross-modulation terms evident in the data of Figure 19, only harmonics of F_1 and F_2. Model 3 provided the best fit to the data of the parallel/cascaded models. The data were better fit when two monocular linear rectifiers fed a binocular compressive (square root) rectifier (Figure 20) than when all three rectifiers were identical linear, compressive (square or third-root) or accelerating (second-power) rectifiers. Even then, the technique was sufficiently selective to show that the Figure 20 model failed to fit the $3F_1$ data.

In applying our approach to binocular interaction we should note that methods for recording EPs to disparity change (Regan and Spekreijse, 1970; Julesz et al., 1980; Norcia et al., 1985), EPs to correlation/anticorrelation (Julesz et al., 1980; Herpers et al., 1981; Lehmann et al., 1978), and interocular suppression of pattern EPs (Spekreijse et al., 1972) all fail when visual acuity is low in one or in both eyes, so that these techniques are of limited effectiveness in assessing binocular interaction in amblyopic children or infants. On the other hand, recording cross-modulation terms produced by dichoptic stimulation with unpatterned-field flicker provide a means of measuring binocularity that does not require good visual acuity or accurate binocular convergence. By using LED stimulators mounted in goggles, the technique might, perhaps, be used to monitor binocularity in young children undergoing therapy for amblyopia.

V. DISCUSSION

This chapter describes an extension of the two-sinewave EP method that allows multineuron models to be tested experimentally by comparing EP data with a library of theoretically obtained predictions for candidate models. In order to use this approach we have developed an ultrahigh-resolution method for wide-bandwidth EP recording and a theoretical procedure for obtaining the predictions of multineuron models.

In brief, the experimental procedure is to stimulate simultaneously with two freqencies F_1 and F_2, holding the amplitude of one input constant while varying the amplitude of the second input. This gives a family of plots of EP amplitude vs. stimulus amplitude, one plot for each of the many ($nF_1 \pm mF_2$) frequency components in the EEG or MEG. As more and more of the frequency components are taken into account, the corresponding family of plots for any given neural model becomes a more and more specific "fingerprint" of that model, thus allowing confident rejection of candidate models. Compared with the well-known Wiener kernel (white-noise) time-domain method of EP analysis (Marmarelis and Marmarelis, 1978), this frequency-domain approach has the advantage that many of the high-order response components (e.g., $n + m = 3$ to 6) can readily be measured physiologically, thus, as just mentioned, allowing sharp discrimination between candidate models. This advantage is empirical and stems from the very narrow bandwidths (e.g., < 0.004 Hz) of the EP components. Because noise is continuously distributed along the frequency axis, comparatively little falls into any given 0.004-Hz frequency bin. But any single bin may contain almost the entire power of an EP component, thus giving high signal-to-noise ratios. In the time domain, however, signal substantially overlaps with noise, and Wiener kernel analysis is a time-domain procedure. Our approach can be used in audition (Regan and Regan, 1988b) as well as in vision (Regan and Regan, 1986, 1986a, 1987, 1988a), and is also convenient for investigating cross-modality interactions. Note that our approach is not a general method. For example, it does not apply to systems whose properties change with time or whose properties depend on the previous history of stimulation. (These limitations also apply to the Wiener kernel method.)

ACKNOWLEDGMENTS

Research sponsored by the Air Force Office of Scientific Research, Air Force Systems Command, USAF, under grant or cooperative agreement number, AFOSR F49620-88-C-0002. The U.S. Government is authorized to reproduce and distribute reprints for Governmental purposes notwithstanding any copyright notation thereon.

This manuscript is submitted for publication with the understanding that the US Government is authorized to reproduce and distribute reprints for Governmental purposes.

REFERENCES

Barlow, H. B. and Levick, W. R. (1965). The mechanism of directionally selective units in the rabbit's retina, *J. Physiol (London)*, **178**, 477–504.

Bennett, W. R. (1933). New results in the calculation of modulation products, *Bell System Tech. J.*, 228–243.

Beverley, K. I. and Regan, D. (1973). Evidence for the existence of neural mechanisms selectively sensitive to the direction of motion in space, *J. Physiol (London)*, **235**, 17–29.

Brugge, J. F. and Reale, R. A. (1985). Auditory cortex, in *Cerebral Cortex*, Vol. **4**, Peters, A. and Jones, E. G., Eds., Plenum Press, New York, 229–272.

Clynes, M., Kohn, M., and Lifshitz, K. (1964). Dynamics and spatial behavior of light evoked potentials, their modification under hypnosis, and on-line correlation in relation to rhythmic components, Ann. N.Y. Acad. Sci., **112**, 468–509.

Cynader, M. and Regan, D. (1978). Neurons in cat parastriate cortex sensitive to the direction of motion in three-dimensional space, *J. Physiol. (London)*, **274**, 549–569.

De Valois, R. L., Albrecht, D. G., and Thorell, L. G. (1982). Spatial frequency selectivity of cells in macaque visual cortex, *Vision Res.*, **22**, 545–599.

Fiorentini, A., Pirchio, M. and Spinelli, D. (1983). Electrophysiological evidence for spatial frequency selective mechanisms in adults and infants, *Vision Res.*, **23**, 119–127.

Foster, K. H., Gaska, J. P., Nagler, M., and Pollon, D. A. (1985). Spatial and temporal frequency selectivity of neurones in visual cortical areas V1 and V2, *J. Physiol. (London)*, **365**, 331–363.

Gabor, D. (1946). Theory of communication, *J. Inst. Elect. Eng. (London)*, **93**, 429–456.

Herpers, M. J., Caberg, H. B., and Mol, J. M. F. (1981). Human cerebral potentials evoked by moving dynamic random dot stereograms, *Electroencephalogr. Clin. Neurophysiol.*, **52**, 50–56.

Hobson, E. W. (1926). *The Theory of Functions of a Real Variable and the Theory of Fourier's Series*, Cambridge University Press, 710.

Hubel, D. H. and Wiesel, T. N. (1962). Receptive fields, binocular interaction and functional architecture in the cat's visual cortex, *J. Physiol. (London),* **160**, 106–154.

Julesz, B., Kropfl, W., and Petrig, B. (1980). Large evoked potentials to dynamic random dot correlograms and stereograms permit quick determination of stereopsis, *Proc. Natl. Acad. Sci. U.S.A.*, **25**, 2348–2351.

Legge, G. E. and Foley, J. M. (1980). Contrast masking in human vision, *J. Opt. Soc. Am.*, **70**, 1458–1471.

Lehmann, D., Skrandies, W., and Lindenmaier, C. (1978). Sustained cortical potentials evoked in humans by binocularly correlated, uncorrelated and disparate dynamic random-dot stimuli, *Neurosci. Lett.*, **10**, 129–134.

Marmarelis, P. Z. and Marmarelis, V. Z. (1978). *Analysis of Physiological Systems. The White Noise Approach,* Plenum Press, New York.

Morgan, M. J. and Regan, D. (1987). Opponent model for line interval discrimination: interval and vernier performance compared, *Vision Res.*, **27**, 107–118.

Movshon, J. A., Thompson, I. D., and Tolhurst, D. J. (1978). Spatial and temporal contrast sensitivity in neurons in areas 17 and 18 of the cat visual cortex, *J. Physiol. (London),* **283**, 79–99.

Norcia, A. M., Sutter, E. E., and Tyler, C. W. (1985). Electrophysiological evidence for the existence of coarse and fine disparity mechanisms in human, *Vision Res.*, **25**, 1603–1611.

Ohzawa, I., Sclar, G., and Freeman, R. D. (1982). Contrast gain control in the cat visual cortex, *Nature (London),* **298**, 266–268.

Orban, G. A. (1985). Velocity tuned cortical cells and human velocity discrimination, in *Brain Mechanisms and Spatial Vision,* Ingle, D., Jeannerod, M., and Lee, D. N., Eds., Martinius Nijhoff, Dordrecht.

Poggio, G. F. and Talbot, W. H. (1981). Mechanisms of static and dynamic stereopsis in foveal cortex of rhesus monkey, *J. Physiol. (London),* **315**, 469–492.

Regan, D. (1983). Spatial frequency mechanisms in human vision investigated by evoked potential recording, *Vision Res.*, **23**, 1401–1408.

Regan, D. (1989). *Human Brain Electrophysiology: Evoked Potentials and Evoked Magnetic Fields in Science and Medicine,* Elsevier, New York.

Regan, D. and Beverley, K. I. (1973). The dissociation of sideways movements from movements in depth: psychophysics, *Vision Res.*, **13**, 2403–2415.

Regan, D. and Regan, M. P. (1986). Spatial tuning and orientational tuning in pattern evoked potentials measured by nonlinear analysis, in *Evoked Potentials III,* Barber, C. and Blum, T., Eds., Boston, Butterworths, 190–197.

Regan, D. and Regan, M. P. (1987). Nonlinearity in human visual responses to two-dimensional patterns, and a limitation of Fourier methods, *Vision Res.*, **27**, 2181–2183.

Regan, D. and Regan, M. P. (1988a). Objective evidence for phase-independent spatial frequency mechanisms in the human visual pathway, *Vision Res.*, **28**, 187–191.

Regan, D. and Regan, M. P. (1988b). The transducer characteristic of hair cells in the human inner ear: a possible objective measure, *Brain Res.*, **438**, 363–365.

Regan, D. and Spekreijse, H. (1970). Electrophysiological correlate of binocular depth perception in man, *Nature (London),* **255**, 92–94.

Regan, M. P. and Regan, D. (1986a). Monocular and binocular nonlinearities in flicker evoked potentials. Spekreijse and van der Tweel's linearization revisited after 21 years, 3rd Int. Evoked Potential Symp., Berlin.

Regan, M. P. and Regan, D. (1988). A frequency domain technique for characterizing nonlinearities in biological systems, *J. Theor. Biol.*, **133**, 293–317.

Regan, M. P. and Regan, D. (1989). Objective investigation of function using a nondestructive zoom-FFT technique for evoked potential analysis, *Can. J. Neurosci.*, **16**, 168–179.

Reichardt, W., Poggio, T., and Hausen, K. (1983). Figure-ground discrimination by relative movement in the visual system of the fly, *Biol. Cybern.*, **46**, 1–15.

Rice, S. O. (1944). Mathematical analysis of random noise, *Bell System Tech. J.*, **23**, 282–332.

Russell, M. H. A., Kulikowski, J. J., and Murray, I. J. (1987). Spatial frequency dependence of the human visual evoked potential, in *Evoked Potentials III,* Barber, C. and Blum, T., Eds., Boston, Butterworths, 231–239.

Schiller, P. H. (1986). The central visual system, *Vision Res.*, **26**, 1351–1386.

Schiller, P. H., Sandell, J. H., and Maunsell, J. H. R. (1986). Functions of the ON and OFF channels of the visual system, *Nature (London),* **322**, 824–825.

Sekuler, R., Pantel, A., and Levinson, E. (1978). Physiological basis of motion perception, in *Perception,* Held, R., Leibowiz, H. W., Teuber, H. -L., Eds., Springer-Verlag, New York, 67–96.

Spekreijse, H. (1966). *Analysis of EEG Responses,* Thesis, D. W. Junk, The Hague.

Spekreijse, H. and Oosting, H. (1970). Linearising: a method for analysing and synthesizing nonlinear systems, *Kybernetik,* **7**, 22–31.

Spekreijse, H. and Reits, D. (1982). Sequential analysis of the visual evoked potential system in man: nonlinear analysis of a sandwich system, *Ann. N.Y. Acad. Sci.*, **388**, 72–97.

Spekreijse, H., van der Tweel, L. H., and Regan, D. (1972). Interocular sustained suppression: correlations with evoked potential amplitude and distribution, *Vision Res.*, **12**, 521–526.

Spekreijse, H., van der Tweel, L. H., and Zuidema, T. (1973). Contrast evoked responses in man, *Vision Res.*, **13**, 1577–1601.
Sternberg, R. L., Shipman, J. S., and Thurston, W. B. (1954). Tables of Bennett functions for the two-frequency modulation product problem for the half-wave linear rectifier, *Q. J. Mech. Appl. Math.*, **7**, 505–511.
Sternberg, R. L., Shipman, J. S., and Kaufman, H. (1955). Tables of Bennett functions for the two-frequency modulation product problem for the half-wave square-law rectifier, *Q. J. Mech. Appl. Math.*, **8**, 457–467.
Thrane, N. (1980). Zoom FFT Tech Rev., **2-1980**, Nærum, Denmark, Bruel & Kjær.
Tyler, C. W., Apkarian, P., Levi, D. M., and Nakayama, K. (1979). Rapid assessment of visual function: an electronic sweep technique for the pattern visual evoked potential, *Invest. Ophthalmol. Vis. Sci.*, **18**, 703–713.
Victor, J. D. and Shapley, R. M. (1980). A method of nonlinear analysis in the frequency domain, *Biophys. J.*, **29**, 459–483.
Volterra, V. (1959). *Theory of Functionals and of Integral and Integro-Differential Equations*, Dover Publications, New York.
Wilson, H. R. (1980). A transducer function for threshold and suprathreshold human vision, *Biol. Cybern.*, **38**, 171–178.

Chapter 10

NONLINEARITIES IN PSYCHOPHYSICAL MODELS OF THE PROCESSING OF SPATIAL FORM AND MOTION

D. Regan

TABLE OF CONTENTS

I. Introduction 294
A. Classic Psychophysics and Animal Physiology 294
B. Linear Systems Analysis and the Wide World of Nonlinear Systems 295

II. Binocular Processing of the Direction of Motion in Depth 296

III. Monocular Processing of Changing Size 298

IV. Line Separation Discrimination 303

Acknowledgments 305

References 306

I. INTRODUCTION

A. CLASSIC PSYCHOPHYSICS AND ANIMAL PHYSIOLOGY

Whatever the limitations of human visual psychophysics may be (and, as we will see, there are several severe limitations), human visual psychophysics has one important advantage compared with studies of animal visual pathways and with studies of computer vision: if one's aim is to understand human vision, there is no substitute for empirical studies of vision in humans. A second feature of visual psychophysics is that it aims to give a functional description of the visual system as a whole rather than a structural description of the physical elements that comprise the visual system. A statement by Marmarelis and Marmarelis (1978, p. 5) addresses this point: " — it may in fact be illusory to think that the smaller the pieces into which a system is dissected the better we will understand it".

From Fechner onward the development of classic psychophysical methods has been motivated by the aim of rendering certain aspects of mental experience amenable to quantitative scientific methods. Rather than dealing with sensation directly, classic psychophysics replaces sensation by a physical scale. The experimenter studies the relation between the stimulus and the subject's behavioral responses and does not attempt to compare the stimulus with the resulting sensation. Thus, psychophysical procedures may be used to apply methods of system analysis to research on human vision.*

To summarize, relevant features of the systems approach include the following:

1. There is a clear distinction between two levels at which a system can be analyzed: the functional level of analysis and the structural (hardware) level of analysis.
2. The systems approach provides a functional identification of the system in the form of a mathematical model that, ideally, allows one to predict the output of the system for any given input. In practice, the experimenter must often be satisfied with a model that is valid for only a restricted range of input conditions.
3. A structural identification of the system cannot, in general, be obtained by comparing the input and output of the system (Marmarelis and Marmarelis, 1978).
4. A system as a whole may exhibit lawful behavior that is not manifest in the properties of any of the individual elements that comprise the system.
5. When treated as a whole, the behavior of a system may be understood at a qualitative level and adequately described at a quantitative level with comparatively straightforward mathematics, even though the description is forbiddingly complex when each individual element is treated separately.
6. In science and technology the fact that the same mathematical model can describe the behavior of quite different physical systems has long been accepted. (A familiar example is that similar differential equations describe oscillations in electrical, mechanical, and acoustical systems.) Such mathematical analogs are important "knowledge multipliers", because they enable knowledge developed in one specialized field to be transferred to another field. The application of systems science in vision research can be regarded as a thorough-going exploitation of this "transfer by analogy" process.

For our present purpose we may regard a quantitative psychophysical model of some aspect of visual function (e.g., spatial vision) as akin to the system analyst's "functional identification of the system in the form of a mathematical model" (point 2 above). And we may regard the findings of cellular-level physiologists, neuroanatomists, and so on, as akin to the system analyst's "structural identification of the system" (point 3 above).

* Textbooks on systems analysis include Aseltine (1958), Bode (1945), Blaquiere (1966), and Hayashi (1964), and analyses of nonlinear biological systems are described in Milsum (1966) and Marmarelis and Marmarelis (1978).

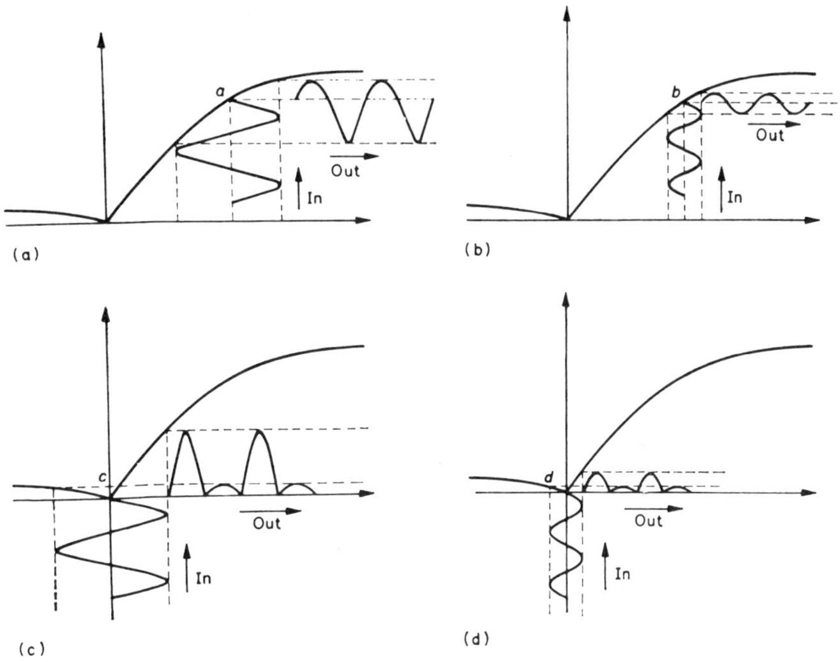

FIGURE 1. Nonessential and essential nonlinearities. Input amplitude is plotted along the abscissa and output along the ordinate. Curvature of the characteristic causes waveform distortion to be strong when the amplitude of the input sinusoid is large (a), but distortion tends to zero as the amplitude of the input is reduced (b). However, when the input waveform is centered on a point on the characteristic where the slope changes discontinuously, waveform distortion is not proportionally reduced by reducing the maplitude of the input (compare c with d). Note that in c and d noise at the input is assumed to be zero. (From Regan, D. (1989). *Human Brain Electrophysiology*, Elsevier, New York. With permission.)

B. LINEAR SYSTEMS ANALYSIS AND THE WIDE WORLD OF NONLINEAR SYSTEMS

As previously mentioned, the first step in systems analysis is to compare the output signal with the input signal. If the relation between the output and input obeys the following requirement, the system is said to behave linearly. If input $I_A(t)$ produces output $O_A(t)$, and input $I_B(t)$ produces output $O_B(t)$, then input $[I_A(t) + I_B(t)]$ produces output $[O_A(t) + O_B(t)]$ irrespective of the choice of $I_A(t)$ and $I_B(t)$. This requirement is sometimes called the superimposition requirement (Bracewell, 1965, p. 185). Time invariance means that the response to I_A (t-T) is O_A (t-T) for all T and $I_A(t)$ (Bracewell, 1965, p. 185). That is, the waveform of the system's response depends on the waveform of the input, but not on the time of application.

An important point about linear behavior is that the same method of linear systems analysis can be used with any linear system. Linear systems analysis has many convenient features. In particular, if the frequency-domain output produced by a sufficient number of different input sinusoids is known, then the time domain output produced by a single input pulse (or any other transient waveform) can be computed by means of, for example, the inverse Fourier transform. The converse calculation can be carried out by means of, for example, the forward Fourier transform. Furthermore, the minimum possible phase shifts produced by any linear system can be calculated if the variation of output amplitude with input oscillation frequency is known (Aseltine, 1958; Bracewell, 1965). Many physical systems whose behavior is quite nonlinear when the input or output amplitude is large behave approximately linearly over a limited range of input amplitudes, and linear analysis can be used within that limitation (e.g., Figure 1A and B). A system which behaves linearly provided that the range of input variation is restricted is called "piecewise linear" (Distefano et al., 1967, pp. 46 and 47).* A case in point is De Lange's research on visual sensitivity to flickering light, described in a Ph.D. thesis that initiated widespread interest in the application of linear systems analysis to visual psychophysics (De Lange, 1952, 1957, 1958). He showed by psychophysical experiment that the human visual system behaves approximately linearly near flicker detection threshold. Therefore, he pointed out, if one has measured detection threshold for sinusoidal flicker at a sufficient number of flicker frequencies, it follows on theoretical grounds that flicker detection threshold for any given flicker waveform can be predicted by means of the inverse Fourier transform. Thus, only one experiment is necessary. This study unified a large amount of apparently unconnected data on flicker detection for different flicker waveforms gathered during previous decades, and at the same time showed these data to be redundant and quite the wrong way to approach the topic of flicker detection. The vision research community was strongly influenced by this electrical engineer's all too brief excursion into vision research. The subsequent application of linear systems analysis to data on contrast detection thresholds for sinewave gratings by Campbell and Robson (1969) was, perhaps, the most influential initiative in spatial vision psychophysics of the last two decades.

On the other hand, although linear systems theory has proved valuable in psychophysics, especially when stimuli are at or near detection threshold, linear systems theory is by no means adequate in visual psychophysics. The profound importance of nonlinearity in our understanding of the human visual system is brought out by the following quotation from Reichardt and Poggio (1981, p. 187) on the topic of neural information processing: " — every nontrivial computation has to be essentially nonlinear, that is, not representable (even approximately) by linear operations". A system is said to behave nonlinearly when the relation between output and input disobeys the superimposition requirement set out above. In practical terms, a major distinction between linear and nonlinear systems is that, for a nonlinear system, the frequency domain behavior cannot be predicted from the time domain behavior unless one already knows the system equation (i.e., the mathematical model) for that particular system. Consequently, the time domain and temporal frequency domain can provide complementary information about the system. For a linear system, on the other hand, the time and temporal frequency characteristics are merely alternative ways of presenting the same information, as already stated.

Rather than review the many examples of nonlinearity that have been encountered in visual psychophysics, I will describe three examples in some detail.

II. BINOCULAR PROCESSING OF THE DIRECTION OF MOTION IN DEPTH

When an object moves along a trajectory that passes between the eyes, the two retinal images move in opposite directions, and when the trajectory passes wide of the head the two retinal images move in the same direction. More particularly, the direction of motion in depth, relative to the head, is uniquely related to the ratio V_L/V_R where V_L and V_R are,

* Figure 1 a to d illustrates the difference between a "nonessential" nonlinearity (which approximates linear behavior more and more closely as input amplitude is progressively reduced) and an "essential" nonlinearity (which continues to behave nonlinearly as input amplitude is progressively reduced). Essentially nonlinear processes of importance in visual psychophysics include rectification (as illustrated in Figure 1c and d), division, multiplication, and thresholding (Beverley and Regan, 1973a; Regan and Beverley, 1980, 1981; Morgan and Regan, 1987). Division, multiplication, and thresholding are discussed in this chapter. Chapter 9 discusses rectification, though in the context of evoked electric and magnetic fields.

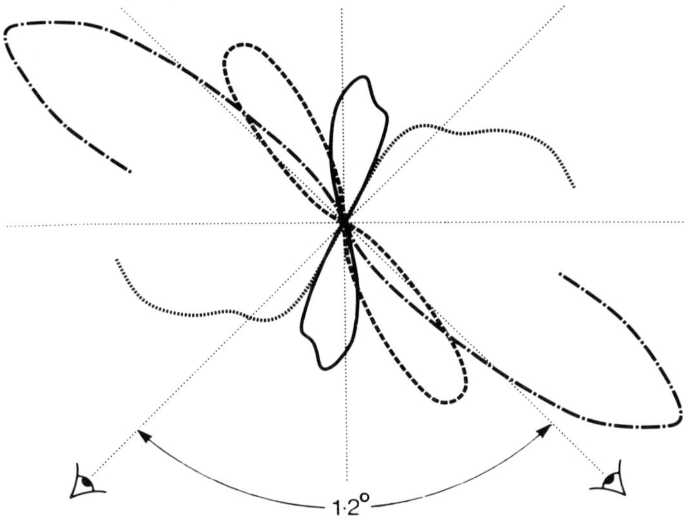

FIGURE 2. Direction tuning curves for the binocular motion-in-depth psychophysical channels of the human visual system. This figure is redrawn in polar coordinates from the data in Beverley and Regan (1973a). Evidence for the existence of neural mechanisms selectively sensitive to the direction of movement in space. [From Beverley, K. I. and Regan, D. (1973a). *J. Physiol. (London)*, **235**, 17–29.]

respectively, the velocities of the left and right retinal images (Beverley and Regan, 1973a, 1975; Regan et al., 1978). (We restrict this discussion to the velocity component contained in a plane that includes the left and right eyes.) Beverley and Regan (1973a, b) raised the question whether the human visual system takes advantage of this geometrical fact and, in particular, whether the human visual system processes retinal image velocity in an essentially nonlinear way so as to create sensitivity to V_L/V_R. The crucial experiment was as follows. Subjects inspected a target that oscillated along a straight line trajectory in depth with a fixed value of $(V_L/V_R)_{ADAPT}$. After a few minutes we found that visual sensitivity to motion in depth was depressed, and this adaptation effect was specific to the direction of motion in depth. For example, after adapting to a trajectory that passed to the right of the nose, visual sensitivity to trajectories passing to the right of the nose was reduced, while sensitivity to trajectories passing to the left of the nose was not. Conversely, after adapting to a trajectory that passed to the left of the nose, visual sensitivity to trajectories passing to the left of the nose was reduced, while sensitivity to trajectories passing to the right of the nose was not (Beverley and Regan, 1973b). This finding can be understood if the human visual pathway contains at least two psychophysical channels tuned to the ratio V_L/V_R; one prefers trajectories that pass to the left of the nose, and the other prefers trajectories that pass to the right of the nose. In a further experiment, subjects adapted to 13 different values of V_L/V_R, i.e., to 13 different directions of motion in depth. The results could be explained in terms of binocular psychophysical channels tuned to only four different values of V_L/V_R. Figure 2 shows the tuning curves of these channels. Sensitivity is proportional to radial distance from the origin. Azimuthal angle is proportional to the ratio V_L/V_R. This plot greatly exaggerates the small range of angles for which V_L/V_R is negative, i.e., trajectories that pass between the eyes.

Searching for a physiological correlate of the psychophysical channels of Figure 2, Cynader and Regan (1978) recorded the firing responses of single neurons in areas 17 and 18 of cat visual cortex and found cells whose binocular preference for a particular direction of motion in depth was sharpened or even created by a nonlinear interaction between signals from left and right eyes. The binocular responses of these cells could not be predicted linearly from their separate responses to monocular stimulation. Figure 3 plots the firing character-

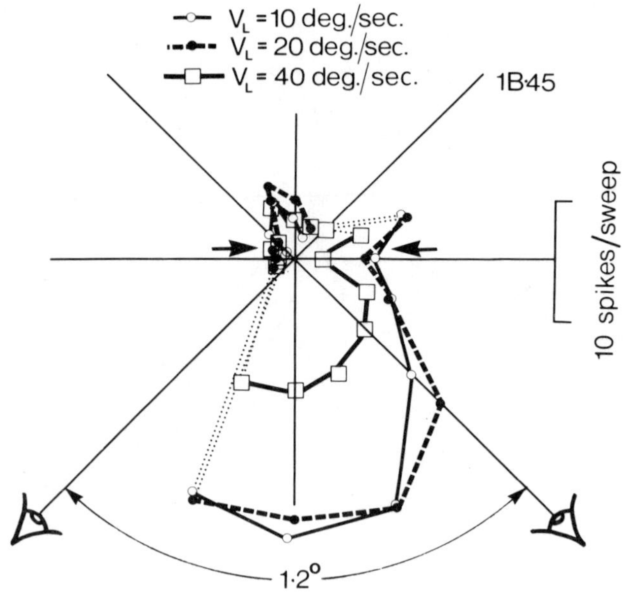

FIGURE 3. A "hitting the head" cell in area 18 of cat visual cortex. The stimulus was a bar that moved along different directions in depth. The number of spikes per sweep is plotted radially. the ration (VL/VR) between the retinal image velocities in the left and right eyes is proportional to azimuthal angle. (The coordinate system is the same as for the psychophysical data of Figure 2). The preferred direction of motion in depth was the same for stimulus speeds of 10, 20, and 40 deg/s. (From Regan, D. and Cynader, M. (1982). *Invest. Ophthalmol. Vision Sci.*, **22**, 525–550.)

istics of a "hitting the head" cell. The two horizontal arrows indicate how binocular inhibition to frontal plane motion created a sharp selectivity for trajectories passing through or close to the head. Tuning could be very narrow: some cells responded to a cone of directions only 1 to 2° wide, the remaining 358 to 359° range of directions producing comparatively weak responses. The main finding illustrated in Figure 3 was that, for some cells, the preferred direction of motion in depth remained constant to within a few tenths of a degree as the absolute speed of the stimulus was doubled, then quadrupled (Regan and Cynader, 1982). This finding can be explained if the neuron was tuned to the ratio V_L/V_R independently of the absolute values of V_L and V_R. In effect, the cell performs the analog of the essentially nonlinear process of computing a ratio.

Classes of "hitting the head" and "missing the head" cells were found in cat visual cortex corresponding to all the psychophysical channels of Figure 2: four classes for approaching motion and four corresponding classes for receding motion (Regan et al., 1979a,b).

III. MONOCULAR PROCESSING OF CHANGING SIZE

If a square object oscillates in depth along the z axis of Figure 4A, its retinal image oscillates in size in such a way that the opposite edge always move at the same speed in opposite directions, as illustrated in Figure 5A. The retinal image of a square object that oscillates in depth along the x axis in Figure 4A moves from side to side so that opposite edges always move at the same speed in the same direction, as illustrated in Figure 5B. And when the pure inphase and pure antiphase oscillations of Figure 5 are mixed in differing proportions, as illustrated in Figure 6, the square appears to oscillate along different directions in depth.

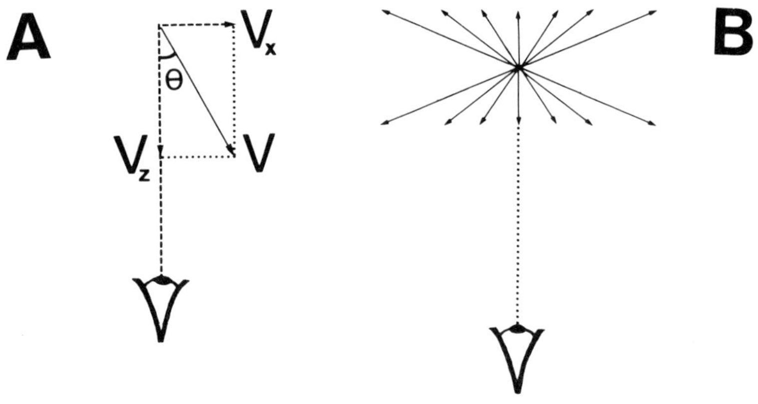

FIGURE 4. (A) Motion along an arbitrary direction θ in depth can be analyzed into an x-axis component V_x, and a z-axis component V_z. (B) The seven trajectories shown have different components of motion in the x direction, but identical components in the z direction. (From Regan, D. and Beverley, K. I., (1980). *J. Opt. Soc. Am.*, **70**, 1289–1296.)

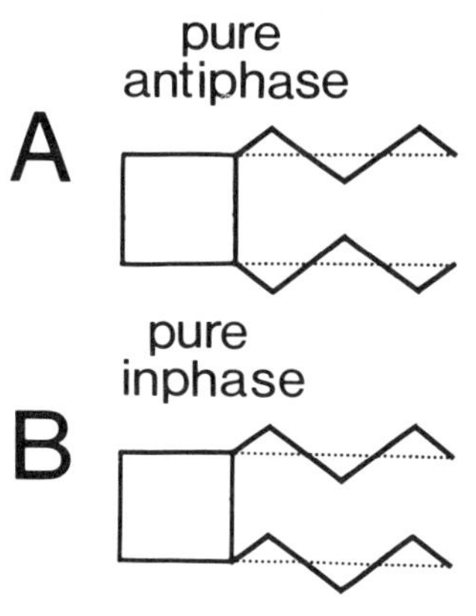

FIGURE 5. (A) Pure oscillations of size. The opposite edges of the square move at the same speed in opposite directions. This is equivalent to motion along the z-axis in Figure 4, i.e., motion along a line through the eye. (B) Pure sideways oscillations. The opposite edges of the square move at the same speed in the same direction. This is equivalent to motion along the x-axis in Figure 4.

The experiment described next was designed to test whether the human visual pathway processes pure antiphase oscillation (corresponding to z-axis motion in Figure 4A) independently of a simultaneously present in-phase oscillation (corresponding to x-axis motion in Figure 4A). In other words, we wished to test whether the visual pathway responds to the z-axis component of the velocity vector V in Figure 4A independently of the direction of angle θ.

We measured the just-detectable (i.e., threshold) pure antiphase (i.e., z-axis) oscillation of a test square before and after adapting to one or other of the adapting oscillations shown in Figure 6. In general, adaptation produced a reduction in visual sensitivity, i.e., the just-detectable oscillation before adaptation was smaller than after adaptation. Threshold elevation was defined as the following ratio: (just detectable oscillation after adaptation)/(just-detectable oscillation before adaptation). The dotted line in Figure 8A shows how threshold elevation for the z-axis test oscillation was affected by the amount of x-axis oscillation in the adapting stimulus. The data points would fall on a horizontal straight line if visual sensitivity to

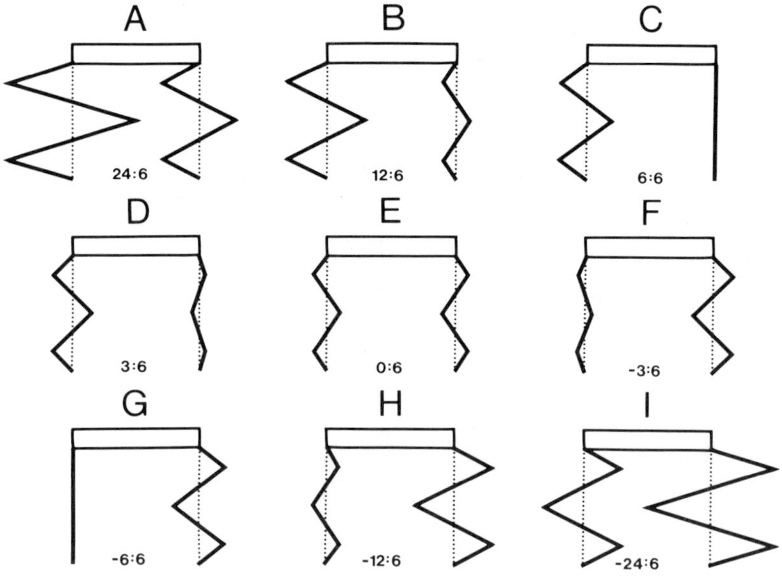

FIGURE 6. Examples of the stimulus oscillations used as adapting stimuli in the experiment whose results are shown here. The amplitudes of the oscillations relative to the size of the rectangle are exaggerated, but the ratios between movements of the left and right edges are to scale. Each oscillation contained the identical antiphase (i.e., z-axis) component of 6-min arc peak to peak. An in-phase (i.e., x-axis) component of amplitude 24-min arc was added in A, 12-min arc in B, 6-min arc in C, and 3-min arc in D. No x-axis component was present in E. Panels F to I show the results of adding an x-axis component to the 6-min arc z-axis oscillation with the phase of the x-axis component inverted with respect to the x-axis component in A to D. The ratio between the instantaneous velocities of the left and right edges of the square is shown in each case (V_L/V_R or V_R/V_L). A negative sign means that the edges always moved in opposite directions. (From Regan, D. and Beverley, K. I. (1980). *J. Opt. Soc. Am.*, **70**, 1289–1296.)

changing size were completely independent of the x-axis component of motion. The rightmost and leftmost points in Figure 8A were, indeed, at the same height as the central point, showing that the threshold elevation produced by a 6-min arc pure z-axis adapting oscillation was the same as when a 48-min arc x-axis oscillation was added to the 6-min arc z-axis oscillation. However, the dotted line in Figure 8A was not quite horizontal: there was a clear dip on either side of the center. Noting that the bottom of the dips occurred at about the 6-min arc on the abscissa, and that this condition corresponds to C or G in Figure 6, we concluded that the bottom of the dips occurred when one or other edge of the adapting stimulus was stationary on the retina. In terms of three-dimensional motion, these are conditions C and G in Figure 7. One possible explanation for the dips in the Figure 8A curve is as follows. Let us model the processing of the z-axis component of motion as shown in Figure 10. We suppose that one of the edges of the square excites the upper motion filter, while the opposite edge excites the lower motion filter. In terms of the Figure 10 model we assume that the dips in Figure 8A are caused by a reduction in the changing-size signal shown as k(x-y) in Figure 10. The dips in Figure 8A can be understood if the output of a motion filter falls off disproportionately steeply at near zero speeds. (For example, a motion filter might give no output until the speed of the exciting edge reached some threshold speed.) Nonlinear behavior that is most evident at low input levels is well known, and there is a wide range of physical examples including the transition from static to sliding friction.

In our next experiment we attempted to reduce the nonlinear interaction between visual responses to the z-axis and x-axis components of motion by adding an ''auxiliary'' oscillaton to all the adapting stimuli shown in Figure 6. The reason for choosing an x-axis (i.e., in-

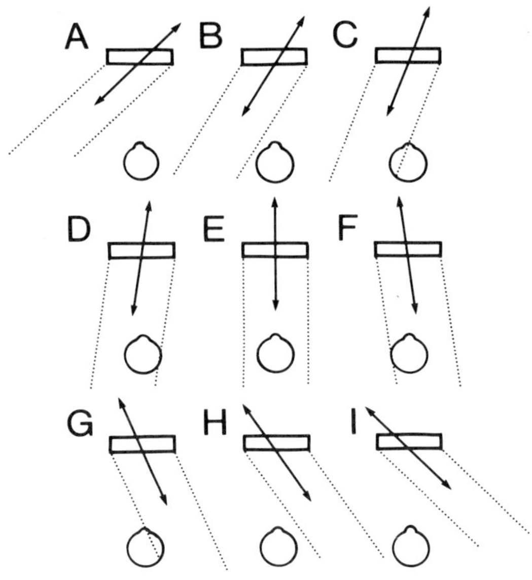

FIGURE 7. Oscillations of the stimulus square in Figure 6 were equivalent to motion along the trajectories illustrated. Directions in A corresponds to those in A in Figure 6, and so on. (From Regan, D. and Beverley, K. I. (1980). *J. Opt. Soc. Am.*, **70**, 1289–1296.)

phase) auxiliary oscillation was to ensure that the auxiliary oscillation did not excite the changing-size filter in Figure 10C. In order to make doubly sure, the frequency of the auxiliary oscillation was set at 8 Hz, a frequency that had previously been shown to be beyond the upper frequency limit of the changing-size filter (Regan and Beverley, 1978). No auxiliary oscillation was added to the test stimulus which remained exactly as in the experiment of Figure 8A.

Figure 8B shows that the result of adding the auxiliary oscillation to the adapting stimuli used in Figure 8A was to remove the two dips in the dotted line. In Figure 8A the visual system processed the z-axis component of motion without any appreciable ''cross-talk'' from the x-axis component of motion, even when the x-axis component was eight times faster than the z-axis component.

The psychophysical effect of an auxiliary signal shown in Figure 8 is closely similar to the linearization phenomenon that is well known in engineering. The concept of linearizing can be understood as follows. In a domestic magnetic audio taperecorder the relation between, on the one hand, the magnetizing (recording) flux produced at the recording head and, on the other hand, the magnetizing field produced by the sound wave becomes markedly nonlinear when the magnetizing field is near zero. This nonlinearity would distort the recording were a high-frequency oscillation not added to the signal to be recorded. The function of the high-frequency oscillation is to reduce the fraction of total recording time for which the magnetizing field is near zero. Thus, the effect of this high-frequency oscillation is to make the recording system behave more linearly. By analogy, we assume that, in our Figure 8B experiment the motion ''jitter'' caused by the auxiliary signal had the effect that, for adapting stimulus (B) in Figure 9, the upper edge was never stationary on the retina so that the failure of a motion filter (Figure 10) to respond for slow excitation speeds was never manifest.

It is well known that if a system consists of a series of linear stages, it is not possible to infer the sequence of processing by comparing the input and output of the system. This is possible, however, for a system that can be modeled as a linear stage followed by a static

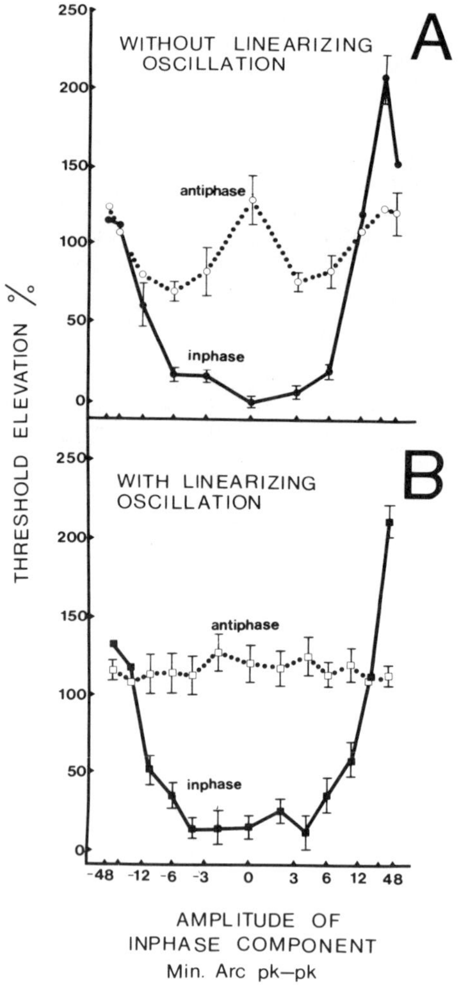

FIGURE 8. (A) Threshold elevations produced by separately adapting to 11 different oscillations in depth, all of which had the same antiphase component of motion as illustrated in Figure 4B. This constant component was a 2-Hz antiphase triangular wave of amplitude 6-min arc (peak to peak) per edge. The peak-to-peak amplitude of the added in-phase component is plotted along the abscissa. The dotted line plots threshold elevations for a pure antiphase test oscillation and indicates that the constant 6-min arc antiphase component of the various adapting oscillations produced a roughly constant threshold elevation independently of the amount of added in-phase component, though elevations were somewhat reduced when one or other edges of the rectangle were stationary (at about 6-min arc on the abscissa). The continuous line plots threshold elevations for a pure in-phase test oscillation. Each point is the mean of eight settings. (B) Repeat of A except that all adapting stimuli had a constant in-phase 8-Hz triangular wave oscillation added to the 2-Hz adapting stimuli (Figure 9). The linearizing oscillation rendered antiphase threshold elevations (dotted line) completely independent of the amount of x-axis motion. Each point is the mean of eight settings. Vertical bars shown 1 ± SE. (From Regan, D. and Beverley, K. I., (1980). *J. Opt. Soc. Am.*, **70**, 1289–1296.)

nonlinearity followed by a second nonlinear stage (Van der Tweel and Spekreijse, 1969; Marmarelis and Marmarelis, 1978). With this point in mind, in a further experiment we attempted to measure the attenuation characteristic peripheral to the nonlinearity in Figure 10C (Regan and Beverley, 1981). The procedure was to vary the amplitude of the auxiliary oscillation until the linearizing effect reached some criterion level. Then the measurement was repeated for a series of different values of auxiliary oscillation frequency. The rationale was that the procedure pitted the input amplitude of the auxiliary oscillation against its

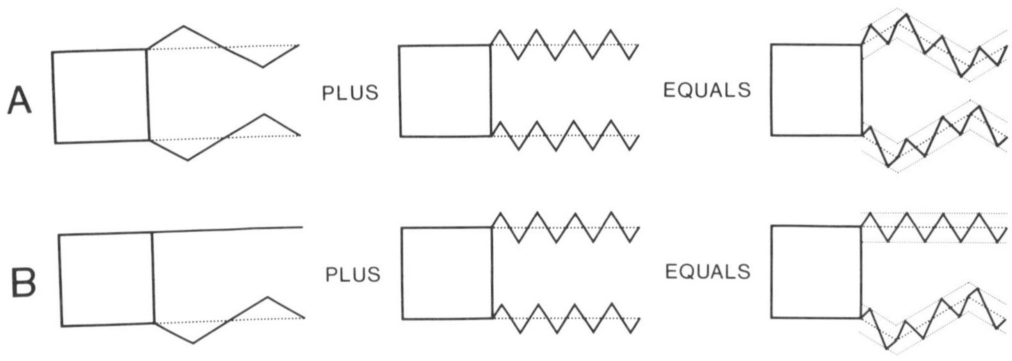

FIGURE 9. The adapting stimulus oscillation used to obtain the data of Figure 8B. An ''auxiliary'' oscillation was added to the 2-Hz adapting oscillation shown in Figure 6. The auxiliary oscillation was a triangular in-phase oscillation of frequency 8 Hz. (A) The pure antiphase (i.e., z-axis) oscillation (E in Figure 6), the auxiliary oscillation, and, at the far right, the sum of the two oscillations. (B) The oscillation (C of Figure 6), the auxiliary oscillation, and the sum of the two oscillations. (From Regan, D. and Beverley, K. I., (1980). *J. Opt. Soc. Am.*, **70**, 1289–1296.)

frequency to achieve a constant amplitude at the site of the nonlinearity. In other words, for every different auxiliary oscillation frequency, the linearizing effect was constant; therefore, the amplitude of the auxiliary signal on arrival at the nonlinearity was also constant. Hence, we had measured the attenuation curve periphral to the nonlinearity. The curve was flat from 1 to about 8 Hz, and then rolled off at about 15 dB per octave.

Finally, we consider the implications of our ''auxiliary signal'' experiments for everyday vision. When a subject's head is restrained and he carefully fixates (as in our experiments), then retinal ''jitter'' velocities are around 14 min arc/s, (Skavenski et al., 1979) or even considerably less (Kowler and Steinman, 1980). In this situation we found the optimal velocity of the auxiliary signal to be about 64 min arc/s [i.e., an 8-Hz triangular wave of amplitude 4 min arc (peak to peak)]. In everyday vision there is, of course, no artificially added auxiliary signal. On the other hand, the head is not restrained in everyday vision, and retinal image ''jitter'' speeds are considerably greater when the head is free than when restrained (Skavenski et al., 1979). Head movements cover the frequency range dc-7 Hz, and the residual retinal image ''jitter'' speeds range from 20 min arc/s (with the head almost still) to more than 180 min arc/s (with head movements) (Skavenski et al., 1979). (Note that our optimal auxiliary signal was roughly in the middle of this velocity range.) Thus, only when a subject carefully fixates with the aid of head restraint can any stimulus edge have a near-zero retinal image velocity as in Figures 5C and 7C. In everyday visual conditions, one consequence of imperfectly compensated head movements will be that whatever the trajectory of a moving object, none of its edges will have a near-zero retinal image velocity. Consequently, in everyday free-head viewing conditions, the changing-size channel can accurately compute the rate of change of size even for trajectories that graze the eye as in Figure 7 C and G.

IV. LINE SEPARATION DISCRIMINATION

It is well established that human observers can discriminate as little as 15-s arc change in the separation of a pair of parallel lines, a change in separation that is less than the 25-s arc distance between the centers of adjacent cones at the foveal center (Osterberg, 1935). Importantly, Morgan and Ward (1985) found that this line separation discrimination threshold was little affected by the presence of flanking lines that were sufficiently close

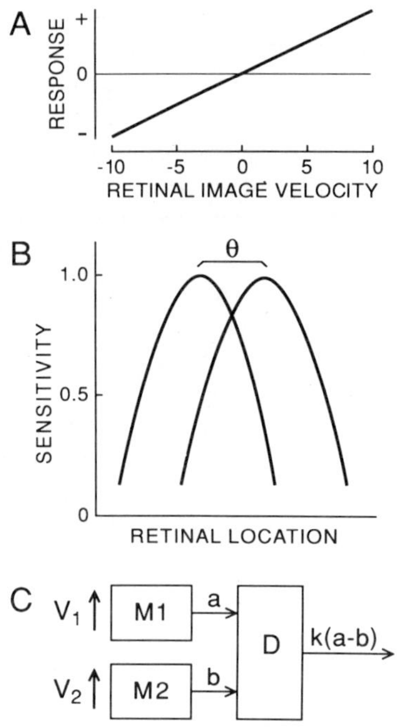

FIGURE 10. Psychophysical model of the processing of changing size (i.e., z-axis motion). The motion filters may overlap, but their sensitivity peaks are some distance apart on the retina. The data of Figure 8 indicate that there is a threshold-type nonlinearity in each of the legs that precede the changing-size filter. (Modified from Beverley, K. I. and Regan, D. (1979). *Vision Res.*, **19**, 1093–1104.)

to the target lines to corrupt the signal from putative neural mechanisms driven from a retinal region that encompassed both the target lines. This finding invalidates hypotheses of line interval discrimination that are based on comparing the responses of many neurons driven from the same retinal location [although hypotheses of this kind can provide a fair account of spatial frequency discrimination and orientation discrimination (Campbell et al., 1970; Regan and Beverley, 1983, 1985; Regan, 1984; Westheimer et al., 1976; Wilson and Gelb, 1984; Wilson and Regan, 1984)]. To explain line separation discrimination we suggested that receptive fields at widely separated locations are coupled together so as to form a kind of nonlinear coincidence detector whose output is maximal when the two receptive fields both receive optimal stimulation from a two-line stimulus that falls precisely on their receptive field centers (Regan and Beverley, 1985, footnote 42; Morgan and Regan, 1987). We suppose that the two receptive fields respond best to lines of the same orientation. By definition, such a coincidence detector is strongly nonlinear, because the output produced by stimulating with one or other line of the pair is negligible compared with the ouptut produced by stimulating with both lines simultaneously. For example, one of the many ways in which such a coincidence detector might operate is for the output to be proportional to the product of the excitations produced by the two lines. A possible physiological basis for this hypothetical coincidence detector is the finding that, in striate cortex, horizontal connections run between columns of similar orientation specificity (Gilbert and Wiesel, 1989). A suggested explanation for the exquisite discrimination of line interval separation is that discrimination is determined by the relative activity of overlapping coincidence detectors (Morgan and Regan, 1987). This idea could also account for the findings that discrimination is almost independent of contrast over a wide range of contrasts, and that discrimination was not affected by varying randomly the contrast of one of the lines (Morgan and Regan, 1987).

Now we turn from the one-dimensional task of line separation discrimination to a two-dimensional task of shape discrimination. In everyday life, one of the most important visual functions is to recognize the shape of an object or a symbol (e.g., a road sign or a letter). This ability requires subjects to discriminate between the shapes of retinal images. Reasoning that a neural mechanism that underlies shape discrimination might operate by encoding the relative dimensions of the retinal image of an object along different azimuths, we designed a simple experiment to investigate this process. The stimulus was a bright rectangle presented on a dim background. The rectangle's vertical sides were of length *a* deg and the horizontal sides were of length *b* deg. Ten rectangles with slightly different values of aspect ratio a/b were presented in random order. Half the rectangles were elongated vertically (i.e., $a/b > 1.0$), and half were elongated horizontally (i.e., $(a/b) < 1.0$). The subject's task was to judge whether the rectangle was elongated vertically or horizontally. To ensure that neither dimension *a* or *b* provided a reliable cue to shape, the area of the rectangle was randomly varied by up to 400% on each presentation. This stratagem forced subjects to base their judgements on the aspect ratio (a/b), i.e., the subject's responses were based on a ratio estimation, an essentially nonlinear process. We found that this nonlinear process could be performed with remarkable precision. For example, the just-noticeable difference in aspect ratio was as low as 1.2% for a rectangle of mean area 0.25 deg^2. Let us assume that the separations of opposite edges (i.e., *a* and *b*) are encoded directly by the hypothetical coincidence detectors described above in the case of line separation discrimination, and then the ratio a/b is encoded at a more central stage of processing. If we assume that the nonlinear process of computing a/b is entirely free from noise, then we must accept that a and b are encoded with a precision of about 10-s arc, considerably less than the 25-s arc separation of the most tightly packed cones. And if we adopt the more plausible position that there is some noise at the a/b stage of processing, then we must assume that *a* and *b* are encoded to even greater precision than 10-s arc. One possible way in which the aspect ratio a/b might be encoded is that the visual system contains two pools of neurons, one sensitive to aspect ratios greater than 1.0, the other pool being sensitive to aspect ratios less than 1.0. Stimulus aspect ratio would be neurally represented by the relative excitation of these two pools. The illusory shifts of aspect ratio produced by adapting to different aspect ratios are consistent with this hypothesis (Regan and Hamstra, 1991, 1992).

ACKNOWLEDGMENTS

I thank Lorna Houston for assistance in preparing this manuscript. This research was partly supported by the NSERC of Canada and also sponsored by the Air Force Office of Scientific Research, Air Force Systems Command, USAF, under grant number AFOSR F49620-88-C-0022. The U.S. Government is authorized to reproduce and distribute reprints for Governmental purposes notwithstanding any copyright notation thereon.

REFERENCES

Aseltine, J. A. (1958). *Transform Methods in Linear System Analysis,* McGraw-Hill, New York.

Beverley, K. I. and Regan, D. (1973a). Evidence for the existence of neural mechanisms selectively sensitive to the direction of movement in space, *J. Physiol. (London),* **235**, 17–29.

Beverley, K. I. and Regan, D. (1973b). Selective adaptation in stereoscopic depth perception, *J. Physiol. (London),* **232**, 40–41.

Beverley, K. I. and Regan, D. (1975). The relation between discrimination and sensitivity in the perception of motion in depth, *J. Physiol. (London),* **249**, 387–398.

Beverley, K. I. and Regan, D. (1979). Visual perception of changing-size: the effect of object size, *Vision Res.,* **19**, 1093–1104.

Blaquiere, A. (1966). *Nonlinear Systems Analysis,* Academic Press, New York.

Bode, H. W. (1945). *Network Analysis and Feedback Amplifier Design,* Van Nostrand, Englewood Cliffs, NJ.

Bracewell, R. (1965). *The Fourier Transform and Its Applications,* McGraw-Hill, New York.

Campbell, F. W. and Robson, J. G. (1969). Application of Fourier analysis to the visibility of gratings, *J. Physiol. (London),* **197**, 551–566.

Campbell, F. W., Nachmias, J., and Jukes, J. (1970). Spatial frequency discrimination in human vision, *J. Opt. Soc. Am.,* **60**, 555–559.

Cynader, M. and Regan, D. (1978). Neurones in cat parastriate cortex sensitive to the direction of motion in three-dimensional space, *J. Physiol. (London),* **274**, 549–569.

De Lange, H. (1952). Experiments on flicker and some calculations of an electrical analogue of the foveal system, *Physica,* **18**, 935–950.

De Lange, H. (1957). Attenuation Characteristic and Phase Shift Characteristic of Human Fovea-Cortex System in Relation to Flicker Fusion Phenomenon, Thesis, Techische Hogeschool, Delft.

De Lange, H. (1958). Research into the dynamic nature of the human fovea-cortex system with intermittent and modulated light, *J. Opt. Soc. Am.,* **48**, 777–784.

Distefano, J. J., Stubberud, A. R., and Williams, I. J. (1967). *Feedback and Control Systems,* McGraw-Hill, New York.

Gilbert, C. D. and Wiesel, T. N. (1989). Columnar specificity of intrinsic horizontal and corticocortical connections in cat visual cortex, *J. Neurosci.,* **9**, 2432–2442.

Hayashi, D. (1964). *Nonlinear Oscillations in Physical Systems,* McGraw-Hill, New York.

Kowler, E. and Steinman, R. M. (1980). Small saccades serve no useful purpose: reply to a letter by R. W. Ditchburn, *Vision Res.,* **20**, 273–276.

Marmarelis, P. Z. and Marmarelis, V. Z. (1978). *Analysis of Physiological Systems; The White-Noise Approach,* Plenum Press, New York.

Morgan, M. J. and Regan, D. (1987). Opponent model for line interval discrimination: interval and vernier performance compared, *Vision Res.,* **27**, 107–118.

Morgan, M. J. and Ward, R. M. (1985). Spatial and spatial frequency primitives in spatial-interval discrimination, *J. Opt. Soc. Am.,* **A2**, 1205–1210.

Milsum, J. H. (1966). *Biological Control Systems Analysis,* McGraw-Hill, New York.

Osterberg, G. (1935). Topography of the layer of rods and cones in the human retina, *Acta Ophthalmol.,* Suppl. 6, 1–103.

Regan, D. (1984). Spatial vision: VEP evidence for mechanisms tuned to spatial frequency, in *Evoked Potentials II,* Nodar, R. H. and Barber, C., Eds., Butterworth, Boston, 287–289.

Regan, D. (1989). *Human Brain Electrophysiology,* Elsevier, New York.

Regan, D. and Beverley, K. I. (1978). Looming detectors in the human visual pathway, *Vision Res.,* **18**, 415–421.

Regan, D. and Beverley, K. I. (1980). Visual responses to changing-size and to sideways motion for different directions of motion in depth: linearization of visual responses, *J. Opt. Soc. Am.,* **70**, 1289–1296.

Regan, D. and Beverley, K. I. (1981). Motion sensitivity measured by a psychophysical linearizing technique, *J. Opt. Soc. Am.,* **71**, 958–965.

Regan, D. and Beverley, K. I. (1983). Spatial frequency discrimination and detection: comparison of postadaptation thresholds, *J. Opt. Soc. Am.,* **73**, 1684–1690.

Regan, D. and Beverley, K. I. (1985). Visual responses to vorticity and the neural analysis of optic flow, *J. Opt. Soc. Am.,* **A2**, 280–283.

Regan, D. and Cynader, M. (1982). Neurons in cat visual cortex tuned to the direction of motion in effect: effect of stimulus speed, *Invest. Ophthalmol. Vision Sci.,* **22**, 525–550.

Regan, D. and Hamstra, S. (1991). Shape discrimination and the judgement of perfect symmetry. ARVO Abstr.

Regan, D. and Hamstra, S. (1992). Shape discrimination and the judgement of perfect symmetry, *Vision Res.,* in press.

Regan, D., Beverley, K. I., and Cynader, M. (1978). Stereoscopic depth channels for position and for motion, in *Frontiers in Visual Science,* Cool, S. J. and Smith, E. L., Eds., Springer-Verlag, New York, 351–372.

Regan, D., Beverley, K. I., and Cynader, M. (1979a). The visual perception of motion in depth, *Sci. Am.,* **241**, 136–151.

Regan, D., Beverley, K. I., and Cynader, M. (1979b). Stereoscopic subsystems for position in depth and for motion in depth, *Proc. R. Soc. London,* **B204**, 485–501.

Reichardt, W. and Poggio, T., Eds. (1981). *Theoretical Approaches in Neurobiology,* MIT Press, Cambridge.

Skavenski, A. A., Hansen, R. M., Steinman, R. M., and Winterson, B. J. (1979). Quality of retinal image stabilization during small natural and artificial body rotations in man, *Vision Res.,* **19**, 675–683.

Van der Tweel, L. H. and Spekreijse, H. (1969). Signal transport and rectification in the human evoked-response system, *Ann. N.Y. Acad. Sci.,* **156**, 678–695.

Westheimer, G., Shimamura, K., and McKee, S. P. (1976). Interference with line-orientation sensitivity, *J. Opt. Soc. Am.,* **66**, 332–338.

Wilson, H. R. and Gelb, D. J. (1984). Modified line element theory for spatial frequency and width discrimination, *J. Opt. Soc. Am.,* **1**, 124–131.

Wilson, H. R. and Regan, D. (1984). Spatial frequency adaptation and grating discrimination: predictions of a line element model, *J. Opt. Soc. Am.,* **A1**, 1091–1096.

Chapter 11

A NONLINEAR MODEL OF FEATURE DETECTION

David C. Burr and M. Concetta Morrone

TABLE OF CONTENTS

I. Introduction 310

II. The Local Energy Model of Feature Detection 310

III. The Model in Two Dimensions 314

IV. Structuring the Image 315

V. Brightness Illusions 318

VI. Discussion 324

References 326

INTRODUCTION

One of the major tasks of the human visual system is to extract quickly and effortlessly the most salient information from an image to form a symbolic representation of the scene. This argument was made most forcefully by Marr (1976, 1982) with the concept of the "primal sketch", an early and grossly simplified abstract representation of the original scene. The primal sketch is not simply a reduced or sampled version of the original image, but a highly nonlinear caricature comprising lines, cylinders, and other basic forms that capture the essential essence of the scene, while ignoring much detail, as well as gradual variation of luminance. Figure 1 depicts a well-known example of how economically an artistic sketch may convey the essential features of an image. A few well-placed strokes are sufficient to represent the image without ambiguity.

Two of the more common visual features are lines and edges, both being rich sources of visual information. The potential of lines to convey information is well illustrated in the sketch of Figure 1, and edges are obviously important in delimiting the boundaries of objects. Indeed, there is a sense in which edges and lines are almost interchangeable, illustrated by the fact that sketch-artists often use lines to symbolize edges (see Figure 1). However, mathematically speaking, the two types of features are quite distinct and, indeed, orthogonal: edges can be described as odd-symmetric functions and lines are even-symmetric functions, orthogonal with respect to each other in L^2.

It is because these two equally important features are orthogonal, defining a two-dimensional feature space, that we propose that a two-dimensional basis set is required for their detection. We further argue that nonlinear combination of the two bases is the most efficient way of detecting lines and edges.

Many models have been proposed for line and edge detection, all similar in one major respect: they convolve the image with simple linear operators of various sizes and search for maxima or zero-crossings in the output. However, this approach runs into several difficulties, particularly when the image features are adjacent and when the features are not simply edges or lines but combinations of both (see Figure 2).

Our approach to line and edge detection differs from that of others in that we have attempted to understand why lines and edges are important to vision, and have been able to provide a robust operational definition of what constitutes a visual feature. Our model, like most others, convolves the image with linear operators of limited bandwidth. But it differs from others in that it requires two sets of linear operators, orthogonal to each other (and related by the Hilbert transform). It also incorporates two major nonlinearities: a second-order (squaring) nonlinearity to combine the output of the matched filters, and a higher-order "nonmaximal suppression" nonlinearity, whereby only the points producing maxima in the combined output are considered as features effective in structuring visual information. This chapter outlines the details of the model and illustrates its application to several interesting images. These demonstrations show how the model successfully predicts human perception, both qualitatively and quantitatively, in conditions where most previous models fail.

II. THE LOCAL ENERGY MODEL OF FEATURE DETECTION

As mentioned above, our model requires two sets of matched operators, one even-symmetric, the other odd-symmetric, hence, one the Hilbert transform of the other. The image is convolved separately by the two sets of operators, and the outputs combined by "Phythagorean sum" (square root of the sum of squares) to give what we term the "local energy" profile. The operation is usually performed separately over several scales and

FIGURE 1. A sketch by Matisse, illustrating how a few well-placed strokes may provide an adequate symbolic representation of an image. Note that the lines have been positioned to correspond to luminance borders, or edges, of the original image.

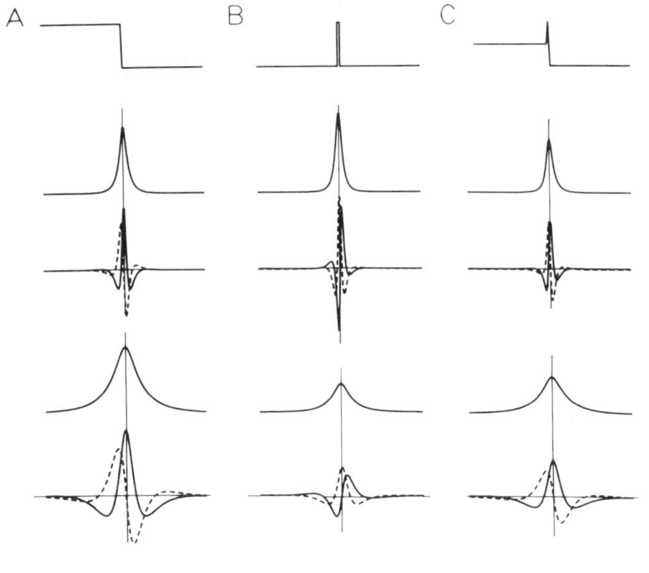

FIGURE 2. Examples of the operation of the model for an edge, a line, and a combination edge-line. Under each profile is shown the output of the even-symmetric (dashed lines) and odd-symmetric (continuous lines) at two spatial scales. Immediately above these traces is the local-energy profile, the square root of the sums of the squared output from the even- and odd-symmetric operators. Whereas the linear operators produce many minima, maxima, and zero-crossings, only some of which fall at the position of the feature, the local-energy functions are always positive, with one clear peak at the position of the feature. The three different features are all marked unambiguously with the same operator. To decide which type of feature caused the local energy peak, it is sufficient to evaluate the response of the linear operators at that point: for the edge, only the odd-symmetric operators respond at that point; for the line, only the even-symmetric operators; for the edge and line, both operators respond, so the feature is recognized veridically.

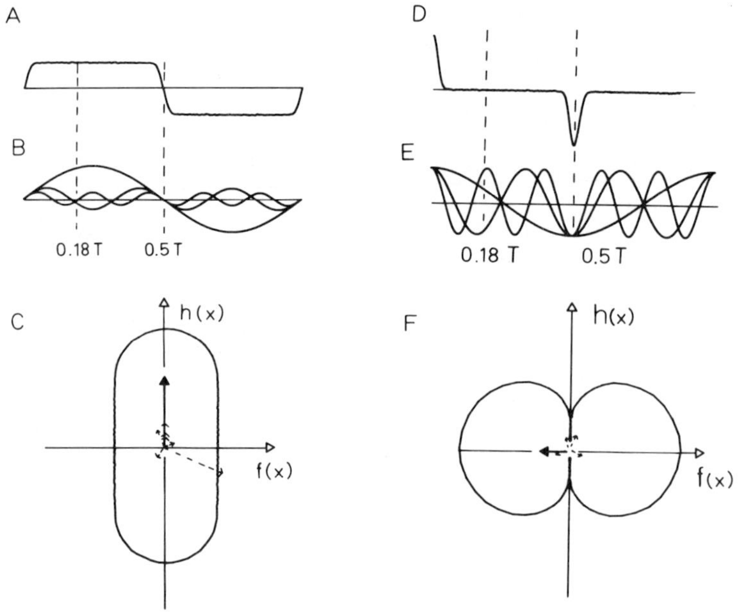

FIGURE 3. Illustration of the relationship between features, phase congruence, and local energy. See text for explanation.

orientations, modeling the action of cortical visual detectors (e.g., Blakemore and Campbell, 1969), but this is, in fact, not an essential feature of the model (see Morrone and Owens, 1987).

Figure 2 illustrates the action of the model in one dimension for three profiles, an edge, a line, and a combination edge and line. For clarity we illustrate the response at only two spatial scales, although at least four are required to model human vision (Wilson and Bergen, 1979). The superimposed curves under the profile show the responses of the even-symmetric operators (dashed curves) and odd-symmetric operators (continuous curves). These linear responses are similar to those predicted by most models of feature detection that use only one class of operator (either even-symmetric: Marr and Hildreth, 1982; Watt and Morgan, 1985; or odd-symmetric: Canny, 1983, 1986) and illustrate some of the difficulties inherent with this approach. Although the odd-symmetric operators peak at the position of the edge, there are also spurious negative peaks at either side of the true edge, and two very large peaks near the bar. The even-symmetric operators face similar problems as line detectors, and the zero-crossings mark several spurious features. The feature that presents most difficulties for models with a single class of detector is the combination edge and line. For this feature, neither the peaks nor the zero-crossings of the even- or the odd-symmetric detectors correspond to the perceived position of the feature. This class of feature, in fact, occurs very frequently in natural scenes, particularly under conditions of oblique lighting (see Horn, 1977; Perona and Malik, 1990; and Figure 4).

Several strategies have been devised to minimize spurious feature marking, usually involving thresholding and comparison across spatial scales (Marr, 1982; Watt and Morgan, 1985; Yuille and Poggio, 1985). However, many of these strategies have been criticized as being computationally expensive, biologically implausible, and often ineffective, particularly with more complicated images with closely adjacent features. And none of these strategies is effective with the combination of line and edge features (like Figure 2C).

The local energy output (Pythagorean sum of even- and odd-symmetric output) is shown directly above the linear outputs. One obvious advantage of this operation is that at each scale the profile has only one peak, and that peak corresponds in position to the feature. Furthermore, the same function signals both types of features with a positive peak (rather than relying on the difficult process of searching for zero-crossings). After the feature has been detected, it can easily be identified by examining the amplitude of the linear operators at that point: a response of odd-symmetric operators signals an edge, and even-symmetric operators a line. If both classes of operators respond at the position of the feature (such as in Figure 2C), then both a line and an edge are signaled.

What were the theoretical considerations that led us to choose local energy as an indicator of visual features? It has long been known that the Fourier phase spectrum is important for vision, indeed, far more important than the Fourier amplitude spectrum. If the phase spectra of two images are interchanged, leaving the amplitude spectra intact, the appearance of the hybrid pictures is determined almost completely by the phase spectra (Oppenheim and Lim, 1981; Piotrowski and Campbell, 1982). We have recently shown that the importance of the phase to vision lies not in the Fourier phase spectrum per se, but in how phase relationships of the various harmonics cause them to interact to create visual features (Morrone and Burr, 1988; Burr and Morrone, 1990). Specifically, we have demonstrated that visually salient features occur at those points of an image where the Fourier components come into phase with each other. Thus, they can be detected by searching for points of congruence in arrival phase.

Figure 3 illustrates the concept of phase congruence of Fourier harmonics and shows how this congruence creates maxima in the local-energy function. The upper traces (A and D) show two periodic waveforms, a squarewave and a series of delta functions, both slightly blurred with a Gaussian filter (to attenuate the infinite series of harmonics). The sinusoids under each waveform represent the first three components in the Fourier expansion. For both waveforms, these harmonics (and all higher harmonics) come into phase periodically, at twice the frequency of the fundamental; and the point where the harmonics come into phase is where the feature is seen, be it edge or line. For an edge, all cosine harmonics have arrival phases of $\pm\ \pi/2$ (depending on the polarity of the edge). For a bar, the arrival phases are all 0 or π. If the feature were a combination edge and line (like Figure 2C), the arrival phases would take on an intermediate value at the point of phase congruence. From these and other observations, we have proposed the generalization that visually salient features always occur at the point of maximum phase congruence (Morrone and Burr, 1988).

The relationship between local energy and phase congruence is illustrated in the lower graphs of Figure 3. These curves, parametric in x, were obtained by plotting the input functions $f(x)$ (Figure 3 A and D) against their Hilbert transforms $h(x)$. In this representation, local energy is given by the distance of the curves from the origin (the norm).*

The arrows inside the graphs depict the vectors associated in this space with the first four individual harmonics of the periodic waveforms, calculated at two positions: $x = 0.5T$ the point where the feature is seen, and $x = 0.18T$, an arbitrary point on the plateau. The norm of the vectors for each harmonic does not vary with position (x) on the waveform [as $\sin^2(2\pi x/T) + \cos^2(2\pi x/T) = 1$], but the argument is proportional to the product of position and spatial frequency ($2\pi x/T$). At $x = 0.5T$ (the position of the edge or line), the arrival phases of all the harmonics are the same ($\pi/2$ in Figure 3C and π in Figure 3F), so the vectors are all aligned. For $x = 0.18T$, however, the arrival phases of the harmonics are very different, so the vectors (indicated by dashed arrows) will all point in different

* The functions f(x) can be considered to be the ouptut of an even-symmetric operator of broad bandwidth, and its Hilbert transform h(x) the output of an odd-symmetric operator of identical bandwidth. In practice the model does not evaluate local energy for the whole pattern, but separately at different scales. However, the concepts illustrated here for the physical image can readily be extended to the output of our linear operators by considering band-pass filtered versions of f(x) and h(x).

directions. Each point of the parametric curves is the resultant of all the vectors associated with the harmonics. Clearly, the norm of the resultant vector will be greatest when the individual vectors are most closely aligned; that is, when the arrival phases of the component harmonics are most similar. Hence, peaks in local energy will mark points of maximal arrival-phase congruence. Although this example uses periodic patterns, the same argument can readily be extended to aperiodic patterns, by considering a patch-wise Fourier analysis.

Operators similar to local energy have been used in several applications, including image enhancement (Granlund, 1978; Wilson et al., 1983), motion detection (Adelson and Bergen, 1985, Heeger, 1987) and in modeling hyperacuity (Klein and Levi, 1985). However, the motivation for using this class of operation has always been to eliminate phase information, rather than to capitalize on it. These seemingly contradictory assertions can be reconciled by the understanding that although local energy is, in fact, insensitive to absolute phase, it is highly sensitive to relative phase organization. For example, the local energy of a sinewave is identical to that of cosinewave; and any L_2 function will have the same local-energy profile as its Hilbert transform, or any other phase-shifted version of it. However, local energy does depend on the relationships between phases of the various Fourier harmonics, as the above example clearly shows. It is for both of these reasons that we find the function so useful. It is sensitive to phase relationships and, therefore, detects points of phase congruence; but it responds equally well, irrespective of the value of the phase at those points of phase congruence, so the same operator can detect both lines and edges. Distinguishing between lines and edges is a simple process, readily established by evaluating the phase (or relative contribution of the even- and odd-symmetric operators) at the maxima of local energy.

III. THE MODEL IN TWO DIMENSIONS

The model has recently been extended to two dimensions (Burr and Morrone, 1990; Morrone and Burr, 1992). This step is not as routine as may be imagined, as the operators must be Hilbert transforms of each other, and the Hilbert transform is not defined in two dimensions. Therefore the operators must be essentially one dimensional and oriented in space. It is interesting that the oriented operators that we were forced to use in order to obtain a Hilbert transform in fact resemble quite closely the receptive field properties of visual neurones. In both cat and monkey, the most striking property of cortical neurons (as distinct from retinal and thalamic neurons) is that they respond very selectively to stimuli of given orientations (Hubel and Wiesel, 1962, 1977). We use four sets of oriented operators, oriented at 0, 90, and $\pm 45°$. Along one axis the operators are modulated with the even- or odd-symmetric profile depicted in Figure 2, and along the other with a simple Gaussian of about the same size.

Another difficulty for the two-dimensional model is in detecting maxima in local energy. If only the absolute maxima of the plane are considered, only a few isolated points would be marked. On the other hand, if maxima are marked line by line (at each orientation), too many features would be marked, producing a tendency for "overshoot", making lines appear longer than they are and corners terminated incorrectly. Our strategy is to establish first the direction of maximal change of energy (within a small window) and search for maxima only along this direction. Again this process finds a neurophysiological analogue in human and animal vision called "cross-orientation inhibition": mutual inhibition between neurons of differing orientation preference (Morrone, Burr, and Maffei, 1982; Morrone and Burr, 1986; Burr and Morrone, 1987).

Finally, there is the difficulty of how information at various scales and various orientations is combined to give an overall feature map of the scene. For this we find little guidance either from physiological or from psychophysical research. Although there is compelling evidence that information is decomposed by filters selective for both orientation and spatial frequency, there is as yet no information of how this information may be resynthesized to yield a unified percept. Indeed, for the human visual system, there may be no need for resynthesis of information at all. However, to gain some insight into how our model performs, we must be able to combine our separate representations to form a single feature map. To achieve this, each map is given a scale-dependent "uncertainty weighting" by blurring the feature maps with Gaussians of space constantly proportional to the scale of the operators, and then summed together. The effect of the uncertainty weighting is to privilege-attach positional information from the higher scales, which we may expect to be more precise.

An example of the operation of the model in two dimensions is shown in Figure 4. The image was convolved separately by even- and odd-symmetric operators of four spatial scales and four orientations. (In practice this operation is achieved using the pyramid technique of Burt and Adelson [1983] for efficiency of computation.) At each scale and at each orientation, the even- and odd-symmetric outputs were combined by Pythagorean sum to produce the local energy profile, from which the local maxima were calculated (along the direction of maximal energy gradient) to yield the independent feature maps. The feature maps were blurred in proportion to scale size (again achieved by "expanding" the pyramid) and summed to give the image of Figure 4B.

The first obvious point to make is that although the feature map may lack some of the artistic merit expressed in Figure 1, it is in fact a quite good sketch of the original image. Furthermore, evey relevant feature, line, edge, and specularity, has been marked, with no glaring false positives. Given that there has been no thresholding or postprocessing, this result is more than acceptable.

The histograms of Figure 4C give an idea of the proportions of various features types by plotting the distributions of phase at the maxima of local energy, separately for three different scales. All three histograms have clear peaks at $\pm\pi/2$, corresponding to positive- and negative-going edges, suggesting that these features are the most common. However, it is important to note that while these phases do predominate, there remains a large proportion of marked features of intermediate phases (like the feature of Figure 2C), features that often result from oblique lighting and shadowing effects. For our model, this class of features presents no difficulty and is veridically marked as combination edge bars. But for any model based on simple linear convolution, these types of features are difficult to localize veridically (see Figure 2C).

IV. STRUCTURING THE IMAGE

One of the clearest illustrations that vision is not strictly linear is provided by Harmon and Julesz's (1973) demonstration of coarse-quantization, illustrated in Figure 5A. The original image of Figure 4A has been "quantized" by setting all the pixels within each square to the mean value of those of the original image. This sampling technique preserves nearly all low-frequency image information (below the Nyquist sampling rate), while introducing the spurious high spatial frequency components that shape the blocks. What is interesting is that although blocking preserves sufficient low frequency information to allow face recognition (readily verified by blurring or distancing the image), by no effort of will can one extract this information from the unfiltered blocked image.

Whatever mechanisms underlie the phenomenon of blocking, they demonstrate a clear nonlinearity in the human visual system. The low-frequency information about the face cannot be extracted when viewed together with the high spurious spatial frequencies, clearly violating the additivity principle of linear systems. The original explanation for the effect was that the high spurious frequencies introduced by blocking mask the lower spatial frequencies that contain the image information, rendering them effectively invisible (Harmon

A
B

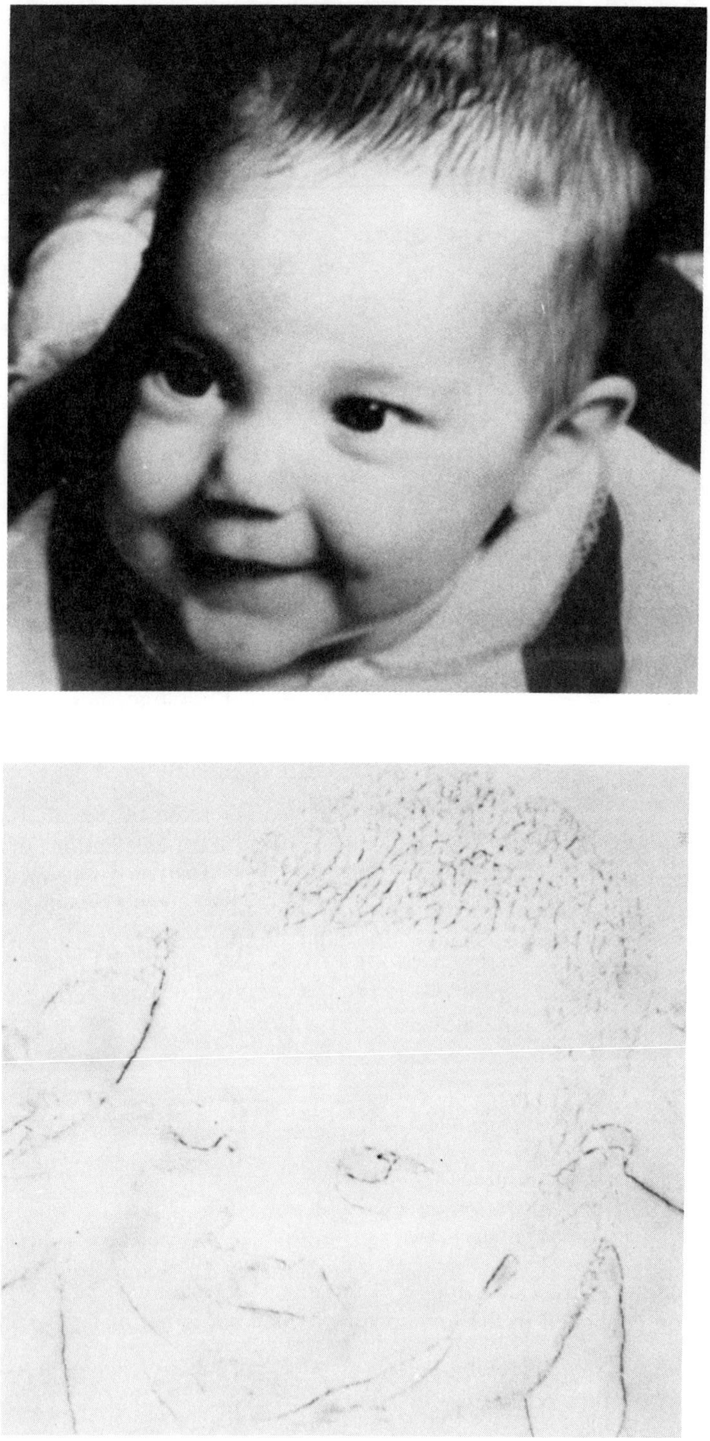

FIGURE 4. An example of results of the model in two dimensions. B is the output of the model applied to A (see text for details). The histograms of C show the distribution of arrival phases at the marked features, for high (upper curve), medium (middle curve), and low (lower curve) scales. Although there is a clear tendency for the phases to group at $+\pi/2$ (corresponding to edges), all phases are represented in the histograms.

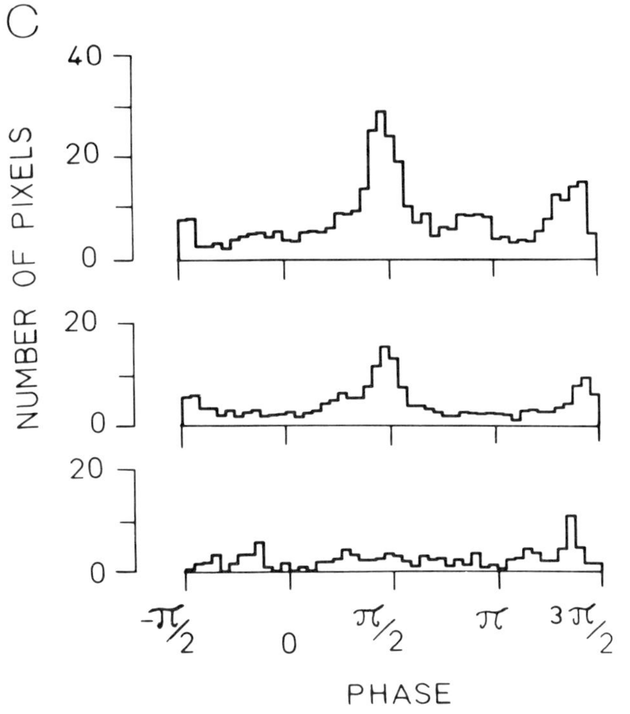

FIGURE 4C.

and Julesz, 1973). However, although nonlinear compression of this sort certainly occurs (e.g., Stromeyer and Julesz, 1972; Legge and Foley, 1980), calculations of the strength of masking (for example, from the data of Legge and Foley, 1980 or Anderson and Burr, 1985) show that the spurious spatial frequencies could not have sufficient power to suppress the low-frequency signals below detection threshold. Perhaps the best demonstration that the low spatial frequencies are not merely "masked" is that addition of further high-frequency energy (in the form of high-pass noise) causes the blocked image to be recognizable (Morrone et al., 1983).

Our explanation for the blocking phenomenon is that the high spatial frequencies do not remove the low spatial frequency information from vision, but structure the way that it is perceived. According to our model, only maxima in local energy are perceived as features, and in this image it is the high spatial frequencies that dictate where the maxima of local energy fall (partly because of the scale-dependent uncertainty weighting). Figure 4C shows the output of the model, produced by the procedure described above. The marked features fall in a grid-like pattern along the borders of the blocks; and according to our model, these features provide an abstract description of the image, causing it to be perceived as blocks. The information contained in the low spatial frequencies is not masked or suppressed, but becomes "compartmentalized" by the structure provided by the maxima of local energy. After the high spatial frequencies of image have been removed by digital blurring (Figure 5B), the low spatial frequencies contribute more to the position of the maxima of local energy, creating a feature map more consistent with the features of the face (Figure 5D), allowing a blurred face to be perceived.

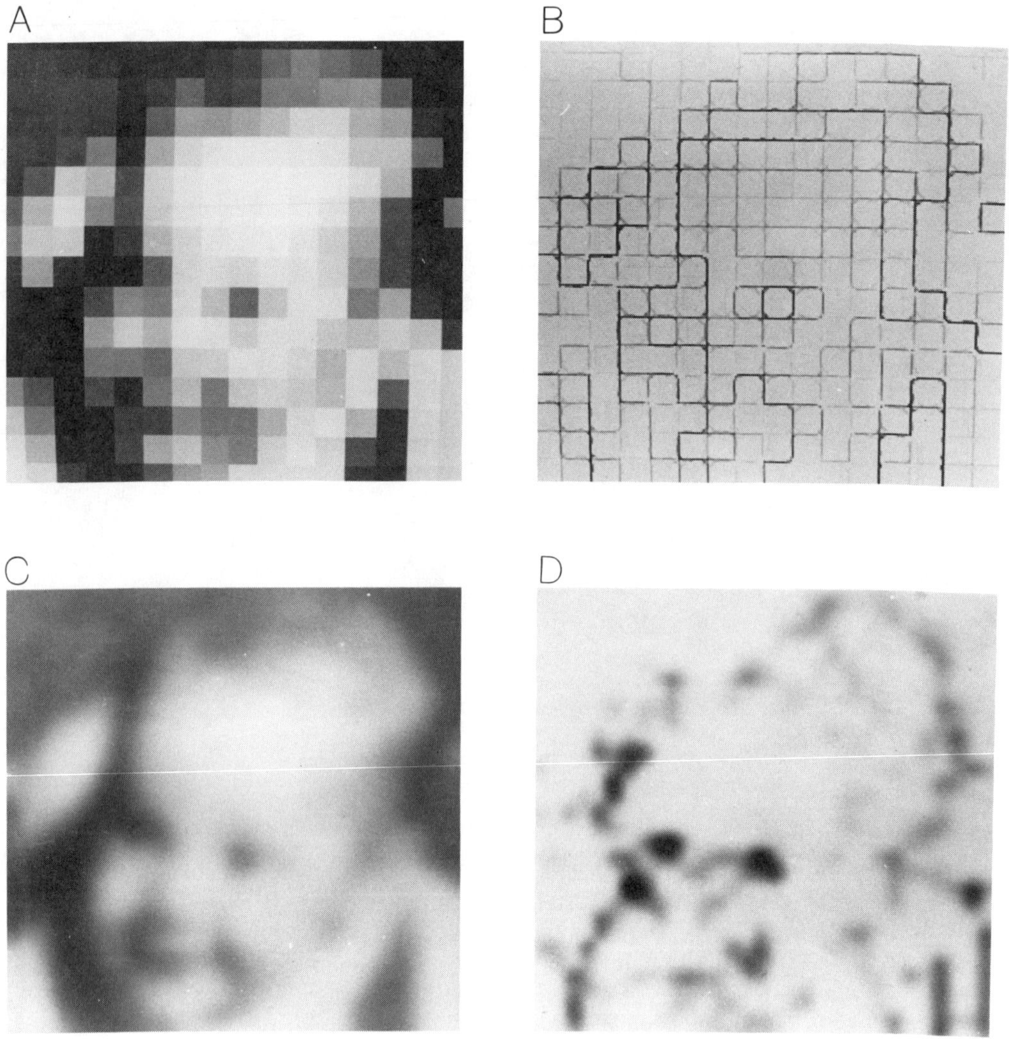

FIGURE 5. Illustration of how the local energy model predicts the "coarse quantization" phenomenon, first described by Harmon and Julesz (1973). The image of 4A has been quantized by setting all pixels within each block to the average level (A). This process preserves most low spatial frequency information (below the *Nyquist* rate, half the sampling frequency), readily verified by viewing the picture from a distance, or blurring the image (C). B and D show the output of the model to the blocked and blurred images (see text for further explanation).

V. BRIGHTNESS ILLUSIONS

Visual illusions have long been invoked to demonstrate that the visual system does not simply transpose the external scene into a veridical internal copy, but encodes important information into a symbolic representation. In doing so it sometimes errs, to produce a so-called visual illusion.

One of the best known and most powerful illusions is the phenomenon of Mach bands, the paradoxical light and dark bands seen where a luminance ramp meets a plateau (Mach, 1865, 1906; Ratliff, 1965). Figure 6 provides two illustrations of the phenomenon, in one and in two dimensions. The figure on the left is a trapezoidal waveform, but is not perceived as such. There are conspicuous spikes of brightness, where there are none in the luminance

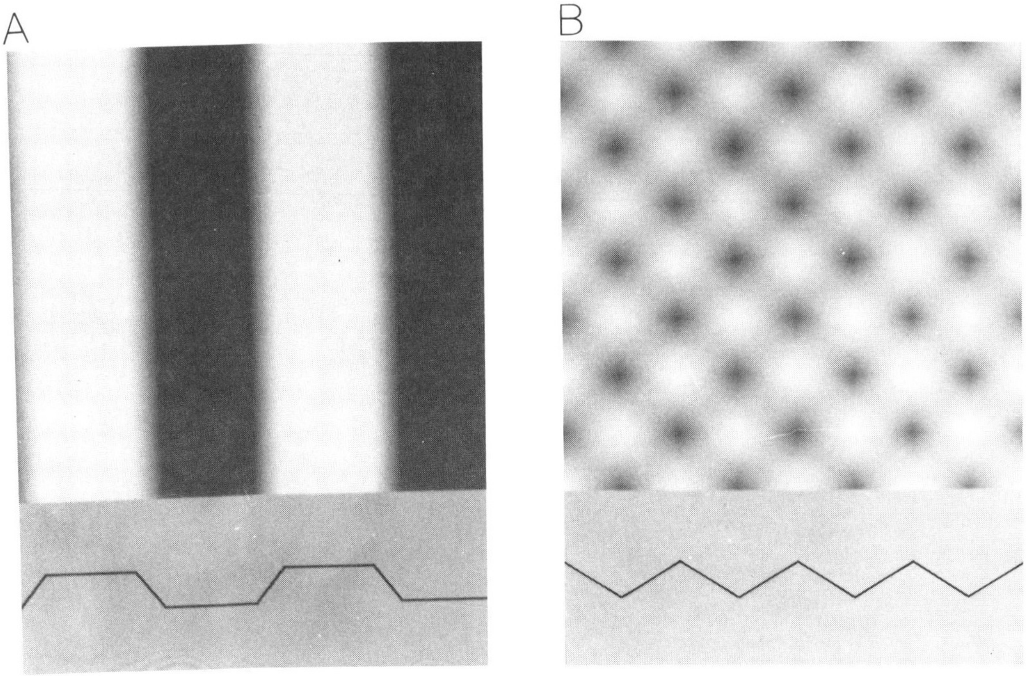

FIGURE 6. Two examples of the well-known brightness illusion of Mach bands. On the left the luminance profile is trapezoidal, but we perceive sharp bright and dark lines where the ramps meet plateaus. The pattern on the right was created by multiplying a vertical with a horizontal triangle wave to form a pyramid-like luminance distribution; yet we perceive sharp bright and dark crosses, not present in the luminance distribution.

profile. Note also that at the point where the bands are seen, there is a change in brightness (partly disguised by the Mach bands), so that the luminance ramp between the bands appears of uniform brightness, with no particular feature at the point where the luminance profile crosses zero. Figure 6B shows the same effect in two dimensions (Morrone et al., 1986). The luminance distribution is, in fact, pyramidal, the product of a horizontal and vertical triangular waveform; yet we perceive the pattern as bright and dark star-like structures.

The explanation for the Mach bands that appears in most textbooks (e.g., Cornsweet, 1970; Ratliff, 1965) was first advanced by Mach himself (Mach, 1906): that lateral inhibition in the retina effectively differentiates the image, producing overshoot and undershoot at the points where the bands are seen. However, a major difficulty with global differentiation (or any other linear high-pass filter operation) is that it implies that as the ramp is made steeper that bands should become stronger and stronger, to be maximal for a steep square edge, which would severely distort our everyday perception. However, the bands, in fact, become weaker as the ramp is steepened and do not occur at all if the ramp extends over less than 30'' (Ross et al., 1981; see also Fiorentini, 1973; Ratliff, 1984).

Our explanation for Mach bands does not rely solely on differentiation or other linear operations, but on the fact that the energy model marks line-like features at the point where the bands are seen (Ross et al., 1989). Figure 7 shows the local energy profile at two scales. At both scales the energy peaks at the points where the bands are seen, so the summed feature map will predict features to be there and there alone. At the higher scale there is a strong response from even-symmetric operators at these peaks, signaling the presence of sharp lines. At lower scales the odd-symmetric operators also respond at the energy peaks, and this response signals that a brightness change should accompany the Mach bands. As

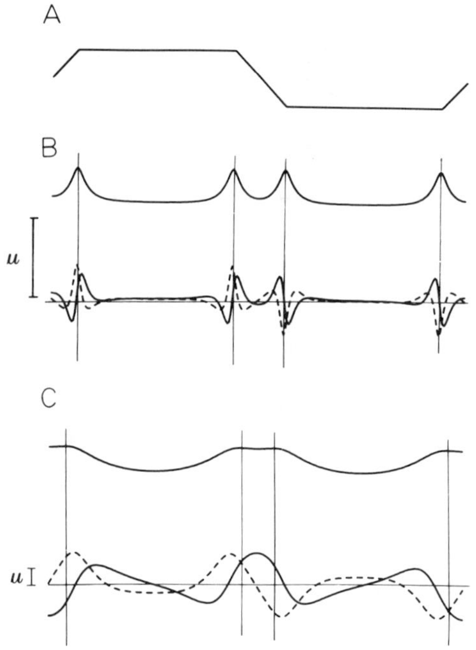

FIGURE 7. Illustration of the local-energy model applied to a trapezoidal waveform. Under the waveform are the local energy profiles of a high (B) and a low scale (C), together with the even-symmetric (dashed lines) and odd-symmetric (continuous lines) output. The energy at the higher scale peaks at the postion where the Mach bands are seen; and at that position there is a strong response from the even-symmetric operators, indicating that the feature should be a line. At the lower scale, the position of the feature is not positioned exactly on the kneepoint of the luminance distribution, but (as discussed elsewhere), the lower scales contribute little to localization of the features. They do, however, contribute to the brightness map; and because the odd-symmetric operators (as well as the even-symmetric operators) respond at the marked feature, they signal a brightness change.

in the previous example, the change in brightness occurs at the feature defined by the local energy peaks, and the high spatial scales dominate in determining the position of these peaks.

Local energy predicts not only the occurrence of Mach bands and where they should occur, but also the perceived strength of the bands. We have measured the strength of the Mach bands, by adjusting the contrast of a trapezoidal waveform until the bands were just visible. The inverse of contrast at threshold gives an estimate of contrast sensitivity, a unitless index greater than 1. It turned out that the contrast thresholds were very easy to set, and three observers agreed closely on the threshold settings. Contrast sensitivity for Mach bands were measured over a wide range of spatial frequencies and for a variety of stimuli filtered in the spatial frequency domain (Morrone et al., 1986; Ross et al., 1989).

Figure 8 shows an example of the results. Contrast thresholds for seeing Mach bands on blurred stimuli were measured as a function of the low-pass filter cutoff frequency (triangular symbols). As the pattern is blurred, and bands become progressively weaker, so more contrast is required to see them. For comparison we also measured contrast thresholds for detecting a high-pass trapezoid stimulus (circular symbols), deliberately chosen to have similar thresholds to those of the Mach bands.

The lines passing through the symbols of Figure 8 are not simply best fits of the data, but quantitative predictions from the local-energy model. To predict the detection thresholds for the high-pass stimulus, we first calculated the local energy at each scale. The energy was then combined by the standard procedure of probability summation across scales and

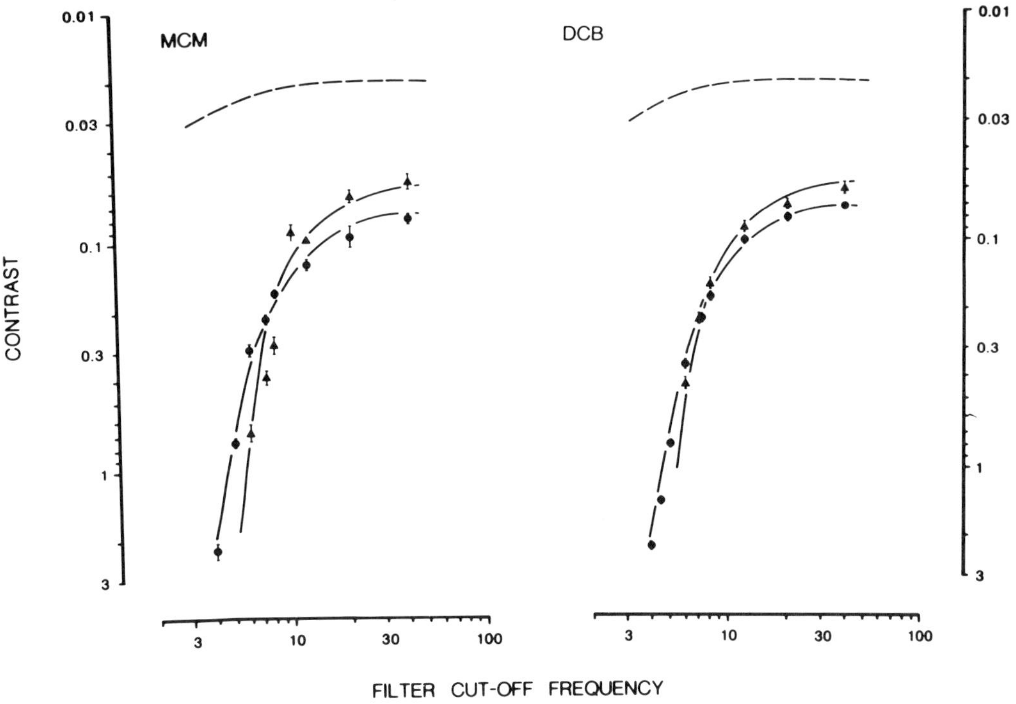

FIGURE 8. Contrast thresholds for seeing Mach bands (triangles) on a low-pass filtered trapezoidal waveform (circle), as a function of the filter cutoff frequency. The circles indicate detection thresholds of a high-pass trapezoid, measured for comparison. The curves passing through the symbols are predictions from the local energy model, while the dashed curves above are predictions derived from lateral inhibition (see Ross et al., 1989 for details).

across space (a form of nonlinear summation that takes account of the fact that increasing the number of visual units that respond to a stimulus increases the probability that at least one will respond and detect it; see Sachs et al., 1971; Ross et al., 1989). The results of these predictions (as a function of filter cutoff frequency) are represented by the continuous curves passing through the circles. To predict the strength of the Mach bands, we considered only energy that contributed to the peaks at the position where the Mach bands are seen. Again, this energy was summed probabilistically across space and across scale to yield the predictions represented by the other set of solid lines in Figure 8.

It shoud be stressed that the predictive curves of Figure 8 were derived with only one degree of freedom: absolute sensitivity determining the height of the curves. Once this was fixed for the detection results, the prediction curve of the Mach band thresholds was entirely determined. Given that there are no other variable parameters (or ''fudge factors'') in the model, the predictions are surprisingly good. Not only do the curves follow the general shape of the data, but they even predict subtle differences in the data, including the crossover around 7 c/deg.

The dashed curves above the figure are predictions of the strength of Mach bands for a model based solely on differentiation or high-pass filtering (assuming even-symmetric filters). It is clear that such a model does not come close to predicting quantitatively the actual data.

Many illusions illustrate how brightness* does not depend entirely on the physical luminance of the stimuli. One of the clearest demonstrations of this is the Craik-Cornsweet-

* A technical term meaning the perceived or apparent luminance of a surface, often misused to refer to physical properties of images.

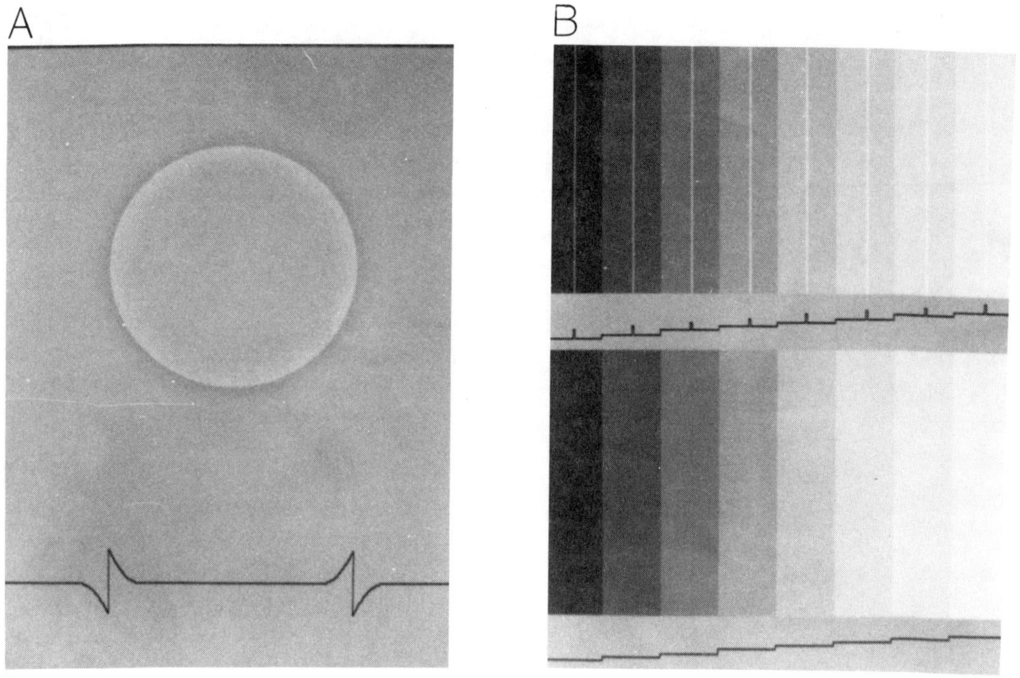

FIGURE 9. (A) An example fo the Craik-O'Brien-Cornsweet illusion. The pattern has identical luminance everywhere except at the border, yet the circle seems to be lighter than the background. (B) The lower figure shows the classic Chevreul illusion: the steps are all of uniform luminance, but do not appear so. When thin lines are added to edge step, they appear to divide each step into two regions of quite different brightness.

O'Brien illusion (Craik, 1966; O'Brien, 1958; Cornsweet, 1970), illustrated in Figure 9A. The luminance is, in fact, identical everywhere except at the border of the perceived circle, yet the circle seems brighter than the background. Original explanations for the phenomenon relied on high-pass filtering of the visual system (Cornsweet, 1970; Campbell et al., 1978), but as for the explanations for Mach bands, the filtering cannot predict quantitatively the results. The illusion holds even at very high spatial frequencies, well outside the range where vision may be expected to perform high-pass filtering (Burr, 1987).

The local-energy model readily accounts for the illusion. Local energy peaks at the borders of the inner circle, so features are seen there. As the profile is odd-symmetric, only odd-symmetric operators respond at the marked feature, so the feature is perceived as an edge, and perceiving an edge means perceiving a brightness step. The step in brightness caused by the edge extends to the whole region, as there are no other features to provide contradictory information.

Other brightness illusions are illustrated in Figure 9B. A staircase luminance profile is not seen veridically, but takes on the "scallopy" appearance of the well-known Chevreul illusion (Chevreul, 1890). For this illusion we do not as yet have a satisfactory explanation, but expect that it reflects a nontransitivity in visual computations. Each edge in the series has the same polarity and should, therefore, signal successive brightness increments. The second region should appear brighter than the first, the third brighter than the second, etc. If complete transitivity were observed, the difference between the brightness of the first and third region should be twice that of the difference between adjacent panels. However, there is good evidence that transitivity is not preserved totally, even for two successive edges of the same polarity (Shapley and Reid, 1985). Failure in preserving transitivity would create

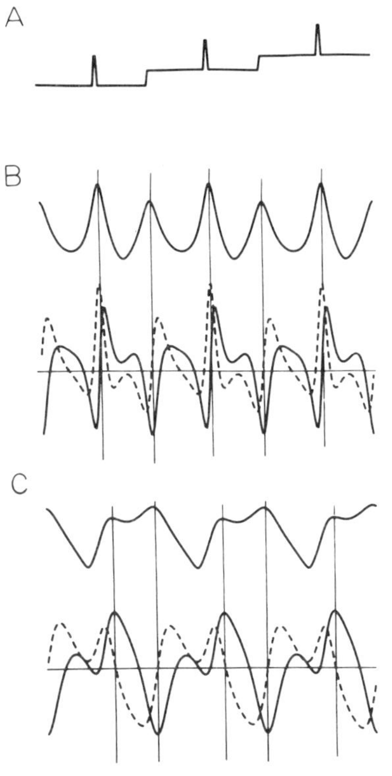

FIGURE 10. Illustration of how local energy predicts the brightness illusion of Figure 9. A shows the input waveform, and B and C the local-energy profiles, together with the even- (dashed line) and odd- (continuous lines) symmetric outputs at two scales. At the high scale (B), the local-energy peaks correspond in position to the lines and edges, acurately predicting the position of these features. At this scale it is mainly the even-symmetric operators that respond at the position of the lines, correctly predicting their appearance. At the lower scale, however, odd-symmetric detectors respond at the peak of local energy caused by the lines, predicting the observed brightness change.

an incompatibility between the local brightness steps and the global brightness of the whole pattern, and this incompatibility may produce the scallopy pattern we observe.

Unfortunately, the above explanation is not yet in a form where it can be tested quantitatively. However, a new illusion, more amenable to quantitative measurement can be created by adding a thin line to each step (Morrone et al., 1991). The series of lines changes the appearance of the pattern completely, causing it to appear ''square-wave-like'' with the regions bounded by lines and edges seen with uniform brighness, with a large brightness step at the position of each line.

Again, the local-energy model can explain readily this illusion. The profiles of Figure 10 show the response of the local energy and of the matched filters at two spatial scales. The energy profile peaks at the positions of the lines and edges, predicting that all features should be seen in those positions. The even-symmetric response at the high spatial scales predicts the presence of lines at those positions. But at the lower scales the odd-symmetric operators also respond at the peaks of local energy, predicting an edge, with an accompanying brightness change. We see both sharp lines and a change in brightness, in agreement with the predictions of the model.

As we saw in the previous example, the real test of a model is how well it predicts quantitatively human performance. We have measured the apparent contrast of the illusion under various conditions. Figure 9B shows how apparent contrast varies as the pattern is

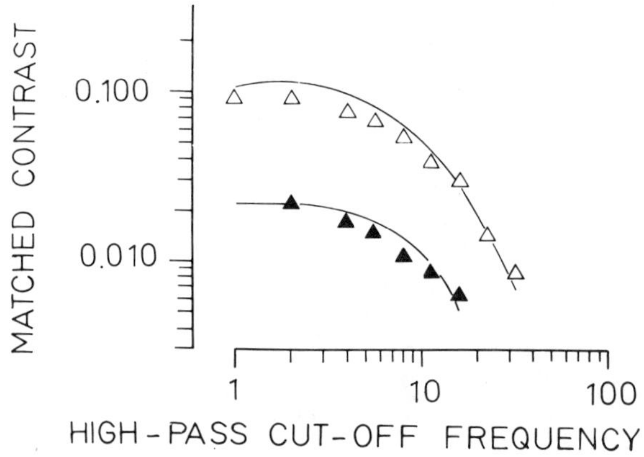

FIGURE 11. Apparent contrast of the illusory edge of Figure 9, measured at high (open triangles) and low (filled triangels) contrast. The lines passing through the symbols are predictions from the local-energy model, obtained from the strength of response of the odd-symmetric operators (see Figure 10).

high-pass filtered. Again, the solid curves are predictions from the model, based on the strength of the odd-symmetric response at the feature. As the low frequencies of the pattern are reduced, the odd-symmetric response becomes progressively diminished, until there should be no illusion at all. And, indeed, the experimental data follow closely the predictions of the model.

VI. DISCUSSION

To conclude, we have presented a model for early human vision whereby the retinal image is not simply transmitted through a system of linear filters, but encoded by nonlinear mechanisms to form an abstract, symbolic representation. In general, this representation reflects the physical reality. However, there exist several examples of visual illusions where humans perceive an image as different from that defined by its physical luminance distribution. In all the examples examined so far, our model performs like human vision in ''seeing'' the illusions and predicts accurately the strength of the illusions under various conditions.

In general, we were guided by the known physiology of mammalian vision in choosing parameters for the model. Our filters all have a limited amplitude spectrum of about 1.5 octaves, agreeing with most physiological and psychophysical data (e.g., Maffei and Fiorentini, 1973; Movshon et al, 1978a,b; Blakemore and Campbell, 1969; Legge and Foley, 1980; Anderson and Burr, 1991). But unlike many models of visual perception (Marr, 1982; Robson, 1980), the limited bandwidth of the operators (forming so-called spatial frequency channels) does not perform a major role in our model. Indeed, we have demonstrated reasonable success by applying the operation with broad-band matched filters of 3 octaves bandwidth (Morrone and Owens, 1987).

However, once the decision has been made for independent analysis of the image at various scales, the problem arises as to how the results of the independent analyses should be combined, and for this problem we find little guidance in the literature. Our strategies of relying more on the higher than lower scales for feature localization (but not in determining brightness) seems intuitively logical and does receive some support from the literature: phase judgments vary little with spatial frequency, implying an improvement in absolute positional judgments at higher scales (Burr, 1980; Tyler and Gorea, 1986). However, we would prefer that this important strategy be based on firmer data and have commenced some studies to this end.

A different strategy for synthesis has been suggested by Perona and Malik (1990). Rather than evaluate local maxima separately over all scales, Perona and Malik search all energy maps (of different scale and orientations) concurrently, and choose the absolute maxima, thereby producing a single feature map. The strategy has the advantage of not requiring reintegration of separate feature maps and is not at all biologically implausible. Hopefully, it will be possible to devise a quantitative test to distinguish between the two strategies.

Another difference between Perona and Malik's approach and ours is that they use a different strategy to evaluate local energy. Rather than nonlinear combination of two orthogonal filters, they convolve the image with many matched filters with differing phase response and choose those that give the greater response. This process is formally equivalent to our approach (consider the parametric plots of Figure 3), but is, of course, computationally much more expensive. Moreover, although the physiological evidence remains somewhat ambiguous (Field and Tolhurst, 1986), there is now clear psychophysical evidence that in human vision, detectors tend to have phase responses of 0, $\pi/2$ and $\pm\pi$, implying that receptive fields are eight even-symmetric or odd-symmetric (Burr et al., 1989). Thus, both the experimental evidence and considerations of computational efficiency imply that the nonlinear technique of taking the Pythagorean sum of orthogonal operators is reasonable.

Most of the operations that we suggest for our model are readily implementable by physiologial ''hardware''. There exist two major classes of neurons in the mammalian primary visual cortex, ''simple cells'' and ''complex cells'' (Hubel and Wiesel, 1962, 1977). Both simple and complex cells respond selectively to stimulus orientation, and the preferred orientation of the stimulus changes progressively with position in the cortex. The major difference between the two cell classes is that simple cells are quasilinear (except for a half-wave rectification resulting from the fact that their firing rate cannot fall below zero), while complex cells exhibit a clear second-order (squaring) nonlinearity (Movshon et al., 1978a,b; Maffei et al., 1979; Spitzer and Hochstein, 1985a,b). Furthermore, the linear simple-cells tend to be grouped so that adjacent cells have similar orientation and spatial frequency tuning, but differ in phase response by $\pi/2$ (Pollen and Ronner, 1981). The simple cells are ideally suited to act like the matched filters of our model. Complex cells, on the other hand, with their second-order nonlinearity, are ideal candidates to extract local energy, either from input supplied by the simple cells or by similar operations performed within their own subunits. If this suggestion were correct, then the complex cells would be primarily responsible for location of visual features, while simple cells were responsible for their identification.

An important nonlinearity in our model is the nonmaximum suppression, where only local peaks in the energy functions are considered to provide brightness information. Although no such mechanism has yet been observed neurophysiologically, it is by no means unreasonable. One of the primary mechanisms of elaboration of sensory information is mutual inhibition between neurons (Eccles, 1969). Lateral inhibition helps shape the receptive-field structure of visual neurons (Hartline, 1949), and inhibition between cone types enhances color sensitivity and helps maintain color constancy (Wiesel and Hubel, 1966). More recently, ''cross-orientation inhibition'' has been demonstrated between orientation-selective cortical neurons, a mechanism which should tend to enhance the response of the strongest firing neurons (Morrone et al., 1982; Morrone and Burr, 1986; Burr and Morrone, 1987; see also Chapter 12 of this volume). Similar sorts of mechanisms could well lead to suppression of weaker responses, ensuring that only local maxima of energy contribute toward the feature map. However, whether this operation is actually implemented neurophysiologically, and if so, at what level, remain open questions.

REFERENCES

Adelson, E. H. and Bergen, J. R. (1985). Spatio-temporal energy models for the perception of motion, *J. Opt. Soc. Am.*, **A2**, 284–299.

Anderson, S. J. and Burr, D. C. (1985). Spatial and temporal selectivity of the human motion detection system, *Vision Res.*, **25**, 1147–1154.

Anderson, S. J. and Burr, D. C., (1991). The Two-dimensional spatial and spatial frequency properties of motion sensitive mechanisms in human vision, *J. Opt. Soc. Am.*, **A8**, 1340–1351.

Blakemore, C. and Campbell, F. W. (1969). On the existence of neurones in the visual system selectively sensitive to the orientation and size of retinal images, *J. Physiol. (London)*, **225**, 437–455.

Burr, D. C. (1980). Sensitivity to spatial phase, *Vision Res.*, **20**, 391–396.

Burr, D. C. (1987). Implications of the Craik-O'Brien illusion for brightness perception, *Vision Res.*, **27**, 1903–1913.

Burr, D. C. and Morrone, M. C. (1987). Inhibitory interactions in the human visual system revealed in pattern visual evoked potentials, *J. Physiol. (London)*, **389**, 1–21.

Burr, D. C. and Morrone, M. C. (1990). Edge detection in biological and artificial visual systems, in *Vision: Coding and Efficiency*, Blakemore, C., Ed., Cambridge, University Press, London.

Burr, D. C., Morrone, M. C., and Spinelli, D. (1989). Evidence for edge and bar detectors in human vision, *Vision Res.*, **29**, 419–431.

Burt, P. J. and Adelson, E. H. (1983). The laplacian pyramid as a compact image code, *IEEE Trans. COM*, **31**, 532–540.

Campbell, F. W., Howell, E. R., and Johnstone, J. R. (1978). A comparison of threshold and suprathreshold appearance of gratings with components in the low and high spatial frequency range, *J. Physiol. (London)*, **284**, 193–201.

Canny, J. F. (1983). Finding edges and lines in images, *M I T AI Lab. Tech. Rep.*, **720**.

Canny, J. (1986). A computational approach to edge detection, *IEEE Trans. PAMI*, **8**, 679–698.

Chevreul, M. E. (1890). *The Principles of Harmony and Contrast of Colours*, (Translated by C. Martel), Bell, London.

Cornsweet, T. N. (1970). *Visual Perception*, Academic Press, New York.

Craik, K. (1966). *The Nature of Psychology*, Cambridge University Press, London.

Eccles, J. (1969). *The Inhibitory Pathways of the Central Nervous System*, Liverpool University Press, Liverpool.

Field, D. J. and Tolhurst, D. J. (1986). The structure and symmetry of simple-cell receptive-field profiles in the cat's visual cortex, *Proc. R. Soc. London*, **B228**, 379–399.

Fiorentini, A. (1972). Mach band phenomena, in *Handbook of Sensory Physiology, Vol VII/4*, Jameson, D. and Huvich, L. M., Eds., Springer-Verlag, Berlin.

Granlund, G. (1978). In search of a general picture operator, *Comput. Graphics Image Process.*, **8**, 155–173.

Harmon, L. D. and Julesz, B. (1973). Masking in visual recognition: effect of two-dimensional filtered noise, *Science*, **180**, 1194–1197.

Hartline, H. K. (1949). Inhibition of activity of visual receptors by illuminating nearby retinal elements in the Limulus eye, *Fed. Proc.*, **8**, 69.

Heeger, D. J. (1987). Model for the extraction of image flow, *J. Opt. Soc. Am.*, **4A**, 1455–1471.

Horn, B. K. P. (1977). Image intensity understanding, *Artif. Intell.*, **8**, 201–231.

Hubel, D. H. and Wiesel, T. N. (1962). Receptive fields, binocular interaction and functional architecture in the cat's visual cortex, *J. Physiol. (London)*, **160**, 106–154.

Hubel, D. H. and Wiesel, T. N. (1977). Architecture of macaque monkey visual cortex, *Proc. R. Soc. London*, **B198**, 1–59.

Klein, S. A. and Levi, D. M. (1985). Hyperacuity thresholds of 1 sec: theoretical predictions and empirical validation, *J. Opt. Soc. Am.*, **A2**, 1170–1190.

Legge, G. E. and Foley, J. M. (1980). Contrast masking in human vision, *J. Opt. Soc. Am.*, **70**, 1458–1471.

Mach, E. (1865). Uber die Wirkung der raumlichen Vertheilung des Lichreizes auf di Neztzhaut, *I.S.-B. Akad. Wiss. Wien Math.*, **54**, 303–322.

Mach, E. (1906). Uber den Einfluss Raumlich und zeitlich variierender Lichtreize auf die Gesichtswahrnehmung, *Sitzungsber. Akad. Wiss. Wien Math.*, **115**, 633–648.

Maffei, L. and Fiorentini, A. (1973). The visual cortex as a spatial frequency analyzer, *Vision Res.*, **13**, 1255–1267.

Maffei, L., Morrone, M. C., Pirchio, M., and Sandini, G. (1979). Response of visual cortical cells to periodic and nonperiodic stimuli, *J. Physiol. (London)*, **296**, 27–47.

Marr, D. (1976). Early processing of visual information, *Philos. Trans. R. Soc. London*, **B275**, 485–526.

Marr, D. (1982). *Vision*, Freeman, San Francisco.

Marr, D. and Hildreth, E. (1980). Theory of edge detection, *Proc. R. Soc. London*, **B207**, 187–217.

Morrone, M. C. and Burr, D. C. (1986). Evidence for the existence and development of visual inhibition in humans, *Nature (London),* **321**, 235–237.

Morrone, M. C. and Burr, D. C. (1988). Feature detection in human vision: a phase dependent energy model, *Proc. R. Soc. (London),* **B235**, 221–245.

Morrone, M. C. and Burr, D. C. (1992). A model of human feature detection based on matched features, in *Robots and Biological Systems,* Dario, P., Sandini, G., and Aebischer, P., Eds., Springer-Verlag, Berlin.

Morrone, M. C., Burr, D. C., and Maffei, L. (1982). Functional significance of cross-orientational inhibition. I. Neurophysiological evidence, *Proc. R. Soc. (London),* **B216**, 335–354.

Morrone, M. C., Burr, D. C., and Ross, J. (1983). Added noise restores recognition of coarse quantised images, *Nature (London),* **305**, 226–228.

Morrone, M. C., Burr, D. C., Ross, J., and Moulden, B. (1991). Illusory squarewaves from staircase plus lines, *Perception,* **20**, A41.

Morrone, M. C. and Owens, R. (1987). Feature detection from local energy, *Pattern Rec. Lett.,* **1**, 103–113.

Morrone, M. C., Ross, J., Burr, D. C., and Owens, R. (1986). Mach bands depend on spatial phase, *Nature (London),* 250–253.

Movshon, J. A., Thompson, I. D., and Tolhurst, D. J. (1978a). Spatial summation in the receptive fields of simple cells in the cat's striate cortex, *J. Physiol. (London),* **283**, 53–77.

Movshon, J. A., Thompson, I. D., and Tolhurst, D. J. (1978b). Receptive field organization of complex cells in the cat's striate cortex, *J. Physiol. (London),* **283**, 79–99.

O'Brien, V. (1958). Contour perception, illusion and reality, *J. Opt. Soc. Am.,* **48**, 112–119.

Oppenheim, A. V. and Lim, J. S. (1981). The importance of phase in signals, *Proc. IEEE,* **69**, 529–541.

Perona, P. and Malik, J. (1990). Detecting and localizing edges composed of steps, peaks and roofs, *Berkeley Tech. Rep.* 90/590.

Piotrowski, L. N. and Campbell, F. W. (1982). A demonstration of the visual importance and flexibility of spatial-frequency amplitude and phase, *Perception,* **11**, 337–346.

Pollen, D. A. and Ronner, S. F. (1981). Phase relationships between adjacent simple cells in the visual cortex, *Science,* **212**, 1409–141.

Ratliff, F. (1965). *Mach Bands: Quantitative Studies on Neural Networks in the Retina,* Holden-Day, San Francisco.

Ratliff, F. (1984). Why Mach bands are not seen at the edges of a step, *Vision. Res.,* **24**, 163–166.

Robson, J. G. (1980). Neural images: the physiological basis of spatial vision, in *Visual Coding and Adaptability,* Harris, L. S. and Erlbaum, L., Eds., Hillsborough, NJ, 177–214.

Ross, J., Holt, J. J., and Johnstone, J. R. (1981). High frequency limitations on Mach bands, *Vision Res.,* **21**, 1165–116.

Ross, J., Morrone, M. C., and Burr, D. C. (1989). The conditions for the appearance of Mach bands, *Vision Res.,* **29**, 699–715.

Sachs, M. B., Nachmias, J., and Robson, J. G. (1971). Spatial frequency channels in human vision, *J. Opt. Soc. Am.,* **61**, 1176–1186.

Shapley, R. and Reid, R. C. (1985). Contrast and assimilation in the perception of brightness, *Proc. Natl. Acad. Sci. U.S.A.,* **82**, 5983–598.

Spitzer, H. and Hochstein, S. (1985a). Simple- and complex-cell response dependences on stimulus parameters, *J. Neurophysiol.,* **53**, 1244–1265.

Spitzer, H. and Hochstein, S. (1985b). A complex-cell receptive field model, *J. Neurophysiol.,* **53**, 1266–1286.

Stromeyer, C. F. and Julesz, B. (1972). Spatial frequency masking in vision: critical bands and spread of masking, *J. Opt. Soc. Am.,* **62**, 1221–1232.

Tyler, C. W. and Gorea, A. (1986). Different encoding mechanisms for phase and contrast, *Vision Res.,* **26**, 1073–1082.

Watt, R. J. and Morgan, M. J. (1985). A theory of the primitive spatial code in human vision, *Vision Res.,* **25**, 1661–167.

Wiesel, T. N. and Hubel, D. H. (1966). Spatial and chromatic interactions in the lateral geniculate body of the rhesus monkey, *J. Neurophysiol.,* **29**, 1115–1156.

Wilson, H. R. and Bergen, J. R. (1979). A four mechanism model for threshold spatial vision, *Vision Res.,* **19**, 19–32.

Wilson, R., Knutsson, H. E., Granlund, G. H. (1983). Anisotropic nonstationary image estimation and its application, IEEE Trans. COM, **31**, 388–397.

Yuille, A. L. and Poggio, T. (1985). Fingerprint theorems for zero-crossings, *J. Opt. Soc. Am.,* **A2**, 683–692.

Chapter 12

SPATIAL AND TEMPORAL NONLINEARITIES IN RECEPTIVE FIELDS OF THE CAT STRIATE CORTEX

A. B. Bonds

TABLE OF CONTENTS

I. Introduction 330

II. The Nature of Cortical Nonlinearities 331

III. Nonlinear Filtering Strategies 333

IV. Orientation-Dependent Response Nonlinearities 334

V. Spatial Frequency-Dependent Response Nonlinearities 339

VI. Contrast-Dependent Response Nonlinearities 342

VII. Temporal Frequency-Dependent Response Nonlinearities 347

VIII. Discussion 349
- A. Global Inhibition 349
- B. Maintenance of Spatial Filter Properties in the Visual Cortex 350

References 351

I. INTRODUCTION

The selectivity of single cells for the spatial and temporal qualities of a stimulus is progressively refined in the hierarchy of the primary visual pathway (e.g., Maffei and Fiorentini, 1973). In the retina of the cat, ganglion cells serve as low-pass spatial frequency filters, with minimal low-frequency attenuation and high-frequency cutoffs ranging from about 0.5 to 2.0 c/deg (Enroth-Cugell and Robson, 1966). Sensitivity for stimulus orientation is limited to a bias averaging about 1.9:1 for best and worst orientations (Levick and Thibos, 1980), and temporal frequency response is broad band, ranging to about 100 Hz (Enroth-Cugell et al., 1983). In the visual cortex of the cat these same characteristics are highly constrained. Low spatial frequency attenuation is very evident in some cells, and spatial frequency bandpass (at half amplitude) averages 1.4 and 1.8 octaves for simple and complex cells, respectively (Movshon et al., 1978b). Orientation selectivity is absolute; there is complete absence of response to orientations orthogonal to the optimal orientation. Finally, the temporal frequency bandpass is limited to 12 to 16 Hz in most cases (Movshon et al., 1978b).

Other, less apparent changes accompany these obvious restrictions on the range of effective stimuli. In the cortex, the maintained discharge is absent (or even "negative"; Movshon et al., 1978a), and peak firing rates are generally lower than those found in either retinal ganglion or LGN cells. The response vs. contrast function is modified from a nearly perfectly logarithmic relationship in the retina (Enroth-Cugell et al., 1983) to a saturating sigmoid (e.g., Albrecht and Hamilton, 1982). Unlike retinal ganglion or LGN cells, cortical cells show strong temporal nonstationarity in the form of contrast adaptation (Albrecht et al., 1984; Ohzawa et al., 1985; Bonds, 1991).

All of the characteristics of striate cortical cells that differentiate them from cells at earlier stages in the visual pathway are a consequence of response reduction of some form or another. This reduction is not a simple, proportional attenuation, but is selective with respect to some aspect of stimulation and is, thus, not simply due to steady-state synaptic inefficiency. A selective decrease in firing rate can result from three possible mechanisms: (1) a reduction of excitation; (2) endogenous properties of the membrane or synapses (e.g., fatigue of the cell due to high sustained firing levels), and (3) an increase in inhibition, which is likely to arise in other cells. The second and third of these alternatives are inherently nonlinear.

Within certain limits, the contributions to cortical selectivity from each of these three mechanisms can be discriminated from one another. The characteristics of the excitatory input from the LGN that feeds the cortex are well known. Development of cortical spatial or temporal filter properties solely by excitatory summation of these signals (e.g., Hubel and Wiesel, 1962) would be a linear process, but this alternative seems unlikely. The spatial and temporal bandwidths of signals from any LGN cell are invariably broader than the bandpass of any cortical cell. There is still no evidence to contradict the notion that connections from LGN to cortical cells are exclusively excitatory (Garey and Powell, 1971). Construction of filters from linear combinations of LGN signals with positive, constant coefficients cannot result in lower bandwidths without the addition of some nonlinearity such as a threshold function or selective inhibition. The time dependence of the response of LGN cells is also much less than that found in cortex (Ohzawa et al., 1985; Bonds, 1991), so modification of temporal response qualities is not dependent on excitatory inputs.

Differentiating between response modification by inherent physiological properties of a given cell and by network properties (i.e., involving the participation of other cells) is more difficult. Physiological modification of response could occur in two forms: anomalies in the firing of the cell itself ("fatigue"), brought on by, e.g., ionic redistribution or conductance modulation, or transmitter depletion at the input synapses. Fatigue can be characterized by its dependence on the response history (prior firing levels) of the recorded cell. It is also likely to affect responsiveness uniformly across all stimulation parameters. Properties such as contrast adaptation might seem to be linked with fatigue since they are associated with high firing rates. However, many response nonlinearities are highly dependent on stimulus configuration (see below), and contrast adaptation can occur independent of the firing rate of the recorded cell (Ohzawa et al., 1985). This would suggest a lesser role for fatigue of the cell or its excitatory input as a mechanism for response limitation. The response history of inhibitory mechanisms is not so easily measured since their activity (and consequently their firing history) is not overtly apparent, so fatigue or even potentiation of these mechanisms and/or synapses cannot be directly ruled out.

The primary hypothesis to be defended here is that inhibition generated by broadly organized networks is responsible in the main for cortical filtering and the accompanying nonlinearities. While the idea that filtering is controlled by intracortical inhibition is not new (e.g., Bishop et al., 1973), the exact role of inhibition has proven difficult to clarify. Since inhibition in isolation is not apparent in extracellularly recorded cortical spike trains, indirect manipulations have been required for its study. The experiments described below have revealed both qualitative and quantitative aspects of the strength and organization of inhibition by observation of the impact of inhibitory stimuli on responses to excitatory stimuli. This interaction is itself a fundamental violation of linearity.

II. THE NATURE OF CORTICAL NONLINEARITIES

Linearity means simply that the whole is equal to the sum of its parts. This can formally be expressed by the concepts of additivity and superposition. Additivity requires that the input/output relationship of a given signal component be defined by multiplication of its strength by some constant scaling factor that is independent from the amplitude of the component. Superposition (or homogeneity) requires that the output from a set of signals be expressed as a linear combination (with fixed coefficients) of the input set. The qualification of fixed coefficients implies that there is no interaction between the signal components, that is, the amplitude of any one signal component cannot modulate that of any other signal component. The application of these rules is nontrivial when considering the nature of receptive fields.

Input signal strength can be characterized within at least two domains, amplitude and area. The measurement of amplitude is in itself not straightforward because a visual image can be measured in terms of either absolute luminance or contrast. Since luminance is essentially normalized by retinal adaptive mechanisms (for review see Enroth-Cugell and Shapley, 1984), it has become customary to express the amplitude of visual stimuli directed toward the cortex in terms of contrast. Contrast, as the name implies, involves a comparison of absolute luminance at one point in the visual field to that at some other point or points (e.g., mean luminance). Thus, even though in the strictest sense the concept of contrast also requires consideration of the spatial distribution of the stimulus, in the usual definition (Michelson contrast) this is ignored.

The spatial coverage and location of the stimulus is also a factor in defining the strength of cortical input. In the simplest case, any receptive field has a finite region over which the distribution of light from the stimulus is integrated. Within these bounds, the size of the stimulus is an important modifier of response amplitude, although this is less important for complex than for simple cells (Hubel and Wiesel, 1962). Input strength can, therefore, depend on both contrast (amplitude) and stimulus size (space) within the limits of the receptive field. In a strictly linear system, increasing contrast and/or field size would predict proportionally increased firing rate (output).

In even simple cells, which are considered the most "linear" of the cortical cells (DeValois et al., 1978; Movshon et al., 1978a), these relationships are tenuous at best. Many studies of the linearity of cortical cells are based on stimulation with drifting one-dimensional, broad-field (at least greater in area than the supposed extent of the receptive field) sinusoidal gratings. These patterns are spectrally pure in luminance, space, and time, so that any departures from linearity introduced by the cells are immediately apparent. In a simple experiment, the spatial qualities of the grating (orientation, spatial frequency) are adjusted so as to be optimal for driving a cell, and contrast is systematically varied. The relationship between contrast and the cell's output (measured most directly as the average firing rate delivered at each contrast) is not linear with contrast, but rather forms a sigmoid (saturating function) against log contrast (e.g., Albrecht and Hamilton, 1982). Doubling contrast, thus, yields less than a doubling of the firing rate; the actual change depends on the initial operating point on the curve. Since this type of nonlinearity results from modification of only stimulus amplitude and spatial characteristics are held constant, it is generally identified as a response amplitude or point nonlinearity as opposed to a nonlinearity of spatial summation (Enroth-Cugell and Robson, 1966).

While a point nonlinearity alone might disqualify these cells from any further application of linear systems analysis (e.g., Fourier transforms, convolution), such manipulations have been usefully applied under limited circumstances (see below). The success of these studies depends on the fact that even though the "point" input/output relationships are strictly nonlinear, they are, within limits, qualitatively linear, as the curves are at least monotonic. The conditions for qualitative linearity are in some cells very limited. There appears to be a real contrast threshold below which cells do not fire (Schumer and Movshon, 1984; Tolhurst and Dean, 1987), and at extremes of high contrast supersaturation, in which the firing rate noticeably decreases with increasing contrast, can arise (Li and Creutzfeldt, 1984; Bonds, 1991).

Point nonlinearities, manifested primarily in the response vs. contrast function, are generally similar across all cortical cells. Nonlinearities of spatial summation can, however, differ markedly between the simple and complex families of cells. Simple cells, as initially described by Hubel and Wiesel (1962), appear to be spatially linear. By their criteria, simple cells are composed of discrete "on" and "off" zones, excitation is usually integrated with stimulus area, and the behavior of a cell can be loosely predicted from its receptive field map. Complex cells generally show "on" and "off" areas uniformly intermixed throughout the receptive field, and response integration as a function of stimulus area is unpredictable at best. This difference in spatial organization is immediately apparent in the form of the responses of these two cell types to drifting gratings. Simple cells respond in bursts of spikes that are modulated at the rate of drift, indicating the ability to discriminate the phase (or position) of the grating, whereas complex cells respond with a steady elevation of firing rate so long as the grating is drifted (e.g., Skottun et al., 1991). Simple cells thus have an appearance of spatial linearity, but complex cells are clearly spatially nonlinear.

Perhaps the strongest argument for the spatial linearity of simple cells is presented by experiments in which the predictability of the relationship between the receptive field profile and the spatial frequency bandpass of the cell is tested (e.g., DeValois et al., 1978; Movshon et al., 1978a). In these experiments, the spatial frequency tuning function of a cell is measured by observing the response amplitude of the cell as a function of stimulus spatial frequency. According to linear systems theory, the spatial arrangement of the receptive field is predicted by the inverse Fourier transform of the spatial frequency response function, or, conversely, the (forward) Fourier transform of the receptive field map predicts the shape of the spatial frequency response function. To confirm this prediction, a one-dimensional receptive field map can be measured by flashing bars of light (aligned on the same axis as the grating) in different positions of the receptive field. A plot of response amplitude vs. bar position (the line weighting function) is then Fourier transformed, and the result is compared with the spatial frequency response function. This relationship is usually not exact, with the predicted tuning curve generally being broader than the measured curve. The threshold (point) nonlinearity shown by most cortical cells has been suggested to be an underlying cause for this discrepancy (Tadmor and Tolhurst, 1989).

Acknowledging these experiments, the argument can be made that some cortical cells can show linear or quasilinear properties. This could mistakenly lead to the conclusion that cortical processes can, at least to a first approximation, be modeled with linear mechanisms (e.g., Legge and Foley, 1980). However, most experiments described in the literature have been conducted with the use of stimuli that are spatially ideal for exciting the cell and are therefore less likely to reveal nonlinearities. This is hardly applicable to real-world vision, in which cells are constantly exposed to broad-spectrum stimuli. It will be shown below that the powerful filtering characteristics of cortical cells, in the domains of orientation, spatial frequency, and contrast gain, result in large part from active nonlinear interactions rather than passive mechanisms. Stimulation with nonoptimal patterns can increase markedly the amount of distortion (nonlinear behavior) exhibited by a cell. This concept can be clarified by examining the possible ways in which nonlinearities interact to implement the filtering process.

III. NONLINEAR FILTERING STRATEGIES

A filter is a device that selectively impedes the transfer of information along some dimension. A filter is not inherently nonlinear, since attenuation (or blockade, in the extreme case) of a signal can still preserve the proportionality and absence of interaction that are characteristic of linearity. However, filters can be constructed in such a way as to violate the principle of superposition. In a strictly linear system, the output that is driven by signal component A cannot be modified by the presence of a (orthogonal) signal component B. The violation of this principle can occur in the visual cortex through interaction of excitation and inhibition. Consider the strategies for restricting the bandpass of a filter. A passive filter achieves its characteristics through selective attenuation. In an electrical model, this results either from series blocking or shunting to ground of signal components of certain frequencies that are selected by the characteristics of reactive elements (capacitors or inductors). Active filters achieve their characteristics through destructive summation of specific signal components. For example, a signal component outside the bandpass of the filter can be eliminated by detection of this component, phase inversion through 180° (negation) and then summation with the primary signal, resulting in cancellation of the unwanted component. With either strategy, the filter action is dependent on selection of the unwanted signal followed by its specific elimination. The amplitude of the signal within the bandpass is usually not acted on. In the visual cortex, there is strong evidence that undesired signal components are often rejected by inhibition that is selectively activated by the presence of that component. Selective activation is similar in principle to the electrical filter strategy. The primary difference between an electrical (linear) filter and a cortical cell is that the electrical filter shunts or cancels only the undesired signal components, whereas inhibition of the cortical cell, even though it may be activated only by the presence of certain stimulus components, attenuates passage of all signals through that cell.

This type of interaction within the neuron leads to reconsideration of the concept of bandpass, since it can clearly differ for excitatory and inhibitory signals. Excitatory bandpass is straightforward; i.e., if a signal component falls within a given parametric range, the cell responds; otherwise it doesn't. Inhibitory bandpass is harder to grasp, since inhibition is not expressed as a response per se. Cortical cells usually have a negligible maintained discharge. In the absence of excitation, activation of inhibition results in no change in firing rate. When recording extracellularly, which is usually the case, inhibition is evident only in terms of its modification of an excitatory response. A cortical inhibitory mechanism may, thus, be associated with two distinct bandpasses, that which activates the inhibitory mechanism and the bandpass of the signals that are affected by it. The overt representation of the latter is equivalent to the excitatory bandpass, since the inhibition is evident through its impact on the excitatory signals passing through the cell. There may also exist excitatory components that are acted on so effectively by inhibition that they never appear. Evidence of their existence can only be revealed through removal of the inhibition. In the event that the excitatory and inhibitory bandpasses differ, the system must be nonlinear since, e.g., a signal component falling outside the excitatory bandpass, and, therefore, ''invisible'' to the cell, but inside the inhibitory bandpass, can, in fact, modify (disproportionately) the output generated by a second signal component falling within the excitatory bandpass. It will be shown that in cortical cells the inhibitory and excitatory bandpasses clearly differ. This not only provides a powerful indicator of nonlinearity, it also addresses the mechanism underlying selective signal attenuation, since in that event the inhibition must be generated by a network that is not directly associated with the recorded cell.

A final consideration of cortical linearity is time dependence of a response. At the visual cortical level, time dependence takes two basic forms: response decay (or adaptation) and nonstationarity. These two characteristics could be considered as the respective deterministic and random components of time-dependent response variability. Adaptation is the stimulus-driven decay of respone amplitude or contrast sensitivity over time. While it may take place across several different time frames, at least within a given cell it is repeatable and predictable. Nonstationarity is a general term used to describe that time-varying component of the response that is not clearly adaptation (e.g., Tomko and Crapper, 1974; Tolhurst et al., 1981). It is not a function of stimulation history (as is adaptation) and may appear as both increases and decreases in response amplitude. Even though a system that varies over time can be perfectly linear at any given moment, here stimulus-driven response variability will be considered a form of serial nonlinearity since it involves interaction between signals that are adjacent in time rather than space.

IV. ORIENTATION-DEPENDENT RESPONSE NONLINEARITIES

Cells in the striate cortex are generally selective for the orientation of a one-dimensional (bar or grating) stimulus. This property can be expressed in an orientation tuning function, in which the response amplitude of the cell is plotted against stimulus orientation. Cells in normal adult cats show a half-width at half-height of this function averaging about 19° (Rose and Blakemore, 1974; Henry et al., 1974). Two general models have been proposed to account for this selectivity, one based primarily on excitatory summation, the other on a combination of excitation and inhibition. Hubel and Weisel (1962) hypothesized that orientation selectivity resulted from structured convergence of excitation from LGN afferent fibers. Linear combination of signals from an ordered row of receptive fields would yield maximal excitation from stimulation with bars oriented in parallel with that row and minimal excitation from orthogonally oriented bars. The absolute orientation selectivity shown by typical cortical orientation tuning functions cannot be accounted for entirely by this scheme. LGN cells do show some intrinsic orientation bias, but this averages only about 2:1 for optimally and orthogonally oriented bars (Vidyasagar and Urbas, 1982; Soodak et al., 1987). Even if the orientation selectivity of a given cortical cell paralleled that of its LGN afferents, in a cell receiving only excitation, bars oriented orthogonally to the preferred orientation would still generate some response, whereas most cortical cells show no response under this condition.

The second model adds some form of inhibition to provide a thresholding function that eliminates the spurious response at the orthogonal orientation. Several variations of this strategy have been suggested, depending on the way in which excitation and inhibition interact. One alternative is to assume that cortical excitation is minimally influenced by stimulus orientation and that selectivity is imposed primarily by inhibition that is itself orientation selective, so-called cross-orientation inhibition. This concept arose from the evidence that stimuli presented at orientations orthogonal to the optimal orientation of a cell caused a reduction in either spontaneous or driven firing rates (e.g., Bishop et al., 1973; Nelson and Frost, 1978; Heggelund, 1981a,b; Morrone et al., 1982). Further support for this hypothesis arises from experiments in which bicuculline, a GABA antagonist, is microiontophoretically injected in the vicinity of a neuron. Blockage of GABA-mediated inhibition is reported to reduce or eliminate orientation selectivity in many, but not all, simple and complex cells (Sillito, 1975, 1979; Tsumoto et al., 1979). These results are interpreted as suggesting that essentially nonorientation selective excitatory input from the LGN is acted on by orientation-dependent inhibition that is strongest when a stimulus is oriented orthogonal to the optimal orientation of a cortical cell.

Another view is that orientation selectivity is defined by both excitatory and inhibitory mechanisms. The elongate shapes of cortical receptive fields (as opposed to the more radially symmetric receptive fields of LGN cells) suggest that there is at least some predisposition for cortical fields to arise from rows of LGN fields (Hubel and Wiesel, 1962). Moreover, as previously mentioned, LGN single units show some response dependence on orientation. The influence of orientation bias originating in the LGN on cortical orientation selectivity is evident from the observation that monosynaptically activated cortical cells show a distribution of orientation preferences similar to that of LGN cells (Vidyasagar and Henry, 1991). Such an orientation bias can easily be further refined by a simple, nonorientation dependent thresholding mechanism. Orientation bandwidth can be decreased to an arbitrarily small value by raising this threshold. By itself this scheme is insufficient, since a fixed threshold will result in an orientation tuning bandwidth that varies as a function of stimulus contrast. Higher stimulus contrasts, resulting in higher mean firing rates, would yield broader orientation tuning, but orientation tuning bandwidth is, in fact, found to be constant with stimulus contrast (Sclar and Freeman, 1982). A fixed threshold would also predict that responses to low-amplitude stimuli, that generate input amplitudes below a criterion cutoff, would be lost, whereas many cortical cells, even with narrow orientation selectivity (that would require a high fixed threshold), have contrast sensitivities as high as those found in LGN cells.

The use of a threshold to refine orientation bias would have merit if the threshold mechanism could adapt to different stimulus contrasts, yielding a slight amount of thresholding at low contrasts and proportionally more at higher contrasts. Response suppression that varies as a function of contrast has been demonstrated in simple cells by testing with two superimposed gratings. In these experiments (Bonds, 1989; Bauman and Bonds, 1991) the excitatory or inhibitory impact of a mask grating of varying geometry was measured by observing how it affected the excitatory response generated by an optimally configured (best orientation, spatial frequency) base grating that was simultaneously presented. The influence of the mask when it was configured to generate mainly inhibition was indicated by the reduction of the base-induced excitation. When both base and mask generated excitation, the relative contribution from each was unclear in the DC response (average impulses per second) because of the response amplitude nonlinearities discussed above. Simple cells are, however, sufficiently spatially linear to yield response modulation that is synchronous with the temporal frequency of drift of periodic stimuli such as gratings. Use of different temporal frequencies of drift for base and mask thus allowed "labeling" of response components so as to permit independent identification of the amplitudes generated by either the base or mask. Assuming that superposition holds (see below), presentation of the base at 2 Hz and the mask at 3 Hz yielded components in the response at these same frequencies detectable via Fourier analysis. The modulatory impact of the mask on the base was then determined (even in the presence of excitation by the mask) by observing the modification of the 2-Hz response amplitude.

Figure 1 shows the results of this manipulation when mask orientation is varied. The base tuning curve (DC response) is measured as a function of the orientation of a single grating in the usual way (open circles, Figure 1A). The mask tuning curve is measured by observing the variation of the DC response while the cell is driven by both the base and superimposed mask; here, the base is fixed and the mask orientation is varied. In this example, the mask is presented at two contrast levels. The solid horizontal line shows the response level obtained from stimulation by the base only. The DC mask tuning curve shows that the mask can contribute both excitation (elevating response above the horizontal line), when the mask is oriented within the confines of the base tuning curve, and inhibition (depressing response below the horizontal line) when it falls outside the base tuning curve. The higher contrast (squares) yields both stronger excitation when inside the base tuning curve and stronger suppression when outside the base tuning curve.

Examination of the 2-Hz response (Figure 1B) leads to a somewhat different interpretation of these results. The curves show the response component generated by the (fixed) base grating as a function of the orientation of the mask grating. The consequence of mask presentation is plotted as negative impulses per second relative to the firing level generated by presentation of the base alone. Here it is clear that the base response is reduced for all mask orientations. There is some orientation selectivity of this depression, since it is stronger at the extremes, but the curve overall is much less dependent on orientation than the excitatory response tuning curve shown in Figure 1A, indicating no direct linkage between excitation and stimulus-mediated depression. In general, the presence of a second grating, in any configuration, was found to reduce the response generated by the first grating. This response reduction is graded in two ways. It may (but does not necessarily) have an orientation dependence. Suppression is usually stronger for mask gratings oriented 20 to 40° from the optimum orientation than it is for either optimally or orthogonally oriented gratings (Bonds, 1989). It is also constrast dependent; higher contrasts yield stronger suppression (filled squares, Figure 1B). In this particular example there appears a third dependence, where suppression is also seen for optimally oriented masks. This may result from response saturation brought on by the high activity of the cell due to higher mask contrast (Figure 1A).

The 3-Hz (mask) response component essentially matches the base tuning curve as mask orientation is varied (Figure 1C). This reflects excitatory contributions from the mask when it is within the excitatory tuning bandpass of the cell. The absence of a 3-Hz response component at the orientation extremes addresses the mechanism responsible for response suppression. Even though the mask clearly suppresses the base-generated response, no 3-Hz component is apparent. The suppressive signal must, therefore, arise from a pathway that is spatially nonlinear, since suppression is not synchronous with the temporal frequency of drift of the mask stimulus. This could arise either from complex cells, which show a steady elevation of firing rate when exposed to drifting gratings, or from a group of simple cells that are arranged with spatially random phase relationships.

Similar behavior is seen for orientations well outside the excitatory orientation bandwidth. Figure 2 shows the orientation tuning function (dotted line) as well as the 2-Hz (base; filled circles) and 3-Hz (mask; squares) response components as a function of mask ori-

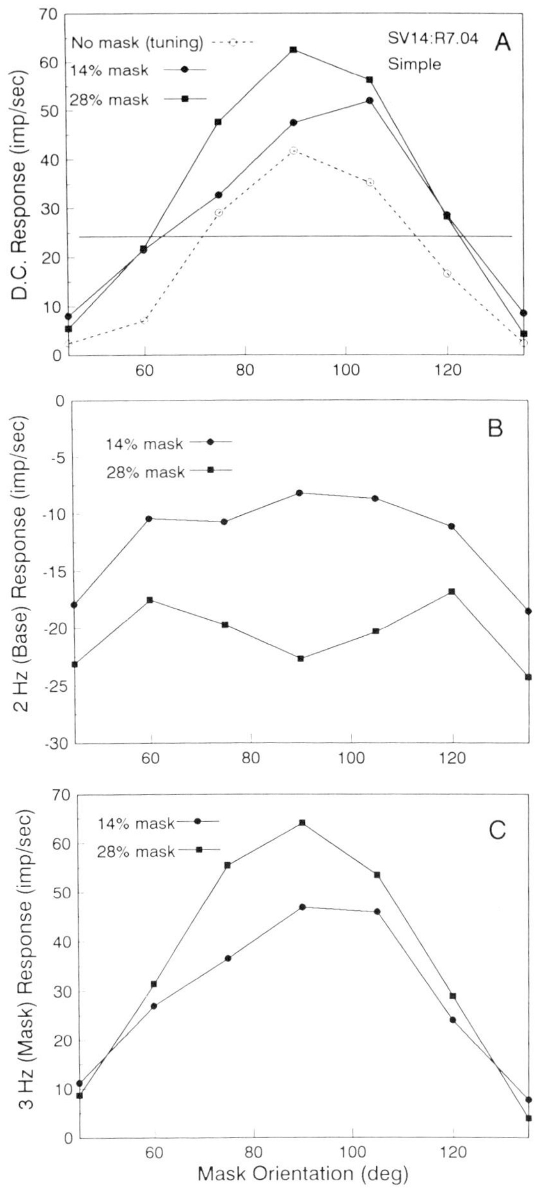

FIGURE 1. Response suppression as a function of mask orientation. (A) DC component of response to single grating of fixed contrast and varying orientation (open circles, dotted line) describes the orientation tuning function. Solid lines show response to simultaneous stimulation with a fixed base grating and mask grating for which orientation was varied. The higher contrast mask (filled squares) yields more excitation within the range of the tuning function and more suppression outside the tuning function. (B) Suppression of 2-Hz (fixed base grating) response component is found at all orientations, but is prominent at the edges of the tuning function. Suppression is also found at the optimal orientation for higher contrast masks. (C) 3-Hz (mask) response component essentially mirrors the excitatory tuning function. (From Bonds, A. B. (1989). *Visual Neurosci.*, **2**, 41–55).

entation through a full 360°. Again, the 2-Hz response is expressed as negative impulses per second relative to the control condition in which the base is presented alone. Here the strongest suppression of the 2-Hz response component is seen for masks 30° to either side of the optimal orientation, as suggested by the results of Figure 1. Beyond this, suppression is reduced and is, in fact, least at orientations orthogonal to the optimal orientation, even

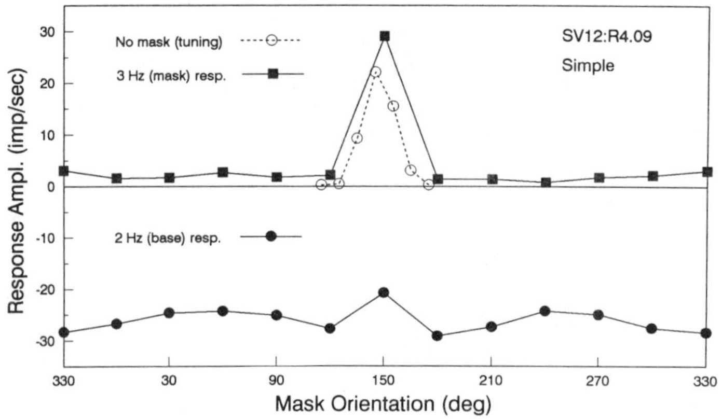

FIGURE 2. Response suppression over the full range of orientations. The upper plot shows both the tuning function of the cell (open circles, bottom line) and the 3-Hz response component which again resembles the excitatory tuning function. The lower plot shows the suppression of the 2-Hz (base) response as a function of mask orientation. Significant suppression is found at all orientations, with local peaks 30° from the tuning optimum of the cell. (From Bonds, A. B. (1989). *Visual Neurosci.*, **2**, 41–55).

though it is still quite strong, with a response reduction of about 22 impulses/s. The absence of a 3-Hz response component for mask orientations outside the excitatory orientation tuning function again suggests an asynchronous suppressive mechanism.

These observations show that mask gratings of any orientation result in response suppression to a greater or lesser extent, which is a clear and general violation of the principle of superposition. The dependence of this effect on orientation is rather different from that predicted by the term "cross-orientation inhibition". In that case, orientation selectivity is presumed to arise from inhibition that is selectively tuned for orientations orthogonal to the optimal orientation. Here we find instead that inhibition is generated by stimuli of all orientations, with local peaks of inhibition often found at orientations 20 to 40° away from the optimum. Such behavior is consistent with the model described above, in which contrast-dependent inhibition acts to impose a variable theshold on inherently orientation-biased signals arriving from the LGN. The local suppression maxima found at the edges of the orientation tuning functions serve to enhance the steepness of these curves.

The reduction of the response from one grating by the presence of another may not only be fundamental in supporting orientation selectivity, but it is also very important for understanding the dynamic nature of the relationship between stimulus contrast and cortical response amplitude. The presence of at least a component of response suppression that is spatially unselective suggests a global inhibitory mechanism, so named because it is activated by the presence of any stimulus that is effective for the visual system of the organism under study. While earlier studies did not explicitly describe this phenomenon, their results are at least consistent with it. The iontophoresis of GABA blockers (Sillito, 1975, 1979) which presumably impaired inhibitory mechanisms not only resulted in a loss of orientation selectivity in most cells, but also produced an elevation in responsiveness at all orientations. This hyperexcitability could be attributed to some pharmacological impact of bicuculline on the cell membrane, but could also have resulted from a release of inhibition generated by all stimuli, suggesting that all stimuli produce a balance of both inhibition, at least some of which is GABA mediated, and excitation. Morrone et al., (1982), using paired superimposed stimuli as in the studies described above, described response suppression for orthogonal masks but did not report results from systematic variation of mask orientation, nor did they separate response components and, thus, were unable to observe suppression for optimally oriented masks.

The separation of response components is performed by Fourier analysis of the response histograms, which implicitly requires response amplitude linearity. Because of the response nonlinearities (primarily thresholding) shown by simple cells, there is some question as to the validity of this manipulation. When the mask is configured to provide excitation, the cell can fire robustly due to stimulation from both the base and the mask. The generation of spectral "cross-talk" (as a result of nonlinearity) between response components could then influence the calculated amplitude of the base- or mask-generated response. If one models the transduction process of a simple cell as a perfect half-wave rectifier, the nature of this nonlinearity is such that the amplitudes of the base and mask response components remain independent. However, when a threshold elevation above the half-wave rectification level is introduced (i.e., the sinusoidal waveform is sliced higher than the midpoint), significant interaction between response components can be seen. For example, if the threshold is raised 20% (of the peak amplitude of the sine wave), the base response amplitude is elevated 16% by cross-talk from a mask component of equal amplitude. The conclusion that optimally configured gratings generate response suppression is supported by two points:

1. The nonlinear interactions occur when the transduction function is not a perfect half-wave rectification. The duty cycle (ratio of the duration of response activity to total duration) of 16 simple cells ranged from 39% to just over 50% (where 50% is perfect half-wave rectification). This corresponds to threshold elevations of 22 to 0% and averaging 8%, respectively, predicting additive cross-talk amounting to at most 18% and averaging 7% (Bauman and Bonds, 1991). This contamination is thus not a major factor.
2. The nature of the contamination is additive, which means that when it occurs the measured base response component is larger than its actual value. This implies that the actual suppression caused by excitatory stimuli is, if anything, greater than indicated by response component separation.

V. SPATIAL FREQUENCY-DEPENDENT RESPONSE NONLINEARITIES

The refinement of spatial frequency selectivity in the visual cortex could result from spatial frequency-dependent inhibition in a way similar to that observed for orientation-dependent inhibition. Inhibition that is selective for spatial frequencies that are outside the excitatory bandpass could serve to narrow the less selective input signals from the LGN. This hypothesis was tested using the paired grating paradigm described above, with the difference that both gratings were of optimal orientation and the spatial frequency of the mask was varied (Bauman and Bonds, 1991). To simplify interpretation of the results, the 42 cells studied were separated into three arbitrary categories based on their excitatory spatial frequency tuning. Cells with response peaks at 0.2 to 0.4 c/deg and little or no low-frequency attenuation were classed as low-pass; those with peaks between 0.4 and 0.9 c/deg and noticeable low-frequency attenuation were classed as bandpass; and those with peaks at or above 1.0 c/deg were classed as high-pass. The population included 15 low-pass, 19 band-pass, and 8 high-pass cells.

Testing with paired gratings showed that 74% of the total sample showed response suppression that was selective for spatial frequency. In all but 2 (of 31) cases, the nature of the suppression was consistent with the spatial frequency tuning of the cell. Figure 3A shows a low-pass simple cell exposed to grating pairs in which the mask was presented at three contrasts and seven spatial frequencies. The excitatory spatial frequency tuning func-

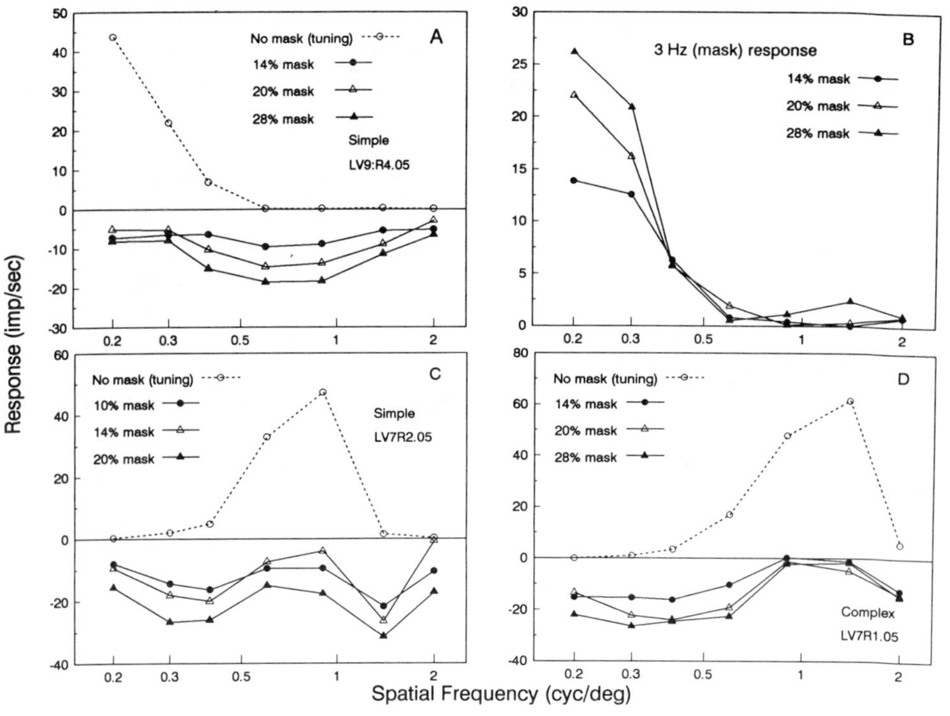

FIGURE 3. Response suppression as a function of mask spatial frequency. (A) Response tuning of a low-pass cell (dotted line, open circles) compared with suppression (in negative impulses per second) resulting from simultaneous presentation of a mask at variable spatial frequencies and three contrasts. Suppression is strongest at about 0.7 cyc/deg, where the excitatory response goes to zero. (B) The 3-Hz response component from the same experiment shown in A essentially mirrors the excitatory response. (C) Response suppression found on both flanks of the tuning curve from a band-pass cell. (D) Low-frequency response suppression found in a high-pass cell. (From Bauman, L. A. and Bonds, A. B., (1991). *Vision Res.*, **31,** 933–944).

tion, measured with a single grating in the usual way (open circles), shows the strongest response at 0.2 c/deg (the lowest spatial frequency tested) and a cutoff at 0.6 c/deg. Suppression of the 2-Hz response is expressed in terms of negative impulses per second relative to the response generated by presentation of the base stimulus alone. Even at the optimal spatial frequency (0.2 c/deg) there is some modest suppression, which is consistent with the observations made in the context of the orientation experiments described above. At higher spatial frequencies, suppression increases to a maximum at about 0.6 to 1.0 c/deg, where the excitatory tuning function cuts off. Any increase of mask contrast results in a monotonic increase of suppression. As found for orientation, the mask (3-Hz) response component at all three contrasts (Figure 3B) resembles the excitatory tuning function of the cell. The low level of 3-Hz activity where suppression is strongest (0.6 to 1.0 c/deg) again suggests a spatially asynchronous (nonlinear) origin for this inhibition.

Spatial frequency-dependent suppression for a bandpass cell is shown in Figure 3C. Mask-induced suppression of this simple cell peaks at or near both of the cutoff spatial frequencies of the tuning function, which serves to restrict total bandwidth. As above, the degree of suppression increases with mask contrast, and suppression is seen even at the optimal spatial frequency. The 3-Hz response function (not shown) mirrored the 2-Hz excitatory tuning function. A high-pass complex cell (Figure 3D) shows suppression that peaks at its cutoff spatial frequency of 0.4 c/deg, in this case below the peak spatial frequency. A small amount of suppression is seen at 2.0 c/deg, suggesting some effort to restrict the

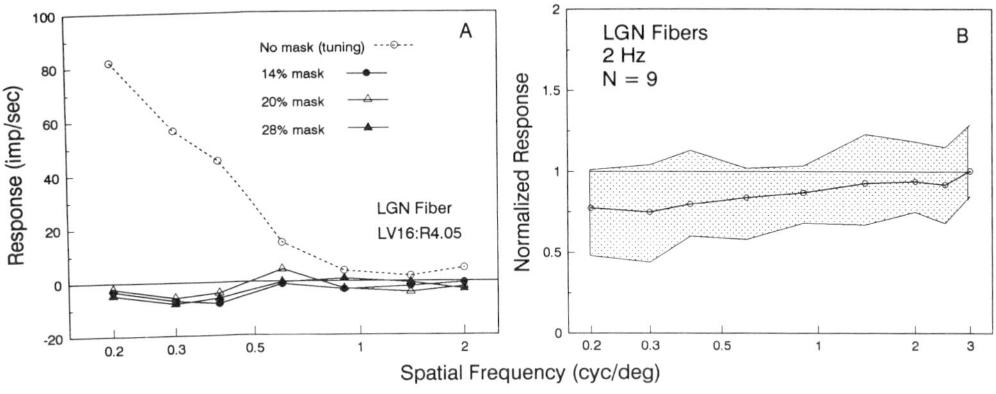

FIGURE 4. (A) Response suppression pattern (as a function of spatial frequency) in an LGN cell. Typically, a slight degree of low-frequency suppression is associated with cells that are generally low-pass. (B) Open circles show the average normalized suppression across a population of 9 LGN cells. The shaded area indicates the entire envelope of normalized suppression found for this population. (From Bauman, L. A. and Bonds, A. B., (1991). *Vision Res.*, in press).

bandwidth here as well. In this cell, suppression at the peak spatial frequency is not prominent.

Of the cells in which response was suppressed by the mask grating, 93% (29/31) showed a clear relationship between the type of suppression and the excitatory tuning of the cell. Cells with low-pass tuning showed high-frequency suppression, and cells with high-pass tuning showed low-frequency suppression. Cells with bandpass tuning showed either low- or high-frequency suppression or a combination of the two (e.g., Figure 3C). In the two contradictory cases, low-frequency suppression was associated with low-pass tuning. On the whole, these correlations suggest that spatial frequency-dependent response suppression is consistent with limitation of the spatial frequency bandpass of cortical cells.

One question that arises from this work is the site of origin of the inhibition. While orientation-dependent inhibition is not likely to be prominent in the LGN (due to the weak orientation dependence of LGN responses), one might expect that, since LGN cells do have spatial frequency bandpasses narrower than those of retinal ganglion cells, spatial frequency-dependent inhibition could be present in LGN responses. The paired grating paradigm was applied to nine LGN cells; results from one cell are shown in Figure 4A. This unit had a low-pass excitatory tuning curve. The small amount of response suppression evident is found at the low-frequency end of the spectrum, coincident with the excitation, which is opposite to the pattern seen in most cortical cells. The average pattern of suppression pooled across all nine LGN cells is shown in Figure 4B. Suppression is weak and dominant mainly at the lower frequencies. The total envelope (not standard deviation) of all suppression seen is denoted by the shaded area. The absence in the LGN of the strength and variety of response suppression seen in the cortex strongly suggests a cortical origin for this effect.

One challenge to the interpretation of these overall results is the form of the stimuli. Superimposition of two gratings could produce interference. Since the gratings are drifted at two different temporal frequencies, there will be periodic cancellations of contrast when the relative phases approach 180°, although this cancellation would not spread across the stimulus field since the gratings are of differing spatial frequencies. A local contrast cancellation would result in response reduction that could be misinterpreted as arising from inhibition. The claim that the response suppression shown here is not an artifact of the stimulation paradigm is supported by three observations.

1. Spatial frequency-dependent response suppression was not seen in 26% of the cortical cells studied and any of the LGN cells. If response suppression were a consequence of contrast cancellation, one would expect that it would be evident in all responses.
2. As shown in Figure 3, the amount of response suppression was monotonically dependent on the contrast of the mask stimulus as its contrast increased from below to above that of the base stimulus. The most effective contrast cancellation would occur when the contrasts were matched, not when the mask contrast was higher than that of the base.
3. The generally complementary behavior of the spatial frequency tuning of the response suppression and the spatial frequency tuning of the cell supports some causal relationship between the two.

VI. CONTRAST-DEPENDENT RESPONSE NONLINEARITIES

As mentioned above, the relationship between the response amplitude (expressed, e.g., in elevation of mean firing rate) of a cortical cell and stimulus contrast is clearly nonlinear. When measured with drifting sinusoidal grating stimuli (e.g., Tolhurst et al., 1981; Dean, 1981; Albrecht and Hamilton, 1982; Li and Creutzfeldt, 1984), the curve usually demonstrates a segment that is linear with contrast near threshold, a suprathreshold segment that is more or less linear with log contrast, and a flattening or inflection (''saturation'' and ''supersaturation'', respectively) at high contrasts. Aside from the case of supersaturation, the response vs. contrast relationship is at least monotonic and affords piecewise linearity. One might, therefore, suspect that the response nonlinearity simply expresses response compression inherent in the cell's firing characteristics. In this case, the logarithmic break in the curve and saturation would result from the cell's increasing inability to perform linear integration of input signals at higher activity levels.

Evidence contrary to this scheme was first cited by Albrecht and Hamilton (1982) and Sclar and Freeman (1982). Both groups found that response saturation was defined not by the firing rate of the cell (as would be expected from response compression) but by the contrast of the stimulus. Saturation was seen at very low firing rates resulting from high-contrast stimulation with nonoptimal spatial frequencies (Albrecht and Hamilton) or orientations (Sclar and Freeman). This result was later confirmed by Li and Creutzfeldt (1984) and Bonds (1991). Two fundamental conclusions can be drawn from this outcome: (1) since the response nonlinearity is controlled by contrast, despite amplitude nonlinearity the bandpass for both spatial frequency and orientation remains essentially constant at different contrasts (Albrecht and Hamilton, 1982; Sclar and Freeman, 1982); (2) since the nonlinearity is not controlled by firing rate, it must originate from some mechanism that is independent of the recorded cell (Bonds, 1991).

Hints as to the nature of this mechanism were offered by a study of the contrast gain control that is characteristic of cortical cells (Ohzawa et al., 1985). When cells are repeatedly driven by contrast values closely grouped about a mean contrast, the origin of the response vs. log contrast function is systematically displaced to the right as the mean contrast level is increased. The response vs. contrast function is, thus, not a static relationship, but is dynamic and is dependent on the recent history of stimulation. Earlier evidence for this conclusion was seen in the cortical response adaptation studies of Vautin and Berkley (1978), Movshon and Lennie (1979), and Albrecht et al., (1984). The reduction of responsiveness as a result of steady stimulation at higher mean contrasts reduces saturation and enhances discriminability of contrast increments at suprathreshold levels.

Most natural viewing is, however, not of a continuously drifting high-contrast grating of one spatial frequency and orientation, but is rather a heterogeneous mix of continually

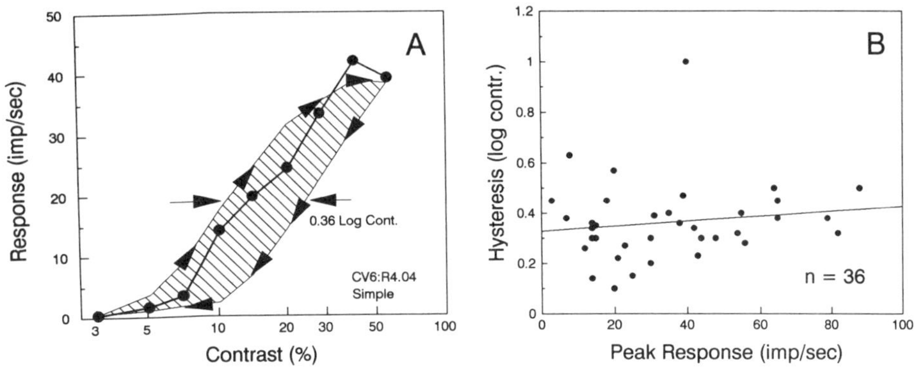

FIGURE 5. (A) The solid line (filled circles) shows the response vs. log contrast relationship of a cortical simple cell measured with randomized presentation of different contrasts (see text). The shaded area is bounded by the response level path followed when contrasts are presented sequentially, first ascending, then descending. Responses on the ascending leg are always larger than those at the same contrast on the descending leg. The resulting dynamic hysteresis is quantified by the difference in contrast required to generate a response of half maximum amplitude on each leg. (B) A scatter plot of the degree of hysteresis expressed in log contrast vs. the maximum firing rate across 36 cells shows no significant correlation. Hysteresis does not result merely from a high firing rate. (Adapted from Bonds, A. B. (1991). *Visual Neurosci.*, **6**, 239–255).

changing contrasts, orientations, and spatial frequencies. The question naturally arises as to whether the above-described gain control, measured under conditions of sustained stimulation, is, in fact, relevant to day-to-day vision, or simply reflects an overload condition that is not of functional significance.

The problem of measuring the short-term impact of contrast on the operating characteristics of a system that is driven by contrast recalls the uncertainty principle of Werner Heisenberg, that a system cannot be measured without perturbing it.One approach is to assume that the adaptive process occurs within a finite amount of time and to measure it as it occurs. As an extension of the adaptation studies cited above, Bonds (1991) presented sequences of ascending, then descending contrasts to cortical cells over fixed time intervals. Here adaptation would be manifested in a decrease in response amplitude at a given contrast on the descending leg of the sequence, after exposure to higher contrasts. For a series of nine contrasts (at 0.15 log intervals) presented over 54 s (ascending from 3.5 to 56% contrast and returning), the response on the descending contrast leg was always less than that for the same contrast on the ascending contrast leg, describing a form of dynamic hysteresis (Figure 5A). A simple figure of merit for this hysteresis is the difference in contrast required to generate a response of half the maximum amplitude on both the upward and downward path. In the case of Figure 5A (horizontal arrows) this figure is 0.36 log unit. Hysteresis measured under similar conditions was found in all of 36 cells so tested, with an average value of 0.36 log unit ($\pm$0.15 log SD).

Adaptation to long-term exposures at high contrasts was tested, but not found, in the LGN (Movshon and Lennie, 1979; Ohzawa et al., 1985). Contrast sequences were similarly applied to four LGN cells whose fibers were recorded in the cortex. Of these, only one showed hysteresis of a magnitude of about 0.1 log unit of contrast. The remaining three showed no noticeable differences in response amplitude for the rising and falling legs of the contrast sequence, further validating the notion that this type of dynamic response modification is inherently cortical.

The maximum firing rate for a given cell during this measurement process varied from a low of 3.2 impulses/s to a high of 88 impulses/s. Substantial hysteresis was found across the entire range of firing levels. Linear regression between the magnitude of hysteresis and the maximum firing rate (Figure 5B) is slightly positive but not significant, indicating that this form of response modulation is independent of firing rate and is thus unlikely to result from simple fatigue.

These hysteresis curves also support the conjecture that saturation is not the result of static response compression. Saturation was evident in the flattening (or downturn) of the response function as contrast was increased (e.g., Figure 5A). In nearly all cases, there was a significant drop in firing rate for the first (and subsequent) contrast decrement(s), indicating that the flattening of the response curve for ascending contrasts is caused by a real, lingering decrease in gain that can be revealed by immediate testing with lower contrasts. The lag in recovery from response saturation suggests that it results from some dynamic process rather than response compression, in which case recovery would be more immediate.

The exact temporal characteristics of gain adjustment were less easy to measure than its existence. Presentation of the same range of contrasts (again, 3.5 to 56%) to one cell over periods ranging from 54 down to 6 s (ascending plus descending contrasts) resulted in monotonic decline of the degree of hysteresis from about 0.4 to 0.2 log unit. Response reduction was still evident after exposure of only 1 to 2 s at higher contrasts. In some cases the response reduction could be seen directly in the histograms of responses to ascending and descending contrast sequences. Figure 6 shows histograms from a simple cell (A) and a complex cell (B). The differences in amplitude at similar contrasts on the ascending and descending legs are obvious, but the paths to these differences are less obvious. The simple cell shows a suggestion of response decay during the 2-s interval of the histograms, as indicated by the negative slope of the regression lines, which were fit to all points in the histograms to indicate general time-dependent response trends. Unexpectedly, however, the overt response decay is more apparent for contrasts on the descending side than those on the ascending side. The complex cell shows behavior even more puzzling. The regression lines all show a positive slope, apparently contradicting the clear reductions in amplitude seen at the same contrast levels on the descending leg.

Gain reduction could occur in the absence of an obvious time-dependent reduction in response amplitude if the gain-setting process were faster than the resolution of the measuring system (i.e., the histograms). In most responses, the events of the first 100 to 150 ms were different from the steady state. These transients appeared as dips (e.g., Figure 6B) as often as peaks, and in many cases were more profound for decreasing contrasts than for increasing contrasts (e.g., Figure 6A). Because of the temporal complexity of these events, their impact on the responsiveness of the cell was not immediately apparent. A direct measurement of the rate at which adaptation occurred using a probe stimulus of varying stimulus-onset asynchrony (e.g., Enroth-Cugell and Shapley, 1973) proved impossible because of the complicated temporal format of cortical responses to brief stimuli. However, it was possible to show that the gain setting process can occur over very brief periods of time by presentation of very brief adaptation stimuli. In these experiments a pedestal stimulus (drifting grating, 14% contrast) of 2-s duration was used to establish a baseline firing rate. Adaptation resulted from the introduction of a brief increase in contrast (to 28%) starting 0.5 s after the appearance of the pedestal. The adapting stimulus duration ranged from 50 to 750 ms. The measure of adaptation was the ratio of spikes generated during the first and last 500 ms of the 2-s pedestal period. Because of the inherently transient nature of the cortical response (presumably reflecting adaptation by the pedestal stimulus itself), this ratio was 0.81 without any adapting stimulus (Figure 7). Even a 50-ms adapting stimulus reduced this figure slightly, and the decline was more or less linear with adapt duration to 750 ms, where the ratio was reduced by about 50%. Exposure to even brief increases in contrast can, thus, reduce the firing rate of the cell noticeably and for some period of time.

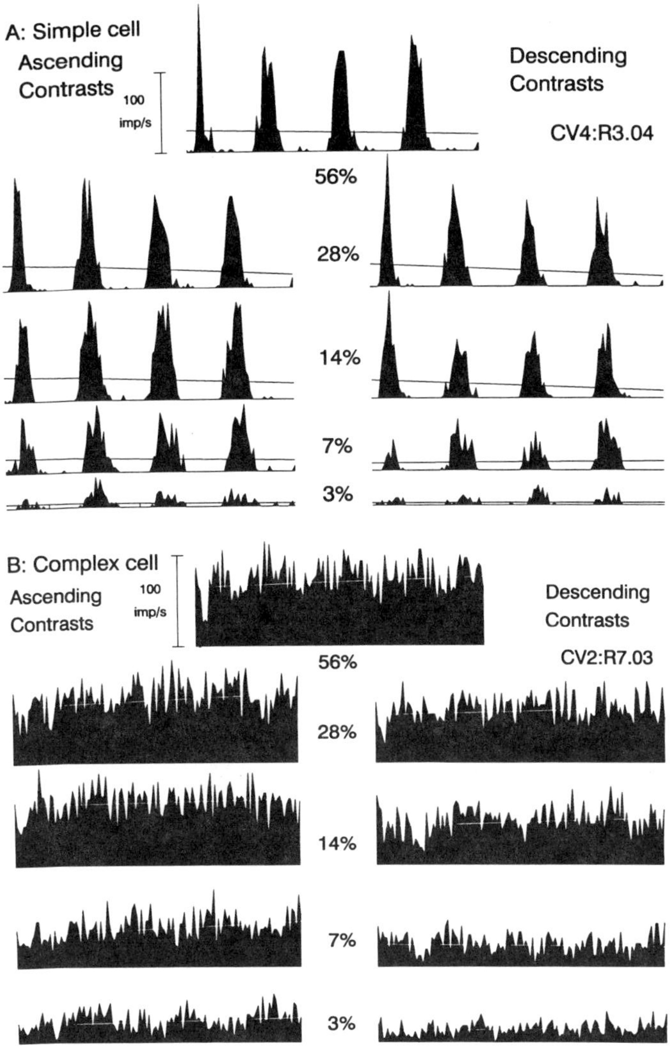

FIGURE 6. Response histograms from presentation of sequentially ascending and descending contrasts. (A) Simple cell. Amplitude differences on the ascending and descending leg are clear. Regression lines fit to the histograms have a negative slope and show an overall tendency for response reduction across time. (B) Complex cell. Even though response amplitude differences at the same contrasts are clear, the regression lines (white) are flat or have a slightly positive slope. (From Bonds, A. B. (1991). *Visual Neurosci.*, **6**, 239–255).

The sensitivity of the gain-setting system to exposure to low levels of contrast was likewise explored. Seven cells were tested with sequences of five contrasts peaking at 14%. In four cases, hysteresis averaging 0.12 log contrast was seen (Figure 8A), although three cells showed negligible adaptation under these circumstances. Cells with high contrast sensitivity demonstrated modest hysteresis even for exposures to near-threshold contrasts. A peak contrast of 5% (shown for 3 s) shows a response differential of nearly 50% at a 3% test contrast (Figure 8B). This same cell showed a slight (but repeatable) response reduction after exposure to a peak contrast of 3% for 3 s. In many cases contrasts that are barely suprathreshold are thus sufficient to alter the gain of a cell.

As were the nonlinearities in the statically measured response vs. contrast function, the dynamic nonlinearities were also found to be dependent on stimulus contrast rather than

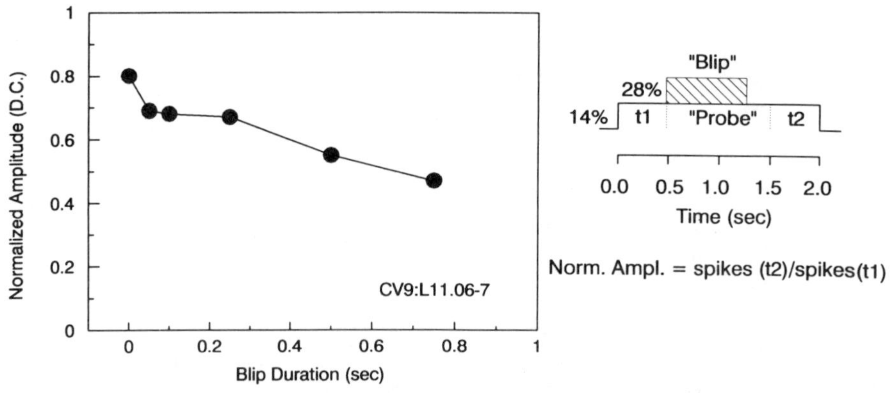

FIGURE 7. Rapid gain change. A pedestal of 14% contrast (''probe'') is introduced for 2 s. At 500 ms, a contrast increment (to 28%) or ''blip'' of variable duration is introduced. The normalized response amplitude is the ratio of spikes during the first and last 500 ms of the pedestal presentation. With no blip, this ratio is 0.81. Even a 50-ms blip causes some response reduction, which monotonically increases for longer blips. (From Bonds, A. B. (1991). *Visual Neurosci.*, **6**, 239–255).

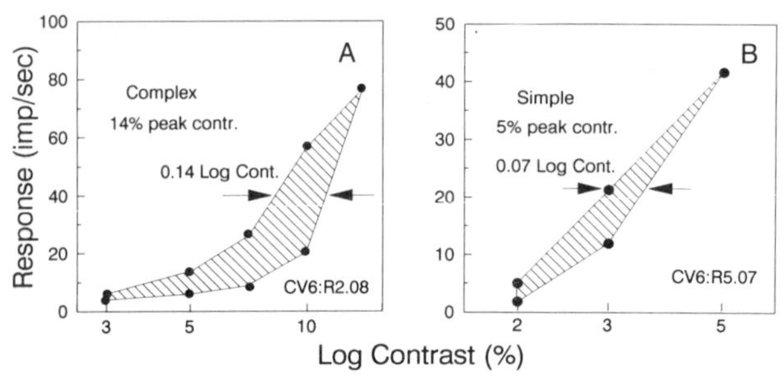

FIGURE 8. Hysteresis at low contrasts. (A) A peak contrast of 14% (3 s per datum) results in hysteresis of 0.14 log contrast. (B) A peak contrast of 5% results in a repeatable loss of amplitude at 3% contrast (3 s per datum). (From Bonds, A. B. (1991). *Visual Neurosci.*, **6**, 239–255).

response level of the recorded cell. The general independence of the degree of hysteresis and firing rate was shown in Figure 5B. Two additional experiments were performed to test explicitly the dependence of hysteresis on contrast. Emulating the paradigm of Sclar and Freeman (1982), hysteresis curves were measured for different peak response levels in the same cell. These levels were defined by selecting different grating orientations for the tests. At each orientation the same set of contrasts was delivered. Over a range of peak amplitudes from 40 to 70 impulses/s, generated at five different orientations, the magnitude of the hysteresis remained essentially constant, ranging from 0.25 to 0.28 log unit. Figure 9A shows the similarity between three such curves normalized to the same peak amplitude. A second approach tested for the existence of hysteresis under conditions of constant contrast by modifying response amplitude through sequential presentations of differing orientations. A control hysteresis curve was first measured under normal conditions (by varying contrast; cross hatches, Figure 9B). Next, response amplitude was varied over the same total range and time frame by varying orientation, while contrast was fixed (shaded shape, Figure 9B). Hysteresis is reduced by about 30%, although not eliminated, when induced with fixed contrast. Both experiments thus indicate that contrast, rather than response level, is of primary importance in setting the operating level of cortical cells.

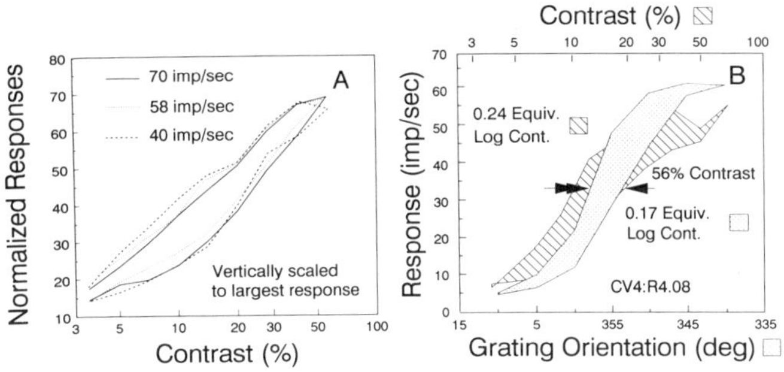

FIGURE 9. Hysteresis appears to be contrast driven, not response driven. (A) Peak response levels were altered across a range of 40 to 70 impulses per second by presenting contrast sequences at three different orientations. When the response curves are normalized, the hysteresis behavior is nearly identical. (B) Here responsiveness was varied either by sequential changes of contrast (at a fixed orientation: diagonal fill) or by sequential changes of orientation (at a fixed contrast: dotted fill). Even though the total range of firing rate was identical, the hysteresis at constant contrast (0.17 log) was markedly less than that resulting from changing contrast (0.24 log). (Adapted from Bonds, A. B. (1991). *Visual Neurosci.*, **6,** 239–255).

VII. TEMPORAL FREQUENCY-DEPENDENT RESPONSE NONLINEARITIES

The concept of temporal frequency "tuning" in striate cortex is not so widely accepted as that of orientation or spatial frequency tuning inasmuch as most cells do not appear to have a clearly optimal temporal frequency. Cortical cells do, however, show limitation of temporal frequency bandwidth. Afferent signals from the LGN demonstrate response modulation to stimuli with frequencies as high as 60 to 70 Hz (e.g., Kaplan et al., 1979) whereas the temporal frequency cutoff in cortical cells rarely exceeds 12 Hz (Movshon et al., 1978a,b). The question remains as to whether cortical cells are inherently insensitive to higher temporal frequencies or whether a form of "tuning" is implemented by active suppression of high-frequency signals. Some evidence for how temporally low-pass behavior comes about is seen in the response histograms to higher temporal frequencies. Figure 10 (Bonds, 1991) shows histograms resulting from exposure to frequencies of 4, 7, and 12 Hz. The dotted line indicates a single exponential process (with positive or negative coefficient) added to a baseline. It is fit (least squares criterion using a Simplex algorithm) to the centers of mass of the individual response cycles. For the 4-Hz response the exponent coefficient is negative and small, indicating a modest response rise over the first cycle or two of stimulation. At 7 Hz, after the first half-cycle, the response gradually drops with a time constant of 1.17 s. Although the 12-Hz response initially shows robust firing and modulation, this diminishes rapidly, with a time constant of only 186 ms. A similar trend was seen across a randomly selected population of ten other cells. The limited bandpass of cortical cells is thus governed by a process that is not instantaneous, but is built up over time.

Indirect evidence for a form of active suppression that acts to limit cortical bandpass can be seen in Figure 11. Both panels show the dependence of mask-induced response suppression on temporal frequency. In Figure 11A, the mask was oriented at 35° from the optimal orientation so as to yield a large suppression and both the temporal frequency of drift and contrast of the mask were varied. Suppression, which is essentially linear with the log of mask contrast, is strong at both 2 and 8 Hz. It is less powerfull but still evident at 16 and even 32 Hz. When mask contrast is fixed and its orientation is varied, a similar pattern emerges (Figure 11B). Suppression is strong at 2 Hz, decreases somewhat at 8 Hz, but is still clear at 16 Hz. As with most cortical cells, neither of these cells generated excitatory responses to single gratings above 12 Hz. The ability of a 16-Hz mask to suppress

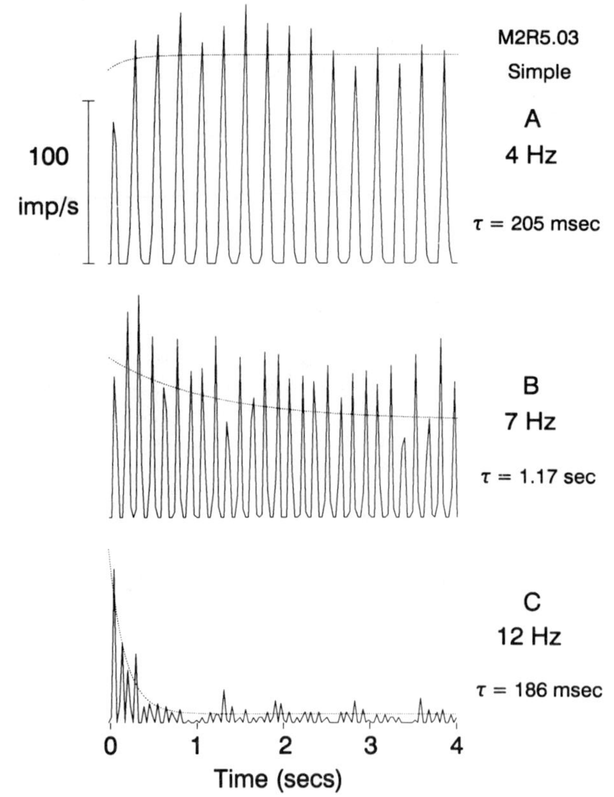

FIGURE 10. Dependence of response decline on temporal frequency of stimulation. Response histograms from stimulation with 4, 7, and 12 Hz (A, B, and C, respectively). The dotted line is a simple exponential fit to the centers of mass of the response bursts. Response decline over time markedly accelerates with higher frequencies of stimulation. (From Bonds, A. B. (1991). *Visual Neurosci.*, **6**, 239–255).

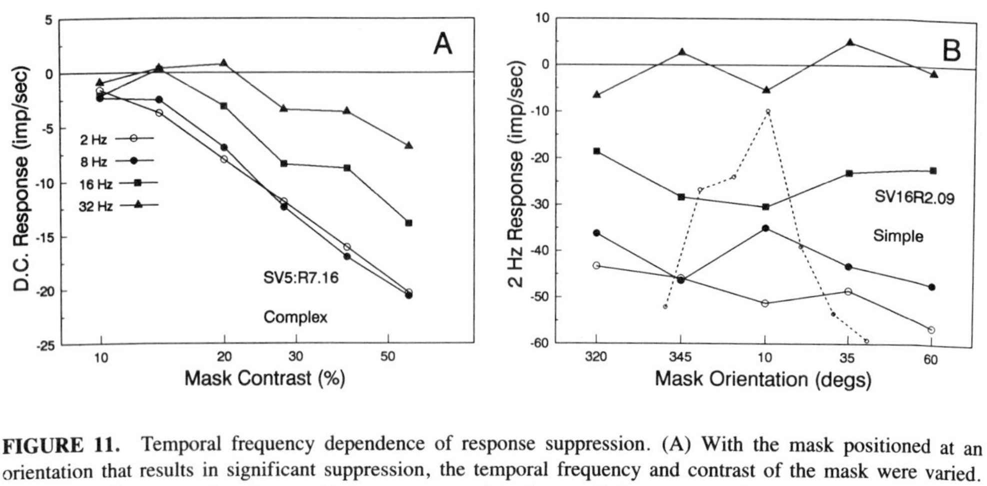

FIGURE 11. Temporal frequency dependence of response suppression. (A) With the mask positioned at an orientation that results in significant suppression, the temporal frequency and contrast of the mask were varied. Suppression is more or less linear with log contrast and is clear at higher contrasts for even a 32-Hz mask. (B) Suppression is essentially uniform with orientation (within the range measured) and is present with a 16-Hz mask (but not a 32-Hz mask). (From Bonds, A. B. (1989). *Visual Neurosci.*, **2**, 41–55).

the response from a 2-Hz base grating demonstrates a nonlinear interaction in the temporal domain that could serve as a substrate for active bandwidth limitation. This result poses something of a mystery: if the interaction is mediated by inhibition, how does it arise? Although intracortical inhibition is abundant, no cortical cells thus far recorded seem to be able to sustain throughput of a 16-Hz (much less 32-Hz) signal. While these frequencies are well within the bandpass of LGN cells, inhibitory synapses between LGN and cortex are yet to be proven (cf. Garey and Powell, 1971). The answer may lie in the intricate feedback circuitry between the LGN and cortex (Sherman and Koch, 1986).

VIII. DISCUSSION

Perhaps the most remarkable quality of cortical cells is the degree to which they don't fire. This is especially impressive when considering the high functional convergence of LGN cells onto cortical cells. Cross-correlation studies estimate convergence ranging from about 10 for simple cells to as high as 30 for complex cells (Tanaka, 1985). Both the maintained discharge and average peak firing rate of cortical cells are, however, lower than found in LGN cells. The elimination of nerve impulses does not result from proportional attenuation or simple thresholding, since cortical cells show exquisite sensitivity, contrast adaptation, and high selectivity to the spatial qualities of stimuli.

The selective elimination of impulses appears to be mediated by inhibitory mechanisms that display two different qualities. One component, that is independent of the spatial characteristics of the stimulus, is primarily associated with the amplitude-dependent (point) response nonlinearities. The other component is spatially dependent and primarily appears to enhance spatial selectivity. The two mechanisms may not be unique and may, in implementation, simply be alternate expressions of a common physiological process. The dichotomy is drawn to simplify analysis of the observations.

A. GLOBAL INHIBITION

Results described above show that even when stimuli are spatially optimized for a cell, their presentation usually results in some degree of suppression of a preexisting response. Moreover, while overall suppression may not be totally independent of the spatial configuration of the stimulus, some degree of suppression occurs as a result of presentation of virtually any stimulus that falls within the range of visibility of the organism under study. This suggests the existence of a global component of inhibition that in the ideal case is activated by all stimuli and, when activated, suppresses all cells. The advantages of this decidedly nonlinear mechanism are several. At the level of the single unit, and when optimal stimuli are used, this mechanism supports contrast adaptation. Modification of the operating point of the cell as a function of input contrast can effectively increase dynamic range. As mean contrast is increased, the response vs. log contrast function moves to the right. This reduces the responsiveness to stimuli that would otherwise cause saturation and positions the response on the sloped area of the curve rather than the flat top (see, e.g., Figure 5A), which improves increment sensitivity at higher contrasts. Contrast gain control in this form is, thus, constructive, since it improves performance in the vicinity of an arbitrary ambient contrast, although information on absolute contrast is lost. This loss may, however, be no more damaging than the loss of information on absolute luminance that occurs in the retina as a result of its own luminance gain control (Enroth-Cugell and Shapley, 1984). In some way the information on absolute contrast must still be available to the system, since it is required to command the neural circuitry that activates the gain control.

The responsiveness of a cell is also reduced by nonoptimal stimuli and even stimuli which do not excite the cell. This is decidedly disadvantageous in the context of a contrast gain control, since sensitivity and gain can be reduced even though the cell is not firing at all. There are at least two possible advantages of this strategy. One is an "attentional" mechanism. Long-term contrast adaptation, in which response to steady-state stimulation is reduced over seconds or even minutes is well-documented (e.g., Vautin and Berkley, 1978; Albrecht et al., 1984; Magnussen and Greenlee, 1985). There is, thus, a tendency for cells to become complacent with the status quo. With the appearance of a novel stimulus, the affected cells will respond with a transient burst. This burst not only rises above the steady-state ambient activity, but actually suppresses the activity from other cells, resulting in a more powerful temporal differentiation. The second, and perhaps more important, advantage lies in control of the signal strength that flows to higher visual centers. Heavy convergence from striate cortical cells could overload subsequent stages with even moderate firing rates from the precursor cortical cells. One role of global inhibition might be maintenance of the total volume of this information at a constant level. This can only occur if the increase in activity at one site is mirrored by a proportional decrease in activity at all others. Such activity could be thought of as another stage of the normalization executed by the visual pathways. Retinal ganglion cells transmit information on relative, not absolute illumination. Single cortical cells transmit information on relative, not absolute contrast. The visual cortex as a whole transmits information in the form of relative, not absolute activity of groups of cells that reexpress the image content as the output of spatial filters.

B. MAINTENANCE OF SPATIAL FILTER PROPERTIES IN THE VISUAL CORTEX

The data presented here can be interpreted as supporting a dual mechanism for derivation of the spatial tuning of cortical cells. The first component is a variable threshold that is mediated by the global inhibitory mechanism discussed above. In this way biases toward selectivity in the signals reaching the cortex can be transformed into absolute selectivity by the elimination of activity that is below some criterion level. The coordinated dependence of both operating point (threshold level) and response amplitude on contrast ensures constant filter properties across all levels of contrast. The second component, found in at least some cells for both orientation and spatial frequency selectivity, is a form of lateral inhibition in the appropriate domain. Inhibitory side-bands that are usually strongest at the orientation or spatial frequency at which the excitatory tuning curve approaches zero act to refine the selectivity that is more crudely defined by the thresholding mechanism.

One puzzling question is why the global inhibitory mechanism, which seems fundamental to defining both the gain and the spatial filter properties of individual cells, is so dependent on network activity. If this gain control served primarily to optimize the contrast increment sensitivity of a given cell, that demand would be most effectively met by a mechanism dependent only on the activity of that cell. However, the definition of spatial selectivity cannot be accomplished with inhibition derived from a simple feedback system that is modulated by the firing rate of the cell. While that configuration could adequately model response amplitude nonlinearities for optimal stimuli, because of these same nonlinearities it is insufficient to maintain constant spatial tuning at different contrasts. A response-set nonlinearity, such as response compression, would result in a relative increase in tuning peaks (sharper tuning) at lower contrasts, since the cell is not driven into the region resulting in compression. However, the shapes of neither orientation nor spatial-frequency tuning functions are dependent on contrast level (Sclar and Freeman, 1982; Albrecht and Hamilton, 1982). The response amplitude nonlinearity must, therefore, be contrast-set and must depend on information available either before or at the site of spatial integration, not after. Such information can only be integrated from a broad, generally organized network.

Another view of the importance of the large-scale integration of information to control the operating point of cortical cells arises when considering that the cortex must function as an integrated entity rather than as a collection of the outputs of individual cells. If each cell were controlled independently so as to optimize its firing level to maximize sensitivity to changes in contrast, then all cells would perpetually generate an output at some median firing level. The modulation of contrast of an image is represented in cortex by the modulation of firing rates, either across time or across cells. Normalization of these firing rates on an individual basis would presumably attenuate severely or eliminate the texture of the visual scene. Homogeneous control of gain across the cellular concert preserves relative activity, thereby preserving the spectrum of the image.

REFERENCES

Albrecht, D. G. and Hamilton, D. B. (1982). Striate cortex of monkey and cat: contrast response functions, *J. Neurophysiol.*, **48**, 217–237.

Albrecht, D. G., Farrar, S. B., and Hamilton, D. B. (1984). Spatial contrast adaptation characteristics of neurones recorded in the cat's visual cortex, *J. Physiol. (London)*, **347**, 713–739.

Bauman, L. A. and Bonds, A. B. (1991). Inhibitory refinement of spatial frequency selectivity in single cells of the cat striate cortex, *Vision Res.*, **31**, 933–944.

Bishop, P. O., Coombs, J. S., and Henry, G. H. (1973). Receptive fields of simple cells in the cat striate cortex, *J. Physiol. (London)*, **231**, 31–60.

Bonds, A. B. (1989). The role of inhibition in the specification of orientation selectivity of cells in the cat striate cortex, *Visual Neurosci.*, **2**, 41–55.

Bonds, A. B. (1991). Temporal dynamics of contrast gain in single cells of the cat striate cortex, *Visual Neurosci.*, **6**, 239–255.

Dean, A. F. (1981). The relationship between response amplitude and contrast for cat striate cortical neurones, *J. Physiol. (London)*, **318**, 413–427.

DeValois, R. L., Albrecht, D. G., and Thorell, L. G. (1978). Cortical cells: bar and edge detectors or spatial frequency analyzers?, in *Frontiers in Visual Science*, Cool, S. J. and Smith, E. L., Eds., Springer Verlag, New York, 544–556.

Enroth-Cugell, C. and Robson, J. G. (1966). The contrast sensitivity of retinal ganglion cells of the cat, *J. Physiol. (London)*, **187**, 517–552.

Enroth-Cugell, C., Robson, J. G., Schweitzer-Tong, D. E., and Watson, A. B. (1983). Spatio-temporal interactions in cat retinal ganglion cells showing linear summation, *J. Physiol. (London)*, **341**, 279–307.

Enroth-Cugell, C. and Shapley, R. M. (1973). Adaptation and dynamics of cat retina ganglion cells, *J. Physiol. (London)*, **233**, 271–309.

Enroth-Cugell, C. and Shapley, R. M. (1984). Visual adaptation and retinal gain controls, *Prog. Retinal Res.*, **3**, 263–346.

Garey, L. J. and Powell, T. P. S. (1971). An experimental study of the termination of the lateral geniculo-cortical pathway in cat and monkey, *Proc. R. Soc. London, Ser. B*, **179**, 41–63.

Heggelund, P. (1981a). Receptive field organization of simple cells in cat striate cortex, *Exp. Brain Res.*, **42**, 89–98.

Heggelund, P. (1981b). Receptive field organization of complex cells in cat striate cortex, *Exp. Brain Res.*, **42**, 99–107.

Henry, G. H., Dreher, B., and Bishop, P. O. (1974). Orientation specificity of cells in cat striate cortex, *J. Neurophysiol.*, **37**, 1394–1409.

Hubel, D. H. and Wiesel, T. N. (1962). Receptive fields, binocular interaction and functional architecture in the cat's visual cortex, *J. Physiol. (London)*, **160**, 106–154.

Kaplan, E., Marcus, S., and So, Y. T. (1979). Effects of dark adaptation on spatial and temporal properties of receptive fields in cat lateral geniculate nucleus, *J. Physiol. (London)*, **294**, 561–580.

Legge, G. E. and Foley, J. M. (1980). Contrast matching in human vision, *J. Opt. Soc. Am.*, **70**, 1458–1471.

Levick, W. R. and Thibos, L. N. (1980). Orientation bias of cat retinal ganglion cells, *Nature (London)*, **286**, 389–390.

Li, C. Y. and Creutzfeldt, O. D. (1984). The representation of contrast and other stimulus parameters by single neurons in area 17 of the cat, *Pfluegers Arch.*, **401**, 304–314.

Maffei, L. and Fiorentini, A. (1973). The visual cortex as a spatial frequency analyzer, *Vision Res.*, **13**, 1255–1267.

Magnussen, S. and Greenlee, M. W. (1985). Marathon adaptation to spatial contrast: saturation in sight, *Vision Res.*, **25**, 1409–1411.

Morrone, M. C., Burr, D. C., and Maffei, L. (1982). Functional implications of cross-orientation inhibition of cortical visual cells, *Proc. R. Soc. London Ser. B*, **216**, 335–354.

Movshon, J. A., Thompson, I. D., and Tolhurst, D. J. (1978a). Spatial summation in the receptive fields of simple cells in the cat's striate cortex, *J. Physiol. (London)*, **283**, 53–77.

Movshon, J. A., Thompson, I. D., and Tolhurst, D. J. (1978b). Spatial and temporal contrast sensitivity of neurones in areas 17 and 18 of the cat's visual cortex, *J. Physiol.*, **283**, 101–120.

Movshon, J. A. and Lennie, P. (1979). Pattern selective adaptation in visual cortical neurones, *Nature (London)*, **278**, 850–852.

Nelson, J. I. and Frost, B. J. (1978). Orientation-selective inhibition from beyond the classic visual receptive field, *Brain Res.*, **139**, 359–364.

Ohzawa, I., Sclar, G., and Freeman, R. D. (1985). Contrast gain control in the cat's visual system, *J. Neurophysiol.*, **54**, 651–667.

Rose, D. and Blakemore, C. B. (1974). An analysis of orientation selectivity in the cat's visual cortex, *Exp. Brain Res.*, **20**, 1–17.

Schumer, R. A. and Movshon, J. A. (1984). Length summation in simple cells of cat striate cortex, *Vision Res.*, **24**, 565–571.

Sclar, G. and Freeman, R. D. (1982). Orientation selectivity in the cat's striate cortex is invariant with stimulus contrast, *Exp. Brain Res.*, **46**, 457–461.

Sherman, S. M. and Koch, C. (1986). The control of retinogeniculate transmission in the mammalian lateral geniculate nucleus, *Exp. Brain Res.*, **63**, 1–20.

Sillito, A. M. (1975). The contribution of inhibitory mechanisms to the receptive field properties of neurones in the striate cortex of the cat, *J. Physiol. (London)*, **250**, 305–329.

Sillito, A. M. (1979). Inhibitory mechanisms influencing complex cell orientation selectivity and their modification at high resting discharge levels, *J. Physiol. (London)*, **289**, 33–53.

Skottun, B. C., DeValois, R. L., Grosof, D. H., Movshon, J. A., Albrecht, D. G., and Bonds, A. B. (1991). On classifying simple and complex cells according to response modulation, *Vision Res.*, **31**, 1079–1088.

Soodak, R. E., Shapley, R. M., and Kaplan, E. (1987). Linear mechanism of orientation tuning in the retina and lateral geniculate of the cat, *J. Neurophysiol.*, **58**, 267–275.

Soodak, R. E., Shapley, R. M., and Kaplan, E. (1987). Linear mechanism of orientation tuning in the retina and lateral geniculate of the cat, *J. Neurophysiol.*, **58**, 267–275.

Tadmor, Y. and Tolhurst, D. J. (1989). The effect of threshold on the relationship between the receptive-field profile and the spatial-frequency tuning curve in simple cells of the cat's striate cortex, *Visual Neurosci.*, **3**, 445–454.

Tanaka, K. (1985). Organization of geniculate inputs to visual cortical cells in the cat, *Vision Res.*, **25**, 357–364.

Tolhurst, D. J. and Dean, A. F. (1987). Spatial summation by simple cells in the striate cortex of the cat, *Exp. Brain Res.*, **66**, 607–620.

Tolhurst, D. J., Movshon, J. A., and Thompson, I. D. (1981). The dependence of response amplitude and variance of cat visual cortical neurones on stimulus contrast, *Exp. Brain Res.*, **41**, 414–419.

Tomko, G. J. and Crapper, D. R. (1974). Neuronal variability: non-stationary responses to identical visual stimuli, *Brain Res.*, **79**, 405–418.

Tsumoto, T., Eckart, W., and Creutzfeldt, O. D. (1979). Modification of orientation sensitivity of cat visual cortex neurons by removal of GABA-mediated inhibition, *Exp. Brain Res.*, **34**, 351–363.

Vautin, R. G. and Berkley, M. A. (1978). Responses of single cells in cat visual cortex to prolonged stimulus movement: neural correlates of visual aftereffects, *J. Neurophysiol.*, **40**, 1051–1065.

Vidyasagar, T. R. and Henry, G. H. (1991). Relationship between preferred orientation and ordinal position in neurones of cat striate cortex, *Visual Neurosci.*, **5**, 565–569.

Vidyasagar, T. R. and Urbas, J. V. (1982). Orientation sensitivity of cat LGN neurones with and without inputs from visual cortical areas 17 and 18, *Exp. Brain Res.*, **46**, 157–169.

Chapter 13

HOW FLYING BEES COMPUTE RANGE FROM OPTICAL FLOW: BEHAVIORAL EXPERIMENTS AND NEURAL MODELS

M. V. Srinivasan

TABLE OF CONTENTS

I. Introduction 354

II. Experimental Methods 354
- A. Analysis of Flight Paths 355

III. Results 357
- A. Dependence of the Centering Response on Contrast Frequency 360
- B. Dependence of the Centering Response on Apparent Angular Speed 362
- C. What Region of the Eye Is Involved in Measuring Apparent Angular Speed? 364
- D. Dependence of the Centering Response on the Intensity Profile of the Pattern 365
- E. Dependence of the Centering Response on the Contrast of the Pattern 365

IV. Discussion 367
- A. The Centering Response vs. the Optomotor Response 367
- B. Directional Motion Detection and the "Obliquity Problem" 369
- C. Relationship of the Centering Response to the Movement Avoidance Response 370
- D. Neural Mechanisms for the Measurement of Angular Speed 371

Acknowledgments 374

References 374

I. INTRODUCTION

The eyes of most insects are comparatively close together and lack focusing mechanisms, so that they must use cues other than stereopsis or accommodation to gauge the distances of objects in the environment. It has been suggested that moving insects perceive the distances of objects in terms of the speeds of their images on the retina, a higher image speed being associated with a smaller range (Wallace, 1959; Collett, 1978; Eriksson, 1980; Goulet et al., 1981; Collett and Harkness, 1982; Horridge, 1986; Cheng et al., 1987; Collett, 1988; Lehrer et al., 1988; Srinivasan et al., 1989). Recently, Sobel (1990a,b) confirmed Wallace's (1959) original finding that locusts use a peering motion of the head to infer object range.

The study conducted by Kirchner and Srinivasan (1989) was inspired by the observation that when bees are made to fly through a small opening (e.g., in a window), they tend to fly through its center, balancing the distances to the left and the right-hand sides of the opening. Kirchner and Srinivasan demonstrated that honeybees flying along a tunnel tend to fly through the middle by balancing the apparent motions of the images of the walls on the two sides. This chapter describes experiments designed to explore this "centering response" quantitatively, in an attempt to discover the parameters of image motion that are used to extract range information, and to probe the mechanisms by which these parameters are measured.

The well-studied optomotor response in insects is mediated by a motion-detecting mechanism that is sensitive to contrast frequency (reviewed by Reichardt, 1987). Such a system is, however, incapable of measuring image speed unambiguously, and independently of the structure of the image (reviewed by Borst and Egelhaaf, 1989). On the other hand, a system that measures the apparent speed of an image independently of its structure would be capable of providing a reliable estimate of range. The results described in this chapter suggest that the estimation of range by the bee's visual system does, indeed, involve movement-sensitive mechanisms that measure the apparent speed of an image largely independently of its structure.

II. EXPERIMENTAL METHODS

The experimental apparatus was a tunnel formed by two walls 40 cm long, 20 cm high, and 12 cm apart (Figure 1). Each wall consisted of a sheet of rubber stretched around a pair of drums. Gratings or other patterns (described below) were affixed to the sheets. A DC motor coupled to one of the drums enabled the pattern on one of the walls to be moved in either direction (arrows, Figure 1) at any desired speed.

Groups of 4 to 6 freely flying honeybees (*Apis mellifera*) were marked and trained to enter the apparatus, fly along the tunnel, and receive a reward of sugar solution in a box at the end of the tunnel (Figure 1). After the bees had received circa 60 rewards, their flight trajectories were videofilmed from above as they flew along the tunnel in either direction (i.e., toward the reward, or homeward after having been rewarded). Depending on the experiment, both walls were stationary or one of them was in motion (Figure 1). In some experiments, sheets of transparent Perspex were placed in front of the walls to prevent the flight trajectories from being inflenced by air currents generated by the moving patterns. However, since experiments conducted without these sheets revealed no obvious differences, data from both kinds of experiments were pooled.

The patterns on the walls usually consisted of a vertically oriented, one-dimensional, black-and-white grating with an intensity profile that was square-wave, random, sinusoidal, or sawtooth-shaped. In some experiments, one of the walls carried a sheet of white (photocopying) paper, or gray paper of the same mean luminance and spectral composition as the gratings.

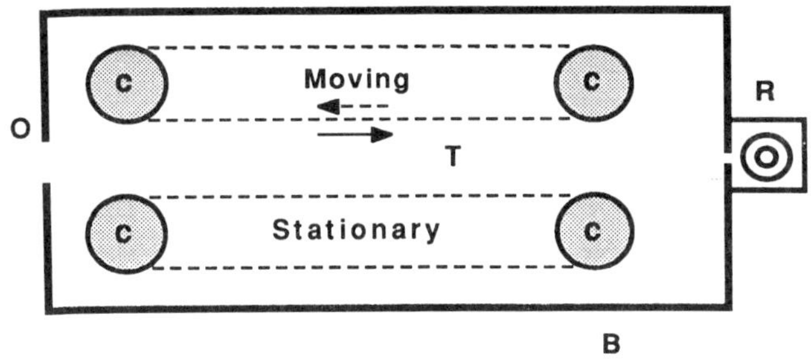

FIGURE 1. Schematic, overhead view of experimental apparatus (not to scale). Bees are trained to enter the Perspex box B through the opening O and to fly along tunnel T, formed by two walls, to receive a reward of sugar water in the box R. Each side wall of the tunnel consists of a sheet of rubber stretched around a pair of drums (C). Gratings or other patterns are affixed to the sheets. A motor coupled to one of the drums enables the pattern on one of the walls to be moved in either direction (arrows) at any desired speed. (Reproduced from Srinivasan, M. V. et al. (1991), with permission of authors and Cambridge University Press, NY.)

The square-wave and random gratings, of contrast 0.85, were constructed by glueing strips of black paper onto a background consisting of white (photocopying) paper. The square-wave gratings had periods of 2.5, 3.5, 5, 7, and 10 cm. The random grating was constructed similarly, except that the widths of successive black bars and intervening white spaces were random, integral multiples of 0.4 cm, determined by the random-number generator on a computer. The sinusoidal and sawtooth-profile gratings were created using a computer-driven laser printer, capable of generating 64 gray levels in square pixels of size 0.7 mm. These patterns were carefully screened for defects arising from toner deficiencies in the laser printer. Screening was done by visual inspection, as well as through intensity-profile measurements made using a video camera coupled to a computer-driven frame grabber. The sinusoidal gratings had periods of 1.5, 3, or 6 cm. The Michelson contrasts used ($[I_{max} - I_{min}]/[I_{max} + I_{min}]$, as derived from the measured intensity profiles) were 0.60, 0.15,or lower than 0.05 (nominally, homogenous gray). The sawtooth-profile gratings had a period of 3.0 cm and a contrast of 0.75.

The (angular) spatial frequencies of the gratings, as seen by a bee flying along the axis of the tunnel, ranged from 0.01 to 0.04 c/deg for the square-wave gratings, 0.017 to 0.07 c/deg for the sinusoidal gratings, and 0.035 c/deg for the sawtooth-profile gratings. The random grating possessed a spectrum of spatial frequencies distributed over the range 0 to 0.13 c/deg. Our prior knowledge of the spatial acuity of the honeybee's visual system (e.g., Srinivasan and Lehrer, 1988) indicates that all of the gratings used in these experiments should have been clearly resolvable.

A. ANALYSIS OF FLIGHT PATHS

Flight trajectories, recorded at 25 frames per second by the camera and video recorder, were subsequently analyzed frame by frame to measure average positions along the width of the tunnel, average flight velocities, etc. This was done either manually — by plotting successive positions of bees on a sheet of transparent acetate paper affixed to the screen of the video monitor — or by superimposing the video image onto a digitizing tablet using a half-reflecting mirror. Detailed histograms of bee positions, as shown in Figures 2 and 3, were obtained by analyzing the digitized data on a computer to determine the relative frequency with which bees were found in each cell of an array of 20 × 10 cells. (This array represented a total area of 26 × 12 cm.) These relative frequencies are depicted as an array of squares, the linear dimension of each square being proportional to frequency. In other instances (as in the data presented in Figures 4 to 7, and Figure 9), each trajectory was assigned a single mean position along the width of the tunnel, by averaging the y coordinates

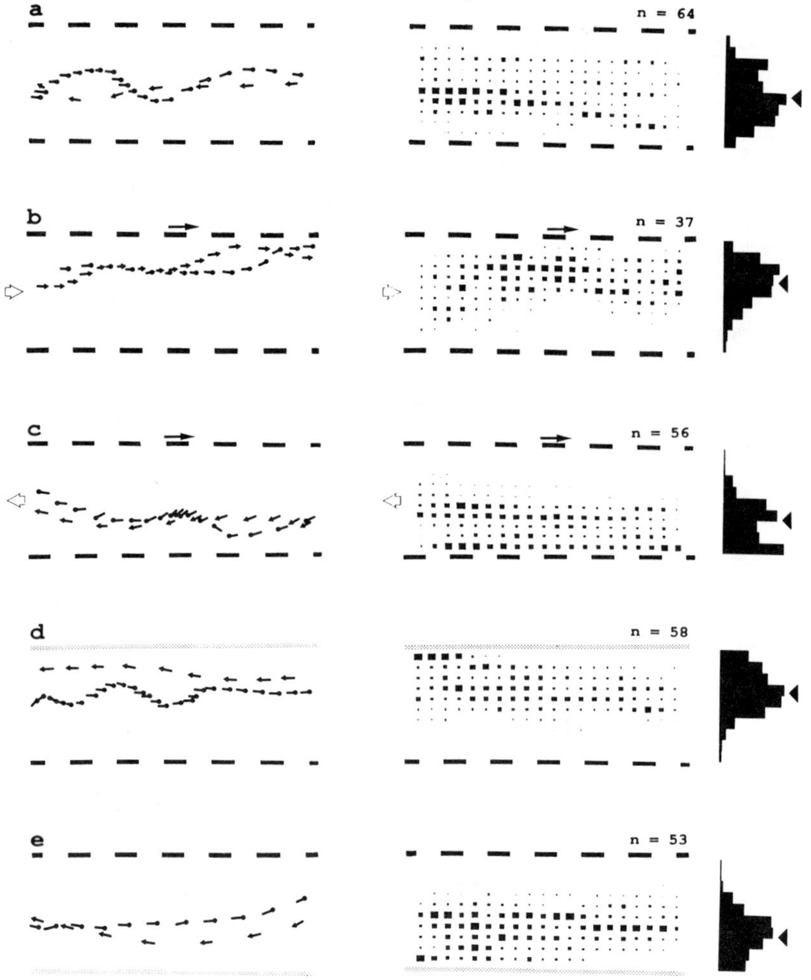

FIGURE 2. The centering response: experiment demonstrating that flying bees use apparent image motion to estimate range. Bees are trained to fly along a tunnel, in which the side walls carried vertically oriented square-wave gratings of period 3.5 cm. Flights are filmed from above at 25 frames per second with (a) stationary gratings on either side, (b) grating moving at a speed of 20 cm/s in the direction of the bees' flight, (c) grating moving at 20 cm/s in the direction opposite to that of the bees' flight, and (d and e) the moving grating replaced by a homogeneously gray sheet. The left-hand column shows two typical trajectories (depicted by the two different bee symbols) in each case. The right-hand column shows the spatial distributions of the bees' positions, obtained by analyzing a number of trajectories in each case. Flights in both directions are included in analyzing a, d, and e, while flights in the directions of the large arrows are analyzed in b and c. The histograms on the extreme right depict the distribution of the bees' trajectories along the width of the tunnel, averaged over a central, 26-cm segment. The arrow next to each histogram depicts the mean value of the distribution. n = number of trajectories analyzed in each case. (Reproduced from Srinivasan, M. V. et al. (1991), with permission of authors and Cambridge University Press, NY.)

of bee positions digitized along a central, 26-cm segment of the trajectory. In some instances (Figures 4 and 9) the mean positions of several trajectories were then averaged to obtain a mean trajectory position, and a standard deviation about this mean. Velocities of bees, and of pattern movement were obtained by measuring the distance traveled over a known number of frames. Knowing these and the position of the trajectory along the tunnel width, the apparent angular speeds of the patterns on the eyes of the flying bee could be calculated using the formulae shown in Figure 5.

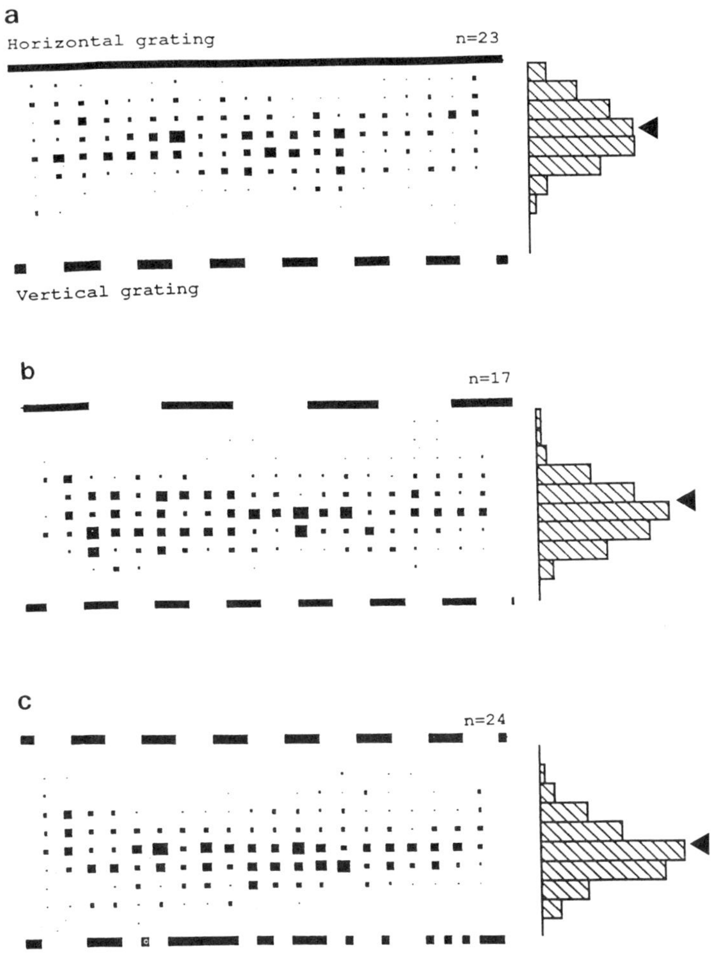

FIGURE 3. Distribution of bees' positions in the tunnel when the two walls carry different, stationary patterns. Flights in both directions are included in the analysis. In all three cases, the pattern on one wall is a vertically oriented square-wave grating of period 3.5 cm (a) The pattern on the other wall is a horizontally oriented square-wave grating of the same period; (b) it is a vertically oriented square-wave grating of period 7.0 cm; (c) it is a spatially random grating with characteristics as described in Section II. In b and c the mean position of the bees' trajectories (arrow) is not displaced significantly from the tunnel axis, while in a, bees fly significantly closer to the wall carrying the horizontal grating. Histograms have same meaning as in Figure 2. n = number of trajectories analyzed in each case. (Reproduced from Srinivasan, M. V. et al. (1991), with permission of authors and Cambridge University Press, NY.)

III. RESULTS

When the two walls of the tunnel were stationary and carried identicial gratings, bees tended to fly through the middle of the tunnel (on average), as shown in Figure 2a. The mean position of the flight trajectories along the width of the tunnel was not significantly displaced from the axis of the tunnel (t test, $p > 0.30$). How were the bees gauging and balancing the distance to the gratings on the two sides? One possibility is that they were balancing the speeds of the retinal images of the gratings on the two eyes. To test this possibility, one can examine flight trajectories when one of the gratings is in motion. When this grating moves at a constant speed in the direction of the bees' flight — thereby reducing the velocity of retinal image motion on that eye relative to the other eye — the bees' trajectories shift toward the side of the moving grating (Figure 2b). However, when the

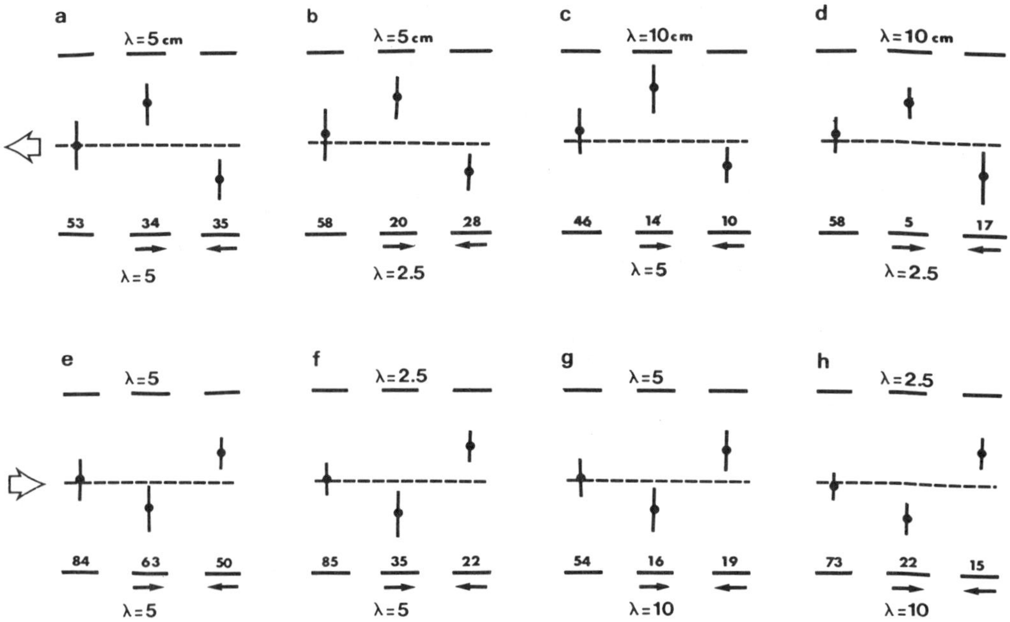

FIGURE 4. Contrast-frequency invariance of the centering response. Analysis of flight trajectories for a number of different combinations (a - h) of vertically oriented square-wave gratings on the two walls. Each of the eight panels shows the mean and the standard deviation of the positions of the flight trajectories for a particular combination of gratings, of period g, when both gratings are stationary (left-hand side of each panel), or when one of the gratings moves at a constant speed of 27.5 cm/s toward the reward box (center) or the exit (right), as shown by the small arrows. The panels a to d represent flights toward the exit, while the panels e to h represent flights toward the reward box (as shown by the large arrows). In all cases, the mean position of the trajectories does not deviate significantly from the tunnel axis when both gratings are stationary, but does so in the appropriate direction (as in Figure 2) when one of the gratings is in motion. The numbers represent the number of flights analyzed in each case. (Reproduced from Srinivasan, M. V. et al. (1991), with permission of authors and Cambridge University Press, NY.)

grating moves in a direction opposite to that of the bees' flight — thereby increasing the speed of retinal image motion on that eye relative to the other eye — the trajectories shift away from the side of the moving grating (Figure 2c). Finally, when one of the gratings is replaced by a homogeneously gray sheet of paper and the other grating is stationary, the trajectories shift toward the homogeneous side (Figure 2d,e). There is a significant difference between the average locations of the flight trajectories, measured along the width of the tunnel, between the experimental conditions corresponding to Figure 2a and b, a and c, b and c, a and d, a and e, and d and e (*t* test, $p < 0.001$ for all cases).

These findings suggest that when the walls are stationary, the bees maintain equidistance by balancing the apparent speeds of the retinal images in the two eyes. A slower motion on one eye is evidently taken to imply that the grating on that side is further away, causing the bee to fly along a trajectory closer to it; a faster motion of the image, on the other hand, has the opposite effect. When one eye sees homogeneous gray, it receives no movement stimulus and the bees consequently fly closer to that side.

What parameters of image motion mediate the centering response? The most likely interpretation of the results described so far is that the flying bees are trying to balance the apparent angular speeds, in the horizontal plane, of the images on the two eyes as they fly along the tunnel. There is, however, another possibility, namely, that they infer the distances to the two walls by moving alternately toward one wall and then the other (i.e., zig-zagging along the tunnel) and comparing the looming (rate of expansion) of the patterns on the two

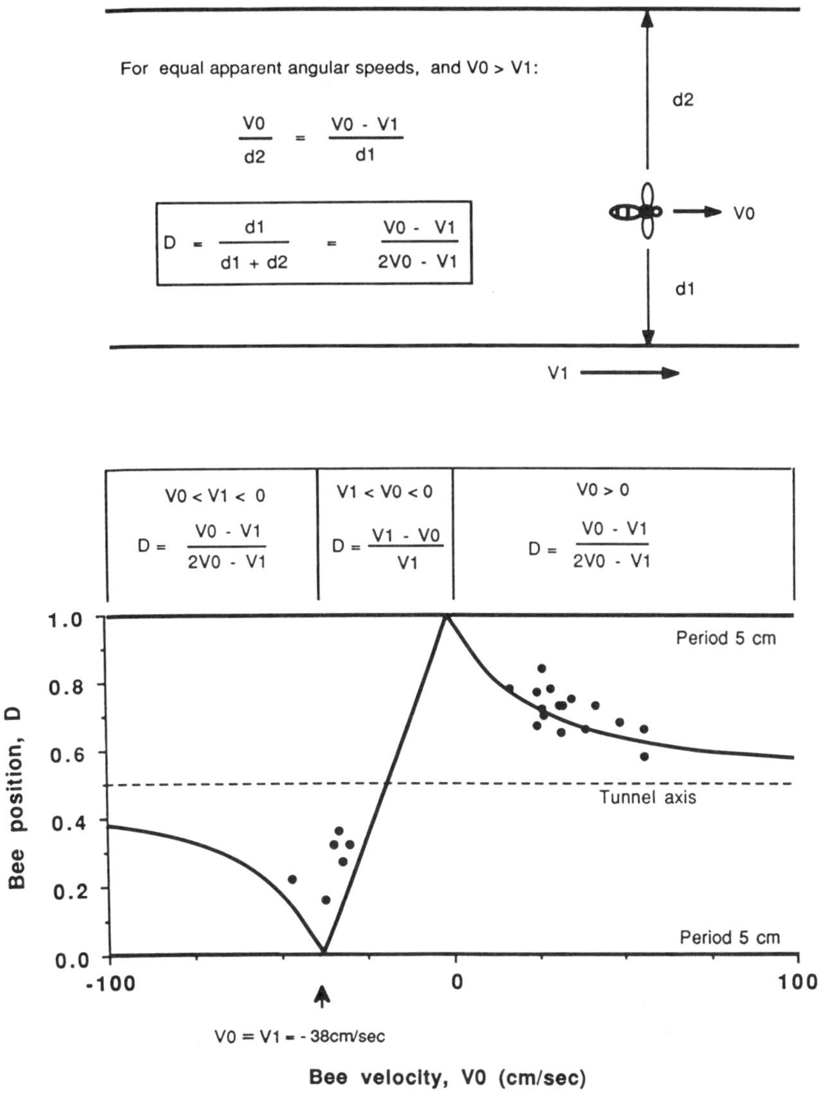

FIGURE 5. Dependence of the centering response on apparent angular velocity. Experiment testing the hypothesis that bees balance the apparent angular speeds of the images on the two eyes as they fly through a tunnel in which one wall carries a stationary, vertically oriented square-wave grating of period 5 cm, and the other an identical, but moving grating. If the hypothesis is true, then for any bee velocity V0 and grating velocity V1 (= −38cm/s in this particular case) it should be possible to predict the position of the trajectory along the width of the tunnel, as explained in the text and specified in the upper panel. The lower panel compares the theoretically predicted positions (continuous curve) with those experimentally observed (dots). The good agreement between theory and experiment supports the hypothesis that the bees are balancing the apparent angular speeds of the images on the two eyes as they fly along the tunnel. (Reproduced from Srinivasan, M. V. et al. (1991), with permission of authors and Cambridge University Press, NY.)

walls. Although this seems unlikely (from the relatively straight profiles of the flight trajectories), this possibility can be tested by analyzing the trajectories of bees flying through the tunnel when both walls are stationary, with one wall carrying a horizontal grating, and the other wall a vertical one. While looming cues can be obtained from the horizontal grating as well as the vertical grating, horizontal motion cues would be provided only by the vertical grating. Therefore, we would expect bees to fly close to the axis of the tunnel if they use

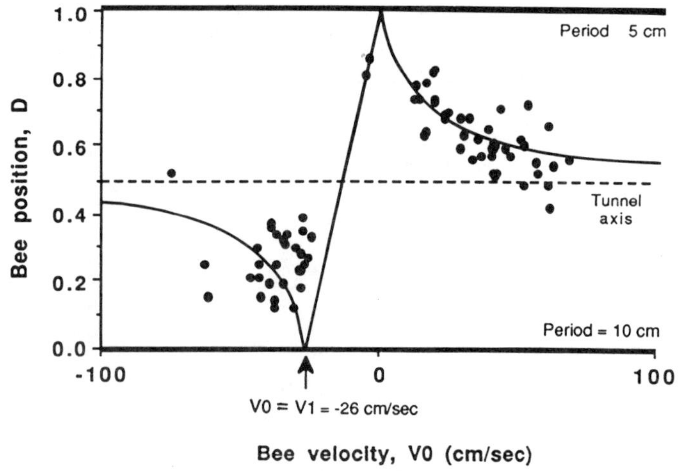

FIGURE 6. Experiment similar to that shown Figure 5, but with gratings of different periods on the two walls. In this case the 5-cm grating is stationary while a 10-cm grating moves at a constant velocity of 26 cm/s. Symbols as in Figure 5. (Reproduced from Srinivasan, M. V. et al. (1991), with permission of authors and Cambridge University Press, NY.)

looming-motion cues, but close to the horizontal grating if they use horizontal-motion cues. The results (Figure 3a) show that the bees fly distinctly closer to the horizontal grating; the mean position of the trajectories deviates significantly from the tunnel axis ($p < 0.0005$). Thus, the importance of looming cues seems negligible. This conclusion is supported by the finding that the position of the flight trajectories is strongly affected by horizontal motion of a vertical grating (Figure 2 b and c). If the centering response were mediated primarily by looming-motion cues, one would not expect the response to be affected by the superposition of a horizontal component of motion, as this would not affect the looming component.

A. DEPENDENCE OF THE CENTERING RESPONSE ON CONTRAST FREQUENCY

The experiments described so far leave open the possibility that the bees were comparing the temporal frequencies of the intensity fluctuations (''contrast frequencies'') produced by the successive dark and light bars of the gratings, as seen by the individual photoreceptors of the two eyes, rather than the apparent angular velocities of the patterns on the two walls. It is important to consider this possibility because it is now well established that the optomotor behavior — the best known motion-driven response in insects — is a response that depends on contrast frequency, rather than angular speed (reviewed by Reichardt, 1987, and Borst and Egelhaaf, 1989).

To test the importance of contrast frequency in the centering response, one can analyze the trajectories of bees flying along the tunnel when the two walls carry stationary gratings, but with the grating on one side having a spatial frequency twice that on the other (Figure 3b). Despite the fact that this configuration produces twice the contrast frequency on one eye as on the other, the mean position of the bees' trajectories do not deviate significantly from the axis of the tunnel (Figure 3b; $p > 0.4$), indicating that contrast frequency is not an important parameter in the centering response. In a further experiment, again performed with stationary patterns on the two walls, one wall carries a periodic grating while the other carries a grating that is spatially random, comprising a broad range of spatial frequencies. Here again, the bees' trajectories are located close to the tunnel axis (Figure 3c; $p > 0.05$), indicating that, as far as the centering response is concerned, the temporal pattern of stimulation of individual photoreceptors is not crucial to the measurement of apparent motion.

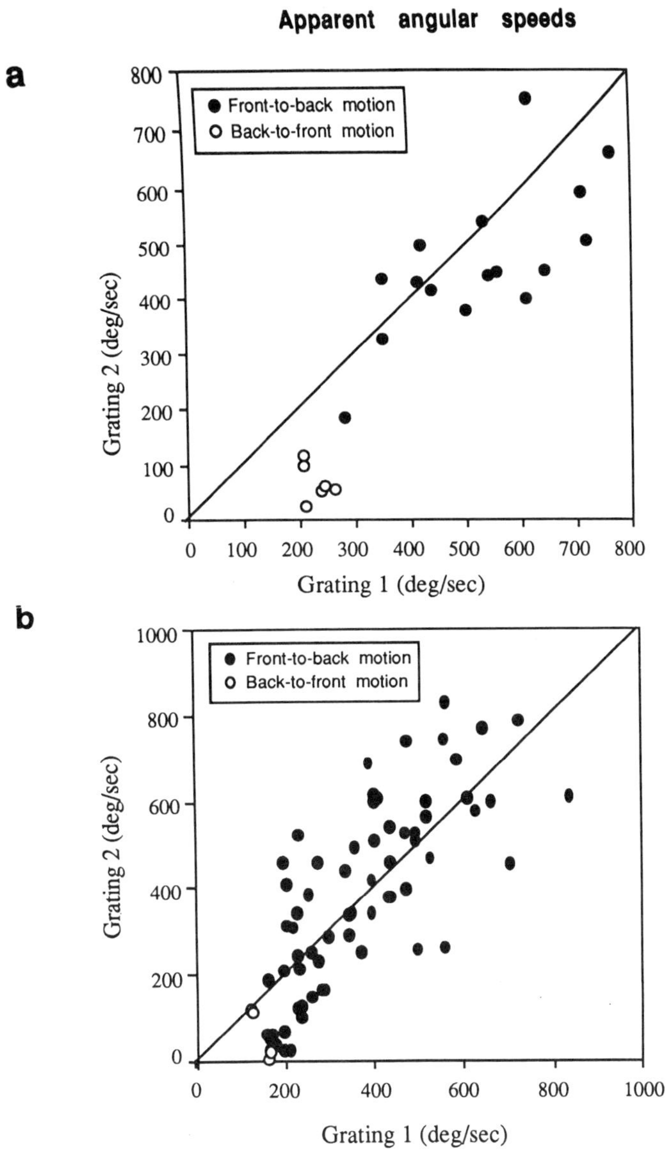

FIGURE 7. Comparison of apparent angular speeds experienced by the two eyes. The upper panel shows a comparison of the apparent angular speeds experienced by the bees' eyes for each of the flights analyzed in Figure 5. The filled circles represent flights in a direction opposite to that in which the grating moves, while the open circles represent flights in the same direction, but at a lower speed, so that one grating appears to move in the back-to-front direction on the eye. The lower panel shows a similar comparison for the flights analyzed in Figure 6. In either case, most of the data points cluster around the 45-degree line, indicating that the bees are, indeed, balancing the apparent angular speeds as seen by the two eyes. (Reproduced from Srinivasan, M. V. et al. (1991), with permission of authors and Cambridge University Press, NY.)

The role of contrast frequency in the measurement of apparent motion can be examined more extensively by using a number of different combinations of spatial frequencies for the gratings on the two walls. The results, shown in Figure 4, indicate that, as long as both gratings are stationary and resolvable by the eye, the bees' trajectories are centered on the axis of the tunnel, regardless of the spatial frequencies of the gratings on the two sides. The insensitivity of the centering response to spatial frequency (and, therefore, contrast frequency) is evident even when the spatial frequencies of the gratings on the two sides differ by a factor of as much as four (Figure 4, see left-hand segments of panels d and h). However, when one of the gratings is in motion (Figure 4, see center and right-hand segments of each panel) the bees' flight trajectories are clearly displaced from the tunnel axis ($p < 0.0025$ in all cases). The direction of displacement then depends consistently on whether the grating moves with or against the direction of the bees' flight. The measurements with the moving gratings constitute an important control, because they demonstrate (1) that the bees are, indeed, attending to the apparent motion of the gratings on the two sides as they fly along the tunnel and (2) when the gratings on the two sides are stationary, the resulting centering response is not simply a consequence of saturation of the responses of the movement-sensitive mechanisms in the two eyes.

In summary, the results of the above experiments suggest that contrast frequency is not an important parameter in the measurement of the apparent motion of the patterns. It is, therefore, very likely that apparent angular speed is, by default, the relevant parameter.

B. DEPENDENCE OF THE CENTERING RESPONSE ON APPARENT ANGULAR SPEED

It is possible to test more directly whether the bees are, indeed, measuring angular speed, as follows. If the flying bees are trying to balance the apparent angular speeds of the patterns as seen by the two eyes, then when one of the patterns is in motion and we know its linear speed as well as that of the bee, we should be able to predict where, along the width of the tunnel, the bee's trajectory should be located so that the apparent angular velocities of the two patterns are equal. With reference to Figure 5, we see that, for a bee flying at a speed V0, and a pattern moving in the same direction with a linear speed V1, equality of apparent angular speeds requires that $V0/d2 = (V1 - V0)/d1$ when $V0 < V1$, $V0/d2 = (V0 - V1)/d1$ when $V0 \geq V1$, and that $V0/d2 = (V0 + V1)/d1$ when the bee flies in a direction opposite to that in which the pattern moves.

The above expressions can be rewritten to evaluate the quantity $d1/(d1 + d2)$, which is the theoretically expected distance of the trajectory from the wall carrying the moving pattern, expressed as a fraction of the width of the tunnel. Denoting this quantity by D, we obtain $D = (V1 - V0)/V1$ when $V0 < V1$, $D = (V0 - V1)/(2.V0 - V1)$ when $V0 \geq V1$, and $D = (V0 + V1)/(2.V0 + V1)$ when V0 is oppositely directed to V1.

The results of an experiment designed to test these predictions are shown in Figure 5. Here each wall carries a square-wave grating of period 5 cm. The grating on one wall is stationary, while that on the other wall (the lower wall in the figure) moves at a constant velocity $V1 = -38$ cm/s. (The negative sign denotes motion toward the entrance.) The continuous curve shows the theoretically expected relationship between bee speed (V0, abscissa) and bee distance (D, ordinate), for bees flying with or against the direction of motion of the moving grating. The dots depict the experimentally measured relationship between the same two variables. Each dot represents one trajectory, where V0 was measured as the average velocity of the bee, in centimeters per second, and D was measured as the mean distance of the trajectory from the moving pattern, expressed as a fraciton of the tunnel width. It is evident that the measurements fit the theory quite well, especially in the case of bees flying in the direction opposite to that of grating motion ($V0 > 0$).

The point $D = 0$ on the theoretical curve corresponds to a situation in which the bee and the pattern are moving in the same direction at the same speed. In this situation, the moving grating does not move relative to the bee's eye. To the bee, therefore, this grating would be indistinguishable from a grating that is stationary and infinitely far away. One would, thus, expect that bees flying at the same speed and in the same direction as the moving grating would fly very close to it; the results (Figure 5) show that this is, indeed, the case.

Consider now a bee flying at different speeds along the axis of the tunnel, while one of the gratings is in motion. As the speed of the bee decreases toward zero, the eye viewing the stationary grating would experience a progressively lower angular speed, and this would be a progressively smaller fraction of the angular speed experienced by the other eye. This is true irrespective of whether the bee flies with or against the moving grating. Therefore, with decreasing flight speed ($|V0| \rightarrow 0$), we would expect the position of the trajectory to shift progressively closer to the stationary grating ($D \rightarrow 1.0$). The data (Figure 5) show that this is, indeed, what occurs. On the other hand, as the speed of a bee flying along the tunnel axis increases ($|V0| \rightarrow$ infinity), the ratio of angular speeds seen by the two eyes would approach unity, irrespective of the bee's direction of flight. Therefore, with increasing flight speed we would expect the position of the trajectory to approach the axis of the tunnel asymptotically ($D \rightarrow 0.5$). The data show that this is, indeed, what occurs in the case of bees flying opposite to the direction of motion of the grating. Corresponding data for bees flying in the same direction as the grating are not available, as very few bees are found to fly at the requisite speeds.

The results of an experiment similar to that described above (see Figure 5) but using gratings of different spatial frequencies on the two walls, is shown in Figure 6. Here one wall carried a stationary grating of period 5 cm, while the other carried a grating of period 10 cm moving at a speed of 26 cm/s. Again, it is evident that the data agree reasonably well with the theoretically expected relationship between D and V0, calculated on the assumption that the flying bee balances the apparent angular speeds of the gratings on the two eyes, irrespective of their spatial structure.

A comparison of the apparent angular speeds seen by the two eyes, for each of the sets of flights analyzed in Figures 5 and 6, is shown in Figure 7 a and b, respectively. For flights in which both gratings appear to move in the front-to-back direction over the eyes (filled circles, Figure 7 a and b), the apparent angular speeds range from circa 200 to circa 800 deg/s, depending on the bee's flight speed and position along the width of the tunnel. It is evident that, in these flights, the bees tend to balance the apparent angular speeds on the two eyes, rarely allowing them to differ by more than 50%.

Until now we have only considered situations in which each grating appears to move in the front-to-back direction over the corresponding eye. Let us now consider the case in which one grating appears to move in the front-to-back direction, while the other appears to move in the back-to-front direction. This situation — which would rarely be encountered during translatory flight in the real world — occurs when a bee flies in the same direction as the moving grating, but at a lower speed. In Figures 5 and 6, this corresponds to the domain of bee velocities for which $V1 < V0 < 0$ (see the center panel above the graph in Figure 5). Not many data points are available in this domain of bee velocities, because bees moving in the same direction as the grating tend to "catch up" with it, as indicated by the clustering of data points near $V0 = V1$ in Figures 5 and 6 (see arrow on the abscissa). Nevertheless, there are several data points pertaining to this condition that are in good agreement with the theoretical prediction (continuous curve) based on the assumption of equal apparent angular speeds on the two eyes. The apparent angular speeds experienced by the two eyes under these conditions are depicted by the open circles in Figure 7 a and b. Some, but not all of the points fall close to the 45-degree line, confirming that there is a tendency to balance apparent angular speeds even under these unnatural conditions. The departure of some of the data points from the 45-degree line (see Figure 7) may well be due to geometrical constraints imposed by the tunnel. For example, considering the data of Figures 5 and 7a, a bee flying in the same direction as the moving grating and at the same speed (38 cm/s) would experience zero angular speed on the eye viewing that grating, but it cannot experience an angular speed of less than 180 deg/s on the eye viewing the other (stationary) grating, owing to the finite width of the tunnel (12 cm).

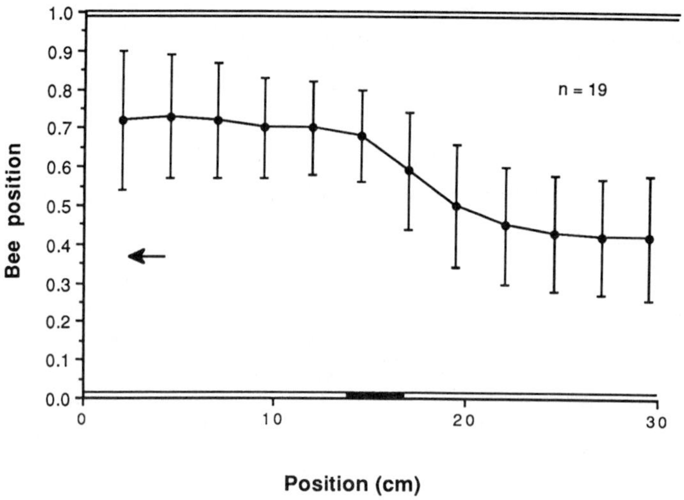

FIGURE 8. Experiment designed to explore the region of the visual field that is involved in monitoring apparent angular speed. The figure shows the trajectory (depicted as the mean and standard deviation of 19 flights) when one wall is white and featureless, while the other carries a single, vertical dark stripe halfway along the tunnel, as shown. There is a sharp kink in the trajectory as the stripe passes the lateral field of view of the eye, suggesting that angular speed is monitored in a direction approximately at right angles to that of flight. (Reproduced from Srinivasan, M. V. et al. (1991), with permission of authors and Cambridge University Press, NY.)

In summary, the above findings indicate that bees flying through the tunnel tend to balance apparent image speeds even when the motions are oppositely directed on the two eyes. In other words, it is possible that the underlying motion-detecting mechanism is insensitive to the direction of movement.

C. WHAT REGION OF THE EYE IS INVOLVED IN MEASURING APPARENT ANGULAR SPEED?

In calculating the theoretical curves for Figures 5 and 6, we have assumed that the bee compares the apparent angular speeds in the lateral regions of the two eyes, i.e., along directions at right angles to the tunnel axis, where the velocity is a maximum. Is this really the case, or do the bees compare average velocities measured over a substantial fraction of the visual field of each eye? To investigate this question, one can analyze flight trajectories of bees when one wall is uniformly white, while the other wall carries a single, dark vertical stripe placed halfway along its length. The results show that the bees deflect sharply away from the stripe, but only when it approaches the lateral direction of view of the eye (Figure 8). Many flights, not included in the analysis shown, actually involve more extreme deflections, ending in crashes against the blank wall and termination of flight. It is clear that the response induced by the stripe is not one of fixation, but, rather, avoidance. The shape of the trajectories suggest that angular speed is monitored chiefly by the lateral fields of view of the eyes, along directions approximately at right angles to that of flight. This concurs with Lehrer's (1990) recent observation that the lateral eye regions are the ones most sensitive to drift of the retinal image. An alternative interpretation of these trajectories would be that the centering response is associated with a fairly high threshold of apparent angular speed that is exceeded only in the lateral fields of view. This possibility, however, is unlikely since the response is driven by apparent angular speeds that span a factor of ten (Figure 7) and is detectable even in walking bees, where these speeds can be as low as 40 deg/s (data not shown).

D. DEPENDENCE OF THE CENTERING RESPONSE ON THE INTENSITY PROFILE OF THE PATTERN

All of the patterns used in the experiments described so far consisted of black and white stripes. These patterns are characterized by regions of uniform intensity, "black" or "white", separated by boundaries where the intensity changes abruptly from one level to the other. Since it is the boundaries, and not the regions of uniform intensity, that provide information on movement, it is of interest to ask whether the two types of boundary, light-to-dark and dark-to-light, are used equally effectively by the bee's visual system to measure apparent angular speed. To examine this question, one can use gratings whose spatial intensity-profile is an asymmetrical sawtooth, as illustrated in Figure 9a. With identical, stationary sawtooth gratings on the two walls, bees always fly close to the axis of the tunnel, regardless of whether the gratings are oriented such that they offer sharp dark-to-light boundaries to both eyes of the flying bee (left panel, Figure 9a) or sharp light-to-dark boundaries on one eye and sharp dark-to-light boundaries on the other (center and right panels, Figure 9a). This finding indicates that sharp boundaries of a single polarity (either one) are sufficient to provide velocity information. However, it does not reveal whether bees are also capable of extracting this information from areas of the pattern where the intensity varies gradually, as in the ramp regions of the sawtooth profiles.

To investigate the latter question, one can analyze the trajectories of bees flying along the tunnel when both walls are lined with sinusoidal gratings (Figure 9b). Again, bees fly close to the axis of the tunnel, even when the spatial periods of the gratings on the two walls are unequal. It is only when the period of one of the gratings is reduced to 1.5 cm (corresponding to an angular period of 14 deg for a bee flying along the axis), that there is a slight but statistically significant tendency for the bees to fly closer to that grating ($p < 0.0005$). In each of the cases shown in Figure 9, motion of a grating causes the bees' flight trajectories to be displaced either toward or away from that grating, depending on the direction of its motion in relation to that of the bee, as in the other experiments (data not shown). This set of results indicates that the bee's visual system can compute the apparent angular speed of a pattern accurately even when there are no sharp boundaries, and irrespective of the slopes of the intensity gradients in the pattern.

E. DEPENDENCE OF THE CENTERING RESPONSE ON THE CONTRAST OF THE PATTERN

The dependence of motion perception on pattern contrast can be investigated by analyzing the trajectories of bees flying through the tunnel while one wall carries a stationary sinusoidal grating of a fixed, relatively high contrast, whereas the other carries a stationary sinusoidal grating of the same period, but whose contrast is varied from one experiment to the other. The results of experiments using two different combinations of contrasts are shown in Figure 9c. The bees tend to fly close to the axis of the tunnel even when the contrasts of the gratings on the two walls are as disparate as 0.60 and 0.15 (Figure 9c). It is only when the contrast of the pattern on one of the walls is very low (<0.05) that the bees' trajectories begin to depart significantly from the axis of the tunnel. In fact, in conducting experiments in which one wall carries a high-contrast pattern and the other a zero-contrast pattern (i.e., a uniformly gray sheet) one finds that considerable care has to be taken in producing and selecting the zero-contrast pattern because even small flaws, originating from spurious residual contrasts, are evidently sufficient to cause the flight trajectories to be located close to the tunnel axis. These findings indicate that the bee's perception of apparent angular speed is accurate even at very low contrast and that the visual system is able to extract the apparent angular speed of a pattern largely independently of its contrast.

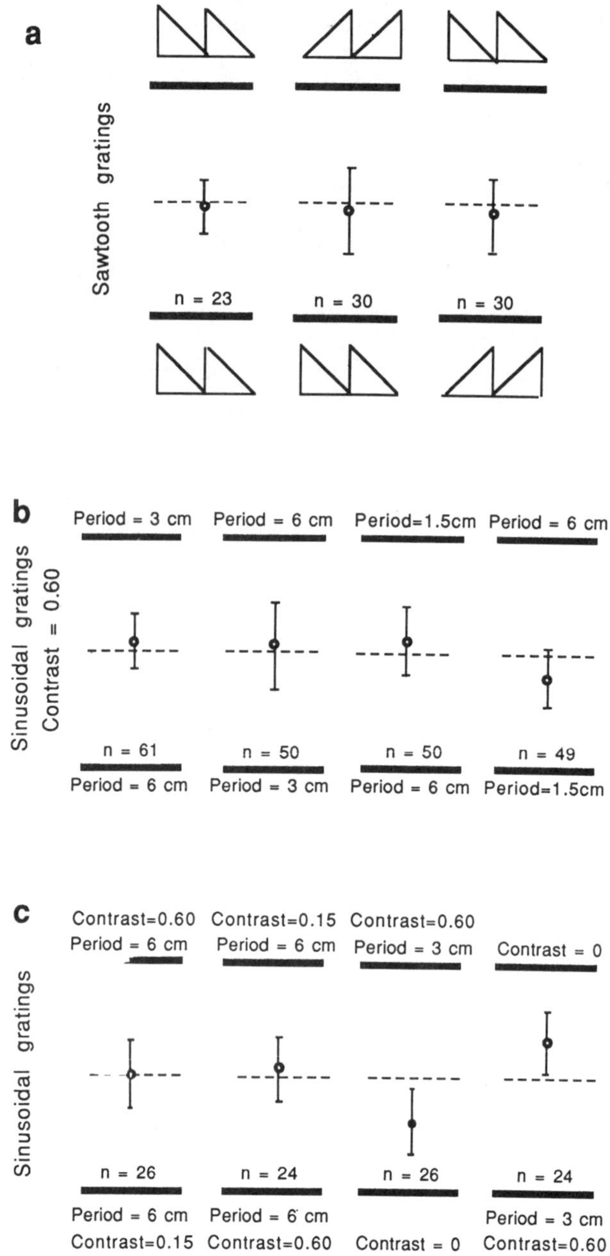

FIGURE 9. Positions of trajectories (mean and standard deviation) of bees flying rightward through the tunnel when the two walls present stationary patterns carrying a variety of intensity profiles and contrasts. (a) Sawtooth-profile gratings, with the abrupt intensity edges facing the same or opposite directions on the two walls; (b) sinusoidal gratings of different periods on the two walls; (c) sinusoidal gratings of equal periods, but different contrasts on the two walls. Bees fly close to the tunnel axis in all cases, except when the contrast of the grating on one side is reduced to zero. n = number of flights, in the direction of the arrow, analyzed in each case. (Reproduced from Srinivasan, M. V. et al. (1991), with permission of authors and Cambridge University Press, NY.)

IV. DISCUSSION

This study, investigating the tendency of freely flying honeybees to fly along the axis of a tunnel, has demonstrated that flying honeybees balance the distances to the two walls by equalizing the apparent angular speeds on the two eyes. Whether image speed is translated into a true percept of range is a question that is not answered by the present investigation. There is, however, much circumstantial evidence to suggest that image speed is perceptually linked to object range in bees and other insects. Flying bees do not go near rapidly moving objects (see Srinivasan and Lehrer, 1984a, and also below), suggesting that rapid image movement is interpreted as being caused by an object that is dangerously close. It has been shown that freely flying bees can be trained to distinguish between objects at different ranges and that this discrimination is most likely mediated by monitoring image speed (Lehrer et al., 1988; Srinivasan et al., 1989). Locusts peer before they jump onto a nearby object and are able to use the apparent motion of the object's image to estimate the object's range and adjust the size of their jump accordingly (Wallace, 1959; Sobel, 1990a,b).

A. THE CENTERING RESPONSE VS. THE OPTOMOTOR RESPONSE

Several lines of evidence suggest that the centering response is mediated by motion-detecting mechanisms that are distinct from those that mediate the optomotor response in insects (reviewed by Reichardt, 1987, and Borst and Egelhaaf, 1989):

1. The response is not one of yaw (i.e., rotation about a vertical axis); rather, it is one in which the flying insect shifts its position in the tunnel laterally. This is seen most clearly in the trajectories shown in Figures 2 d and e and 8, where one of the walls carries no pattern. If these trajectories were governed by a classical yaw-controlling optomotor response, we would expect them to curve toward the wall carrying the pattern; what we find, instead, is that the trajectories are consistently displaced away from that wall. von Buddenbrock and Moller-Racke (1952) investigated the behavior of flies walking toward a light source along a tunnel lined with gratings on the side walls. When one wall carried a vertical grating and the other wall was blank, the flies turned toward the patterned wall, exhibiting a distinctly optomotor-like behavior. This is in contradiction to our findings with flying honeybees. Interestingly, however, we have observed that naive bees, just learning to enter the tunnel and find the reward, occasionally show a tendency to turn toward the wall carrying the pattern. It is possible that these bees are exhibiting an optomotor response, just like the flies, which is suppressed, as training proceeds, to reveal another visual motor mechanism.
2. In many insects, including bees and flies, it is well established that the optomotor response elicited by a moving grating depends primarily on the contrast frequency induced by the grating, and not on its apparent angular speed (reviewed by Reichardt, 1987; and Borst and Egelhaaf, 1989). A contrast-frequency-dependent response has also been observed in studies of motion detection and ocular tracking in man (e.g., Kelly, 1979; Burr and Ross, 1982; Gellman et al., 1990). Behavioral measurements of the optomotor response of the bee (Kunze, 1961) and measurements of the response of large-field motion-sensitive descending neurons in the bee (Ibbotson and Goodman, 1990) reveal a bell-shaped response curve with a peak at a relatively low contrast frequency of 10 Hz and a half-width of approximately 10 Hz. If the centering response were driven by the same motion-sensitive mechanisms as the optomotor response, then the data of Kunze (1961) and of Ibbotson and Goodman (1990) would predict that, in the tunnel, stationary gratings inducing contrast frequencies of, say, 2 Hz on one eye and 8 Hz on the other would produce substantially different motion signals in the

two eyes, impairing the operation of the centering response. An analysis of the contrast frequencies experienced by bees flying through the tunnel under the various experimental conditions indicates, however, that, contrary to this expectation, the centering response operates perfectly even when the contrast frequencies experienced by the two eyes are very disparate (details in Srinivasan et al., 1991). This demonstrates that the visual mechanisms mediating the centering response do not confound angular speed with contrast frequency. All of the indications are that the centering response is driven by the apparent angular speed of the moving pattern and is largely independent of the spatial structure of the pattern or the contrast frequencies that it induces, provided that the pattern carries a few vertically oriented features that can be used to measure angular speed in the horizontal plane. The visual mechanisms mediating the centering response must be capable of encoding unambiguously angular speeds ranging from 100 to 800 deg/s, even if the contrast frequencies on the two sides differ by a factor of 10 (Srinivasan et al., 1991). This is true for a domain of contrast frequencies ranging from a fraction of a cycle per second to at least 30 Hz. David (1982) has encountered mechanisms of movement perception that are qualitatively similar. He finds that fruitflies flying in a wind tunnel regulate their ground speed according to the apparent angular speed of the environment, which they are able to measure accurately irrespective of its spatial-frequency content. Land and Collett (1974) inferred the presence of neural mechanisms sensitive to angular velocity in their study of chasing behavior in flies.

3. Further evidence that contrast frequency does not play an important role in the centering response comes from the following consideration. We have already seen that motion of the pattern on one of the walls causes the bees' trajectories to shift away from the axis of the tunnel (Figures 2, 4, 5, 6). This shift alters the apparent angular speeds on the two eyes, but not the contrast frequencies: the latter depend only on the speeds of the bee and the pattern, and are independent of the bee's position along the width of the tunnel. It, therefore, follows that the centering response cannot be mediated by monitoring the contrast frequencies on the two eyes, because these do not change when the bee shifts its lateral position.
4. There is some evidence (as already described in Section III) to suggest that the centering response is mediated by movement-responsive mechanisms that are insensitive to the direction of apparent motion as seen by the eye. That is, apparent motion in the front-to-back direction is not distinguished from apparent motion in the back-to-front direction at the same angular speed. This is different from the optomotor response, which is clearly direction sensitive, and which needs to be so if it is to fulfill its role of visual stabilization. This difference is, in itself, not compelling evidence against the participation of optomotor-like mechanisms in the centering response because a direction-insensitive response can readily be generated by summing the outputs of a pair of directionally sensitive movement detectors with opposite preferred directions. However, it is interesting in the light of Sobel's (1990a,b) recent study of range estimation by locusts, which provides compelling experimental evidence that peering locusts use image speed, and not velocity, to estimate object range. They appear not to distinguish between motions of images in and against the direction of motion of the head, as long as the image speeds are identical. It may well be that direction-insensitive mechanisms of movement detection offer a simple, as yet undiscovered means of measuring apparent angular speed independently of image structure, which direction-sensitive mechanisms do not.

It is likely that the centering response reflects an innate behavior that many locomoting insects use to maneuver through narrow gaps. To execute such maneuvers successfully, it

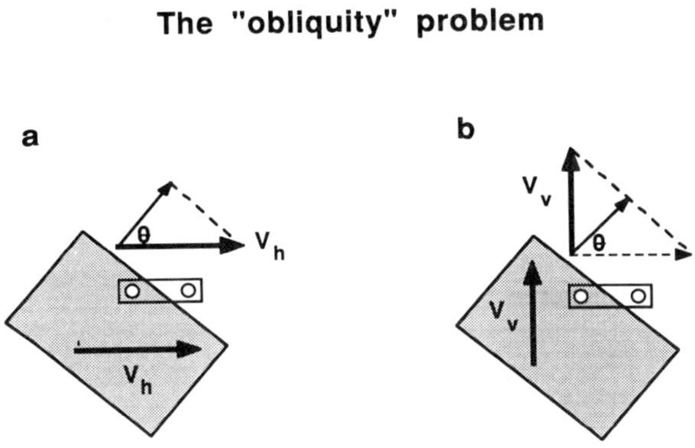

FIGURE 10. Illustration of the "obliquity" problem. The figure compares the responses evoked by a moving, oblique edge, in a nondirectional speed detector and in a one-dimensional speed detector with a horizontally oriented axis (horizontal rectangular box). (a) When the edge moves at a physical velocity of V_h in the horizontal direction (heavy arrow), the one-dimensional detector will register the correct speed (V_h), while the nondirectional detector will register the speed perpendicular to the edge, $V_h \cdot \cos\theta$ (thin arrow); (b) when the edge moves at a physical velocity of V_v in the vertical direction (heavy arrow), the one-dimensional detector will register a "cross-talk" of $V_v \cdot \tan\theta$ (dashed arrow), which can become very large as $\theta \rightarrow 90$ deg. The nondirectional detector, on the other hand, will register the speed perpendicular to the edge, $V_v \cdot \sin\theta$ (thin arrow), which can never be greater than V_v. Thus, as the orientation of the edge approaches the horizontal, the one-dimensional detector is very much more susceptible than the nondirectional detector to cross-talk from the vertical component of motion.

would seem desirable to use a motion-perception system that is sensitive to angular speed, rather than contrast frequency or spatial structure. Such a system would enable an insect to fly through the middle of a gap between two branches, for example, even if the textures of the bark on the two branches are substantially different.

B. DIRECTIONAL MOTION DETECTION AND THE "OBLIQUITY PROBLEM"

As we have already noted, there is evidence that range is estimated from the outputs of motion detectors which register apparent angular speed, but which are largely insensitive to the direction of motion of the image. Considering that the optical flow during straight-ahead motion consists mainly of horizontal motion from front to back, why do insects not use one-dimensional velocity detectors with this preferred direction? The answer may lie in the problem of "obliquity" that can occur when one-dimensional detectors are used to measure the speed of a two-dimensional image. This is illustrated in Figure 10, where we compare the responses evoked by an oblique edge, moving at a physical velocity of V_h in the horizontal direction, in a nondirectional speed detector and in a one-dimensional speed detctor with a horizontally oriented axis (shown as the horizontal rectangular box). It is clear from Figure 10a that the one-dimensional detector will register the correct speed, while the nondirectional speed detector will underestimate the speed because it senses only the component of motion perpendicular to the edge (this is the so-called "aperture problem"; reviewed by Hildreth and Koch, 1987). Thus, at first sight, it might appear that the one-dimensional detector performs more accurately than the nondirectional detector. However, the situation is very different when we consider the effect of noise in the motion of the edge, that is, when a small component of vertical motion, V_v, is added to the horizontal motion. In reality, noise of this kind is inevitable: it can be produced by fluctuations in flight altitude, for example, or by a slight roll of the head. Introduction of this vertical component would cause the one-dimensional detector to register an additional horizontal component of speed, as shown in Figure 10b. This "cross-talk" can assume very large values as the orientation of the edge approaches the horizontal, because the time taken for the edge to cross two neighboring points along the horizontal axis then becomes infinitesimally small. As the orientation of the edge crosses horizontal, the apparent velocity reverses polarity abruptly and now assumes very large values in the opposite direction. Errors arising from the obliquity problem could be reduced by low-pass filtering the image severely along the axis perpendicular to the detector's axis (in this case, along the vertical); however, this would reduce the spatial resolution of speed along that axis.

The nondirectional detector, on the other hand, would be relatively immune to the obliquity problem. Introduction of V_v would cause the detector to register some change in speed (Figure 10b), but the magnitude of this change can never be greater than V_v.

It is important to note that the obliquity problem is different from the "aperture" problem: it arises because of the one-dimensionality of the motion detector, and not because of the finite viewing aperture. The obliquity problem would be experienced only by a one-dimensional detector that measures speed; a detector whose response depends on contrast frequency (such as the insect optomotor neuron) would not be affected in this way (see van Hateren, 1990). The obliquity problem can be avoided by the use of a two-dimensional velocity detector (e.g., Horn and Schunck, 1981; Nagel, 1987; Srinivasan, 1990); however, the neural circuitry required to realize such schemes can be complex.

The susceptibility of the one-dimensional detector to the obliquity problem may be one reason why the measurement of range from optical flow involves nondirectional, rather than one-dimensional mechanisms for measuring speed. Given that most natural objects contain textures with variously oriented edges, it should be possible to measure apparent speed quite accurately by monitoring the maximum or the average value of the outputs of a group of nondirectional speed detectors.

From the above considerations we may predict that, under real (i.e., noisy) conditions, as the orientation of the edge approaches the direction of motion of the eye, the apparent speed of the edge would be underestimated by a nondirectional detector and overestimated by a one-dimensional detector. Bees flying through a tunnel lined with horizontal stripes on one wall and vertical stripes on the other tend to fly very close to the wall carrying the horizontal stripes (Figure 3a), suggesting that the speed of the horizontal stripes is underestimated severely and that the measurement of speed is therefore, indeed, nondirectional. Whether peering locusts over- or underestimate the range of an oblique stripe, however, remains to be examined.

C. RELATIONSHIP OF THE CENTERING RESPONSE TO THE MOVEMENT AVOIDANCE RESPONSE

It is known that bees, trained to collect a reward at a grating, find it difficult to approach the grating when it is in motion (Srinivasan and Lehrer, 1984a). It has also been shown previously that bees avoid movement, but not flicker (Srinivasan and Lehrer, 1984b, 1985). Thus, the neural mechanisms underlying this "movement avoidance response" (MAR) must respond to true movement and not to mere modulations of intensity in time. The MAR operates over a wide domain of contrast frequencies ranging from a few to 200 Hz (Srinivasan and Lehrer, 1984a). It is conceivable that the centering response of bees flying through the tunnel in the present experiments is a consequence of balanced, oppositely directed MARs elicited by gratings on the two sides. At high contrast frequencies (60 to 200 Hz) the MAR is dependent on contrast frequency, rather than angular speed. At lower contrast frequencies (0 to 20 Hz), however, the MAR appears to depend on angular speed rather than contrast frequency (see Figure 5, Srinivasan and Lehrer, 1984a), a property which meshes well with that of the centering response.

D. NEURAL MECHANISMS FOR THE MEASUREMENT OF ANGULAR SPEED

We are yet to discover the neurons and processing mechanisms that are involved in the measurement of the apparent angular speed of the image. The so-called "optomotor" neurons are the most widely studied movement-sensitive neurons in the insect optic lobe (reviewed by Hausen and Egelhaaf, 1989). The response of these neurons to the transient movement of spots or bars is approximately proportional to angular speed, and roughly independent of the size or shape of the object (Collett and King, 1975; Olberg, 1981; but see Horridge and Marcelja, 1990). However, the steady-state response to a moving, periodic grating depends strongly on the spatial structure of the grating (e.g., spatial frequency and contrast) and the temporal frequency that it induces, rather than on its angular velocity (Eckert, 1980; Borst and Egelhaaf, 1989, review; Horridge and Marcelja, 1990). What is needed are neurons, with input field sizes of around 10 deg, that measure local image speed independently of the structure or contrast of the image, over an angular-speed range of 100 to 800 deg/s, and a contrast-frequency range of 1 to 30 Hz. To this date, there appear to be no published accounts of neurons satisfying all of these requirements.

A simple neural scheme for the measurement of local image speed is suggested in Figure 11. The moving image (level A) is first converted to a binary image composed of two levels ("black" and "white"), by an array of neurons which possess high sensitivity to contrast and saturate at low contrasts. The binary neural image (level B), which moves at the same velocity as the original image, is then spatially low-pass-filtered by a subsequent array of neurons, resulting in a moving neural image (level C) in which the abrupt edges of the binary image have been converted to ramps of constant slope. The speed of the image can then be monitored by measuring the rate of change of response at the ramps. Accordingly, the neural image at this level is temporally differentiated by an array of phasically responding neurons, giving a moving neural image composed of a train of pulses (level D), one located at each edge of the binary image. The amplitude of each pulse will then be proportional to the rate of change of intensity at the corresponding ramp and, therefore, to the instantaneous speed of the image at that location. A subsequent stage of rectification ensures that the response is positive, regardless of the polarity of the edge or the direction of movement (level E). Thus, we have a scheme which measures the local speed of the image, independently of structure, contrast, or direction of movement. A subsequent stage of lateral inhibition (not shown) would ensure that the system responds only to movement and not to flicker.

The scheme proposed here is, in effect, a modified version of the well-known gradient scheme which computes velocity as the ratio of the local temporal gradient to the local spatial gradient (Marr and Ullman, 1981; Horn and Schunck, 1981; reviewed by Buchner, 1984; Bouzerdoum and Pinter, 1989; Fei Jin and Srinivasan, 1990; Srinivasan, 1990). However, the present scheme has the advantage that it avoids the mathematical operation of division (and the attendant problems at low gradients, where the computation approaches the indeterminate value of zero/zero). It achieves this by parsing the image into edges and "standardizing" the spatial gradient to a constant, finite value at each edge.

The results of a quantitative simulation of the responses of such a scheme are shown in Figure 12. It is assumed that at the front end of the system (i.e., between A and B in Figure 11) the signals are transmitted linearly for contrasts ranging from 0 to 0.10. Hard saturation is assumed to occur at higher contrasts. The subsequent low-pass filtering stage (between B and C in Figure 11) is assumed to be performed by neurons with Gaussian receptive-field profiles, with an angular half-width of 10 degrees. The temporal differentiator following this stage (between C and D in Figure 11) is assumed to have an impulse response similar to that of a lamina monopolar cell (LMC). Following Laughlin et al. (1987) and Fei Jin and Srinivasan (1990), the impulse response h(t) of the LMC is modeled by the function

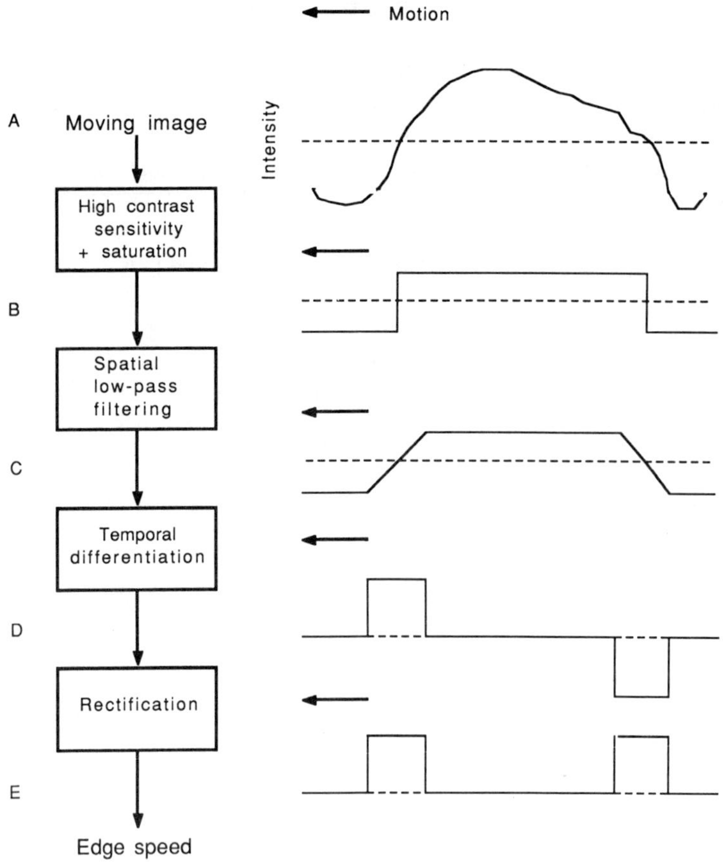

FIGURE 11. A simple neural scheme proposed for measurement of the angular speed of an image, independently of its structure or contrast. The moving image (A), whose spatial intensity profile at a given instant is as shown (top right), is converted to a moving binary neural image (B). The binary image is then spatially low-pass-filtered to yield a moving neural image in which the abrupt edges have been converted to ramps of constant spatial slope (C). This neural image is then differentiated in time by an array of phasically responding neurons to yield a neural image composed of pulses (D). The amplitudes of these pulses encode the local angular speed of the image. A subsequent stage of rectification ensures that angular speed is encoded independently of edge polarity or direction of motion. (Reproduced from Srinivasan, M. V. et al. (1991), with permission of authors and Cambridge University Press, NY.)

$$h(t) = a_d.\exp(-t/t_d) - \exp(-t/t_h)$$
$$\text{with } a_d = 0.2,\ t_d = 0.003 \text{ s, and } t_h = 0.0006 \text{ s}$$

The variation of peak response with angular speed is shown in Figure 12a, for moving gratings of a number of different spatial periods. It is evident that the response depends primarily on the angular speed of the grating and not on its spatial period (or on its contrast frequency). The variation of the peak response with grating contrast is shown in Figure 12b for gratings of two different periods, each moving at two different angular speeds. It is evident that, at contrasts above saturation, the response depends primarily on angular speed and not on contrast or grating period. All of these properties are manifest because the scheme parses the moving pattern into its constituent edges, and the relatively small visual field of each neuron allows it to respond to a single edge.

The LMCs in the lamina ganglion of the insect visual system are highly sensitive to moving edges, giving a response whose amplitude increases monotonically with edge speed

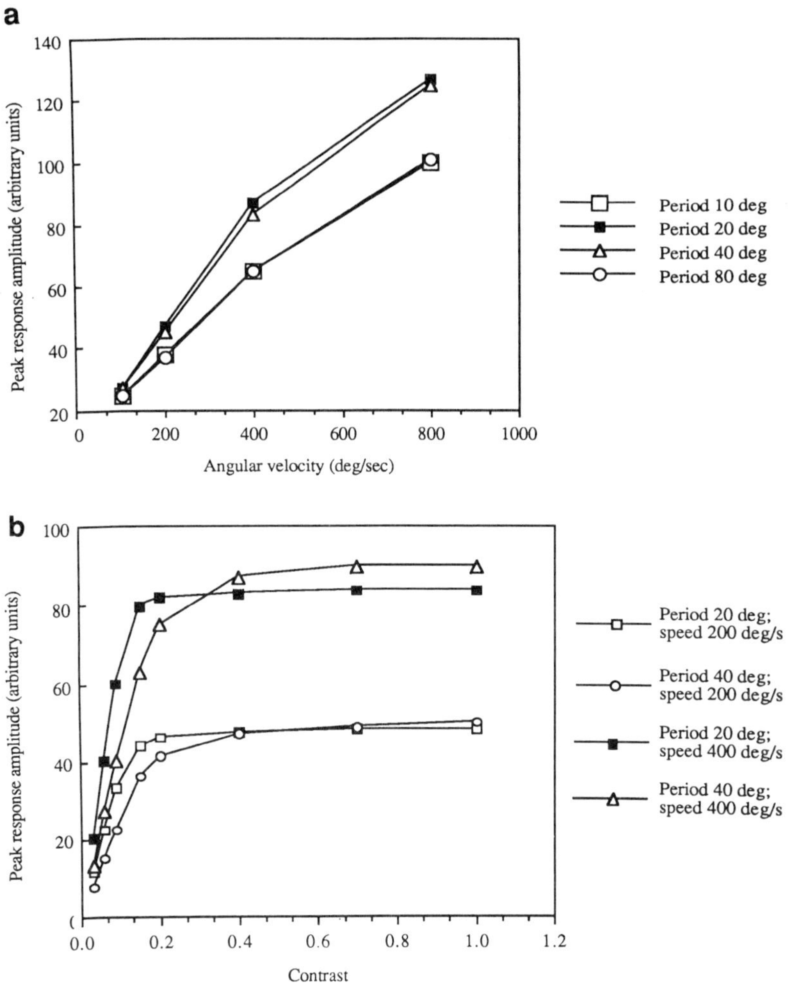

FIGURE 12. (a) Variation of response with angular speed for a simulated neural speed detector, for gratings of a number of different spatial periods. It is evident that the amplitude of the response depends primarily on the angular speed of the grating, and not on its spatial period. (b) Response-vs.-contrast curves for a simulated neural speed detector, for gratings of two different spatial periods moving at each of two different angular speeds. It is evident that, at contrasts higher than 0.1, the amplitude of the response depends primarily on the angular speed of the grating and is independent of its contrast or spatial period.

over a range of circa 0 to 1000 deg/s (Srinivasan et al., 1990) and with sinusoidal temporal frequency over a range of 0 to 30 Hz (Coombe et al., 1989). This property of the temporal frequency-response is mirrored closely by the MAR (Srinivasan and Lehrer, 1984a). However, the responses of the LMCs are not entirely independent of the contrast of the edge. Neurons with dynamic properties similar to those of the LMCs, but which are less sensitive to variations of edge contrast, have been encountered recently by Osorio (see chapter in this book) in the medulla of the locust. Further work is needed to determine how accurately these medullary neurons encode image speed and to ascertain whether they do, indeed, play a speed-encoding role.

ACKNOWLEDGMENTS

Much of the work reviewed in this chapter was done in collaboration with Wolfgang Kirchner of the Zoologisches Institut, University of Würzburg, Miriam Lehrer of the Zoologisches Institut, University of Zürich, and Zhang Shao-Wu of the Academia Sinica in Beijing, all of whom were Visiting Fellows at the Centre for Visual Sciences at the Research School of Biological Sciences of the Australian National University, Canberra. Kym Witney's asistance in digitizing and analyzing bee flights was invaluable, as was Alex Saeck's generosity in providing and maintaining the bee colonies.

REFERENCES

Borst, A. and Egelhaaf, M. (1989). Principles of visual motion detection, *Trends in Neurosci.*, **12**, 297–306.

Bouzerdoum, A. and Pinter, R. B. (1989). Image motion processing in biological and computer vision systems. In Visual Communications and Image Processing IV, *Proc. SPIE*, **1199**, 1229–1240.

Buchner, E. (1984). Behavioral analysis of spatial vision in insects, in *Photoreception and Vision in Invertebrates*, Ali, M. A., Ed., Plenum Press, New York, 561–621.

Burr, D. C. and Ross, J. (1982). Contrast sensitivity at high velocities, *Vision Res.*, **22**, 479–484.

von Buddenbrock, W. and Moller-Racke, L. (1952). Beitrag zum Lichtsinn der Fliege *Eristalomyia tenax, Zool. Anz.*, **149**, 51–61.

Cheng, K., Collett, T. S.,Pickhard, A., and Wehner, R. (1987). The use of visual landmarks by honeybees: bees weight landmarks according to their distance from the goal, *J. Comp. Physiol. A.*, **161**, 469–475.

Collett, T. S. (1978). Peering — a locust behaviour pattern for obtaining motion parallax information, *J. Exp. Biol.*, **76**, 237–241.

Collett, T. S. (1988). How ladybirds approach nearby stalks: a study of visual selectivity and attention, *J. Comp. Physiol. A*, **163**, 355–363.

Collett, T. S. and Harkness, L. I. K. (1982). Depth vision in animals, in *Analysis of Visual Behavior*, Ingle, D. J., Goodale, M. A., and Mansfield, R. J. W., Eds., M. I. T. Press, Cambridge, 111–176.

Collett, T. and King, A. J. (1975). Vision during flight, in *The Compound Eye and Vision in Insects*, Horridge, G. A., Ed., Clarendon Press, Oxford, 437–466.

Coombe, P. E., Srinivasan, M. V., and Guy, R. G. (1989). Are the large monopolar cells of the insect lamina on the optomotor pathway?, *J. Comp. Physiol. A.*, **166**, 23–35.

David, C. T. (1982). Compensation for height in the control of groundspeed by Drosophila in a new, 'Barber's Pole' wind tunnel, *J. Comp. Physiol.*, **147**, 485–493.

Eckert, H. (1980). Functional properties of the H1-neurone in the third optic ganglion of the blowfly, *Phaenicia, J. Comp. Physiol.*, **135**, 29–39.

Eriksson, E. S. (1980). Movement parallax and distance perception in the grasshopper Phaulacridium vittatum (Sjostedt), *J. Exp. Biol.*, **86**, 337–340.

Fei Jin, Z. and Srinivasan, M. V. (1990). Neural gradient models for the measurement of image velocity, *Visual Neurosci.*, **5**, 261–271.

Gellman, R. S., Carl, J. R., and Miles, F. A. (1990). Short latency ocular following responses in man, *Visual Neurosci.*, **5**, 107–122.

Goulet, M., Campan, R., and Lambin, M. (1981). The visual perception of relative distances in the wood-cricket, *Nemobius sylvestris, Physiol. Entomol.*, **6**, 357–367.

van Hateren, J. H. (1990). Directional tuning curves, elementary movement detectors, and the estimation of the direction of visual movement, *Vision Res.*, **4**, 603–614.

Hausen, K. and Egelhaaf, M. (1989). Neural mechanisms of visual course control in insects, in *Facets of Vision*, Stavenga, D. C. and Hardie, R. C., Eds., Springer-Verlag, Berlin, 391–424.

Hildreth, E. C. and Koch, C. (1987). The analysis of visual motion: from computational theory to neuronal mechanisms, *Ann. Rev. Neurosci.*, **10,** 477–533.

Horn, B. K. P. and Schunck, B. (1981). Determining optical flow, *Artif. Intell*, **17**, 185–203.

Horridge, G. A. (1986). A theory of insect vision: velocity parallax, *Proc. R. Soc. London Ser. B*, **229**, 13–27.

Horridge, G. A. and Marcelja, L. (1990). Responses of the H1 neuron of the fly to contrast and moving bars, *Philos. Trans. R. Soc. London Ser. B*, **329**, 75–80.

Ibbotson, M. R. and Goodman, L. J. (1990). Response characteristics of four wide-field motion-sensitive descending interneurones in *Apis mellifera, J. Exp. Biol.*, **148**, 255–279.

Kelly, D. H. (1979). Motion and vision. II. Stabilized spatio-temporal threshold surface, *J. Opt. Soc. Am.*, **69**, 1340–1349.

Kirchner, W. and Srinivasan, M. V. (1989). Freely flying honeybees use image motion to estimate object distance, *Naturwissenschaften*, **76**, 281–282.

Kunze, P. (1961). Untersuchung des Bewegungssehens fixiert fliegenden Bienen, *Z. Vgl. Physiol.*, **44**, 656–684.

Land, M. F. and Collett, T. S. (1974). Chasing behaviour of houseflies (Fannial canicularis), *J. Comp. Physiol.*, **89**, 331–357.

Laughlin, S. B., Howard, J., and Blakeslee, B. (1987). Synaptic limitations to contrast coding in the retina of blowfly *Calliphora, Proc. R. Soc. London Ser. B*, **231**, 437–467.

Lehrer, M. (1990). How bees use peripheral eye regions to localize a frontally-positioned target, *J. Comp. Physiol.*, **167**, 173–185.

Lehrer, M., Srinivasan, M. V., Zhang, S. W., and Horridge, G. A. (1988). Motion cues provide the bee's visual world with a third dimension, *Nature (London)*, **332**, 356–357.

Marr, D. and Ullman, S. (1981). Directional selectivity and its use in early visual processing, *Proc. R. Soc. London Ser. B*, **211**, 151–180.

Nagel, H.-H. (1987). On the estimation of optical flow: relations between different approaches and some new results, *Artif. Intell.*, **33**, 299–324.

Olberg, R. M. (1981). Object and self-movement detectors in the ventral nerve cord of the dragonfly, *J. Comp. Physiol.*, **141**, 327–334.

Reichardt, W. (1987). Evaluation of optical motion information by motion detectors, *J. Comp. Physiol.*, **161**,. 533–547.

Sobel, E. C. (1990a). Depth perception by motion parallax and paradoxical parallax in the locust, *Naturwissenschaften*, **77**, 241–243.

Sobel, E. C. (1990b). The locust's use of motion parallax to measure distance, *J. Comp. Physiol.*, **167**, 579–588.

Srinivasan, M. V. (1990). Generalized gradient schemes for the measurement of two-dimensional image motion, *Biol. Cybern.*, **63**, 421–431.

Srinivasan, M. V. and Lehrer, M. (1984a). Temporal acuity of honeybee vision: behavioural studies using moving stimuli, *J. Comp. Physiol. A*, **155**, 297–312.

Srinivasan, M. V. and Lehrer, M. (1984b). Temporal acuity of honeybee vision: behavioural studies using flickering stimuli, *Physiol. Entomol.*, **9**, 447–457.

Srinivasan, M. V. and Lehrer, M. (1985). Temporal resolution of colour vision in the honeybee, *J. Comp. Physiol.*, **157**, 579–586.

Srinivasan, M. V. and Lehrer, M. (1988). Spatial acuity of honeybee vision and its spectral properties, *J. Comp. Physiol. A*, **162**, 159–172.

Srinivasan, M. V., Lehrer, M., Zhang, S. W., and Horridge, G. A. (1989). How honeybees measure their distance from objects of unknown size, *J. Comp. Physiol. A*, **165**, 605–613.

Srinivasan, M. V., Lehrer, M., Kirchner, W., and Zhang, S. W. (1991). Range perception through apparent image speed in freely-flying honeybees, *Vision Neurosci.*, **6**, 519–535.

Srinivasan, M. V., Pinter, R. B., and Osorio, D. (1990). Matched filtering in the visual system of the fly: large monopolar cells of the lamina are optimized to detect moving edges and blobs, *Proc. R. Soc. London Ser. B*, **240**, 279–293.

Wallace, G. K. (1959). Visual scanning in the desert locust *Schistocerca gregaria*, Forskal, *J. Exp. Biol.*, **36**, 512–525.

Chapter 14

RETINOTOPIC VISION IN THE LOCUST

D. Osorio

TABLE OF CONTENTS

I. Introduction 378

II. Medulla Physiology 378
- A. Preparation and Recording 379
- B. Sustaining Cells 380
- C. Transient Cells 382
- D. Adaptation 384
- E. Spatial Interactions and Nondirectional Motion Detection 385

III. Medulla Cell Properties and the Organization of Retinotopic Visual Processing 386
- A. Transient Cell Properties and Function 387
- B. Comparison of Medulla Transient Cells with Retinal Ganglion Y Cells 388

IV. Concluding Remarks: The Role of Linear and Nonlinear Transformations in Early Visual Processing 389

References 389

I. INTRODUCTION

The visual systems of insects and vertebrates have evolved independently, yet both groups live in the same world and have comparable repertoires of visual behavior. Unfortunately, few records have been made from the cells that carry out retinotopic visual processing in insects, and so we have few points of comparison with the vertebrate visual ganglia. This paucity of records is especially unfortunate because the anatomy of the insect optic lobe has been well known and its regular structure admired for many years (Cajal and Sanchez, 1915; Strausfeld, 1976). The neural anatomy of the insect optic lobes is rather conservative (Strausfeld, 1976; Osorio, 1991b; Chapter 16, this volume), and so although most anatomical studies are from the fly, these are, in general, applicable to the locust, whose physiology is described here. The outer two of the insect's three (retinotopically ordered) optic ganglia, the lamina and the medulla, are columnar with a set of 40 to 50 morphologically distinct cells repeated behind each facet (ommatidium) in the compound eye and, hence, each sampling point in visual space (Figure 1; Strausfeld, 1976; Fischbach and Dittrich, 1989). The third optic ganglion, the lobula, is also columnar, but has a lower periodicity with one set of cells for every four ommatidia (Braitenberg, 1972). Beyond the lobula retinotopic visual processing is absent.

Whereas the nature of various intermediate "neural representations" of the retinal input has been a principal subject of vertebrate research (Marr, 1982; DeYoe and Van Essen, 1988), with few exceptions this aspect of insect visual processing has been overlooked. Instead the role of a given "identifiable" cell in controlling a particular behavior is a common focus of interest for invertebrate neurobiologists (Hoyle, 1977), an approach which may exaggerate the impression that insects are capable only of a limited repertoire of stereotyped reflexes, each controlled by its own dedicated neural circuitry. This article takes a different view of the insect visual system to look at some of the cells in the medulla of the optic lobe. The medulla, with its orderly structure and dense neuropil, plainly invites comparison with retinotopically arranged visual ganglia in vertebrates. Just as visual areas occupy much of the vertebrate brain, retinotopic vision appears to be an expensive process for insects; the medulla contains over half of the cells in the insect brain (Strausfeld, 1976).

The medulla contains cells with a variety of structures (Strausfeld, 1976), but the columnar neurons with fields 1 to 5 receptors across that form a core of afferent pathways are the subject of this work. Some properties of the neurons in the insect medulla are outlined, focusing on temporal properties and local spatiotemporal interactions within the receptive fields. Some of these findings are documented in more detail elsewhere (Osorio, 1991a). Spatial receptive fields are currently under investigation (along with more quantitative modeling of temporal responses) and are not described; however, all the cells mentioned here had receptive fields of about 2 to 10° across (1 to 7 receptors). To conclude, the neurophysiology and the organization of visual processing in the insect medulla are briefly compared with their vertebrate counterparts.

II. MEDULLA PHYSIOLOGY

Unsurprisingly, medulla cells have diverse responses. For example, some are directionally selective and others spectrally opponent (Osorio, 1986a,b); the majority, however, fall into two main physiological classes, as was noted by Honegger (1978, 1980), who following a classification of cat retinal ganglion cells (e.g., Levick and Thibos, 1983) called them "sustaining" and "transient". Like the sustaining and transient (or X and Y) classes of retinal ganglion cell, the insect's medulla cells are, respectively, quasilinear and highly nonlinear. The functional analogy between the insect and vertebrate ganglia implied by this

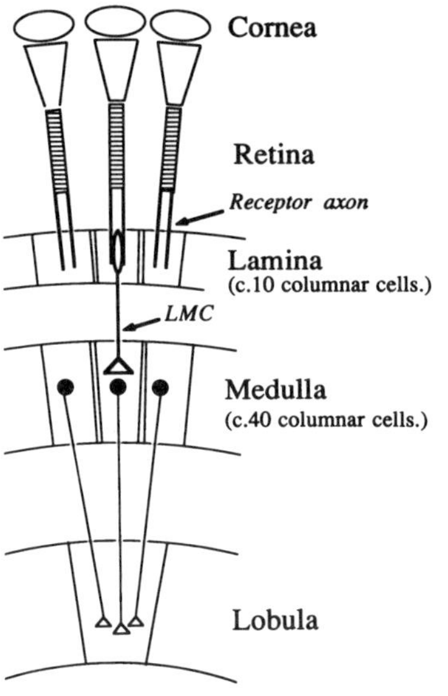

FIGURE 1. Diagram of the insect peripheral visual system, showing the optics, retina, and three main retinotopic ganglia of the optic lobe. A set of eight coaxial receptors lies behind each facet lens in the compound eye cornea (in the fly, the receptors are not coaxial, but each shares its visual field with one receptor in each of six neighboring facets [Kirschfeld, 1967]). Six coaxial receptors project to the lamina where they terminate on the large monopolar cells (LMCs). The LMCs, of a single lamina column together with two further photoreceptors (which in locust are UV sensitive), and (about) three second-order lamina cells terminate in a corresponding column of the medulla. Thus, each column of cells in the medulla receives its primary inputs from the light focused by a single facet lens and, hence, a single point in visual space. Many medulla cells end in the lobula, which has a coarser structure than the lamina and medulla. Beyond the lobula on the ''afferent'' visual pathway, there is no evidence for a columnar anatomy or retinotopic visual processing. Note that there is no constriction in the visual pathway equivalent to the vertebrate optic nerve. It should be emphasized that this diagram does not illustrate the full range of neural anatomies or likely processing pathways that may exist. For example, there are wide-field afferent and efferent cells at every level in the optic lobe, and many ''columnar cells'' have dendritic fields that extend beyond the confines of a single medulla column and so may receive inputs from several points. Finally, there are many amacrines with dendritic fields of varying extent. A good summary of optic lobe anatomy can be found in Strausfeld's recent (1989) article and references therein.

nomenclature remains speculative (see below), and to emphasize this we use the terms ''sustaining'' and ''transient'' for the insect medulla cells and X and Y for the two best characterized classes of (cat) retinal ganglion cell (Enroth-Cugell and Robson, 1966).

A. PREPARATION AND RECORDING

Responses in the medulla, here recorded intracellularly, are complex, especially those of the transient cells, and describing and modeling the underlying processing presents a formidable challenge. This article outlines responses to some simple stimuli that provide a basis for describing some aspects of medulla physiology, in particular local spatiotemporal interactions underlying the responses of the transient cells. Stimuli were produced by ultra-bright light emitting diodes (LEDs), used within the linear part of their operating range. These LEDs were arranged in a hexagonal pattern, with the distance between centers approximately equal to the interreceptor angle of the locust (1.5°) and centered in the visual field using a Cardan-arm perimeter.

Both (5 ms) pulse and step, and also sinusoidal stimuli were used. Compound sinusoids were used to measure contrast-response functions and to analyze local spatiotemporal interactions (Figure 2): two sinusoids whose frequencies differed by 1/64th or 1/128th of a

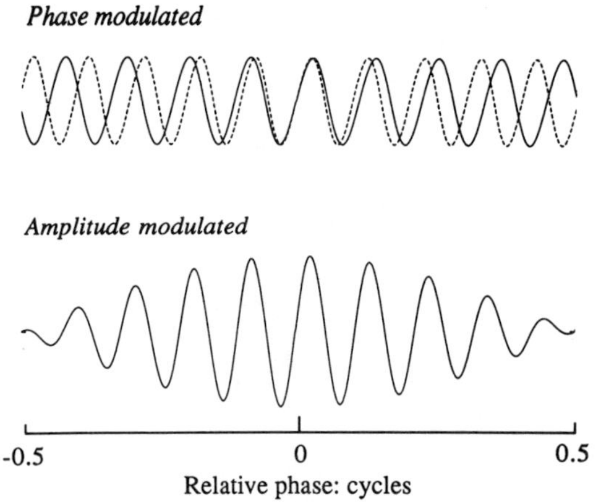

FIGURE 2. Diagram of the compound sinusoidal stimuli used. Two frequencies differing by a fraction of a period were used. If the signals are presented separately at two different positions, there is a shift in relative phases of the two signals over one beat period (above). Alternatively, the two sinusoids can be added, in which case the amplitude of the signal is modulated (below). In this diagram the two signals complete a full cycle of phase shifts after 9 and 10 cycles, whereas in our experiments the beat period was 64/65 or 128/129 cycles. With a long beat period, segments of stimulus approximate fragments of a pure sinusoid (or pair of sinusoids with a fixed phase shift).

period were either added together to give a signal which approximated a single frequency modulated in amplitude/contrast with a beat period of 64 or 128 cycles or, alternatively, the two different frequencies were presented at separate points in the receptive field. In the latter case the gradual phase-shifting between the two signals gave a scan of the time-phase dependent interactions between the outputs of neighboring receptors. If the receptive field is (locally) uniform (e.g., impulse responses have the same form) and the outputs of a pair of receptors are summed linearly, then the responses to these two types of sinusoidal stimuli will be the same. Thus, the former signal, which we call *amplitude modulated,* can be used as a control for spatial interactions underlying responses to the latter *phase modulated* signals. Responses to these stimuli were analyzed by dividing the response to the 64 or 128 cycles of a single beat period into 16 segments and measuring the frequency components in each segment by fast Fourier analysis. Corrections were made for distortions in the responses caused by the use of compound stimuli, and an estimate of the noise due to spontaneous activity was subtracted from the measured (rms power) response at each harmonic. Further details of the stimuli and methods of analysis are given elsewhere (Osorio, 1991a).

B. SUSTAINING CELLS

The sustaining cells of the medulla have a range of spatial, temporal, and spectral properties (Osorio, 1986a, 1987b), but their responses are essentially linear. In particular, the temporal tuning measured using sinusoids is close to that predicted by taking a Fourier transform of the impulse response, and integration within the receptive field is linear (Osorio, 1987b; unpublished observations).

One type of sustaining cell has temporal properties similar to those of the inputs to the medulla, the lamina large monopolar cells (LMCs; see James this volume), but inverted (Figure 3). When light adapted, both cells are phasic, being tuned to frequencies of about 15 Hz; when dark adapted they are tonic. The temporal tuning of the locust's LMCs and of these sustaining cells is surprisingly high, given that flicker fusion in the receptors occurs at about 60 Hz (Pinter, 1972; unpublished observations). By comparison, flies' receptors can transmit frequencies of over 200 Hz, but like the locust their lamina cells are tuned to

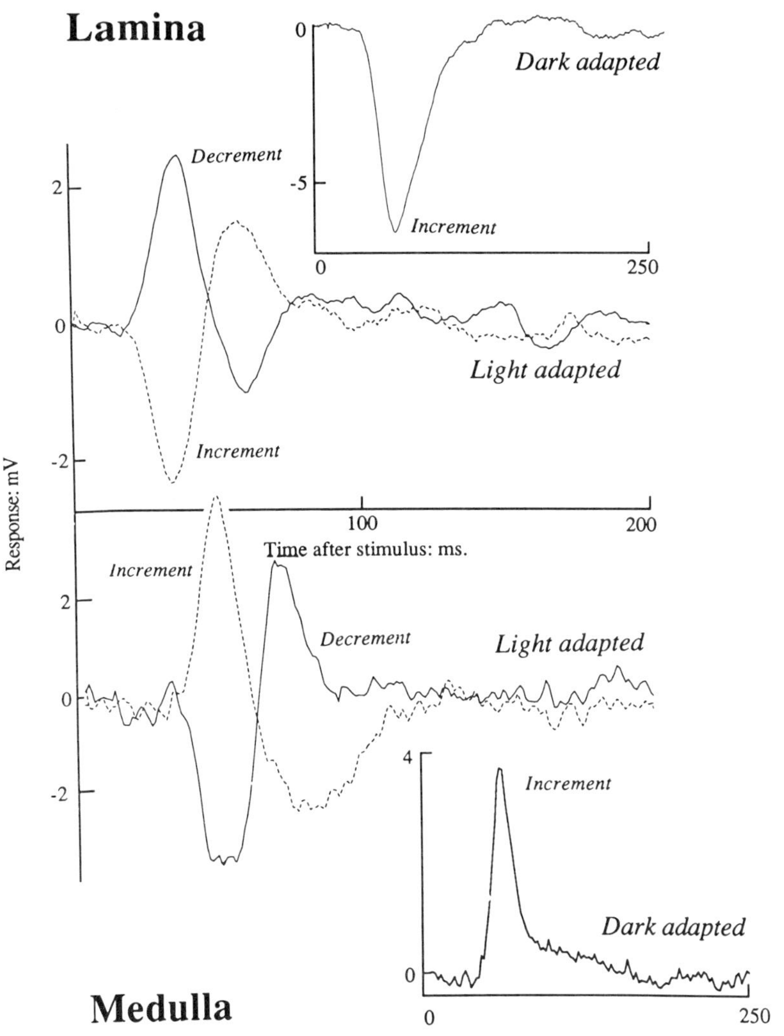

FIGURE 3. Signal-averaged (30 stimulus cycles) responses of quasilinear cells of the lamina and medulla to brief (5-ms) pulses. Main panels show light-adapted, and insets show dark-adapted, responses. The lamina unit is an input to the medulla, and the responses of the two cells are similar save for an inversion. As in other illustrations (except where stated), responses were digitized at 1 kHz. When light adapted, both cells are tuned to a frequency of around 15 Hz. (Above) Responses of a lamina large monopolar cell (LMC). While locust lamina cells have not been described in detail, they are similar to those of the fly (see James, this volume), giving graded potential hyperpolarizations to illumination. Responses are tonic when the cells are dark adapted and become much more phasic on light adaptation. The LMC receptive field is about 2° across with weak surround antagonism (James, this volume; unpublished observations). (Below) Responses of a medulla sustaining cell with a time course similar to that of the LMCs but with a larger (5°) receptive field. Unlike LMCs the medulla cells produce spikes (which were relatively small in this record).

frequencies of 10 to 20 Hz at the same light intensity(James, Chapter 2 this volume). Thus, temporal integration of receptor signals is significantly more pronounced behind the ''fast'' eye of the fly than the ''slow'' eye of the locust.

Apart from the change in tuning that occurs as a function of mean light intensity, the sustaining cells show some marked deviations from linearity at high contrasts (Figure 4). Most obviously, responses tend to saturate, particularly to hyperpolarizing stimuli. In addition, there is an indication that the saturation (of the first harmonic response component)

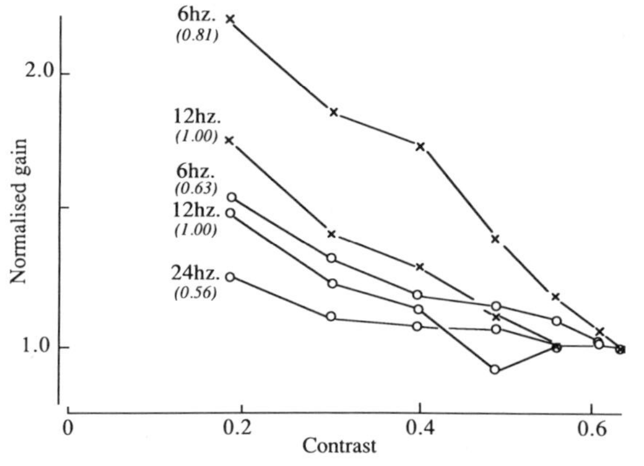

FIGURE 4. Contrast-response functions of two sustaining cells (o and x) of the kind shown in Figure 3, plotted as response divided by stimulus contrast and normalized to the response at the maximum contrast. A linear contrast-response function in each case would give a horizontal line, and a compressive nonlinearity is evident in the sustaining cells' responses. This compression may be more marked at low than at high frequencies, an effect that is not attributable to simple response saturation: in both cells the maximum response at 6 Hz is less than the response at 12 Hz, but the contrast response function shows a greater compressive nonlinearity. (The number in brackets beside each curve gives the amplitude of the maximum response normalized to the response at 12 Hz for the cell in question.) The values of stimulus contrast given may not accurately portray the contrast-response function of the cell, because sustaining cells show linear spatial summation and the stimuli used (three 1° diameter LEDs) occupied less than a quarter of the 5° diameter receptive fields. The functions were obtained using amplitude-modulated sinusoids, with responses to increasing and decreasing phases of the beat period averaged; there were no systematic differences between responses to the two phases.

is more marked at low than at high frequencies (in five cells studied including the two illustrated in Figure 4). This frequency dependence, which is not a simple consequence of saturation, will effectively elevate the frequency tuning of the sustaining cells to high contrast signals. These properties are reminiscent of, but smaller than, the contrast gain-control seen in cat retinal ganglion cells (Shapley and Victor, 1978).

C. TRANSIENT CELLS

Like the sustaining cells, transient cells are a diverse physiological class; for example, receptive fields vary from below 5° to over 20° across, and some cells respond to extended stimuli, whereas others do not. Nonetheless these cells share a number of features, indicating that they have common physiological properties and input pathways. It is the local processing on the pathways between photoreceptors and transient cells which are described here.

Transient cells characteristically give a graded potential response and one or two spikes at fixed latencies to any (suprathreshold) stimulus irrespective of its amplitude (Osorio, 1987a, 1991a). Responses to increments and to decrements and those to pulse and to step intensity changes are similar, although the sensitivity to stimuli of different polarities may differ (Figure 5a,b). The rectifying character of the cells is clearly seen in the graded potential components of responses to sinusoids which show a strong second harmonic component (Figures 6 and 9). Unfortunately there is no clear relationship between the amplitudes of the first and second harmonic components. Usually the amplitude (rms power) of the second harmonic is roughly proportional to that of the first harmonic (Figure 6a), but under some conditions response power appears to "drain" from the fundamental so that as contrast is increased the amplitude of this response component at first increases roughly linearly, but then drops sharply once a certain threshold is passed (Figure 6b). This behavior is roughly what might be expected if the unit were changing from half-wave to full-wave rectification,

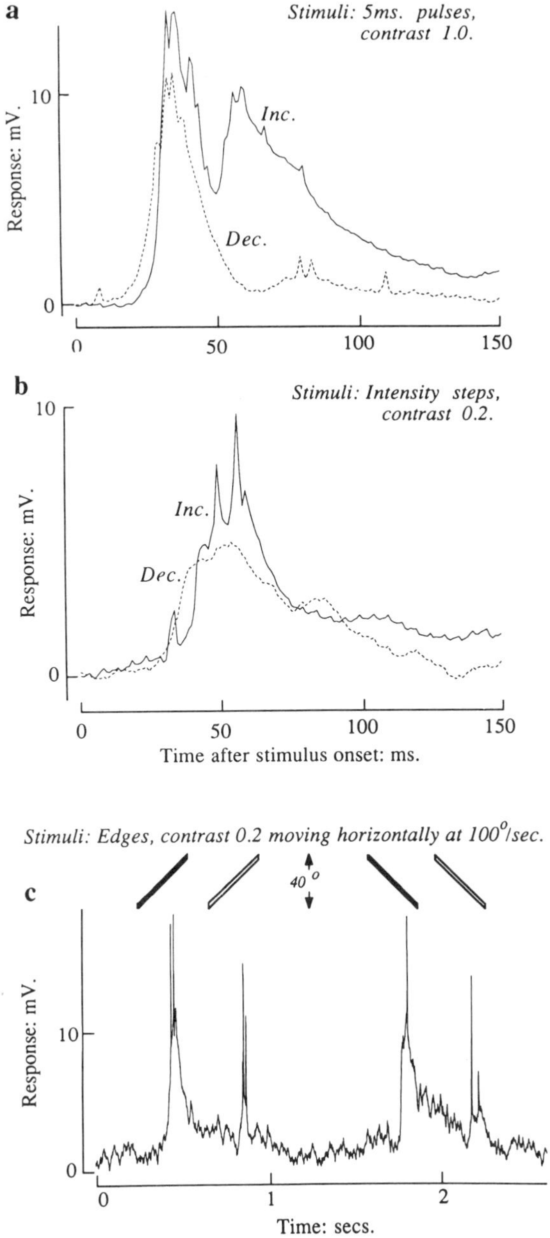

FIGURE 5. Signal-averaged responses of a transient cell to static (a,b) and to moving (c) stimuli. The response latency in this particular record was more variable than in others (Osorio, 1987a, 1991a). (a) Responses (20 stimulus cycles averaged) to brief pulses filling the receptive field. These stimuli were provided using a CRT display mean luminance circa- 500 cd.m^{-2}. The latencies of the responses to increments and decrements are similar, but there is a secondary excitatory input from the incremental pulse which is inhibitory to decrements. (b) Responses to step intensity changes. The response waveforms to step and to pulse intensity changes are quite similar. As with pulse stimuli the cell is more responsive to increases than to decreases in intensity. (c) Responses to moving edges. Movement of a line indicated that the receptive field was about 5° across. Here edges of both polarities were moved in both horizontal directions through the receptive field. Solid lines denote the movement of a dark edge, and open lines above the response trace indicate the movement of a light edge. The cell shows no sign of directional selectivity. Note that, in contrast to static stimuli, the cell is more responsive to dark than to light edges.

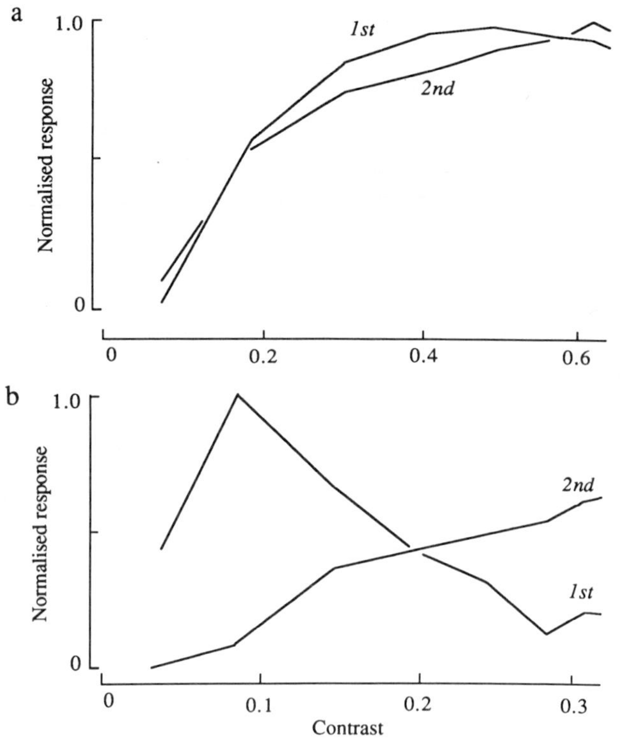

FIGURE 6. Normalized responses (rms power) of two transient cells to 6-Hz static sinusoidal flicker of varying amplitude. Normalization is to the maximum fundamental response (but in **a** the second harmonic rms power has been multiplied by 2.5 times). In **a** (cell illustrated in Figure 9) responses are "well behaved", approximately proportional to contrast with a marked compression (cf. Figure 4). This behavior is seen in many transient cells and also in medulla sustaining cells and lamina cells. Note that the second harmonic component is proportional to the fundamental. The response illustrated in **b** is anomalous but quite frequently encountered; here the fundamental response collapses above a certain threshold contrast, and, in certain cases, over 95% of response power is at the second harmonic. The effects of adaptation in the records illustrated were small after the first few stimulus cycles.

but it is not easy to make generalizations about the relationship between stimulus contrast and the amplitude of the various response components.

The transient cells receive inputs in the distal part of the medulla close to the termination sites of the LMCs (Osorio, 1987a), and their responses could conceivably arise by combining the outputs of two cells with responses of opposite sign such as those illustrated in Figure 3, with a simple synaptic nonlinearity like half-wave rectification. However, significant further signal processing seems to occur between the receptors and the transient cells.

D. ADAPTATION

When successive stimuli are presented at one point transient cell responses decline (or the probability of spiking drops). This adaptation decays after about 100ms (Figure 7a). Adaptation is restricted to the outputs of a single ommatidium (Figure 7b) and so can be attributed to self-inhibition in the pathway between the receptors and transient cells. The strength of the adaptation is dependent on the strength of the stimulus so that when a cell is adapted with a train of stimuli of one contrast, an increase in contrast will elicit an unadapted response (Figure 8b). Inhibition must occur before (full) signal rectification because it is polarity specific — decrements do not inhibit increment responses and vice versa (Figure 8a) — but is not attributable to processes in the receptors as these are close

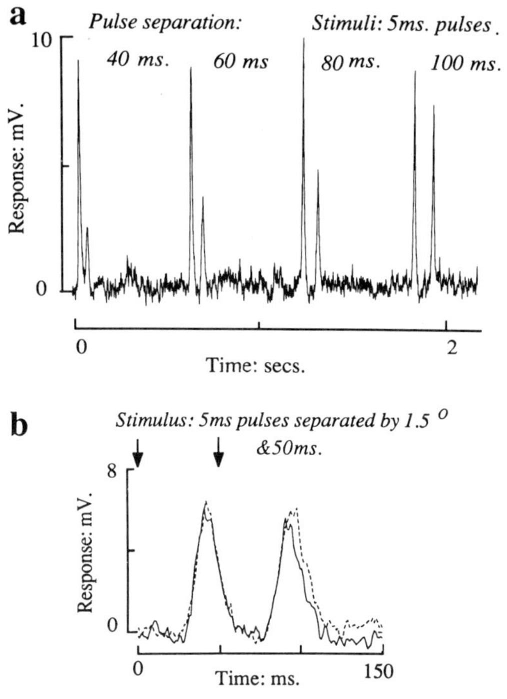

FIGURE 7. The time course of local adaptation in a transient cell (a) and (b) evidence this adaptation is localized to the outputs of a single ommatidium (i.e., point in visual space) (a) A series of pulse pairs presented with varying separations between pulses. The inhibition of the response to the second stimulus wanes over about 100 ms (Osorio, 1991a). Averaged responses to 20 stimulus cycles. (b) Responses of the cell to consecutive stimuli (arrowed) to neighboring receptors. The continuous and broken lines represent reversed orders of stimulus presentation, giving apparent upward and apparent downward motion, respectively. There is no evidence for inhibition or for nonlinear spatiotemporal interactions between the outputs of neighboring receptors (see also Figure 9).

to linear in the conditions of these experiments (James, Chapter 2 of this volume; Pinter, 1972; personal observation) and is absent from medulla sustaining cells. (Note that the sustaining cells are tuned to frequencies of between 10 and 20 Hz where there is marked adaptation of transient cells.)

E. SPATIAL INTERACTIONS AND NONDIRECTIONAL MOTION DETECTION

Transient cells are responsive to movement, giving a train of spikes as a line or an edge is moved through the receptive field (Figure 5c, Osorio 1991a). This responsivity might arise because the outputs of separate receptors give independent flicker responses, or it may be due to local spatiotemporal interactions of the kind that underlie directional selectivity (Borst and Egelhaaf, 1988; Egelhaaf et al., 1989). In fact there is no evidence for nonlinear interactions between the outputs of neighboring (or nearby) photoreceptors when successive stimuli are given with intervals ranging from 10 to 200 ms (e.g., Figure 7b). Similarly, if interactions between the outputs of nearby receptors are probed using sinusoidal flicker at various frequencies (3 to 24 Hz) and a full cycle of phase shifts, there is no evidence for spatial interactions other than linear pooling (Figure 9; Osorio, 1991a). (Motion-sensitive interactions are clearly visible in directional cells under the same conditions [Osorio, unpublished observations]).

The transient cells do not seem to resemble directional units, but it might be argued that local adaptation acts as a mechanism of nondirectional motion detection, allowing the cells to distinguish motion in any direction from static flicker. Self-inhibition suppresses

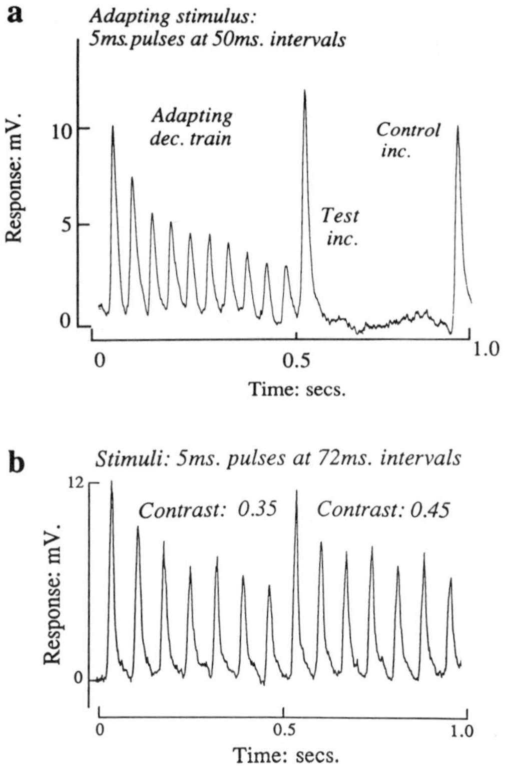

FIGURE 8. Demonstration of the polarity (a) and contrast (b) sensitivity of adaptation in a transient cell. (a) The response is adapted with a train of ten decrement pulses (contrast: 0.5; duration: 5 ms) presented at 50-ms intervals (20 stimulus cycles averaged). After a further 50 ms a test increment is presented and after a further 400-ms a control increment. The amplitudes of the test and the control are approximately equal, indicating that the increment response is unaffected (or slightly enhanced) by adaptation with decrements, and that the self-inhibition occurs before rectification. (b) The cell is adapted with a train of seven decrements at a single point, presented at 72-ms intervals. After a further 72 ms, a second train of higher contrast signals is presented. The initial response and adaptation to this second train is similar to that to the first (20 stimulus cycles averaged with 400 ms intervals between cycles).

responses to consecutive stimuli in the outputs from a single ommatidium, but does not affect those of its neighbors. This veto of responses to static flicker is reminiscent of the veto said to underlie directional selectivity in the rabbit retinal ganglion (Barlow and Levick, 1964). In the rabbit's directional cells receptors (or subunits) respond to static flicker. Motion in the preferred direction gives responses from adjacent receptors which resemble a series of independent flicker responses. These responses are inhibited if the motion is in the null direction. Interestingly, the time course of the local adaptation in transient cells (decaying in about 100 ms) is similar to that for the inhibition that underlies directional selectivity in the locust's medulla (Osorio, 1986b, 1991a).

III. MEDULLA CELL PROPERTIES AND THE ORGANIZATION OF RETINOTOPIC VISUAL PROCESSING

In the insect's medulla a set of over 40 anatomically distinct cells is repeated behind each facet in the insect's eye, and each one probably has a distinct input circuitry and physiology. We are a long way from making sense of why visual information should be divided into 40 separate channels at the finest scale permitted by the retinal sampling array.

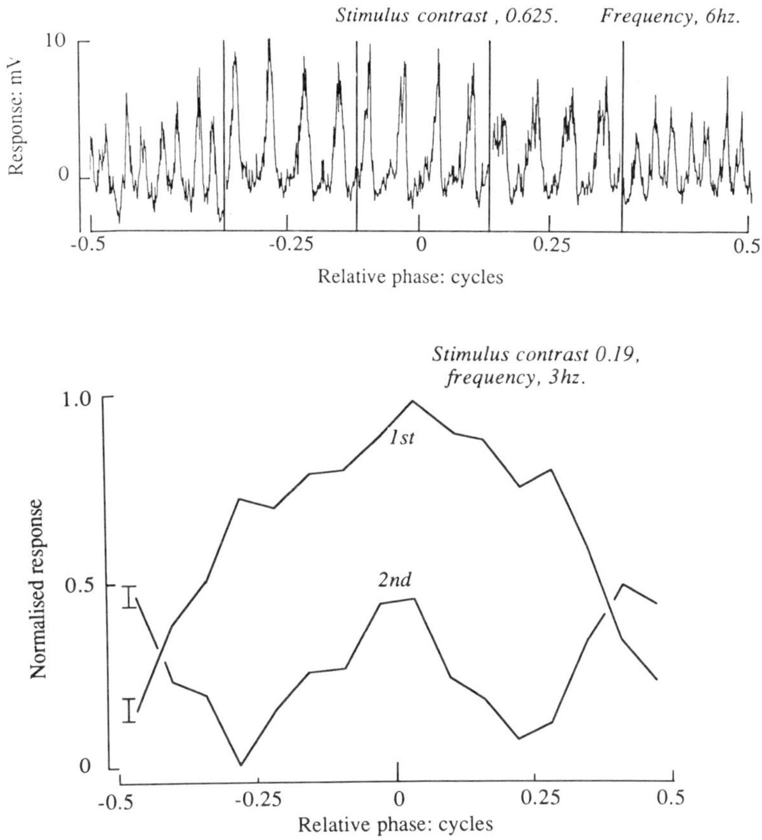

FIGURE 9. Transient cell responses as a function of relative phase between sinusoidally flickering stimuli at two points 3° apart in the receptive field. Data were obtained using phase-modulated signals (above). The contrast-response function of the cell is shown in Figure 6a. (Above) Responses to segments of four "periods" at five different (approximate) phase shifts. The stimuli shifted through a full cycle of phase shifts in 128 periods. When the stimuli are in antiphase, the frequency-doubled response is clearly visible; at other relative phases the response is dominated by the first harmonic component. The high stimulus contrast allows the graded-potential response to be clearly visible above the noise (5 stimulus cycles averaged). Responses to five beat periods averaged, sampling frequency circa 200 Hz. (Below) First and second harmonic response components as a function of relative phase between the stimuli. The stimuli were placed at points of equal sensitivity so that there is a minimal fundamental response component when signals are in antiphase. The vertical bars at the left end of each curve indicate the estimated noise at each frequency. This estimate has been subtracted from the crude measurement of response (rms power) at each phase shift. In this record the second harmonic component has approximately equal maxima when the stimuli are in phase and in antiphase, and nulls when the phase shifts are ± 0.25 cycles; this is to be expected if rectification occurs before spatial pooling. In practice, transient cells vary in this respect; in some units the second harmonic shows smaller maxima to antiphase than to in-phase signals and minima at shift of greater than 0.25 cycles. In these latter units there appears to be some spatial pooling of receptor outputs before rectification. The degree of pooling may be frequency dependent; pooling is stronger at high than at low frequencies. The reasons for the variability between cells and frequency dependence are unclear.

Nonetheless, it seems worthwhile to consider some general questions about the function of the cells and to compare the main retinotopic visual ganglion in the locust with the vertebrate brain and, in particular, the retinal ganglion.

A. TRANSIENT CELL PROPERTIES AND FUNCTION

The most striking feature of our findings is the diversity of responses and the extent to which the responses of the transient cells are transformed by comparison with their (possibly indirect) inputs from the LMCs. In particular, transient cells show three important nonlinearities; these, roughly speaking, resemble adaptation, rectification, and thresholding. Adaptation is sensitive to stimulus polarity (Figure 8a) and so occurs before full rectification, but might occur in independent "on" and "off" channels (Franceschini et al., 1989). Thresholding, to produce a fixed spiking response to all suprathreshold stimuli, probably occurs in the transient cells themselves.

The combined effects of thresholding and adaptation suggest that inputs from any one ommatidium will give a single transient cell spike to the highest contrast increment and another to the highest contrast decrement occurring in each 100-ms period. Analysis of local spatiotemporal interactions within the receptive field (Figures 7b and 9; Osorio, 1991a), and movement of lines and edges through the receptive field, indicate that transient cell responses are nondirectional. While adaptation appears to be mediated by self-inhibition localized to the outputs of single ommatidia, there is some spatial pooling of receptor outputs before rectification, especially at high temporal frequencies (Figure 9; Osorio, 1991a). This pooling will render the cells more sensitive to extended stimuli than to incoherent signals. Object boundaries are extended and tend to have a higher contrast than internal texture, suggesting that they will preferentially excite the transient cells. In an earlier paper (Osorio et al., 1987) we suggested that self-inhibition will act to reset the cells' threshold to respond to object boundaries, but to ignore textural detail and so provide a basis for image segmentation uncluttered by structural details.

B. COMPARISON OF MEDULLA TRANSIENT CELLS WITH RETINAL GANGLION Y CELLS

Do the similarities between cells in the optic medulla and those in the cat retinal ganglion, which prompted the division of the insect's cells into "sustaining" and "transient" categories (Honegger, 1978), reflect fundamental functional analogies? The nonlinearities in transient (or Y) cells produce refined and simplified versions of the retinal image. In future we hope to build models of (nonlinear) neural properties to allow us to predict what image features are encoded by any unit and so to assess whether the insects and the vertebrates cells are, indeed, functionally analogous. Here we can only make a start to compare cell physiologies.

The cat's Y cells, like the locust's transient cells, have a rectification-like process in their input pathways so that responses to increments and decrements are similar. In the cat, rectification occurs locally, prior to spatial integration of the signals, and is attributed to subunits within the receptive field (Hochstein and Shapley, 1976). Similar subunits probably exist in the locust, but their size is uncertain; adaptation is localized to the outputs of single ommatidia (Figure 7b), but if defined on the basis of the summation area for local rectification, the size of the subunits seems to be frequency dependent. At high frequencies there is more spatial integration before rectification than at low frequencies (Osorio, 1991a). Unlike the vertebrate Y cells there is no sign of a linear envelope surrounding the subunits in the locust's transient cells (unpublished observations). In this respect they are more similar to some of the W cells in the retinal ganglion, which also lack a linear envelope (Troy et al., 1988).

Medulla transient cells and also retinal Y cells adapt to stimulus contrast, but probably in differing ways. Contrast gain control in the vertebrate retina is associated with a change in the temporal properties of the responses so that ganglion cells are tuned to higher temporal frequencies at high than at low contrasts (Shapley and Victor, 1978). There is some indication that a comparable contrast-dependent change in temporal-frequency tuning occurs in the sustaining cells of the locust medulla (Figure 4), but in medulla transient cells gain changes are marked, whereas response timing seems to be unaffected by adaptation (Figures 7 and 8; Osorio, 1991a).

IV. CONCLUDING REMARKS: THE ROLE OF LINEAR AND NONLINEAR TRANSFORMATIONS IN EARLY VISUAL PROCESSING

The anatomy of the insect medulla is different from that of the vertebrate retinal ganglion (Figure 1). In the insect there is no bottleneck equivalent to the optic nerve, and the medulla is clearly the main center of retinotopic visual processing; beyond the medulla there is spatial integration of visual information. By comparison there are many centers of retinotopic vision beyond the retinal ganglion. Nonetheless, there are similarities between the two systems, and it is tempting to ask whether the qualitative division of neural responses into linear and highly nonlinear sets are due to a convergence of function.

Linear transformations are, in general, reversible, provided that there is no loss in the number of degrees of freedom, the input can be recovered by back transformation. This is not the case for nonlinear transformations; stimulus polarity or amplitude, for instance, are not encoded by medulla transient cells. Nonlinear cells are needed for analysis, abstracting particular features or stimulus parameters from the input. Only the simplest of behaviors, for example, phototaxis, could be controlled by purely linear processing, and "essential" nonlinearities must underlie the neural control of most behaviors (Poggio and Reichardt, 1976).

Partly because their transformations are reversible, it is attractive to describe linear cells (with band-pass frequency tuning) not in terms of the features that they detect, but rather as "unprejudiced" transmitters of all information that have evolved to remove redundancy from the retinal input and so to make best use of the limited "neural bandwidth" available. Cells in the insect lamina (Laughlin, 1982; Srinivasan et al., 1982), the vertebrate retinal ganglion, and visual cortex (Sakitt and Barlow, 1982; Bossomaier and Snyder, 1986; Atick and Redlich, 1990a,b) have been described in this way, and this view has considerable appeal, not least as a "null hypothesis" against which to test any theory of function. If a set of units encodes all information in the retinal image, there may be little point in arguing that they are adapted to abstract any particular feature.

Unfortunately, the observation of "parallel" quasilinear and nonlinear units does not fit easily with the view that the cells are involved in transmission and analysis, respectively. This parallel organization is seen both in the insect medulla and in the mammalian retinal ganglion (and also the visual cortex). Perhaps to understand the functions either of the transient or of the sustaining cell classes, we need to consider how their neural representations might be combined in later processing. For example, the locust's transient cells could abstract a set of object border primitives suited to making a primary segmentation of the image. This segmentation might structure (capture) or expedite the analysis of internal detail, for example, texture and color, based on the outputs of the more linear cells (Osorio et al., 1987).

REFERENCES

Atick, J. J. and Redlich, A. N. (1990a). Predicting ganglion and simple cell receptive field organizations from information theory, Preprint No. IASSNS-HEP-89/55.

Atick, J. J. and Redlich, A. N. (1990b). Towards a theory of early visual processing, Preprint No. IASSNS-HEP-90/10.

Borst, A. and Egelhaaf, M. (1989). Principles of visual motion detection, *Trends Neurosci.*, **12**, 297–306.

Bossomaier, T. R. J. and Snyder, A. W. (1986). Why spatial frequency processing in the visual cortex? *Vision Res.*, **26**, 1307–1309.

Braitenberg, V. (1972). Periodic structures in the visual system of the fly, in *Information Processing in the Visual Systems of Arthropods*, Wehner, R., Ed., Springer-Verlag, Berlin, 3–15.

Cajal, S. R. and Sanchez y Sanchez D. (1915). Contribucion al conocimiento de los centros nerviosos de los insectos, *Tab. Lab. Invest. Biol. Madr.*, **13**, 1–168.

DeYoe, E. A. and Van Essen, D. C. (1988). Concurrent processing streams in monkey visual cortex, *Trends Neurosci.*, **11**, 219–226.

Egelhaaf, M., Borst, A., and Reichardt, W. (1989). Computational structure of a biological motion detection system as revealed by local detector analysis in the fly's nervous system, *J. Opt. Soc. Am. A*, **6**, 1070–1087.

Enroth-Cugell, C. and Robson, J. G. (1966). The contrast sensitivity of retinal ganglion cells of the cat, *J. Physiol. (London)*, **187**, 517–552.

Fischbach, K. -F. and Dittrich, A. P. M. (1989). The optic lobe of *Drosophila melanogaster*. I. A Golgi analysis of wild-type structure, *Cell Tissue Res.*, **258**, 441–475.

Franceschini, N., Riehle, A., and le Nestour, A. (1989). Directionally selective motion detection by insect neurons, in *Facets of Vision*, Stavenga, D. G. and Hardie, R. C., Eds., Springer-Verlag, Berlin, 360–390.

Hochstein, S. and Shapley, R. M. (1976). Linear and nonlinear spatial subunits in Y cat retinal ganglion cells, *J. Physiol. (London)*, **262**, 265–284.

Honegger, H. -W. (1978). Sustained and transient responding units in the medulla of the cricket, *Gryllus campestris*, *J. Comp. Physiol.*, **125**, 259–266.

Honegger, H. -W. (1980). Receptive fields of sustained medulla cells in crickets, *J. Comp. Physiol.*, **136**, 191–201.

Hoyle, G. (1977). *Identified Neurons and Behaviour of Arthropods*, Plenum Press, New York.

James, A. C. (1991). Nonlinear operator network models of processing in the fly lamina. (this volume).

Kirschfeld, K. (1967). Die projektion der optischen Umwelt auf das Raster der Rhabdomeren im Komplexauge von *Musca*, *Exp. Brain Res.*, **3**, 248–270.

Laughlin, S. B. (1982). Matching coding to scenes in order to enhance efficiency, in *The Physical and Biological Processing of Images*, Braddick, O. J. and Sleigh, A. C., Eds., Springer-Verlag, Berlin, 42–52.

Levick, W. R. and Thibos, L. N. (1983). Receptive fields of cat ganglion cells: classification and construction, in *Progress in Retinal Research, Vol. 2*, Osborne, N. and Chader, G., Eds., Pergamon Press, Oxford, 267–319.

Marr, D. (1982). *Vison*, Freeman, San Francisco.

Osorio, D., (1986a). Directionally selective cells in the locust medulla, *J. Comp. Physiol. A*, **159**, 841–847.

Osorio, D. (1986b). Ultraviolet sensitivity and spectral opponency in the locust, *J. Exp. Biol.*, **122**, 193–208.

Osorio, D. (1987a). The temporal properties of non-linear transient cells in the locust medulla, *J. Comp. Physiol. A*, **161**, 431–440.

Osorio, D. (1987b). Temporal and spectral properties of sustaining cells in the medulla of the locust, *J. Comp. Physiol. A*, **161**, 441–448.

Osorio, D. (1991a). Mechanisms of early visual processing in the medulla of the locust optic lobe: how self-inhibition, spatial-pooling and signal rectification contribute to the properties of transient cells, *Visual Neurosci.*, **7**, 345–355.

Osorio, D. (1991b). Patterns of function and evolution in the arthropod optic lobe, in *Vision and Visual Dysfunction, Vol. II. Evolution of the Eye and Visual System*, Cronly-Dillon, J. R. and Gregory, R., Eds., Macmillan, London, 203–228.

Osorio, D., Snyder, A. W., and Srinivasan, M. V. (1987). Bi-partitioning and boundary detection in natural scenes, *Spatial Vision*, **2**, 191–198.

Pinter, R. B. (1972). Frequency and time domain properties of retinular cells of the desert locust (*Schistocerca gregaria*) and the house cricket (*Acheta domesticus*), *J. Comp. Physiol.*, **77**, 383–397.

Poggio, T. and Reichardt, W. (1976). Visual control of orientation behaviour in the fly. II. Towards the underlying neural interactions, *Q. Rev. Biophys.*, **9**, 377–438.

Sakitt, B. and Barlow, H. B. (1982). A model for economical encoding of the visual image in cerebral cortex, *Bio. Cybern.*, **43**, 97–108.

Shapley, R. M. and Victor, J. D. (1978). The effect of contrast on the transfer properties of cat retinal ganglion cells, *J. Physiol. (London)*, **285**, 275–298.

Srinivasan, M. V., Laughlin, S. B., and Dubs, A. (1982). Predictive coding: a fresh view of inhibition in the retina, *Proc. R. Soc. London Ser. B*, **216**, 427–459.

Strausfeld, N. J. (1976). Atlas of an Insect Brain, *Springer-Verlag, Berlin.*

Strausfeld, N. J. (1989). Beneath the compound eye: neuroanatomical analysis and physiological correlates in the study of insect vision, in *Facets of Vision*, Stavenga, D. G. and Hardie, R. C., Eds., Springer-Verlag, Berlin. 317–359.

Troy, J. B., Einstein, G., Schuurmans, R. P., Robson, J. G., and Enroth-Cugell, Ch. (1989). Responses to sinusoidal gratings of two types of very non-linear retinal ganglion cells of cat, *Visual Neurosci.*, **3**, 213–223.

Chapter 15

THE TRADE-OFF BETWEEN RESOLUTION AND SENSITIVITY IN COMPOUND EYES

Eric J. Warrant and Peter D. McIntyre

TABLE OF CONTENTS

I. Introduction 392
- A. The Visual Environment 392
- B. Adequate Resolution and Sensitivity: Fundamental Requirements in Any Eye Design 392
- C. Specializations of Eyes for Different Ranges of Environmental Light Intensity 394

II. Resolution, Sensitivity, and Eye Design in Compound Eyes 395
- A. The Two Broad Design Divisions: Apposition and Superposition Eyes 395
  - 1. Apposition Eyes 396
  - 2. Superposition Eyes 397
- B. The Design Parameters of Compound Eyes and the Trade-Off between Resolution and Sensitivity 398

III. The Alteration of Resolution and Sensitivity by Light Adaptation 406
- A. Light Adaptation and the Dynamic Response Range of Photoreceptors 406
- B. How Is Light Adaptation Achieved, and What Are Its Effects on Resolution and Sensitivity? 409
  - 1. The Photoreceptor 410
  - 2. The Sensors 411
  - 3. The Controller 412
    - a. The Optical Density of the Aperture 412
    - b. The Optical Density of the Photoreceptor 415

IV. Conclusions 417

Acknowledgments 418

References 418

I. INTRODUCTION

1. THE VISUAL ENVIRONMENT

Light is one of the most important sensory stimuli in the natural habitat. From the deepest ocean depths to the highest mountain ranges, there are few habitable places on Earth which escape its presence (Figure 1). In some places the light may be so dim that only a few photons arrive each minute. In other places hundreds of millions of photons may arrive each microsecond. However, despite this enormous variation in light intensity (20 log units of intensity from nocturnal deep sea to bright sunshine; Figure 1), each of these places is uniquely suited to a myriad of different animals, almost all of which, to some extent, rely on the available light for survival.

Even though such huge variations in light intensity exist on the Earth, visual systems are only required to deal with comparatively narrow ranges of intensity. For instance, a terrestrial animal which has a strictly nocturnal activity would most likely possess a visual system optimized to detect visual information available within the range of intensities normally encountered in its habitat at night time. In addition, the visual system may have mechanisms in place which allow the animal to cope with sudden unexpected exposures to bright sunlight during the daylight hours when the animal is sheltering. Such considerations lead one to conclude that visual systems have evolved to operate in restricted windows of light intensity within the vast range of light intensities encountered on the Earth.

The specializations demanded by a particular window of light intensities are often most obvious at the extreme peripheral level of the visual system, the eyes. These organs have a construction and physiology which is a direct reflection of the visual demands placed on an animal by its habitat. In effect, the eye represents the first transformation of the visual information available within the range of intensities the visual system normally experiences. The eye's physical limitations determine the quantity and type of information which is accepted for higher visual processing in the brain, and, therefore, an appreciation of these limitations is a necessary prerequisite for a proper understanding of the visual system as a whole. This is the purpose of this chapter. We will examine the limitations of the first visual transformation by considering the trade-off made between two very important requirements of an eye intended to operate efficiently within a particular range of light intensities: sufficient *sensitivity* to the available light and adequate *resolution* of spatial information inherent in that light. The way in which these two requirements interact and are optimized is determined by the physical structure of the eye and by its physiology. In this respect, some of the greatest variations in eye design and function are found within the compound eyes of insects and crustaceans (Nilsson, 1989), whose typical habitats span almost the full range of intensities found on the Earth. Compound eyes of insects and crustaceans are, therefore, ideal eyes with which we can explore the theme of this chapter, and we will restrict our discussion to them. Occasionally, however, we shall also refer to the somewhat different compound eye of the horseshoe crab, *Limulus*.

B. ADEQUATE RESOLUTION AND SENSITIVITY: FUNDAMENTAL REQUIREMENTS IN ANY EYE DESIGN

Photons arrive at the eye surface at completely random intervals. The intensity of the stimulus is measured as the average rate of arrival of these photons. If only a few photons arrive over a long period of time, it is very difficult for a photoreceptor to determine accurately the intensity of the stimulus. In other words, if the photon arrival rate is too low, there is considerable uncertainty or ''noise'', which limits the reliable detection of the stimulus. As the rate of photon arrival increases, the relative magnitude of the noise becomes less, and the photoreceptor can more accurately determine the true stimulus intensity. If an eye can

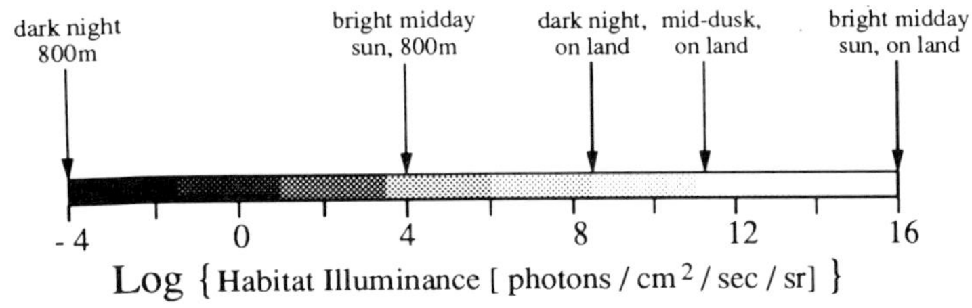

FIGURE 1. Habitat illuminances on Earth span a total range of at least 20 logarithmic units from darkest deep sea to bright sunshine. The quoted units of illuminance assume an incident wavelength of 550 nm. Terrestrial values were measured directly over a 12-h period with an International Light IL-700 research radiometer, spanning late morning to midevening (Warrant, unpublished results). The marine values were adapted from Clarke and Denton, 1962.

somehow capture more photons, relative noise can be reduced. This is, in fact, the first requirement in eye design: the attainment of adequate *sensitivity* to light.

The second requirement of eye design is the ability to determine from where in space light is incident. This is achieved by having photoreceptors viewing separate directions in space: each photoreceptor within a retina is a sampling station which is responsible for determining, among other things, the average intensity of light falling from a particular direction. The amount of spatial detail that can be extracted from a visual scene, therefore, depends on the density of photoreceptors viewing it. If the density is low (i.e., few photoreceptors per unit angle in space), little spatial detail may be extracted. In other words, an eye design must provide adequate spatial *resolution*.

It is the dual requirement of adequate sensitivity and adequate resolution that provides the first compromise in the design of a successful eye. For reasonable resolution, one requires that photoreceptors sample the visual world as densely as possible. This means that each photoreceptor must collect light from as small a solid angle of visual space as possible. Obviously, as this solid angle becomes smaller, the number of photons arriving at a photoreceptor from this direction must become fewer (eye size remaining constant). This means that the statistical reliability of the signal generated in the photoreceptor must degenerate for denser and denser sampling (Barlow, 1952; Synder et al., 1977). If the light levels are dim, one could imagine that the sampling density would not need to be very great before the sensitivity of each photoreceptor to light became unacceptably low. From these simple considerations alone, it would seem that there is a trade-off between resolution and sensitivity, and that whatever compromise is found somehow depends on the mean environmental luminance experienced by the eye in question.

This simple argument based on photoreceptor sampling density is but one indicator of the trade-off between resolution and sensitivity. Even in the brightest intensitics, where the requirement for an eye to have high sensitivity is somewhat unnecessary, resolution is still limited by the wave nature of light. When parallel wavefronts of light pass through an aperture, they suffer from the effects of diffraction. This causes an angular spread of the advancing wavefronts which leads to an uncertainty concerning their direction of origin. The uncertainty is most clearly demonstrated if we place a lens in the aperture. Instead of the parallel wavefronts being focused to an infinitesimally small point on the image plane, diffraction ensures that the image will be a blurred spot, or Airy disk. The severity of this blurring is inversely proportional to the diameter of the lens aperture, and so a simple solution to the problem is to make the lens aperture wider. This solution has the added advantage of increasing sensitivity because a wider aperture allows greater capture of photons. Unfortunately, as an aperture is made wider (while maintaining a constant focal distance), a second group of optical limitations comes into play: spherical and chromatic aberrations.

These effects are most pronounced when the ratio of the focal length to the width of a lens is low. They tend to blur the image and, therefore, reduce resolution. Of course, these aberrations can, in turn, be counteracted by allowing the focal length to become longer, but then the total length of the optical system is increased. If the optical system was an eye, such a length increase would be reflected as an increase in eye size. In other words, maximization of both resolution and sensitivity can be achieved by an increase in eye size (Land, 1981), but of course there must be some limit to this argument. Otherwise houseflies could have eyes the size of soccer balls. Clearly this is absurd on the basis of aerodynamics alone! We may conclude, therefore, that optimum eye size depends not only on resolution and sensitivity, but also on other factors important to the animal's survival.

Resolution and sensitivity also depend on motion (both self and object motion). If an animal turns with a particular angular velocity, the number of photons that can be captured by each photoreceptor is limited by its integration time: the faster the velocity, the fewer the photons that can be captured from each point in space as the eye moves (Snyder, 1977). Motion also reduces resolution (Srinivasan and Bernard, 1975; Snyder, 1977).

In other words, there are many different optical effects which compete with each other in the successful design of an eye. All of them, to some extent, have a bearing on both the resolution and the sensitivity. Exactly which effects are allowed to dominate in the design of an eye is very much a reflection of the luminous conditions under which the eye must operate. In this respect the eyes of insects and crustaceans provide probably the richest insights into eye design one can find in nature.

C. SPECIALIZATIONS OF EYES FOR DIFFERENT RANGES OF ENVIRONMENTAL LIGHT INTENSITY

There are two main features of the window of light intensity in which an animal is normally active. First, it has a total *range*, that is to say, the window of intensities is bounded by a minimum intensity and a maximum intensity that the animal is likely to experience. As an example, an animal active only in daylight hours is likely to experience intensities within a range of 2 to 3 log units (Compton, 1978). The second feature of the window is that it contains an *average* intensity the animal is likely to encounter. The range of intensities experienced, therefore, extends above and below this average. Of course, for the sake of simplicity, the word "average" is used here rather loosely: the average intensity is the one most commonly experienced by an animal, and it is not our intention to imply that it lies precisely in the middle of the range. Returning to our example above, the 2 or 3 log units of intensity experienced by a day-active animal are located along the environmental luminance bar of Figure 1 at a position which is set by its average diurnal light intensity. In fact, we may continue this reasoning to suppose that every animal adapted for life at a particular average intensity, I, along this luminance bar has its own luminance window which is a range of intensities positioned at I.

What do these windows mean for the evolution of an eye's optimum design and physiology? First, it can be generally said that the average intensity an animal experiences is reflected in the type of optical design that its eye possesses. For instance, animals from very dim habitats usually have an optical design evolved to maximize sensitivity to light. Thus, as a first approximation, variations in the physical design of an eye allow different animals to live at very different average intensities, and this provides the first major compromise in the trade-off between resolution and sensitivity; in order to maximize photon capture, eyes adapted to vision at low light levels must, by necessity, sacrifice those features which also endow the animal with high spatial resolution.

If the optical design of an eye is a reflection of the average light intensity an animal has evolved to exist in, then how does the eye reflect the range of intensities normally

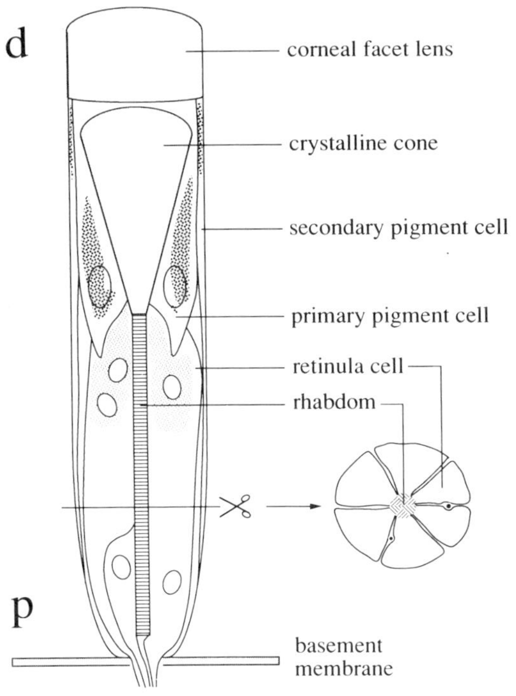

FIGURE 2. A schematic longitudinal section through an ommatidium of an apposition eye showing its main features. The corneal facet lens is at the distal (d) end of the ommatidium and the basement membrane (through which the axon passes to the optic lobe of the brain) is at the proximal (p) end. The horizontal line passing through the ommatidium shows the position from where the transverse schematic (bottom right) is taken.

experienced around this average? The first thing to note is that photoreceptors only have a finite response range, that is, photoreceptors can only respond without saturation to a finite range of light intensities. This operating range of light intensities is often, for certain animals, considerably smaller than the total range of intensities that the animal normally experiences. Somehow, then, the eye must be able to shift the position of its operating range within the total range so that all of its normally experienced light intensities may be perceived. The visual processes by which the operating range is shifted are collectively referred to as *adaptation*. Adaptation provides the second major compromise between resolution and sensitivity in the eye. If the intensity of light being encountered by an eye suddenly drops, the processes of adaptation invoke changes in both the structure and physiology of the eye which shifts the operating range of the photoreceptors by increasing the sensitivity of the eye to light. The very changes which bring about this improvement in sensitivity almost invariably reduce the ability of the eye to resolve spatial details.

We will now examine how the optical design of compound eyes and the process of adaptation both affect the trade-off between resolution and sensitivity.

II. RESOLUTION, SENSITIVITY, AND EYE DESIGN IN COMPOUND EYES

A. THE TWO BROAD DESIGN DIVISIONS: APPOSITION AND SUPERPOSITION EYES

All compound eyes are constructed of single optical units called ommatidia. In the simplest description, these consist of two major components (Figure 2): a lens unit which collects and focuses incoming light (consisting of an external corneal facet lens and, im-

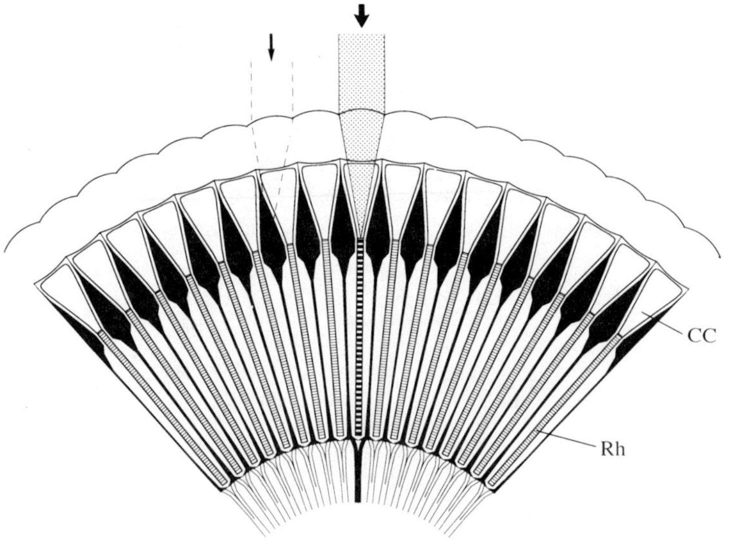

FIGURE 3. A schematic focal apposition eye. Axial rays (heavy arrow) are focused by the corneal facet lens and crystalline cone (CC) on the entrance of the rhabdom (Rh). Nonaxial rays (light arrow) are absorbed by screening pigments between the crystalline cones (black regions). Diagram courtesy of Dr. D.-E. Nilsson.

mediately proximal, a crystalline cone lens), and a photoreceptor (or rhabdom) which absorbs and transduces focused light (a fuller description can be found in Nilsson, 1989). The size and structure of these components varies widely between species, and these variations provide the basis for the enormous range of designs found among compound eyes.

Each ommatidium is a conical tube which tapers distally to proximally. This shape allows for many ommatidia to be packed tightly together, thereby creating a compound eye. The number of ommatidia in a compound eye may be greater than several thousand, as in some dragonfly species, or fewer than ten, as in some ant species.

Compound eyes fall into two broad optical categories: the apposition and superposition types (Exner, 1891). The essential difference between the two designs lies in the number of lens units that provide light to a single rhabdom. In apposition eyes, only one is involved; light rays entering a corneal facet lens are focused exclusively onto the rhabdom within the same ommatidium. Rays are prevented from passing to neighboring ommatidia by a sleeve of light-absorbing screening pigment (Figure 2) which entirely encases each ommatidium. In superposition eyes, on the other hand, no such optical isolation of ommatidia is present (except in the extreme light-adapted state in some species), and light rays entering many corneal lenses are focused onto a single rhabdom.

For our purposes only a description of the salient features of apposition and superposition eyes is required. For a fuller description of the variations and subtleties inherent in the structure of different compound eyes, the interested reader is referred to the excellent reviews which exist on this topic (Land, 1981, 1989; Nilsson, 1989, 1990).

1. Apposition Eyes

Apposition eyes are common both in insects and crustaceans, and are especially common in those which are diurnal. In the most common type of apposition eye, the focal type, parallel light rays from a point source are focused by the axial corneal facet lens onto the distal tip of the rhabdom, which is looking in the direction from which the light is incident (Figure 3). Rays which enter neighboring facets are absorbed by the sleeve of screening pigment which surrounds each ommatidium. The rhabdom acts as a light guide, capturing

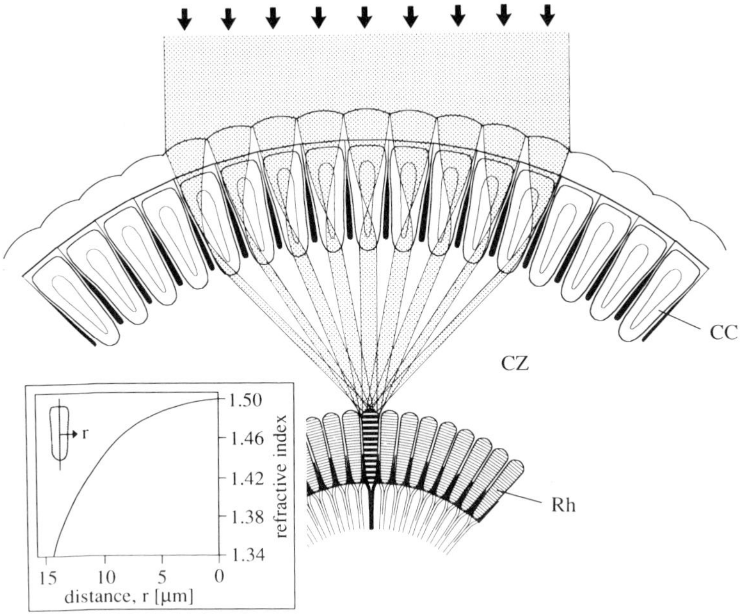

FIGURE 4. A schematic refracting superposition eye. Parallel rays (arrows) are refracted by the corneal facet lenses and crystalline cones (CC) of many ommatidia and focused across the clear zone (CZ) onto the target rhabdom (heavy hatching). Diagram courtesy of Dr. D.-E. Nilsson. (Inset) The refractive-index gradient of a crystalline cone (waist region) from the refracting superposition eye of the dung beetle, *Onitis aygulus*. The gradient is measured perpendicular to the axis of the crystalline cone (as illustrated), where r = 0 corresponds to a point on the axis.

the incident cone of light by total internal reflection. In some species such as the common housefly, the rhabdom is so thin that its diameter is comparable to the wavelengths of light that it absorbs. The rhabdom then acts as a waveguide and propagates the incoming light in distinct modes (reviews: Land, 1981; van Hateren, 1989). A second type of apposition eye recently found in butterflies (Nilsson et al., 1984) is the afocal type, which differs from the focal type by having a strong gradient of refractive index in the extreme proximal end of the crystalline cone. For reasons explained later, this design is superior to the focal design from the point of view of both resolution and sensitivity (van Hateren and Nilsson, 1987).

2. Superposition Eyes

Like apposition eyes, superposition eyes are common in both insects and crustaceans, and are especially common in those which live in dim habitats. However, unlike in apposition eyes, image formation relies on the cooperative action of many ommatidia. This is facilitated by the presence of a wide, optically homogeneous region called the clear zone, which separates the lenses from the retina. The inclusion of a clear zone is a marked structural departure from the apposition design where retina and lenses directly abut. In refracting superposition eyes (Figure 4), a design common in insects and some groups of crustaceans, the lens pair of each ommatidium acts as a low-power telescope. A significant proportion of the lens refractive power is due to a gradient of refractive index present in each corneal lens and crystalline cone (Figure 4 inset). Incident parallel light rays (from a point source) are brought to an intermediate focus within each contributing crystalline cone and recollimated. The resulting pencils of parallel rays are then directed toward the retina across the clear zone; each pencil from each cooperating ommatidium is aimed (ideally) at the same target rhabdom (Cleary et al., 1977; Caveney and McIntyre, 1981; McIntyre and Caveney,

1985). Whether or not these pencils have exact superposition on the retina depends on the precision of the optics (Land, 1984).

Besides the refracting type, two other types have also been described: (1) the reflecting type, found in many decapod crustaceans, focuses light, as its name suggests, by reflection rather than refraction (Vogt, 1975; Land, 1976), and (2) the parabolic type, found in many swimming and hermit crabs (and possibly also mayflies), which focuses light by a combination of reflection and light-guide action (Nilsson, 1988).

B. THE DESIGN PARAMETERS OF COMPOUND EYES AND THE TRADE-OFF BETWEEN RESOLUTION AND SENSITIVITY

What is immediately obvious from our description of apposition and superposition eyes is that the aperture through which light enters the eye is considerably different in each design: in apposition eyes the aperture is a single facet lens, whereas in superposition eyes it is a large patch of facet lenses, the number of which can, in some species, be a couple of thousand. It is easy to see that the superposition design has a vastly superior potential to collect light, and so, from the point of view of sensitivity, the size of the aperture is one of the important design parameters of a compound eye. As we shall see, there are several such parameters, and all of them have a bearing on the resolution and sensitivity attainable by the eye. Exactly which values of these parameters evolve in a given species is a reflection of the visual needs of the animal and the luminous intensity of its habitat.

Let's begin by considering the sensitivity of a given eye design to light. As we mentioned earlier, a fundamental limitation to vision is the ability of an eye to capture photons. If the random arrival of photons is too low, there is considerable uncertainty, or noise, which limits the reliable detection of the stimulus. Photon arrival, and consequent absorption by the photoreceptors, can be described by Poisson statistics (Scholes, 1964, 1965; Lillywhite, 1977). In this statistic, if the mean number of photons counted by a photoreceptor in a given sample is N, the standard deviation in the count (i.e., the noise), is $\sqrt{N}$. Thus as the photon capture (N) increases, the relative magnitude of the noise $(\sqrt{N}/N = 1/\sqrt{N})$ decreases. To see what this means for vision at these threshold levels of light intensity, let us consider two neighboring photoreceptors, one of which is receiving photons from a source with a brightness I, and the other of which is receiving photons from a source of brightness $(I + \Delta I)$. We may now ask, what is the minimum number, N_{min}, of counted photons required for the two neighboring photoreceptors to discern that the two sources have different brightnesses, with (say) 95% reliability? If we approximate the Poisson distribution by the normal distribution (which we can if the average number of photons counted in a sample is greater than about 20; Land, 1981), N_{min} is approximately $8/(\Delta I/I)^2$ for 95% reliability, where $\Delta I/I$ is the brightness difference between the two sources (Land, 1981). Thus, for example, to distinguish two sources which differ in brightness by just 5%, the minimum average photon count required is approximately 3200 photons. The important point is that more and more photons are required by the eye to detect finer and finer brightness differences. These arguments suggest that an eye can discern more fine spatial details in a scene (i.e., subtle localized intensity fluctuations) if it can somehow increase photon capture. There is a limit to this argument because ultimately, at much higher suprathreshold intensities, the physical structure of the eye and the limitations of its optics determine how much spatial detail can be resolved. Between these extremes of light intensity both structural limitations and photon noise limit spatial resolution, as we shall see below.

An extremely useful and widely quoted quantifier of the sensitivity, S, of eyes to extended sources is the Kirschfeld-Land sensitivity equation (Kirschfeld, 1974; Land, 1981):

$$S = (\pi/4)^2(A/f)^2d^2[1 - \exp(-kl_{rh})] \qquad (1)$$

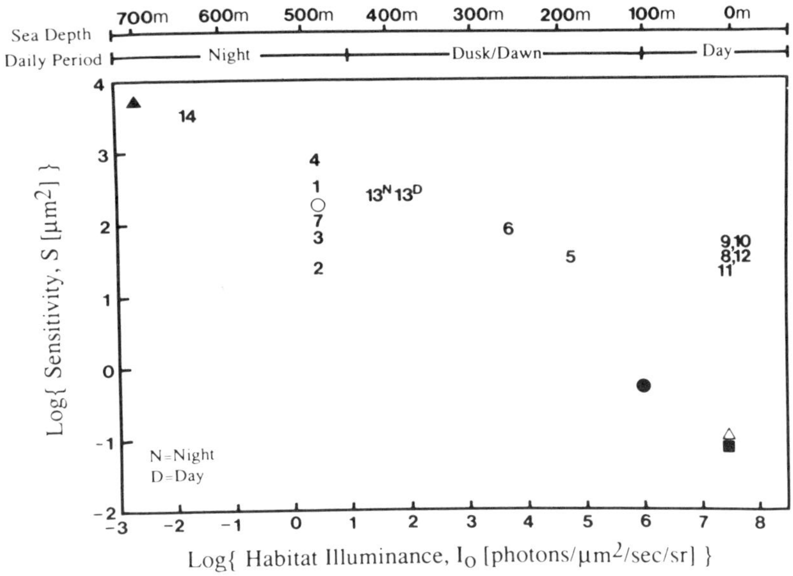

FIGURE 5. Comparative sensitivities (S) of arthropod eyes to an extended source, calculated with the aid of Equation 1 (see text). The arthropods chosen exist in marine and terrestrial habitats, and the calculated sensitivity for each is plotted with respect to the average environmental light intensity it normally experiences (I_o). Sensitivity is calculated for superposition eyes (numbers) and apposition eyes (solid symbols). Sensitivities of two spider simple eyes are also shown (open symbols). The terrestrial (time of day) and marine (sea depth) light intensity equivalents are also indicated. 1 = *Cherax* (crayfish); 2 = *Acetes* (shrimp); 3 = *Ephestia* (moth); 4 = *Anoplognathus* (scarab beetle); 5-7 = *Onitis westermanni, O. alexis,* and *O. aygulus* (dung beetles); 8-11 = *Trapezites, Toxidia, Netrocoryne,* and *Ocybadistes* (skipper butterflies); 12 = *Phalaenoides* (moth); 13 = *Meganyctiphanes* (euphausiid), which migrates vertically in the ocean depending on whether it is during the night (N) or during the day (D); 14 = *Oplophorus* (deep-sea crustacean); open circle = *Dinopis* (spider); open triangle = *Phidippus* (spider); closed triangle = *Cirolana* (isopod); closed circle = *Echinosquilla* (stomatopod); closed square = *Ciulfina* (mantid). From Warrant and McIntyre, 1990b, where references for the data points can be found.

where A is the diameter of the aperture through which light is entering (equal to the diameter of a single facet lens in apposition eyes and the diameter of the patch of contributing facet lenses [or superposition aperture] in superposition eyes), f is the posterior nodal distance, d is the diameter of the rhabdom, k is the natural extinction coefficient of the rhabdom (usually taken as 0.0067 μm^{-1}, the value found in lobster photoreceptors by Bruno et al., 1977) and l_{rh} is the length of the rhabdom. The sensitivity relates the number of photons absorbed by a rhabdom to the number emitted per unit area of an extended source.

The term $[1 - \exp(-k\, l_{rh})]$, which has values between 0 and 1, describes what proportion of light incident on a rhabdom of length l_{rh} is absorbed by it (assuming all light incident is retained by total internal reflection). With the value of k defined above, 95% of the incident light will be absorbed by a rhabdon of length 450 μm. Thus, all other factors ignored, a longer rhabdom will absorb more light. More light can also be accepted by a rhabdom for absorption if it is wider (i.e., larger value of d). Longer, wider rhabdoms, therefore, improve sensitivity, as calculated according to Equation 1. The other way of increasing sensitivity is to increase the ratio of the aperture diameter to the posterior nodal distance (A/f), and this is precisely the sensitivity advantage that superposition eyes have over apposition eyes, as we alluded to earlier: A/f is between 2 and 10 times larger in the former design than in the latter (Warrant and McIntyre, 1991).

The relationship between habitat intensity (I_o) and S (calculated with Equation 1) for a number of different insects and crustaceans with superposition and apposition eyes is shown in Figure 5. What is immediately noticeable is that, in general, eyes existing in dimmer

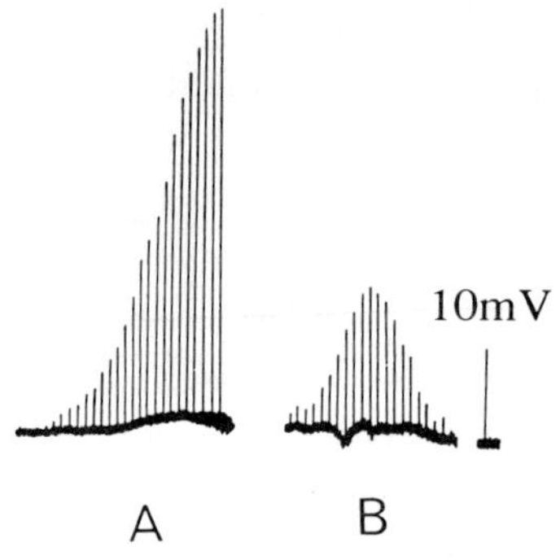

FIGURE 6. Electrical responses of a photoreceptor from the superposition eye of the dung beetle, *Onitis alexis*. (A) Responses to 100-ms flashes of light of increasing intensity (in steps of 0.2 logarithmic units) at $\lambda = 550$ nm. The largest of these responses resulted from stimulation at the maximum test flash intensity of 5.6×10^{11} photons/cm^2/s measured at the cornea. (B) Responses to constant, moderately intense 100-ms flashes of light (550 nm), each flash being delivered to the eye following a 0.5° displacement of the pinhole source in the horizontal direction through the predetermined axis of the cell. A calibration step of 10 mV is also shown.

habitats have higher sensitivity, as one may intuitively expect. A surprising eye is the apposition eye of the deep sea isopod *Cirolana borealis* (black triangle, Nilsson and Nilsson, 1981) which has a higher sensitivity than any other eye on the graph, including all the superposition eyes. This eye has a value of *A/f* comparable to superposition eyes and also has an enormous rhabdom ($d = 100$ μm), which proves that if properly constructed, an apposition eye can also have very high sensitivity.

The sensitivity of an eye to a point source differs from that of an extended source in that only the diameter of the aperture *(A)* and the length of the rhabdom (l_{rh}) is important; for the same rhabdom length, a larger aperture will collect more photons from the point source. Thus, one might predict that for the same point source stimulation, superposition eyes would demonstrate higher sensitivity than apposition eyes. This is easy to test by intracellularly recording the graded-potential responses of a photoreceptor to stimulation by brief flashes of light of increasing intensity from a point source aligned with the photoreceptor's visual axis. The resulting stimulus-response function (or V-LogI curve) is an indication of a photoreceptor's sensitivity and operating range (Figure 6A). A very convenient measure of point source sensitivity can be obtained from the V-LogI curve, the PAQ_{50}; the flux of axial photons (from a centred point source) required to elicit a 50% maximum response from the photoreceptor (Laughlin and Hardie, 1978). The lower the PAQ_{50} the higher the sensitivity. PAQ_{50}s for different species (measured in the dark-adapted state) are listed in Table 1. Examination of the table indicates that our general prediction is correct; i.e., superposition eyes are more sensitive to point source stimulation than apposition eyes. We may further surmise that this is mostly because superposition eyes have considerably larger apertures, but, of course, we cannot rule out the possibility that diurnal animals (with apposition eyes) have an intrinsically reduced response to each absorbed photon (i.e., a reduced gain) than nocturnal animals (with superposition eyes).

Our electrophysiological experiment with a point source (Figure 6) can be extended to our next topic of interest: resolution. Suppose we choose an intensity for our point source that gives a response of 40 or 50% maximum. If we now move the point source in small angular steps through the receptive field of the photoreceptor, delivering a flash at each step, we record its angular response (Figure 6B). When these responses are converted to sensitivity values via transformation through the V-LogI curve, we obtain a measurement of the spatial properties of the photoreceptor known as the angular-sensitivity function, the

TABLE 1
Values of PAQ_{50} Measured in Compound Eyes

Species	Animal	Eye	A (μm)	l_{rh} (μm)	PAQ_{50}	Ref.
Cherax (night DA)	Crayfish	N/DSup	800	95	2×10^8	3
Onitis (night DA)	Dung beetle	CSup	513	86	7.7×10^8	1
Tenodera (fovea)	Mantid	DApp	50	530	4.0×10^9	4
Phalaenoides	Moth	DSup	312	120	4.2×10^9	1
Tenodera (dorsal eye)	Mantid	DApp	35	320	7.1×10^9	4
Onitis (day DA)	Dung beetle	CSup	440	86	1.7×10^{10}	1
Cherax (day DA)	Crayfish	N/DSup	800	95	3.2×10^{10}	3
Calliphora	Fly	DApp	37	200	4.2×10^{10}	2
Hemicordulia	Dragonfly	DApp	40	800	1.4×10^{11}	1

Note: The PAQ_{50} is the number of axial photons at peak wavelength (photons/cm²/s) required to elicit a response from a photoreceptor which is 50% of its maximum. NSup = nocturnal superposition eye; CSup = crepuscular superposition eye; DSup = diurnal superposition eye; DApp = diurnal apposition eye; DA = dark adapted; A = diameter of aperture; l_{rh} = length of the rhabdom; 1 = Warrant and McIntyre, 1990b; 2 = Laughlin and Hardie, 1978; 3 = Bryceson, 1986; 4 = Rossel, 1979.

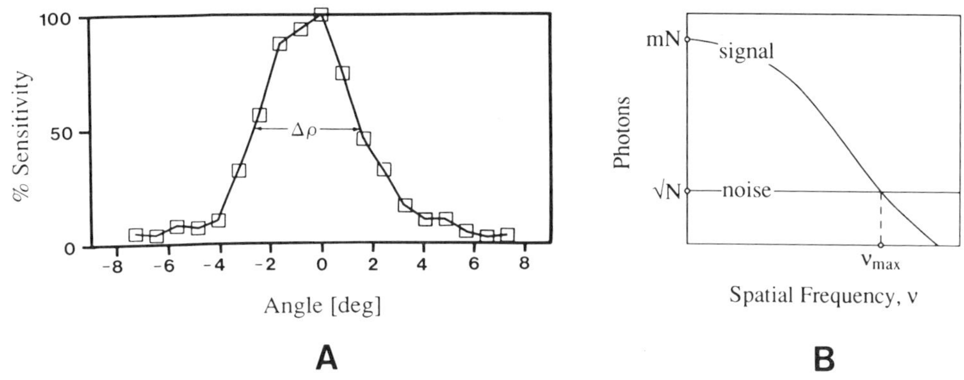

FIGURE 7. A. A typical angular-sensitivity function from a single photoreceptor of the dung beetle *Onitis alexis* in the night dark-adapted state. Δρ is the acceptance angle, defined as the half-width of the angular-sensitivity function (here equal to 4.1°). The narrower the angular-sensitivity function (i.e., the smaller Δρ) and the lower the amplitude of its flanks, the greater the potential range of spatial frequencies resolvable by the eye (see text). (B) A schematic graph of signal (Equation 2) and noise (Equation 3) in a photoreceptor. The signal is calculated using an MTF which is derived from the Fourier transform of the photoreceptor's angular-sensitivity function. ν_{max} is the maximum spatial frequency detectable by the eye, that is, the spatial frequency at which signal and noise become equal (found at the intersection of the signal and noise curves). N is the number of photons absorbed by the photoreceptor during its integration time, and m is the mean contrast of the scene.

half-width of which is referred to as the acceptance angle, Δρ (Figure 7A). The shape and width of the angular-sensitivity function contains nearly all the information one needs to know concerning the limits of resolution in compound eyes. It reflects (1) the positions and angles of incidence of all rays focused to the small spot (or blur-circle) on the retina, and this in turn reflects the factors which limit the optics, and (2) the extent to which rays that are aimed at the target rhabdom are retained by it alone (Warrant and McIntyre, 1990a). Generally speaking, good resolution results from a narrow angular-sensitivity function (i.e., small Δρ) with low-amplitude flanks (i.e., those parts of the function measured at angles well away from 0°, for example, at angles greater than 3° or less than −4° in Figure 7A).

The photoreceptor behaves approximately as a linear system (Pinter, 1972), and this leads to another wonderful feature of the angular-sensitivity function; i.e., its Fourier transform is the spatial modulation transfer function (MTF) of the eye, and this accurately describes the range of spatial frequencies available in a scene, that, in the absence of photon noise, the eye should be able to detect (Synder, 1977, 1979; Dubs, 1982; Warrant and McIntyre, 1990a). The MTF is a function of the spatial frequency v [i.e, M(v)] and has a value which lies between 0 and 1 [i.e., $0 \leq M(v) \leq 1$], the actual value depending on v. Generally, the higher v the lower M(v). The value of v at which the MTF becomes zero defines the range of spatial frequencies that the eye can potentially detect (i.e., all spatial frequencies up to v cycles per degree). A qualitative property of the MTF is that narrower, less-flanked angular-sensitivity functions result in wider ranges of detectable spatial frequencies (Dubs, 1982; Warrant and McIntyre, 1990a). At any average environmental luminance being viewed by an eye, the signal produced in the photoreceptor at each spatial frequency v, from a scene of mean contrast *m*, which produces a mean photon capture *N*, is (Snyder, 1977):

$$\text{Signal} = m\,N\,M\,(v) \tag{2}$$

Nevertheless, photon noise is always present, no matter how high the environmental light intensity, and its effects are always such as to reduce the range of perceivable spatial frequencies (Snyder, 1977, 1979; Snyder et al., 1977):

$$\text{Noise} = \sqrt{N} \tag{3}$$

and is independent of spatial frequency.

Signal and noise are plotted together schematically in Figure 7B. The spatial frequency at which the signal crosses the noise is the maximum detectable spatial frequency, v_{max} (Warrant and McIntyre, 1990a). Spatial frequencies greater than v_{max} are not perceived by the eye because they are lost in the noise. In fact, at every average environmental luminance which produces a mean photon capture *N*, there will be a particular value of v_{max} which limits the resolution of a given eye with a certain construction (Warrant and McIntyre, 1990a).

In general, if an eye is able to increase the signal (Equation 2), it can thereby decrease the relative magnitude of the noise (Equation 3) and, thus, extend the total range of perceivable spatial frequencies by elevating v_{max}. And herein lies the great paradox in eye design: the very factors which allow an increase in the signal by an elevation of *N* are the same factors which tend to decrease it through a broadening of the angular-sensitivity function. To see which of the competing effects wins, we need to reexamine Equation 2. If we assume the angular-sensitivity function (with acceptance angle, $\Delta\rho$) is Gaussian (which is a reasonable approximation in some eyes), Equation 2 can be rewritten as (Snyder, 1979):

$$\text{Signal} = m\,N\,\exp(-3.56\,v^2 \Delta\rho^2) \tag{4}$$

The problem is most obvious for animals which live in dim habitats where the allowable photon catch *(N)* is already low. It might seem a reasonable strategy to have an eye with the design parameters of aperture and rhabdom diameter pushed to maximum size in order to maximize photon catch and increase signal by increasing *N*. We saw such a strategy in the apposition eye of the deep-sea isopod *Cirolana* mentioned earlier. However, an increase in aperture size (*f* remaining constant) results in two effects which alter the image on the

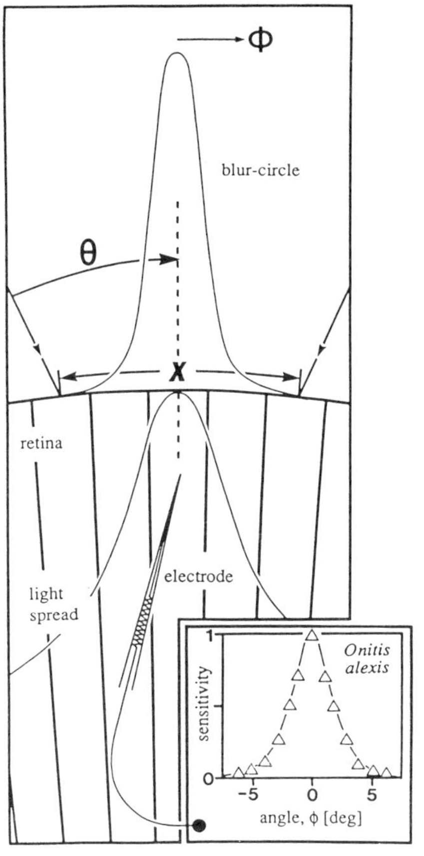

FIGURE 8. Image formation and light spread in compound eyes. Incident parallel light rays (on the eye surface) are focused by the lens(es) to form a blur-circle of light on the retina. The rays forming the blur-circle are spread spatially over the retina (x) and are incident over a range of angles (θ) to the target rhabdom axis. As the blur-circle is moved across the retina in small angular steps (φ), an electrode records the response of the target rhabdom as an angular sensitivity function (example for *Onitis alexis*, Figure 7), whose shape is partly determined by the extent of the light-spread.

retina (Figure 8): (1) an increase in the range of angles (θ) with which rays, forming the blur-circle, are incident on the retina, and (2) an increase in optical aberrations, which spatially spread the blur-circle *(x)* over the retina. These two effects lead to a spreading of light in the retina, which, in turn, results in a widening of the angular-sensitivity function and an increase in the extent of its flanks (Warrant and McIntyre, 1991). Generally speaking, a widening of the angular-sensitivity function means an increase in Δρ, and an examination of Equation 4 shows that increasing Δρ decreases signal. Thus, the trade-off between resolution and sensitivity now becomes obvious: there must be several sets of optimum eye parameters, for each average habitat luminance, which maximize signal and thereby maximize v_{max}. Each optimum set represents an eye construction which has found the best trade-off (at least in mathematical terms) between capture of the available photons and attainment of the narrowest-possible angular-sensitivity function.

There is one further thing we should mention about apposition eyes with regard to the trade-off between resolution and sensitivity. In apposition eyes only, Δρ is related to the half-width of the blur-circle (if diffraction is limiting, this is the ratio of the stimulating wavelength to the corneal facet diameter: λ/D) and the angular size of the rhabdom (the ratio of its diameter to the posterior nodal distance; d/f) by (Snyder, 1977, 1979):

TABLE 2
Values of ν_{max} in Three Superposition Eyes

Species	Animal	Activity	A/f	Δρ	ν_{max}	ν_s	Sampling
Anoplognathus	Beetle	Nocturnal	2.1	5	0.12	0.38	Oversampling
Onitis	Beetle	Crepuscular	1.2	4.3	0.26	0.23	~Matched
Phalaenoides	Moth	Diurnal	0.7	1.7	0.86	0.29	Undersampling

Note: ν_{max} is the maximum detectable spatial frequency in the eye (cycles/deg); ν_s is the sampling frequency of the photoreceptor array (cycles/deg); A/f is the relative aperture (see text); Δρ is the acceptance angle (deg); Oversampling occurs for $\nu_{max} < \nu_s$, matching for $\nu_{max} = \nu_s$, and undersampling for $\nu_{max} > \nu_s$. Data from Warrant and McIntyre 1990a.

$$\Delta\rho^2 = (\lambda/D)^2 + (d/f)^2 \tag{5}$$

An improvement in resolution (i.e., a decrease in Δρ) will, therefore, occur for a decrease in rhabdom diameter *(d)* or an increase in posterior nodal distance *(f)*. Also, an inspection of Equation 1 shows that sensitivity to an extended source is proportional to $(d/f)^2$. Thus, sensitivity is increased when there is an increase in *d* or a decrease in *f*. Thus, in apposition eyes, any attempt to improve sensitivity by, say, an increase in *d* will result in a decrease in resolution, and vice versa.

Of course, our discussions so far have been somewhat hypothetical because the processes of evolution do not work according to equations. Whatever design evolves is not only a reflection of the visual requirements of the animal, but is also due to other constraints, such as the ancestral history of the animal, developmental constraints, and constraints due to the materials and design of the organism (Dumont and Robertson, 1986). For instance, it may be more important for an animal to sacrifice a wide range of perceivable spatial frequencies in favor of a higher ratio of signal to noise at all spatial frequencies within a much narrower range. In other words, it really depends on what a particular animal needs to see in order to survive. There is no point having a visual system constructed to perceive spatial details that the animal never uses. It is better instead to heighten sensitivity to those spatial details which are meaningful to the animal's normal behavior.

Nevertheless, our general conclusion is still borne out in nature. Animals which live in dim habitats have eye constructions which maximize sensitivity to a narrow range of spatial frequencies (i.e., they have a low ν_{max}). Thus, if one can calculate ν_{max} for a particular animal, one has gone a long way in predicting the compromises that have been made between resolution and sensitivity during the evolution of its eye. Values of ν_{max} for three insects with superposition eyes that live at very different light intensities are given in Table 2. What is immediately obvious is that the dimmer the habitat of the animal, the lower its value of ν_{max}. Both the decrease in habitat luminance and the accompanying increase in Δρ (concomitant with the strive for greater photon capture through an increase in A/f) are probably responsible for the decline in ν_{max}.

So far we have neglected the effects of the packing of photoreceptors in the retina. Their angular separation influences how densely they can sample the information available in a scene. The maximum spatial frequency theoretically attainable by a hexagonal sampling array of photoreceptors in the retina (ν_s) is equal to $1/(\sqrt{3}\,\Delta\phi)$ where Δφ is the angular separation of the photoreceptors, better known as the interommatidial angle. How does ν_{max} compare with ν_s? For an ideal eye construction we would predict that ν_{max} should be matched to ν_s, that is, they should be equal. If the sampling array is coarser than ν_{max} (i.e., $\nu_{max} > \nu_s$) the eye is said to undersample, if denser (i.e., $\nu_{max} < \nu_s$) the eye is said to oversample

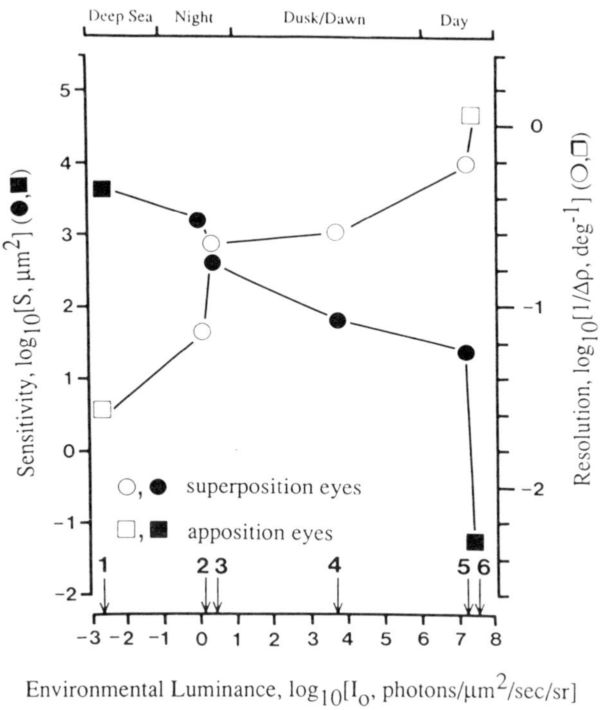

FIGURE 9. The trade-off between resolution (open symbols) and sensitivity (closed symbols) in six different arthropods with superposition eyes (circles) and apposition eyes (squares), which exist at different environmental luminances, I_o (assuming λ = 550 nm), spanning an intensity range of 11 logarithmic units from dark deep sea to bright daylight. Resolution has been defined as the inverse of the acceptance angle, $\Delta\rho$. Sensitivity has been calculated using Equation 1. Numbered arrows along the abscissa refer to each data pair (resolution and sensitivity) from each species. 1 = the isopod *Cirolana* (Nilsson and Nilsson, 1981); 2 = the crayfish *Cherax* (Bryceson and McIntyre, 1983); 3 = the Christmas beetle *Anoplognathus* (Meyer-Rochow and Horridge, 1975); 4 = the dung beetle *Onitis* (Warrant and McIntyre, 1990a); 5 = the day moth *Phalaenoides* (Horridge et al., 1977; Horridge et al., 1983b); 6 = the mantid *Ciulfina* (Horridge and Duelli, 1979).

(Table 2). The relative advantages of the different sampling strategies are beyond the scope of this review, and a discussion of these is found elsewhere (Snyder et al., 1986; Warrant and McIntyre, 1990a).

We can gain further insight into the trade-off between resolution and sensitivity if, for a certain group of animals (each of whom live at very different light intensities I_o), we plot together both their sensitivity S (calculated with Equation 1) and their "resolution" ($1/\Delta\rho$). The results (Figure 9) indicate that for the actual eye design solutions that have evolved for different animals, there is, indeed, a rough trade-off: if resolution is low, sensitivity is high, and vice versa. Of course, the correlation between resolution and sensitivity is not perfect because of differences in eye size, and in addition there may be specialized examples of eye designs where *both* resolution and sensitivity have been cleverly maximized in a way that our simple trade-off relationship (Figure 9) does not predict. For instance, if we compare the superposition eye of the moth *Phalaenoides* (5 in Figure 9) with the apposition eye of the mantid *Ciulfina* (6 in Figure 9), which both live in bright daylight, the former has almost the same narrowness of angular sensitivity as the latter, but 1000 times greater sensitivity (as is also evident in Figure 5). No doubt, the calculated value of ν_{max} for *Phalaenoides* may be somewhat larger than for *Ciulfina*. Such conclusions may also be reached if the afocal apposition eyes of butterflies are compared with their normal focal counterparts. Due to their unique optics (Nilsson et al., 1984; van Hateren and Nilsson, 1987) the afocal design performs considerably better than the focal design in terms of both resolution and sensitivity, even though both designs are used at the same intensities.

Our discussions thus far have only considered stationary photoreceptors acting in isolation. If, however, an animal is required to move rapidly, both resolution and sensitivity suffer, as we mentioned in Section I. This is regardless of habitat luminance. A useful strategy to offset the problem is to have eyes (or eye regions) as big as possible (Land, 1981), and this is often the strategy adopted by animals which rely on the motion of themselves and/or the environment to survive. Second, we have not covered the possibility of photoreceptors pooling their outputs for improved photon perception at low light levels. This is a particularly useful strategy and one which allows a behavioral response from the animal that would not be possible had pooling been absent (Doujak, 1985). Nevertheless, pooling always occurs at the expense of resolution and, so, does not contradict our reasoning presented above.

III. THE ALTERATION OF RESOLUTION AND SENSITIVITY BY LIGHT ADAPTATION

We saw in the last section how the environmental light intensity is a major factor in the evolution of a compound eye design. We also explored the reasons why, at any given intensity, a particular eye design might be favored because of its optimal balance between resolution and sensitivity. For instance, we noted that animals which live in dimmer habitats generally possess eye designs which maximize photon capture. How, then, with its given eye design, can an animal cope with excursions in habitat luminance away from the average level it normally experiences? Provided that the excursions are not too great, the eye can cope by invoking within itself certain physiological and anatomical changes which adjust its operating range to match the new luminance. These changes are collectively termed adaptation, and they perform their task by dynamically controlling the sensitivity of the eye to light. In so doing they can simultaneously alter, intentionally or otherwise, the ability of the eye to resolve spatial detail. We will now discuss how adaptation provides a second, dynamic, determinant of the trade-off between resolution and sensitivity.

A. LIGHT ADAPTATION AND THE DYNAMIC RESPONSE RANGE OF PHOTORECEPTORS

As we saw in Section II.B, arthropod photoreceptors have sigmoidal stimulus-response functions referred to as V-LogI curves (Figure 6A). As their name suggests, these functions are not linear, and this is because of self-shunting and, in addition, at higher intensities, declining photon effectiveness in the transduction processes within the photoreceptor (Lillywhite and Laughlin, 1979; Laughlin and Hardie, 1978; Laughlin, 1981). Rather, they are logarithmic transforms of stimulus intensity. Let us consider the V-LogI curves shown in Figure 10, which were obtained from a single intracellular recording over a 24-h period from the photoreceptor of a crepuscular dung beetle, *Onitis alexis,* in states of both light and dark adaptation. At this stage we will consider only the left-most curve, that was obtained in dark-adapted conditions at night. The first point to note is the proportionality of the transform between about 20 and 80% of its maximum voltage response (53 mV). In this region, the response increment ΔV is the same for identical stimulus increments $\Delta LogI$. This range of intensities, which corresponds to responses within the "linear" region of the V-LogI curve, is sometimes referred to as the operating range of the photoreceptor. In *Onitis alexis,* which has a superposition eye, the operating range is approximately 3 log units of intensity. In many apposition eyes, it is often around half or two thirds of this value, as in flies and dragonflies (Laughlin and Hardie, 1978). Apart from this, there are two other

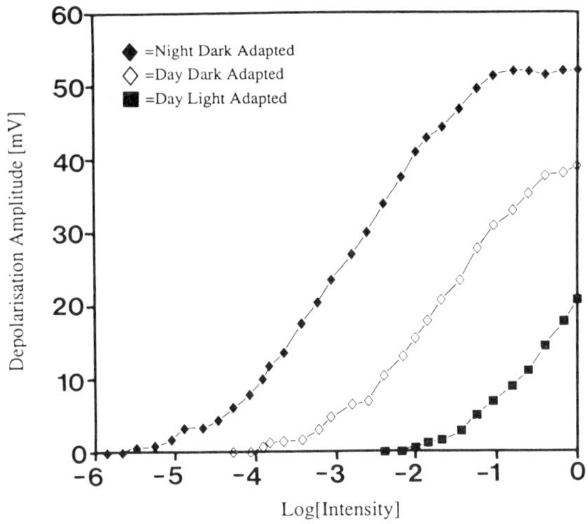

FIGURE 10. Intensity-response (V-LogI) functions measured from a single retinula cell of *Onitis alexis* over a 24-h period during different states of adaptation (dark adapted during nighttime and daytime, light adapted during daytime).

qualities of the V-LogI curve worth noting. First, the ''linear'' region has a certain slope, which has been referred to as the contrast efficiency (Laughlin and Hardie, 1978; Laughlin, 1981). This name has been chosen because a constant logarithmic intensity increment, ΔLogI, corresponds to a constant contrast increment. Thus, the slope of the V-LogI curve is a reflection of the efficiency with which contrasts are coded within the retina. Second, the V-LogI curve has a certain position along the LogI axis which indicates the sensitivity of the photoreceptor to the absolute light intensity. This position is conveniently quantified by the PAQ_{50}, which we introduced earlier, that is, the axial flux of photons required to elicit a 50% maximum response from the photoreceptor. The reciprocal of this value has been termed the range sensitivity (Laughlin and Hardie, 1978). Simply put, those photoreceptors which have a high range sensitivity are those which require fewer photons to elicit a half-maximum response. Such photoreceptors would have V-Log I curves shifted in position toward dimmer intensities on the LogI axis. An inspection of PAQ_{50}s in Table 1 shows, for instance, that *O. alexis* has a higher range sensitivity during both day and night than the dragonfly *Hemicordulia tau*.

So far, we have been dealing with the characteristics of the V-LogI curve in the dark-adapted state. What happens if we allow the photoreceptor to light adapt? In carefully controlled experiments, Laughlin and Hardie (1978) measured the response of fly and dragonfly photoreceptors to point-source flashes (intensity I_t) superimposed on a number of different sustained-background adapting-intensities (I_a). Of course, the photoreceptor cannot distinguish the difference between the energy from the point source and that from the adapting background, so that during the flash, the effective stimulus intensity is, therefore, simply the sum of the two ($I_a + I_t$). A brief summary of their results is (Figure 11) as follows:

1. When an adapting background is introduced to a previously dark adapted photoreceptor, it suddenly depolarizes, then immediately repolarizes through, first, a ''fast'' and, then, a ''slow'' mechanism of light adaptation, so that after some seconds a new sustained level of receptor potential is reached. The magnitude of this sustained level above the dark-adapted resting potential gets larger for greater adapting intensities, but after a certain adapting intensity it begins to level off. Any subsequent flash will produce a response superimposed on the sustained level.

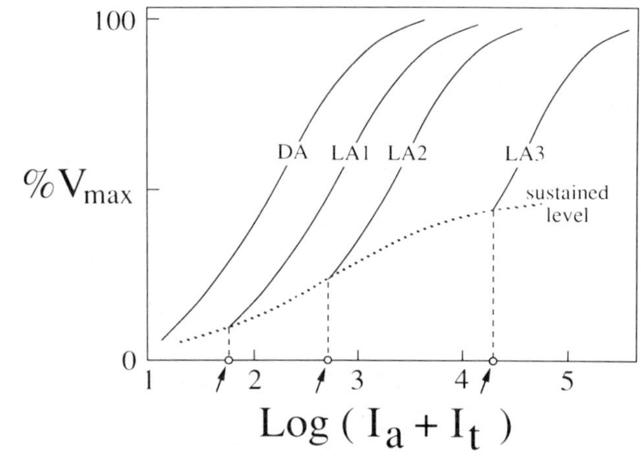

FIGURE 11. Schematic V-LogI curves during dark adaptation (DA) and three states of light adaptation (in order of increasing intensity [arrows]: LA1, LA2, LA3) from a single photoreceptor. The intensity of light which the photoreceptor experiences results from the sum of the adapting intensity (I_a) and the test flash intensity (I_t). The dotted line represents the steady-state sustained response of the photoreceptor to continuous background illumination. As light adaptation increases, the V-LogI curve shifts to higher intensities while maintaining contrast constancy (i.e., constant slope) and approximately the same operating range. (Adapted from Laughlin and Hardie, 1978).

2. Measured V-LogI functions have the same slope (i.e., contrast efficiency) and roughly the same operating range, but differ in their position along the LogI axis. For higher adapting intensities, the processes of light adaptation shift the V-LogI function to higher intensities; that is, the range sensitivity of the photoreceptor is decreased. The mechanisms which bring this about are the same as those responsible for the logarithmic transform: self-shunting and declining photon effectiveness in the transduction mechanisms (see above).

In addition to preservation of the operating range and contrast efficiency, recent results indicate that light adaptation also preserves a maximum ratio of signal to noise in the photoreceptor, which, in the absence of these mechanisms, would have been severely diminished due to saturation (Howard et al., 1987).

These are very important results because they indicate that the way the photoreceptor responds to fluctuations around a particular mean luminance is unaltered by a shift in the mean luminance, i.e., the response to a particular contrast increment is the same at all the mean luminances the visual system is expected to cope with. Another way of saying this is that the eye adheres to the principles of contrast constancy, and this is one of the great unifying features of vertebrate and invertebrate vision. The principles of contrast constancy also operate in both the photoreceptors and retinal ganglion cells of vertebrates (Normann and Werblin, 1974; Sakmann and Creutzfeldt, 1969).

The fact that the position of the V-LogI curve is shifted along the LogI axis is indicative of the ability of light adapting mechanisms to alter dynamically the sensitivity of the eye to light. The magnitude of the shift is termed the range shift, and as the adapting background intensity is increased, the light adaptation becomes more effective until eventually the range shift becomes proportional to the background intensity (Laughlin and Hardie, 1978). The total range shift differs between species and probably reflects the total range of environmental intensities normally experienced by an animal; generally, it seems that superposition eyes have both a larger operating range and a greater range shift than apposition eyes (Warrant and McIntyre, 1990b). For instance, the dung beetle *O. alexis* (superposition eye) has an operating range of 3 log units which can be shifted through a total range shift of at least 4 log units via light adaptation. On the other hand, the apposition eye of the fly *Calliphora stygia* (Laughlin and Hardie, 1978), has an operating range of about 1.5 log units and a total range shift of about 2 log units. Interestingly, the ratio of operating range to range shift is approximately the same in both species.

B. HOW IS LIGHT ADAPTATION ACHIEVED, AND WHAT ARE ITS EFFECTS ON RESOLUTION AND SENSITIVITY?

There are basically two ways compound eyes adapt to light, and both have parallels in vertebrate eyes. First, if the eye experiences a sudden and sustained increase in light intensity, pupil mechanisms are activated which decrease the amount of light incident on the eye which reaches the retina. In compound eyes the pupil is usually a screen of migrating light-absorbent pigment granules which reside within specialized pigment cells. Second, the eye can respond by altering the amount of light which reaches the retina that is actually absorbed and transduced by the photoreceptors. This can be achieved by altering the amount of light actually captured by each photoreceptor by (1) altering its optical properties (e.g., its length and width; its refractive index relative to its surrounds) and/or (2) altering its transduction gain (i.e., the magnitude of the membrane potential generated by each absorbed photon). Pupil mechanisms and photon-capture mechanisms both combine to shift the position of the V-LogI curve in response to a change in adaptation state. Exactly what the relative contributions of the two mechanisms are to this range shift is a subject of considerable interest and has so far only been determined in a few species. The discussion on adaptation mechanisms which follows is brief by necessity, and for a fuller description, the interested reader is referred to the superb review by Autrum (1981).

It is easier to understand the mechanisms of adaptation if we construct a very simple model of the eye (Figure 12). In this model system, we can see the purpose of adaptation, which is to change the sensitivity of the eye by shifting the V-LogI curve while simultaneously preserving its contrast efficiency and operating range. The range of membrane potentials (V) which constitutes the response is, therefore, a regulated output of the system. Adaptation is the process by which changes in this regulated output are counteracted when a sustained change in light intensity (the input) is experienced by the eye. The inputs to the system are of two types:

1. Disturbance inputs: sudden *unpredictable* changes in the intensity of light incident on the eye, e.g., due to an animal moving into shade
2. Circadian inputs: show *predictable* changes in the intensity of light incident on the eye due to the daily light cycle (in many species, the eye preadapts in anticipation of the cyclic change)

The model illustrates the main steps in adaptation: (1) a change in the input is detected (e.g., by feedforward and/or feedback mechanisms) by one or more sensors; then (2), after certain delays, the output signals of these sensors alert controlling mechanisms to alter the amount of light incident on, absorbed by, and transduced in the photoreceptors; and, finally, (3), the output membrane potentials from the controlled system, the photoreceptor, are then altered, so that they return to the regulated range. In addition to setting an appropriate sensitivity for the eye by a suitable shift in the V-LogI curve, the processes of adaptation also, invariably, alter the spatial properties of the eye. This is reflected in differences in the measured angular-sensitivity function (Figures 6B and 7A) at different states of adaptation. Such differences imply that adaptation also affects the resolution of the eye.

We will now consider each section of the model in more detail, beginning with the controlled system, the photoreceptor.

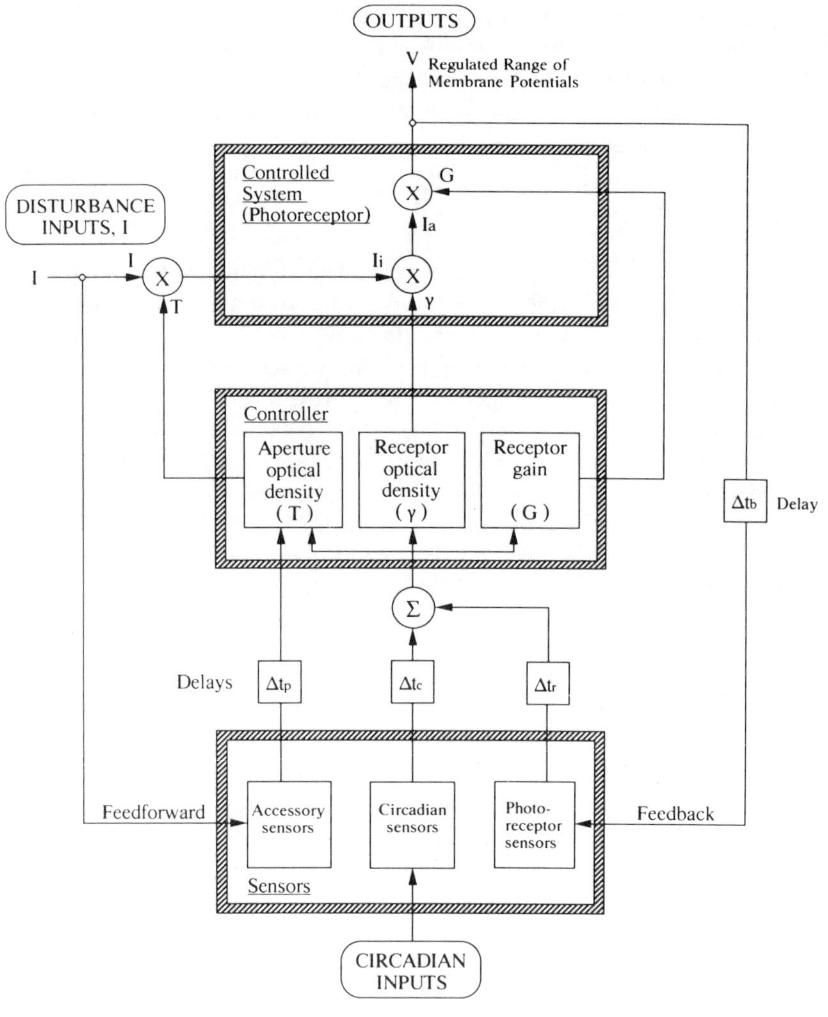

FIGURE 12. A simple model of the compound eye showing the main ways in which inputs (disturbance and circadian) are sensed (by accessory, circadian, and photoreceptor sensors) and controlled (by controller mechanisms for aperture [T] and photoreceptor [γ] optical densities as well as for photoreceptor gain [G]). A regulated output (a range of membrane potentials, V) is produced by the photoreceptor (the Controlled System). The symbols X and Σ denote multiplication and addition, respectively. Other symbols are defined in the text.

1. The Photoreceptor

The photoreceptor contains visual pigment which absorbs photons and causes appropriate conductances through the phototransductive membrane. This results in an appropriate current and membrane potential *(V)*, and the mechanism seems to be mediated by transmembrane fluxes of calcium ions (Sandler and Kirschfeld, 1988). The photoreceptor has two properties which can be controlled by the ''controlling'' mechanisms in the model. First, the photoreceptor has an intrinsic conductance gain (*G*), which is the ratio of activated depolarizing conductances to effective photons. Both the magnitude and duration of currents produced by single photons can be altered (Fein and Charlton, 1977). Second, the photoreceptor has a particular size and shape and, in addition, a refractive-index difference relative to its surrounding medium. These properties determine its optical density, that is, the fraction (γ) of photons incident on the retinal surface that are absorbed effectively by the photoreceptor.

Thus, very crudely, we can say that the output range of membrane potentials *(V)* is determined by the gain (*G*) and the absorbed photon signal (I_a) and that, in turn, I_a is determined by the photon intensity incident on the retina (I_i) and the absorption fraction (γ).

2. The Sensors

The sensors detect changes to the inputs of the system (disturbance and circadian) and then, after certain delays, invoke changes in the controller mechanisms. There are three main sensors: those which are accessory and lie within extrarhabdomal structures, that detect disturbances directly (feedforward); those which detect circadian changes; and those which detect changes in the output potentials of the photoreceptor (feedback). The first two have been implicated as specific sensor organs. The last is merely a convenient way to categorize the actions of the photoreceptor itself; a change in membrane potential changes the directions and fluxes of certain ions across the membrane, and this, in turn, can affect the gain *(G)* and the optical densities of the photoreceptor (γ) and aperture (*T* = the fraction of light incident on the eye that is transmitted to the retina). Each of the three sensors invoke changes in one or more of the control mechanisms after delays of Δt_p (accessory sensors), Δt_c (circadian sensors), and Δt_r (photoreceptor sensors).

Accessory sensors have been found in moth superposition eyes and are specialized light-sensitive regions located just proximal to the crystalline cones, but exactly in which cell(s) is not yet known. They could be within the retinula cell extensions, the cone cells, or either of the primary or secondary screening pigment cells (Land, 1987; Nilsson et al., 1989, 1992). Their only effect is to alter the optical density of the aperture *(T)* by inducing the migration of screening pigment.

Circadian sensors in invertebrates have been the subject of intense study (Page, 1982). Specific extraocular light-sensitive organs, capable of entraining and resetting circadian oscillator circuits in the brain, have been either found or implicated in both insects (reviews: Dushay et al., 1989; Tomioka et al., 1990) and crustaceans (Fuentes-Pardo and Inclan-Rubio, 1987; Inclan-Rubio and Fuentes-Pardo, 1987; Sandeman et al., 1990). The sensors feed directly into circadian oscillators (or "pacemaker") circuits which, via direct neurosecretory efferent input to photoreceptors and pigment cells, alter the optical densities of both the aperture and photoreceptor and also alter the photoreceptor gain (Fleissner, 1986; Barlow et al., 1989). Thus, as night approaches, the circadian system causes the photoreceptor gain to increase gradually and both optical densities to decrease similarly; these changes can improve the sensitivity of the eye (through a shift in the V-LogI curve) by a factor of at least 10. The approach of day causes the opposite procedure. It has been suggested that the circadian input preadapts the eye and could (during the daytime) protect a sheltering nocturnal animal from sudden and unexpected exposure to bright daylight (Leggett and Stavenga, 1981; Kier and Chamberlain, 1990; Warrant and McIntyre, in preparation).

Photoreceptor "sensors" receive feedback (with delay Δt_b) from the output potential (V), and this invokes changes in photoreceptor gain (G) and optical density (γ). The optical density can be altered by (1) photoreceptive microvillar breakdown and synthesis, which alters its shape and cross-sectional area for photon capture (review: Blest, 1980) and (2) changing the refractive-index of the medium surrounding the rhabdom by the migration of screening pigment granules or mitochondria toward (light adaptation) or away (dark adaptation) from the rhabdom, a process common in apposition eyes. In many superposition eyes, photoreceptor "sensors" also invoke changes in the optical density of the aperture (*T*) by causing the migration of screening pigment granules within the pigment cells (White et al., 1983; Bernard et al., 1984; Weyrauther, 1986). As we mentioned before, this can also be invoked by accessory sensors. Which of the two sensor types is responsible in a given eye depends on the species (Nilsson et al., 1989).

Of course, in a given eye at a given time of day, more than one sensor may be receiving input and giving output. Their combined effect on the controller mechanisms will, therefore, reflect the relative magnitudes of their outputs. For instance, during the daytime, *but in darkness,* the circadian sensor of a nocturnal animal may be providing such a strong output (i.e., that the eye should be prelight-adapted) that the other sensors, signaling that the eye should dark adapt, are simply overruled.

3. The Controller

The controller contains the mechanisms which alter the optical densities of the aperture *(T)* and the photoreceptor (γ) and also alter the photoreceptor gain *(G)*. We have already introduced some of the ways in which these mechanisms operate, especially in relation to the gain, a discussion of which we will omit here. We will now examine them in a little more detail, especially in reference to their effects on resolution (via the shape and width of the angular-sensitivity function) and sensitivity (via the position of the V-LogI curve for point-source-stimulation and via the parameters of Equation 1 for extended source stimulation).

a. The Optical Density of the Aperture

The optical density of the aperture is the fraction (*T*) of light striking the surface of the eye (*I*) which is transmitted to the retina. The most common mechanism of attenuation is to absorb incoming light by light-absorbent screening pigment granules (located within specialized pigment cells) which, in response to a change in light intensity, migrate to an appropriate position within the eye. The greater an increase in light intensity, the further advanced the migration and the greater the light attenuation.

In some apposition eyes the primary pigment cells (with their pigment granules) form a migrating iris over the rhabdom entrance (Figure 13A). During light adaptation the cells migrate inward, simultaneously pressing into the soft cone cells and pushing the rhabdom down proximally. This action produces a thin cone thread and reduces the diameter of the rhabdom entrance *(d)*. During dark adaptation the migration is outward, and the full rhabdom diameter is exposed. Thus, because of an increase in *d*, dark adaptation brings a decrease in resolution (Equation 5) and an increase in sensitivity (Equation 1). For precisely the opposite reason, light adaptation brings an increase in resolution and a decrease in sensitivity. This trend is seen in all the species which possess this type of eye. For instance, in *Limulus*, $\Delta\rho$ increases from 6 to 13° on going from light adaptation to dark adaptation (Barlow et al., 1989), and in the beetle *Creophilus* the increase is 5 to 7° (Meyer-Rochow, 1974). In these species and in several species of ants (Felisberti and Ventura, 1989; Ventura et al., 1976; Menzi, 1987), sensitivity improves considerably on dark adaptation. In addition, in continuous darkness, most species display the influence of a circadian rhythm: sensitivity is higher (by over 10 times), and resolution lower, at night than during the day.

In other apposition eyes, adaptation alters the posterior nodal distance of the lens (*f*) by altering the length of the crystalline cone (Figure 13B), as in the shrimp *Artemia* (Nilsson and Odselius, 1981). During light adaptation the crystalline cone becomes longer (and *f* bigger) than during dark adaptation. Examination of Equations 1 and 5 show that, just as before, sensitivity is higher and resolution lower during dark adaptation than during light adaptation. No circadian rhythm operates in the eye of *Artemia*.

In superposition eyes, light adaptation can cause migrations of distal screening pigment in one of two ways (Figure 13C): (1) restricted migrations (LA:A) only around the proximal tips of the crystalline cones (as is common in species which experience only a limited range of light intensities) or (2) large proximal migrations (LA:B) into the clear zone (as is common in species which experience a wide range of light intensities). This latter migration mechanism

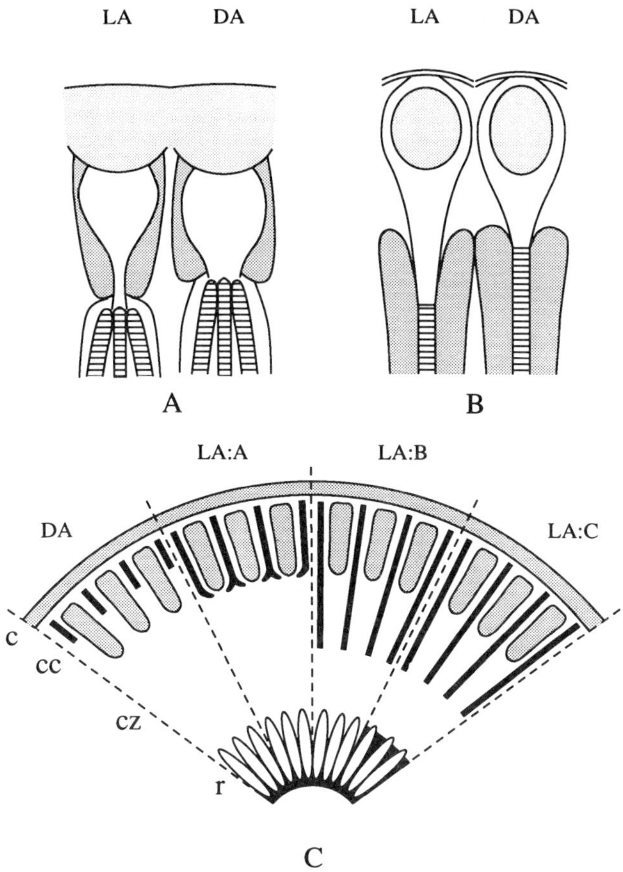

FIGURE 13. Mechanisms of controlling the optical density of the aperture (T) in compound eyes in response to dark (DA) and light (LA) adaptation. (A) Changing the diameter of a variable pigment pupil in front of the rhabdom (apposition eyes); (B) changing the focal length of the ommatidium (apposition eyes); (C) changing the positions of screening pigment granules (superposition eyes). In the DA state, distal pigment (black) lies between the crystalline cones (cc) just proximal to the cornea (c). In many species, proximal pigment lies near the basement membrane. In the cone pigment migration mechanism (LA:A), light adaptation induces migration of the distal pigment around the proximal ends of each crystalline cone so as to occlude all but a small exit aperture. In the longitudinal pigment migration mechanism (LA:B and LA:C), distal pigment migrates proximally into the clear zone (cz). This may also be accompanied (LA:C) by distal migrations of the proximal pigments between the rhabdoms (r). (A and B courtesy of Dr. D.-E. Nilsson.)

can also be accompanied by distal light-adapted migrations of proximal screening pigments (LA:C) between the rhabdoms, and this has the effect of reducing light spread in the retina (Figure 8) and improving resolution (Bryceson and McIntyre, 1983; Warrant and McIntyre, in preparation).

In the first, *cone,* mechanism (Warrant and McIntyre, in preparation) pigment migration subserving light adaptation results in an attenuation in the transmitted light, but *not* in a reduction (except in extreme light adaptation) of the diameter of the superposition aperture through which light enters. Because the diameter of this aperture (A) is large relative to f, the considerable number of spherically aberrated rays from the periphery of the aperture is, therefore, not significantly reduced on light adaptation. Thus, image quality (i.e., the quality of the blur-circle and the extent to which it spreads in the retina [Figure 8] remains nearly unchanged: $\Delta\rho$ is found to be invariant with the state of adaptation, as in the dung beetle

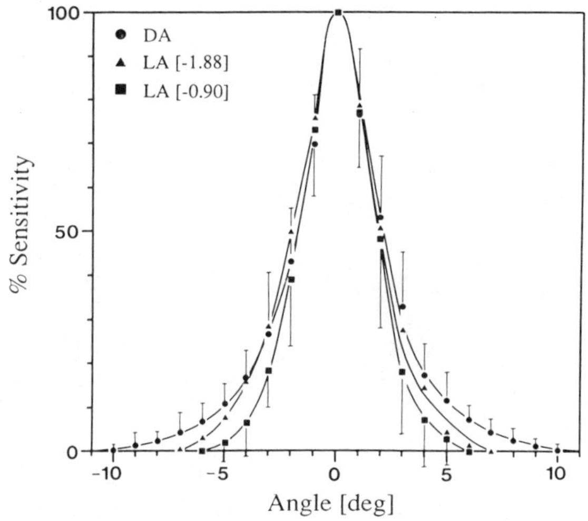

FIGURE 14. Averaged angular-sensitivity functions from several photoreceptors of many dung beetles (*Onitis alexis*) at night in states of dark adaptation (DA: circles), partial light adaptation (LA; −1.88 log units adapting intensity: triangles) and stronger light adaptation (LA; −0.90 log units of adapting intensity: squares). Solid curves are fitted by eye to the data. Error bars: standard deviations on the data points.

O. alexis, where $\Delta\rho$ is constant at 4° (Warrant and McIntyre, 1990a). Nevertheless, light adaptation does affect the shape of the angular-sensitivity function: as light adaptation proceeds, the flanks in the function are reduced in magnitude (Figure 14). From our discussions in Section II.B, this will improve the MTF and thereby improve resolution. Such a flank reduction may indicate that the rays exiting from each crystalline cone are aberrated (i.e., not parallel, as is optimal) and that the cone mechanism can remove the more severely aberrated rays from each cone during light adaptation.

The situation, however, is quite different in the second, *longitudinal,* mechanism where proximal migrations of screening pigment quickly reduce both the superposition aperture and the peripheral aberrated rays. This results in a considerable improvement in image quality on light adaptation. In the crayfish *Cherax* (Walcott, 1974; Bryceson and McIntyre, 1983) and in the water beetle *Macrogyrus* (Horridge et al., 1983a), refinement of both the width and shape of the angular-sensitivity function improves resolution on light adaptation ($\Delta\rho$ reduces from 13 to 5° in *Cherax* and from 12 to 3° in *Macrogyrus*).

Both mechanisms of pigment migration can be under the control of a circadian rhythm; i.e., pigment is pushed into a light-adapted position during daytime, regardless of whether the animal is in total darkness. In *Onitis* this has been used to investigate the effects of pigment migration on sensitivity (Figure 10) and, in particular, to determine what its contribution is to the range shift of the V-LogI curve and what is due to changes in the transduction gain (Warrant and McIntyre, 1990b). Between the extremes of daytime light adaptation and night-time dark adaptation, the total range shift (sensitivity change) in the eye is at least 4 log units, of which 2 log units can be attributed to changes in transduction gain. Similar shifts due to transduction gain, but much larger shifts due to pigment migration (3 to 4 log units), have been obtained in moths and crayfish, which have the longitudinal migration mechanism (Bernhard and Ottoson, 1960, 1964; Bernhard et al., 1963; Rodriguez-Sosa and Aréchiga, 1982). The 6 or more log units of total sensitivity change one finds in superposition eyes is enormous when compared with the 2 log unit change typical in apposition eyes (Srinivasan and Bernard, 1980). The reason for the difference is possibly that superposition eyes are more likely to experience a wider range of light intensities (Warrant and McIntyre, 1990b).

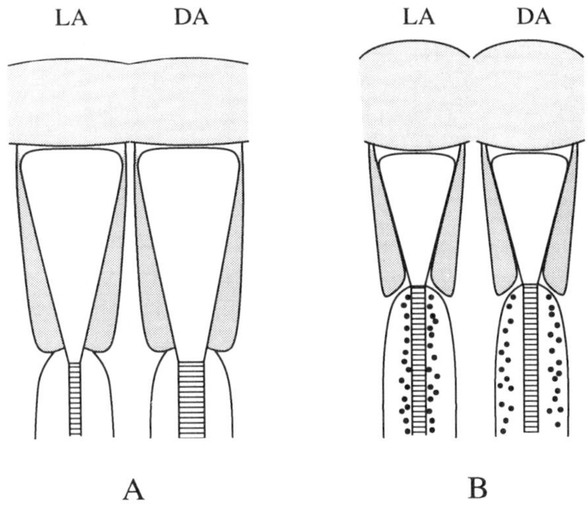

FIGURE 15. Mechanisms of controlling the optical density of the photoreceptor (γ) in compound eyes in response to dark (DA) and light (LA) adaptation (apposition eyes). (A) Changing the diameter of the rhabdom; (B) changing the radial position of screening pigment granules in the retinula cells. (Diagram courtesy of Dr. D.-E. Nilsson.)

But as we can see, our trade-off is just as evident in superposition eyes as it is in apposition eyes. An improvement in sensitivity during dark adaptation is accompanied by a loss in resolution, and vice versa during light adaptation.

b. The Optical Density of the Photoreceptor

In addition to its intrinsic absorption, the rhabdom has an optical density set by its shape, its size, its relative refractive index and by the properties of the medium which immediately surrounds it. These factors determine what fraction (γ) of the light incident on the retina, that is intended for a particular rhabdom, is absorbed by that rhabdom alone. These factors change during changes in adaptation state. This not only results in changes in the number of absorbed photons (sensitivity), but also in changes in resolution (Δρ). Another possible determinant of γ is the concentration of unbleached photopigment in the rhabdom. While this is an important factor in photopic (cone-driven) sensitivity in vertebrate eyes (Valeton and van Norren, 1983), it is considered to be of negligible importance in invertebrate eyes (Hamdorf, 1979) where photoreconversion is rapid.

Adaptational changes in γ via changes in rhabdom diameter (through synthesis and breakdown of photoreceptive microvilli; Blest, 1980) is a strategy so far only described in apposition eyes (Figure 15A). Apart from one exceptional family of moths (the Pyralids; Horridge and Giddings, 1971; Bernard et al., 1984), no examples are known in compound eyes where adaptation changes the *length* of the rhabdoms. In contrast, distal-proximal movements of the entire rhabdom (unaltered length) are well known (Autrum, 1981). An increase in rhabdom diameter *(d)* is common in response to dark adaptation, when photon capture, and therefore sensitivity (Equation 1), is required. Inspection of Equation 5 shows that resolution will be reduced during dark adaptation. Light adaptation brings a reduction in *d* and a corresponding increase in resolution and decrease in sensitivity. In addition, in many species, a circadian rhythm provides an extra control over rhabdom diameter. During the day the rhabdom is narrower than during the night, but at either of these times a change in adaptation state can override the circadian rhythm (Williams, 1982). Examples of this mechanism are found in the following groups (during transition from day light adaptation to night dark adaptation):

1. In locusts (Williams, 1982, 1983; Horridge et al., 1981) which demonstrate an increase in *d* from 2 to 5 μm, an increase in Δρ from 2 to 5°, and an increase in (extended source) sensitivity of 100 times
2. In the mantids *Tenodera* and *Orthodera* (Rossel, 1979; Horridge et al., 1981) which demonstrate an increase in *d* from 1.5 to 5 μm (*Orthodera*), an increase in Δρ from 0.7 to 2° (*Tenodera fovea*), and an increase in (extended source) sensitivity of 10 times (*Tenodera*)
3. In crabs *Ocypode, Scylla,* and *Hemigrapsus* (Nässel and Waterman, 1979; Leggett and Stavenga, 1981; Arikawa et al., 1987) which demonstrate increases in *d* of up to 3 times, an increase in Δρ from 4 to 11° (*Scylla*), and an increase in (extended source) sensitivity of at least 10 times. In crabs, some of the observed changes may also be due to screening pigment migration (Leggett and Stavenga, 1981).

Adaptational alterations of the refractive index difference between the rhabdom and its surrounding medium will also alter γ. When this refractive index difference is maximized, so too is light capture within the rhabdom (Land, 1981). This gain in sensitivity during dark adaptation is achieved by a decrease in the refractive index of the medium which immediately surrounds the rhabdom. More incident light rays are thereby captured by total internal reflection within the rhabdom. During light adaptation, the refractive index increases again, and this results in a decrease in sensitivity. In compound eyes, the refractive index immediately outside the rhabdom (within the retinula cell) can be altered by two main mechanisms. First, increases and decreases in refractive index can, respectively, occur when mitochondria migrate toward the rhabdom during light adaptation and away from it during dark adaptation. This is common in locusts (Horridge and Barnard, 1965). In special cases the migration of pigment granules can achieve the same result. A second mechanism involves the formation of a lower refractive-index palisade around the rhabdom during dark adaptation and its dissolution during light adaptation. This is found in many species, notably locusts, dragonflies, and cockroaches. How do these mechanisms affect resolution? During light adaptation when mitochondria coat the rhabdom, the cone of light which is actually absorbed by the rhabdom is effectively narrower (i.e., smaller θ in Figure 8) because the steeper rays cannot be internally reflected. This results in a narrower angular sensitivity and improved resolution (Autrum, 1981). The opposite is true during dark adaptation, especially when a palisade is present. In locusts, Tunstall and Horridge (1967) found that increases in Δρ can be correlated with the outward migration, during dark adaptation, of mitochondria and the appearance of a palisade, but changes in rhabdom diameter (not known at that time) may also be partly responsible. Nevertheless, an increase in Δρ following palisade formation has been predicted theoretically in cockroaches, and the results agree well with experiment (Snyder and Horridge, 1972).

Another common mechanism by which γ can be altered is by the migration of pigment granules within the retinula cells (apposition eyes, Figure 15B). Pigment granules, lying alongside the rhabdom, migrate radially away from the rhabdom during dark adaptation and toward it again during light adaptation. This is frequently found in apposition eyes with very thin rhabdoms functioning as waveguides (e.g., in flies, bees, and butterflies) and in this respect is different from the mitochondrial mechanism operating around the larger rhabdoms of locusts. Instead of strictly mediating changes in relative refractive index, these pigment granules, in their light-adapted position, absorb some of the energy of higher-order waveguide modes propagating along the outside of the rhabdom (Snyder, 1975; van Hateren, 1989). This affects not only the sensitivity, but also the resolution. Due to the small distances of migration (a few microns), this radial pupil (Kirschfeld and Franceschini, 1969; Stavenga, 1979) is both fast (typically 5 to 60 s) and effective (attenuation of light by 1 to 2 log units of intensity; Srinivasan and Bernard, 1980). In thin rhabdoms, the waveguide modes are clearly visible in an ophthalmoscope (Nilsson et al., 1988; Land and Osorio, 1990); in the thinnest rhabdoms, such as those found in small butterflies (circa 1 μm wide) only the lowest (first) mode is present. In slightly larger rhabdoms both the first and second modes may propagate. In even larger rhabdoms (circa 4 μm wide), such as in the butterfly *Melanitis*, the third and fourth may also be present. Higher-order modes accept light from larger angles, and the presence of a higher-order mode, therefore, results in a broadening of the angular sensitivity. In addition, the higher-order modes have a greater proportion of their energy propagating outside the rhabdom compared with lower-order modes. Thus, during light adaptation, inwardly migrating pigment granules successively absorb these exterior higher modes, starting from the highest, and this not only reduces sensitivity, but due to the loss of higher modes, narrows the angular sensitivity and improves resolution (Smakman et al., 1984; Nilsson et al., 1988; Land and Osorio, 1990). The opposite is true during dark adaptation. This has recently been demonstrated in *Melanitis*, where impressive optical and electrophysiological investigations have shown that $\Delta\rho$ decreases from 2.9 to 1.5° during light adaptation (Land and Osorio, 1990) and does so very quickly (5 s). In a previous study in dragonflies, the time course of $\Delta\rho$ change from one state of adaptation to another was so fast (1.5° [dark adapted] to 0.7° [light adapted] in less than 10 s) that a radial pupil was implicated, but not proved (Warrant and Pinter, 1990). Despite this however, in those thinnest rhabdoms propagating only the lowest mode, no such improvement in resolution takes place (Hardie, 1979; Nilsson et al., 1988; Land and Osorio, 1990).

We should finally mention two other ways in which γ can be altered. The first, found in superposition eyes, we have already mentioned in connection with aperture screening pigments: stationary or migrating screening pigment granules located between the rhabdoms. In light-adapted eyes, or those normally experiencing only daylight, this pigment shield cuts out stray light passing between rhabdoms and thereby improves resolution, but reduces the total amount of light absorbed in the retina (Warrant and McIntyre, 1991). The second mechanism involves a light reflecting layer (the tapetum) found at the base of the eye in many species (mostly in superposition eyes, but also in the apposition eyes of amphipods and isopods). The tapetum allows light that has passed proximally through the rhabdom to be reflected backward for a second chance of absorption in the rhabdom (Land, 1981). This *effectively doubles the length of the rhabdom* (l_{rh}), and inspection of Equation 1 shows that this will improve sensitivity. However, particularly in unshielded retinas, reflection of steeply incident rays will degrade resolution. In some eyes, particularly in crustaceans, the tapetum can be exposed during dark adaptation or obscured (by pigment), or even removed, during light adaptation (Debaisieux, 1944).

IV. CONCLUSIONS

We have discussed the trade-off between resolution and sensitivity in compound eyes from two perspectives. First, we have shown that evolution tends to establish a particular *static* trade-off on the basis of habitat luminance. Animals which have evolved to exist in dim habitats have eyes constructed to maximize photon capture, but in doing so have sacrificed the ability to resolve a wide range of spatial frequencies. Second, we have shown that the range of intensities that an animal normally experiences can, through adaptation, provide a *dynamic* fine-tuning of the trade-off. Light adaptation almost always brings with it a decrease in sensitivity and an increase in resolution, and dark adaptation, the opposite. Thus, through both the static and dynamic trade-offs between resolution and sensitivity, each animal is uniquely suited to the range of light intensities it normally experiences.

ACKNOWLEDGMENTS

The authors would like to express their sincere gratitude to Dr. Dan-E. Nilsson for stimulating discussions and for critically reading the manuscript.

REFERENCES

Arikawa, K., Kawamata, K., Suzuki, T., and Eguchi, E. (1987). Daily changes of structure, function and rhodopsin content in the compound eye of the crab *Hemigrapsis sanguineus, J. Comp. Physiol. A,* **161**, 161–174.

Autrum, H. (1981). Light and dark adaptation in invertebrates, in *Vision in Invertebrates (Handbook of Sensory Physiology,* Vol. VII/6C), Autrum, H. Ed., Springer-Verlag, Berlin, pp 1–91.

Barlow, H. B. (1952). The size of ommatidia in apposition eyes, *J. Exp. Biol.,* **29**, 667–674.

Barlow, R. B., Chamberlain, S. C., and Lehman, H. K. (1989). Circadian rhythms in the invertebrate retina, in *Facets of Vision,* Stavenga, D. G. and Hardie, R. C., Eds., Springer-Verlag, Berlin, 257–280.

Bernard, G. D., Owens, D. O., and Hurley, A. V. (1984). Intracellular optical physiology of the eye of the pyralid moth *Amyelois, J. Exp. Zool.,* **229**, 173–187.

Bernhard, C. G. and Ottoson, D. (1960). Studies on the relation between the pigment migration and the sensitivity changes during dark adaptation in diurnal and nocturnal Lepidoptera, *J. Gen. Physiol.,* **44**, 205–215.

Bernhard, C. G., Höglund, G., and Ottoson, D. (1963). On the relation between pigment position and light sensitivity of the compound eye in different nocturnal insects, *J. Insect Physiol.,* **9**, 573–586.

Bernhard, C. G. and Ottoson, D. (1964). Quantitative studies on pigment migration and light sensitivity in the compound eye at different light intensities, *J. Gen. Physiol.,* **47**, 465–478.

Blest, A. D. (1980). Photoreceptor membrane turnover in arthropods: comparative studies of breakdown processes and their implications, in *The Effects of Constant Light on Visual Processes,* Williams, T. P. and Baker, B. N., Eds., Plenum Press, New York, 217–245.

Bruno, M. S., Barnes, S. N., and Goldsmith, T. H. (1977). The visual pigment and visual cycle of the lobster *Homarus, J. Comp. Physiol.,* **120**, 123–142.

Bryceson, K. P. (1986). Diurnal changes in photoreceptor sensitivity in a reflecting superposition eye, *J. Comp. Physiol. A.,* **158**, 573–582.

Bryceson, K. P. and McIntyre, P. (1983). Image quality and acceptance angle in a reflecting superposition eye, *J. Comp. Physiol.,* **151**, 367–380.

Caveney, S. and McIntyre, P. (1981). Design of graded-index lenses in the superposition eyes of scarab beetles, *Philos. Trans. R. Soc. London Ser. B,* **294**, 589–632.

Clarke, G. L. and Denton, E. J. (1962). Light and animal life, in *The Sea,* Hill, M. N., Ed., Wiley-Interscience, New York, 456–468.

Cleary, P., Deichsel, G., and Kunze, P. (1977). The superposition image in the eye of *Ephestia kühniella, J. Comp. Physiol.,* **119**, 73–84.

Compton, J. C. (1978). Facts behind exposure determination, *J. Biol. Photogr. Assoc.,* **46**, 159–166.

Debaisieux, P. (1944). Les yeux des Crustacés: structure, développement, réactions à l'éclairement, *Cellule,* **50**, 9–122.

Doujak, F. E. (1985). Can a shore crab see a star?, *J. Exp. Biol.,* **166**, 385–393.

Dubs, A. (1982). The spatial integration of signals in the retina and lamina of the fly compound eye under different conditions of luminance, *J. Comp. Physiol.,* **146**, 321–343.

Dumont, J. P. C. and Robertson, R. M. (1986). Neuronal circuits: an evolutionary perspective, *Nature (London),* **233**, 849–853.

Dushay, M. S., Rosbash, M., and Hall, J. C. (1989). The disconnected visual system mutations in *Drosophila melanogaster* drastically disrupt circadian rhythms, *J. Biol. Rhythms,* **4**, 1–27.

Exner, S. (1981). *Die Physiologie der facettirten Augen von Krebsen und Insecten,* Franz Deuticke, Leipzig; English translation: Hardie, R. C. (1988). *The Physiology of the Compound Eyes of Insects and Crustaceans,* Springer-Verlag, Berlin.

Fein, A. and Charlton, J. S. (1977). A quantitative comparison of the effects of intracellular calcium injection and light adaptation on the photoresponse of *Limulus* ventral photoreceptors, *J. Gen. Physiol.,* **70**, 591–600.

Felisberti, F. and Ventura, D. F. (1989). Morphological changes in the ommatidia of an ant in the day and night states, *Braz. J. Med. Biol. Res.,* **22**, 1159–1162.

Fleissner, G. (1986). Die innere Uhr und der Lichtsinn von Skorpionen und Käfern, *Naturwissenschaften,* **73**, 78–88.

Fuentes-Pardo, B. and Inclan-Rubio, V. (1987). Caudal photoreceptors synchronize the circadian rhythms in crayfish. I. Synchronization of ERG and locomotor circadian rhythms, *Comp. Biochem. Physiol. A*, **86**, 523–527.

Hamdorf, K. (1979). Physiology of invertebrate visual pigments, in *Vision in Invertebrates (Handbook of sensory physiology,* **Vol. VII/6A**), Autrum, H., Ed., Springer-Verlag, Berlin, 145–224.

Hardie, R. C. (1979). Electrophysiological analysis of fly retina. I. Comparative properties of R1-6 and R7 and R8, *J. Comp. Physiol.*, **129**, 19–33.

Horridge, G. A. and Barnard, P. B. T., (1965). Movement of palisade in locust retinula cells when illuminated, *Q. J. Microsc. Sci.*, **106**, 131–135.

Horridge, G. A. and Duelli, P. (1979). Anatomy of the regional differences in the eye of the mantis *Ciulfina*, *J. Exp. Biol.*, **80**, 165–190.

Horridge, G. A., Duniec, J., and Marçelja, L. (1981). A 24-hour cycle in single locust and mantis photoreceptors, *J. Exp. Biol.*, **91**, 307–322.

Horridge, G. A. and Giddings, C. (1971). The retina of *Ephestia* (Lepidoptera), *Proc. R. Soc. London Ser. B*, **179**, 87–95.

Horridge, G. A., Marçelja, L., Jahnke, R., and McIntyre, P. (1983a). Daily changes in the compound eye of a beetle *(Macrogyrus), Proc. R. Soc. London Ser. B*, **217**, 265–285.

Horridge, G. A., Marçelja, L., and Jahnke, R. (1983b). Retinula cell responses in a moth superposition eye, *Proc. R. Soc. London Ser. B*, **220**, 47–68.

Horridge, G. A., McLean, M., Stange, G., and Lillywhite, P. G. (1977). A diurnal moth superposition eye with high resolution *Phalaenoides tristifica* (Agaristidae), *Proc. R. Soc. London Ser. B*, **196**, 233–250.

Howard, J., Blakeslee, B., and Laughlin, S. B. (1987). The intracellular pupil mechanism and photoreceptor signal:noise ratios in the fly *Lucilia cuprina, Proc. R. Soc. London Ser. B*, **231**, 415–435.

Inclan-Rubio, V. and Fuentes-Pardo, B. (1987). Caudal photoreceptors synchronize the circadian rhythms in crayfish. II. Functional relationships between caudal and visual photoreceptors, *Comp. Biochem. Physiol. A*, **86**, 529–536.

Kier, C. K. and Chamberlain, S. C. (1990). Dual controls for screening pigment movement in photoreceptors of the *Limulus* lateral eye: circadian efferent input and light, *Vision Neurosci.*, **4**, 237–255.

Kirschfeld, K. (1974). The absolute sensitivity of lens and compound eyes, *Z. Naturforsch.*, **29C**, 592–596.

Kirschfeld, K. and Franceschini, N. (1969). Ein Mechanismus zur Steuerung des Lichtflusses in den Rhabdomeren des Komplexauges von *Musca, Kybernetik*, **6**, 13–22.

Land, M. F. (1976). Superposition images are formed by reflection in the eyes of some oceanic decapod crustacea, *Nature (London)*, **263**, 764–765.

Land, M. F. (1981). Optics and vision in invertebrates, in *Handbook of Sensory Physiology*, **Vol VII/6B,** Autrum, H., Ed., Springer-Verlag, Berlin, 471–592.

Land, M. F. (1984). The resolving power of diurnal superposition eyes measured with an ophthalmoscope, *J. Comp. Physiol.*, **154**, 515–533.

Land, M. F. (1987). Screening pigment migration in a sphingid moth is triggered by light near the cornea, *J. Comp. Physiol. A*, **160**, 355–357.

Land, M. F. (1989). Variations in the structure and design of compound eyes, in *Facets of Vision*, Stavenga, D. G. and Hardie, R. C., Eds., Springer-Verlag, Berlin, 90–111.

Land, M. F. and Osorio, D. C. (1990). Waveguide modes and pupil action in the eyes of butterflies, *Proc. R. Soc. London Ser. B*, **241**, 93–100.

Laughlin, S. B. (1981). Neural principles in the peripheral visual systems of invertebrates, in *Vision in Invertebrates (Handbook of Sensory Physiology,* **Vol. VII/6B)**, Autrum, H., Ed., Springer-Verlag, Berlin, 133–280.

Laughlin, S. B. and Hardie, R. C. (1978). Common strategies for light adaptation in the peripheral visual systems of fly and dragonfly, *J. Comp. Physiol.*, **128**, 319–340.

Leggett, L. M. W. and Stavenga, D. G. (1981). Diurnal changes in angular sensitivity in crab photoreceptors, *J. Comp. Physiol.*, **144**, 99–109.

Lillywhite, P. G. (1977). Single photon signals and transduction in an insect eye, *J. Comp. Physiol.*, **122**, 189–200.

Lillywhite, P. G. and Laughlin, S. B. (1979). Transducer noise in a photoreceptor, *Nature (London)*, **277**, 569–572.

McIntyre, P. and Caveney, S. (1985). Graded-index optics are matched to optical geometry in the superposition eyes of scarab beetles, *Philos. Trans. R. Soc. London B*, **311**, 237–269.

Menzi, U. (1987). Visual adaptation in nocturnal and diurnal ants, *J. Comp. Physiol. A*, **160**, 11–22.

Meyer-Rochow, V. B. (1974). Fine structural changes in dark-light adaptation in relation to unit studies of an insect compound eye with a crustacean-like rhabdom, *J. Insect. Physiol.*, **20**, 573–589.

Meyer-Rochow, V. B. and Horridge, G. A. (1975). The eye of *Anoplognathus* (Coleopter: Scarabaeidae), *Proc. R. Soc. London Ser. B*, **188**, 1–30.

Nässel, D. R. and Waterman, T. H. (1979). Massive diurnally modulated photoreceptor membrane turnover in crab during light and dark adaptation, *J. Comp. Physiol.*, **131**, 205–216.

Nilsson, D.-E. (1988). A new type of imaging optics in compound eyes, *Nature (London)*, **332**, 76–78.

Nilsson, D.-E. (1989). Optics and evolution of the compound eye, in *Facets of Vision*, Stavenga, D. G. and Hardie, R. C., Eds., Springer-Verlag, Berlin, 30–73.

Nilsson, D.-E. (1990). From cornea to retinal image in invertebrate eyes, *Trends Neurosci.*, **13**, 55–64.

Nilsson, D.-E. and Nilsson, H. L. (1981). A crustacean compound eye adapted for low light intensities (Isopoda), *J. Comp. Physiol.*, **143**, 503–510.

Nilsson, D.-E. and Odselius, R. (1981). A new mechanism for light-dark adaptation in the *Artemia* compound eye (Anostraca, Crustacea), *J. Comp. Physiol.*, **143**, 389–399.

Nilsson, D.-E., Land, M. F., and Howard, J. (1984). Afocal apposition optics in butterfly eyes, *Nature (London)*, **312**, 561–563.

Nilsson, D.-E., Land, M. F., and Howard, J. (1988). Optics of the butterfly eye, *J. Comp. Physiol. A*, **162**, 341–366.

Nilsson, D.-E., Henrekson, I., and Järemo, A.-C. (1989). Pupil control in compound eyes: more than one mechanism in moths, in *Neurobiology of Sensory Systems*, Singh, R. N. and Strausfeld, N. J., Eds., Plenum Press, New York, 17–22.

Nilsson, D.-E., Hamdorf, K., and Höglund, G. (1992). Localization of the pupil trigger in insect superposition eyes, *J. Comp. Physiol. A*, **170**, 217–226.

Normann, R. A. and Werblin, F. S. (1974). Control of retinal sensitivity. I. Light and dark-adaptation of vertebrate rods and cones, *J. Gen. Physiol.*, **63**, 37–61.

Page, T. L. (1982). Extraretinal photoreception in entrainment and photoperiodism in invertebrates, *Experientia*, **38**, 1007–1013.

Pinter, R. B. (1972). Frequency and time domain properties of retinula cells of the desert locust *(Schistocera gregaria)* and the house cricket *(Acheta domesticus)*, *J. Comp. Physiol.*, **77**, 383–397.

Rodriguez-Sosa, L. and Arêchiga, H. (1982). Range of modulation of light sensitivity by accessory pigments in the crayfish compound eye, *Vision Res.*, **22**, 1515–1524.

Rossel, S. (1979). Regional differences in photoreceptor performance in the eye of the praying mantis, *J. Comp. Physiol.*, **131**, 95–112.

Sakmann, B. and Creutzfeldt, O. D. (1969). Scotopic and mesopic light adaptation in the cat's retina, *Pfleugers Arch.*, **313**, 168–185.

Sandeman, D. C., Sandeman, R. E., and de Couet, H. G. (1990). Extraretinal photoreceptors in the brain of the crayfish *Cherax destructor*, *J. Neurobiol.*, **21**, 619–629.

Sandler, C. and Kirschfeld, K. (1988). Light intensity controls extracellular Ca^{2+} concentration in the blowfly retina, *Naturwissenschaften*, **75**, 256–258.

Scholes, J. H. (1964). Discrete subthreshold potentials from the dimly lit insect eye, *Nature (London)*, **202**, 572–573.

Scholes, J. H. (1965). Discontinuity of the excitation process in locust visual cells, *Cold Spring Harbor Symp. Quant. Biol.*, **30**, 517–527.

Smakman, J. G. J., van Hateren, J. H., and Stavenga, D. G. (1984). Angular sensitivity of blowfly photoreceptors: intracellular predictions and wave-optical predictions, *J. Comp. Physiol. A*, **155**, 239–247.

Snyder, A. W. (1975). Photoreceptor optics — theoretical principles, in *Photoreceptor Optics*, Synder, A. W. and Menzel, R., Eds., Springer-Verlag, Berlin, 38–55.

Snyder, A. W. (1977). Acuity of compound eyes: physical limitations and design, *J. Comp. Physiol.*, **116**, 161–182.

Snyder, A. W. (1979). Physics of vision in compound eyes, in *Handbook of Sensory Physiology*, **Vol VII/6A**, Autrum, H., Ed., Springer-Verlag, Berlin, 225–313.

Snyder, A. W. and Horridge, G. A. (1972). The optical function of changes in the medium surrounding the cockroach rhabdom, *J. Comp. Physiol.*, **81**, 1–8.

Snyder, A. W., Stavenga, D. G., and Laughlin, S. B. (1977). Spatial information capacity of compound eyes, *J. Comp. Physiol.*, **116**, 183–207.

Snyder, A. W., Bossomaier, T. R. J., and Hughes, A. (1986). Optical image quality and the cone mosaic, *Science*, **231**, 499–501.

Srinivasan, M. V. and Bernard, G. D. (1975). The effect of motion on visual acuity of the compound eye: a theoretical analysis, *Vision Res.*, **15**, 515–525.

Srinivasan, M. V. and Bernard, G. D. (1980). A technique for estimating the contribution of photomechanical responses to visual adaptation, *Vision Res.*, **20**, 511–521.

Stavenga, D. G. (1979). Pseudopupils of compound eyes, in *Handbook of Sensory Physiology*, **Vol VII/6A**, Autrum, H., Ed., Springer-Verlag, Berlin, 357–439.

Tomioka, K., Okada, Y., and Chiba, Y. (1990). Distribution of circadian photoreceptors in the compound eye of the cricket *Gryllus bimaculatus*, *J. Biol. Rhythms*, **5**, 303–313.

Tunstall, J. and Horridge, G. A. (1967). Electrophysiological investigations of the optics of the locust retina, *Z. Vergl. Physiol.*, **55**, 167–182.

Valeton, M. J. and van Norren, D. (1983). Light adaptation of primate cones: an analysis based on extracellular data, *Vision Res.*, **23**, 1539–1547.
van Hateren, J. H. (1989). Photoreceptor optics, theory and practice, in *Facets of Vision,* Stavenga, D. G. and Hardie, R. C., Eds., Springer-Verlag, Berlin, 74–89.
van Hateren, J. H. and Nilsson, D.-E. (1987). Butterfly optics exceed the theoretical limits of conventional apposition eyes, *Biol. Cybern.*, **57**, 159–168.
Ventura, D. F., Martinoya, C., Bloch, S., and Puglia, N. M. (1976). Visual sensitivity and the state of adaptation in the ant *Atta sexdens* (Hymenoptera: Formicoidea), *J. Comp. Physiol.*, **110**, 333–342.
Vogt, K. (1975). Zur Optik des Flußkrebsauges, *Z. Naturforsch.*, **30**, 691.
Walcott, B. (1974). Unit studies on light-adaptation in the retina of the crayfish, *Cherax destructor, J. Comp. Physiol.*, **94**, 207–218.
Warrant, E. J. and Pinter, R. B. (1990). Changes of acuity during light and dark adaptation in the dragonfly compound eye, *Z. Naturforsch.*, **45C**, 137–141.
Warrant, E. J. and McIntyre, P. D. (1990a). Limitations to resolution in superposition eyes, *J. Comp. Physiol. A*, **167**, 785–803.
Warrant, E. J. and McIntyre, P. D. (1990b). Screening pigment, aperture and sensitivity in the dung beetle superposition eye, *J. Comp. Physiol. A*, **167**, 805–815.
Warrant, E. J. and McIntyre, P. D. (1991). Strategies for retinal design in arthropod eyes of low F-number, *J. Comp. Physiol. A*, **168**, 499–512.
Weyrauther, E. (1986). Do retinula cells trigger the screening pigment migration in the eye of the moth *Ephestia kühniella? J. Comp. Physiol. A*, **159**, 55–60.
White, R. H., Banister, M. J., and Bennett, R. R. (1983). Spectral sensitivity of screening pigment migration in the compound eye of *Manduca sexta, J. Comp. Physiol. A*, **153**, 59–66.
Williams, D. S. (1982). Ommatidial structure in relation to turnover of photoreceptor membrane in the locust, *Cell Tissue Res.*, **225**, 595–617.
Williams, D. S. (1983). Changes of photoreceptor performance associated with the daily turnover of photoreceptor membrane in the locust, *J. Comp. Physiol.*, **150**, 509–519.

Chapter 16

NONLINEAR LATERAL INHIBITION APPLIED TO MOTION DETECTION IN THE FLY VISUAL SYSTEM

Abdesselam Bouzerdoum and Robert B. Pinter

TABLE OF CONTENTS

I. Introduction 424

II. The Fly Visual System 425
  A. The Compound Eye 425
  B. The Lamina Ganglionaris 426
  C. The Medulla 428
  D. The Lobula Complex 429
    1. Lobula 429
    2. Lobula Plate 430

III. Elementary Movement Detection 432

IV. Shunting Lateral Inhibition 435
  A. The Movement Detector 436

V. Response Characteristics 437
  A. Response to Low-Contrast Stimuli 437
  B. Response to Drifting Gratings 438
  C. Contrast Frequency Dependence 439
  D. Spatial Frequency Dependence 441
  E. Adaptation of Contrast Sensitivity Functions 443
  F. Dependence on Mean Luminance Level 444

VI. Discussion 444

Acknowledgment 446

References 446

I. INTRODUCTION

Vision is one of the most important senses to animals, for to interact with the world, almost all but a few simple animals need to know what is present at a distance in the world and where it is. Seeing is particularly vital to fast flying animals like insects; they rely on their sense of sight for landing on a target, chasing a mate, or escaping a predator. Vision is a complex function which comprises many information processing tasks. One early and fundamental visual task is the processing of motion information. This peripheral process is prerequisite to many other brain functions such as, e.g., separating a figure from its surround, detecting moving objects, avoiding obstacles, or stabilizing the flight course (see Nakayama, 1985; Wehner, 1981). Sometimes objects and targets can be identified by features independent of movement, such as shape and color; often, however, movement is the definitive feature.

It is, hence, not surprising to find movement-detecting units in the peripheral visual systems of animals. The vertebrate retinae, for example, contain ganglion cells that are highly sensitive to motion (Maturana et al., 1960; Maturana and Frenk, 1963; Barlow and Hill, 1963; Michael, 1968; Lipetz and Hill, 1970; Werblin, 1970; Marchiafava, 1979). Some of these cells are *directionally selective;* they respond vigorously to motion in a particular direction termed the preferred direction and give a weak or no response at all to motion in the opposite, or null, direction (Maturana and Frenk, 1963; Barlow et al., 1964; Michael, 1968; Werblin, 1970; Marchiafava, 1983). Movement appears even more important to insects; movement-sensitive units have been discovered in both the second and third optic ganglia as well as in the protocerebrum (Horridge et al., 1965; Collett and Blest, 1966; Bishop and Keehn, 1967; Bishop et al., 1968; McCann and Dill, 1969; Mimura, 1971; Hausen, 1976; DeVoe, 1980; DeVoe et al., 1982). Furthermore, all tangential cells identified so far in the lobula plate, the posterior part of the fly's third visual ganglion, are directionally selective movement-detecting (DSMD) neurons (Dvorak et al., 1975; Eckert and Bishop, 1978; Hausen, 1976, 1981, 1982a, b). This tangential system has been extensively investigated because of its primary involvement in the neural control of the insect's optomotor response (Hausen and Egelhaaf, 1989; Hausen, 1984; Wehner, 1981).

The tangential cells of the lobula plate are wide-field neurons which receive their inputs from large retinotopic arrays of small-field movement detectors (Hausen, 1981, 1982a, b; Egelhaaf, 1985). It is widely believed that the small-field movement detectors which feed the large-field tangential cells operate on the principle of a nonlinear asymmetric interaction of signals mediated by adjacent visual elements (Buchner 1976; Franceschini et al. 1989; Zaagman et al. 1978). Up till now, the nonlinear interaction that has most widely been invoked is of the correlation type, i.e. a time averaging operation following the multiplication of two appropriately filtered signals (Buchner, 1976; Egelhaaf and Reichardt, 1987; Reichardt, 1987, 1961). The correlation type interaction was first proposed by Hassenstein and Reichardt (1956) to account for the behavioral optomotor response of insects. It has subsequently been applied in the description of the activity of movement detecting neurons of the fly's lobula plate (Egelhaaf and Borst, 1989; Egelhaaf et al., 1989; Egelhaaf and Reichardt, 1987; Zaagman et al., 1978); but in the mean time it has been shown to account for certain aspects in motion vision in other species including man (Foster, 1971; van Santen and Sperling, 1984).

A simple correlation detector may account for a large number of functional properties of movement-sensitive interneurons in the lobula plate, but it cannot explain most of their nonlinear characteristics. Being formulated, in their early stages, as linear time-invariant spatiotemporal filters, the correlation detector and its variants cannot account for the saturation and adaptation properties exhibited by these neurons (Lenting et al., 1984; Dvorak et al., 1980; de Ruyter van Steveninck et al., 1986; Borst and Egelhaaf, 1987; Egelhaaf and Borst, 1989). Recently, we have proposed a motion detector that shares the same topological structure and is, to second order, functionally equivalent to the correlation model; it was, however, based on a nonlinear interaction of the shunting inhibitory type (Bouzerdoum and Pinter, 1991, 1989a). Unlike the correlation model, our motion detector, which was named multiplicative inhibitory motion detector (MIMD), can account for many nonlinear characteristics of the fly's DSMD interneurons (preliminary results on the applicability of the MIMD to the activity of these neurons can be found in Bouzerdoum and Pinter, 1990).

The purpose of this chapter is to investigate the biological plausibility of the MIMD and its ability to account for the functional characteristics of the DSMD neurons. After a somewhat detailed description of the fly's primary visual system, we discuss the concept of elementary movement detection and its underlying mechanism in the fly visual system. Then, we propose shunting lateral inhibition as the mechanism mediating directional selectivity at the level of the small-field movement detectors. This gives rise to the MIMD, which will be used to predict the transient and steady-state response properties as well as the spatial characteristics of the fly's DSMD neurons. The chapter is concluded with a discussion section.

II. THE FLY VISUAL SYSTEM

The primary visual system of insects consists of a pair of multifaceted eyes known as the compound eyes, the visual ganglia in the two optic lobes, and the tracts and projection centers of the visual interneurons in the protocerebrum (Strausfeld, 1976a). A secondary visual subsystem is composed of three ocelli and the ocellar tract. The high dipteran visual system is the best known and perhaps the most thoroughly studied insect visual system to date. In this visual system, each optic lobe comprises three visual ganglia; they are commonly called, from the periphery inward, the lamina ganglionaris,* the medulla, and the lobula complex, which contains two synaptic layers: the lobula and the lobula plate (Figure 1).

Each of these optic ganglia is strictly organized into columns and strata, and consists of an inner plexiform region of synaptic neuropil flanked or faced with granular layers which contain the neuron somata. The columnar neuropils are retinotopically connected by anteroposteriorly crossed projections of fibers, i.e., the external and internal optic chiasma. The columnar fibers projecting retinotopically through the chiasmata form the major visual pathway from the eye to the anterior and posterior lateral protocerebrum, where the main output regions of the optic lobes reside. Here, we introduce in some detail the primary visual system of flies with emphasis on the movement detection system; a more comprehensive neuroanatomical description is found in the *Atlas of an Insect Brain* (Strausfeld, 1976a).

A. THE COMPOUND EYE

The fly compound eye is composed of approximately 3000 to 4000 ommatidia separated from each other by screening pigment cells. Each ommatidium is a functional unit comprising a dioptric apparatus and a retinula, which contains eight receptor or retinular cells labeled R1-8. The dioptric apparatus funnels light to the caps of the retinular cells through a lens system formed by a facet or corneal lens and a crystalline cone. It is the hexagonal array of the corneal lenses that gives the compound eye its faceted appearance.

The photosensitive structure of the compound eye, i.e., the rhabdom, is formed by microvillar contributions (rhabdomeres) of the retinular cells. In higher Diptera, a pair of central rhabdomeres is surrounded by six peripheral ones in an asymmetric trapezoidal pattern

* In modern terminology, the lamina ganglionaris is simply referred to as the lamina; here, the two terms are used interchangeably.

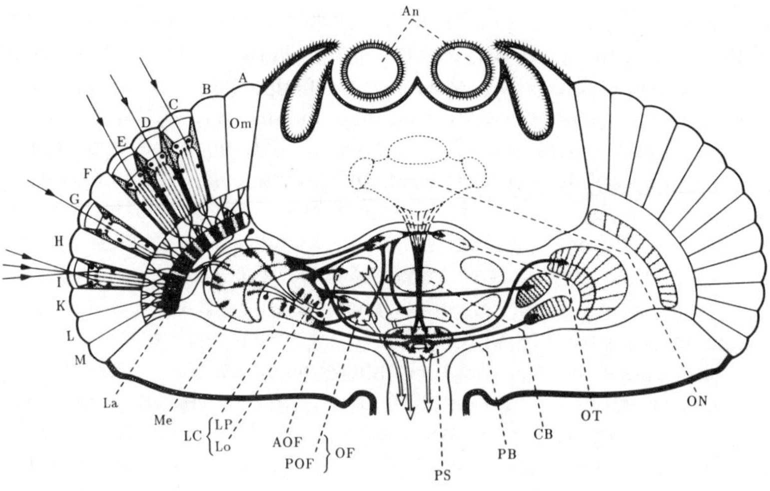

FIGURE 1. Schematic representation of the fly visual system. Corneal lens, crystalline cone, retinular cells and their optical axes are illustrated in ommatidia C-E, G, and I, respectively. In particular, ommatidium I illustrates the divergence of the optical axes of retinular cells lying in a single ommatidium; ommatidium G shows the common optical axis of the central receptors; and ommatidia C-E show three of the seven parallel optical axes lying in neighboring ommatidia (see text for more details). Projections from the optic ganglia to higher-order centers in the midbrain are indicated by black arrows, while the output projections to the thoracic ganglia are indicated by white or open arrows. Abbreviations: An, antennae; Om, ommatidium; La, lamina; Me, medulla; LC, lobula complex; LP, lobula plate; Lo, lobula; AOF, anterior optic focus; POF, posterior optic focus; OF, optic focus; PS, posterior slope; PB, protocerebral bridge; CB, central body; OT, optic tubercle; ON, ocellar neuropil. Modified from (Kirschfeld 1979).

(Nilsson, 1989; Shaw, 1989). The peripheral rhabdomeres extend over the whole ommatidium length and belong to one receptor cell each (R1-6). The central rhabdomeres, on the other hand, are shorter and belong to two retinular cells: the superior and the inferior central cells (R7/8). Unlike in most insects, the rhabdomeres in higher dipteran flies are unfused and, hence, act as separate lightguides, except for the central rhabdomeres: arranged in tandem, one on top of the other, they form a single lightguide. As a result, the peripheral rhabdomeres of an ommatidium possess divergent optical axes, and the central rhabdomeres share a common one which represents the optical axis of the ommatidium (Figure 1). However, the optical axes of six different peripheral lightguides, lying in a characteristic group of six ommatidia surrounding a central one, are parallel. The rhabdomeres are arranged in a way such that one of the peripheral rhabdomeres in each of the six neighboring ommatidia is "looking" in the same direction in space as the central pair of rhabdomeres lying in the central ommatidium. Another common characteristic of these rhabdomeres is that the axons of their receptor cells all converge to the lamina cartridge beneath the central ommatidium, where their signals are pooled (see next section on the lamina ganglionaris). The central cells, however, differ from the peripheral ones in their synaptic connections. The latter, having short axons, terminate in the first optic ganglion (the lamina ganglionaris), whereas the former, equipped with long axons, course through the lamina to reach synaptic endings in the medulla. (For further details on the compound eyes of flies and other insects, see Nilsson, 1989; or Goldsmith and Bernard, 1974.)

B. THE LAMINA GANGLIONARIS

In Diptera, the first optic ganglion, namely the lamina ganglionaris, is a curved plate lying beneath the retina. It is composed of structurally identical columnar subunits known as optic cartridges, which exactly match in number the overlying ommatidia. The cylindrical structure of each cartridge is surrounded by three epithelial glial cells isolating it from adjacent cartridges and preventing intercartridge connectivities, except at the proximal and distal borders of the lamina. As already mentioned, the axons of a set of retinular cells looking at the same point in space project to the same optic cartridge, where the short axons form a hollow column of terminals and the long axons pass through to the medulla without making any synaptic contacts at the lamina level. This organization increases the amount of light gathered from a point in space at least sixfold, since a cartridge receives inputs from six different facets instead of one. This enhances the signal-to-noise ratio without degrading visual resolution (Nilsson, 1989; Shaw, 1989). The orderly structure of the lamina is very complex and several aspects of its circuitry are still debatable or, as yet, inadequately understood (Shaw, 1984, 1989; Strausfeld, 1989). However, studies starting early this century (Vigier, 1908; Cajal and Sanchez, 1915) and spanning over eight decades have revealed a great deal about the structure of this neuropil. It is one of the most comprehensively described ganglia of any nervous system today. The interested reader is referred to one of many comprehensive reviews (Laughlin, 1981; Shaw, 1981, 1984, 1989; Strausfeld and Nässel, 1981; Strausfeld, 1984, 1989) for further details, because in this section we give only a brief description to the lamina ganglionaris.

The most recent accounts (Shaw, 1989; Strausfeld, 1984, 1989) indicate that neurons inhabiting the lamina include, in addition to peripheral retinular cell endings and central receptor axons, four types of centrifugal cells, two types of amacrine cells, and six types of relay neurons: five monopolar cells designated L1-5 according to Braitenberg (1967), and one T-cell, T1. The centrifugal cells are so named because their cell bodies are located in the medulla or just beneath the lamina and their fibers project centrifugally into the synaptic region in the lamina, where they make different kinds of synaptic contacts. The centrifugal species include narrow-field neurons, C2 and C3 which are confined to a single cartridge, and wide-field tangential cells, TAN1/2 and TAN3. The only anaxonal neurons in the lamina are the amacrine cells, which possess the largest variety of synaptic connections of any lamina neuron. Each cartridge receives six α-processes orginating from several different amacrine cells. The T-cell has a basket-shaped dendritic arborization climbing the length of the cartridge, and a cell body originating above the medulla, with a long axon across the external chiasma to the lamina and a short axon into the medulla. The monopolar cells, on the other hand, have their somata lying in a granular layer overlying the external plexiform layer, where the synaptic connections take place; each cell body sends a single axon through the cartridge terminating in the medulla.

The monopolar cells do not, however, form a homogeneous population because they differ in size, shape, and synaptic contacts. The cells L1 and L2 are the largest relay neurons in the lamina and, hence, termed the large monopolar cells (LMCs); whereas L5 is the smallest and has minute fibers, thus called the midget cell. The LMCs are very similar in shape and form the central elements in each cartridge. They are closely associated with the short axon endings and receive in equal numbers the majority of their synapses. The similarities between the two LMCs end here because each cell terminates at a different level in the medulla: L2 ends superficially while L1 terminates deep in the medulla. Another major difference between L1 and L2 is that L2 is presynaptic to L1 and to the receptors R1-6 at the base of the lamina. The L3 monopolar, which runs parallel to L1 and L2, receives from the peripheral photoreceptors about a third as many inputs as an LMC. The remaining monopolars, L4 and L5, being third-order interneurons, get no direct input from the photoreceptors. The L4 cell derives its input indirectly from the photoreceptors via the amacrine cells; furthermore, its axon sends collaterals which synapse onto LMCs in its own and two adjacent cartridges. Another network of lateral connections involves the amacrines,

L4 and T1 whose dendrites are both pre- and postsynaptic to α-processes of the amacrine cells. For T1 there are still conflicting reports as to whether it is a second- or a third-order interneuron with respect to the photoreceptors (Shaw, 1984, 1989; Strausfeld, 1989), and its role is still somewhat enigmatic. The role of the LMCs is, however, much clearer. Since their primary input is provided by the photoreceptors, the LMCs are considered a major channel for transmitting information about the intensity received by a single sampling station from the retina to the medulla. Their functional role, however, is coding of contrast, which allows the visual system to cope better with large variations in background intensities (Laughlin, 1981, 1989a,b). Laminar units appear to be mainly involved in adaptational processes to enhance visual acuity (for a review see, e.g., Laughlin, 1989b). Although searched for, responses to motion have not been recorded in the lamina (Horridge et al., 1965; Mimura, 1974). The notion that motion-sensitive neurons exist at the lamina level is now virtually discounted.

C. THE MEDULLA

The second visual ganglion of the optic lobe, the medulla, is organized into columns. There are as many columns in the medulla as there are optic cartridges in the lamina or ommatidia in the retina. The outer two thirds of the neuropil is segregated from the inner third by a serpentine layer of tangential cell axons. The medulla is linked to the first and third optic ganglia via the external and internal chiasmata, respectively. The most distinguishing feature of the medulla is the large number and elaborate shapes of its cell species. A population of 120 distinct cell types have been identified, 43 of which are relay neurons (Campos-Ortega and Strausfeld, 1972; Strausfeld, 1976a). Each kind of relay neuron has a characteristic shape; moreover, every neuron projects to either one or both neuropils in the lobula complex. The medulla is also characterized by a large variety of amacrine cells, which may comprise as many as 35 different morphological types (Strausfeld, 1976a, 1989). Furthermore, it is the most peripheral neuropil at which binocular interactions are established. The two medullae in the fly optic lobes are connected by the posterior optic tract, which is chiefly composed of tangential cell processes. The medulla also conveys signals ipsilaterally to the optic tubercle via the anterior optic tract and to the posterior protocerebrum (Figure 1).

The many thousands of neurons contained in the medulla can be readily classified into two general categories: the tangential cells and the perpendicular (columnar) neurons. The perpendicular neurons are further subdivided into long and short axonal cells, and diffuse anaxonal neurons. The short axoned perpendiculars all serve to link the outer and inner medulla layers lying above and beneath the serpentine layer. The long axoned cells projecting to the lobula complex include three main types: transmedullary (TM) neurons to the lobula, T-cells to the lobula (T2, T3) or lobula plate (T4, T5), and Y-cells to both the lobula and lobula plate (Strausfeld, 1976a, b). The T- and some of the Y-cells have their somata located beneath the inner medulla layer or behind the lobula plate, while transmedullary neurons originate from perikarya above the outer medulla layer. Most medullary units are aperiodic in the sense that they do not appear regularly in every column. In fact, there are only 34 periodic elements besides monopolar and receptor cell endings and medulla-lamina neurons: they include 13 amacrines, 13 relays, and 8 tangentials. The amacrine and the tangential cell processes contribute to the formation of at least 24 discrete strata of lateral connections. Each species of lamina neurons terminates at a different stratum in the medulla.

Although not much is known about synaptic interconnections within the medulla, two major retinotopic projection modes directly involving laminar units have been recognized: one involved in color coding, the other in motion information processing. In the first, wide-field transmedullary neurons get inputs from R7-8 and L3, which have no interactions in the lamina, and output to a large variety of retinotopic columnar neurons in the lobula. In contrast, the second projection mode involves small-field medullary relays; they derive their input from R1-6 via the LMCs and synapse onto the bushy T4 and T5 cells, which provide inputs to wide-field colorblind motion-sensitive neurons in the lobula plate. This suggests that at the level of the lamina there is already segregation of retinotopic projections to the lobula and lobula plate. (See also Srinivasan, this volume).

There are only three types of small-field medullary neurons — SUB, TM1, and TM5 — that receive input solely from a set of terminals arising from a single optic cartridge, thus preserving retinotopically the initial sampling of the visual field by the photoreceptor mosaic. Two of them, TM1 and SUB, have dendrites at the same level as the L2 cell endings. The two cell types, however, terminate in two different regions: TM1 cells end at a superficial level of the lobula containing only T5 cells; whereas, the SUB neurons terminate in the deepest stratum of the medulla, where the dendrites of the bushy T4 cells reside (Strausfeld, 1984, 1989). In the medulla, each column is represented by four axons belonging to the bushy T4 and T5 cells, which exist as identical pairs in each retinotopic column. The four axons project to the lobula plate and terminate at four different levels hosting dendrites of the wide-field motion-sensitive neurons (Buchner and Buchner, 1984; Strausfeld, 1984). This suggests that the T4 and T5 twin cells are involved in motion detection, although their functional properties are still unknown. Motion-sensitive units have been electrophysiologically recorded in the medulla (Bishop et al., 1968; McCann and Dill, 1969; Mimura, 1971, 1972; DeVoe and Ockleford, 1976; DeVoe, 1980), but they have not been identified histologically. Anatomically, the medulla is the most complex structured neuropil of the optic lobe. This complexity is reflected in the variety of the functionally different units encountered in this region. They range from simple contrast detectors, On-units, to directional and nondirectional motion-sensitive neurons. (A detailed account of the functional units and their responses in the medulla can be found in Mimura, 1975; see also Osorio, this volume.)

D. THE LOBULA COMPLEX

The third optic ganglion in Diptera, i.e., the lobula complex, is divided into an anterior part, the lobula, and a posterior part, the lobula plate (Figure 1). Each neuropil possesses a columnar structure. However, the columnar cells coarsen the retinotopic periodicity of the lamina and medulla; there is only one lobular column for every six medullary ones. The lobula complex is connected to the medulla by the internal optic chiasma. There is a chiasma between the medulla and lobula, but none between the medulla and lobula plate. The lobula and lobula plate are facing each other, i.e., their outer surfaces lie opposite each other. The equivalent columns of the lobula and lobula plate are also opposite and are linked by small-field neurons which pass into the optic foci of the posterior protocerebrum and into the posterior slope. The columnar cells leave the lobula complex as dense bundles and end in the visual centers of the ventrolateral protocerebrum (Strausfeld, 1976a, b; Strausfeld and Nässel, 1981; see also Figure 1). The output elements of the optic foci of the ventrolateral protocerebrum, namely the descending neurons, project through the cervical connective farther to the motor neuropils of the thoracic ganglion, where connections diverge onto motor neurons of various muscle groups.

1. Lobula

As discussed in Section II.C, the lobula receives inputs from two types of medullary-neurons: the TM cells and the Y-cells, which also project into the lobula plate. The output neurons of the lobula, further, mediate signals to the ipsilateral optic tubercle via the anterior optic tract and to the optic foci of the anterior protocerebrum (Figure 1). The contralateral lobulae in both optic lobes are directly connected by the central commissure, the most dominant optic tract of the protocerebrum (Strausfeld, 1976a). The lobula neuropil is composed of several classes of regularly spaced small-field columnar neurons, which convey signals to the anterior optic foci of the ventrolateral protocerebrum. The most prominent of these classes are Col A neurons studied anatomically by Hausen and Strausfeld (1980) in the lobula of *Calliphora*. These neurons are subperiodic columnar elements forming a very regular neuronal lattice coarsened 1 to 9 with respect to the lobular columns. Their dendrites form a plexus of arborizations at the posterior surface of the neuropil, and their axon endings converge onto the ipsilateral giant descending neuron of the fly (for further details on the synaptic organization of Col A neurons, see Hausen et al., 1980; Hausen and Strausfeld, 1980). As far as we know, the physiological and functional properties of Col A neurons are unknown. However, it is speculated (Hausen, 1984) from the structural features of the lobula, that these neurons constitute an array of small-field on-off units and are involved in the jump- and flight-initiation circuitry.

2. Lobula Plate

This is the primary motion computation center in the fly visual system. To our knowledge, 50 large interneurons have been identified so far in the lobula plate (Hausen, 1987, cited in Hausen and Egelhaaf, 1989). They comprise anatomically two types of neurons: anaxonal amacrine cells and axonal tangential cells. The large tangential cells, which constitute the principal anatomical feature of the lobula plate, have been the subject of intensive studies for the past two decades (Borst and Egelhaaf, 1987; Braitenberg, 1972; Dvorak et al., 1975; Eckert and Bishop, 1978; Eckert, 1980, 1982; Egelhaaf, 1985; Egelhaaf and Borst, 1989; Hausen, 1976, 1981, 1982a, b, 1984, 1987, 1989; Hengstenberg et al., 1982; Hengstenberg, 1982; Pierantoni, 1976; de Ruyter van Steveninck et al., 1986; Strausfeld and Bassemir, 1985a, b; Zaagman et al., 1978). This group of neurons comprises 25 to 30 different classes of wide-field DSMD neurons. Although they all respond to whole-field motion, the DSMD neurons differ in their preferred directions (Hausen, 1981). They also differ in their receptive fields and response properties owing to the interactions with other tangential cells of the ipsilateral and contralateral optic lobes. According to their functional characteristics, these neurons occupy different layers in the lobula plate. Neurons sensitive to horizontal motion are located in the outer layers, while those sensitive to vertical motion reside in the neuropil inner layers. Anatomically, the large tangential cells include centripetal output cells conveying signals to the brain, centrifugal feedback cells projecting from the brain into the lobula plate, heterolateral visual connection cells linking the lobula plate with the contralateral optic lobe, and some other complex cells possessing several axonal destinations (Hausen, 1981, 1984; Hausen and Egelhaaf, 1989).

The lobula plate receives a complex network of columnar inputs from the medulla and lobula, though this input network is not yet fully understood. The Y-cells and the T-cells are the best known columnar neurons that provide input to the lobula plate. The unique columnar twin elements of the medulla (T4) and lobula (T5) have been, in particular, well investigated (Buchner and Buchner, 1984; Strausfeld, 1984, 1989). The four twin-cell axons associated with each retinotopic column terminate at four discrete strata hosting dendrites that map horizontally or vertically into the retinotopic mosaic, thus forming the so-called "horizontal" and "vertical" systems. Both the horizontal and vertical systems (HS and VS, respectively) are composed of giant centripetal cells, which terminate in the ipsilateral parts of the brain (Hausen, 1976; Pierantoni, 1976). These cells are the most prominent in the tangential system; they respond favorably to horizontal or vertical motion (Hausen, 1981). The HS consists of three tangential cells residing near the anterior surface of the neuropil and projecting into the ventrolateral brain (Pierantoni, 1976; Hausen, 1981; Strausfeld and Bassemir 1985a). Having a constant pattern of dendritic spread covering the dorsal, medial, and ventral regions of the neuropil, the HS neurons have been accordingly named the HS north (HSN), equatorial (HSE), and south (HSS) horizontal cells (Pierantoni, 1976). They all respond optimally to ipsilateral progressive (front-to-back) horizontal motion over the upper, middle, and lower parts of the visual field, and are inhibited by motions in the reverse directions (Dvorak et al., 1975; Pierantoni, 1976; Hausen, 1976, 1981, 1982a, b; Eckert, 1981). In contrast, only HSN and HSE are sensitive to binocular rotatory wide-field motion, which suggests that information about regressive motion in the contralateral visual field is conveyed to them by heterolateral horizontal-motion-sensitive neurons (Hausen, 1982b). HSE, in addition, is activated by contralateral regressive motion.

The VS, on the other hand, contains 11 cells (VS1-11) possessing dendrites that are serially arranged at the posterior surface of the lobula plate and axons that terminate in the posterior perioesophageal region of the protocerebrum (Pierantoni, 1976; Eckert and Bishop, 1978; Hausen, 1976, 1981; Hengstenberg et al., 1982; Strausfeld and Bassemir, 1985b). A detailed investigation of the synaptic connectivity of the VS cells has revealed two main output pathways to the motor system (Strausfeld and Bassemir, 1985b). In one pathway VS2-3 are connected to the motor neurons of particular neck muscles; in the other, VS4-9 converge onto the descending neurons leading to the thoracic ganglion. The vertical system is anatomically and functionally inhomogeneous. All VS cells respond, in general, to ipsilateral downward motion; but only VS2-6, which have their dendrites restricted to the posterior surface of the neuropil, show this preferred direction throughout their receptive fields (Dvorak et al., 1975; Eckert and Bishop, 1978; Hausen, 1976, 1981; Hengstenberg, 1982). The other cells, VS1 and VS7-9, which possess dorsal dendritic aborizations in the anterior layer of the neuropil, share the same preferred direction with VS2-6 in the ventral parts of their receptive fields only. Dorsally, they respond to horizontal motion: VS1, the most lateral unit, responds to regressive motion, whereas VS7-9 respond to progressive motion. The remainder of these cells, VS10-11, are very thin elements; their responses are supposed to resemble those of VS7-9. Additional responses have been recorded for vertical motion in the contralateral visual field; the preferred directions are downward for VS1 and upward for VS7-9 (Hausen, 1981).

The HS and VS cells are examples of the homolateral centripetal elements of the lobula plate, which contains also heterolateral centripetal cells. The best-known heterolateral centripetals are H2, H3, and H4, which have dendritic arborizations lying deep in the neuropil and axons projecting to the contralateral protocerebrum or to both sides of the brain. While H2 and H3 are selectively activated by regressive and progressive motion, respectively, the H4 cell responds to regressive as well as to ipsilateral upward motion (Hausen, 1976, 1981; Eckert, 1981). Other heterolateral output elements such as the figure detection (FD) cells discovered by Egelhaaf (1985b) and the vCal1 cell (Hausen and Egelhaaf, 1989) also exist in the lobula plate, but the details of their synaptic connectivity are still lacking.

Centrifugal cells projecting from the brain back into the lobula plate also comprise homolateral and heterolateral elements. Homolateral units are represented by a set of two centrifugal horizontal (CH) cells: the dorsal (DCH) and the ventral (VCH) cells. The DCH and the VCH cells possess small dendritic arborizations surrounding the terminals of the HS cells and axons projecting into the dorsal and ventral half of the ipsilateral lobula plate, respectively. They both respond to ipsilateral progressive and binocularly to contralateral regressive motion (Hausen, 1976, 1981). The V1-cell is a centrifugal heterolateral element. It projects from the VS-cell termination area into the posterior surface of the contralateral lobula plate. This cell responds mainly to ipsilateral downward motion (Hausen, 1976, 1981; Eckert and Bishop, 1978).

Heterolateral units connecting both lobula plates directly include H1, V2, and V3. The cells V2 and V3 are sensitive to ipsilateral upward and downward motion, respectively, whereas the giant H1 neuron is sensitive to ipsilateral regressive motion (Hausen, 1976, 1981; Eckert, 1980). H1 is the only directionally selective cell in the optic lobe that has been consistently investigated for more than two decades (Bishop et al., 1968; Hausen, 1976, 1981; Zaagman et al., 1978; Eckert, 1980; Dvorak et al., 1980; Lenting et al., 1984; Maddess and Laughlin, 1985; de Ruyter van Steveninck et al., 1986; Borst and Egelhaaf, 1987; Franceschini et al., 1989); the data accumulated has made a significant contribution to the understanding of the functional characteristics of DSMD neurons in the fly visual system. There is only one H1 neuron in each lobula plate; it appears to be involved in the control of the optomotor torque response. The two bilaterally symmetric H1 cells exert mutual inhibition, thereby each cell is particularly sensitive to either clockwise or anticlockwise rotatory (yaw) motion of the visual field.

In summary, the primary visual system of the fly is dominated by a retinotopic organization in which the lobula plate represents the last orderly projection of the visual field. The medulla is the most peripheral structure where movement detection takes place. The lobula plate, however, is characterized by large DSMD neurons, which seem to share a common network of presynaptic elements derived from the medulla. It is not known whether these presynaptic units are motion sensitive or motion insensitive. It is, though, common knowledge that the wide-field DSMD neurons of the lobula plate integrate signals from large retinotopic arrays of small-field elementary movement detectors, and that the actual motion detection process does not take place at their dendrites (Hausen, 1982a, b; Egelhaaf, 1985; Hausen and Engelhaaf, 1989). Nevertheless, it cannot be decided yet, whether these small-field movement detectors reside in the medulla or lobula plate. Possible candidates are the bushy T4 and T5 cells, but it is not known if they are sensitive to motion. Next, we examine the concept of elementary movement detection, then discuss some functional properties of the DSMD neurons.

III. ELEMENTARY MOVEMENT DETECTION

The elementary movement detector (EMD) is the minimum prerequisite for directionally selective detection of motion in the visual field. It is based on the principle of asymmetrical interaction between two adjacent channels (Figure 2). Asymmetry is necessary for the detector to acquire direction selectivity. For if the system were symmetric, the two input channels could be interchanged without altering its response. This would be equivalent to reversing the direction of motion but still obtaining the same response. Therefore, without an asymmetrical interaction, the movement detector loses its ability to respond differentially to motion in opposite directions. There exist mainly two general schemes for realization of the asymmetrical interaction. One works by rejecting the null stimulus by a veto operation (Figure 2a). The other selects the preferred direction by detecting a specific conjunction of excitation (Figure 2b). The best known conjunctive scheme is the correlation model suggested by Hassenstein and Reichardt (1956) to account for the characteristics of insect optomotor response. In this model, the interaction between adjacent channels is implemented by a multiplication followed by an infinite time averaging operation, i.e., a first-order correlation (for a review see, e.g., Bouzerdoum and Pinter, 1991, 1989a).

The first veto scheme was proposed by Barlow and Levick (1965), who discovered that inhibition is the mechanism underlying directional selectivity of ganglion cells in the rabbit retina. They suggested that inhibition is triggered selectively in such a way that at each subunit of the receptive field a delayed inhibitory mechanism vetoes the excitatory response in the null direction, but appears too late to cancel the response in the preferred direction. Barlow and Levick demonstrated that directional selectivity of retinal ganglion cells is based on sequence discrimination within small-field synaptic subunits, or EMDs. More specifically,

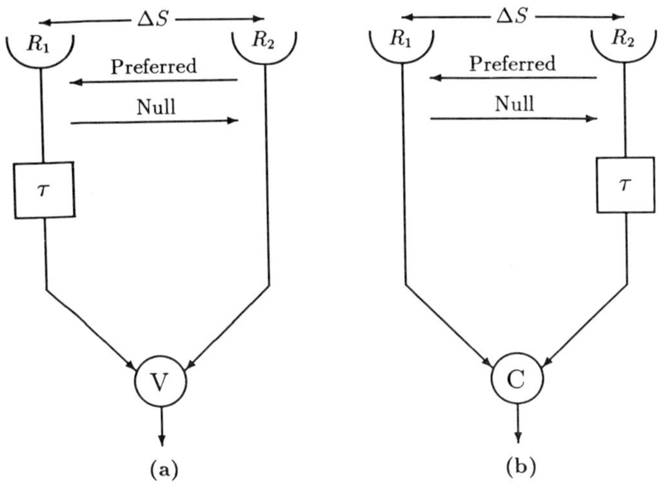

FIGURE 2. Elementary movement detector (EMD). The visual field is sampled at two receptor regions, R_1 and R_2, separated by a distance ΔS. The signal from one receptor is passed through an ''appropriate'' time delay, such as a low-pass filter of time constant τ, before interacting with the signal from the adjacent channel. (a) Veto scheme: If the two signals reach ⓥ concurrently (null direction), they cancel each other and no motion is signaled. However, if the two signals arrive at ⓥ separately (preferred direction), the veto signal is unable to suppress the other signal, which indicates motion. (b) Conjunctive scheme: If the signals from the two adjacent channels arrive simultaneously at ⓒ (preferred direction), then a conjunction of excitation is detected signaling motion, whereas, if the two signals arrive separately, the unit ⓒ stays quiescent.

they showed that over the whole receptive field, successive stimulation of two subunits close to each other caused a response that depends on whether the order corresponds to motion in the preferred or null direction, but the effect decreased at greater separations. Michael (1968) confirmed this in the ground squirrel by concurrently moving two spots in opposite directions; he found that the spot moving in the null direction inhibited the response to the spot moving in the preferred direction, but there was no interaction when the distance separating the two dots was appreciable.

The initial stages of movement detection in insects also appear to be based on sequence discrimination by EMDs. Both behavioral and electrophysiological experiments on flies indicate that movement detection takes place between neighboring points of the sampling lattice of the eye (Kirschfeld, 1972; Buchner, 1976; Buchner et al., 1978; Bishop et al., 1968; McCann, 1973; Zaagman et al., 1978; Srinivasan and Dvorak, 1980; Riehle and Franceschini, 1984; Franceschini et al., 1989). For example, sequential stimulation confined to pairs of identified photoreceptors in single ommatidia, induced optomotor turning reactions in the fly (Kirschfeld, 1972). Franceschini and co-workers recorded a sequence-dependent response in the DSMD neuron, H1, by activating two adjacent cartridges with stimulation of the photoreceptor pair (R1,R6) within a single ommatidium. In particular, they found that the sequence R1→R6 evoked an excitatory response, whereas the sequence R6→R1 induced an inhibitory or no response, which was in accordance with the preferred and nonpreferred directions of the H1 neuron, respectively. There have been, however, conflicting reports regarding the nature of the mechanism mediating direction selectivity in the fly visual system. Franceschini et al. (1989), based on electrophysiological experiments, concluded that facilitation rather than inhibition is responsible for directional selectivity; whereas, Schmid and Bülthoff (1988), based on neuropharmacological coupled with electrophysiological experiments, suggested that inhibition is the mechanism underlying directional selectivity in the fly visual system. Yet an earlier report (Mimura, 1972) indicates that both mechanisms are involved: facilitation in the preferred direction and inhibition in the null direction.

Nevertheless, the conflict may be resolved if one considers the spatial arrangement and nature of synaptic contacts of the input elements conveying motion signals to the wide-field tangential cells. There exists convincing evidence that these DSMD neurons receive both excitatory and inhibitory inputs from large retinotopic arrays of small-field EMDs, which possess opposite preferred directions (Hausen, 1981, 1982a; Egelhaaf, 1985; Hausen and Egelhaaf, 1989; Franceschini et al., 1989). Consequently, movement in a particular direction (preferred direction) activates mainly the excitatory inputs, whereas movement in the reverse direction excites the inhibitory ones. This confirms Mimura's report since we believe that the units he recorded were mostly large-field DSMD neurons. Although his report is accurate, it does not concern the mechanism underlying direction selectivity at the level of the small-field EMDs. Franceschini and his colleagues discovered that stimulation of a single photoreceptor (R1 or R6) does not elicit any response, but when stimulation of R6 is preceded by that of R1, the H1 neuron fires; furthermore, the evoked response is always time-locked with the second flash (Riehle and Franceschini, 1984; Franceschini et al., 1989). They concluded, therefore, that the priming flash on R1 must have exerted a facilitatory influence rendering the secondary flash on R6 more effective. This would have been the case if only one EMD feeding the H1 neuron were activated by this type of stimulation (see Figure 2). It is unlikely, however, that only one EMD is activated when stimulating a single or a pair of receptors projecting to adjacent cartridges, for synchronous flashes on R1 and R6 failed to elicit a response. This, along with the fact that the null sequence (R6→R1) caused an inhibition in the resting discharge of H1, led Franceschini and colleagues to conclude that the H1 neuron is fed antagonistic signals by coexisting EMDs which have opposite orientations.

Figure 3 illustrates the basic functional structure describing the flow of motion information from the lamina to a DSMD neuron. In this structure, the interaction between the EMD input channels is of the veto type; nevertheless, it can still explain the results obtained by Franceschini and co-workers. For example, stimulating two photoreceptors, like R1 and R6, separately or simultaneously evokes no response because the activated EMDs convey antagonistic signals which cancel out at the level of the DSMD neuron. In contrast, sequential stimulation of a pair of neighboring receptors upsets the balance between the excitatory and inhibitory signals delivered to the DSMD neuron, thereby inducing either an excitatory or an inhibitory response — depending on whether the flash sequence mimics motion in the preferred or null direction (see Figure 3). Since the signals generated by the first flash are opposite and cancel each other, the response of the DSMD neuron is always time-locked with the onset of the second flash. This shows that lateral interactions of the veto type could also mediate direction selectivity in the fly visual system. Furthermore, the functional structure in Figure 3 is compatible with the results obtained by Schmid and Bülthoff (1988) from neuropharmacological experiments. They demonstrated that the inhibitory neurotransmitter γ-amino-butyric acid (GABA) is responsible for the veto mechanism; by blocking the inhibitory interactions with picrotoxin, an antagonist of GABA, the direction selectivity of H1 was abolished. Therefore, despite conflicting reports, experimental evidence to date favors the veto scheme.

In conclusion, elementary motion detection is based on sequence discrimination between adjacent subunits. In flies, the mechanism of sequence discrimination is most likely caused by lateral interactions of the inhibitory type. This phenomenon is known simply as lateral inhibition. In the remainder of this chapter we examine lateral inhibition as a description of the functional characteristics of the DSMD neurons in the lobula plate.

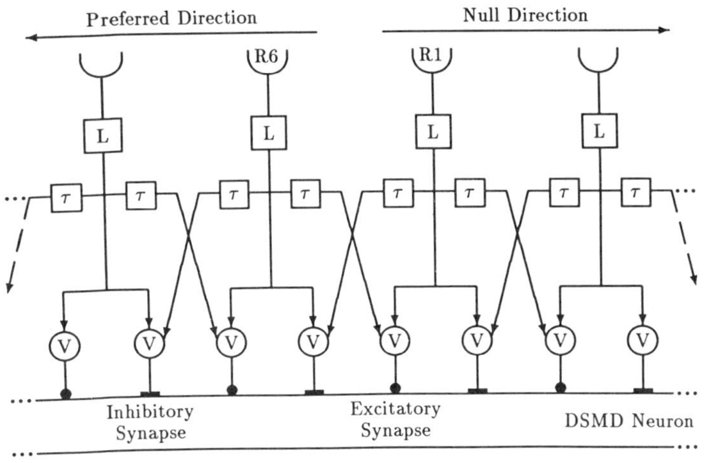

FIGURE 3. Schematic functional representation of a DSMD neuron. The DSMD neuron receives excitatory and inhibitory signals from an array of functionally identical EMDs, which differ only with respect to the orientation of their sampling bases. The basic information flow from a lamina cartridge (box L) to the DSMD neuron is represented by a puppet. The legs of the puppet transmit signals to its knees where they interact with the delayed signals (box τ), carried laterally by the puppet arms of the neighboring columns. The interaction at the knees is of the veto type. Stimulating photoreceptors, singly or in paired combination, activates synchronously an equal number of EMDs conveying opposite signals which lead to mutual cancellation and, thus, no response is evoked in the DSMD neuron. For the same reason, sequential flashing of a pair of adjacent photoreceptors (e.g., R1 and R6) does not elicit a response, until after the onset of the second flash when a lateral signal from the first flash vetoes one of the signals generated by the second stimulus. This upsets the balance between the activated excitatory and inhibitory synapses, thereby resulting in excitation or inhibition of the DSMD neuron. Excitation is evoked by a sequence mimicking motion in the preferred direction (R1→R6), while inhibition is induced by the reverse sequence (R6→R1).

IV. SHUNTING LATERAL INHIBITION

The concept of lateral inhibition arose in the extensive studies of Hartline and Ratliff on the compound eye of the ''horseshoe crab,'' *Limulus*. These studies revealed that inhibitory interactions are exerted mutually among neighboring ommatidia in the lateral eye of *Limulus* (Hartline and Ratliff, 1957; Ratliff, 1961). Lateral inhibition as originally described by Hartline and Ratliff is linear; that is, the phenomenon is describable by a system of linear equations, once a threshold has been reached (Hartline and Ratliff, 1957). Linear lateral inhibition may be sufficient to account for some visual functions like contrast enhancement; it is not, however, sufficient for others. Motion detection, for example, requires a nonlinear asymmetrical interaction if the system is to have an average response that is directional (Buchner, 1976; Poggio and Reichardt, 1973). Thus, nonlinear rather than linear lateral inhibition must be the mechanism underlying the asymmetrical interaction in the veto scheme.

Numerous studies suggest that the dominant factor in GABA-mediated postsynaptic inhibition is the shunt of excitatory potential due to an increased membrane conductance to chloride or potassium ions. In the fly visual system, the GABA-mediated effects at the EMD level may also be of the shunting type since the GABA action is blocked by picrotoxin, an antagonist of GABA whose site is thought to reside on chloride channels. However, since the equilibrium potential for chloride ions is close to the resting potential in some neurons, the shunting inhibition, also called silent inhibition, is effective only in the presence of an excitation. If we further postulate that the shunting conductance is modulated by the activity of the inhibiting unit, then shunting inhibition translates into multiplicative interaction between the activities of the pre- and postsynaptic units. Therefore, shunting lateral inhibition

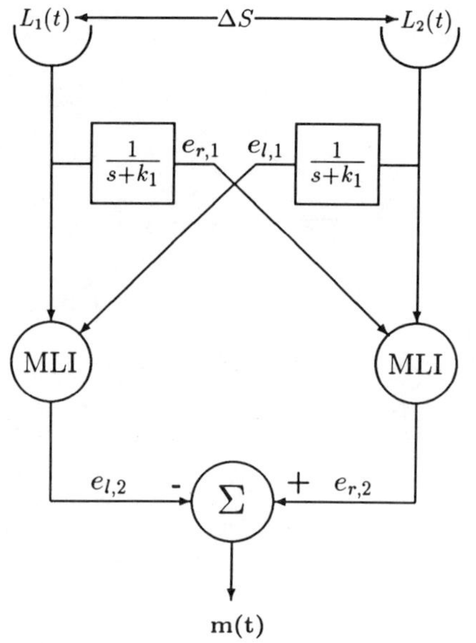

FIGURE 4. Multiplicative inhibitory motion detector (MIMD). The outputs of the MLI boxes are $\dot{e}_{r,2} = L_2(t) - k_2 e_{r,2}\,[1 + k_{12} f(e_{r,1})]$, and $\dot{e}_{l,2} = L_1(t) - k_2 e_{l,2}\,[1 + k_{12} f(e_{l,1})]$. After (Bouzerdoum and Pinter 1991, 1990).

is termed multiplicative lateral inhibition (MLI). A multiplicative lateral inhibitory neural network (MLINN) is described by

$$\frac{de_i}{dt} = L_i(t) - k_i e_i \left[1 + \sum_{j \neq i} k_{ij} f_{ij}(e_j) \right] \tag{1}$$

where $e_i \geq 0$ represents the excitation of the ith cell or cell subunit, interpretable as the deviation of the membrane voltage from the resting potential; $L_i \geq 0$ is the external input to the ith neuron; $k_i > 0$ is the rate constant for the ith neuron; $k_{ij} \geq 0$, represents the synaptic efficacy (or coupling strength) of the jth neuron potential transmitted to the ith neuron; and f_{ij} is the lateral inhibition function. It is a positive monotonic increasing function for positive arguments and represents the dependence of the shunting conductance in the ith neuron on the voltage of the jth neuron (for a derivation see, e.g., Pinter, 1983; Bouzerdoum, 1991; Bouzerdoum et al., 1991). It has been shown that unidirectional MLINNs, or networks with one-sided coupling, can subserve directional selectivity (Bouzerdoum and Pinter, 1989b; Bouzerdoum and Pinter, 1991; Bouzerdoum et al., 1991). (For more on the properties of these networks see Bouzerdoum, 1991.)

A. THE MOVEMENT DETECTOR

A movement detector consists of two mirror symmetrical subunits or EMDs with outputs having opposite polarities. Previously, we have defined a multiplicative inhibitory motion detector (MIMD) as a detector where the nonlinear interaction of the EMD is mediated by multiplicative lateral inhibition (Bouzerdoum and Pinter, 1991, 1989a). Figure 4 illustrates an MIMD composed of two mirror symmetrical unidirectional MLINNs tuned to motion in opposite directions; its final output is given by the difference between the outputs of the two subunits. Each leg or subunit of this MIMD is described by a pair of coupled ordinary differential equations, that is,

$$\dot{e}_{\xi,1} = L_p(t) - k_1 e_{\xi,1}$$

$$\dot{e}_{\xi,2} = L_q(t) - k_2 e_{\xi,2}[1 + k_{12} f(e_{\xi,1})] \tag{2}$$

where $\xi = l,r$ depending on which leg (left or right) is being described; if $\xi = r$, $p = 1$, $q = 2$; otherwise $p = 2$, $q = 1$. Next, we will discuss the functional characteristics of this detector and compare them with those of fly DSMD neurons.

V. RESPONSE CHARACTERISTICS

Due to its structural symmetry, the MIMD in Figure 4 does not respond to a uniform field of illumination and responds to leftward and rightward motion with the same strength and opposite sign (Bouzerdoum and Pinter, 1989a, 1991). In this section we investigate both the transient and steady-state characteristics of the MIMD response to drifting sinusoidal gratings commonly used for studying the response of the fly's DSMD neurons. To conduct an analysis of the MIMD functional properties, an approximation to its response is in order.

A. RESPONSE TO LOW-CONTRAST STIMULI

In general, the system of equations (Equation 2) is not amenable to an elementary treatment to yield an explicit solution. To exhibit the solution, one often has recourse to a method of numerical integration. However, the solution may be approximated, especially for low-contrast stimuli, if certain rather mild conditions are met. Here, we apply the perturbation methods to obtain an approximate solution for inputs, $L_i(t)$ $(i = 1,2)$, of the form

$$L_i(t) = L_0 + c l_i(t)$$

where the mean visual field input strength $L_0 = (L_{max} + L_{min})/2, 0 \leq \|l_i(t)\| \leq L_0$, and the contrast $c = (L_{max} - L_{min})/(L_{max} + L_{min})$.

For a fairly broad class of this type of input, the system of differential equations (Equation 2) admits a unique solution that is continuous in both t and c (see Bouzerdoum, 1991 for a proof). Therefore, for a stimulus in this class, $e_{\xi,1}$ and $e_{\xi,2}$ can be written as

$$e_{\xi,1} = x_0 + c x_{\xi,1},$$

$$e_{\xi,2} = y_0 + c y_{\xi,1} + c^2 y_{\xi,2} + \cdots = y_0 + \sum_{n=1} c^n y_{\xi,n} \tag{3}$$

Note that the expression of $e_{\xi,1}$ does not contain terms higher than the first power of c. This is because $e_{\xi,1}$ is the output of a linear time-invariant system, i.e., it is described by an ordinary linear differential equation of constant coefficients. Now the lateral inhibition function $f(e_{\xi,1}) = f(x_0 + c x_{\xi,1})$ can be expanded into a Taylor series form around x_0.

$$f(x_0 + c x_{\xi,1}) = f(x_0) + c x_{\xi,1} f'(x_0) + \frac{c^2 f''(x_0)}{2!} x_{\xi,1}^2 + \cdots$$

$$= f(x_0) + \sum_{n=1}^{\infty} \frac{c^n f^{(n)}(x_0)}{n!} x_{\xi,n}^n \tag{4}$$

Differentiating Equation 3 and substituting $f(e_{\xi,1})$ for its Taylor series into Equation 2 yields, after equating the coefficients of powers of c to zero, the following equations:

$$\dot{x}_{\xi,1} = l_p(t) - k_1 x_{\xi,1}$$

$$\dot{y}_{\xi,1} = l_q(t) - k_2 k_{12} f'(x_0) x_{\xi,1} y_0 - \alpha y_{\xi,1}$$

$$\dot{y}_{\xi,2} = -\frac{k_2 k_{12} f''(x_0)}{2!} x_{\xi,1}^2 y_0 - k_2 k_{12} f'(x_0) x_{\xi,1} y_{\xi,1} - \alpha y_{\xi,2}$$

$$\vdots$$

$$\dot{y}_{\xi,n} = -k_2 k_{12} \sum_{j=1}^{n} \frac{f^j(x_0)}{j!} x_{\xi,1}^i y_{\xi,n-j} - \alpha y_{\xi,n}, \text{ for } n \geq 2 \tag{5}$$

where, $x_0 = L_0/k_1$, $y_0 = L_0/\alpha$, and $\alpha = k_2[1 + k_{12} f(x_0)]$.

For values of c such that $0 < c < 1$, a pth order approximation $m_p(t)$ of the MIMD response, $m(t)$, is given by

$$m_p(t) = \sum_{j=1}^{P} c^j (y_{r,j} - y_{l,j}) \tag{6}$$

Note that the smaller is c, the more accurate is the approximation. Thus, for low-contrast stimuli one could always get a fairly good approximation to the MIMD response by solving the set of differential equations in Equation 5.

B. RESPONSE TO DRIFTING GRATINGS

Sine-wave gratings are commonly used in vision to evoke the spatial and temporal frequency responses of the visual systems. In the case of the MIMD, the response to a drifting sine-wave grating usually oscillates, after the transient dies out, around an average response which depends strongly on contrast and spatial and temporal frequency contents of the moving pattern. If only nonlinearities up to second order are considered, then the response of the MIMD is approximated by

$$m_2(t) = c(y_{r,1} - y_{l,1}) + c^2(y_{r,2} - y_{l,2}) \tag{7}$$

Let Mr denotes the time-averaged $m_2(t)$, i.e. the MIMD mean steady-state response associated with second-order nonlinearities. It can be easily shown, by solving Equation 5, that Mr due to a moving sine-wave grating, $L(s,t) = L_0 + cL_0 \cos(2\pi\mu f_s s + 2\pi f_t t + \varphi)$, satisfies the relation

$$Mr = \frac{k_2 k_{12} c^2 L_0^2 f'(x_0)(\alpha - k_1)\omega}{\alpha(k_1^2 + \omega^2)(\alpha^2 + \omega^2)} \sin(2\pi\mu f_s \Delta S) \tag{8}$$

where c is the grating contrast, $\omega = 2\pi f_t$ is the angular frequency, f_s is the grating spatial frequency, L_0 is the mean luminance level, and μ represents the direction of grating motion (1 for leftward motion and -1 for rightward motion).

From Equation 8 we see that the MIMD mean steady-state response, due to second-order nonlinearities, is insensitive to contrast reversal and depends on it quadratically, c^2. Moreover, Mr is fully directional, i.e., its sign depends on the direction of grating motion

μ. However, since the MIMD employs spatial sampling, it is necessarily subject to the limitations on spatial resolution set by the Shannon sampling theorem; that is, the MIMD can best resolve a grating whose spatial wavelength λ is at least twice as long as ΔS, the distance separating the two sampling channels (i.e., $2\Delta S < \lambda$). For a spatial wavelength λ such that, $\Delta S < \lambda < 2\Delta S$, the sign of $\sin(2\pi\mu f_s \Delta S)$ becomes opposite that of μ. Direction selectivity then reverses sign, and the detector signals the wrong direction of motion. This phenomenon, known as *spatial aliasing,* is well known for insect visual systems. It has been explained by the concept of geometrical interference between the array of receptors and the moving pattern causing a reversal of the direction of movement of the interference pattern by 180° (Götz, 1964). Eckert (1980) found, by extracellular recordings from the H1 neuron, that when the spatial wavelength of the moving pattern is smaller than twice the interommatidial angle, the response properties with regard to the direction of pattern movement are reversed: regressive motion causes inhibition and progressive motion causes excitation.

Marmarelis and McCann (1973), when studying movement detection in class II neurons or giant tangential cells of the fly's lobula plate, found that nonlinearities higher than second order do not contribute significantly to the response. In order to apply the MIMD to the activity of the DSMD neurons, we chose a lateral inhibition function of the form $f(e) = e$; thus $\alpha = k_2(k_1 + k_{12}L_0)/k_1$ and $f'(x_0) = 1$. For this particular function the second-order approximation is justified since odd-order nonlinearities do not contribute to the mean response, and a nonlinearity of order $2p$ adds a term in c^{2p}, which can be neglected for $p \geq 2$, especially for low contrast values.

If we assume that the spatial integration of local movement signals at the level of the wide-field DSMD neurons is linear, then the effects of the excitatory and inhibitory synaptic contacts of the individual EMDs with the DSMD neurons (Figure 3) are, respectively, additive and subtractive. Thus, to first order, the response of a DSMD neuron may be obtained by spatially summing the outputs of a one-dimensional array of uniformly spaced MIMDs that share the input channels with neighboring detectors. The response of such an array of MIMDs (ARMIMD) to both stationary and moving patterns was simulated. Not surprisingly, it was found that, like the MIMD, an ARMIMD does not respond to uniform stationary patterns and responds with equal strength and opposite sign to a pattern moving in opposite directions. Examination of the ARMIMD response to moving gratings, however, revealed that for long enough arrays the steady-state response settles to a constant level while the transient response oscillates with a frequency corresponding to the grating contrast frequency (i.e., temporal frequency). Figure 5a shows the response of an ARMIMD to a sinewave grating moving abruptly from left to right with constant speed, after it was at rest for some time. Figure 5b illustrates the response to a drifting sine-wave grating obtained by intracellular recordings from the fly's large-field DSMD neurons (for more, see Egelhaaf and Borst, 1989; Borst and Egelhaaf, 1989). Note the close similarity between the simulated and the recorded response. At the onset of pattern motion the response oscillates with a frequency that reflects exactly the contrast frequency of the grating, then settles to its steady-state level. Note also that both the transient and the steady-state responses depend on the temporal or contrast frequency. This contrast frequency dependence is investigated in the following subsection.

C. CONTRAST FREQUENCY DEPENDENCE

The MIMD is, in fact, not a velocity but a contrast frequency detector. In the previous section (Figure 5) we saw that, in addition to exhibiting a phasic response at stimulus onset, a large enough ARMIMD responds tonically to motion of a periodic pattern. The dependence of both response components on the contrast frequency was determined for an array of uniformly spaced, identical MIMDs. Figure 6 depicts the dependence of the transient response on the grating contrast frequency for different background (mean field) luminance levels as

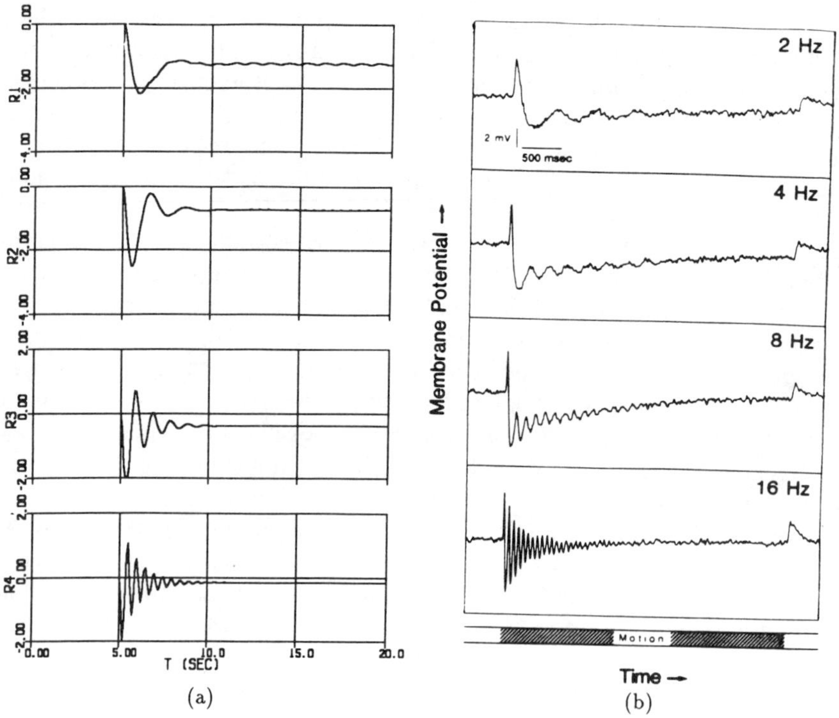

FIGURE 5. (a) Response of an ARMIMD to a sine-wave grating for temporal frequencies of (from top to bottom) 0.25, 0.5, 1, and 2 Hz. The grating starts drifting from left to right after it was at rest for 5 s. Here, $f(e) = e$. (b) Intracellular response of an HS fly DSMD neuron to a sine-wave grating moving abruptly from back to front at different temporal frequencies as indicated in the figure. Adapted from (Egelhaaf and Borst 1989).

indicated. The curves in this figure were obtained by recording the peak amplitude in response to a grating of fixed contrast and spatial wavelength, moving across the ARMIMD. Note that all the curves peak at a certain frequency that depends on mean luminance level. In general, the peak amplitude response of the ARMIMD increases, and the peak frequency shifts to higher temporal frequencies on increasing the mean luminance level (see Figure 6); however, saturation is expected for a sufficiently high mean luminance level.

Figure 7 illustrates the mean steady-state response as a function of contrast frequency normalized with respect to contrast and spatial wavelength. The curves were obtained from Equation 8 with parameters $k_1 = k_2 = k_{12} = 1$ and lateral inhibition function $f(e) = e$. The mean steady-state response increases with contrast frequency until it reaches a maximum; then it falls off at higher frequencies. The frequency of maximum steady-state response is increased only slightly on increasing the mean luminance level; it is given, from Equation 8 by

$$f_t = \left[\frac{-(k_1^2 + \alpha^2) + \sqrt{(k_1^2 + \alpha^2)^2 + 12k_1^2\alpha^2}}{24\pi^2} \right]^{1/2} \tag{9}$$

which may be approximated by $f_t \approx k_1/2\pi$ for sufficiently high mean luminance levels. The fact that this peak frequency does not change with spatial wavelength indicates that the steady-state response of the MIMD does not depend on grating velocity.

The curves in Figures 6 and 7 demonstrate the ARMIMD and, hence, the MIMD ability to respond to a broad range of pattern velocities. The response characteristics in these curves

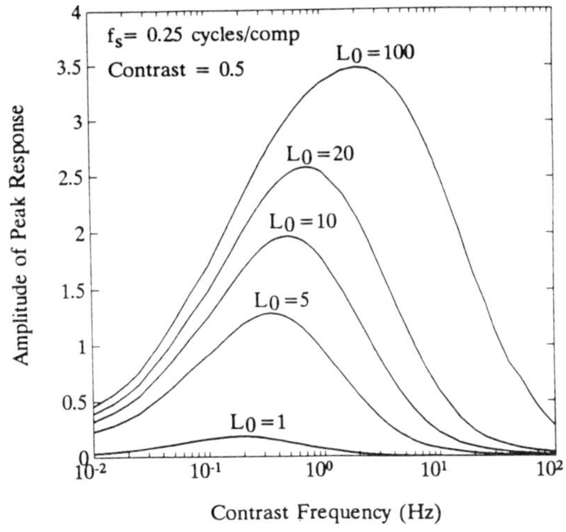

FIGURE 6. Transient peak amplitude response of an ARMIMD as a function of contrast frequency for different mean luminance levels: L_0 = 100, 20, 10, 5, and 1 (arbitrary units).

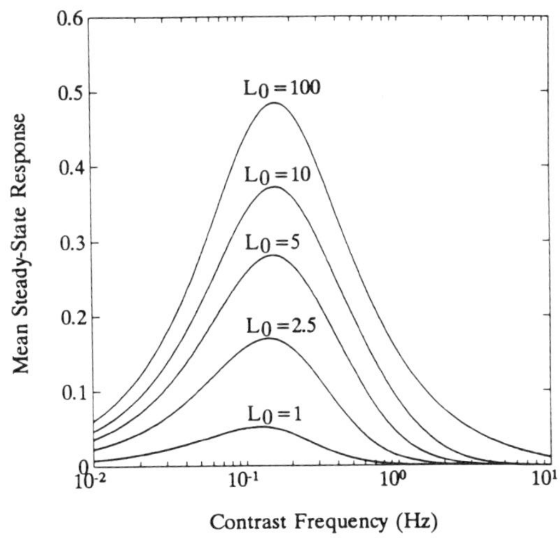

FIGURE 7. The mean steady-state response computed from Equation 8 as a function of contrast frequency for different mean luminance levels: L_0 = 100, 10, 5, 2.5, and 1 (arbitrary units). The computed responses are normalized with respect to contrast and spatial wavelength.

are in full agreement with those of tangential cells in the lobula plate. All DSMD neurons of the lobula plate tested so far exhibit similar response characteristics. The response does not depend on pattern velocity, but rather on contrast frequency; the response range covers about 3 log units of contrast frequency (0.01-0.05 Hz to 20-50 Hz); the response amplitudes increase from lower threshold to peak and fall off sharply above this value; the response peaks are consistently found at 1-5 Hz (Eckert, 1980, 1982; Hausen, 1981, 1982b; Hengstenberg, 1982; Zaagman et al., 1978).

D. SPATIAL FREQUENCY DEPENDENCE

In addition to the contrast frequency dependence, the MIMD or an ARMIMD mean steady-state response shows a strong dependence on the spatial frequency content of moving

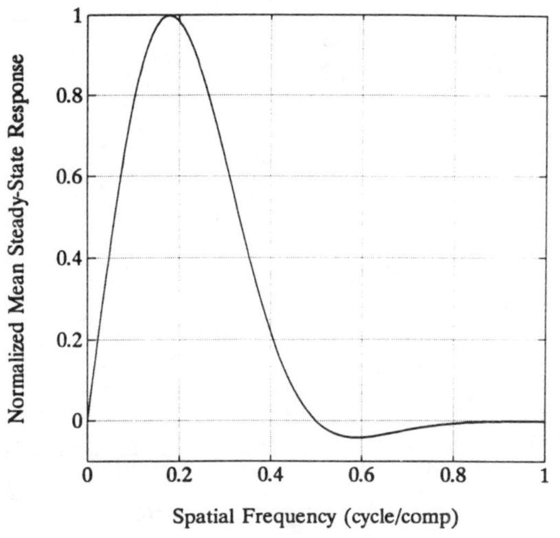

FIGURE 8. The effect of the contrast transfer factor of Equation 8 on the MIMD mean steady-state response as a function of spatial frequency. The response is normalized with respect to its peak value. The attenuation at high frequencies is evident.

stimuli. Equation 8 predicts a sinusoidal response with respect to spatial frequency. Since the period of this sinusoid is equal to $(1/\Delta S)$ cycles per degree or 1 cycle per compartment,* responses in the range $(1/2\Delta S) < f_s < (1/\Delta S)$ are equal and opposite in sign to responses in the range $f_s < (1/2\Delta S)$. Contrary to this prediction, the behavioral responses of flies are not periodic; the negative response measured for $\Delta\varphi^* < \lambda = 1/f_s < 2\Delta\varphi^*$, where $\Delta\varphi^*$ represents the effective interommatidial angle, are smaller in magnitude than the positive responses measured for $\lambda > 2\Delta\varphi^*$ (Buchner, 1976; Heisenberg and Buchner, 1977).

In our theoretical analysis, we have so far always assumed that the receptors had a needle-shaped spatial sensitivity distribution; i.e., each receptor samples the visual field at a single point. However, real photoreceptors have instead an approximately Gaussian (bell-shaped) spatial or angular sensitivity distribution. They get their input by spatially integrating the luminance distribution located within their range and, hence, act as a spatial low-pass filter. The cutoff frequency of the low-pass filter is determined by the width of the spatial distribution at half maximum or the acceptance angle, $\Delta\rho$. If a sine-wave grating is presented to the photoreceptors, its contrast, c, is attenuated by a factor (see, e.g., Götz, 1964)

$$\frac{c_{rec}}{c} = e^{-\frac{\pi^2}{4\ln 2}(\Delta\rho/\lambda)^2} = e^{-3.56(\Delta\rho f_s)^2} \tag{10}$$

where c_{rec} is the effective contrast in the receptors. Since low-pass filtering severely limits the transfer of high frequencies, we can expect the response to lose its periodicity with respect to spatial frequency. The effect of contrast attenuation on the MIMD mean steady-state response is presented in Figure 8. The curve in this figure has been plotted using a contrast transfer parameter $\Delta\rho = \beta\Delta S$, with $\beta = 1.09$. The constant β has been inferred from values of $\Delta\rho$ and $\Delta\varphi^*$ ($\beta = \Delta\rho/\Delta\varphi^*$) as they were determined for the compound eyes of fruitfly *Drosophila melanogaster* ($\Delta\rho = 5°$, $\Delta\varphi^* = 4.6°$; Heisenberg and Buchner, 1977) and of the blowfly *Calliphora erythrocephala* ($\Delta\rho = 1.59°$, $\Delta\varphi^* = 1.45°$; Zaagman et al., 1978).

* Here a compartmental unit (comp) is equivalent to a visual angle of ΔS degrees.

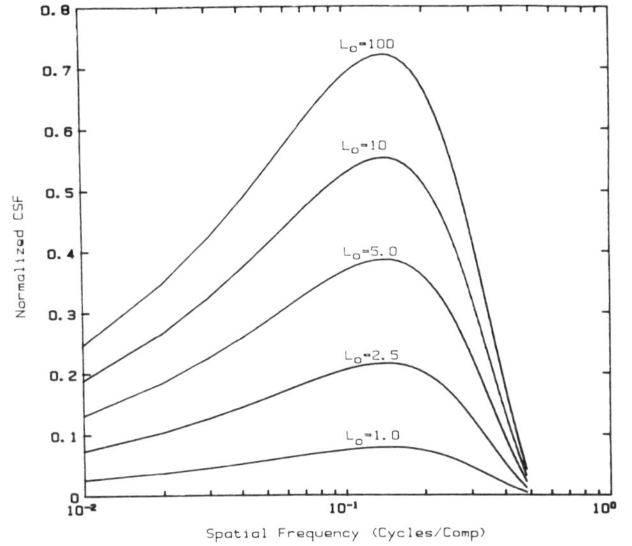

FIGURE 9. CSFs of a MIMD as computed from Equation 8 for different mean luminance levels: $L_0 = 100$, 10, 5, 2.5 and 1 (arbitrary units).

E. ADAPTATION OF CONTRAST SENSITIVITY FUNCTIONS

In spatial vision, sine-wave gratings are frequently used to describe the perceptual spatio-temporal frequency response, which is commonly known as the contrast sensitivity function (CSF). The CSFs of visual systems are obtained by determining the inverse of the threshold contrast (i.e., the contrast sensitivity) at a set of points in the spatial frequency domain. Dvorak et al. (1980) measured at different mean luminance levels the spatial CSFs for the H1 DSMD neuron in the fly lobula plate. They found that the form of the CSF varies markedly with mean luminance; especially, the CSF increases as the mean luminance level of the stimulus is raised. At high mean luminance levels, the CSF peaks at a certain spatial frequency and falls off at higher and lower frequencies. Moreover, the high frequency roll-off becomes less steep and the peak frequency shifts toward lower frequencies as the illumination level is decreased. (See also Srinivasan and Dvorak, 1980).

The adaptation, or change of CSF on change of mean luminance level can be accounted for by considering the response of an MIMD to a moving sinusoidal grating. Equation 8 with $k_1 = k_2 = k_{12} = 1$ was used to compute normalized CSFs at different values of L_0. The CSFs in Figure 9 were obtained by including the effect of spatial filtering that takes place at the receptor level, i.e., by multiplying the contrast with the term $e^{-3.56(\Delta\rho f_s)^2}$. Using an effective contrast transfer parameter $\Delta\rho = 1.6\ \Delta S$ resulted in pushing the peak spatial frequency of the CSFs to a lower value (from 0.25 cycle/comp for $\Delta\rho = 0°$ to 0.14 cycle/comp). Figure 9 clearly demonstrates that the CSF of the MIMD adapts to mean luminance changes in the same way the CSF of DSMD neurons does, except for the peak which occurs at the same spatial frequency, $f_s = 0.14$ cycles/comp, in the case of the MIMD CSFs. However, in insect compound eyes the effective contrast transfer parameter, $\Delta\rho$, increases on lowering the mean luminance level L_0 (Buchner, 1976; Heisenberg and Buchner, 1977; Srinivasan and Dvorak, 1980; Pinter, 1985; Warrant and Pinter, 1989). This increase of $\Delta\rho$ is due to a mechanism of adaptation to low light levels that results in widening the angle subtended by the photoreceptive waveguides, hence, increasing absolute sensitivity by sacrificing spatial acuity (Nilsson, 1989; see also Laughlin, 1989b). If this $\Delta\rho$ change with mean luminance had been taken into account, then the peaks of the MIMD CSFs would also be shifted to lower frequencies for lower values of L_0 (see Bouzerdoum, 1991).

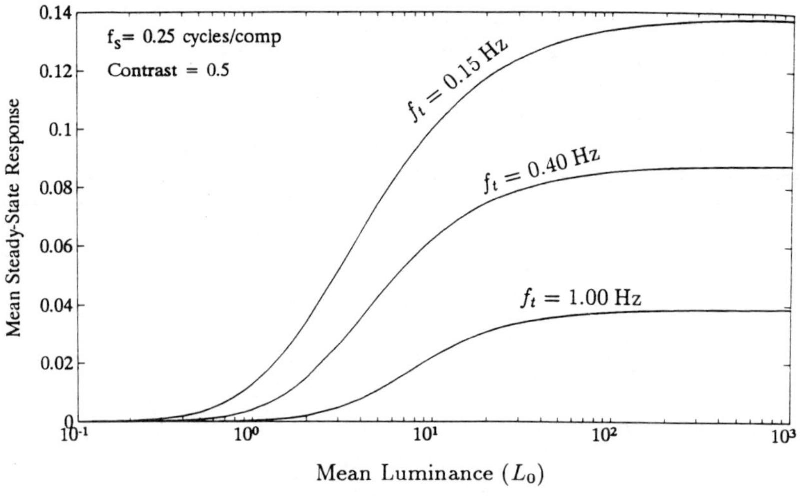

FIGURE 10. The ARMIMD steady-state response as a function of mean luminance for the contrast frequencies indicated in the figure. The responses are normalized with respect to the number of detectors in the array.

F. DEPENDENCE ON MEAN LUMINANCE LEVEL

Figure 10 depicts the steady-state response of an ARMIMD to drifting sine-wave gratings of fixed contrast as a function of mean luminance level. The variations of the curves in this figure agree well with those of the H1 neuron response. They are characterized by slow increase at low levels, saturation at high levels, and a rapid increase spanning about 2 log units of mean luminance at intermediate levels. The range of the steady-state response of H1 does also cover about 2 log units of mean luminance from threshold to saturation (Eckert, 1980; Hausen, 1981, 1984). This is different from the saturation at the photoreceptor level which spans a range of 5 log units of mean luminance. However, since the saturation phenomenon may occur at all levels of the complex architecture of the fly visual system, we cannot know for sure if the saturation of the DSMD neurons with respect to mean luminance is caused by saturation of EMDs feeding them, as suggested here.

VI. DISCUSSION

Insect optomotor responses are based on directional movement detection (Hassenstein and Reichardt, 1956; Reichardt, 1961; Götz, 1964; Kirschfeld, 1972; Buchner, 1976; Buchner et al., 1978; Reichardt and Egelhaaf, 1988). In flying and walking insects, the reactions evoked by the perceived apparent motion of a stationary milieu serve to minimize unvoluntary deviations and stabilize the course. In short, apparent vertical translatory motions are transformed into lift and thrust responses; whereas rotatory motions are transformed into torque around the vertical, longitudinal, and transverse body axes (yaw, roll, and pitch responses). The well-studied torque and thrust responses in flies are chiefly controlled by two independent movement detecting systems sensitive to horizontal and vertical motion, respectively (Kirschfeld, 1972; Buchner, 1976; Buchner et al., 1978). There exists now ample evidence indicating that the optic lobes of the fly contain two groups of neurons which respond preferentially to horizontal and vertical motion (Bishop and Keehn, 1967; Bishop et al., 1968; Mimura, 1971; Hausen, 1981). It has been repeatedly suggested that the giant tangential cells of the lobula plate are intimately involved with coding the information prerequisite to the optomotor responses (Bishop and Keehn, 1967; Bishop et al., 1968; McCann and Dill, 1969; Eckert, 1980; Hausen, 1981, 1982a, b, 1984; Hausen and Egelhaaf, 1989; Hengstenberg, 1982).

The anatomical and functional characteristics of these neurons were discussed in Section II, which describes in some detail the primary visual system of the fly.

The Reichardt correlation detector has been successfully and very extensively applied to account for behavioral and electrophysiological responses of the fly (Buchner, 1976; Heisenberg and Buchner, 1977; Buchner et al., 1978; Zaagman et al., 1978; Reichardt, 1987; Reichardt and Egelhaaf, 1988; Egelhaaf and Borst, 1989; Egelhaaf et al., 1989). Despite its predictive power, the correlation detector still suffers from some limitations. Based on linear time-invariant filtering, this detector cannot account for the saturation and adaptation phenomena exhibited by the responses of the large tangential cells (Dvorak et al., 1980; Lenting et al., 1984; Maddess and Laughlin, 1985; de Ruyter van Steveninck et al., 1986; Borst and Egelhaaf, 1987; Egelhaaf and Borst, 1989). Here we have departed from the tradition of employing the correlation detector to account for the response characteristics of the large DSMD neurons. We have instead applied a detector, namely the multiplicative inhibitory motion detector (MIMD), that is structurally and to second order functionally equivalent to the former detector; however, this detector is biologically more plausible: it is based on the nonlinear mechanism of shunting lateral inhibition (Bouzerdoum and Pinter, 1991).

The applicability of the MIMD to the activity of motion sensitive interneurons in the lobula plate has been demonstrated. The activity of these neurons was simulated by linearly summing the outputs of a one-dimensional array of MIMDs. Although the comparisons were qualitative, the MIMD responses showed a close similarity to those of DSMD neurons in both transient and steady-state characteristics. The MIMD also showed the potential to account even for the nonlinear characteristics of these neurons, such as saturation and adaptation to changes in mean luminance (see Figures 9 and 10). The discrepancies observed are due mostly to a choice of parameters. For example, the steady-state responses of the lobula plate DSMD neurons peak at frequencies in the range 1 to 5 Hz, but the MIMD responses, as depicted in Figure 7, peak at approximately 0.16 Hz. Though, we have shown that at sufficiently high mean luminance levels the MIMD response peaks at approximately $k_1/2\pi$ Hz. So, with the appropriate choice of the parameter k_1 the MIMD can be made to peak at the same temporal frequency as a DSMD neuron does.

Throughout this study, we have used a first-order low-pass filter in the lateral branch of the motion detector, but a higher-order low-pass or a high-pass filter could have been used as well. Franceschini and colleagues (1989) suggested that the lateral branches of the EMDs feeding the H1 DSMD neuron in the fly lobula plate contain, to a first approximation, low-pass filters of higher order, whereas the direct branches contain first-order high-pass filters. Note, however, that Franceschini and colleagues have assumed a facilitatory rather than an inhibitory mechanism for the interaction between the lateral and the direct input channels of the EMD circuit. By contrast, if the interactions are of the inhibitory type or veto type, then the direct branches contain higher-order low-pass filters and the lateral branches contain first-order high-pass filters. From Equation 5 it is clear that the direct input branch of a MIMD acts as a higher-order low-pass filter; it is composed of a cascade of low-pass filters. Thus, it is perhaps more appropriate to choose a first-order high-pass filter in the lateral branch of each MIMD.

In applying our theory to motion detection in the fly visual system, we have assumed that the spatial integration process which takes place at the level of the DSMD neurons is linear, while it is, in fact, nonlinear, mediated possibly by shunting inhibition (Hengstenberg, 1982; Reichardt et al., 1989). In principle, this process is almost linear if the activation of single input channels produces only minute voltage changes at the output sites of the dendrites (Hengstenberg, 1982). This nonlinear summation is a gain control mechanism by which a lobula plate DSMD neuron maintains a high gain for singly activated input channels and prevents being overloaded when many input elements are active (Hausen, 1984). Two basic approaches have been proposed for the implementation of the gain-control mechanism. One approach is based on presynaptic inhibition where a large field element, so-called pool cell, summates the outputs of local movement detectors, then inhibits either the individual movement detectors or directly the output cell (for more details, see Reichardt et al., 1989). The second approach, on the other hand, is an intracellular shunting model where a voltage-dependent potassium channel shunts depolarizing membrane voltages (Hengstenberg, 1982). As of now, it is not possible to distinguish between the two models; however, the second approach is more appealing, in the sense that it essentially relies on the intrinsic properties of neurons rather than on a circuit composed of different cells whose existence is, as yet, to be determined.

ACKNOWLEDGMENT

This work is supported in part by NSF grants BNS 8510188 and MIP 8822121.

REFERENCES

Ariel, M. and Adolph, A. R. (1985). Neurotransmitter inputs to directionally sensitive turtle ganglion cells, *J. Neurophysiol.*, **54**, 1123–1143.

Ariel, M. and Daw, N. W. (1982). Pharmacological analysis of directionally sensitive rabbit retinal ganglion cells, *J. Physiol. (London)*, **324**, 161–185.

Barlow, H. B. and Levick, W. R. (1965). The mechanism of directionally selective units in the rabbit's retina, *J. Physiol. (London)*, **178**, 477–504.

Barlow, H. B., Hill, R. M., and Levick, W. R. (1964). Retinal ganglion cells responding selectively to direction and speed of image motion in the rabbit, *J. Physiol. (London)*, **173**, 377–407.

Barlow, H. B. and Hill, R. M. (1963). Selective sensitivity to direction of motion in ganglion cells of the rabbit's retina, *Science*, **139**, 412–441.

Bishop, L. G., Keehn, D. G., and McCann, G. D. (1968). Motion detection by interneurons of the optic lobes and brain of the flies *Calliphora phaenicia* and *Musca domestica*, *J. Neurophysiol.*, **31**, 509–525.

Bishop, L. G. and Keehn, D. G. (1967). Neural correlates of the optomotor response in the fly, *Kybernetik*, **3**, 288–295.

Borst, A. and Egelhaaf, M. (1989). Principles of visual motion detection, *Trends Neurosci.*, **12**, (8), 297–306.

Borst, A. and Egelhaaf, M. (1987). Temporal modulation of luminance adapts time constant of fly movement detectors, *Biol. Cybern.*, **56**, 209–215.

Bouzerdoum, A. (1991). Nonlinear Lateral Inhibitory Neural Networks: Analysis and Application to Motion Detection, Ph.D. dissertation, University of Washington, Seattle.

Bouzerdoum, A. and Pinter, R. B. (1991). Modeling the processing and perception of visual motion, in *Sensory Neural Networks: Lateral Inhibition*, Nabet, B. and Pinter, R. B., Eds., CRC Press, Boca Raton, FL, 47–68.

Bouzerdoum, A., Nabet, B., and Pinter, R. B. (1991). Analysis and analog implementation of directionally sensitive shunting inhibitory neural networks, in *Visual Information Processing: From Neurons to Chips*, SPIE-1473, 29–38.

Bouzerdoum, A. and Pinter, R. B. (1990). A shunting inhibitory motion detector that can account for the functional characteristics of fly motion-sensitive interneurons, in Proc. Int. Joint. Conf. Neural Nets, 17–21, San Diego, CA, pp I-149-I-153.

Bouzerdoum, A. and Pinter, R. B. (1989a). Image motion processing in biological and computer vision systems, in *Visual Communications and Image Processing IV*, **SPIE-1199**, 1229–1240.

Bouzerdoum, A. and Pinter, R. B. (1989b). Biophysical basis, stability, and directional response characteristics of multiplicative lateral inhibitory neural networks, Int. Joint Conf. Neural Nets, 18–22 June, Washington, D.C.

Braitenberg, V. (1972). Periodic structures and structural gradients in the visual ganglia of the fly, in *Information Processing in the Visual Systems of Arthropods*, Wehner, R., Ed., Springer-Verlag, Berlin, 3–15.

Braitenberg, V. (1967). Patterns of projection in the visual system of the fly. I. Retina-lamina projections, *Exp. Brain Res.*, **3**, 271–298.

Buchner, E. and Buchner, S. (1984). Neuroanatomical mapping of visually induced nervous activity in insects by H-deoxyglucose, in *Photoreception and Vision in Invertebrates,* Ali, M. A., Ed., Plenum Press, London, 623–634.

Buchner, E., Götz, K. G., and Straub, C. (1978). Elementary detectors for vertical movement in the visual system of *Drosophila, Biol. Cybern.,* **31**, 235–242.

Buchner, E. (1976). Elementary movement detectors in an insect visual system, *Biol. Cybern.,* **24**, 85–101.

Cajal, S. R. and Sánchez, D. (1915). Contribucion al Conocimiento de los Centros Nerviosos de los Insectos. I. Retina y Centros Opticos, *Trab. Lab. Invest. Biol. Univ. Madrid,* **13**, 1–168.

Caldwell, J. H., Daw, N. W., and Wyatt, H. J. (1978). Effects of picrotoxin and strychnine on rabbit retinal ganglion cells: lateral interactions for cells with more complex receptive fields, *J. Physiol. (London),* **276**, 277–298.

Campos-Ortega, J. A. and Strausfeld, N. J. (1972). Column and layers in the second synaptic region of the fly's visual system: the case for two superimposed neural architectures, in *Information Processing in the Visual Systems of Arthropods,* Wehner, R., Ed., Springer-Verlag, Berlin, 31–36.

Collet, T. and Blest, D. A. (1966). Binocular, directionally-selective neurons, possibly involved in the optomotor response of insects, *Nature (London),* **212**, 1330–1333.

DeVoe, R. D., Kaiser, W., Ohm, J., and Stone, L. S. (1982). Horizontal movement detectors of honeybees: directionally-selective visual neurons in the lobula and brain, *J. Comp. Physiol.,* **A-147**, 155–170.

DeVoe, R. D. (1980). Movement sensitivities of cells in the fly's medulla, *J. Comp. Physiol.,* **A-138**, 93–119.

DeVoe, R. D. and Ockleford, E. M. (1976). Intracellular responses from cells of the medulla of the fly, *Calliphora erythrocephala, Biol. Cybern.,* **23**, 13–24.

Dvorak, D., Srinivasan, M. V., and French, A. S. (1980). The contrast sensitivity of fly movement-detection neurons, *Vision Res.,* **20**, 397–407.

Dvorak, D. R., Bishop, L. G., and Eckert, H. E. (1975). On the identification of movement detectors in the optic lobe, *J. Comp. Physiol.,* **A-100**, 5–23.

Eckert, H. (1982). The vertical-horizontal neurone (VH) in the lobula plate of the blowfly, *Phaenicia, J. Comp. Physiol.,* **A-149**, 195–205.

Eckert, H. (1981). The horizontal cells in the lobula plate of the blowfly, *Phaenicia sericata, J. Comp. Physiol.,* **A-143**, 511–526.

Eckert, H. (1980). Functional properties of the H1-neurone in the third optic ganglion of the blowfly, *Phaenicia, J. Comp. Physiol.,* **A-135**, 29–39.

Eckert, H. and Bishop, L. G. (1978). Anatomical and physiological properties of the vertical cells in the third optic ganglion of *Phaenicia sericata* (Dipt. Call.), *J. Comp. Physiol.,* **A-126**, 57–86.

Egelhaaf, M., Borst, A., and Reichardt, W. (1989). Computational structure of a biological motion-detection system as revealed by local detector analysis in the fly's nervous sytem, *J. Opt. Soc. Am.,* **A-6**, 1070–1087.

Egelhaaf, M. and Borst, A. (1989). Transient and steady-state response properties of movement detectors, *J. Opt. Soc. Am.,* **A-6** 116–127.

Egelhaaf, M. and Reichardt, W. (1987). Dynamic response properties of movement detectors: theoretical analysis and electrophysiological investigation in the visual system of the fly, *Biol. Cybern.,* **56**, 69–87.

Egelhaaf, M. (1985). On the neuronal basis of figure-ground discrimination by relative motion in the visual system of the fly. II. Figure-detection cells, a new class of visual interneurones, *Biol. Cybern.,* **52**, 195–209.

Foster, D. H. (1971). A model of the human visual system in its response to certain classes of moving stimuli, *Kybernetik,* **8**, 69–84.

Franceschini, N., Riehle, A., and Le Nestour, A. (1989). Directionally selective motion detection by insect neurons, in *Facets of Vision,* Stavenga, D. G. and Hardie, R. C., Eds., Springer-Verlag, Berlin, 360–389.

Furman, G. G. and Frischkopf, L. S. (1964). Model of neural inhibition in the mammalian cochlea, *J. Acoust. Soc. Am.,* **36**, 2194–2201.

Goldsmith, T. H. and Bernard, G. D. (1974). The visual system of insects, in *The Physiology of Insecta,* **Vol. 2**, 2nd ed., Morris, R., Ed., Academic Press, London, 165–272.

Götz, K. G. (1964). Optomotorische Untersuchung des visuellen Systems einiger Augenmutanten der Fruchtfliege *Drosophila, Kybernetik,* **2**, 77–92.

Hartline, H. K. and Ratliff, F. (1957). Inhibitory interaction of receptor units in the eye of *Limulus, J. Gen. Physiol.,* **40**, 357–376.

Hassenstein, B. and Reichardt, W. (1956). Functional structure of a mechanism of perception of optical movement, in Proc 1st Int. Congr. Cybern., Namur, 797–801.

Hausen, K. and Egelhaaf, M. (1989). Neural mechanisms of visual course control in insects, in *Facets of Vision,* Stavenga, D. G. and Hardie, R. C., Eds., Springer-Verlag, Berlin, 30–73.

Hausen, K. (1987). The neural architecture of the lobula plate of the blowfly, *Calliphora erythrocephala, Cell Tissue Res.,* submitted.

Hausen, K. (1984). The lobula-complex of the fly: structure, function and significance in visual behavior, in *Photoreception and Vision in Invertebrates,* Ali, M. A., Ed., Plenum Press, London, 523–559.

Hausen, K. (1982a). Motion sensitive interneurons in the optomotor system of the fly. I. The horizontal cells: structure and signals, *Biol. Cybern.,* **45**, 143–156.

Hausen, K. (1982b). Motion sensitive interneurons in the optomotor system of the fly. II. The horizontal cells: receptive field organization and response characteristics, *Biol. Cybern.,* **46**, 67–79.

Hausen, K. (1981). Monocular and binocular computation of motion in the lobula plate of the fly, *Dtsch. Zool. Ges.,* **74**, 49–70.

Hausen, K., Wolburg-Buchholz, K., and Ribi, W. A. (1980). The synaptic organization of visual interneurons in the lobula complex of flies, *Cell Tissue Res.,* **208**, 371–387.

Hausen, K. and Strausfeld, N. J. (1980). Sexually dimorphic interneuron arrangements in the fly visual system, *Proc. R. Soc. London Ser.,* **B-208**, 57–71.

Hausen, K. (1976). Functional characterization and anatomical identification of motion sensitive neurons in the lobula plate of the blowfly *Calliphora erythrocephala, Z. Naturforsch.,* **C-31**, 629–633.

Heisenberg, M. and Buchner, E. (1977). The rôle of retinula cell types in visual behavior of *Drosophila melanogaster, J. Comp. Physiol.,* **A-117**, 127–162.

Hengstenberg, R., Hausen, K., and Hengstenberg, B. (1982). The number and structure of giant vertical cells (VS) in the lobula plate of the blowfly *Calliphora erythrocephala, J. Comp. Physiol.,* **A-149**, 163–177.

Hengstenberg, R. (1982). Common visual properties of giant vertical cells (VS) in the lobula plate of the blowfly *Calliphora, J. Comp. Physiol.,* **A-149**, 179–193.

Horridge, G. A., Scholes, J. H., Shaw, S., and Tunstall, J. (1965). Extracellular recordings from single neurons in the optic lobe and brain of the locust, in *The Physiology of the Insect Central Nervous System,* Treherne, J. E. and Beament, J. W. L., Eds., Academic Press, New York, 165–202.

Kirschfeld, K. (1979). The visual system of the fly: physiological optics and functional anatomy as related to behavior, in *The Neuroscience: Fourth Study Program,* Schmitt, F. O. and Worden, F. G., Eds., MIT Press, Cambridge, 297–316.

Kirschfeld, K. (1972). The visual system of musca: studies on optics, structure and fusion, in *Information Processing in the Visual Systems of Arthropods,* Wehner, R., Ed., Springer-Verlag, Berlin, 61–74.

Laughlin, S. (1989a). Coding efficiency and design in visual processing, in *Facets of Vision,* Stavenga, D. G. and Hardie, R. C., Eds., Springer-Verlag, Berlin, 213–234.

Laughlin, S. (1989b). The role of sensory adaptation in the retina, *J. Exp. Biol.,* **146**, 39–62.

Laughlin, S. (1984). The role of parallel channels in early visual processing by the arthropod compound eye, in *Photoreception and Vision in Invertebrates,* Ali, M. A., Ed., Plenum Press, London, 457–481.

Laughlin, S. (1981). Neural principles in the peripheral visual systems of invertebrates, in *Handbook of Sensory Physiology,* **Vol. VII/6B** (Invertebrates Visual Centers and Behavior I), Autrum, H., Ed., Springer-Verlag, Berlin, 133–280.

Lenting, B. P. M., Mastebroek, H. A. K., and Zaagman, W. H. (1984). Saturation in a wide-field, directionally selective movement detection system in fly vision, *Vision Res.,* **24**, 1341–1347.

Lipetz, L. E. and Hill, R. M. (1970). Discrimination characteristics of the turtle's retinal cells, *Experientia,* **26**, 373–374.

Maddess, T. and Laughlin, S. B. (1985). Adaptation of the motion-sensitive neuron H1 is generated locally and governed by contrast frequency, *Proc. R. Soc. London Ser.,* **B-225**, 251–275.

Marchiafava, P. L. (1983). The organization of inputs establishes two functional and morphologically identifiable classes of ganglion cells in the retina of the turtle, *Vision Res.,* **23**, 325–338.

Marchiafava, P. L. (1979). The response of retinal ganglion cells to stationary and moving visual stimuli, *Vision Res.,* **19**, 1203–1211.

Marmarelis, P. Z. and McCann, G. D. (1973). Development and application of white-noise modeling techniques for studies of insect visual nervous system, *Kybernetik,* **12**, 74–90.

Maturana, H. R. and Frenk, S. (1963). Directional movement and horizontal edge detectors in the pigeon eye, *Science,* **142**, 977–979.

Maturana, H. R., Lettvin, J. Y., McCulloch, W. S., and Pitts, W. H. (1960). Anatomy and physiology of vision in the frog *(Rana Pipiens), J. Gen. Physiol.,* **142**, (Suppl. 2), 129–171.

McCann, G. D. (1973). The fundamental mechanism of motion detection in the insect visual system, *Kybernetik,* **12**, 64–73.

McCann, G. D. and Dill, S. F. (1969). Fundamental properties of intensity, form and motion perception in the visual nervous system of *Calliphora phaenicia* and *Musca domestica, J. Gen. Physiol.,* **53**, 385–413.

Michael, C. R. (1968). Receptive fields of single optic nerve fibers in a mammal with an all-cone retina. II. Directionally sensitive units, *J. Neurophysiol.,* **31**, 257–2617.

Mimura, K. (1975). Units of the optic lobe, especially movement, in *The Compound Eye and Vision in Insects,* Horridge, G. A., Ed., Clarendon Press, Oxford, 423–436.

Mimura, K. (1974). Analysis of visual information in lamina neurones of the fly, *J. Comp. Physiol.*, **A-88**, 335–372.

Mimura, K. (1972). Neural mechanisms, subserving directional selectivity of movement in the optic lobe of the fly, *J. Comp. Physiol.*, **80**, 409–437.

Mimura, K. (1971). Movement discrimination by the visual system of flies, *Z. Vergl. Physiol.*, **73**, 105–138.

Nakayama, K. (1985). Biological image motion processing: a review, *Vision Res.*, **25**, 625–660.

Nilsson, D.-E. (1989). Optics and evolution of the compound eye, in *Facets of Vision,* Stavenga, D. G. and Hardie, R. C., Eds., Springer-Verlag, Berlin, 30–73.

Pierantoni, T. (1976). A look into the cock-pit of the fly. The architecture of the lobular plate, *Cell Tissue Res.*, **171**, 101–122.

Pinter, R. B. (1985). Adaptation of spatial modulation transfer functions via nonlinear lateral inhibition, *Biol. Cybern.*, **51**, 285–291.

Pinter, R. B. (1983). The electrophysiological bases for linear and for nonlinear product term lateral inhibition and the consequences for wide field textured stimuli, *J. Theor. Biol.*, **105**, 233–243.

Poggio, T. and Reichardt, W. (1973). Considerations on models of movement detection, *Kybernetik,* **13**, 223–227.

Ratliff, F. (1961). Inhibitory interaction and the detection and enhancement of contours, in *Sensory Communication,* Rosenblith, W. A., Ed., MIT Press, Cambridge, 183–203.

Reichardt, W., Egelhaaf, M., and Guo, A.-K. (1989). Processing of figure and background motion in the visual system of the fly, *Biol. Cybern.*, **61**, 327–345.

Reichardt, W. and Egelhaaf, M. (1988). Properties of individual movement detectors as derived from behavioural experiments on the visual system of the fly, *Biol. Cybern.*, **58**, 287–294.

Reichardt, W. (1987). Evaluation of optical motion information by movement detectors, *J. Comp. Physiol.*, **A-161**, 533–547.

Reichardt, W. (1961). Autocorrelation, a principle for the evaluation of sensory information by the nervous system, in *Sensory Communication,* Rosenblith, W. A., Ed., MIT Press, Cambridge, 303–317.

Riehle, A. and Franceschini, N. (1984). Motion detection in flies: parametric control over ON-OFF pathways, *Exp. Brain Res.*, **54**, 390–394.

de Ruyter van Steveninck, R. R., Zaagman, W. H., and Mastebroek, H. A. K. (1986). Adaptation of transient responses of a movement-sensitive neuron in the visual system of the blowfly *Calliphora erythrocephala, Biol. Cybern.*, **54**, 223–236.

van Santen, J. P. H. and Sperling, G. (1984). Temporal covariance model of human motion perception, *J. Opt. Soc. Am.*, **A-1**, 451–473.

Schmid, A. and Bülthoff, H. (1988). Using neuropharmacology to distinguish between excitatory and inhibitory movement detection mechanisms in the fly *Calliphora erythrocephala, Biol. Cybern.*, **59**, 71–80.

Shaw, S. R. (1989). The retina-lamina pathway in insects, particularly Diptera, viewed from an evolutionary perspective, in *Facets of Vision,* Stavenga, D. G. and Hardie, R. C., Eds., Springer-Verlag, Berlin, 186–212.

Shaw, S. R. (1984). Early visual processing in insects, *J. Exp. Biol.*, **112**, 225–251.

Shaw, S. R. (1981). Anatomy and physiology of identified non-spiking cells in the photoreceptor-lamina complex of the compound eye of insects, especially Diptera, in *Neurons without Impulses,* Roberts, A. and Bush, B. M. H., Eds., Cambridge University Press, London, 61–116.

Srinivasan, M. V. and Dvorak, D. R. (1980). Spatial processing of visual information in the movement-detecting pathway of the fly: characteristics and functional significance, *J. Comp. Physiol.*, **A-140**, 1–23.

Strausfeld, N. J. (1989). Beneath the compound eye: neuroanatomical analysis and physiological correlates in the study of insect vision, in *Facets of Vision,* Stavenga, D. G. and Hardie, R. C., Eds., Springer-Verlag, Berlin, 317–359.

Strausfeld, N. J. and Bassemir, U. K. (1985a). The organization of giant horizontal-motion-sensitive neurons and their synaptic relationships in the lateral deutocerebrum of *Calliphora erythrocephala* and *Musca domestica, Cell Tissue Res.*, **242**, 531–550.

Strausfeld, N. J. and Bassemir, U. K. (1985b). Lobula plate and ocellar interneurons converge onto a cluster of descending neurons leading to neck and leg motor neuropil in *Calliphora erythrocephala, Cell Tissue Res.*, **240**, 617–640.

Strausfeld, N. J. (1984). Functional Neuroanatomy of the blowfly's visual system, in *Photoreception and Vision in Invertebrates,* Ali, M. A., Ed., Plenum Press, London, 483–522.

Strausfeld, N. J. and Nässel, D. R. (1981). Neuroarchitectures serving compound eyes of crustacea and insects, in *Handbook of Sensory Physiology,* **Vol. VII/6B** (Invertebrates Visual Centers and Behavior I), Autrum, H., Ed., Springer-Verlag, Berlin, 1–132.

Strausfeld, N. J. and Campos-Ortega, J. A. (1977). Vision in insects: pathways possibly underlying neural adaptation and lateral inhibition, *Science,* **195**, 894–897.

Strausfeld, N. J. (1976a). *Atlas of an Insect Brain,* Springer-Verlag, Berlin.

Strausfeld, N. J. (1976b). Mosaic organizations, layers and visual pathways in the insect brain, in *Neural Principles in Vision,* Zettler, F. and Weiler, R., Eds., Springer-Verlag, Berlin, 245–279.

Vigier, P. (1908). Sur l'existence réelle et le rôle des appendices piriform des neurones. La neurone périoptique des diptère, *C. R. Soc. Biol. Paris,* **64**, 959–961.

Warrant, E. J. and Pinter, R. B. (1990). Changes of acuity during light and dark adaptation in the dragonfly compound eye, *Z. Naturforsch.*, **45c**, 137–141.

Wehner, R. (1981). Spatial vision in arthropods, in *Handbook of Sensory Physiology,* **Vol. VII/6C** (Invertebrates Visual Centers and Behavior II), Autrum, H., Ed., Springer-Verlag, Berlin, 287–616.

Werblin, F. S. (1970). Response of retinal cells to moving spots: intracellular recording in *Necturus maculosus, J. Neurophysiol.*, **33**, 342–350.

Wyatt, H. R. and Daw, N. (1975). Directionally sensitive ganglion cells in the rabbit retina: specificity for stimulus, direction, size, and speed, *J. Neurophysiol.*, **38**, 613–626.

Zaagman, W. H., Mastebroek, H. A. K., and Kuiper, J. W. (1978). On the correlation model: performance of a movement detecting neural element in the fly visual system, *Biol. Cybern.*, **31**, 163–168.

Chapter 17

LATERAL INHIBITION AND IMAGE PROCESSING

M. E. Jernigan and G. F. McLean

The fundamental problem of vision is the extraction of significant information from images under a tremendously wide range of imaging conditions and consequent variations in image quality. Because of imperfections in the imaging system (limited resolution, transmission and detection errors, relative motion, etc.), variations in illumination, and obscuration of the optical pathway (cloud, smoke, haze, etc.), images may be blurred or noisy, have excessive dynamic range or inadequate contrast, or all of these, perhaps in different regions of a single image. How is it that natural vision systems address these problems, and how might artificial vision systems be designed to address them?

Early efforts to understand the image processing effected in the retinas of biological vision systems consisted of simple linear models of spatial summation and lateral inhibition (Barlow, 1953; Hartline and Ratliff, 1957). Viewed as discrete signal processing systems, these two processes form linear combinations of input picture element intensities with positive (summation) or negative (inhibition) weights, which may be written for a simplified one-dimensional case:

$$y(n) = \sum_{k \in N} h(k)x(n - k) \tag{1}$$

where the weights, $h(k)$, can be interpreted as a discrete system impulse response, and the sum is over a neighborhood N around a picture element at n. Figure 1 shows typical summation and lateral inhibition impulse responses.

In digital filter design, these weights can be chosen to achieve spatial smoothing or low-pass filtering to suppress high spatial frequency noise and local sharpening or high-pass filtering to enhance edges. For example, the simplest edge-enhancement schemes are unsharp masking and subtracting the Laplacian (second difference). In unsharp masking

$$y(n) = 2x(n) - \bar{x}(n) \tag{2}$$

where the coefficient 2 achieves unit gain for constant input and $\bar{x}(n)$ is the local mean, estimated with a local average. For the simplest 3-point average

$$\bar{x}(n) \simeq \frac{1}{3}\,[x(n - 1) + x(n) + x(n + 1)] \tag{3}$$

which results in an unsharp mask impulse response of

$$h(n) = \left[-\frac{1}{3} \;\; \frac{5}{3} \;\; -\frac{1}{3} \right] \tag{4}$$

The second difference scheme is

$$\begin{aligned} y(n) &= x(n) - [x(n - 1) - 2x(n) + x(n + 1)] \\ &= 3x(n) - x(n - 1) - x(n + 1) \end{aligned} \tag{5}$$

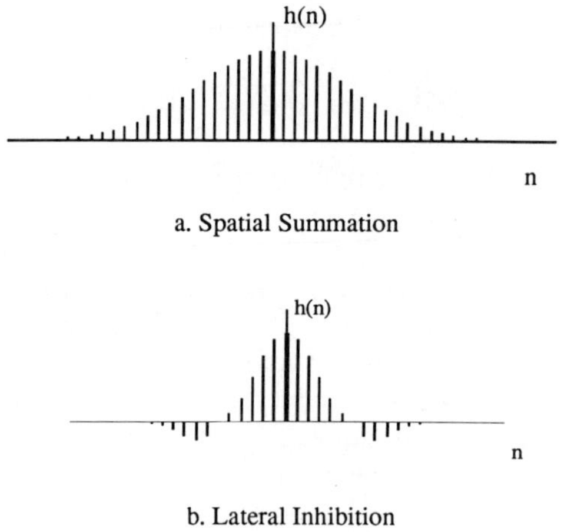

FIGURE 1. Linear models of summation and inhibition.

with impulse response

$$h(n) = [-1 \quad 3 \quad -1] \tag{6}$$

In a broader context, both these schemes are instances of an inverse filter which deblurs an image degraded by an exponential blur process whose discrete impulse response is

$$h_d(n) = \frac{1 - a}{1 + a} a^{|n|} \tag{7}$$

Taking the z transform

$$H_d(z) = \frac{(1 - a)^2}{1 + a^2 - az - az^{-1}} \tag{8}$$

With the inverse system

$$\begin{aligned} H(z) &= H_d^{-1}(z) \\ &= \frac{1 + a^2}{(1 - a)^2} - \frac{a}{(1 - a)^2} z - \frac{a}{(1 - a)^2} z^{-1} \end{aligned} \tag{9}$$

which has the difference equation

$$y(n) = \frac{1 + a^2}{(1 - a)^2} x(n) - \frac{a}{(1 - a)^2} [x(n - 1) + x(n + 1)] \tag{10}$$

Unsharp masking is, thus, the inverse of a blur process whose impluse response is given by Equation 4 with $a \simeq 0.21$, while subtracting the second difference compensates for a blur with parameter $a \simeq 0.382$. The form of these edge enhancement processes, shown in Figure 2, is the simplest possible symmetric linear lateral inhibition model.

More sophisticated filters combine smoothing and sharpening effects and have impulse responses which closely model observed neurophysiological responses. Difference of Gaus-

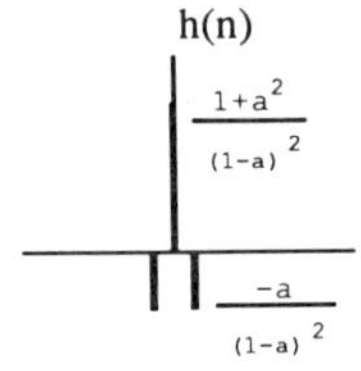

FIGURE 2. Lateral inhibition as a deblurring inverse filter (unsharp masking: $a \simeq 0.21$; subtracting the second difference: $a \simeq 0.382$).

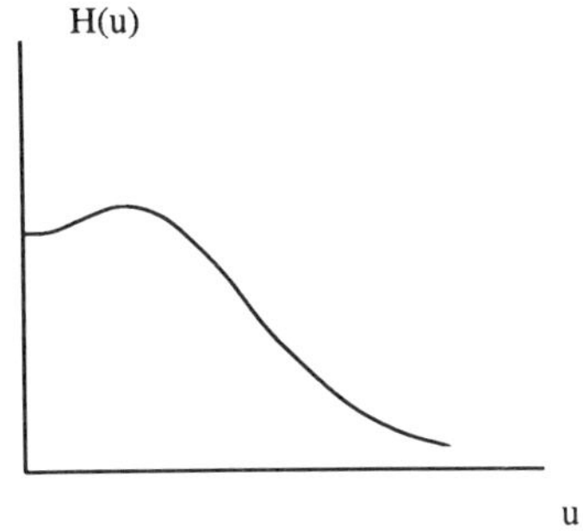

FIGURE 3. Typical optimal image restoration filter frequency response.

sians, Gabor filters, and the Gaussian smoothed Laplacians have all been advocated both as models and as image enhancement operators (Jernigan and Wardell, 1981; Marr, 1982; Wilson and Bergen, 1979). Similar processes emerge from an optimal restoration derivation. The classical Wiener deconvolution filter for a minimum mean-squared error estimate of a linearly degraded, noise-corrupted signal has a frequency response whose general form, under reasonable models of the linear degradation, and signal and noise spectra, are shown in Figure 3. The response is reminiscent of empirical measurements of the modulation transfer function in human vision and is easily approximated with a difference of Gaussians or Gabor filter. It is worth noting the essential similarity of the difference of Gaussians, Gabor filters, and the Gaussian smoothed Laplacian. All have impulse responses qualitatively similar to that shown in Figure 1b. Differences relate to the size and distribution of the weights in the central and surrounding regions. The Gabor filter is distinct in its secondary excitatory and inhibitory surrounding regions.

All of these classic linear models fall short, both as models of lateral interaction in the retina and as effective mechanisms of image enhancement or restoration. The difficulty lies in the nature of images, or more precisely, the nature of significant information in images. Images are fundamentally multiplicative, being formed as the product of an illumination and a reflectance component. Images are also nonstationary; indeed, it is exactly the change in local gray-level statistics that conveys the significant information for perception and interpretation of images. Changes in local intensity or in the pattern of spatial variation in intensity (texture) correspond to edges which make up the boundaries of objects whose shape and texture lead to identification and recognition. Classical linear, shift invariant processes are unable to treat separately multiplicative input components and are unable to respond differently to different local image characteristics. Further, the empirical evidence has long since demonstrated significant nonlinear and adaptive behavior on the part of neurophysiological mechanisms.

In the early 1970s, the image processing community began to explore one aspect of the problem: the intrinsically multiplicative nature of images as a product of illumination (whose

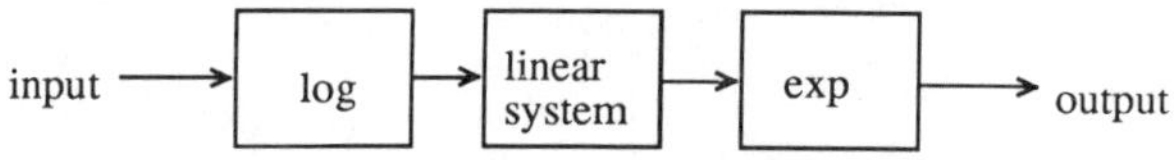

FIGURE 4. Canonical form of a multiplicative homomorphic system.

variation is often undesired) and reflectance (generally taken as signal). Stockham observed that a logarithmic non-linearity in the response of retinal receptors is consistent with a multiplicative image model in the sense of Oppenheim's canonical homomorphic systems (Oppenheim and Schafer, 1975; Stockham, 1972). That is, a system such as that in Figure 4 obeys a generalized superposition principle with respect to multiplication:

$$T[r(x,y)i(x,y)] = T[r(x,y)]T[i(x,y)] \tag{11}$$

So-called homomorphic, or density domain, image processing systems are effective for local contrast enhancement (edges as signal) and global dynamic range reduction (illumination variation as noise). Such systems can also be viewed directly in terms of discrete picture elements as local product interactions, where input pixels are weighted with exponents which are the impulse response weights of the linear filter in Figure 4 (Moloney and Jernigan, 1990).

Specifically, in one dimension,

$$y(n) = \prod_{k \in N} x^{h(k)}(n - k) \tag{12}$$

Applying a high-pass filter in the density domain — the textbook homomorphic filtering example — is a form of unsharp masking, or deblurring, which in the simplest case yields

$$\begin{aligned} y(n) &= \frac{x^{5/3}(n)}{[x(n - 1)x(n + 1)]^{1/3}} \\ &= \frac{x^2(n)}{\bar{x}_G(n)} \end{aligned} \tag{13}$$

where $\bar{x}_G(n)$ is the geometric mean of $x(n)$, or $[x(n - 1)x(n)x(n + 1)]^{1/3}$ for a 3-point estimate. This nonlinear edge enhancement scheme does relatively more enhancement where the local mean is less, and vice versa.

Except for the initial logarithmic nonlinearity, this scheme is still based on a simple linear model of lateral inhibition. Beginning in 1980, Pinter began to work with a local nonlinear model of lateral inhibition based on multiplicative lateral interaction (Pinter, 1984). The model has both nonlinear and recurrent characteristics which result in local gain adaptation, adaptation of the local impulse response, and luminance adaptive edge enhancement. Its image enhancement potential is described by Jernigan et al. (1991). The steady-state behavior takes the form

$$y(n) = \frac{x(n)}{1 + \sum_{k \neq 0, k \in N} f_k[y(n - k)]} \tag{14}$$

with the functions $\{f_k[y(n - k)]\}$ describing the nature of the local interaction. An alternate expression more clearly manifests the multiplicative inhibition:

$$y(n) = x(n) - y(n)\left\{\sum_{k\neq 0, k\in N} f_k[y(n - k)]\right\} \tag{15}$$

Several forms of these functions have been discussed, including polynomial (Pinter, 1984), exponential, and linear nearest neighbor (Nabet and Pinter, 1991; Pinter, 1984). The luminance adaptive behavior emerges as long as the inhibition is the product of the output and some function (diminishing with distance) of neighboring outputs. An interesting interpretation emerges if the solution of the model can be written in the form

$$y(n) = \frac{x(n)}{\Pi_{k\neq 0, k\in N}\, y^{a_k}(n - k)} \tag{16}$$

which assumes the function of neighboring outputs can be expressed as a product of terms with exponents $\{a_k\}$. Taking the log on both sides of this expression:

$$\log y(n) = \log x(n) - \sum_{k\neq 0, k\in N} a_k \log y(n - k) \tag{17}$$

or

$$\sum_{k\neq N} a_k \log y(n - k) = \log x(n) \tag{18}$$

where $a_0 = 1$. This expression is a recursive linear system operating in the density domain. Referring to Figure 4, the linear difference equation of the linear filter is

$$\sum_{k\in N} a_k \tilde{y}(n - k) = \tilde{x}(n) \tag{19}$$

where $\bar{x}(n) = \log x(n)$ and the weights $\{a_k\}$ are coefficients in the difference equation and can be chosen to accomplish any desired recursive digital filtering, including low pass (smoothing) and high pass (sharpening). Greater flexibility and specificity in filter design may be achieved by introducing zeroes in the transfer function by allowing a linear combination of neighboring inputs on the right-hand side of Equation 19.

As a simple example, suppose the following second-order recursive system is applied in the density domain:

$$\tilde{y}(n) + \frac{a}{1 + a^2}[\tilde{y}(n - 1) + \tilde{y}(n + 1)] = b\left[\frac{1 - a^2}{1 + a^2}\right] x(n) \tag{20}$$

The symmetric, noncausal impulse response of this system is

$$\tilde{h}(n) = b(-a)^{|n|} \tag{21}$$

which is shown in Figure 5.

This system shows secondary excitatory and inhibitory behavior of diminishing significance (recall the characteristic Gabor filter response). With the effects of the logarithmic nonlinearity, both primary and secondary inhibition show an implicit luminance dependence, with relatively less edge enhancement and ringing as luminance decreases. The multiplicative lateral inhibition model (Equation 14) is characterized by a nonlinear mapping of constant inputs which follows a power law. That is, for a constant input L, the output is L^{α}, where

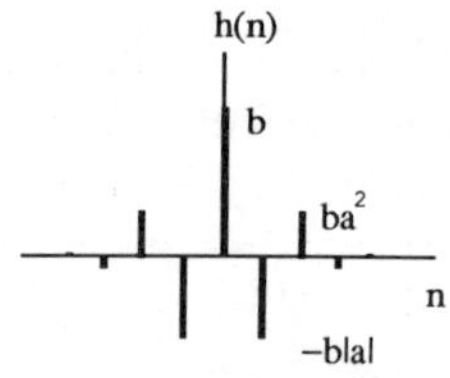

FIGURE 5. Exponential high pass filter ($|a| < 1$).

α is slightly less than 1/2 for the nearest neighbor implementation and greater than 1/2 for the full exponential model (Pinter, 1984). In addition, the relative edge enhancement, defined as the peak-to-peak difference divided by the change in background luminance increases with luminance at low luminance values and then levels off to a constant (near 1.4 for the nearest neighbor and near 1.7 for the exponential versions). The recursive system of Equation 15 has similar behavior with a constant response of

$$y(n) = L^{b\frac{1 - a^2}{(1 - a)^2}} \quad for\ x(n) = L \tag{22}$$

and a relative edge enhancement of

$$\begin{aligned} REE &= \frac{Peak\ to\ Peak}{\Delta\ background} \\ &= \frac{s^b - 1}{s^{\frac{b}{1+a}} - s^{\frac{ba}{1+a}}} \end{aligned} \tag{23}$$

where S is the ratio of the input luminance at the edge [i.e., $x(n) = L + LSu(n)$, where $u(n)$ is a unit step]. The edge enhancement is approximately

$$REE \simeq \frac{1 + a}{1 - a} \tag{24}$$

for $1 < S < 2$. With Equations 22 and 24, the recursive homomorphic system can be made to match both DC gain and relative edge enhancement behavior of either nearest neighbor or exponential forms of multiplicative lateral inhibition.

This simple example demonstrates the essential similarity of multiplicative lateral inhibition and homomorphic recursive high frequency enhancement filters. While achieving an implicit luminance dependent behavior that appears to model the empirical behavior of neurophysiological mechanisms in biological retinas, these processes are limited in their applicability to the range of problems encountered in digital image processing. We conclude this article with a brief description of recent explicitly adaptive mechanisms better suited to many such problems.

Over the last 10 years, image processors have become increasingly interested in adaptive systems for image processing, those which adapt to local characteristics of the input image to achieve edge dependent smoothing, luminance dependent sharpening, or both. In the case of edge dependent smoothing, the reduced sensitivity of human vision to random noise in the vicinity of strong edges has been exploited by Anderson and Netravali (1976), Nagao and Matsuyama (1979), Lee (1980), Mahesh et al. (1990), and others to devise processes which smooth furthest from edges while allowing sharp edges to pass unblurred. Because the perceptually important edges are less affected than by simple linear shift invariant filters, subjective image quality is better. On the other hand, luminance-dependent sharpening combined with dynamic range reduction has been described by Peli & Lim (1982) and deVries (1990). This adaptive edge enhancement competes favorably with homomorphic or density domain high frequency emphasis on similar images.

While most of these adaptive filtering strategies have taken nonrecursive forms, recursive forms are also possible (Jernigan and Erler, 1991). Recursive adaptive filters provide for more flexible adaptation of the form and extent of the local impulse response and allow for an effectively greater extent of local interaction. Their implementation, particularly by simulation on conventional computers, is generally more complicated than their nonrecursive counterparts, but may be more efficient in the use of lateral interaction in neural networks, both synthetic and natural. In implementations of both Lee's local linear minimum-squared error estimator and a recursive adaptive version, better image quality is achieved more readily with the recursive implementation (Jernigan and Erler, 1991).

An adaptive difference equation in one dimension takes the form:

$$y(n) = \sum_{l \in N_i} b_l x(n - l) - \sum_{k \in N_0, k \neq 0} a_k y(n - k) \tag{25}$$

where N_i and N_0 are the input and output neighborhoods of interaction, and the weights $\{b_l\}$ and $\{a_k\}$ are dependent on local image characteristics. For example, Lee's nonrecursive additive noise filter has coefficients

$$\mathrm{b}_0 = \frac{\sigma_s^2 + \frac{1}{N}\sigma_n^2}{\sigma_x^2}$$

$$b_l = \frac{1}{N}\frac{\sigma_n^2}{\sigma_x^2} \qquad l \in N_i,\ l \neq 0 \tag{26}$$

where σ_s^2 and σ_x^2 are local estimates of signal and input variance, respectively, and σ_n^2 is the noise variance. The recursive form is achieved by choosing $\{a_k\}$ such that the corresponding pole locations approach the unit circle on the real axis as the local signal-to-noise variance ratio decreases, thereby achieving maximum smoothing furthest from strong edges. A simple first-order, causal recursive adaptive filter would have

$$a_1 = \mathrm{c}_1 \exp^{-c_2\frac{\sigma_s^2}{\sigma_x^2}} \tag{27}$$

where c_1 controls the extreme position of the pole and thus the maximum smoothing ($c_1 = 0.9$ corresponds roughly to a ten-point average) and c_2 may be chosen to determine the rate at which the pole moves toward the origin as the local signal-to-noise variance ratio increases. Figures 6 through 9 show the effect of both nonrecursive and recursive adaptive filters on an image corrupted with known noise.

To achieve edge enhancement, the weights must effect lateral inhibition, either nonrecursively as with unsharp masking and subtracting the second difference, or recursively for which the coefficients $\{a_k\}$ determine poles which tend toward -1 in the complex z plane as the degree of enhancement (overshoot as well as secondary ringing) increases.

FIGURE 6. Original image of San Francisco.

The relative success of recent adaptive image-processing strategies derives from their explicit estimation of significant local attributes and the consequent appropriate adjustment of local parameters. However, it appears that the input dependent behavior of certain mechanisms in biological visual systems is achieved differently, by virtue of their nonlinear nature. The luminance-dependent behavior of the nonlinear lateral inhibition model is achieved implicitly by the nonlinearity rather than with any explicitly adaptive mechanism. Both homomorphic filters and multiplicative lateral inhibition achieve a relative edge enhancement which is largely independent of local contrast. The introduction of the nonlinearity whether in the form of a logarithmic point operation or as a multiplicative lateral interaction allows the separation of the illumination and reflectance components and, hence, the simultaneous

FIGURE 7. Noisy version of image. Additive Gaussian distributed noise is used, with $\sigma_n^2 = 100$.

reduction of dynamic range and enhancement of local contrast. To achieve more versatile, spatially adaptive, edge-dependent smoothing and sharpening typically desired in digital image processing, interaction among neighboring inputs and outputs may be made to depend explicitly on input characteristics including local mean, variance, and gradient measures.

ACKNOWLEDGMENT

We gratefully acknowledge Mr. Kevin Erler's contributions to the recursive adaptive filter discussion, including sample images; Mr. Richard Belshaw's work with the multiplicative lateral inhibition model; and the support of the Natural Science and Engineering Research Council of Canada (Grant A4427).

FIGURE 8. Image enhanced using the Lee filter.

FIGURE 9. Image enhanced using the recursive adaptive filter.

REFERENCES

Anderson, G. L. and Netravali, A. N. (1976). Image restoration based on a subjective criterion, *IEEE Trans. Syst. Man Cybern.*, **6(12)**, 845–853.

Barlow, H. B. (1953). Summation and inhibition in the frog's retina, *J. Physiol. (London),* **119**, 69–88.

DeVries, F. P. Ph. (1990). Automatic, adaptive, brightness independent contrast enhancement, *Signal Process.*, **21**, 169–182.

Hartline, H. K. and Ratliff, F. (1957). Inhibitory interaction of receptor units in the eye of *Limulus, J. Gen. Physiol.*, 40.

Jernigan, M. E., Belshaw, R. J., and McLean, G. F. (1991). Nonlinear lateral inhibition and image processing, in *Sensory Neural Networks: Lateral Inhibition,* Pinter, R. B. and Nabet, B., Eds., CRC Press, Boca Raton, FL.

Jernigan, M. E. and Erler, K. (1991). Adaptive processing of images, in Pacific Rim Conf. on Communications, Computers and Signal Processing, Victoria, Canada, May 1991. IEEE.

Jernigan, M. E. and Wardell, R. W. (1981). Does the eye contain optimal edge detection mechanisms? *IEEE Trans. Syst. Man Cybern.*, **11**, 441–444.

Lee, J.- S. (1980). Digital image enhancement and noise filtering by use of local statistics, *IEEE Trans. Pattern Anal. Machine Intell.*, **2(2)**, 165–168.

Mahesh, B., Song, W. J., and Pearlman, W. A. (1990). Adaptive estimators for filtering noisy images, *Opt. Eng.*, **29**, 488–494.

Marr, D. (1982). *Vision,* W. H. Freeman, San Francisco.

Moloney, C. R. and Jernigan, M. E. (1990). Adaptive image estimation based on multiplicative superposition, *Opt. Eng.*, **29**, 478–487.

Nagao, M. and Matsuyama, T. (1979). Edge preserving smoothing, *Comput. Vision Graphics Image Process.*, **9**, 394–407.

Nabet, B. and Pinter, R. B. (1991). *Sensory Neural Networks,* CRC Press, Boca Raton, FL.

Oppenheim, A. V. and Schafer, R. W. (1975). *Digital Signal Processing,* Prentice-Hall, Engelwood Cliffs, NJ, chap. 10.

Peli, T. and Lim, J. S. (1982). Adaptive filtering for image enhancement, *Opt. Eng.*, **21**, 108–112.

Pinter, R. B. (1984). Adaptation of receptive field spatial organization via multiplicative lateral inhibition, *J. Theor. Biol.*, **110**, 435–444.

Stockham, T. G. (1972). Image processing in the context of a visual model, *Proc. IEEE,* **60**, 828–842.

Wilson, H. R. and Bergen, J. R. (1979). A four mechanism model for threshold spatial vision, *Vision Res.* **19**, 19–32.

Chapter 18

ELECTRONIC HARDWARE FOR VISION MODELING

Bahram Nabet

TABLE OF CONTENTS

I. Introduction 464

II. Feedback Neural Networks 464

III. Feedforward Networks 468

IV. Alternative Methods 470

V. Conclusions 473

Acknowledgments 473

References 474

I. INTRODUCTION

Many computational models of neuronal interactions can be traced to the seminal work of Hodgkin and Huxley (1952), who described the variation of cell membrane potential as a function of ionic currents in conductive channels

$$C \frac{dV}{dt} = (V^{+} - V)\, g^{+} + (V^{-} - V) g^{-} + (V^{p} - V) g^{p} \tag{1}$$

where C is the membrane capacitance; V^{+},V^{-}, and VP are excitatory, inhibitory, and passive saturation points, respectively; and g^{+},g^{-}, and g^{p} are excitatory, inhibitory, and passive membrane conductances, respectively. For example, simple assignment of variables by choosing the state variable $x_i = V$, setting $g^{p} = A$, $V^{+} = B$, $V^{-} = V^{p} = 0$, and considering the case where the excitatory conductance is proportional to the input current to the cell ($g^{+} = I_i$) and inhibitory conductance is controlled by input from other cells of the network $g^{-} = \Sigma_k I_K$, transforms Equation 1 into

$$\frac{dx_i}{dt} = -A\, x_i + (B - x_i)\, I_i - x_i \sum_{k \neq i} I_k \tag{2}$$

which is a one-layer feedforward, or nonrecurrent, neural network model with self-excitation and multiplicative, or shunting, inhibition, the computational properties of which have been extensively studied by Grossberg (for a review see Grossberg, 1989). It is a straightforward matter to show that feedback, or recurrent, versions of neuronal interactions are obtained when the conductances are modulated by the state variables (voltages, in this case) of other cells, rather than inputs as is the case in Equation 2.

The Hodgkin-Huxley equation, thus, provides the crucial insight for the electronic circuit designers that models of neural networks can be constructed from electronic devices or circuits whose conductance may be varied as a function of voltage or current values. The simplest such device is a single field effect transistor (FET) which, when operated below pinchoff, acts as a voltage-controlled conductance. Sections II and III of this chapter show feedback and feedforward networks which can be built from these elements. Section IV describes alternative methods for hardware implementation of neural network models.

II. FEEDBACK NEURAL NETWORKS

Consider the equivalent circuit of an isopotential patch of membrane shown in Figure 1a, where g^{+}, g^{-}, and g^{p} are as defined above, which is used for modeling synaptic-mediated excitation and inhibition (Hodgkin and Huxley, 1952; Rall, 1962). Assuming, without loss of generality, Nernst potential (modeled by the batteries) of zero, ignoring the excitatory ionic channel g^{+}, and considering the case where the inhibitory conductance is controlled by the voltage of other cells, that is, $g^{-} = \Sigma_j f_j(x_j)$, we obtain

$$\frac{dx_i}{dt} = I_i - Ax_i - x_i \sum_{j \neq i} f_j(x_j) \tag{3}$$

where the capacitance value is set to 1, I_i is (excitatory) current inputs to the cell, and A is the passive membrane conductance. This is a multiplicative lateral inhibitory feedback neural network.

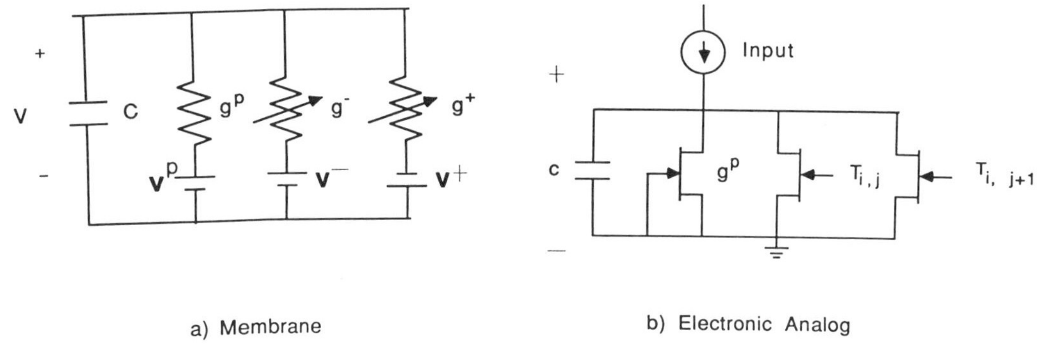

FIGURE 1. (a) An isopotential patch of membrane. Batteries show Nernst potentials. g^+, g^-, and g^p are, respectively, passive, excitatory, and inhibitory conductances. (b) Electronic analog of a. The voltage at the gate of transistor T_{ij} controls its conductance. This circuit is approximately described by Equation 3 of the text.

Figure 1b shows a possible electronic analog of the same biological model. The inhibitory conductance is an FET whose conductance is modulated by the voltage present at its gate. Each interconnection is implemented by one transistor. Such implementations are truly analog in that the physics of the device performs the "computation", and no digital-to-analog conversion, or vice versa, is required.

For small values of node voltage x_i, the simple circuit of Figure 1b is described by Nabet et al. (1989) as:

$$\frac{dx_i}{dt} = I_i - Ax_i - x_i \sum_{j \neq i} f_j(x_j) + K_i x_i^2 \tag{4}$$

which is the desired neural network Equation (2) with the addition of a quadratic self-excitatory nonlinearity $K_i x_i^2$. This term is due to the nonlinearity of the transconductance of the transistor and can be eliminated by proper designs, but it is fortuitous in this case since it provides self-excitatory feedback.

Figure 2 shows the response of a network comprising 32 cells, connected to their nearest neighbors, to a point source stimulus located on cell number 17. In this figure the ratio of the stimulus to its surround has been kept at 2:1, but the mean value has been changed to study the effect of varying mean light intensity on reflectance processing. The characteristic Mexican hat response is clearly observed at high mean intensities, but the inhibitory influence decreases with decreasing available light. This can be a technologically desired effect in the form of a sensor for an imaging device which uses the available ambient light to better process the input image. Inset of this figure shows similar variation of the receptive field in the LMC cells of the fly *Lucilia sericata*. The upper curve of the inset shows the receptive field variation under high light conditions, while the lower is taken under reduced mean luminance levels. These results are in agreement with theoretical analysis based on Equation 3 (Pinter, 1984). A detailed description of field adaptation as a function of mean luminance variation can be found in Pinter and Nabet (1991).

On the other hand, variation of contrast with mean intensity may be undesired since it opposes the tendency to see constant contrast under varying illumination conditions. The circuit of Figure 1 can be altered to allow for an almost constant lightness profile, that is, discounting the illuminant, with changing lighting conditions. This circuit design can be motivated by examining the adaptation level, or steady state form, of Equation 3

$$x_i = \frac{I_i}{A + \Sigma f_j(x_j)} \tag{5}$$

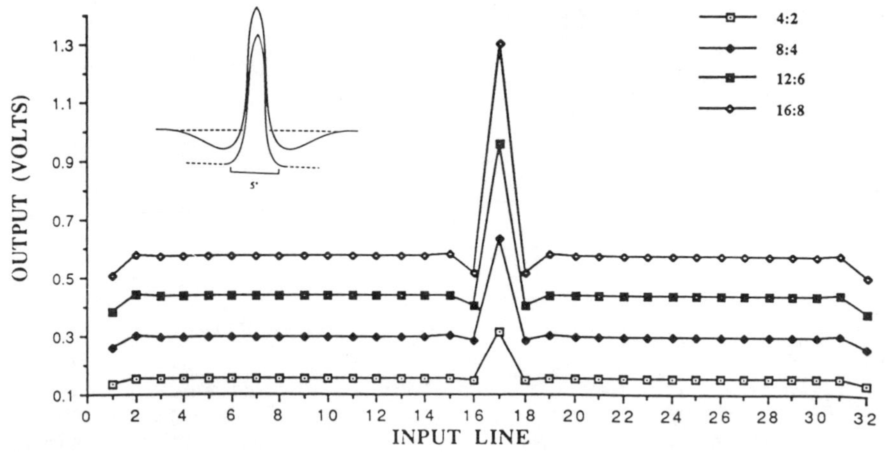

FIGURE 2. Response of the network to a point source input. The ratio of the contrast of the point source with respect to its surround was kept constant while mean intensity was increased from top to bottom. This figure, thus, shows rearrangement of the receptive field due to change of intensity. Inset of the figure shows similar behavior in the receptive field of the LMC cells of the fly *Lucilia sericata*. Horizontal axis of the inset, 5 degrees visual angle; vertical axis, relative voltage units of hyperpolarization.

which shows the automatic gain control property, an important property afforded the network by the presence of the multiplicative term. The second term of the denominator is a monotonic nondecreasing function; hence, as input increases this term also increases, thus reducing x_i and allowing a large dynamic range of input to be processed by the cell. The first term of the denominator can also contribute to this gain control by varying with mean intensity. In other words, if $A = g(I_i)$ or $A = g(I_{total})$, where I_{total} is total input excitation, then the passive membrane conductance contributes to light adaptation. The first case is a local dependency, while the second case is a universal dependency. Both cases are biologically plausible.

Figure 3 shows this effect implemented in hardware. The curve marked as output 1 shows the response to a point source input similar to the response of the bottom curves of Figure 2. The lower curve, marked as output 2, is the response when the passive membrane resistance is changed to a higher value. This lower curve is similar to the top curve of Figure 2 which was obtained for high input intensities; hence, this change of membrane resistance mimics operation under high mean intensity while the actual intensity is low. Note that both the gain, noted on the graph by the normalization value, and the receptive field profile change in this figure in accordance with Equation 5. This result is consistent with the experiments of Barlow et al. (1957) in cat retinal ganglion cells and Ramoa et al. (1985) in cat cortical cells, which showed that light adaptation increases the prominence of the antagonistic surround relative to the center.

Response of the feedback network to a bar pattern is shown in Figure 4. Each cell in the network of this figure is connected to its nearest and next-to-nearest cells with connection strength reduced by one half as the distance between cells increases; edges are enhanced and uniform areas are suppressed. The amount of enhancement is proportional to the available ambient (light) intensity, similar to the response to a point source shown in Figure 2. As mentioned above, change of passive membrane resistance can counter the effect of varying illumination on processing of contrast.

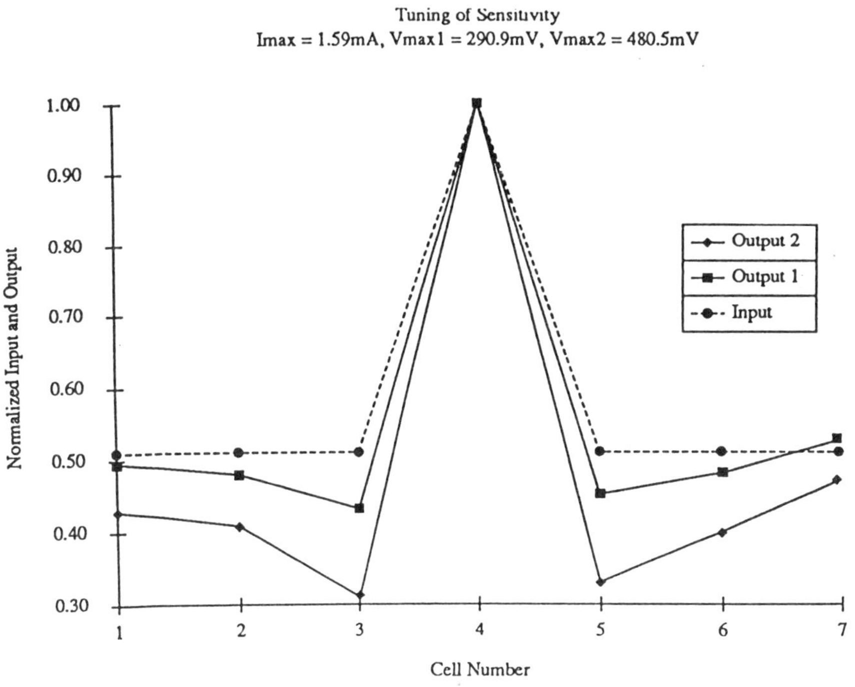

FIGURE 3. Adaptation by change of membrane resistance. The same input pattern can elicit different outputs if the passive membrane resistance is changed. Upper output curve is for lower resistance.

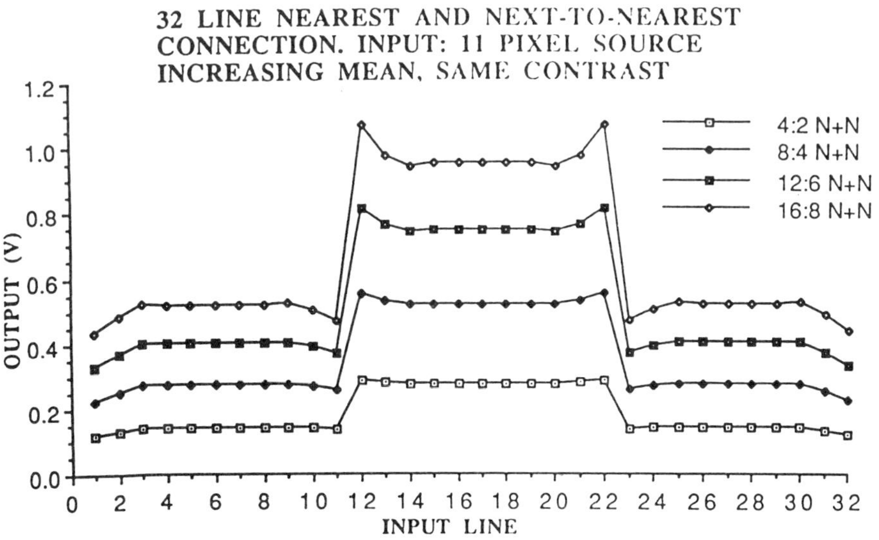

FIGURE 4. Response of the feedback network to a bar pattern of input. Enhancement at the edges and attenuation of the uniform areas is notable.

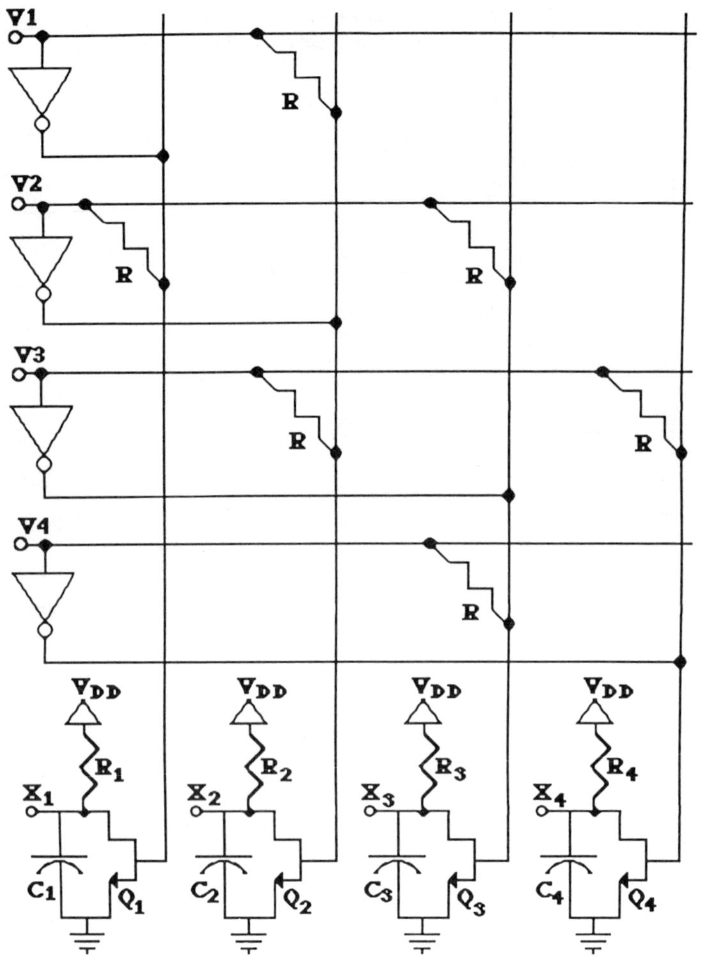

FIGURE 5. Circuit implementation of a shunting feedforward network. Nearest-neighbor connections and self-excitation is shown for four cells.

III. FEEDFORWARD NETWORKS

As mentioned in Section I, feedforward networks of type similar to Equation 2 can be derived from the Hodgkin-Huxley equations when membrane channel conductances become dependent on input values rather than activity of other cells in the network. A circuit implementing this relationship is shown in Figure 4. The top portion of this circuit is the familiar crossbar architecture which defines the connectivity pattern of the network. The input lines are shown as voltages and are also available in inverted form to allow both excitation and inhibition. The lower array of transistors performs multiplication and nonlinear transformation of the input.

The circuit of Figure 5 can be approximately described by

$$\frac{dx_i}{dt} = a_i - b_i x_i - x_i \left(\sum_{j \neq i} V_j / R_j\right) + c_i x_i V_i^2 \qquad (6)$$

where the capacitance is taken as 1, V_j is input, X_i is output of cell i, and a_i, b_i, and c_i are

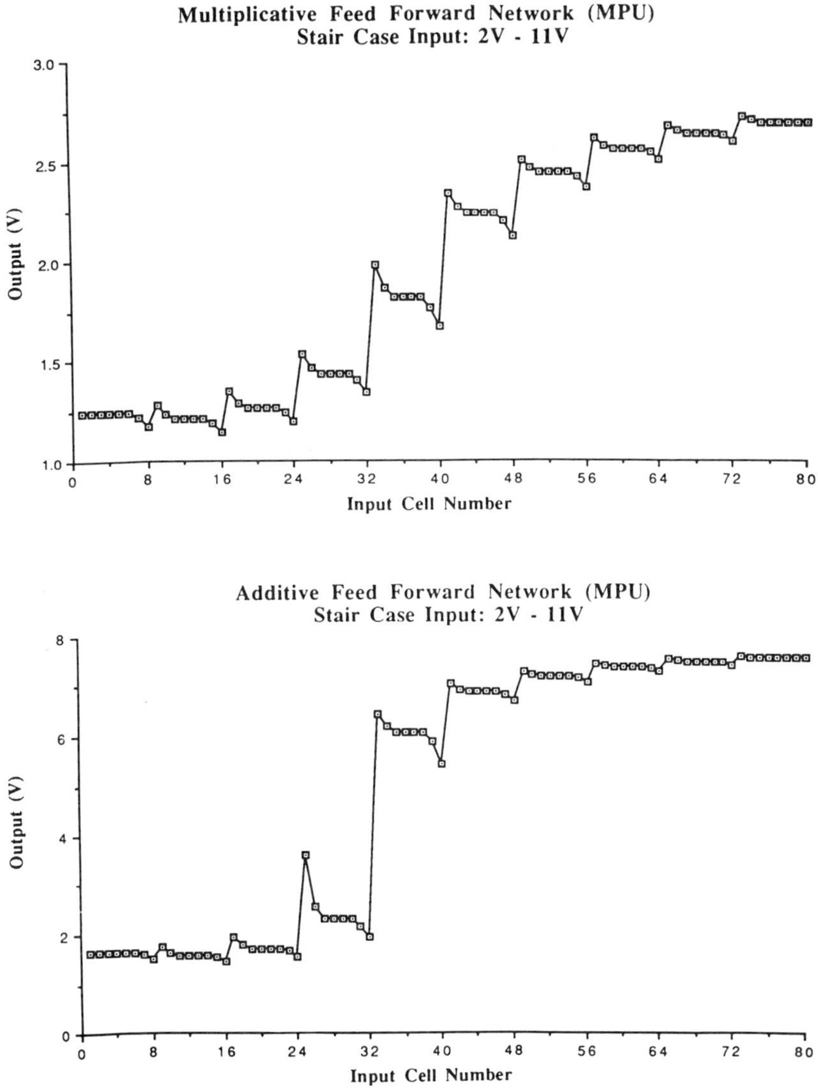

FIGURE 6. (Top) Response of a feedforward shunting network to a staircase input pattern with steps of 1 V. (Bottom) Response of a feedforward network without the multiplicative effect to the same input pattern.

constants dependent on device characteristics and circuit parameters. The term in parentheses in the right-hand side of Equation 6 shows the desired multiplicative effect which causes an automatic gain control that allows the large input dynamic range of input to be processed by the network.

A network of 80 cells with self-excitatory and two nearest-neighbor inhibitory connections was simulated in PSPICE. A staircase input pattern ranging from 2 to 12 V with 1-V steps was presented to the network. The output of each neuron was limited to almost one tenth of the input range. The results of the implementation are shown in Figure 6 (top). The output has the distinctive pattern of lightness perception in psychophysical experiments (see, for example, Levine, 1985) wherein edges are enhanced while uniform areas are attenuated. The network has also represented an input range which is an order of magnitude larger than the dynamic range of each neuron and at the same time has preserved and enhanced the edges which contain the most information in the image. It is noteworthy that this is a postsensory but preattentive effect, that is, the input to the network can be construed as the output of the photoreceptors which is proportional to the logarithm of the actual light intensity.

Figure 6 (bottom) shows an identical network which is operated in a range that does not have the multiplicative effect. Although both networks are allowed similar dynamic range of activity, at lower and higher ranges of input assigned output levels are barely distinguishable in the network without gain control; the multiplicative network has allocated its dynamic range more suitably.

Response of the network to a bar pattern is shown in Figure 7. The top figure shows a response similar to the feedback case. In the lower figures the relative strengths of inhibition to excitation are changed by changing the resistance values while the excitatory or inhibitory nature of connections is unaltered. It is seen that this change can cause the network to detect only the edges, that is, find the zero crossings, as the middle plot suggests, or even exhibit an off-center on-surround, or off-unit, response as in the lower figure. These types of on- and off-units are found at early levels of a visual system and are used in an architecture developed by Grossberg and Todorovic (1988) which describes many properties of preattentive vision. Specifically, level 2 of this architecture (see Grossberg and Wyse, Chapter 21 of this volume, Figure 5) is comprised of cells with properties similar to retinal ganglion cells or lateral geniculate cells with receptive fields and shunting interactions similar to those implemented above.

The receptive field defined by the connection strengths of the feedforward network is symmetric in the above cases, but it can be designed to be antisymmetric. Figure 8 shows a bar pattern processed by a network where each cell is excited by itself and the cell to its right, but is inhibited by the two cells at its left. The response, as expected, is asymmetric and shows preference of one orientation of contrast compared with the opposite one. This type of receptive field is often used for determination of the direction of motion (Bouzerdoum et al., 1991). Combinations of different contrast polarity-sensitive cells have been used (Grossberg et al., 1989) to construct network architectures which are sensitive to direction of motion but insensitive to polarity of contrast. In that architecture direction-of-contrast-sensitive units appear at level 3 (see Grossberg and Wyse, Chapter 21 of this volume, Figure 5) and receive input from on- and off-units similar to those shown in Figure 7. Figure 8 (bottom) shows the response of the network to the same pattern as in Figure 8 (top) but with a connection strength profile where the ratio of inhibition to excitation is opposite to the previous case. It is seen that the network is now an off-center on-surround as opposed to the previous case.

IV. ALTERNATIVE METHODS

The circuits described above are economical in terms of device count since one transistor performs a multiplication operation, and addition is done by current summing. However, a shortcoming is the limited dynamic range of resistance (typically 200 to 3000 Ω) that can be modulated in single transistors. This problem can be alleviated by using current mirror circuits with a linear voltage input/current output relationship. Scaling of the transistors will provide different "resistance" values. Combination of several resistors and their control by digital circuitry can provide programmable connection weights. Such control also allows for possible "learning" of the connection strengths (Mueller et al., 1990).

Power consumption can also become prohibitive in chips where transistors operate in either linear or saturation modes. A highly successful alternative is presented in the pioneering work of Mead (1989) where transistors are operated in the subthreshold, or weak inversion, region of operation. This mode of operation is usually discounted in digital design as a

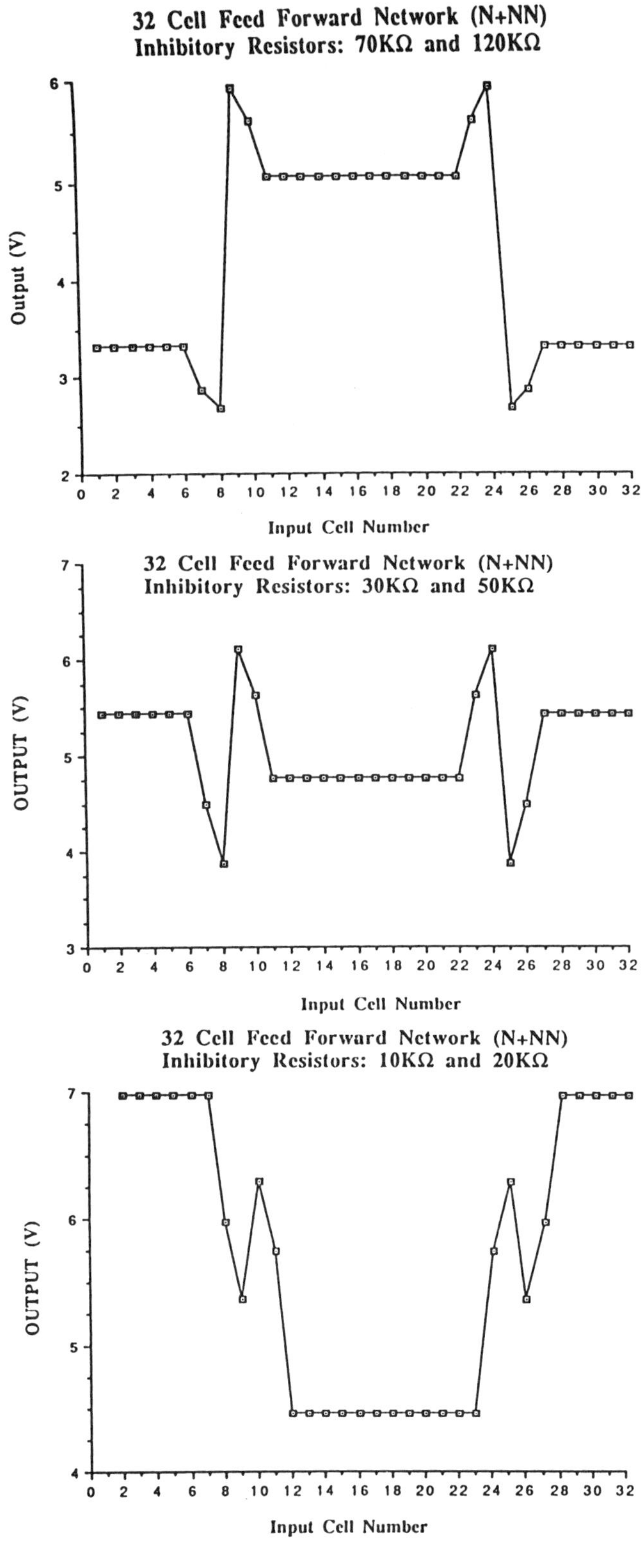

FIGURE 7. Response of a feedforward network to a bar pattern of input. The relative weight of inhibition to excitation is changed from top to bottom. The middle figure computes the zero crossings of the intensity pattern while the bottom figure shows a response (off-center on-surround) opposite that of the top figure.

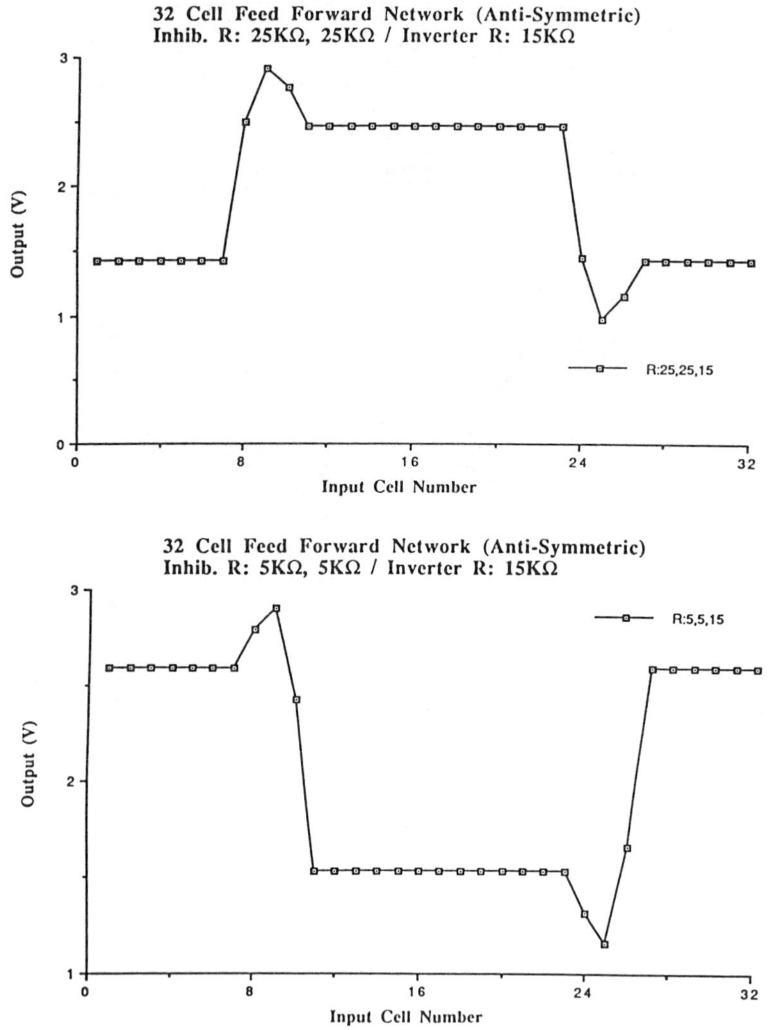

FIGURE 8. (Top) Response of an asymmetrically connected network to a bar pattern. Preference for polarity of contrast is noticeable. (Bottom) Response to the same input when the relative weight of inhibition to excitation is changed.

parasitic effect, but Mead has employed the well-defined, but hard to control, subthreshold current conduction in a series of ''silicon retinas'' (Mead and Mahowald, 1988). These models utilize a photodetector which is a ''parasitic'' by-product of CMOS processing as the photosensor (Mead, 1985). In addition to low power consumption, these circuits have a very large dynamic range, well over six orders of magnitude. As silicon processing advances, more uniformity will be achieved, and subthreshold design can become standard for neural network implementations which do not require strict control of variables. Examples of the cases requiring stricter precision than presently possible with subthreshold circuits include most learning algorithms such as the error back-propagation.

Another notable implementation of circuits for peripheral processing of visual information is the use of resistive arrays for, for example, estimation of velocity of moving images (Hutchinson et al., 1988). In this approach a visual task is formulated in terms of energy functions that need to be minimized. Rather than building biological analogs, a resistive network is constructed whose energy maps this energy function. The network then naturally settles into the state of least power dissipation; the associated voltages are the solutions sought.

V. CONCLUSIONS

Hardware circuits described here are inspired by the basic membrane equation which suggests synaptic conductance modulation as a mechanism for achieving shunting feedback and feedforward network interaction. The circuits use the basic property of single field-effect transistors which, in below pinch-off operation, are voltage-controlled conductances.

The implementations serve the obvious purpose of capturing technologically desirable features such as edge and contrast enhancement and dynamic range compression. The results reported here showed, as expected, similarity to biological and psychophysical experiments. The inherent fault tolerance due to parallelism and the analog, rather than digital, nature of operations allow these tasks to be performed with few devices and at high speeds. In addition, the hardware models are abstraction tools which help study the behavior of the networks; mean intensity dependence and the effect of variation of inhibition to excitation strengths which were shown in this work are much more easily performed on electronic hardware than physiological wetware. Since the networks are mathematically described, an important application will be the study of dynamic behavior of the networks and its comparison with mathematical models.

Shortcomings of designs suggested above were also mentioned to be the limited dynamic range of resistance that can be modulated in single transistors operating in linear region, and power consumption. Circuits and technologies to overcome these deficiencies were mentioned.

Early stages of hardware circuits of neural networks for vision have produced some of the most tangible products of neural network technology, and new complicated systems based on simple components are the logical next step of development. One such system is being developed at MIT Lincoln Laboratory (Waxman, 1991) which, among other tasks, determines the velocity of moving images. This chip integrates photodetectors, photoemitter, processing circuitry, and lenslets on the same substrate.

ACKNOWLEDGMENTS

This work was supported by Drexel University Research Scholar Award. PSPICE simulation results are due to L. Hojda.

REFERENCES

Barlow, H. B., Fitzhugh, R., and Kuffler, S. W. (1957). Change in organization of the receptive fields of the cat's retina during dark adaptation, *J. Physiol. (London)*, **137**, 338–354.

Bouzerdoum, A., Nabet, B., and Pinter, R. B. (1991). Analysis and analog implementation of directionally sensitive shunting inhibitory neural networks, in Proc. SPIE Int. Symp. on Optical Eng. and Photonics in Aerospace Sensing, April 1–5, Orlando, FL, 29–37.

Grossberg, S. (1988). Nonlinear neural networks: principles, mechanisms, and architectures, *Neural Networks*, **1**, 17–61.

Grossberg, S. and Todorovic, D. (1988). Neural dynamics of 1-D and 2-D brightness perception: a unified model of classical and recent phenomena, *Percept. and Psychophys.*, **43**, 241–277.

Grossberg, S., Mingolla, E., and Todorovic, D. (1989). A neural network architecture for preattentive vision, *IEEE Trans. Biomed. Eng.*, **36(1)**, 65–84.

Hodgkin, A. L. and Huxley, A. F. (1952). A quantitative description of membrane current and its application to conduction and excitation in nerve, *J. Physiol. (London)*, **117**, 500–544.

Hutchinson, J., Koch, C., Luo, J., and Mead, C. (1988). Computing motion using analog and binary resistive networks, *IEEE Comput. Mag.*, **21**, 52–64.

Levine, M. D. (1985). *Vision in Man and Machine*, McGraw-Hill, New York.

Mead, C. A. (1989). *Analog VLSI and Neural Systems*, Addison-Wesley, Reading, MA.

Mead, C. A. (1985). A Sensitive Electronic Photoreceptor, Chapel Hill Conf. on VLSI, 463–471.

Mead, C. A. and Mahowald, M. (1988). A silicon model of early visual processing, *Neural Networks*, **1**, 91–97.

Mueller, P., Blackman, D., and Furman, R. (1990). Neural computation of visual images, in *An Introduction to Neural and Electronic Networks*, Zorentzer, S. F., Davis, J. L., and Lau, C., Eds., Academic Press, San Diego.

Nabet, B. (1991). Adaptation by change of passive membrane resistance, in *Proc. Int. Joint Conf. Neural Networks*, Seattle, July 8–12, in press.

Nabet, B. and Pinter, R. B. (1991). *Sensory Neural Networks: Lateral Inhibition*, CRC Press, Boca Raton, FL.

Nabet, B., Darling, R. B., and Pinter, R. B. (1989). Analog implementation of shunting neural networks, in *Advances in Neural Information Processing*, Vol. 1, Touretzky, D., Ed., 695–702.

Pinter, R. B. (1984). Adaptation of receptive field spatial organization by multiplicative lateral inhibition, *J. Theor. Biol.*, **110**, 435–444.

Pinter, R. B. and Nabet, B. (1991). Field adaptation in visual systems is a function of nonlinear lateral inhibition, in *Neural Networks: Concepts, Applications and Implementations*, Vol. 3, Antognetti, P. and Milutinovic, V., Eds., 21–48.

Rall, (1962). Theory of physiological properties of dendrites, *Ann. N.Y. Acad. Sci.*, **96**, 1071–1092.

Ramoa, A. S., Freeman, R. D., and Macy, A. (1985). Comparison of response properties of cells in cat visual cortex at high and low luminance levels, *J. Neurophysiol.*, **54**, 61–72.

Waxman, A. M. (1991). Neurodynamics of Real Time Image Velocity Extraction, presentation given at Neural Networks for Vision and Image Processing, Wang Institute, Boston University, May 10–12, 1991.

Chapter 19

MULTIPLICATIVE INHIBITION AND VOLTERRA SERIES EXPANSION

Bahram Nabet and Robert B. Pinter

TABLE OF CONTENTS

I. Introduction 476

II. Temporal Volterra-Wiener Series Expansion 476
  A. First-Order Temporal Kernel 479
  B. Second-Order Temporal Kernel 481

III. Spatial Kernels 482
  A. Zero-Order Kernel 483
  B. First-Order Spatial Kernel 483
  C. Second-Order Spatial Kernel 484
  D. Higher-Order Spatial Kernels 485
  E. Classification Properties 486

IV. Conclusions 488

Acknowledgment 488

Appendix 488

References 491

I. INTRODUCTION

Some of the models for description of aspects of visual systems discussed in the various chapters of this volume are based on nonlinear lateral inhibition. Bouzerdoum and Pinter (Chapter 16) have developed motion detection techniques based on asymmetrically connected multiplicative lateral inhibitory interactions. Grossberg and Wyse (Chapter 21) have used shunting inhibition to discount the illuminant in a layer of a boundary contour system which is then embedded in an architecture that is capable of figure-ground separation. Nabet (Chapter 18) has discussed electronic circuits which can implement multiplicative inhibitory interaction as variations of the Hodgkin-Huxley equation and demonstrate interesting image-processing capabilities. In this chapter nonlinear system synthesis and identification methods of Volterra-Wiener series expansion are used to analyze the behavior of these systems. A variational analysis technique is shown to establish the connection between the differential equations describing the network dynamics and the Volterra series directly. Stability of the networks under study is shown in the appendix.

II. TEMPORAL VOLTERRA-WIENER SERIES EXPANSION

Linear mathematical methods of system identification are incapable of satisfactorily analyzing nonlinear systems.

The nonlinear analysis technique of Volterra-Wiener expansions has been successfully applied to, among others, the analysis of wave propagation in random media, scattering by random surfaces, Brownian motion, theory of turbulence, and biological systems (for references see Ogura, 1986). In this section this method is applied to the analysis of temporal and spatial characteristics of the class of neural networks under study.

Consider the simple neural network system defined by

$$\frac{dx_i}{dt} = I_i - a_i x_i + K_i x_i^2 - x_i\left(\sum_j K_{ij} x_j\right), \tag{1}$$

where x_i is the (voltage) state variable, I_i is the (current) input, a_i is a passive decay term corresponding to the passive membrane conductance in a biological network, and K_i and K_{ij} are connection strength. This equation then describes a recurrent neural network with a self-excitatory on-center and multiplicative inhibitory off-surround. Despite its simplicity the system described by Equation 1 exhibits the most important properties of multiplicative inhibition (see Nabet and Pinter, 1991 for details).

The objective is to write the solution to Equation 1 in the Wiener-Volterra expansion form

$$\begin{aligned} x_i = H_{i0} &+ \int_{-\infty}^{\infty} \mathbf{H}_{i1}(\tau)\, l(t-\tau)\, d\tau + \\ &\int_{-\infty}^{\infty}\int_{-\infty}^{\infty} \mathbf{H}_{i2}(\tau_1,\tau_2)\, l(t-\tau_1)\otimes l(t-\tau_2)\, d\tau_1\, d\tau_2 + \cdots \end{aligned} \tag{2}$$

In Equation 2, $\mathbf{H}_{ij}(\tau_1,\tau_2, \ldots, \tau_j)$ is the ith row of the jth Volterra kernel and $\otimes$ is the Kronecker Product used to denote multidimensional dot products, and is defined by

$$\begin{bmatrix} a_{11}B & \cdots & a_{1m_a}\mathrm{B} \\ \vdots & & \vdots \\ a_{n_a1}\mathrm{B} & & a_{n_am_a}\mathrm{B} \end{bmatrix} \tag{3}$$

for matrices $A = [a_{ij}]$ and $B = [b_{ij}]$, of dimension $n_a \times m_a$ and $n_b \times m_b$, respectively.

The first integral of Equation 2 is the familiar convolution operator, and the higher-order terms show the inclusion of higher-order nonlinearities in analysis and identification of nonlinear systems.

The existence of the convergent Volterra series is shown for a large class of systems in Boyd and Chua (1985). Study of Wiener-Volterra systems with application to identification and synthesis of physiological systems is done in a text by Marmarelis and Marmarelis (1978). Schetzen (1980) and Rugh (1981) have provided classic treatments of the Volterra-Wiener approach from a system theoretical point of view. The Wiener kernels can be found by the familiar Gram-Schmidt orthogonalization procedure, the formulas for which are discussed in all the above three texts.

A variational analysis technique (Rugh, 1981; Pinter, 1987) can be applied to Equation 1 to produce a solution in the desired form. The kernels can then be extracted by observation, and the input-output relationship is modeled more accurately compared with a linearized model. This technique, which is also extensively used in quantum mechanical analysis to find, for instance, higher-order Hamiltonians of Schrödinger's equation, is outlined below.

1. Let the input $I_i = L_0 + \lambda l_i$ where λ is the perturbation parameter.
2. Assume a solution of the form $x = \bar{x} + \lambda x_1 + \lambda^2 x_2 + \ldots$, and differentiate it.
3. Insert this solution into the original differential equation.
4. Equate like powers of λ in equations which result from steps 2 and 3 above.

The following derivation proceeds according to the above outline.

Let the input

$$I_i(t) = L_0 + \lambda l_i(t) \tag{4}$$

where L_0 is the input background, uniform across the array and independent of time, and λ is the perturbation parameter, be applied to the system of Equation 1. The response of the network is

$$\frac{dx_i}{dt} = L_0 + \lambda l_i(t) - a_i x_i + K_i x_i^2 - x_i \left(\sum_j K_{ij} x_j\right). \tag{5}$$

The total input is then the vector $\mathbf{I} = [L_0 + l_1, L_0 + l_2, \ldots, L_0 + L_n]'$, and the output vector is $x = [x_1, x_2, \ldots, x_n]'$.

Next, the standard assumption is made that the response to the perturbed input can be written in the form

$$x_i = \lambda^0 \bar{x}_i + \lambda^1 x_{i1} + \lambda^2 x_{i2} + \lambda^3 x_{i3} + \cdots \tag{6}$$

where x_i is the ith term of the output vector, and the second index of the subscript notates the order of the term which is the order of approximation.

Differentiating this equation leads to

$$\dot{x}_i = \lambda^0 \times 0 + \lambda^1 \dot{x}_{i1} + \lambda^2 \dot{x}_{i2} + \lambda^3 \dot{x}_{i3} + \cdots \tag{7}$$

Substituting the assumed form of solution to Equation 6 in the perturbed differential Equation 5 results in

$$\dot{x}_i = L_0 + \lambda l_i - a_i[\lambda^0 \bar{x}_i + \lambda^1 x_{i1} + \lambda^2 x_{i2} + \lambda^3 x_{i3} + \cdots] +$$
$$K_i[\lambda^0 \bar{x}_i + \lambda^1 x_{i1} + \lambda^2 x_{i2} + \lambda^3 x_{i3} + \cdots]^2 -$$
$$[\lambda^0 \bar{x}_i + \lambda^1 x_{i1} + \lambda^2 x_{i2} + \lambda^3 x_{i3} + \cdots] \sum_j K_{ij} [\lambda^0 \bar{x}_j + \lambda^1 x_{j1} + \lambda^2 x_{j2} + \cdots] \tag{8}$$

Equations 7 and 8 should hold for all values of λ, which means that like powers of λ should have the same coefficients. Equating the terms that are multiplied by λ^0 in Equations 7 and 8 results in

$$L_0 - a_i\bar{x}_i + K_i\bar{x}_i^2 - \bar{x}_i \sum_j K_{ij}\bar{x}_j = 0 \tag{9}$$

which is the steady-state response of the network to the uniform input, L_0.

One solution to Equation 9 is $\bar{x}_i = \bar{x}_j$ for all j, that is, the response of the network to a uniform input is uniform across the array. This transforms Equation 9 into

$$\bar{x}_i^2[K_i - \sum_{j \neq i} K_{ij}] - a_i\bar{x}_i + L_0 = 0 \text{ for all } i \tag{10}$$

which has the solution

$$\bar{x}_i = \frac{-a_i + \sqrt{a_i^2 + 4L_0 \sum_{j=1}^{n} K_{ij}}}{2 \sum_{j=1}^{n} K_{ij}} \tag{11}$$

where $K_{ii} = -K_i$ and only the solution which satisfies the positivity constraint $\bar{x}_i \geq 0$ has been chosen. This response to background intensity is the zero-order kernel, namely

$$H_{0i} = \bar{x}_i \tag{12}$$

Equation 11 clearly demonstrates that the output is proportional to the square root of the uniform input. This is the dynamic range compression property of multiplicative inhibition and has been extensively studied by Pinter (1987) and Pinter and Nabet (1991).

From this treatment it is seen that, in general, if the order of $x_i f_i(x_i)$ is n, i.e., if

$$x_i f_i(x_i) \doteq \alpha x_i^n + \beta x_i^{n-1} + \cdots + \gamma \tag{13}$$

then the input is compressed by an order of $1/n$, or

$$\bar{x}_i \propto [L_0]^{\frac{1}{n}} \tag{14}$$

Specifically, if the feedback nonlinearity is exponential within a range, the output will be approximately proportional to the logarithm of the input within that range.

A. FIRST-ORDER TEMPORAL KERNEL

The first-order Volterra kernel which performs the familiar (linear) convolution operation is found by equating the coefficients of λ^1 to obtain

$$\dot{x}_{i1} = l_i(t) + x_{i1}(-a_i + \bar{x}_i 2K_i - \sum_j K_{ij}\bar{x}_j) - \bar{x}_i \sum_j K_{ij}x_{j1} \tag{15}$$

For simplicity let

$$b_i = a_i - \bar{x}_i 2K_i + \sum_{j\neq i} K_{ij}\bar{x}_j \tag{16}$$

In matrix notation Equation 15 can be written as

$$\begin{bmatrix} \dot{x}_{11} \\ \dot{x}_{21} \\ \vdots \\ \dot{x}_{n1} \end{bmatrix} = - \begin{bmatrix} b_1 & \bar{x}_1 K_{12} & \bar{x}_1 K_{13} & \cdots & \bar{x}_1 K_{1n} \\ \bar{x}_2 K_{21} & b_2 & \bar{x}_2 K_{23} & \cdots & \bar{x}_2 K_{2n} \\ \vdots & & \ddots & & \vdots \\ \bar{x}_n K_{n1} & \cdots & & & b_n \end{bmatrix} \begin{bmatrix} x_{11} \\ x_{21} \\ \vdots \\ x_{n1} \end{bmatrix} + \begin{bmatrix} l_1 \\ l_2 \\ \vdots \\ l_n \end{bmatrix} \tag{17}$$

and can further be simplified by noting that $\bar{x}_i = \bar{x}_j$, but is left in the more general form to allow further refinement by, for example, using linear programming techniques to analyze the zero-order kernel.

Equation 17 is in the form

$$\dot{\mathrm{x}}_1 = \mathbf{A}\mathrm{x}_1 + 1 \tag{18}$$

and can be solved for x by using techniques of linear algebra. It is important to note that matrix **A** is dependent on the zeroth order kernel, which is, in turn, dependent on the mean intensity of the input. The Volterra expansion technique is, thus, capable of propagating the intensity dependence of the model to higher-order kernels.

The solution to Equation 18 is

$$\mathrm{x}_1(t) = e^{\mathbf{A}(t-t_0)}\mathrm{x}_1(t_0) + \int_{t_0}^{t} e^{\mathbf{A}(\tau)}1(t - \tau)\, d\tau \tag{19}$$

Assuming, without loss of generality, a time origin of zero and zero initial conditions, i.e., $t_0 = 0$ and $\mathrm{x}_1(t_0) = 0$, the generating solution Equation 19 can be written as

$$\mathrm{x}_1(t) = \int_{t_0}^{t} e^{\mathbf{A}(\tau)}1(t - \tau)\, d\tau \tag{20}$$

The state transition matrix $e^{\mathbf{A}(t)}$ can be found by taking the inverse Laplace transform of $(S\mathbf{I} - \mathbf{A})^{-1}$, i.e.,

$$e^{\mathbf{A}(t)} = L^{-1}(S\mathbf{I} - \mathbf{A})^{-1} \tag{21}$$

The roots of the characteristic functions $|S\mathbf{I} - \mathbf{A}| = 0$ are the eigenvalues and can, in general, be repeated or nondistinct, but since the coefficient matrix is symmetric ($K_{ij} = K_{ji}$), from Equation 17 it is observed that **A** is Hermitian and, hence, has real eigenvalues and can be diagonalized even if the eigenvalues are not distinct (Franklin, 1968). Proof of global asymptotic stability in the appendix is also equivalent to proving that the eigenvalues of **A** are negative. From these considerations the state transition matrix can be written as

$$e^{\mathbf{A}(t)} = \begin{bmatrix} \sum_{k=1}^{n} a_{11k} e^{a_k t} & \sum_{k=1}^{n} a_{12k} e^{a_k t} & \cdots & \sum_{k=1}^{n} a_{1nk} e^{a_k t} \\ \vdots & \vdots & & \vdots \\ \vdots & \vdots & & \vdots \\ \sum_{k=1}^{n} a_{n1k} e^{a_k t} & \sum_{k=1}^{n} a_{n2k} e^{a_k t} & \cdots & \sum_{k=1}^{n} a_{nnk} e^{a_k t} \end{bmatrix} \tag{22}$$

where the eigenvalues a_k are negative. Had the **A** matrix not been Hermitian, possible nondistinct eigenvalues would produce terms of the form $\Sigma_k b_k t^k e^{a_k}$ which would considerably complicate the derivation. Furthermore, the proof of stability is necessary in order to argue that **A** is negative definite.

Finding the state transition matrix, in general, is not a trivial problem, and often numerical methods should be employed for its determination. The above derivation shows the existence of a closed form solution, the general shape of that matrix, and, specifically, its relevance to Volterra series expansion.

The first-order generating solution is

$$\begin{bmatrix} x_{11} \\ x_{21} \\ \vdots \\ x_{n1} \end{bmatrix} = \begin{bmatrix} \sum_{j=1}^{n} \int_0^t \sum_{k=1}^{n} a_{1jk} e^{a_k \tau} l_j(t - \tau)\, d\tau \\ \vdots \\ \sum_{j=1}^{n} \int_0^t \sum_{k=1}^{n} a_{njk} e^{a_k \tau} l_j(t - \tau)\, d\tau \end{bmatrix} \tag{23}$$

If Equation 6 is rewritten, by observing that the perturbation parameter may be implicit as a scaling parameter in the kernels, as

$$x = \bar{x} + x_1 + x_2 + \cdots \tag{24}$$

then

$$x = \bar{x} + \int_0^t e^{\mathbf{A}\tau} l(t - \tau)\, d\tau + x_2 + \cdots \tag{25}$$

it is observed, by comparing Equations 25 and 2, that $e^{\mathbf{A}\tau}$ is, indeed, the desired first-order kernel which, as previously mentioned, is dependent on the zeroth kernel and ultimately background intensity. Hence,

$$\mathbf{H}_1(\tau) = e^{\mathbf{A}\tau} \tag{26}$$

where each term of the Volterra kernel is

$$\mathbf{H}_{i1}(\tau) = \sum_k a_{ijk} e^{a_k \tau},\ j = 1, 2, \cdots \tag{27}$$

B. SECOND-ORDER TEMPORAL KERNEL

The second-order kernel is found by equating coefficients of λ^2 to get

$$\dot{x}_{i2} = x_{i2}(-a_i + \bar{x}_i 2K_i - \sum_j K_{ij}\bar{x}_j) - \bar{x}_i \sum_j K_{ij}x_{j2} + K_i x_{i1}^2 - x_{i1} \sum_{j \neq i} K_{ij}x_{j1} \quad (28)$$

and is seen to be of the same form as Equation 15 but with the input l_i replaced by $K_i x_{i1}^2 - x_{i1}\Sigma_{j\neq i}K_{ij}x_{j1}$. Equation 28 can be written in the matrix form

$$\begin{bmatrix} \dot{x}_{12} \\ \dot{x}_{22} \\ \vdots \\ \dot{x}_{n2} \end{bmatrix} = \mathbf{A} \begin{bmatrix} x_{12} \\ x_{22} \\ \vdots \\ x_{n2} \end{bmatrix} + \begin{bmatrix} m_1 \\ m_2 \\ \vdots \\ m_n \end{bmatrix} \quad (29)$$

where

$$m_i = x_{i1} \sum_{j=1}^{n} K_j' x_{j1} \quad (30)$$

is a function of the first-order solution and

$$K_j' = \begin{cases} -K_{ij} & \text{for } j \neq i \\ K_i & \text{for } j = i \end{cases} \quad (31)$$

Again, assuming that $x(0) = 0$, the second-order generating solution is

$$x_2(t) = \int_{t_0}^{t} e^{\mathbf{A}(\tau)}\, m(t - \tau)\, d\tau \quad (32)$$

The ith term of $x_2(t)$ is found from Equation 22 to be

$$x_{i2}(t) = \int_0^t \sum_l \sum_k a_{ilk} e^{a_k \tau} m_l(t - \tau)\, d\tau \quad (33)$$

If $m_l(t)$ is replaced by its definition (Equation 30) with the first-order solution explicitly substituted from Equation 23, this equation becomes

$$x_{i2}(t) = \int_0^t \sum_l \sum_k a_{ilk} e^{a_k \tau} \left[\sum_j \int_0^{t-\tau} \sum_m a_{ljm} e^{a_m \tau_1} l_j(t - \tau - \tau_1) \right] d\tau_1$$
$$\left[\sum_p K_p' \sum_q \int_0^{t-\tau} \sum_r a_{pqr} e^{a_r \tau_2} l_q(t - \tau - \tau_2) \right] d\tau_2 d\tau \quad (34)$$

In order to put this equation in a form from which the second-order Volterra kernel can be extracted by observation, the following change of variables should be made

$$\mu = \tau + \tau_1$$

$$\gamma = \tau + \tau_2$$

and for simplicity of notation let

$$c_i = \sum_k \sum_l \sum_m \sum_p \sum_r a_{ilk} a_{ljm} a_{pqr} K'_p \tag{35}$$

Making these substitutions in Equation 34 while carefully treating the limits of the integrals results in

$$x_{i2}(t) = \sum_j \sum_q c_i \int_0^t \int_0^t \left[\int_0^t e^{\tau(a_k - a_m - a_r)} e^{a_m \mu} e^{a_r \gamma} u(\mu - \tau) u(\gamma - \tau)\, d\tau \right]$$
$$l_j(t - \mu) l_q(t - \gamma)\, d\mu d\gamma \tag{36}$$

where $u(\alpha)$ is the step function defined in the usual manner.

Comparison of Equations 36 and 2 shows that the quantity in brackets is the second-order temporal Volterra kernel and can be integrated to produce

$$h_{jq}(\tau_1, \tau_2) = \sum_m \sum_r \sum_k c'_i \frac{e^{a_m \tau 1} e^{a_r \tau 2}}{(a_k - a_m - a_r)} \left[e^{(a_k - a_m - a_r) \min(\tau_1, \tau_2)} - 1 \right] \tag{37}$$

where

$$c'_i = \sum_l \sum_p a_{ilk} a_{ljm} a_{pqr} K'_p \tag{38}$$

This is the same form as that derived by Pinter (1987), even though the model defined by Pinter does not have the quadratic self-excitatory term. The reason for similarity lies in the fact that equations with the self-excitatory term are still globally asymptotically stable, as proven in the appendix, and the **A** matrix remains Hermitian. These were the constraints that led to the general form of the solution in Equation 23; hence, the final result should be similar in form, although the eigenvalues are different in the two cases.

It is seen that the second-order temporal kernel consists of permutations of the products of exponential terms with eigenvalues of the **A** matrix as rate constants. In both the first- and the second-order temporal kernels, Equations 25 and 37, the eigenvalues become larger as the mean intensity, L_0, increases. Thus, temporal behavior of the kernels becomes faster and moves closer to the time origin, and kernels become smaller in magnitude. This is the basic temporal adaptation (Rushtonian transformation in vision science) and is propagated to all higher-order kernels (Pinter and Nabet, 1991).

It is noteworthy that in physical implementations, even in the solid-state integrated circuit implementations which produce very uniform device characteristics, it is hard to maintain the absolute symmetry of the interconnections. This may seem to invalidate the symmetry constraint which was repeatedly used in this chapter, but a theorem by Franklin (1968) can be invoked which states that such a matrix is arbitrarily close to a diagonalizable matrix, and, hence, the results of this chapter hold with arbitrary accuracy for a physically implemented case.

III. SPATIAL KERNELS

The same technique as described above can be applied to the steady-state solution of Equation 1, the existence of which is due to the global stability of the network, to find the solution in the spatial Volterra-Wiener expansion form:

$$x_i = H_{0i} + \sum_j H_1(i,j)l_j + \sum_j \sum_k H_2(i,j,k)l_j l_k + \cdots \tag{39}$$

where the kernel subscript denotes the order of the kernel, which is the order of approximation, and the kernels are written with spatial indices as an argument to stress the spatial dependence.

Extraction of the spatial Volterra kernels, however, is simpler if the results of the temporal case are used, as shown below.

A. ZERO-ORDER KERNEL

The zeroth spatial kernel has already been derived as the response to the (uniform) background intensity in steady state and is shown in Equation 11. This response, as expected, is invariant with time and across space as indexed by the cell number, assuming boundary effects are compensated by, for example, imposing periodic boundary conditions; also $a_i = a_j$ for every i and j. Both the spatial uniformity of the response and its dependence on background intensity level can be experimentally verified (Nabet and Pinter, 1991).

B. FIRST-ORDER SPATIAL KERNEL

The first-order spatial response is found by solving the first-order network Equation 1, which is the linearized model of the network, in the steady state (Pinter, 1987). Thus,

$$\mathrm{x}_1 = -\mathbf{A}^{-1}(L_0)\mathrm{l} \tag{40}$$

where **A** and l are defined as previously and the dependence of matrix **A** on the mean intensity via the zero-order kernel is explicitly shown. The *i*th term of the first-order response is simply

$$x_{i1} = \sum_j b_{ij} l_j \tag{41}$$

from which the *i*th term of the first-order kernel is observed to be

$$H_1(i,j) = b_{ij} \tag{42}$$

where b_{ij} is the term of the negative of the inverted matrix $\mathbf{A}^{-1}$, the existence of which is a result of the fact that **A** is positive definite.

Contrary to the temporal kernel case, a closed form analytical expression for $\mathbf{A}^{-1}$ cannot be written.* For the case where **A** is a Toeplitz matrix, which is applicable here, a direct relationship between **A** and its inverse exists, but that relationship which involves a Fourier transform is not directly related to the present discussion.

From Equation 40 it is seen that $\mathbf{A}^{-1}$ (L_0) specifies the receptive field of the network. A column of the inverted matrix has been plotted by Pinter (1987) for different values of the mean intensity L_0 and is reproduced here (in Figure 1). The on-center off-surround shape of the response and its dependence on the mean intensity level is very similar to the experimental results presented by Nabet (Chapter 18). This shows the suitability of the Volterra-Wiener expansion method for analyzing the intensity dependent behavior of the network.

It should be emphasized that although the first-order kernel is derived from a linearized system model, its intensity dependence is due to the multiplicative nonlinearity which gave rise to the intensity dependent zero-order kernel. A comparable additive model would not produce intensity dependent kernels.

* Common matrix inversion methods such as the cofactor method are algorithms rather than closed form analytical expressions.

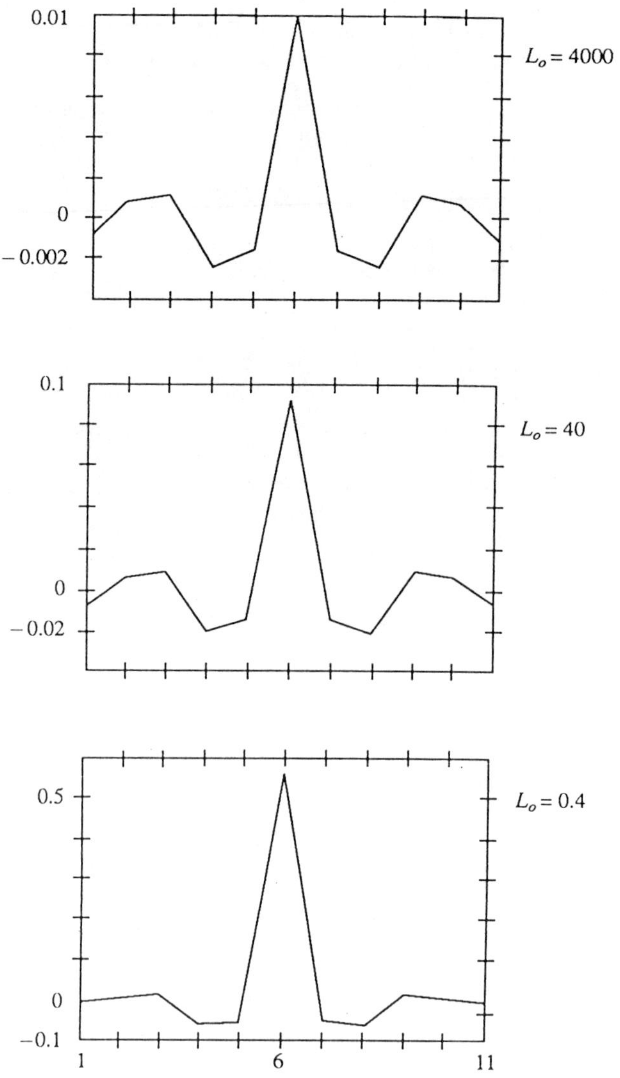

FIGURE 1. Receptive field of the linearized network plotted with different mean intensity values. From Pinter (1987).

C. SECOND-ORDER SPATIAL KERNEL

The second-order spatial Volterra kernel is determined, in a similar manner to the first-order kernel, by finding the steady-state solution of Equation 29, namely,

$$x_2 = -\mathbf{A}^{-1}m \tag{43}$$

where m is defined as in Equation 30. The dependence of the higher-order kernels on the mean intensity via the zero-order kernel is apparent from this equation. Furthermore, it is notable that the matrix inversion is performed only once, and the results are propagated through the higher-order kernels.

Each term of x_2 is

$$x_{i2} = \sum_j b_{ij} m_j \tag{44}$$

where, as before, the second index of the response specifies the order of the kernel.

Substituting the definition of m_j from Equation 30, inserting the expression for the first-order response from Equation 41, and reordering the terms results in

$$x_{i2} = \sum_j \sum_k \left[\sum_p \sum_q K'_q b_{ip} b_{pj} b_{qk} \right] l_j l_k \tag{45}$$

The term within the brackets is the desired ith term of the second-order spatial kernel, namely,

$$H_2(i,j,k) = \sum_p \sum_q K'_q b_{ip} b_{pj} b_{qk} \tag{46}$$

D. HIGHER-ORDER SPATIAL KERNELS

The same procedure can be applied to find any order kernel, assuming that in the steady state the derivative of each of the terms is finite, an assumption which is plausible due to the convergence of the series and asymptotic global stability of the network. In general the mth order spatial kernel is found by equating the coefficients of λ^m in Equation 8 and observing that the mth order response is the steady-state form of the resulting equation, namely,

$$x_{im} = \sum_{j=1}^{n} b_{ij} p_{jm} \tag{47}$$

where, as before, b_{ij} is the term of the negative of the inverted matrix A, and p_{im} is a perturbation input term which depends on all the previous responses in the following manner:

$$p_{im} = x_{i1} \sum_{j=1}^{n} K_{ij} x_{j,m-1} + x_{i2} \sum_{j=1}^{n} K_{ij} x_{j,m-2} + \cdots + x_{i,m-1} \sum_{j=1}^{n} K_{ij} x_{j1} \tag{48}$$

This identity can be more compactly written as

$$p_{im} = \sum_{k=1}^{m-1} \sum_{j=1}^{n} K_{ij} x_{i,k} x_{j,m-k} \tag{49}$$

which, from Equation 47, leads to a steady-state mth order spatial response of

$$x_{im} = \sum_{l=1}^{n} \sum_{k=1}^{m-1} \sum_{j=1}^{n} K_{ij} b_{il} x_{ik} x_{j,m-k} \tag{50}$$

Equation 50 is a recursive formula from which the mth-order response can be derived based on the previous responses.

The third-order response, for example, is found by simply letting m = 3 in Equation 50 to get

$$x_{i3} = \sum_{l=1}^{n} \sum_{j=1}^{n} K_{ij} b_{il} x_{i1} x_{j2} + \sum_{l=1}^{n} \sum_{j=1}^{n} K_{ij} b_{il} x_{i2} x_{j1} \tag{51}$$

Substituting the expressions for first- and second-order responses (Equations 41 and 45), and reordering the terms results in

$$x_{i3} = \sum_{j} \sum_{p} \sum_{q} \left[\sum_{p_2} \sum_{p_3} \sum_{p_4} K_{ij} K_{jp_3} b_{ip_2} b_{iq} b_{jp} b_{jp_4} b_{p_3,p_4} \right] l_j l_p l_q \tag{52}$$

It is observed that the term within the brackets is the third-order kernel, namely,

$$H_i(i,j,p,q) = \left[\sum_{p_2} \sum_{p_3} \sum_{p_4} K_{ij} K_{jp_3} b_{ip_2} b_{iq} b_{jp} b_{jp_4} b_{p_3,p_4} \right] \tag{53}$$

The recursive nature of Equation 50 is especially suited for numeric calculation by a digital computer and is facilitated by the fact that the most computationally intensive operation of inversion of a matrix is done only once and recalled later. On the other hand, recursion disallows writing of an independent closed-form formula for the *m*th-order kernel, the necessity for which is dependent on the rate of the convergence of the Volterra series. Since the significance of the contribution of kernels higher than the third order to the series convergence is unknown in many practical cases of interest, Equation 50 is sufficient for present purposes.

The major aim of this treatment was to show that the one layer network consisting of only multiplicative inhibitory, self excitatory, and linear feedback terms, can be represented as in Equation 39. If this output is further transformed by placing a nonlinearity in front of each cell, such as a hard limiting circuit or a sigmoidal nonlinearity, then the output is

$$x_i = S \left[H_0 + \sum_{j} H_1(i,j) l_j + \sum_{j} \sum_{k} H_2(i,j,k) l_j l_k + \cdots \right] \tag{54}$$

Each unit of the network, thus, not only calculates the weighted sum of the input, but is also sensitive to the higher-order correlations as demonstrated by the fact that the third term of the Equation 54 is the definition of a weighted (spatial) auto correlation function of the input. A unit defined by Equation 54 has special classification properties which are discussed next.

E. CLASSIFICATION PROPERTIES

Typical units of (additive) neural network architectures transform a weighted sum of the input as in the classic threshold logic units (TLU) of McCulloch and Pitts (1943), which were shown to be able to realize *any* logical function. A TLU is defined by

$$x_i = S \left[\sum_{j=1}^{n} H(i,j) l_j \right] \tag{55}$$

where x_i is the output of the unit, S is a nonlinearity, l_j is the input, and $H(i,j)$ are weighting terms, also called connection strengths, written in this form to underline similarity with the Volterra kernel expansion.

The quantity inside the brackets is a discriminant function which specifies the decision boundaries for the classification of the input pattern (Nilsson, 1965; Minsky, 1967). If the nonlinearity S is a threshold element, as is the case for a TLU, then the response x_i can classify the input into two categories which are separated by the decision boundary defined by the discriminant function. A two-category classifier is termed a pattern dichotomizer. Operation of Equation 55 is shown in Figure 2.

Since the discriminant function of Equation 55 is a linear weighted sum of the input, it defines a linear decision boundary, a hyperplane in general, and can, thus, correctly classify

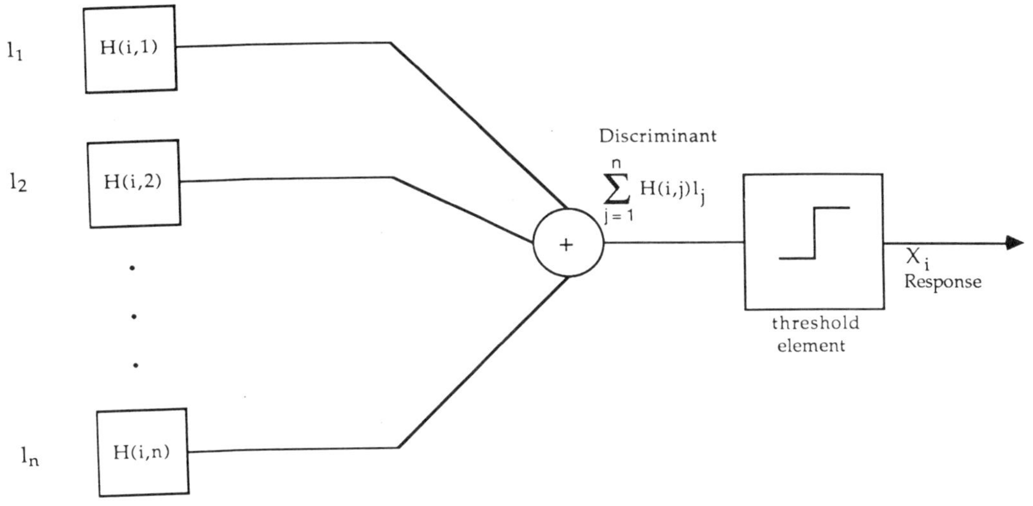

FIGURE 2. The threshold logic unit as a pattern classifier.

only linearly separable input vectors. More importantly, it was shown by Minsky and Papperl in their influential book *Perceptrons* (1969) that single feedforward slabs of TLUs, as in Rosenblatt's perceptron (1958), can only perform linearly separable mappings. It was also suggested in that book that cascades of slabs of first-order units will be of no more practical interest than single layers. The latter was an incorrect generalization as proven by the immense popularity of the error back-propagation algorithm (Rumelhart et al., 1986) which trains multilayer networks to classify both linearly and nonlinearly separable input patterns. This linear separability constraint is often cited as one of the most important reasons for the previous decline of interest in neural network theory and application.

Inspection of Equation 54, which is an expansion of the simplest shunting network with only lateral inhibition, shows that the term inside the brackets is an *n*th-order discriminant function, and, hence, a shunting net is an nth-order classifier. This result is interesting, especially in light of the fact that as shown elsewhere in this book (Nabet, this volume) only one transistor per interconnection is necessary for implementing such a network.

The limitations of shunting networks for classification should also be mentioned. The decision boundaries defined by Equation 54 are rigid since the weights are constant. More importantly, the higher-order weights are not independent in the Volterra expansion case and cannot be individually set. This limits the shape of the decision boundaries and, hence, the type of nonlinearly separable functions that can be correctly classified by this class of one-layer neural networks. The mean input intensity dependence of the kernels, however, can be used to alter, and, hence, control, the shape of the decision regions. This may extend the results from the analysis of neural networks to their synthesis.

Units of the form in Equation 54 have been termed high-order neurons (Giles and Maxwell, 1987) and have been shown to possess very desirable learning, generalization, and knowledge representation capabilities and can be hand-crafted to solve many interesting high-order problems.

Finally, Minsky and Pappert (1969) were well aware of high-order discriminant functions, but correctly observed that implementation of such units would suffer from combinatorial explosion of weights. This treatment shows that all high-order weights are implicit in the closed form of a simple network with only multiplicative inhibition.

IV. CONCLUSIONS

The Volterra-Wiener series expansion can be applied to one-layer neural network models to allow for systematic inclusion of higher-order nonlinearities in the description of temporal and steady-state behavior of the network. This technique was shown to be especially suitable for modeling the adaptation of temporal response and the spatial receptive field of the networks to variation in mean input levels.

In this chapter, temporal and spatial Volterra kernels for a neural network model whose activity is governed by multiplicative lateral inhibition were derived. The spatial expansion showed that a network with first-order weights can exhibit higher-order properties such as classification of nonlinearly separable input functions. Formulas derived for high-order kernels (weights) show their dependence on lower-order terms, and, ultimately, on mean value of the input. Since these weights define the shape of the decision regions, this limits the functions that can be classified, but also shows that changing the mean input (or the total activity, or the background or ambient light in an optical input case) is one mechanism through which some control over the decision regions can be exerted.

The mathematical analysis, relation to vision science, neural network theory, image processing, motion detection, and technological applications discussed in this book all have a common base, namely, the study of the global behavior arising from local activity of simple units. The final conclusion is so simple as to be redundant: a number of simple units, be they neurons or transistors, if connected correctly, can perform very complicated tasks.

ACKNOWLEDGMENT

This work was supported in part by a Drexel University Research Scholar Award, Engineering Foundation Research Initiation Award, and NSF grant MIP8822121.

APPENDIX

The emergent global properties of a network, rather than the behavior of the individual units and the local computation performed by them, describe the behavior of the network. Global stability analysis techniques, such as Liapunov energy functions, show the conditions under which a system approaches an equilibrium point in response to an input pattern. The equilibrium point is then the stored representation of the input. This property can be termed content-addressable memory (CAM) property.

Local stability, by contrast, involves the analysis of network behavior around individual equilibrium points. This appendix shows the global stability of the simple multiplicative inhibitory network analyzed in this chapter.

The stability analysis is greatly facilitated by a theorem by Cohen and Grossberg (1983), who showed that a system of the general form

$$\frac{dx_i}{dt} = a_i(x_i)\left[b_i(x_i) - \sum_{j=1}^{n} c_{ij}d_j(x_j)\right] \tag{56}$$

which includes many of the most popular neural network models as shown by Grossberg (1988), admits the global Liapunov function

$$V(x) = -\sum_{i=1}^{n}\int_0^{x_i} b_i(\xi_i)d_i'(\xi_i)d\xi_i + \frac{1}{2}\sum_{j,k=1}^{n} c_{jk}d_j(x_j)d_k(x_k) \tag{57}$$

under the constraints

$$c_{ij} = c_{ji} \tag{58}$$

$$a_i(x_i) \geq 0 \tag{59}$$

$$d_j'(x_j) \geq 0 \tag{60}$$

If the following substitutions are made in Equation 56

$$c_{ij} = k_{ij} \tag{61}$$

$$d_j(x_j) = f_j(x_j) \qquad \text{with } f_j'(x_j) \geq 0 \tag{62}$$

$$b_i(x_i) = \left(\frac{I_i}{x_i} - \frac{1}{R_i}\right) \tag{63}$$

$$a_i(x_i) = \frac{x_i}{C_i} \qquad \text{with } x_i \geq 0 \tag{64}$$

which satisfy the symmetry, positivity, and monotonicity requirements as expressed in Equations 58-60, then Equation 56 is transformed into

$$\frac{dx_i}{dt} = \frac{x_i}{C_i}\left[\left(\frac{I_i}{x_i} - \frac{1}{R_i}\right) - \sum_{j=1}^{n} k_{ij} f_j(x_j)\right] \tag{65}$$

and can be rewritten as

$$C_i \frac{dx_i}{dt} = I_i - \frac{x_i}{R} - x_i\left[\sum_{j \neq i} k_{ij} f_j(x_j)\right] \tag{66}$$

This is the original multiplicative lateral inhibitory Equation 1 of this chapter without the self-excitatory term $K_i x_i^2$ and is, thus, shown to be globally stable.

The global Liapunov function for Equation 66 can be found by substituting Equations 61-64 in Equation 57 to obtain

$$V(x) = -\sum_{i=1}^{n} \int_0^{x_i} \left(\frac{I_i}{\xi_i} - \frac{1}{R_i}\right) f_i'(\xi_i) d\xi_i + \frac{1}{2} \sum_{j,k=1}^{n} k_{jk} f_j(x_j) f_k(x_k) \tag{67}$$

Since $f_j(x_j)$ is a nonnegative monotone nondecreasing function of x_j according to the constraint in Equation 62, the second term under the integral sign can be easily integrated and assuming, without loss of generality, that $f_j(0) = 0$, the Liapunov function is rewritten as

$$V(x) = -\sum_{i=1}^{n} \int_0^{x_i} \frac{I_i}{\xi_i} f_i'(\xi_i) d\xi_i + \sum_{i=1}^{n} \frac{1}{R_i} f_i(x_i) + \frac{1}{2} \sum_{j,k=1}^{n} k_{jk} f_j(x_j) f_k(x_k) \tag{68}$$

A problem is immediately apparent with this function; a constant term in $f'_j(\xi_i)$ will produce $\ln(\xi_i)$ when integrated, which equals $-\infty$ when the lower bound of the integral, zero, is substituted for ξ_i.

This problem can be solved by replacing the lower bound of the integral by a positive constant λ_i. It should then be shown that the new function

$$V(x) = -\sum_{i=1}^{n}\int_{\lambda i}^{xi}\frac{I_i}{\xi_i}f'_i(\xi_i)d\xi_i + \sum_{i=1}^{n}\frac{1}{R_i}f_i(x_i) + \frac{1}{2}\sum_{j,k=1}^{n}k_{jk}f_j(x_j)f_k(x_k) \tag{69}$$

has the Liapunov function property, i.e., it satisfies

$$\frac{d}{dt}V\left(x(t)\right) \le 0 \tag{70}$$

on all admissible trajectories.

This proposition is verified by noting that by chain rule

$$\frac{d}{dt}V\left(x(t)\right) = \sum_i \frac{\partial V[x(t)]}{\partial x_i(t)}\frac{dx_i(t)}{dt} \tag{71}$$

Directly differentiating $V(x)$ with respect to x_i will result in

$$\frac{\partial V(x(t))}{\partial x_i(t)} = -\frac{\partial}{\partial x_i}\left(\int_{\lambda_i}^{xi}\frac{I_i}{\xi_i}f'_i(\xi_i)d\xi_i\right) + \frac{f'_i(x_i)}{R_i} + \frac{1}{2}f'_i(x_i)\sum_j k_{ij}f_j(x_j)$$
$$+ \frac{1}{2}f'_i(x_i)\sum_k k_{ki}f_k(x_k) \tag{72}$$

Using the symmetricity requirement in Equation 58 of $k_{ij} = k_{ji}$, and letting

$$g'_i(\xi_i) = \frac{I_i}{\xi_i}f'_i(\xi_i) \tag{73}$$

Equation 72 can be simplified to

$$\frac{\partial V(x(t))}{\partial x_i(t)} = -\frac{\partial}{\partial x_i}[g_i(x_i) - g_i(\lambda_i)] + \frac{f'_i(x_i)}{R_i} + f'_i\sum_j k_{ij}f_j(x_j) \tag{74}$$

Since $\partial g_i(\lambda_i)/\partial x_i = 0$, Equation 74 can be substituted in Equation 71, with dx_i/dt replaced by its definition Equation 65, to yield

$$\frac{d}{dt}V(x(t)) = -\sum_{i=1}^{n}\frac{x_i f'_i(x_i)}{C_i}\left[\frac{I_i}{\xi_i} - \frac{1}{R_i} - \sum_{j=1}^{n}k_{ij}f_j(x_j)\right]^2 \tag{75}$$

Using constraints of Equations 62 and 64 which specify admissible trajectories, it is seen that

$$\frac{d}{dt}V(x(t)) \le 0, \tag{76}$$

which is the desired result.

It should be noted that the above derivation is basically a plausibility argument and for that reason has not been treated as a proof to a proposition. The proof can be completed by showing stability of *all* the trajectories. A rigorous proof that the integral $\int_0^{x_i}$ can be replaced by the an integral $\int_{\lambda_i}^{x_i}$ for the general case where

$$\lim_{\xi \to 0+} |b_i(\xi) d_i'(\xi)| = \infty \quad (77)$$

in Equation 56 has been shown by Cohen and Grossberg (1983).

It is a straight-forward matter to show that a network with self-excitation, such as the one described by Equation 1 of this chapter, is also globally stable. If the extent of excitation reaches to neighbors, however, stability cannot be assured. Examples have been given (Cohen, 1988) where a system satisfies all the conditions of Equations 58-60, but the function *b(x)* extends farther than the cell *i*; the network then is shown to oscillate consistently.

REFERENCES

Boyd, S. and Chua, L. O. (1985). Fading memory and the problem of approximating non-linear operators with Volterra series, *IEEE Trans. Circuits Syst.*, **32**, 1150–1160.

Cohen, M. A. (1988). Sustained oscillations in a symmetric cooperative-competitive neural network: disproof of a conjecture about content-addressable memory property, *Neural Networks,* **1**, 217–221.

Cohen, M. A. and Grossberg, S. (1983). Absolute stability of global pattern formation and parallel memory storage by competitive neural networks, *IEEE Trans. Syst. Man Cybern.*, **13**, 815–826.

Franklin, J. N. (1968). *Matrix Theory,* Prentice-Hall, Engelwood Cliffs, NJ.

Giles, C. L. and Maxwell, T. (1987). Learning invariance and generalization in high-order neural networks, *Appl. Opt.*, **26(23)**, 4972–4977.

Grossberg, S. (1988). Nonlinear neural networks: principles, mechanisms and architectures, *Neural Networks,* **1**, 17–61.

Marmaralis, P. Z. and Marmaralis, V. Z. (1978). *Analysis of Physiological Systems,* Plenum Press, New York.

McCulloch, W. S. and Pitts, W. (1943). A logical calculus of the ideas immenant in nervous activity, *Bull. Math. Biophys.*, **5**, 115–133.

Minsky, M. L. (1967). *Computation: Finite and Infinite Machines,* Prentice-Hall, Englewood Cliffs, NJ.

Minsky, M. L. and Pappert, S. (1969). *Perceptrons,* MIT Press, Cambridge.

Nabet, B. and Pinter, R. B. (1991). *Sensory Neural Networks: Lateral Inhibition,* CRC Press, Boca Raton, FL.

Nilsson, N. J. (1965). *Learning Machines,* McGraw-Hill, New York.

Ogura, H. (1986). Estimation of Wiener kernels of a nonlinear system and fast algorithm using digital Laguerre filters, in *The 15th NIBB Conf. Information Processing in Neuron Networks,* Naka, K. and Ando, Y., Eds., 14–63.

Pinter, R. B. (1987). Kernel synthesis from non-linear multiplicative lateral inhibition, in *Advanced Methods of Physiological System Modelling,* **Vol. 1**, Marmaralis, V. Z., Ed., Biomedical Simulations Resource, University of Southern California, Los Angeles, 258–277.

Pinter, R. B. and Nabet, B. (1991). Field adaptation in visual systems is a function of nonlinear lateral inhibition, in *Neural Networks: Concepts, Applications and Implementation,* **Vol. 3**, Antognetti, P. and Milutinovic, V., Eds., 21–48.

Rosenblatt, F. (1958). The perceptron: a probabilistic model of information storage and organization in the brain, *Psychol. Rev.*, **65**, 386–408.

Rugh, W. J. (1981). *Non-linear system theory — The Volterra-Wiener approach,* John Hopkins University Press, Baltimore.

Rumelhart, D. E., Hinton, G. E., and Williams, R. J. (1986). Learning internal representation by error propagation, in *Parallel Distributed Processing,* Rumelhart, D. E. and McClelland, J. L., Eds., MIT Press, Cambridge.

Schetzen, M. (1980). *The Volterra and Wiener Theory of Non-Linear Systems,* John Wiley & Sons, New York.

Chapter 20

INFANT NEURAL CONTROLLER FOR ADAPTIVE SENSORY-MOTOR COORDINATION*

Michael Kuperstein

TABLE OF CONTENTS

I. Introduction 494

II. Locating Stationary Targets 495
 A. Reaching Stationary Targets with Multiple Joints 503

III. Dynamic Movement Control for Unforeseen Payloads 507

Acknowledgments 511

References 512

Appendix 513

* Reprinted with permission from *Neural Networks,* **4**, M. Kuperstein, INFANT Neural Controller for Adaptive Sensory-Motor Coordination, © 1991, Pergamon Press plc.

I. INTRODUCTION

The human brain develops accurate sensory–motor coordination in the face of many unforeseen changes in the dimensions of the body, strength of the muscles, and placements of the sensory organs. This is accomplished for the most part without a teacher.

Autonomous robots of the future will have to confront similar constraints if they are to be effective in uncertain environments. They will have to locate, reach, and pick up novel payloads, all in real time and be expandable to accommodate many joints and sensors. They will have to learn and maintain accurate performance even after unpredictable changes are made in either the geometrical, mechanical, torque, or sensing parameters or from internal parameter changes or damage.

Traditional autonomous robots are controlled by either direct program control, teaching pendants, model reference control, or conventional adaptive control techniques. The first two methods are not adaptive. The last two require a model of the robot plant and actuators, which may be difficult to obtain beforehand (Bolles and Roth, 1988). Also, conventional multijoint dynamic operations require intensive computations that slow down the control process considerably.

The neural network approach appears to be the most promising for addressing the problem of adaptive control in uncertain environments. Progress with this approach will not only become the basis for more adaptive robots, but will also guide experimental analysis and understanding of animal and human movement control. Most of the work to date has focused on dynamic motor control and sensory–motor coordination which has recently led to successful implementations in robotics. Some past key milestones include the first neural controller by Widrow and Smith (1963), who used an adaptive linear element (ADALINE) to balance an inverted broom stick. The controller applied a positive or negative fixed impulse depending on the broom state. Later, Albus (1975) presented the cerebellar model articulation controller (CMAC), which is essentially an adaptive, distributed table lookup method. Combining the CMAC model with a conventional adaptive controller, Miller (1987, 1989) has implemented a system that learns associations between sensor measurements and desired command signals. Also working on adaptive trajectory tracking. Miyamoto et al. (1988) and Kawato et al. (1988) have implemented a neural network which generates adaptive inverse kinematics. From a different direction, Grossberg and Kuperstein (1986/1989) showed how ego-centered coordinates, which can be represented by neural topographies, allow unsupervised adaptive sensory-motor calibrations. Extending this work, Kuperstein (1987, 1988b) and Kuperstein and Rubinstein (1989) have implemented a neural controller in a 5 degree-of-freedom robot to grasp objects in arbitrary positions in three-dimensional space. This controller, called INFANT, learns visual-motor coordination without any knowledge of the geometry of the mechanical system and without a teacher. The INFANT controller is the subject of this review. New work in motor control is now underway in adaptive sequence generation, such as that of Jordan (1989) who uses a recurrent back-propagation method.

In the following review, I present a theory and implementation that suggest how three types of adaptive sensory–motor coordinations might be learned and maintained by animals as well as robot controllers. They are:

1. Locating stationary targets with movable sensors
2. Reaching arbitrarily positioned and oriented targets in three-dimensional space with multijoint arms
3. Positioning an unforeseen payload with accurate and stable movements despite unknown sensor feedback delay

FIGURE 1. Display of the problem of grasping an object.

These three abilities can be tied together by focusing on the overall problem that a human faces in grabbing an object (shown in Figure 1). Here, two moving eyes first have to fixate the object in three-dimensional space. Then, what the eyes see has to generate a plan to reach the cylinder. Finally, the plan has to be performed with inertial dynamics to grab the cylinder. Keep in mind, the human does not know the spacing of his eyes, nor the length of his arms, nor the strength of his muscles, nor the feedback delay of his reflexes.

This review describes neural network architectures consisting of large arrays of neuronal units. The networks combine the constraints of *self-consistency* and *topography* to achieve adaptive sensory-motor coordination without supervision. Together they comprise the INFANT model. The self-consistency hypothesis is an extension of results from developmental studies in coordination behavior. Studies in the kitten (Hein, 1970, 1972, 1974; Hein and Held, 1962; Held and Hein, 1963) show that visually guided behavior develops only when changes in visual stimulation are systematically related to self-produced movement. The hypothesis is also consistent with the motor theory of speech perception (Liberman et al., 1967; Williams and Nottebohm, 1985).

The INFANT controller's architecture relies on the topography of neural units in a network. Topography is the ordered contiguous representation of inputs or outputs across a surface with possible overlap of neighboring representations. Topographic mappings have been found in most sensory and motor brain structures (Kandel and Schwartz, 1985). In the following neural networks, topography is the universal format for representing, combining, and transforming all sensations, plans, and movements. The major theme of the following studies is how various geometries of interacting topographic neural fields can satisfy the constraints of adaptive behavior in complete sensory-motor circuits.

In the following sections, I present the motivations, constraints, experiments, results, and interpretations of the INFANT controller. Since all the neural networks are designed to use essentially the same processing and learning rules, they will be summarized in the Appendix. Detailed equations can be seen in referenced papers. Alternate processing and learning rules could have been used with similar results. The ones I chose appeared to be the simplest and most convenient for computer processing.

II. LOCATING STATIONARY TARGETS

When a target object is perceived in space, its registration on our senses will move as we move. How, then, do we perceive it to be stationary and know where to reach it as we move? Put in different words, how can objects be perceived in their own coordinates when they are sensed in an egocentric frame of reference? For this perception to occur, the representation of the spatial location of the object must somehow be invariant with respect to many combinations of the sensory registrations of the object. Consider the case of perceiving a target with an eye that is constantly shifting its gaze. There are many combinations of the retinal position of the target and the orbital position of the eye that can be used to view the target. And yet it is perceived as stationary.

An invariant target position map of this kind can be very beneficial for object recognition. When the perceptual system attempts to recognize an object, the association of where an object is, with what an object is, can be more easily made with an invariant target map. Otherwise, the perceptual system would have to reassign the location properties of an object whenever that object is moved relative to the viewer.

What constitutes invariant perception in a neural network? Whatever the internal representation of invariant perception, it is evident that a system has it when its actions are invariant in response to different configurations of an object. But by introspection, a system does not have to respond overtly to an object for the system to have an invariant perception of the object. To satisfy both conditions, an invariant representation can be defined as one that has the potential for invariant action. This definition suggests that at least the representations for action in response to varying presentations of an object are the same. Of course, other prior stages of representation between sensation and action can also be invariant. But the essential element of this definition is that it always ties invariant perception to invariant action.

This definition constrains mechanisms of invariant perception to be concerned with motor performance. In particular, this section shows how spatial targets can be represented by correlating the signals that orient the eye toward those targets, with the visual feedback of those targets. More generally, a space can be represented by signals used to orient in that space or self-consistency.

I present a model for the representation of target positions in the brain that is invariant with respect to combinations of any retinal registration and eye posture (detailed in Kuperstein, 1988a). This model adapts to many sets of parameters that describe the sensory and motor plants as well as variabilities in the internal structure of the model. For focus, the model will be concerned with orientation positions of one eye to targets presented to the peripheral visual field, with the head fixed. Targets used by this model are defined as salient points of light contrast. The extension of this definition from point targets to real object targets will be addressed later.

The model is designed to be extended to any number of sensory-motor maps used to indicate the position of a target in space by any orienting motor plant. This includes orienting the eye with the head free and orienting the limbs to a visual target.

The goal of the model is to represent the final muscle activation that will orient an eye so that a target will be focused on the fovea. Within this context, invariant representation is defined as the potential for orienting to a stationary target independent of its sensory registration on the retina and the initial posture of the eye in its orbit. Over performance trials, the accuracy of the system in computing target position will improve through a simple learning rule which uses the perception of target position error after each attempted orientation.

Physical constraints on the eye and the retina are the basis for the input representations. The orientation to a visual target by one eye can be computed from the measures of the retinal position of a target and the posture of the eye in its orbit. The dimensionality of these two measures are very different. The retinal coordinates are determined relative to the fovea, while eye position coordinates are determined by either absolute or relative muscle contractions of the six eye muscles. Not only must a target map transcend the difference between

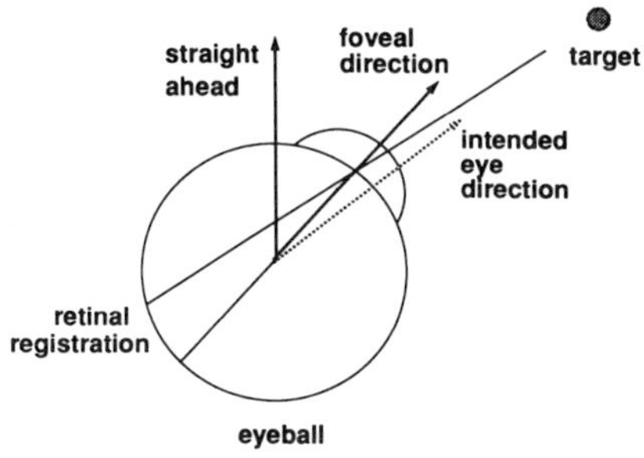

FIGURE 2. Geometry of the eyeball and target fixation.

coordinate types, but it must also yield an invariant representation. Its output for a given target position must be independent of the combination of eye position and retinal position used to generate an orientation to that target.

Adaptability of the target computation is crucial. Measures of target position made external and internal to the system surely change over time. These include changes in the eye optics and muscle characteristics, as well as potential neural damage and developmental changes. The system must be able to recalibrate itself to maintain accurate performance.

Some functional constraints that the eyeball and retina place on orienting to new targets help shape the way retinal and eye position coordinates will be combined. The eye rotates around a point close to the center of the eyeball. The angle of eyeball rotation is monotonically related to eye muscle contraction. As seen in Figure 2, the visual angle of a target is made between a line from the eye lens to the fovea, and a line from the lens to where the target projects on the retina. As a target becomes more eccentric from straight ahead, both measures of visual angle and eye position become monotonically more eccentric, when each measure is made while the other is held constant.

Because the physical geometry of the eyeball and eye muscle innervation is an unforeseen target, position is not simply a linear combination of visual angle and eye orbital angle. However, target position is a monotonic function of retinal and eye position. The model crucially depends on this relationship in order to be viable.

Now let's consider how these measures can be represented. By considering the computational benefits and drawbacks of both retinal and motor coordinates, I will present a representation that accommodates both coordinates and satisfies all of the constraints for an invariant target position map discussed thus far.

Targets in the model (points of light contrast) are represented as unimodal distributions of neural activity in a visual map. Such a visual map, of course, requires considerable preprocessing of visual inputs to the eye. Details of this preprocessing are outside the scope of the model.

The visual distributions are peaked at locations corresponding to the visual angle of the targets. Thus, the visual map has a typical retinal topography. This includes two hemifields (left-right) of neural units arrayed in two dimensions and spatially ordered by visual angle.

Eye position is determined by the distribution of muscle lengths for the six extraocular eye muscles. In the model, six neural populations, called motor maps, are activated by the contraction of each muscle, respectively. These maps will represent the eye posture during the visual registration of the target. Just like the visual map, the motor maps require considerable preprocessing of proprioceptive inputs from the muscles or corollary discharges of motor outputs to the muscles. Again, details of this preprocessing are outside the scope of the model.

With the sensory representations defined this way, the task of combining representations that yield an invariant target representation is further constrained. Retinal coordinates are

FIGURE 3. Interleaving, columnar architecture of the target map. One column is shown by the rectangle. The distribution of muscle representations for the motor inputs are schematized by arrows that show which directions the muscles pull the eye. The circles of varying levels of boldness represent three overlapping visual inputs evoked by spot light targets on three different trials. The target map has a global retinal topography.

in dimensions of relative topographic position, while the motor coordinates are in dimensions of relative activation amplitude.

The visual and motor maps, defined thus far, have complementary ways of representing position information. The visual map is sensitive to the location of a target on a two-dimensional plane and insensitive to the intensity amplitude of the target. On the other hand, each motor map is sensitive to the amplitude of muscle length. Defined this way, neither type of map uses the full information potential of a neural network. Activity in a network can be made most informative by using both its amplitude and location in the network.

The system can benefit from the complementary ways in which the two types of maps are used by superimposing the two together into a third type of map, called the target map. In this configuration, muscle representations are interleaved into columns that are spaced periodically across a retinal topography. This architecture is similar to orientation columns in the visual cortex (Hubel and Wiesel, 1962). There, columns spaced periodically across a retinal topography include all orientation-specific responses to light contrasts. Here, shown in Figure 3, the columns include all muscle representations. Within each column, a distribution of activity amplitudes, gated by visual input, locally represent the muscle lengths. Each column is a self-similar structure to the muscles. By interleaving two dimensions of muscle representations, namely, three agonist-antagonist pairs, and superimposing them with two dimensions of visual representation, four dimensions are packed into two.

There are two kinds of rules that can be used to combine visual and motor coordinates. The first uses gating. At every position of the target map, visual inputs gate the motor inputs either on or off. This isolates different visual pockets of motor inputs for different retinal measures of a target because the visual inputs are spread topographically. Within a visual pocket, the amplitude of the motor inputs represent eye position.

The second rule uses the first rule plus addition. This addition rule comes from the relationships in the measures of target position. As a target becomes more eccentric, either retinal position or the agonist muscle length alone will change monotonically. Thus, by adding visual inputs of the target to agonist motor inputs of initial eye position, elements in the target map have an activity that is crudely proportional to target position. Visual subtraction occurs for antagonist muscle representations.

By using modifiable input weights together with a local addition/subtraction rule, it becomes possible to generate the appropriate global nonlinearity that is required to describe the relationships of retinal and eye positions to target position. With nonuniform weights each target map column locally adds or subtracts visual inputs to motor inputs slightly differently. The architecture becomes piece-wise linear but as a whole, nonlinear. Of the two rules, the second is more accurate according to computer simulations.

Close inspection of this map shows that it has two types of topographies: retinal and muscle. Note that adjacent elements of the target map represent agonist-antagonist muscles. That gives the map its muscle topography. But visual gating of activity also gives it a broad retinal topography. How, then, do elements of this map get transformed to achieve target position? The real global invariance property is determined by the collective outputs of all active columns. Each muscle representation in every column of the target map outputs to a separate muscle network that corresponds to one of the six muscles. This disentangles the interleaved architecture of muscle representation columns in the target map. The mass convergence of outputs from the target map to the six muscle networks generates the global distribution of motor activity that represents the target position.

Because each muscle network in this final output stage receives an input from many columns, the total activity is relatively insensitive to neuronal noise. The pattern of global activity across the six muscle networks at this stage will be almost identical for the same target, as measured from many combinations of retinal and eye positions. This global pattern determines the invariant orientation to the same target as measured with any combination of retinal and eye positions.

The output of the model must then be further processed to control the dynamics of muscle contraction. This is presumably done by a saccade generator. For a discussion of a model for the saccade generator, see the study by Grossberg and Kuperstein (1986/1989, chap. 7). The present model is designed to interface to a saccade generator by generating a distribution of activity in muscle coordinates that represents the target position.

Learning is crucial to achieving and maintaining the accuracy of the target map. Successful representation of invariant target position depends on the right distribution of weight values for the inputs to the target map. How are these weights to be changed? Fundamentally, the only way the system can be sure whether a target map is correct is to use it for orienting to a target. Because the goal of orienting to a target is to move the eye so that the target lies on the fovea, the truest measure of error is the retinal position of the target relative to the fovea after an attempted orientation.

Other measures of error can come from either proprioceptive feedback or corollary discharge of motor outputs to the muscles. If either of these alternatives are used to correct the modifiable weights directly, then they would have to be first calibrated to the retinal measure of error in order to be reliable.

Once a reliable measure of error is obtained, the system must modify the weights of the inputs to the target map in order to decrease any performance error. Determining which weights to change and by how much presents a problem. Many combinations of retinal and orbital positions can be used for the same target on different trials. Therefore, changing the weights corresponding to one set of retinal and eye positions to improve the target position for one target affects many other target positions used on later trials. In particular, there is no guarantee that improving one combination of retinal and eye positions will not hurt another combination and end up in unstable performance. In view of the overlapping nature of the target map, will any learning scheme converge so that all combinations of retinal and eye positions become accurate?

In order to minimize any negative impact of learning across combinations of different retinal and eye positions, I use an incremental learning scheme. In this scheme, the weights of the current combination of visual and motor inputs to the target map are not modified to achieve perfect accuracy for each trial. Rather, they are modified to achieve accuracy a little at a time. The hope with this scheme is that, over many trials, all the incremental changes will accumulate to give the most parsimonious accuracy for all possible combinations.

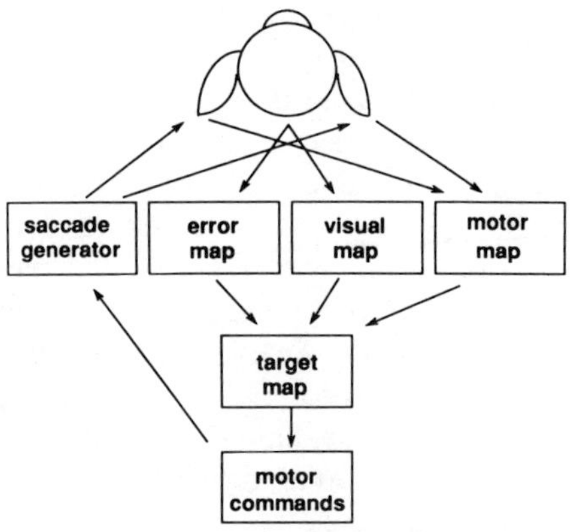

FIGURE 4. The overall flow of information for two of the six muscles in the model. First, the retinal position of the target and muscle contractions of the initial eye position are registered in a visual map and motor map. The two kinds of maps are combined in the target map with a columnar organization of muscle representations in a global retinal topography. The visual map inputs gate the motor inputs to the target map. The outputs from the elements of each target map column separate and aggregate into their respective motor command structure. These, in turn, activate muscle contraction in each eye muscle for orientation to the target, via a saccade generator. After an attempted orientation, a visual error registered on the retina activates an error map which modifies the input weights in the target map for the combination of retinal and eye positions just used.

Modifications of the input weights for specific target map elements can be mediated by an error map. It first transforms an error registered as the retinal position of a target into motor coordinates after each orientation attempt. This transformation is necessary in order to modify the input weights of the target map, which are organized in columns of muscle representations. One such transformation was presented in the discussion of saccadic error correction by Grossberg and Kuperstein (1986/1989, chap. 3). Component sectors of the error map correspond to the six muscle directions. The inputs of the error map to the target map follow the interleaved columnar architecture of the target map. The inputs are timed to become active after a measure of performance error is registered. Each of the six inputs of the map reaches the corresponding muscle representation in all columns for both hemifields of the target map.

The overall flow of information for eye orientations in one dimension in the model is shown in Figure 4. It schematizes the processing of only two of the six muscles in bilaterally symmetric hemifields. Figure 5 shows the activity distributions of all neural maps in the full model for one trial. Figure 6 shows the generation of target map errors from a visual error in one inaccurate orientation trial.

Computer simulations of the model were run under various parameter conditions to determine its accuracy, convergence, and stability. The final average performance error for all combinations of retinal and eye positions is about 1% of the total visual field. The model converged in 100,000 trials. Errors made in the model compare favorably with errors made in human saccades (Weber and Daroff, 1972), which range from 3 to 5% of the visual field.

A model that can map targets invariantly must show that all combinations of retinal and eye positions for the same target will, indeed, yield a similar target position. The ability of the current model to do this is best exemplified by comparing two trials with the same targets, but starting from different combinations of retinal and eye positions. Sampled comparisons of this type yielded an average difference between the intended and actual target positions of less than 0.8% of the visual field.

To test the stability of the model, changes in the following parameters were analyzed: the shape, width, scaling, and mirror imaging of the visual map; the shape of the motor

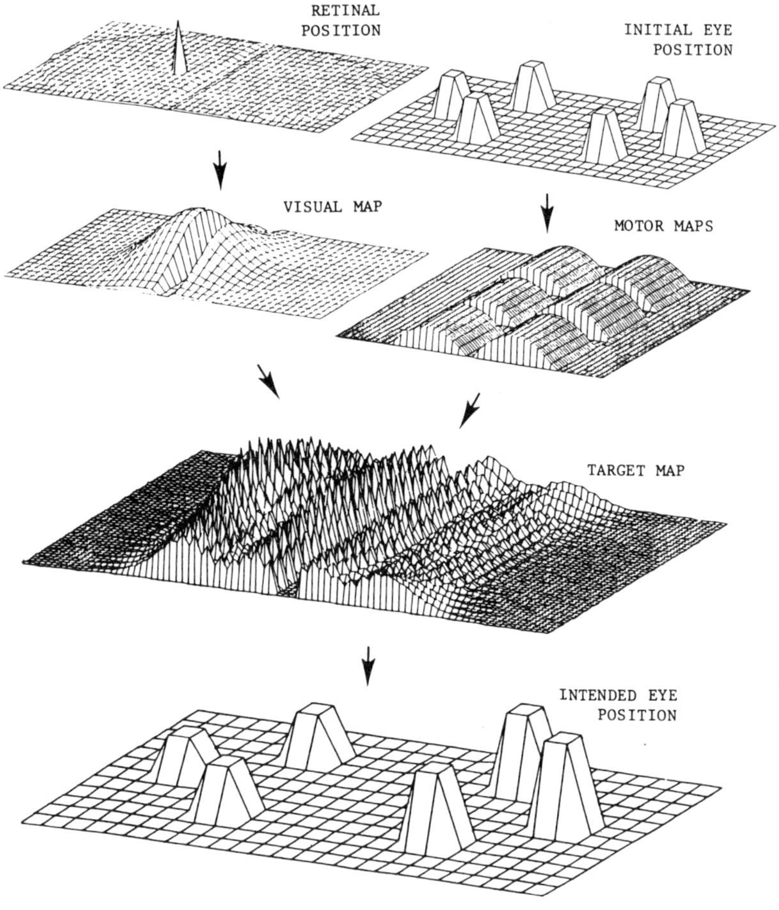

FIGURE 5. Activity distributions in the model for one trial shown in bilateral hemifields. Information flows from the registration of retinal and eye positions to the visual and motor maps to the target map and, finally, to the intended eye position for the target. Values are in relative units. The visual map hemifield and the six motor maps have a population of 20 × 20 elements each. The target map hemifields have 40 × 60 elements each. In this trial, the initial eye position is 0.2 unit to the right of center and 0 unit down from middle (in the range 0-1). The light target is 0.1 unit to the left of center and 0 unit down from middle. Note that the optics invert the image. The registration of initial eye position is represented as levels of contraction for the six eye muscles. The six corresponding motor maps show the recruitment distributions for each contraction level. The motor-visual transformation here is linear so the intended target eye position should be 0.3 unit to the right. The computed intended eye position for the target generated by the target map is 0.306 to the right and 0.08 down, which is in error by 0.6% of the total visual field.

maps; the rate of learning; noise in the target map; asymmetric muscle directions; weakened muscles; different muscle scaling; different network size; and some network damage. Within a wide range of values for each parameter the performance of the model did not significantly change.

The brain structure most closely related to the present model is the posterior parietal cortex. This brain structure "is the most likely location of cells coding the location of visual stimuli in space, since clinical and experimental studies implicate it as essential for accurate spatial orientation and perception" (Critchley, 1982). Recordings of single neurons in this area from behaving monkeys show visual responses that are dependent on the angle of gaze (Sakata et al., 1980; Anderson et al., 1985). In the Anderson et al. experiment, monkeys

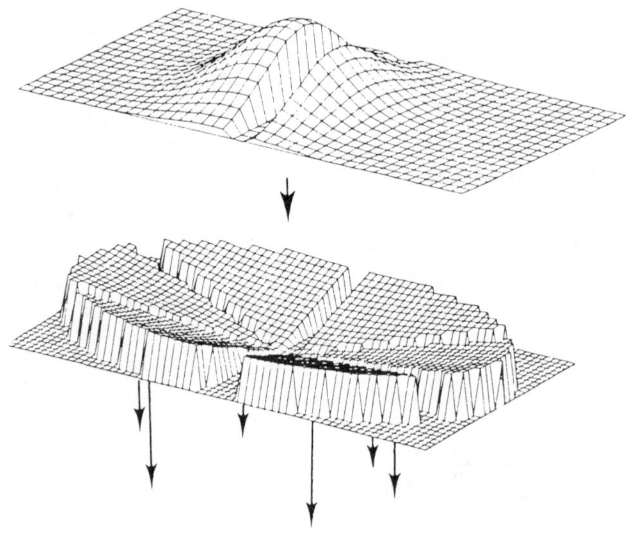

FIGURE 6. The generation of target map errors from a visual error in one inaccurate orientation trial. The visual activity distribution shown on top results from the target being off fovea after an attempted target orientation. This distribution is multiplied by error gradients for each muscle shown on the bottom. The resulting error components are projected to corresponding motor elements in the target map in order to modify the input weights.

performed a task in which they oriented their eyes to targets at different visual maps starting from various orbital positions of the eye, with head fixed. Visually sensitive neurons in the posterior parietal cortex showed response tuning to target coordinates.

Anderson's group claims that for many neurons this angle-of-gaze effect can largely be described by gain factors, which are a function of eye position, multiplied by the response profile of the retinal receptive fields. This characterization is similar to the formulation of the model where the motor inputs to the target map of the model are the product of eye position and visual input.

Moreover, the model predicts columns of visually gated muscle representations in an overall retinal topography. Each column should contain all the eye muscle representations measured in terms of optimal gaze directions. It is important to note that even though the model depicts columns of motor representations, it predicts that only visual responses will be measured because all the motor representations are visually gated. This means that without visual input, the motor inputs will not cause any measurable responses.

All these predictions can be tested experimentally. If a microelectrode is penetrated fairly parallel to the cortical layers, one should find visually gated, gaze-related neural responses that change in two ways through the penetration. First, neighboring neurons should represent components of gaze in the direction of the six eye muscles. This might occur in some order. Second, this order should cycle over some distance. There should be differences in the relative activity observed across cycles. This is due to changes in the topographic position of the visual gating.

So far, discussion of the model has dealt with point targets. How will the model react to a distribution light contrast? First, the visual inputs that reach the target map have already been extensively preprocessed. Not only are they transformed from intensity information to contrast information, but they may have already been gated by habituation and processed for subjective saliency. As a result, the original distribution of light intensity may be reduced to a small number of targets. Perhaps the number of targets may range from one up to, potentially, the limit of short-term memory, around seven.

With no further processing, the model will yield the "center of mass" of parallel targets. This is a weighted average of all the target positions, where the weights are determined by the saliency of the contrast information in the visual field. Such a result may have some value in exploring the visual environment, where it is important to orient to new clusters of light contrast that make up new objects.

However, a different use of a target map becomes very beneficial when there is more than just one target to orient to. Consider the problem of orienting to a number of light targets in sequence from a presentation of targets that turns off before any movement. Hallet and Lightstone (1976) presented human subjects with two light targets in sequence before any eye movement. With both lights off, subjects correctly oriented to the spatial locations of the targets in sequence. If only retinal information is used to orient to each light, a problem arises after the first orientation. Because the retina moves in orienting to the first light, the retinal registration for the second light can no longer be used to compute the orientation to that light.

Somehow, appropriate vector combinations of the orbital eye position and the visual maps of the lights are computed for orienting to consecutive lights. The target map yields such a vector computation because it gives an invariant representation of a target independent of the combination of retinal and eye positions used to achieve a target.

There is still the problem of how to manipulate more than one light at a time. The model suggests a way to resolve this problem. If the outputs of the target map could be gated by all the visual distributions of the targets independently, then target positions of each target could be read out one at a time.

The model presented here is designed to be extended in two ways. One extension is to represent targets by combining more sensory maps, and the other is to represent targets for more than one joint. The first extension is easily made. The case of more sensory inputs to the target map is exemplified when targets are represented with the freely moving head. Here the invariant representation of a target depends on the retinal position, eye orbital position, and head position when the target is registered. The head position is determined either by vestibular or neck-muscle inputs.

The target map, as fomulated here, can accommodate as many input maps as are needed to determine target position, as long as each input measure is a monotonic function of the target location. Each input map would still be interleaved with all other inputs. The topography would still be globally retinal. In the case of a free-moving head, each column in the target map would contain representations from not only all eye muscles but also all neck muscles and/or all vestibular canals. The system would likely converge through learning for two reasons. First, all individual sensory inputs still have a monotonic relation to target displacement. This ensures that there will not be any ambiguous or contradictory target representation for any combination of inputs. Second, the principle of shared incremental learning, previously discussed, applies not only to two inputs but to any number of inputs. The usefulness of this target map may be extended to arm and leg joints as I will describe in the next section.

A. REACHING STATIONARY TARGETS WITH MULTIPLE JOINTS

This section addresses the problem controlling a multijoint arm to reach objects in three-dimensional space (detailed in Kuperstein, 1988b; Kuperstein and Rubinstein, 1989).

The design of the neural controller called INFANT that adaptively controls the arm has a number of constraints. First, the controller should contain only minimal information about the physical plant or actuator characteristics. To the extent it can be independent of any physical plant, it can be generically applied to all plants.

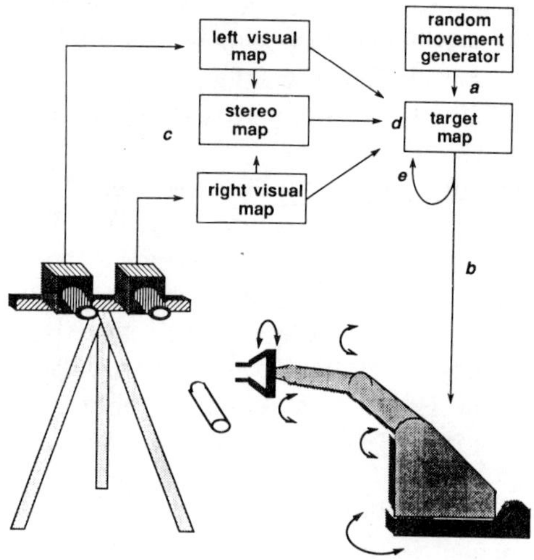

FIGURE 7. The circular reaction. Self-produced motor signals that manipulate an object target are correlated with target sensation signals. The sequence for training is a, b, c, d, e. Correlated learning is done in step e. After the correlation is achieved, target sensation signals alone can evoke the associated motor signals to reach the target accurately. The sequence for performance is c, d, b.

The controller must learn the association between the stereo image of an object and the pattern of motor signals to all the joints. Because an object can be at any location in three-dimensional space and at any orientation, the controller must have some way of processing three-dimensional information from stereo views of the object.

The controller computation must be parallel so that it can be generalized to many joints without increasing processing time.

The INFANT controller is designed to learn self-consistency between sensory and motor signals without supervision. In general, it operates in two phases in a process called *circular reaction*. The current use of circular reaction comes from one of the developmental stages of infant development (Piaget, 1952). In the first phase, sensory-motor relations are learned via correlations between object sensation and self-produced movement signals. In the second phase, the system uses the learned correlations to transform the sensation of an object into a movement that reaches the object.

The schematic diagram in Figure 7 shows the major modules that embody the INFANT controller. During learning in the first phase, a random movement generator first produces random postures of the multijoint arm in space, while the gripper holds an object such as a cylinder. Then the stereo cameras snap an image of the arm holding the object. These images are transformed into neural input maps (left and right visual maps). A third input map (stereo map) is produced as a disparity between the left and right visual maps. The signals coming from the input maps are modulated by a set of weight signals to produce target signals. The weights constitute the global association between all possible images of an object and the arm motor signals.

The target map outputs are compared with the random outputs that generated the arm posture to produce error signals. These error signals are then used to update the weight signals, which improves the reaching performance.

During performance in the second phase, the object is located somewhere free in space. The sequence of transformations from the stereo cameras to the target map is reiterated as in the learning phase. The resulting target map outputs then drive the arm actuators. Because of the learned associations between the object images and the arm motor signals, the hand will superimpose on the object.

Control performance is achieved without sensory feedback and with almost no knowledge of the spatial relations of the mechanical system. The controller only knows how many arm joints there are and the limits of their ranges.

There are a number of methods of transforming visual scenes to neural inputs that will yield similar results. One method is to transform visual scenes into topographic maps like the discussion in the last section. Then a stereo disparity map can be computed by the difference between the left and right visual maps that is also gated by both maps. This method is most brain-like but requires extensive computer power to implement (Kuperstein, 1988b).

Another method is to transform visual scenes into parameters that summarize the location for the center of mass of visual contrast in each scene along with their disparity. Such parameters cannot be used effectively as direct neural inputs because simply attaching neural weights to them would not allow sufficient nonlinearity in the global sensory-motor mapping. Therefore, the parameters can be further transformed into topographic maps in which the peak of a unimodal activity distribution moves across a topographic field in proportion to the amplitude of the parameter. Such a transformation was successfully used to achieve accurate reaching performance (Kuperstein and Rubinstein, 1989).

Even after extensive learning and accurate performance, environmental parameters can change in the real world. To maintain accurate performance in the face of unpredictable changes in the geometry of the physical plant, additional learning can occur. This learning is based on errors derived from visual feedback after an attempted grasp.

To illustrate this type of learning, suppose the gripper misses the target in an attempt to grasp a ball. In this case, the visual mismatch between the position of the ball and the position of the gripper is the source of an error. How does a visual source of error generate a motor correction without any prewired sensory-motor transformation? One way is for the visual position of the gripper to get transformed by the current learned sensory-motor map into a new motor target. The difference between the motor target for the ball and the motor target for the gripper results in a motor error which can be used to modify the weights affected by the ball visual input. Such an error changes the weights in the correct direction. The updated weights represent an improved correspondence between where the ball is seen with where the gripper should move. After the weights are changed, the system can attempt to reach the ball again. This cycle of grasp attempt and weight change repeats until the visual position error between the ball and the gripper lies within a specified accuracy. In practical experiments, the average number of improvement cycles is two or three trials.

Such a sensory-based learning scheme was also considered and then rejected as an alternate method for learning the fundamental sensory-motor coordination from the beginning. It would require prior knowledge of the correspondence between sensory errors and motor errors which does not satisfy one of the initial constraints. The self-consistent learning scheme developed from the circular reaction avoids this requirement and, thus, allows motor learning from any relation between the cameras and the arm. Once a unified, global association between sensation and posture is learned, the association can be used to define the correspondence between sensory errors and motor errors.

An earlier version of the INFANT controller was simulated for the physical plant shown in Figure 1 (Kuperstein, 1988b). This included not only a 5 degree-of-freedom arm, but also two movable eyes on a stationary head. In this case, the activity from six muscles each on two eyes was included as another neural input to the target map. Reaching accuracy averaged 4% of the arm's length in position and 4° of solid angle in orientation.

The INFANT controller has been implemented on an industrial robot arm and stereo cameras (shown in Figure 8), and a commercial image processing system. The implementation was achieved by developing software that embodies the neural network transformations

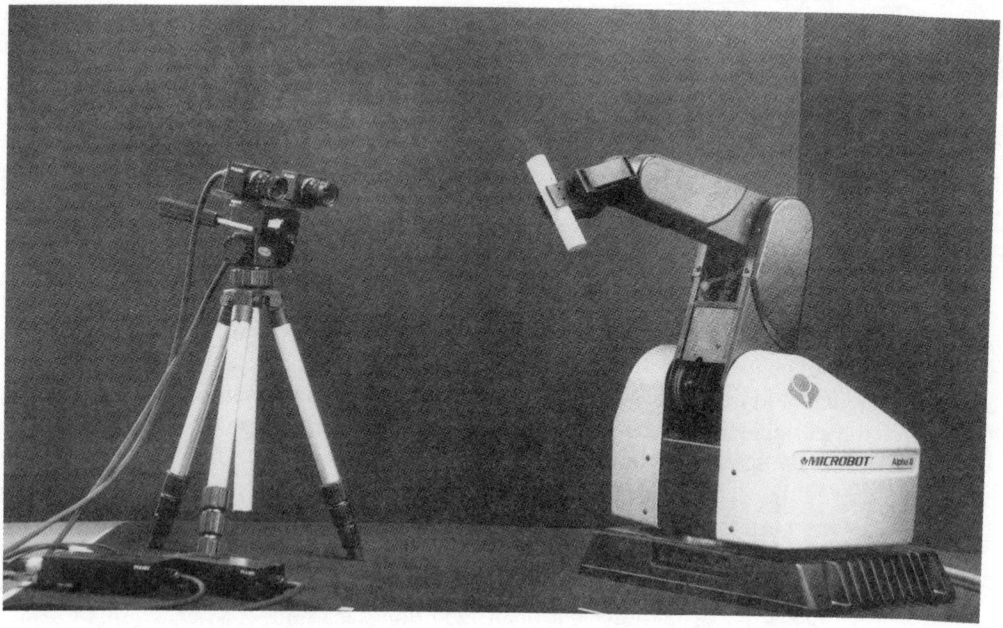

FIGURE 8. The 5 degree-of-freedom arm and stereo camera system.

into an image processor (detailed in Kuperstein and Rubinstein, 1989). The controller was able to reach accurately a cylinder that was arbitrarily positioned anywhere in space within arm's reach. It could not only reach the cylinder that it was trained on, but could also accurately reach many other elongated objects, including cylinders of different diameters, lengths, and visual contrasts, as well as a piece of paper that was rolled into an irregular elongated form. Performance accuracy for reaching a cylinder target after 1200 learning trials averaged 3% of the length of the arm. The average angular deviation between intended gripper wrist orientation and the actual cylinder orientation was 6° of solid angle.

Stability of the system was tested by relearning whole runs with different arm and camera positions and a range of values for neural parameters, including neural map width, learning error rate, and normalization constants. The controller showed very similar error convergence for a large range of parameter choices.

There are five key conclusions from this section:

1. The INFANT controller learns about space by associating how the arm moves with what the cameras see. It uses almost no knowledge of the geometry of the mechanical system, and there are no inverse computations nor control transfer functions.
2. Learning is achieved without any external source of error correction. This allows the controller to be completely independent of supervision. The learning for one elongated object transfers to any elongated object. Moreover, no calibration for binocular parallax is required.
3. The implementation degrades gracefully to hardware errors.
4. The INFANT controller adapts to unforeseen changes in the geometry of the motor plant.
5. The algorithmic flow of control is both feedforward and parallel during performance, which will ultimately make it fast with parallel hardware.

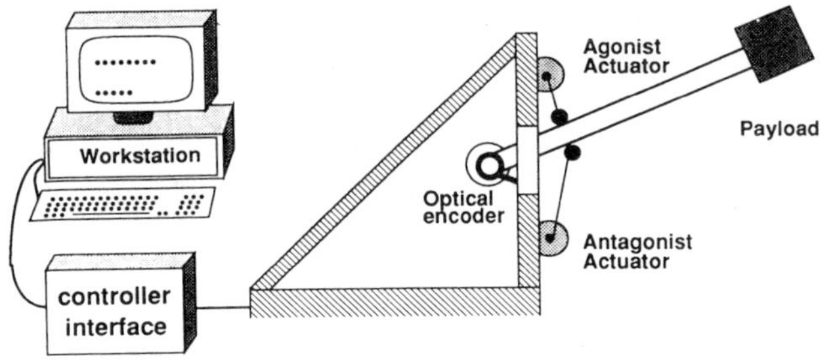

FIGURE 9. Schematic of the single joint plant.

Further work needs to done in the following areas: perceiving one object from many in a scene; perceiving occluded objects; adaptively reaching moving objects; and extending the INFANT controller to multiple limbs and human-like hands.

III. DYNAMIC MOVEMENT CONTROL FOR UNFORSEEN PAYLOADS

In this section, I describe a neural controller that learns to move and position accurately a single joint link carrying an unforeseen payload (extended from Kuperstein and Wang, 1990). This problem was physically implemented with two antagonistic torque motors pulling on a link with tendons. A schematic of the setup is shown in Figure 9.

The standard approach for a controller to move a novel payload is to program a velocity or acceleration ramp at the beginning and end of each movement. The ramp is calibrated to move a worst-case payload. This means that movements for most payloads are much slower than they need to be. Such inefficiency is exacerbated by arbitrarily directed movements in three-dimensional space. A neural controller learns near optimal dynamics for moving unforeseen payloads in three-dimensional space.

There are a number of constraints in this problem that should be satisfied for a generic controller to be useful. First, a lightweight robot should be able to handle the dynamic inertial interactions that it will encounter with real world payloads, especially when the payloads are larger than the mass of the robot. Payloads can vary considerably between successive movements. A generic controller should be able to measure payload mass indirectly by some uncalibrated means.

To improve positioning performance, kinematic feedback sensors need to measure the length, velocity, and acceleration of each actuator tendon. A generic controller should not need these measures to be calibrated, even when they can change over time.

In animals and some robot systems, there is some fixed arbitrary delay between torque onset and the effect of sensor feedback on torque generation. In some cases, this feedback delay can be a considerable fraction of the movement duration. These delays can be caused by either slow sensor feedback transmission or slow movement generation. If this delay is not adaptively anticipated, the controller will become unstable. In general, a generic controller should not need to know the phase relationships between output torque dynamics and input sensor dynamics. Although feedback measurements should have accurate repeatability over short time ranges, such as hours, they are allowed to drift over longer time ranges, such as days, and should not need to be calibrated to any objective metric.

A generic controller should learn and maintain accurate performance, even after unpredictable changes are made in either the geometrical, mechanical, torque, or sensing parameters or from partial internal damage. Learning should be achieved without operator assistance. Changes in the following parameters, which could occur during extended wear, should be accommodated accurately:

1. Transformation of actuator signals to actuator movement
2. Sensing feedback calibrations from the actuators and payload
3. Sensor feedback delays
4. Link lengths
5. Signal noise
6. Partial internal processor faults

The design of the neural network that adaptively controls an actuator to move unforeseen payloads has a number of features that satisfy the constraints of a generic controller. First, it does not contain information about the plant or actuator characteristics, except for the number of inputs, outputs, and their range of activation. The less specific information the controller requires about any one plant, the more useful and flexible it will be for many plants.

The neural controller learns to complete a stable movement by modifying the amplitudes of the thrusting burst and braking burst so that there are no oscillations when posture is achieved.

The neural network computation is parallel so that it can be generalized to many joints without increasing processing time.

The implementation of the dynamic neural controller is embodied in a single joint link that is moved by two strings connected to antagonistic torque motors. This muscle analogy was chosen for the implementation to allow variable stiffness during posture. The link is 18 in. long and the motors provide 85 oz-in. of torque at 3000 rpm. An optical encoder is attached to the link pivot shaft and is used to provide position and velocity information. A commercial PC-AT controller computer board outputs controls to the motor amplifiers and reads the optical encoder. It resides in a PC-AT computer which hosts the operations and the neural network.

The basic design for the neural controller is shown in Figure 10. It is made to interface with the output of the INFANT controller for adaptive sensory-motor coordination discussed in the last section. The input control commands of initial position, intended position step, and measure of payload are first transformed into neural input maps. They are laid out in two-dimensional topographic input representations. The locations of the peaks of Gaussian distributions in each dimension of the representation vary according to the amplitude of the commands. Thus, for input N_{ij}, the index i' is proportional to the initial position, and peak index j' is proportional to the desired position step:

$$N_{ij} = \exp\{-\lambda \cdot [(i - i')^2 + (j - j')^2]\} \text{ for } i = 1, 2 \cdots I; j = 1, 2 \cdots J \quad (1)$$

where λ is a constant for the Gaussian width and I and J are the network population size.

Each neural output torque signal T_k is computed to be the sum of its inputs times their neural weights W_{ijk}:

$$T_k = \Sigma_{ij} N_{ij} \cdot W_{ijk} \text{ for } i = 1, 2 \cdots I; j = 1, 2 \cdots J; k = 1,2 \quad (2)$$

where k is the actuator index. The rectified torques operate on the link over time t by combining the neural output T_k, desired position D_k, current position $C_k(t)$ and torque constant α, after some sampling delay Δ:

$$\tau_k(t + \Delta) = \max[a \cdot (T_k + D_k - C_k(t)), 0] \quad (3)$$

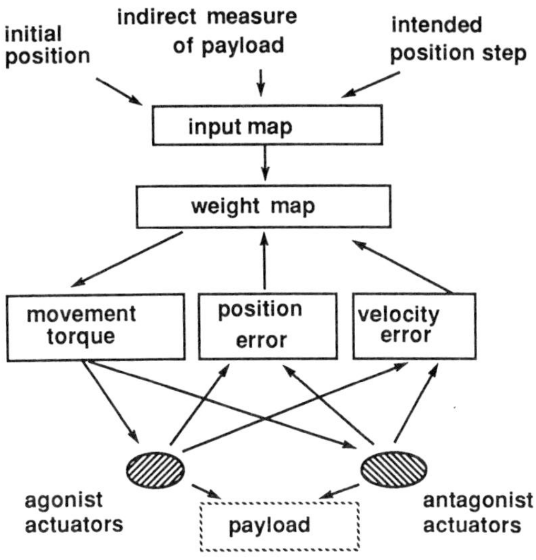

FIGURE 10. Connection scheme of the neural controller. The input command maps are transformed into outputs of actuator control. Resulting performance errors in joint angular position and velocity are used to modify the weight maps, gated by the input maps.

The antagonistic torques are applied to the link until one of two conditions is reached. If the desired position is reached, the velocity error for each actuator is sensed. If the desired position is not reached within a time limit, the position error for each actuator is sensed. These errors, E_{kn}, (k is actuator index and n is error type) update the modifiable weights. When the movement is finished, antagonistic postural signals P_k are instated:

$$\tau_k(t + \Delta) = \max[\beta \cdot (P_k + D_k - C_k(t)), 0] \tag{4}$$

where β is a measure of isometric tension at posture. The weights W_{ijk} associated with the inputs N_{ij} are then changed by the errors E_{kn} gated by a learning rate δ:

$$\Delta W_{ijk} = \delta \cdot N_{ij} \cdot \sum_n E_{kn} \tag{5}$$

Even though learning is piece-wise linear, the network yields the required emergent nonlinear properties of the transformation as a population.

The adaptive feedforward control process is achieved without knowledge of the actuator characteristics, except for its activation range. Note that in more complicated problems where additional parameters of speed, power, or isometric tension are required, more feedforward gains can be applied in parallel. Also, in some situations where adaptive compliance is required, more force-feedback circuits would be used.

The payload is assumed to be constant within a single movement, but variable between movements. The measure of payload only needs to be a monotonic function of actual payload weight which can be determined by a number of means. In this implementation, the agonist actuator gradually increased its torque until the unknown payload began to move. The final torque is a measure of the payload.

The neural controller is tested for stability by analyzing its performance to unforeseen parameter changes. Unexpected changes in payload and joint viscosity are applied, as well as changes in the actuator response characteristics and in the internal parameters of learning.

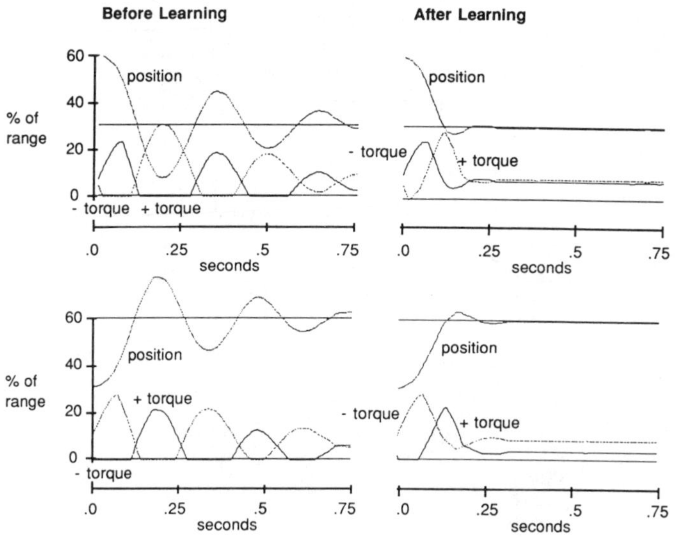

FIGURE 11. Performance trials for dynamic positioning of arm before and after learning. The horizontal lines show the intended position. The source of the difference is that the braking signal occurs earlier after learning. One sample per trial of angular position and velocity errors is used to improve subsequent performance by changing the neural weights.

The errors determine the adaptive goal of the control, which in this case is to end the movement at the intended position with no velocity. It is not necessary to calibrate exactly the movement error. The effects of the error signals are incremental and, therefore, progressively improve the accuracy of the movements. It is only required that the error signal should change the gain in the right direction after a given movement.

Figure 11 shows examples of performance trials for different initial positions, different intended positions, before and after learning. Note that for this problem, both the agonistic and antagonistic torque signals could have flowed to one actuator with a stiff push-pull link. However, the two-actuator design was created to prepare for future problems of adaptive compliance and stiffness during posture. Those problems would require different levels of isometric tension from antagonistic actuators.

As shown in Figure 12, the average movement achieved its intended position within 5% of the total joint angle range without significant oscillations after about 600 trials. This performance was achieved where each trial had an arbitrary initial position and arbitrary intended position.

Whole learning runs that were done with different link masses, direction of gravity, joint viscosity, and nonlinear torque-to-actuator transformations showed very similar error convergence.

This design concept for the neural controller is generic because it only requires minimal knowledge of the physical plant or actuator characteristics and is designed to be extended to an arbitrarily large number of joints. The overall flow of control is fedforward during movement, which makes it fast. Each new joint in the robot system will require adding new neural input and weight maps to the architecture. The next challenge is to understand how this type of generic architecture will perform with the highly nonlinear coupling that occurs between multiple joints.

The population of units in each weight map is arbitrary, and increasing the number should increase the resolution of performance.

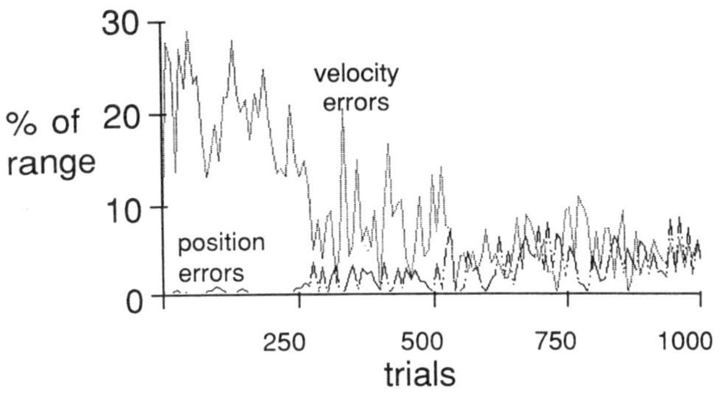

FIGURE 12. Performance errors during learning. Error asymptote of 5% of full range is reached after 600 trials.

This neural controller has three major properties that other dynamic neural controllers do not:

1. It is designed to move accurately an unforeseen payload to arbitrary targets without end-point oscillations, using only one error sample per movement.
2. It can accurately anticipate the effects of large payload inertias and/or delayed sensory feedback. Without anticipation, these effects can cause inappropriate deviations or oscillations during both movement and posture.
3. It is designed to be the basis of incorporating dynamic control for an arbitrary number of joints without changing the structure or slowing the speed of the computation.

All three neural controllers discussed in this review have a number of application benefits. They will deal effectively in novel working environments, such as in space, because of their ability to deal with unforeseen changes in the mechanical plant and actuators. Their adaptability will allow continuous selfcalibration in the face of mechanical wear, and its generic design will allow it to be implemented in many different types of robots. The parallel feedforward control architecture will make robot control very fast and the modifiable neural weights with overlapping functionality will allow fault tolerance. This will greatly reduce tooling costs, setup time, and task failure in autonomous robot operations.

Hypotheses on neural controllers put forward here can also be tested by researchers for their validity in the brain. Combining both the approaches of brain analysis and machine synthesis will bring us more understanding about human sensory-motor coordination than either approach alone.

ACKNOWLEDGMENTS

This work was made possible by grants from the National Science Foundation, Air Force Office of Scientific Research, and the Office of Naval Research.

REFERENCES

Albus, J. (1975). A new approach to manipulator control: the cerebellar model articulation controller (CMAC), *J. Dyn. Syst. Meas. Control,* **37(3)**, 220–227.

Anderson, R. A., Essick, G. K., and Siegel, R. M. (1985). Encoding of spatial location by posterior parietal neurons, *Science,* **230**, 456–458.

Bolles, R. and Roth, B. (Eds.) (1988). *Robotics research: the fourth international symposium,* MIT Press, Cambridge.

Critchley, M. (1982). *The parietal lobes,* Hafner Press, New York.

Grossberg, S. and Kuperstein, M. (1986/1989). *Neural Dynamics of Adaptive Sensory-Motor Control,* Elsevier/North Holland, Amsterdam (Reprinted by Pergamon Press, New York, 1989).

Hallet, P. E. and Lightstone, A. D. (1976). Saccadic eye movements to flashed targets, *Vision Res.,* **16**, 107–114.

Hein, A. (1970). Recovering spatial motor coordination after visual cortex lesion, in *Perception and its Disorders,* Hamburg, D. A., Pribram, K. H., and Stunkard, A. J., Eds., Williams and Wilkins, Baltimore, 163–175.

Hein, A. (1972). Acquiring components of visually guided behavior, in *Minnesota symposia on child development,* **6**, Pick, A., Ed., University of Minnesota, Minneapolis, 53–68.

Hein, A. (1974). Prerequisite for development of visually guided reaching in the kitten, *Brain Res.,* **71**, 259–263.

Hein, A. and Held, R. (1962). A neural model for labile sensorimotor coordinations, in *Biological Prototypes and Synthetic Systems 1.* Bernard, E. E. and Kare, M. R., Eds., Plenum Press, New York, 71–74.

Held, R. and Hein, A. (1963). Movement-produced stimulation in the development of visually guided behavior, *J. Comp. Physiol. Psychol.,* **56**, 872.

Hubel, D. H. and Wiesel, T. N. (1962). Receptive fields, binocular interaction and functional architecture in the cat's visual cortex, *J. Physiol.,* **160**, 106–154.

Jordan, M. I. (1989, June). Generic constraints on underspecified target trajectories. International Joint Conference on Neural Networks, Washington, D.C.

Kandel, E. R. and Schwartz, J. H. (1985). *Principles of Neural Science,* Elsevier, New York.

Kawato, M., Uno, Y., Isobe, M., and Suzuki, R. (1988). Hierarchical neural network model for voluntary movement with application to robotics, *IEEE Control Syst. Mag.,* **8**(2), 8–16.

Kuperstein, M. (1987, March). Adaptive visual-motor coordination in multijoint robots using parallel architecture, in Proc. IEEE Int. Conf. Automat. Robotics, Raleigh, NC, 1595–1602.

Kuperstein, M. (1988a). An adaptive neural model for mapping invariant target position, Behav. Neurosci., **102**(1), 148–162.

Kuperstein, M. (1988b). Neural network model for adaptive hand-eye coordination for single postures, *Science,* **239**, 1308–1311.

Kuperstein, M. and Rubinstein, J. (1989). Implementation of an adaptive neural controller for sensory motor coordination, *IEEE Control Syst. Mag.,* **V9**(3), 25–30.

Kuperstein, M. and Wang, J. (1990). Neural controller for adaptive movements with unforeseen payloads, *IEEE Trans. Neural Networks,* **VI**(1), 137–142.

Liberman, A. M., Cooper, F. S., Shankweiler, D. P., and Studdert-Kennedy, M. (1967). Perception of the speech code, *Psychol. Rev.,* **74**, 431.

Luschei, E. S. and Fuchs, A. F. (1972). Activity of brain stem neurons during eye movements of alert monkeys, *J. Neurophysiol.,* **35**, 455–461.

Miller, W. T. (1987). Sensor-based control of robotic manipulators using a general learning algorithm, *IEEE J. Robotics Automation,* **RA-3**(2), 157–165.

Miller, W. T. (1989). Real-time application of neural networks for sensor-based control of robots with vision, *IEEE Trans. Syst. Man Cybern.,* **19**(4), 825–831.

Miyamoto, H., Kawato, M., Setoyama, T., and Suzuki, R. (1988). Feedback-error-learning neural network for trajectory control of a robotic manipulator, *Neural Networks,* **1**, 251–265.

Piaget, J. (1952). *The Origins of Intelligence in Children,* (M. Cook, Trans.). International University Press, New York.

Sakata, H., Shibutani, H., and Kawano, K. (1980). Spatial properties of visual fixation neurons in posterior parietal association cortex of the monkey, *J. Neurophysiol.,* **43**, 1654–1671.

Weber, R. and Daroff, R. (1972). Corrective movements following refixation saccades, types, and control system analysis, *Vision Res.,* **12**, 467–475.

Widrow, B. and Smith, F. (1963). Pattern recognizing control systems, in *Computer and Information Sciences (COINS) Symposium,* Spartan Brooks, Washington, D.C.

Williams, G. and Nottebohm, F. (1985). Auditory responses in avian vocal motor neurons: a motor theory for song perception in birds, *Science,* **229**, 279.

APPENDIX

For robustness, all the neural controllers discussed in this review use essentially the same processing and learning rules. Each individual neural output C_j is computed to be either the sum or the multiplication of its inputs N_i via their neural weights W_{ij}. For example:

$$C_j = \Sigma_i N_i \cdot W_{ij} \text{ for } i = 1, 2 \cdots I; j = 1, 2 \cdots J \quad (1)$$

Errors in the system, E_j, are typically the difference between the desired or actual outputs, D_j, and the computed outputs C_j:

$$E_j = D_j - C_j \quad (2)$$

The weights W_{ij} associated with the inputs N_i are then changed by the errors E_j gated by a learning rate r:

$$\Delta W_{ij} = r \cdot N_i \cdot E_j \quad (3)$$

Chapter 21

A NEURAL NETWORK ARCHITECTURE FOR FIGURE-GROUND SEPARATION OF CONNECTED SCENIC FIGURES*

Stephen Grossberg and Lonce Wyse

TABLE OF CONTENTS

I. Introduction: Automatic Figure-Ground Separation and General-Purpose Vision 516

II. Theoretical Background 516

III. Figure-Ground Separation by Artificial Detectors: Laser Radar Arrays 518

IV. Figure-Ground Separation in FACADE Theory 519

V. Discounting Variable Illumination and Filling In 519

VI. Multidimensional Fusion 521

VII. Figure-Ground Separation by a Monochromatic FBF Network: the Dye-Injected FBF 525
- A. Step 1: Discount the Illuminant 525
- B. Step 2: CORT-X Filter 525
- C. Step 3: Filling In 530
- D. Step 4: Figure-Ground Separation 532

VIII. FBF Network Equations 535
- A. Input Images 535
  - 1. Step 1: Discount the Illuminant 535
    - a. ON-C Network 535
    - b. OFF-C Network 536
  - 2. Step 2: CORT-X Filter 537
    - a. Simple Cells 538
    - b. Complex Cells 538
    - c. Hypercomplex Cells: First Competitive Stage 539
    - d. Hypercomplex Cells: Second Competitive Stage 539
    - e. Multiple Scale Interaction: Boundary Localization and Noise Suppression 539
    - f. Long-Range Cooperation: Boundary Completion 540
    - g. CORT-X Output 540
  - 3. Step 3: Filling In 540
  - 4. Step 4: Figure-Ground Separation 541

IX. Concluding Remarks: Recognition of Conjunctive Features 541

* Reprinted with permission from *Neural Networks*, S. Grossberg and L. Wyse, A Neural Network Architecture for Figure-Ground Separation of Connected Scenic Figures, © Pergamon Press plc.

Acknowledgments 542

References 542

Appendix 543

I. INTRODUCTION: AUTOMATIC FIGURE-GROUND SEPARATION AND GENERAL-PURPOSE VISION

An important stage in the perception and recognition of objects is the process whereby a figure, or object, in a scene is separated from other figures and background clutter. This is called the stage of figure-ground separation. Whereas knowledge about a figure may facilitate its separation, such knowledge is clearly not *necessary* for biological vision systems to carry out figure-ground separation. Experiences abound of unfamiliar figures that "pop out" from their backgrounds before they ever enter our corpus of learned knowledge about the world. The fact that figure-ground separation can occur even for unfamiliar figures contributes to the general-purpose nature of biological vision, which can process both unfamiliar and familiar scenes, and does not require prior instruction about an environment in order to operate effectively.

The present article describes a new type of system that is capable of automatic figure-ground separation. This process separates scenic figures whose emergent boundary segmentations (defined below) surround a connected region. As a result of this property, such a system can automatically distinguish between connected and disconnected spirals (Figure 1), a benchmark that gained fame through its emphasis in the book by Minsky and Papert (1969, 1988) on perceptrons. Why the present biologically motivated algorithm can distinguish these figures in a way that humans cannot is discussed in Section IX.

II. THEORETICAL BACKGROUND

This article contributes to the development of a self-organizing neural network architecture for invariant pattern recognition in a cluttered environment. Carpenter and Grossberg (1987a) described a version of this architecture (Figure 2). A related scheme is considered herein, whose primary functional stages are as follows:

Stage 1. Discount the illuminant.
Stage 2. Detect, regularize, and complete figure boundary; suppress interior and exterior image noise.
Stage 3. Detach figure from ground.
Stage 4. Filter to give invariance under translation, rotation, and contraction; for example, use a log-polar-Fourier filter (Cavanagh, 1978, 1984; Casasent and Psaltis, 1976; Szu, 1986).
Stage 5. Let invariant spectra of the boundary-enhanced, noise-suppressed, detached figures be the input patterns to an ART 2 architecture for stable self-organization of recognition categories (Carpenter and Grossberg, 1987b, 1988).

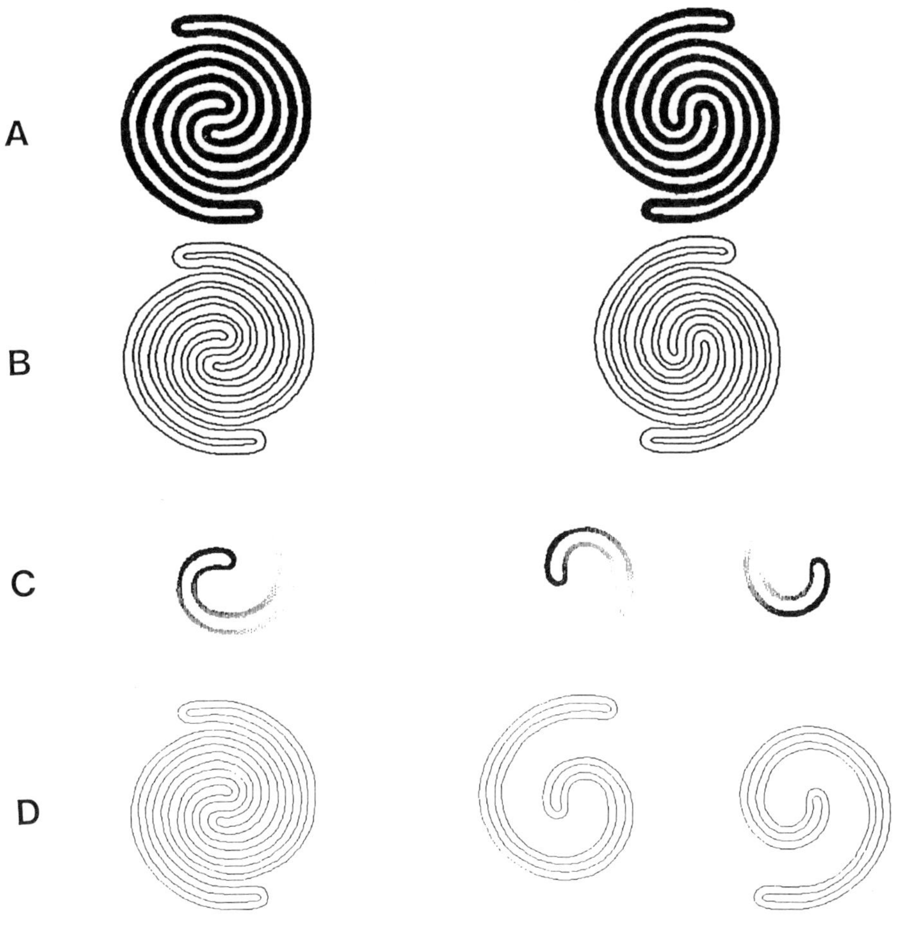

FIGURE 1. (A) Input images. The left image is connected. The right image consists of two disconnected components. (B) Boundary segmentations. (C) Beginning of filling in. (D) FBF outputs. The left boundary lies on a single FBF slab. Each connected component of the right boundary lies on a different slab.

In Carpenter et al. (1989), a neural network preprocessor for the second stage — the boundary segmentation stage — of such an architecture was described. This boundary preprocessor, called the CORT-X filter, detects, regularizes, and completes sharp (even one pixel wide) boundaries, while simultaneously suppressing the noise. The CORT-X filter is based on the biologically derived boundary contour system (BCS) of Grossberg and Mingolla (1985a, b). The BCS uses nonlinear feedback interactions to select and complete sharp boundaries over long gaps in image contours. The CORT-X model uses only feedforward interactions that are faster to simulate and easier to implement in hardware. Their ability to complete boundaries is also more limited, but is adequate for many applications. The CORT-X filter is herein modified to deal with at least 50% analog noise. This modified filter is called CORT-X 2 to distinguish it from the original model.

The present article uses this modification to develop a new approach to designing the third stage — the figure-ground separation stage — of such an architecture. At least three methods exist for solving this problem. Section III reviews a method for figure-ground separation that uses combinations of laser radars, or related artificial detectors (Figure 2b). Sections IV to VI review a method that arises in a neural network model of biological vision that has been called FACADE theory (Grossberg, 1987, 1990; Grossberg et al., 1989).

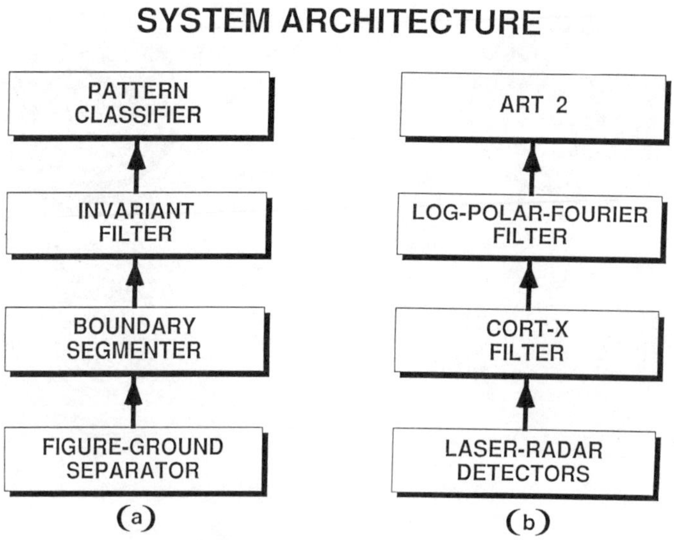

FIGURE 2. System architecture: (a) functional stages; (b) one computational realization. The present article describes a new model of figure-ground separation that uses a modified CORT-X filter boundary segmenter.

Section VII begins the exposition of the new model that we have developed, which is called an FBF model. The FBF model is capable of separating connected figures from their backgrounds in response either to monochromatic images, such as a gray-scale photograph, or from images derived from multiple detectors. Its mechanisms are based on those described in FACADE theory, which are adapted for use in a setting where only a single detector may be available. Readers may skip on a first reading directly to Section VII for an exposition of the new model.

III. FIGURE-GROUND SEPARATION BY ARTIFICIAL DETECTORS: LASER RADAR ARRAYS

Consider architectures wherein an invariant filter, such as a log-polar-Fourier filter, is used to achieve the capacity of the architecture for invariant pattern recognition. Then the filter output is unchanged by translations, rotations, and dilations of its input pattern. In such an architecture, the figure to be recognized must be separated from its background at a processing stage prior to the stage at which the invariant filter acts, as in Figure 2. If a fixed background were also processed along with a changing figure, then the output from the filter would also change, hence, would not be invariant. A conceptually simple technique is to utilize a detector which is itself capable of automatically separating figure from ground. Such an approach motivated the original architecture of Carpenter and Grossberg (1987a). There it was assumed that the detector consists of pairs of laser radars whose outputs are and-gated to separate figure from ground.

For example, a range detector focused at the distance of the figure can extract the figure and a contiguous piece of the ground. The figure can be detached from the ground by spatially intersecting the range pattern with a pattern from another detector that is capable of differentiating figure from ground. A Doppler image can be intersected with the range image when the figure is moving. The intensity of laser return can be intersected with the range image when the figure is stationary (Gschwendtner et al., 1983; Harney, 1980, 1981; Harney and Hull, 1980; Hull and Marcus, 1980; Kolodzy, 1987; Sullivan, 1980, 1981; Sullivan et al., 1979).

More generally, arrays of laser radar detectors may be used to separate image figures into distinct network levels, or slabs. This can be accomplished by intersecting the output signals from multiple detectors that simultaneously inspect the image. For example, a series of range detectors can register all objects at a regular series D, $2D$, $3D$, . . . , ND of distances, within some tolerance ΔD; a series of Doppler detectors can register all objects at a regular series S, $2S$, $3S$, . . . , MS of speeds, within some tolerance ΔS; and an $N \times M$ matrix of intersection images can be generated which extracts the figure at each combination of distance iD and speed jS within this tolerance. Then, $N \times M$ copies of the architecture illustrated in Figure 2 can work simultaneously in parallel to recognize these figures independent of their position, orientation, and size, by preprocessing each of the $N \times M$ laser radar images with CORT-X and invariant filters before they activate a parallel array of ART 2 architectures.

IV. FIGURE-GROUND SEPARATION IN FACADE THEORY

In biological vision, the retinal detectors do not, in themselves, separate figure from ground. One task of neural network research is to suggest how subsequent network processes which are activated by the retinal detectors may generate this competence. Grossberg (1987; reprinted in Grossberg, 1988) has, for example, introduced a neural theory of binocular vision in which figural components of an image are separated from one another into distinct network levels, or slabs. A macrocircuit of this theory is shown in Figure 3, where the vertically hatched boxes form part of the BCS and the dotted boxes form part of the feature contour system (FCS).

The theory describes how parallel and hierarchical interactions between the BCS and FCS generate a multiplexed, multiple-scale representation, called a FACADE representation, of the scene's Form-And-Color-And-DEpth. Within this representation, figural components which encode distinctive combinations of features, such as prescribed combinations of color, depth, and size, are segregated from one another into different network levels. These levels, in turn, activate subsequent stages of network processing that are designed for visual object recognition (Figure 4).

For present purposes, the main insight that may be derived from FACADE theory is that a properly designed sequence of FCS-BCS-FCS operations can separate figure from ground. Henceforth, all networks that use this strategy will be called *FBF networks*. To arrive at this insight, we consider two different competences of FACADE theory: discounting variable illumination and multidimensional fusion.

V. DISCOUNTING VARIABLE ILLUMINATION AND FILLING IN

The theory provides an explanation of how variable illumination conditions are automatically discounted and used to trigger a filling-in process that completes a surface representation over image regions which are suppressed by the discounting process. A monocular version of this process was modeled by Cohen and Grossberg (1984) and Grossberg and Todorović (1988) to explain data about monocular brightness perception. This monocular model is schematized in Figure 5.

In this model, variable illumination conditions are discounted by a shunting on-center off-surround network (Level 2), which constitutes the first FCS stage. Image regions of high relative contrast are amplified and regions of low relative contrast are attenuated as a consequence of the discounting process. The shunting network, in turn, topographically activates a filling-in network (Level 6) which constitutes the second FCS state. This filling-in network uses a nonlinear diffusion process to complete a brightness representation over both the amplified and attenuated image regions.

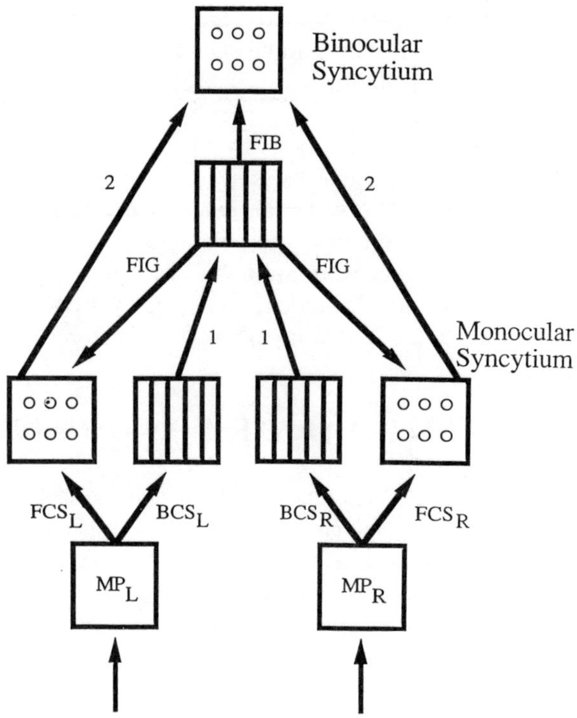

FIGURE 3. Macrocircuit of monocular and binocular interactions within the boundary contour system (BCS) and the feature contour system (FCS). Left and right monocular preprocessing stages (MP_L and MP_R) send parallel monocular inputs to the BCS (boxes with vertical lines) and the FCS (boxes with three pairs of circles). The monocular BCS_L and BCS_R interact via bottom-up pathways labeled 1 to generate a coherent binocular boundary segmentation. This segmentation generates output signals called filling-in generators (FIGs) and filling-in barriers (FIBs). The FIGs input to the monocular filling-in domains, or syncytia, of the FCS. The FIBs input to the binocular filling-in domains, or syncytia, of the FCS. Inputs from the MP stages interact with FIGs at the monocular syncytia where they select those monocular FC signals that are binocularly consistent. The selected FC signals are carried by the pathways labeled 2 to the binocular syncytia, where they interact with FIB signals from the BCS to generate a multiple-scale representation of form and color in depth within the binocular syncytia. The present article describes some monocular properties of the interactions from an MP stage through the first few BCS and FCS stages, namely, those symbolized by the pathways labeled 1 and FIG.

Filling in is restricted to compartments whose boundaries are defined by topographic signals from the BCS (Levels 2 to 5). The BCS converts signals from the first FCS stage (Level 2) into a boundary segmentation, one of whose functions is to trigger a BCS-FCS interaction that contains the filling-in process at the second FCS stage. The result of this FBF interaction is a surface representation of featural quality, such as brightness or color, that is relatively uncontaminated by illumination conditions.

Figures 6 and 7 summarize computer simulations of Grossberg and Todorović (1988) that illustrate how the illuminant is discounted in FCS Level 2, and how the subsequent BCS-FCS interaction at Level 6 controls the filling-in process that completes the brightness representation. The image schematized in Figure 6a is called a McCann-Mondrian (Land, 1977). It is a patchwork of rectangular regions, each with a different luminance. The image is uniformly illuminated. In Figure 6a, the radius of each circle is proportional to the luminance registered by a network node located at the center of the circle.

Figure 6b represents the activation pattern of the shunting on-center off-surround network at Level 2, Figure 6c represents the boundary representation at Level 5, and Figure 6d

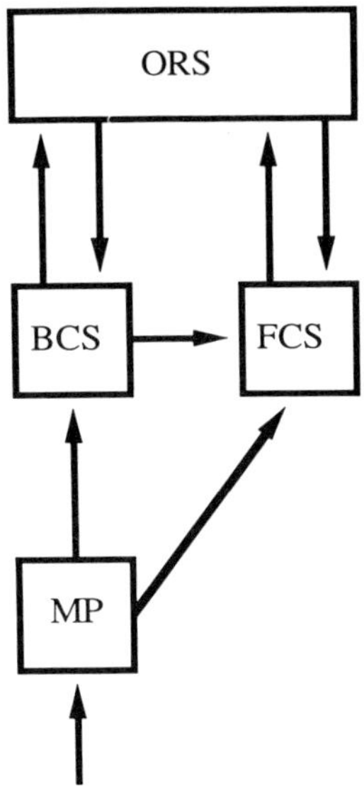

FIGURE 4. A macrocircuit of processing stages. Monocular preprocessed signals (MP) are sent independently to both the boundary contour system (BCS) and the feature contour system (FCS). The BCS preattentively generates coherent boundary structures from these MP signals. These structures send outputs to both the FCS and the object recognition system (ORS). The ORS, in turn, rapidly sends top-down learned template signals, or expectations, to the BCS. These template signals can modify the preattentively completed boundary structures using learned, attentive information. The BCS passes these modifications along to the FCS. The signals from the BCS organize the FCS into perceptual regions wherein filling in of visible brightnesses and colors can occur. This filling-in process is activated by signals from the MP stage. The completed FCS representation, in turn, also interacts with the ORS.

represents the filled-in representation at Level 6. The diffusion spatially averages the activation differences of Figure 6b within the compartments defined in Figure 6c.

In Figure 7a, the same image depicted in Figure 6a is illuminated from the lower right corner. Because the shunting on-center off-surround network at Level 2 effectively discounts the illuminant, the Level 2 activation patterns in Figures 7b and 6b are essentially identical. Hence, the subsequent boundary patterns (Figures 7c and 6c) and filled-in patterns (Figures 7d and 6d) are also essentially identical.

Note in Figure 7d that the brightness, or activation level, of the square region in the upper left corner is larger than that of the square region in the lower right corner. In contrast, in Figure 7a, the luminance, or activation level, of the upper left corner is smaller than that of the square region in the lower right corner. This luminance-to-brightness reversal compensates for the larger intensities of illumination in the lower right region.

VI. MULTIDIMENSIONAL FUSION

An FBF interaction may also be used to represent scenic form, notably scenic surface properties, and to separate figure from ground. This type of form and color fusion is achieved by suitably embedding an FBF interaction into a binocular version of the theory. In the binocular theory, depth is also encoded into the representation; hence the mnemonic Form-And-Color-And-DEpth, or FACADE.

The binocular version of the theory suggests how monocular image data from both eyes can be selectively processed so that only the binocularly consistent monocular data from each eye is allowed to influence the FACADE representation. Figure 3 schematizes the

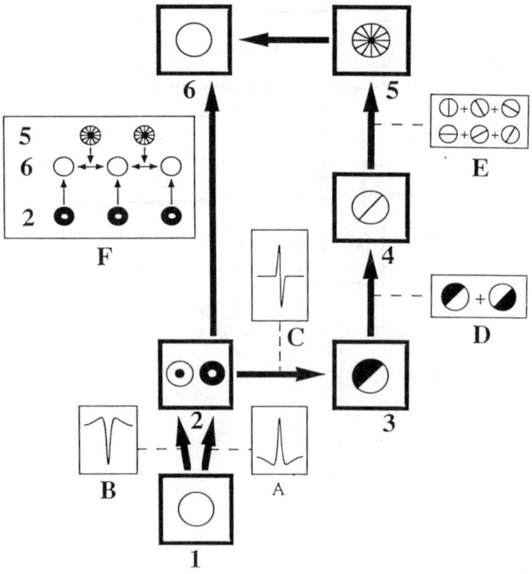

FIGURE 5. Model of how the feature contour system discounts variable illuminants and regulates featural filling-in. The thick-bordered rectangles numbered 1 to 6 correspond to the levels of the system. The symbols inside the rectangles are graphical mnemonics for the types of computational units residing at the corresponding model level. The arrows depict the interconnections between the levels. The thin-bordered rectangles coded by letters A through E represent the type of processing between pairs of levels. Inset F illustrates how the activity at Level 6 is modulated by outputs from Level 2 and Level 5. This simplified model directly extracts boundaries from image contrasts, rather than generating emergent segmentations from image contrasts. The key elements of the model concern how the Level 2 network of shunting on-center off-surround interactions discounts variable illuminants while extracting feature contour signals, and how Level 5 fills in these signals via a nonlinear diffusion process within the compartments defined by boundary contour system output signals.

network that is used. In it, pairs of monocular BCS signals are derived from monocular FCS patterns that discount the illuminant. These monocular BCS signals interact to form the binocular boundary segmentation along the pathways labeled 1. The boundary segmentation regularizes and completes all the boundary data, across multiple spatial scales, that are capable of being binocularly fused. This binocular boundary segmentation sends topographic signals, called filling-in barriers (FIBS), to the monocular filling-in networks within the FCS that are generated by the left eye (FCS_L) and the right eye (FCS_R). This BCS-FCS interaction allows only those monocular featural data from FCS_L and FCS_R that are consistent with the binocular boundary segmentation to fill in and generate topographic output signals, labeled 2, to the binocular *FCS* stage. This BCS-FCS interaction carries out a type of figure-ground separation, since only those FCS regions can generate output signals that are surrounded by binocular filling-in generators (FIGS) from the BCS.

The binocular FCS stage is called the binocular syncytium. In the binocular syncytium, the selected monocular FCS signals from both eyes interact once again with the binocular BCS signals. Here the FCS signals again activate a filling-in process within the compartments that are defined by the BCS signals. These BCS signals are thus called FIBS. The FACADE representation that is generated within the binocular syncytium groups distinctive combinations of features into figures within separate network levels, or slabs. These slabs, thereupon, send adaptively filtered signals to subsequent processing levels for purposes of visual object recognition.

Within such a biological theory of vision, the process of separating figures into different slabs exploits the fact that the retina contains photodetectors with different spectral sensi-

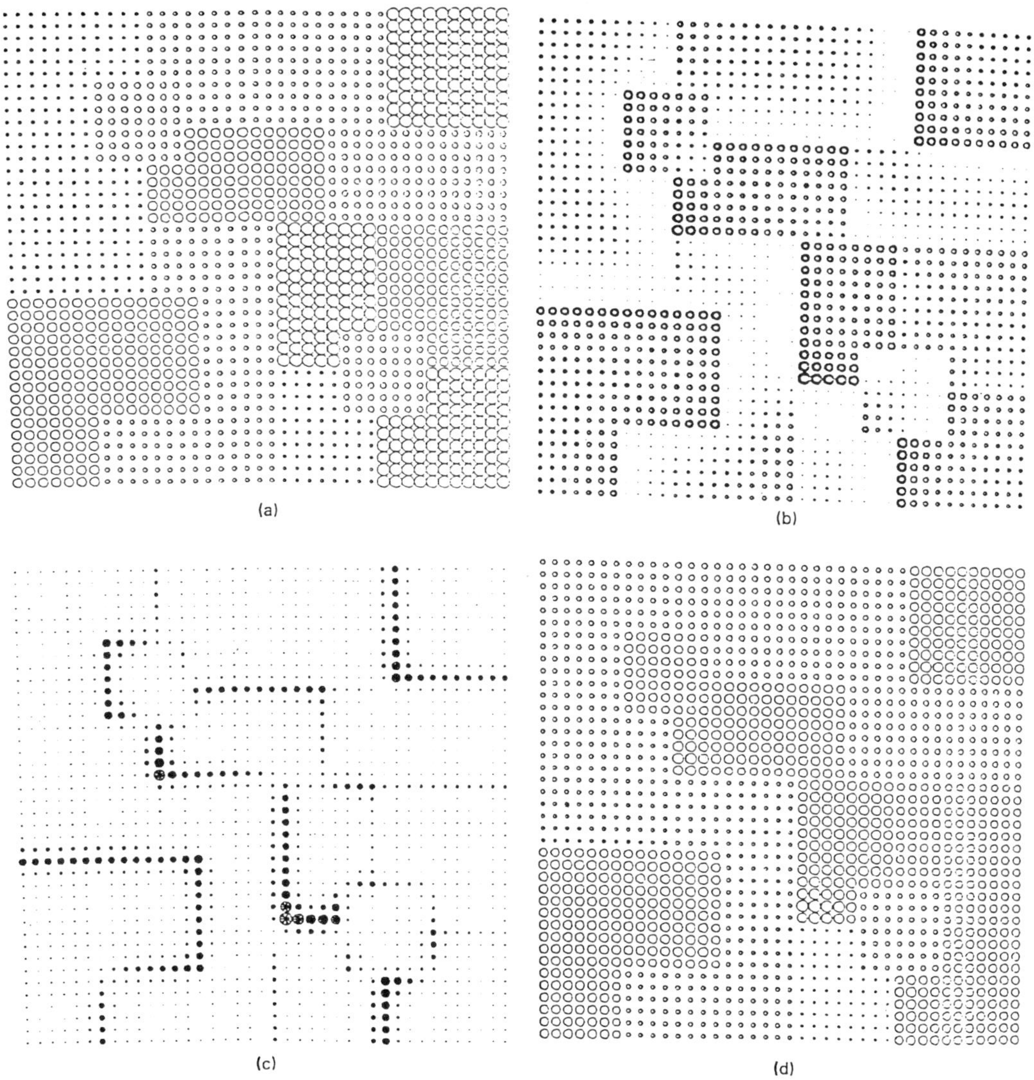

FIGURE 6. The evenly illuminated Mondrian. (a) The stimulus distribution consists of 13 homogeneous polygons with 4 luminance levels. Note that the square in the upper left portion of the stimulus has the same luminance as the square in the lower right portion. However, the average luminance of the regions surrounding the lower square is higher than the corresponding average luminance for the upper square. (b) The on-cell distribution. The amount of on-cell activity within the upper square is higher than within the lower square. (c) The boundary contour output. (d) The filled-in syncytium. The upper square is correctly predicted to look brighter than the lower square.

tivities; for example, three types of retinal cones and one type of retinal rod. The theory suggests how figures may be spatially parsed into separate slabs based, in part, on the distinctive colors that are derived from these detectors. In addition, there exist multiple spatial scales and multiple binocular disparity computations within the theory that further parse figural components into separate slabs based on different size-disparity correlations (Grossberg, 1987b; Grossberg and Marshall, 1989).

Thus, although FACADE theory uses an FBF network — actually an FBFBF network — to achieve figure-ground separation, this network exploits the existence of multiple detector types and multiple-scale reactions to these detector types to carry out the separation.

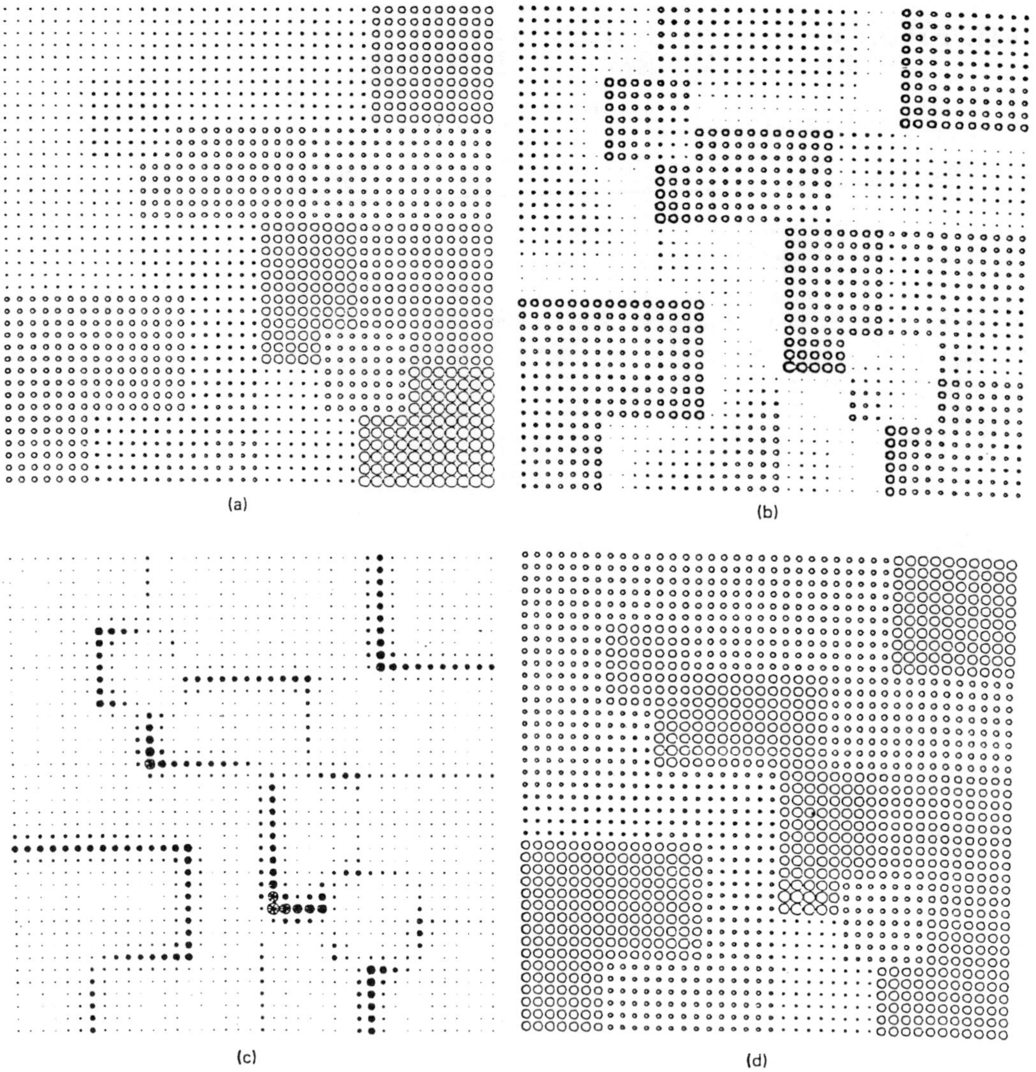

FIGURE 7. The unevenly illuminated Mondrian. (a) The stimulus distribution simulates the transformation of Figure 6a caused by the presence of a light source whose intensity decreases linearly from the lower right corner toward the upper left corner of the stimulus. The lower square is now more luminant than the upper square. (b) The on-cell distribution. (c) The boundary contour output. (d) The filled-in syncytium. Figures 7b to d are very similar to the corresponding figuresfor the evenly illuminated Mondrian (Figure 6). This illustrates the model's discounting of the illuminant. In addition, the upper square is still predicted to appear brighter than the lower square.

It remains to consider how well figural components may be separated into separate slabs even if only a single general-purpose detector is used that, in itself, cannot separate figure from ground. For example, how can individual figures be separated from the cluttered ground of a picture taken with a camera that uses monochromatic film? We now show how a suitably designed FBF network can accomplish this task for at least certain classes of images. Section VII provides an intuitive description of network stages and their effects. Section VIII describes network equations and parameters. These sections can be read in either order.

VII. FIGURE-GROUND SEPARATION BY A MONOCHROMATIC FBF NETWORK: THE DYE-INJECTED FBF

Because the multiple detectors are not available in the monochromatic case, we assume that the filling-in process is activated by *internally* generated input sources. In particular, the network "paints" each connected figure of the image by using an internally generated "dye" that triggers the filling-in of that figure. This heuristic is realized by using the following procedure, which was briefly reported in Grossberg and Wyse (1989).

A. STEP 1: DISCOUNT THE ILLUMINANT

At the first FCS stage, variable illumination conditions are discounted both by a shunting on-center/off-surround network (ON-C) and an off-center/on-surround network (OFF-C), operating in parallel. The ON-C network has a zero baseline activity (decays to zero if there is no signal within its entire receptive field), while the OFF-C network has a positive baseline activity. Because the OFF-C filter has a positive baseline activity and is inhibited by positive signal values, the network performs an image inversion which is normalized by the shunting interactions (Figure 8).

The ON-C and OFF-C networks operate in a complementary fashion. Along a straight boundary between a region of strong signal and one of no signal, both types of networks respond similarly by enhancing the contrast. At a concave corner of a high signal region, the ON-C network responds more strongly than the OFF-C network, while at a convex corner the converse is true (Grossberg and Todorović, 1988, Figure 6). This ON-OFF complementarity property also plays an important role in noise suppression when it interacts with the CORT-X filter, as the next section explains.

B. STEP 2: CORT-X FILTER

The ON-C and OFF-C shunted images are transformed by a CORT-X filter into a boundary representation. The CORT-X boundary filter as developed for binary monochromatic images has been described in detail elsewhere (Carpenter et al., 1989). Some modifications have been made herein so that analog monochromatic images can be processed (Section VIII). This modified filter is called CORT-X 2. The rho-space model of Walters (1986) is an alternative oriented filter that shares a number of properties with the CORT-X filter. A brief description and motivation for the filter stages follows.

Each processing layer has the same number of cells as pixels in the image. The architecture is completely feedforward. Cells at a given layer have input fields (IFs) that integrate over an area in the previous layer local to its position in the field. Two separate scales (input field sizes) are used in parallel in the early stages of processing and are subsequently combined to take advantage of the best of their respective processing capabilities. The term input field, or in-field, is used instead of receptive field because the latter term from neurophysiology typically refers to the region at the first processing layer that influences the activity of a cell at any subsequent layer. Our layer-by-layer analysis of scale sizes requires a more microscopic analysis of network geometry.

The first stage of the model, called the simple cell layer, is an oriented contrast detector that is sensitive to the orientation, amount, direction, and spatial scale of image contrast at a given image location. The orientation sensitivity is the result of an elliptically shaped IF. Several fields of different orientations each operate in parallel at each position. The sensitivity to amount and direction of contrast is produced by exciting the cell with signal present in one half of its IF, inhibiting its output with some proportion of the signal in the other half, and then thresholding the result (Figure 9). The result is a half-wave rectification of the signal.

A

B C

FIGURE 8. (A) The original figure in 50% noise (half the pixels are random); (B) the result of the ON-C shunting filter; (C) the result of the OFF-C shunting filter. The Appendix explains the way in which image intensities were normalized.

The ON-C shunting network and the OFF-C shunting network each input to separate networks of simple cells at each receptive field size. Thus, the ON-C and OFF-C networks together activate four networks of simple cells.

The outputs of these parallel simple-cell networks are next combined at each position (Figure 9). This second stage of the CORT-X filter, called the complex cell layer, is sensitive to the orientation, amount, and spatial scale of the contrast of the image at a given point, but not to the direction of contrast. This is achieved by summing the outputs of all like-oriented simple cells at each position, including cells that are sensitive to opposite direction of contrast and that receive inputs from either the ON-C or OFF-C shunting networks. Two networks of complex cells sensitive to the two scale sizes are generated in this way. Adding the half-wave rectified outputs from pairs of simple cells that are sensitive to opposite direction of contrast has the same net effect as full-wave rectification.

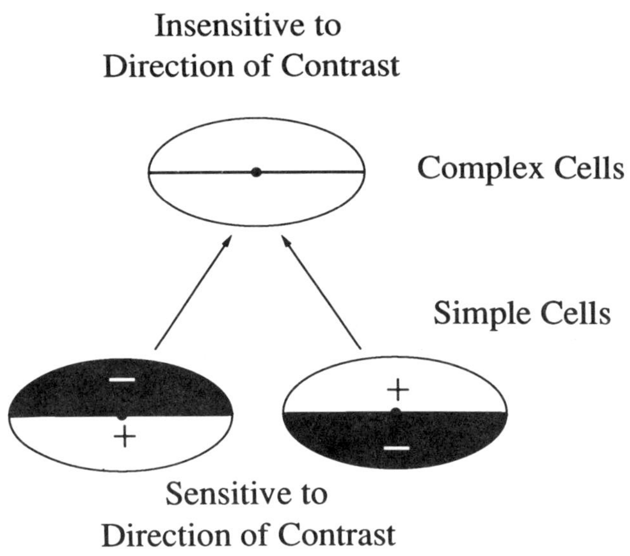

FIGURE 9. The simple cell and complex cell layers that process each of the two shunting network images in parallel. A horizontally oriented set of cells is shown. In Figures 10 and 11, only the outputs of horizontally oriented cells are displayed.

The complementarity property helps to suppress noise when the ON-C and OFF-C outputs are processed separately by their own parallel networks of simple cells and complex cells. To a first approximation, the contrast detectors respond to the ratio of signal between each half of an oriented receptive field. As such, a small amount of noise signal against a background of no signal would effect the contrast detectors more than the same amount of ''dropout'' noise against a background of strong signal. The inversion of the image performed by the OFF-C filter changes the direction of contrast between any noise and its signal background. Thus, the noise will be disruptive in only one of the two parallel networks of simple cells and complex cells, while actual region boundaries will be strongly detected in both (Figure 10).

The output from both networks of contrast detectors are then summed, an operation which takes advantage of the ON-OFF complementarity property. First, it yields roughly equal responses to both concave and convex curvatures. Second, since noise in a given region is suppressed in one or the other contrast-filtered shunting outputs, while boundaries one are strong in both, the summation strengthens the boundary signals relative to the noise (Figure 11).

Figure 11 demonstrates that the two scales also exhibit another type of complementary processing capabilities (Carpenter et al., 1989). The smaller-scale filter does a better job of boundary localization than the larger-scale filter, especially at positions of high boundary curvature, whereas the larger filter does a better job of noise suppression and boundary completion. Canny (1986) has suggested how a single spatial scale can trade off between these virtues, but notes that ''we cannot improve both simultaneously'' (p. 684). The CORT-X family of models suggests a strategy whereby two or more scales can be combined to realize the best features of each. In particular, the large-scale filter achieves good noise suppression far from the image boundaries, but not within the radius of large scale IFs near these boundaries. The small-scale filter is relatively poor at noise suppression anywhere. The next stage of filtering is designed to control this noise near boundaries. Each complex cell excites the hypercomplex cell at its position and orientation in the next level, called the

FIGURE 10. (A) Small-scale horizontal complex-cell output from the ON-C image of Figure 8B; (B) small-scale horizontal complex-cell output from the OFF-C image of Figure 8C; (C) large-scale horizontal complex-cell output from the ON-C image of Figure 8B; (D) large-scale horizontal complex-cell output from the OFF-C image of Figure 8C.

first competitive stage, while inhibiting hypercomplex cells at nearby positions that are not colinear with its axis of symmetry (Figure 12).

The next level, the second competitive stage, sharpens the activation pattern across orientations at each position and scale. In particular, at each position and scale, that cell is chosen whose orientation receives the maximal input from the first competitive stage (Figure 13).

The final operations include cooperative interactions between both filter sizes that select their desirable properties and eliminate their undesirable ones. The ability of the small scale to localize boundaries and the ability of the large scale to suppress noise and complete gaps in the boundaries are maintained by these cooperative interactions (Figure 14A).

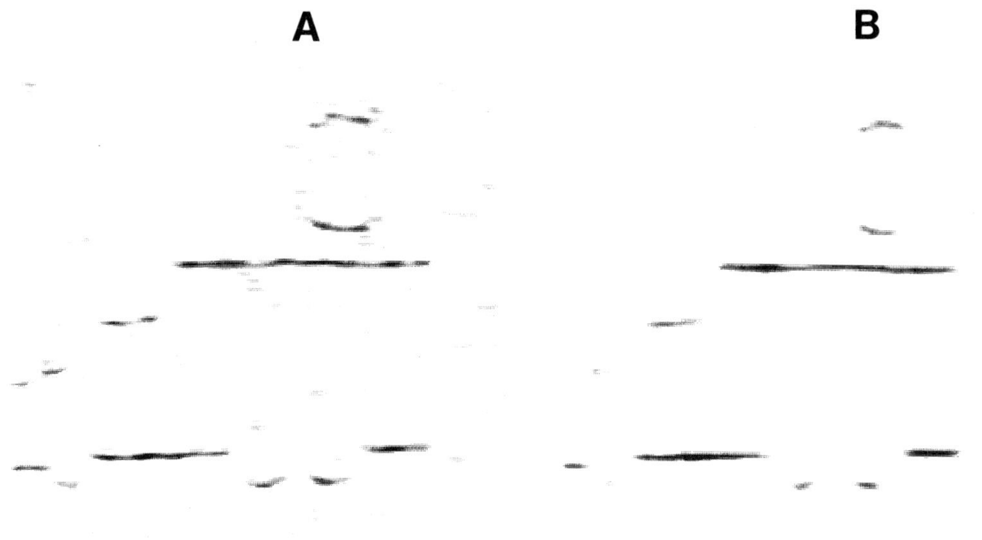

FIGURE 11. (A) ON-OFF small-scale horizontal complex-cell response derived by adding Figures 10A and B. (B) ON-OFF large-scale horizontal complex-cell response derived by adding Figures 10C and D.

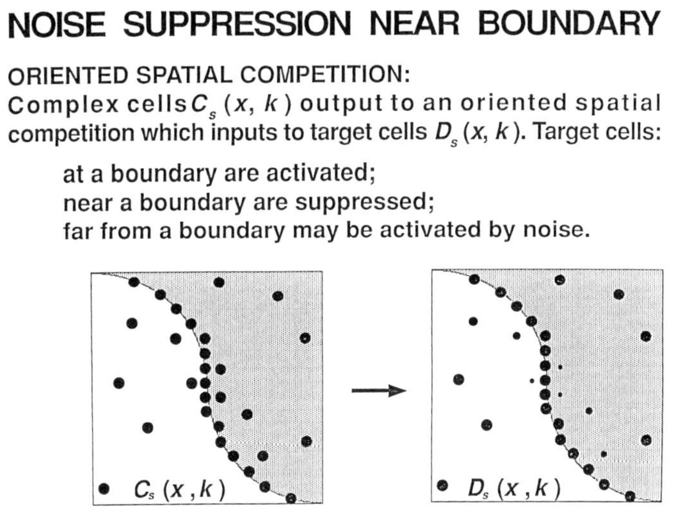

FIGURE 12. Oriented spatial competition inhibits noise pixels near the boundary. (From Carpenter, G. A., Grossberg, S., and Mehanian, C. [1989]. *Neural Networks,* **2,** 169–181. With permission.)

Gaps in the boundary become more likely as the noise level of the original image increases. To overcome this problem, the cooperative level also includes oriented long-range interactions among the selected maximal responses in Figure 13. Due to these long-range interactions, an inactive cell can become active if enough cells of a particular orientation at the previous level are active on both sides of the cell along the oriented axis of this cell (Figure 14B).

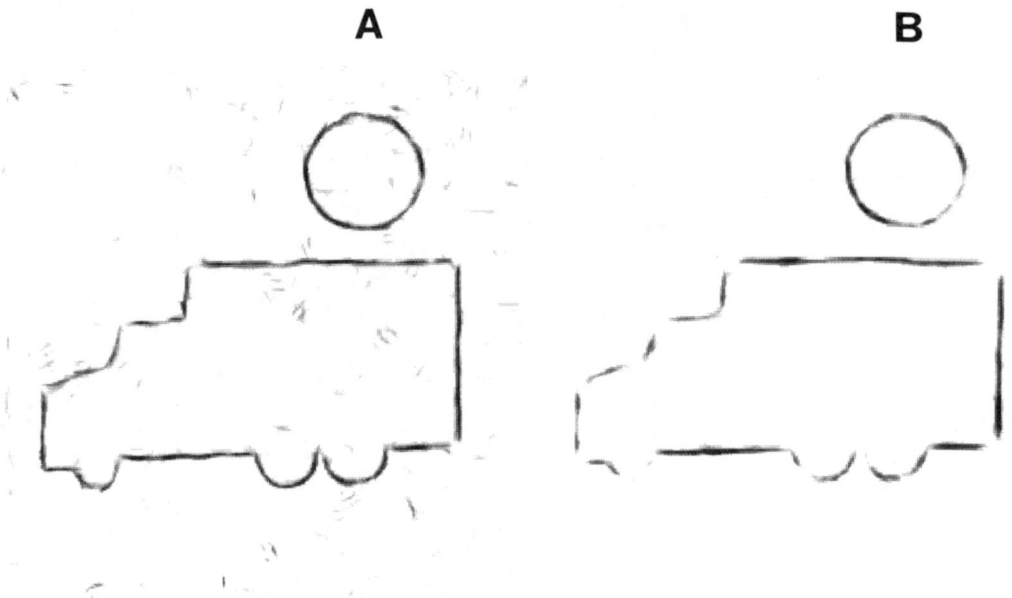

FIGURE 13. Output from the second competitive stage due to competition between orientations at each location. (A) Small scale; (B) large scale.

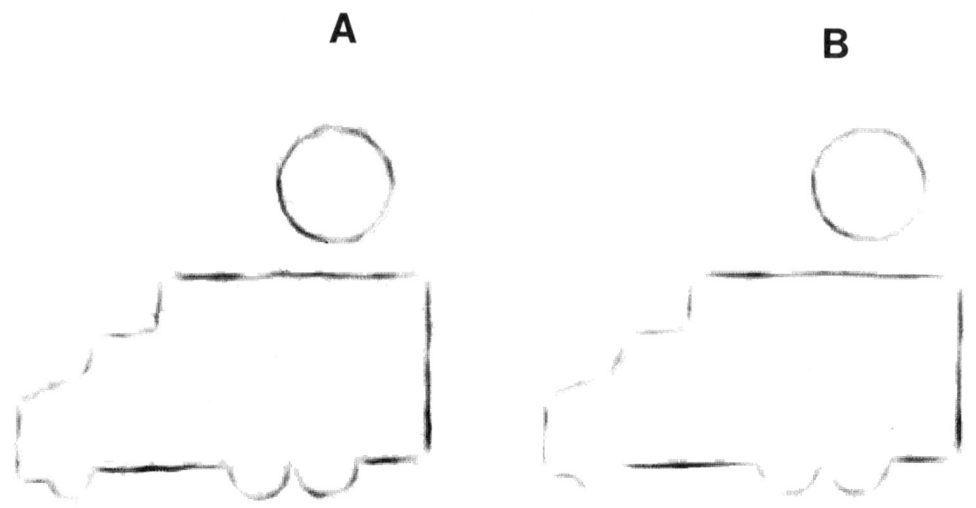

FIGURE 14. (A) Unoriented cooperation between both scales; (B) oriented cooperation within the large scale.

The final output of the CORT-X filter is the sum of the combined-scales image and the completed-gap image (Figure 15).

C. STEP 3: FILLING IN

The output from the CORT-X filter is topographically mapped into M filling-in networks F_m, $i = 1, 2, \ldots, M$. In the Grossberg and Todorović (1988) article, the signals that trigger filling in are generated by the image (Section V). In the present application, they

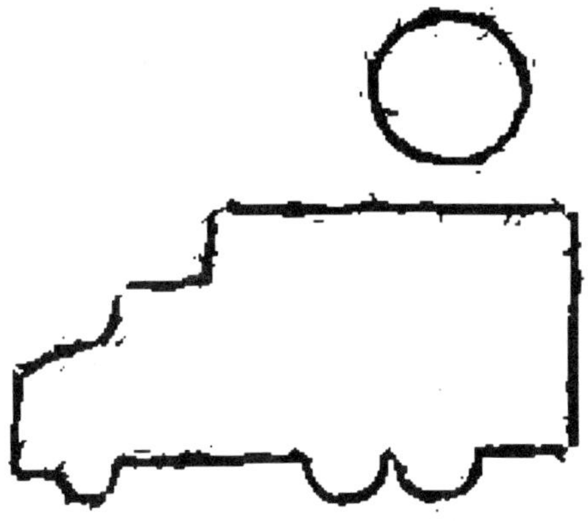

FIGURE 15. Final CORT-X filter output.

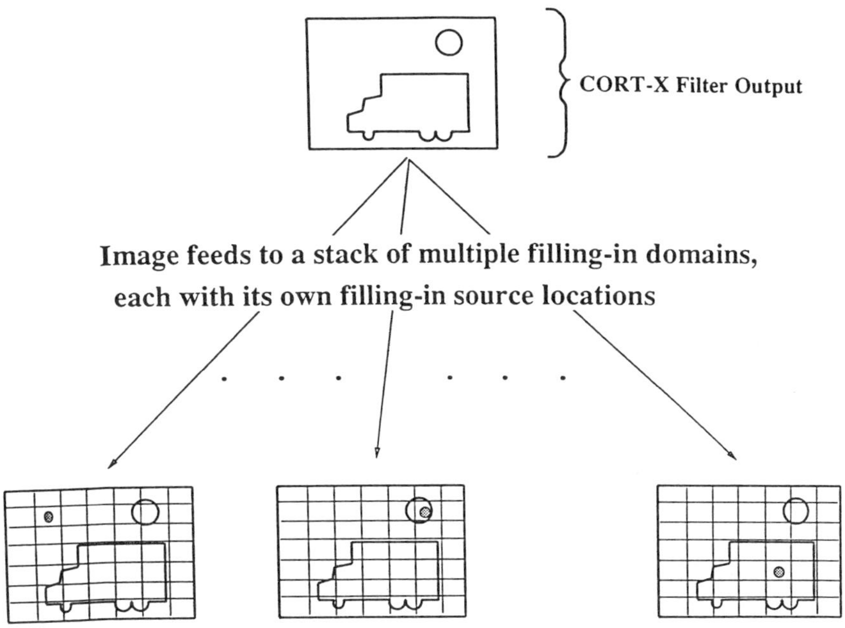

FIGURE 16. Copies of the CORT-X filter output are sent to M filling-in networks (three of which are pictured here). Activity is injected into a different place in each (gray disks) and begins to spread.

are generated by input sources that lie within the network. Moreover, each internally generated input is delivered to its filling-in network at a different position.

Imagine, for definiteness, an $n \times n$ grid of $M = n^2$ nodes laid out over the boundary image generated by the CORT-X filter. Each filling-in network, F_m, is associated with a different grid point where it will receive a featural "dye" injection into its copy of the boundary image. The injection then spreads unimpeded where there is no boundary signal, but does not spread through points where a boundary signal exists (Section VIII). Thus, each injection fills in the connected figure that surrounds the injection point (Figure 16). If the grid is dense enough, then all connected figural components will receive an injection within its boundary in at least one filling-in network F_m. This process is easy to replicate in large numbers because all the networks are identical except for the different, but regular, locations of the injected inputs, and all can operate independently and asychronously in parallel.

D. STEP 4: FIGURE-GROUND SEPARATION

Each filling-in network feeds its activation pattern in parallel to another pair of shunting networks, one ON-C and one OFF-C. This completes the second ''F'' operation in the FBF model. Because of their contrast-enhancing and ratio-processing properties, these filters amplify the filled-in activity near figural boundaries while tending to suppress the low-contrast regions generated by the spreading of activation across the interiors of figures and background regions.

Consider first the case where there are no holes in a boundary at which the spreading activation could leak through. If the filling-in process has been given enough time to reach all of the enclosing boundary signals, then the ON-C filter produces an output only inside the enclosing boundaries, due to the injected activity (Figure 17E). The OFF-C filter produces signal only outside the enclosing boundaries, due to the spontaneous baseline activity which has not been quenched by the input injection (Figure 17H). In this hole-free case, the ON-C filtered images effectively separate each connected region from all others in the original image. However, because of the Gaussian shapes of the kernels used in these filters, the edges in the ON-C patterns also exhibit a Gaussian spatial spread. Also, smooth gradients, such as those generated by activation spreading across regions where there are no boundary signals, may not be quenched to zero if the center and surround kernels are of unequal area. If the pattern recognition stage is based on image boundary signals, then better boundaries must be reconstructed. One way is to subject each image to another pass through the CORT-X filter. Such a step would be useful if the second CORT-X filter used a longer-range cooperative kernel to complete boundaries across any remaining gaps. An alternative approach is to multiplicatively gate each ON-C pattern with the original CORT-X output. The resulting M images are then ready to be passed along and contain among them all the separated figural boundaries of the original image (Figure 17 J to L). This method can be run in real-time until the output from each copy generates a recognition event at its ART network, or the input image is removed.

If the original image contains so much noise that the CORT-X filter is unable to produce boundaries without weak points or small gaps, then significant activity could leak out of figural components during the filling-in process (Figure 17A). On the other hand, the ART recognition process, operating in real-time, could recognize the figure before the equilibrium state of equal activation on both sides of the boundary is reached.

Leakage of diffusing activity causes no problem if there are no other nearby object boundaries in the original scene. The spread of activation would produce a smooth gradient through the gap, and the shunting operation would not detect any contrast at the point of leakage or outside the object until well after the recognition event occurred (Figure 17J). If, however, the boundary of another object were near a boundary gap, then leaking activation followed by the ON-C shunting operation could detect the spurious boundary if the injection site were closer to the spurious boundary than to other boundaries of the figure.

In this case, the OFF-C filtered image is helpful. The ON-C signal at boundary regions exterior to the desired object is not as strong as the OFF-C filter signal at these points unless the boundary and the injection site are both proximal to the boundary gap. This property is due to the fact that very little injected activity would have spread there to excite the ON-C field, and the OFF-C field is tonically active, as shown in Equation 16, Section VIII. Subtracting the OFF-C image from the ON-C image before CORT-X gating, therefore, helps to correct this pathological situation. The OFF-C subtraction does not, moreover, distort the

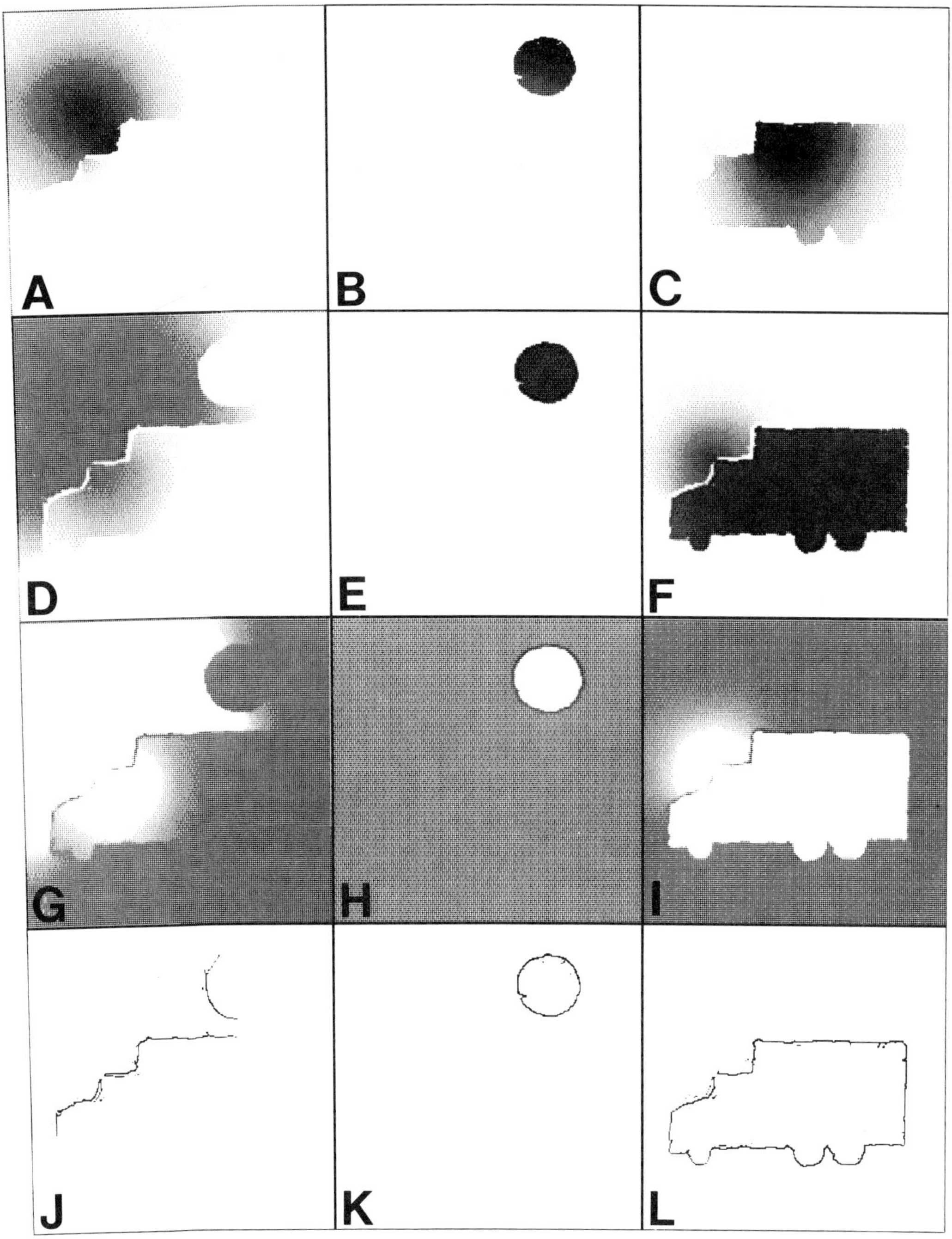

FIGURE 17. Row 1: filling in of activity is initiated at three different places in parallel filling-in networks. Due to the quantization of the gray scale, small filled-in activations do not print, even though they are detected by the shunting networks, as noted in rows 2 through 4. Row 2: the ON-C filter output of the respective filled-in regions. Row 3: the OFF-C filter output of the respective filled-in regions. Row 4: the final "separated figure" outputs to be passed along to a pattern recognizer such as an ART network.

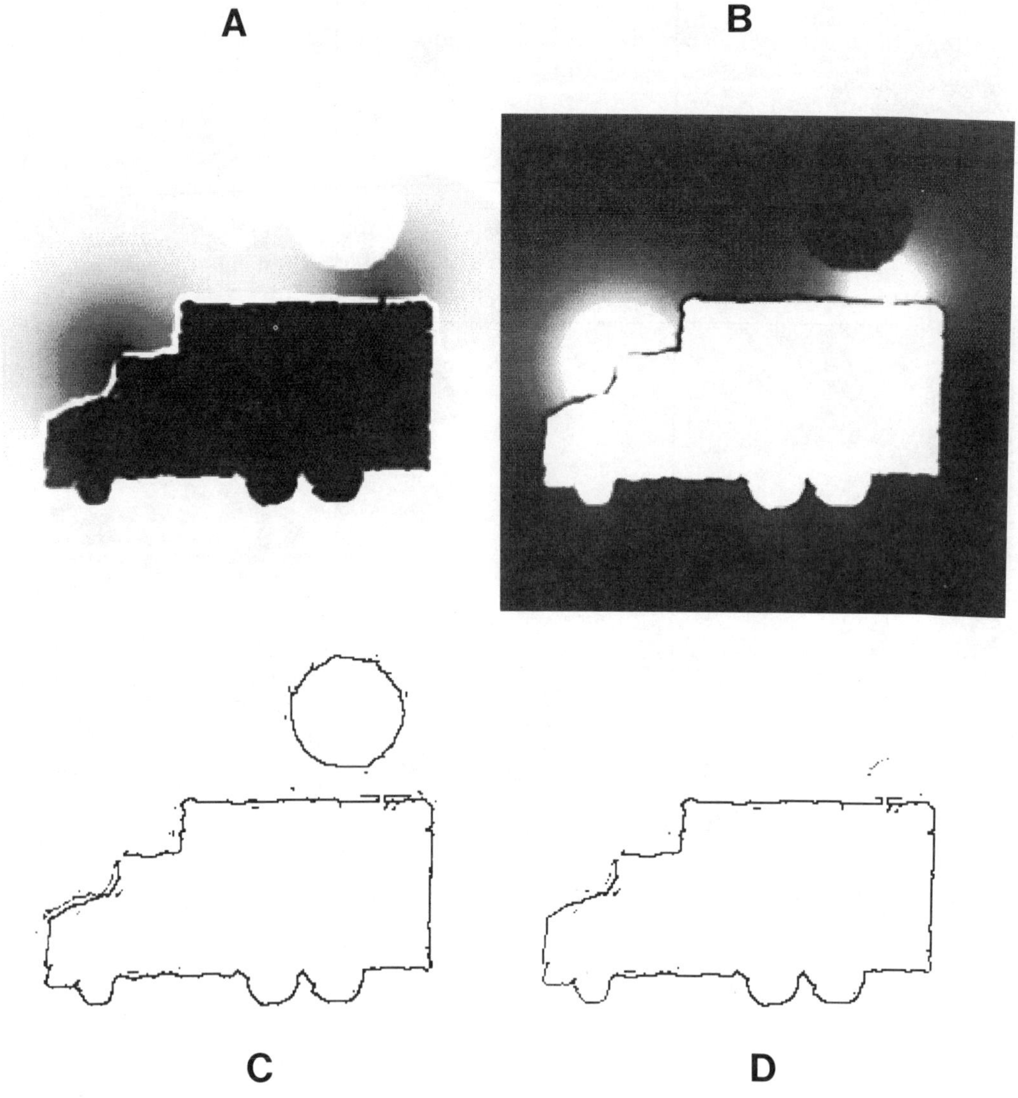

FIGURE 18. (A) The ON-C filter output of the filled in region with a hole in the boundary proximal to another figure; (B) the OFF-C filter of the filled in region; (C) the output as it would look without using the complementary OFF-C image; (D) the final image which uses the OFF-C filter to control the effects of the leakage, leaving only the desired boundary signal.

desired object boundary when the ON-C signal is stronger than the OFF-C signal. In Figure 18, for example, a hole was made in the boundary of the truck near the boundary of another figure before filling in occurs. The ON-C network clearly detects the outer boundary of the moon (Figure 18A). The OFF-C filtered signal is, however, stronger at these points (Figure 18B). Without taking the OFF-C network information into account, the output could cause difficulties for a pattern recognizer since the objects would not be separated (Figure 18C). Combining the two shunting network filters produces the desired separation (Figure 18D). The output of such an operation computes a double opponent receptive field (Grossberg, 1987b).

A single figure is typically large enough to enclose several injection sites across the set of filling-in networks F_m. Thus, even if the injection site is close to a boundary gap and to a nearby spurious boundary in one network F_m, the injection sites of other filling-in networks will be further from the boundary gap. In some of these networks, double opponent processing can compensate for the gap and trigger a correct recognition of the figure from the corresponding ART network. If the spurious boundaries are strong enough, the corresponding ART network will remain silent, as in Figure 17J, because the combination of partial figure and background is not similar enough to a previously learned recognition code.

In summary, successful parallel separation of multiple connected figures from other figures and background can be performed with this architecture under noisy conditions by exploiting the possibility of efficient parallel recognition by ART networks of boundary segmentations that are generated by sampling multiple filling-in perspectives.

VIII. FBF NETWORK EQUATIONS

A. INPUT IMAGES

In our computer simulations, the images are 256*256 arrays with signal values in the interval [0,1]. The simulations pictured herein represent maximum signal strength by black and minimum signal strength by white. Noise was generated by randomly choosing a percentage of pixels and setting their values to a random number, or gray level, in the interval [0,1]. For the simulation pictured herein, 50% of the pixels were randomized. The input pattern $\{I_{ij}\}$ is, thus, represented as gray levels on a set of square pixels $\{P_{ij}\}$. Pixel P_{ij} attains the value I_{ij} at the set of image points $\{(u,v){:}i \leq u < i + 1, j \leq v < j + 1\}$.

1. Step 1: Discount the Illuminant

a. ON-C Network

Each node v_{ij} is placed, for notational convenience, at the center of the corresponding pixel P_{ij} where it receives input I_{ij}. The activity x_{ij} of the node v_{ij} at lattice position (i,j) obeys the shunting on-center off-surround equation:

$$\frac{d}{dt}x_{ij} = -Ax_{ij} + (B - x_{ij})C_{ij} - (x_{ij} + D)\,E_{ij} \tag{1}$$

where C_{ij} is the on-center interaction and E_{ij} is the off-surround interaction. Each C_{ij} and E_{ij} is a discrete convolution of the input pattern $\{I_{ij}\}$ with a Gaussian kernel (Figure 19A). Thus,

$$C_{ij} = \sum_{p,q} I_{pq}C_{pqij} \tag{2}$$

and

$$E_{ij} = \sum_{p,q} I_{pq}E_{pqij} \tag{3}$$

where

$$C_{pqij} = C\exp\{-\alpha^{-2}\log 2[(p - i)^2 + (q - j)^2]\} \tag{4}$$

and

$$E_{pqij} = E\exp\{-\beta^{-2}\log 2[(p - i)^2 + (q - j)^2]\} \tag{5}$$

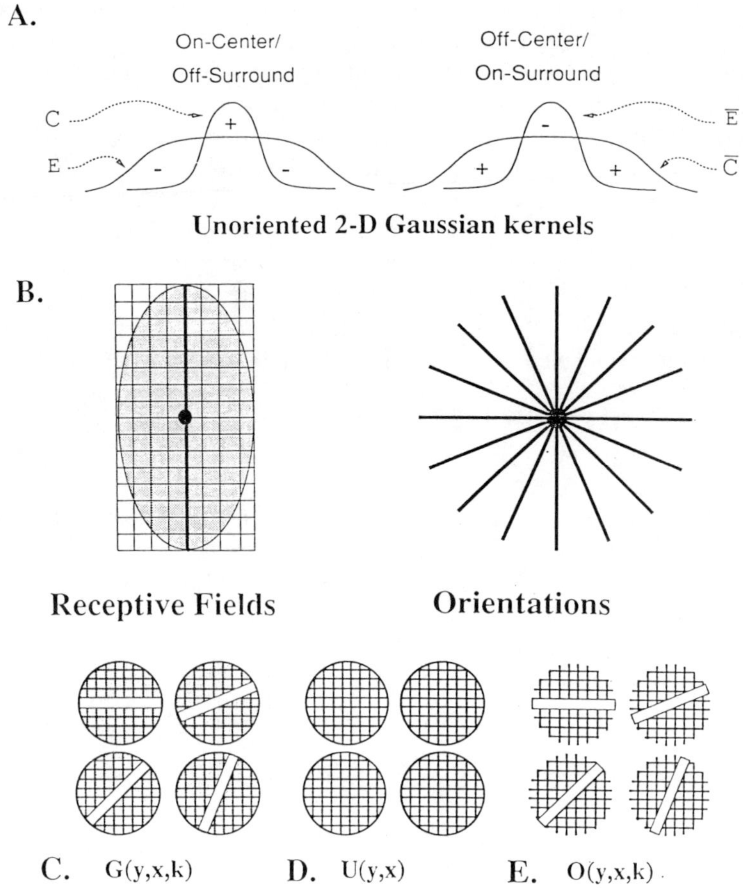

FIGURE 19. The kernels used in the CORT-X filter. All oriented kernels have orientations at every $\pi/8$ radians.

In our simulations, $A = 134$, $B = 1$, $C = 7$, $D = .5$, $E = 3.333$, $\alpha = 1.3$, and $\beta = 1.875$. For this choice of parameters, the ON-C and OFF-C Gaussians are of equal area. At equilibrium ($dx_{ij}/dt = 0$), Equation 1 yields

$$x_{ij} = \frac{\Sigma_{(p,q)}(BC_{pqij} - DE_{pqij})I_{pq}}{A + \Sigma_{(p,q)}(C_{pqij} + E_{pqij})I_{pq}} \tag{6}$$

b. OFF-C Network

The activity $\bar{x}_{ij}$ of the node v_{ij} at lattice position (i,j) obeys the shunting off-center on-surround equation

$$\frac{d}{dt}\bar{x}_{ij} = -A(\bar{x}_{ij} - S) + (\overline{B} - \bar{x}_{ij})\overline{C}_{ij} - (\bar{x}_{ij} + \overline{D})\overline{E}_{ij} \tag{7}$$

where the on-center kernel of $\bar{x}_{ij}$ is the off-surround kernel of x_{ij} and the off-surround kernel of $\bar{x}_{ij}$ is the on-center kernel of x_{ij}. In particular,

$$\overline{B} = D \tag{8}$$

$$\overline{C}_{ij} = E_{ij} \tag{9}$$

$$\overline{D} = B, \tag{10}$$

and

$$\overline{E}_{ij} = C_{ij}. \tag{11}$$

By Equations 8 to 11, Equation 7 may be written as

$$\frac{d}{dt}\overline{x}_{ij} = - A(\overline{x}_{ij} - S) + (D - \overline{x}_{ij})E_{ij} - (\overline{x}_{ij} + B)C_{ij}. \tag{12}$$

At equilibrium,

$$\overline{x}_{ij} = \frac{AS + \Sigma_{p,q)}(DE_{pqij} - BC_{pqij})I_{pq}}{A + \Sigma_{(p,q)} (C_{pqij} + E_{pqij})I_{pq}} \tag{13}$$

It follows by summing Equations 6 and 13 that

$$x_{ij} + \overline{x}_{ij} = \frac{AS}{A + \Sigma_{(p,q)}(C_{pqij}) + E_{pqij})I_{pq}} \tag{14}$$

which shows that for images $\{I_{pq}\}$ with constant Gaussianly filtered total activity

$$\sum_{p,q} (C_{pqij} + E_{pqij})I_{pq} \tag{15}$$

the sum $x_{ij} + \overline{x}_{ij}$ is conserved and maintained at a positive value that increases with the tonic activity level S of $\overline{x}_{ij}$ in the dark. In our simulations, $S = 0.2$. The value for the parameter A (given the other parameter values) was chosen so that the activation of cell $\overline{x}_{ij}$ takes on values between S and O under spatially uniform illumination between 0 and 1 within its receptive field. Under such spatially uniform illumination conditions, $I_{pp} = I$ for all (p,q). Since for our choice of parameters the two Gaussian kernels are of equal area, let $\Phi = \Sigma_{p,q}E_{pqij} = \Sigma_{p,q}C_{pqij}$. Then Φ factors out so that the equilibrium Equation 13 becomes

$$\overline{x}_{ij} = \frac{AS + I(D - B)\Phi}{A + 2I\Phi} \tag{16}$$

We derive the value for A by

$$A = \frac{(B - D)\Phi}{S} \tag{17}$$

so that when $I = 1$, the numerator vanishes. When $I = 0$, $\overline{x}_{ij} = S$ independent of A.

2. Step 2: CORT-X Filter

All simple-cell input fields are elliptical. Two sizes of input fields were used, indexed by the subscript s. The smaller scale, $s = 1$, had a major axis of 12 pixels and a minor axis of 6 pixels. The larger scale, $s = 2$, had a major axis of 20 pixels and a minor axis of 10 pixels. Orientations were chosen around the clock spaced by $\pi/8$ degrees. They are indexed below by the subscript k.

a. Simple Cells

A simple cell with index (i,j) is centered at the lower left-hand corner of pixel P_{ij}. By this convention, the nodes v_{pq} of the shunting variables x_{pq} do not lie on the oriented axes that separate the excitatory and inhibitory halves of vertically and horizontally oriented receptive fields. The output of the pair of simple cells of scale s centered at index (i,j) with receptive field orientation k is defined by

$$S_{sL}(i,j,k) = \max[L_s(i,j,k) - \alpha_s R_s(i,j,k) - \beta_s, 0] \tag{18}$$

and

$$S_{sR}(i,j,k) = \max[R_s(i,j,k) - \alpha_s L_s(i,j,k) - \beta_s, 0] \tag{19}$$

where $L_s(i,j,k)$ and $R_s(i,j,k)$ are the contributions of the left-half $l_s(i,j,k)$ and right-half $r_s(i,j,k)$, respectively, of the oriented input field; that is,

$$L_s(i,j,k) = \frac{\sum_{(p,q)\in l_s(i,j,k)} x_{pq} w_{pq}}{\sum_{(p,q)\in l_s(i,j,k)} w_{pq}} \tag{20}$$

and

$$R_s(i,j,k) = \frac{\sum_{(p,q)\in r_s(i,j,k)} x_{pq} w_{pq}}{\sum_{(p,q)\in r_s(i,j,k)} w_{pq}} \tag{21}$$

where w_{pq} is a weighting factor equal to the proportion of the area of pixel P_{pq} (taken to be one square unit) covered by the receptive field. An activity x_{pq} was included in L_s or R_s if its pixel had a nonzero intersection with the corresponding half of the receptive field. Parameters α_s are threshold contrast parameters and parameters β_s are threshold noise parameters. We chose $\alpha_1 = 1.4$, $\alpha_2 = 2.0$, and $\beta_1 = \beta_2 = \beta = 0.012$. Each simple cell in Equations 18 and 19 is sensitive to the opposite direction of contrast than its companion, as indicated by the indices L and R in S_{sL} and S_{sR}, respectively.

The ON-C and OFF-C networks each input to a different network of simple cells. The simple cells that receive signals from the ON-C network are denoted by S^+_{sL} and S^+_{sR}. The simple cells that receive signals from the OFF-C network are denoted by S^-_{sL} and S^-_{sR}.

The complex cells pool inputs from all simple cells of like orientation and scale that are centered at the same location, as described below.

b. Complex Cells

The complex cell output $C_s(i,j,k)$ is defined by

$$C_s(i,j,k) = F[S^+_{sL}(i,j,k) + S^+_{sR}(i,j,k) + S^-_{sL}(i,j,k) + S^-_{sR}(i,j,k)] \tag{22}$$

Such a cell is sensitive to spatial scale s and amount of contrast centered at cell x with orientation k, but it is insensitive to direction of contrast. In our simulations, $F = 0.5$.

c. *Hypercomplex Cells, First Competitive Stage*

The hypercomplex cells $D_s(i,j,k)$ at the first competitive stage receive input from the spatial competition among the complex cells; that is,

$$D_s(i,j,k) = \max\left[\frac{C_s(i,j,k)}{\epsilon + \mu\,\Sigma_m \Sigma_{p,q} C_s(p,q,m) G_s(p,q,i,j,k)} - \tau,\ 0\right] \quad (23)$$

where $\epsilon = 0.1$, $\mu = 5$, $\tau = 0.01$. The oriented competition kernel $G_s(y,x,k)$ is normalized so that

$$\sum_{p,q} G_s(p,q,i,j,k) = 1 \quad (24)$$

As in Figure 19C, they are circular. Complex cells at the kernel periphery are weighted by the proportion of their area (taken to be one square unit) that is covered by the kernel. The gray areas in Figure 19C are inhibitory. Any cells whose defining pixel location lies within the one unit wide band through the middle of the kernel do not contribute to the inhibition. In our simulations, the small-scale kernel is 8 units in diameter, and the large-scale is 16 units in diameter.

d. *Hypercomplex Cells: Second Competitive Stage*

The hypercomplex cells $D_s(i,j)$ at the second competitive stage realize a competition among the oriented activities $D_s(i,j,k)$ at each position x. For simplicity, the process is modeled as a winner-take-all competition; namely,

$$D_s(i,j) = D_2(i,j,K) = \max_k D_s(i,j,k) \quad (25)$$

where index K denotes the orientation of the maximally activated cell.

e. *Multiple Scale Interaction: Boundary Localization and Noise Suppression*

The interaction between scales is defined by the equation

$$B_{12}(i,j) = D_1(i,j) \sum_{p,q} D_2(p,q) U(p,q,i,j) \quad (26)$$

The unoriented excitatory kernel $U(p,q,i,j)$ is circular (Figure 19D) and normalized so that

$$\sum_{p,q} U(p,q,i,j,k) = 1 \quad (27)$$

and had a diameter of 8 units. All cells covered by the kernel contribute to the excitation to the extent that their area (taken to be one square unit) is covered by the kernel. The small-scale hypercomplex cell $D_i(i,j)$ accurately localizes boundary segments and suppresses noise near the boundary. The large-scale hypercomplex cell $D_2(p,q)$ suppresses noise far from the boundary. The product $D_1(i,j)D_2(p,q)$ would simultaneously realize both constraints except that, due to the poor spatial localization of D_2, this term may be zero at boundary points of high curvature, thereby canceling the good localization properties of $D_1(i,j)$. The effect of $D_2(p,q)$ in the equation is made more spatially diffuse by the kernel $U(i,j,p,q)$. The size of $U(i,j,p,q)$ is chosen to scale with that of $D_2(p,q)$ in order to compensate for the positional uncertainty of $D_2(p,q)$; a larger choice of $D_2(p,q)$ would necessitate a larger choice of $U(i,j,p,q)$. Although term $\Sigma_y D_2(p,q)U(p,q,i,j)$ in Equation 26 localizes the boundary even less accurately than $D_2(p,q)$ does, the *product* of $D_1(i,j)$ with $\Sigma_{p,q}D_2(p,q)U(p,q,i,j)$ in Equation 26 restores this loss of boundary localization. Moreover, kernel $U(p,q,i,j)$ causes no harm at locations p,q that are far from the boundary, since $D_2(p,q) = 0$ there (Carpenter et al., 1989).

f. Long-Range Cooperation: Boundary Completion

The function $B_{12}(i,j)$ represents the image boundary well except where boundary pixels are missing due to noise. More and larger boundary gaps are generated as the noise level increases.

The large detectors $D_2(i,j)$ can be used to overcome this problem partially. Because of the spatial uncertainty of the large detectors $D_2(i,f)$, they are capable of responding at locations where pixel signal strength has been reduced by noise. Such boundary signals may, however, be poorly localized. To overcome this tradeoff between boundary completion and localization, cooperative interactions among the large scale cells are defined by

$$B_2(i,j) = D_2(i,j) \max \left[\sum_{p,q} D_2(p,q,K)O(p,q,i,j,K) - \delta, \mathrm{O}\right] \tag{28}$$

The oriented kernel $O(p,q,i,j,k)$ is defined by the one-unit-wide white strips in Figure 19E. Any cells with centers that lie within the one-unit-wide band contribute to the cooperative process. The kernel is normalized so that

$$\sum_{(p,q)\text{ in kernel}} O(p,q,i,j,K) = 1 \tag{29}$$

In our simulations, the length of the kernel was 12 units, and $\delta = 0.001$.

g. CORT-X Output

The final output of the CORT-X filter is the rectified sum of the multiple-scale interaction and the cooperative interaction:

$$B(i,j) = 1[B_{12}(i,j) + B_2(i,j)] \tag{30}$$

where

$$1[w] = \begin{cases} 1 & \text{if } w > 0 \\ 0 & \text{otherwise} \end{cases} \tag{31}$$

is the Heaviside function.

3. Step 3: Filling In

In each filling-in network F_m, an input is injected into a different area, with the shape of either a narrow Gaussian or a single node. For example, let $I = I(m)$ and $J = J(m)$ define the injection indices (I,J) of F_m. Then the injected input pattern to F_m was chosen to be

$$X_{ij}^{(m)} = X \exp\{-\gamma^{-2} \log 2\, [(I - i)^2 + (J - j)^2]\} \tag{32}$$

This input pattern triggers filling in within F_m via the nonlinear diffusion equation (Grossberg and Todorvić, 1988)

$$\frac{d}{dt} S_{ij}^{(m)} = -MS_{ij}^{(m)} + \sum_{p,q\in Nij} (S_{pq}^{(m)} - S_{ij}^{(m)})P_{pqij} + X_{ij}^{(m)} \tag{33}$$

where $S_{ij}^{(m)}$ is the activity of the (i,j) node of F_m, and the index set N_{ij} of the summation contains the nearest neighbors of (i,j). The permeability coefficient P_{pqij} is defined by

$$P_{pqij} = \frac{\delta}{1 + \epsilon[B(p,q) + B(i,j)]} \tag{34}$$

where $B(p,q)$ and $B(i,j)$ are the outputs (Equation 30) from the CORT-X filter at the positions (p,q) and (i,j), respectively. Thus, activity spreads poorly, if at all, between cells where boundary signals are large, and easily where boundary signals do not exist.

In the simulations, the parameters $M = 0.0001$, $X = 50$, $\gamma = .5$, $\epsilon = 100{,}000$, $\delta = 10$.

4. Step 4: Figure-Ground Separation

Each filling-in network F_m inputs its filled-in image to shunting ON-C and OFF-C networks using the same equations (1 and 7) as above, with parameters $A = 1$, $B_{ij} = \overline{D}_{ij} = 1$, $C = 18$, $D_{ij} = \overline{B}_{ij} = 0.5$, $E = 3.333$, $\alpha = 2.96$, $\beta = 7$, and $S = 0.2$. Here the on-activations $x_{ij}^{(m)}$ and off-activations $\overline{x}_{ij}^{(m)}$ are parameterized by the filling-in network F_m from which they are derived. The boundary representation at position (i,j) of a figure derived from network F_m is defined by

$$R_{ij}^{(m)} = 1\left[(x_{ij}^{(m)} - \overline{x}_{ij}^{(m)})B(i,j)\right] \tag{35}$$

where $1[w]$ is the Heaviside function. In other words, the figural boundary that is separated by network F_m is computed from the double opponent filter $[x_{ij}^{(m)} - \overline{x}_{ij}^{(m)}]$ of filled-in F_m activation, gated by the CORT-X boundary segmentation $B(i,j)$.

IX. CONCLUDING REMARKS: RECOGNITION OF CONJUNCTIVE FEATURES

The FBF network separates figure from ground by using regular arrays of feedforward networks to discount the illuminant and to generate boundary segmentations, nearest-neighbor feedback signals for filling-in, and a proliferation of these circuits in parallel copies that input to parallel arrays of ART pattern recognition networks. The FBF networks, thus, seem to be appropriate designs for implementation in parallel hardware capable of operating at high rates in real time.

It remains to be discussed why the biologically motivated FBF network can automatically distinguish between the pair of connected and disconnected Minsky-Papert figures in Figure 1 that humans cannot distinguish. The main reason is that FBF networks use internally generated "dye injections". These inputs are topographically distributed across the entire perceptual space such that each dye injection is delivered to a different filling-in domain, or slab. In human perception, by contrast, the same feature-contour signals that initiate Boundary contour formation also act as input sources that trigger featural filling-in (Grossberg, 1987a, 1987b; Grossberg and Mingolla, 1985a; Grossberg and Todorović, 1988). The regions used by Minsky and Papert (1969) were all of the same color, of similar overall shape, and occupied essentially the same region of their respective images. They would, therefore, tend to fill in the same slab, or set of slabs, when they are being perceived by humans. Hence, they could not be distinguished by filling in different slabs as a function of their connectivity.

This observation clarifies a recent controversy about human perception, namely, whether target figures that differ from distractor figures by more than one type of feature can be separated from them by rapid parallel processing that does not require serial search. Treisman and colleagues (Treisman and Gelade, 1980; Treisman and Souther, 1985) have suggested that such parallel processing can occur only if the target is distinguished from distractors along a single stimulus dimension, whereas if a target is defined by the conjunction of two or more stimulus dimensions, then it can only be separated from the distractors by a serial search process.

An exception to this rule was discovered by Nakayama and Silverman (1986), who showed that targets which differ from distractors by a combination of disparity and color, or of disparity and motion, can be rapidly separated without serial search. This result is explained by the theory of three-dimensional vision, called FACADE theory (Section IV), which motivated the design of FBF networks. In Grossberg (1987b), FACADE representations of different disparity and color combinations activate different combinations of filling-in domains. They are *structurally* separated in the representation and, hence, can be rapidly detected.

In summary, the difficulty of distinguishing the connected and disconnected Minsky-Papert displays can now be explained by the same mechanisms that explain rapid search of Nakayama-Silverman displays, and that provide the heuristics for designing an FBF network for automatic figure-ground separation.

ACKNOWLEDGMENTS

This research was supported in part by the Air Force Office of Scientific Research (AFOSR 90-0175), the Army Research Office (ARO DAAL-03-88-K0088), DARPA (AFOSR 90-0083), and Hughes Research Laboratories (S1-804481-D and S1-903136); and by the American Society for Engineering Education and Hughes Research Laboratories (S1-804481-D).

The authors wish to thank Cynthia E. Bradford, Carol Yanakakis Jefferson, and Diana Meyers for their valuable assistance in the preparation of the manuscript.

REFERENCES

Canny, J. (1986). A computational approach to edge detection, *IEEE Trans. Pattern Anal. Machine Intell.*, **8**, 679–698.

Carpenter, G. A. and Grossberg, S. (1987a). Invariant pattern recognition and recall by an attentive self-organizing ART architecture in a nonstationary world, in Caudill, M. and Butler, C., Eds., Proc. IEEE Int. Conf. Neural Networks, **Vol. II,** 737–746.

Carpenter, G. A. and Grossberg, S. (1987b). ART 2: Stable self-organization of pattern recognition codes for analog input patterns, *Appl. Opt.*, **26**, 4919–4930.

Carpenter, G. A. and Grossberg, S. (1988). The ART of adaptive pattern recognition by a self-organizing neural network, *Computer,* **21**, 77–88.

Carpenter G. A., Grossberg, S., and Mehanian, C. (1989). Invariant recognition of cluttered scenes by a self-organizing ART architecture: CORT-X boundary segmentation, *Neural Networks,* **2**, 169–181.

Casasent, D. and Psaltis, D. (1976). Position, rotation, and scale invariant optical correlations, *Appl. Opt.*, **15**, 1793–1799.

Cavanagh, P. (1978). Size and position invariance in the visual system, *Perception,* **7**, 167–177.

Cavanagh, P. (1984). Image transforms in the visual system, in Dodwell, P. C. and Caelli, T., Eds., *Figural Synthesis,* Erlbaum Associates, Hillsdale, NJ.

Cohen, M. A. and Grossberg, S. (1984). Neural dynamics of brightness perception: features, boundaries, diffusion, and resonance, *Percept. Psychophys.*, **36**, 428–456.

Grossberg, S. (1987). Cortical dynamics of three-dimensional form, color, and brightness perception. II. Binocular theory, *Percept. Psychophys.*, **41**, 117–158.

Grossberg, S. (1988). *Neural Networks and Natural Intelligence,* MIT Press, Cambridge.

Grossberg, S. (1990). 3-D vision and figure-ground separation by visual cortex, submitted.

Grossberg, S. and Marshall, J. (1989). Stereo boundary fusion by cortical complex cells: a system of maps, filters, and feedback networks for multiplexing distributed data, *Neural Networks,* **2**, 29–51.

Grossberg, S. and Mingolla, E. (1985a). Neural dynamics of form perception: boundary completion, illusory figures, and neon color spreading, *Psychol. Rev.,* **92**, 173–211.

Grossberg, S. and Mingolla, E. (1985b). Neural dynamics of perceptual grouping: textures, boundaries, and emergent segmentations, *Percept. Psychophys.,* **38**, 141–171.

Grossberg, S., Mingolla, E., and Todorović, D. (1989). A neural network architecture for preattentive vision, *IEEE Trans. Biomed. Eng.,* **36**, 65–84.

Grossberg, S. and Todorović, D. (1988). Neural dynamics of 1-D and 2-D brightness perception: a unified model of classical and recent phenomena, *Percept. Psychophys.,* **43**, 241–277.

Grossberg, S. and Wyse, L. (1990). Automatic figure-ground separation by an iterated boundary segmentation and filling-in network, in *Proc. Conf.* Neural Networks for Automatic Target Recognition, Wang Institute of Boston University, Boston, MA, May 11–13, 1990, 27.

Gschwendtner, A. B., Harney, R. C., and Hull, R. J. (1983). Coherent IR radar technology, in *Optical and Laser Remote Sensing,* Killinger, D. K. and Mooradian, A., Eds., Springer-Verlag, New York.

Harney, R. C. (1980). Infrared airborne radar, in *Proc. IEEE 1980 Electronic and Aerospace Systems Conf.,* 462–471.

Harney, R. C. and Hull, R. J. (1980). Compact infrared radar technology, in CO_2 laser devices and applications, *Proc. SPIE,* **227**, 162–170.

Hull, R. J. and Marcus, S. (1980). A tactical 10.6 micrometer imaging radar, *IEEE Proc. Natl. Aerosp. Electron. Conf.,* **2**, 662–668.

Kolodzy, P. (1987). Multidimensional machine vision using neural networks, in Caudill, M. and Butler, C., Eds., *Proc. IEEE Int. Conf. Neural Networks,* **2**, 747–758.

Land, E. H. (1977). The retinex theory of color vision, *Sci. Am.,* **237**, 108–128.

Minsky, M. L. and Papert, S. A. (1969). *Perceptrons,* MIT Press, Cambridge.

Minsky, M. L. and Papert, S. A. (1988). *Perceptrons: Expanded Edition,* MIT Press, Cambridge.

Nakayama, K. and Silverman, G. H. (1986). Serial and parallel processing of visual feature conjunctions, *Nature (London),* **320**, 264–265.

Sullivan, L. J. (1980). Infrared coherent radar, in CO_2 laser devices and applications. *Proc. SPIE,* **227**, 148–161.

Sullivan, L. J. (1981). Firepond laser radar, *Electro/81 Conf. Rec.,* Session 34.

Sullivan, D. R., Harney, R. C., and Martin, J. B. (1979). Real-time quasi-3-dimensional display of infrared radar images. Real-time signal processing II, *Proc. SPIE,* **180**, 57–65.

Szu, H. (1986). Holographic coordinate transformations and optical computing, in *Optical and Hybrid Computing,* SPIE **Vol. 634,** Szu, H., Ed., International Society for Optical Engineering, Bellingham, WA.

Treisman, A. and Gelade, G. (1980). A feature integration theory of attention, Cognit. *Psychol.,* **12**, 97–136.

Treisman, A. and Souther, J. (1985). Search asymmetry: a diagnostic for preattentive processing of separable features, *J. Exp. Gen. Psychol.,* **114**, 285–310.

Walters, D. K. W. (1987). Rho-space: a neural network for the detection and representation of oriented edges, in *Proc. 9th Annu. Conf. Cognitive Science Society,* Erlbaum Press, Hillsdale, NJ, 455–460.

APPENDIX

IMAGE RENDERING

Each image has been scaled so that the maximum signals strength is mapped to black, and the minimum signal strength is mapped to white. Intermediate values map linearly onto a gray scale. The maximum signal strengths for the images are as follows: Figure 8: (A) 1.0, (B) 0.125, (C) 0.197. Figure 10: (A) 0.041, (B) 0.067, (C) 0.018, (D) 0.058. Figure 11: (A) 0.054, (B) 0.035. Figure 13: (A) 0.261, (B) 0.219. Figure 14: (A) 0.013, (B) 0.038. Figure 15: 1.0. Figure 17: Row 1 (col 1) 143.2, (col 2) 753.3, (col 3) 195.3; Row 2 (col 1) 0.324, (col 2) 0.330, (col 3) 0.325; Row 3 (col 1) 0.287, (col 2) 0.396, (col 3) 0.333; Row 4 (col 1) 1.0, (col 2) 1.0, (col 3) 1.0. Figure 18: (A) 0.325 (B) 0.332 (C) 1.0 (D) 1.0.

INDEX

A

Adaptive linear element controller (ADALINE), 494
Adaptive systems, 129–131
  dark background, 160–161
Angular speed estimation in bees, 354–373, see also Bees
Aperture optical density and light adaptation, 412–415
Appearance/disappearance (AP/DIS) response, 244–262, see also Isodipole textures
Artifacts in binary M-sequences, 200–202, 210–211
Asymmetrical response: frequency domain analysis, 266–280
Autonomous robot design, 494, see also Figure-ground separation

B

Bees
  centering response
    apparent angular speed and, 362–364
    contrast frequency and, 360–362
    neural mechanisms of anguLar speed measurement, 371–373
    obliquity problem in motion detection, 369–370
    vs. optomotor response, 367–369
    pattern contrast dependency of, 365–366
    pattern intensity profile and, 365
    region of eye used in angular speed estimation, 364
    relationship to movement avoidance, 370
  range computation study
    experimental design, 354–355
    flight path analysis, 355–357
    results, 357–367
Binary M-sequences, see M-sequences
Binary stimuli advantages and disadvantages
  of asymmetric (Poisson), 152–153
  of symmetric, 151–152
Binocular neurons, 287–288
Binocular processing: motion direction in depth, 296–298
Blocking, 317
Brightness, 321
Brightness illusions, 318–324

C

Cat models
  conditioned kernel response in ganglion cell, 136–144
  directional selectivity in striate cortex, 75–86
  spatial and temporal nonlinearities of striate cortex, 330–351
Centering response in bees, 354–373, see also Bees
Cerebellar model articulation controller (CMAC), 494
Chevreul illusion, 322–323 Cholesky factorization, 224–225
Compound eyes
  apposition, 395–396
  fly model, 424–446, see also Fly model
  light adaptation
    in fly model, 443–444
    mechanisms and effects, 409
      controller, 412–417
      photoreceptor, 410–411
      sensors, 411–412
    photoreceptor dynamic response range and, 406–409
  resolution/sensitivity design trade-off, 398–406
  superposition, 397–398
Computational models
  alternative methods, 470–473
  feedback neural networks, 464–467
  feedforward networks, 468–470
Content-addressable memory (CAM), 488–491
Contrast-dependent cortical nonlinearities, 342–347
Contrast frequency response, 439–441, 443
Cortical nonlinearities
  cat model, 330–351
  contrast-dependent, 342–347
  cortical maintenance of spatial filter properties, 350–351
  discussion: global inhibition, 349–350
  frequency domain analysis, 283–288
  filtering strategies, 333–334
  nature of, 331–333
  orientation-dependent, 334–339
  spatial frequency-dependent, 339–342
  temporal frequency-dependent, 347–349
CORT-X filter, 517–518, 525–535, see also Figure-ground separation
Craik-Cornsweet-O'Brien illusion, 321–322
Cross-kernels in motion detection, 95–104
Crustacean compound eye design, 392–417, see also Compound eyes

D

Dark background adaptive systems, 160–161
Depth perception
  direction of motion, 296–298
  size change, 298–303
Deterministic methods
  binary of nonlinear systems, 177–182
    vs. Lee-Schetzen technique, 182–184
  completion of incomplete data sets, 184–188
  general considerations, 173–174
  linear systems
    testing strategies, 175–176
    two-kernel computation, 176–177

multiinput systems, 202–203
general considerations, 203–204
M-sequence approach, 205–213, see also M-sequences
partial orthogonalization, 204–205
overcoming limitations of binary testing, 188–189
multilevel analysis, 189
quaternary stimuli, 189–190
signal-to-noise considerations, 190–191
vs. stochastic, 172
testing efficiency
binary M-sequences, 192–202
complete sequences, 191–192
Diagonal elements, 145, 161–162
Directional selectivity: simple cells of cat striate cortex, 75–86
DSMD neurons of fly, 429–432, 433–435

E

Edge discrimination, 310, 311, see also Feature detection
Electronic hardware, see Computational models
Elementary motion detection of fly, 432–437
Estimation, see Wiener kernel analysis
Eyes
compound, 391–417, see also Compound eyes
resolution and sensitivity as design requirements, 392–394
specialization for light intensity, 394–395
visual environment and, 392

F

FACADE (Form-And-Color-And-DEpth) representation, 516–542, see also Figure-ground separation
Fast Fourier transform: zoom mode, 280–283
Fast orthogonal search method, 224–227
Fast Walsh transform, 177–182
binary M-sequences, 194–197
vs. Lee-Schetzen technique, 182–184
FBF networks, 516–542, see also Figure-ground separation
Feature detection
brightness illusions, 318–324
discussion, 324–325
image structuring, 315–318
local energy model, 310–314
two-dimensional model, 314–315
Feedback models, 464–467
Feedforward models, 468–470
Figure-ground separation
by FACADE theory, 519
FBF network equations, 535–541
filling and, 520–521
by laser radar arrays, 518–519
by monochromatic FBF network, 525–535
multidimensional fusion, 521–524
recognition of conjunctive features, 541–542
theoretical background, 516–518
variable illumination and, 519–520
Filtering
in cortical nonlinearity, 333–334
in image processing, 451–459
Fly lamina model
anatomy/physiologic considerations, 44–46
LMC kernels, 47–50
receptor kernels, 46–47
operator network models, 40–42
polynomial time-invariant operators, 42–44
receptor-LMC transmission, 58–61
dynamics, 61–62
parallel LNL models, 64–68
receptor nonlinearity transmission, 62–64
structural modeling, 50–51
cascading of monomial time-invariant operators, 51–52
linear-nonlinear-linear (LNL) cascade models, 52–55
$L + LNL$ models, 55–57
surround effects
lateral antagonism, 69–71
optical cross-talk, 68–69
Fly model
discussion, 444–445
elementary movement detection, 432–437
multiplicative inhibitory motion detector concept, 425
response characteristics
adaptation, 443–444
contrast frequency dependence, 439–441
drifting gratings, 438–439
low contrast stimuli, 437–438
mean luminance level dependence, 444
spatial frequency dependence, 441–443
shunting lateral inhibition, 435–437
visual system
compound eye, 425–426
lamina ganglionaris, 426–428
lobula complex, 429–432
medulla, 428–429
Fourier coefficiencts, 29
Fourier phase spectrum: in feature detection, 313–314
Frequency domain methods, 28–30, 172, 267–268
experimental results, 283–289
sum-of-two-sinusoids, 267–280, see also Sum of sinusoids method
zoom-FFT and ultra-high resolution spectra, 280–282
Full History Walsh transform method, 120–123, 168

G

GABA-mediated motion detection mechanisms, 434–436
Gabor filter, 453, 455
Gaussian filter, 453
Gaussian white noise, 17–21
advantages and disadvantages as stimulus, 156
Wiener kernel analysis, 148–151
Global inhibition, 349–351

Goussard/Korenberg least squares kernel estimation, 119–122
Gram-Schmidt orthogonalization, 223, 230

H

Hadamard sets/Hadamard transforms, 176–177
Hermite polynomials, 20
Honeybee range computation study, 354–374, see also Bees
Horizontal systems of fly, 430, 431–432
Horseshoe crab: lateral inhibition in, 435, see also Compound eyes

I

Illusions and feature detection, 318–324
Image processing and lateral inhibition, 451–459
INFANT neural controller
  dynamic movement control for unforseen payloads, 507–511
  learning problems in robot design, 494
  locating stationary objects, 495–503
  reaching stationary targets with multiple joints, 503–507
  self-consistency and topography constraints, 495
Insects
  compound eye design, 392–417
  range conputation in bees, 354–373
  resolution vs. sensitivity in compound eyes, 391–417
  retinotopic vision in the locust, 378–389
  see also Compound eyes and individual insect models
Isodipole textures
  analytic methods
    color incorporation, 250–251
    visual evoked potential measurement, 251
    visual stimuli, 249
  constraints and working model, 247–248
  discussion, 255–262
  results, 251–255
  stochastic analysis, 244–247

K

Kernel method
  measurement methods, 16
    frequency domain methods, 28–30
    Gaussian white noise as stimulus, 17–21
    M-sequence approach, 25–28
    multiple-input systems, 30–33
    non-Gaussian orthogonal expressions, 21–24
    non-Gaussian white noise, 22
    random impulse train, 22
    related approaches, 30
    stochastic systems, 24–25
  in motion detection, 91–104, see also Motion detection
  theoretical background
    functional expansions, 9
    Volterra approach, 9–11
    Wiener and Volterra systems compared, 13–16
  see also Volterra-Wiener series expansion; Wiener kernel analysis
Kernels
  spatial, 482–487
  temporal, 476–481
Kernel variance, 157–164
Korenberg/Goussard least squares kernel estimation, 119–122

L

Laplacian filter, 453
Lateral inhibition, 435–437
  image processing and, 451–459
Least squares kernel estimation method, 119–122
Lee-Schetzen method, 34, 119, 167, 199
  vs. fast Walsh transform, 182–184
  vs. Full History Walsh Transform method, 121–122
Light adaptation in compound eyes
  fly model, 443–444
  mechanisms and effects, 409
    controller, 412–417
    photoreceptor, 410–411
    sensors, 411–412
  photoreceptor dynamic response range and, 406–409
Light intensity and eye design, 392, 394–395
Limulus, see Horseshoe crab
Linear systems
  deterministic analysis, 175–177
  impulse response, 3–4
  multiple-input systems, 7–8
  superposition property, 2–3
  transfer function, 4–7
Line discrimination, 303–305, 310, 311, see also Feature detection
LNL (linear-nonlinear-linear) cascades, 55–57, 230–231
Lobula complex of fly, 429–432
Local motion detection mechanisms, 92–93
Locust model
  linear and nonlinear transformations in early visual processing, 387–389
    medullary cell properties, 386–388
    medullary physiology, 378–379
      adaptation, 384–385
      experimental protocol, 379–380
      spatial interaction and nondirectional movement detection, 385
    sustaining cells, 380–382
    transient cells, 382–384
  retinotopic vision in, 378
Low contrast: fly visual system response, 437–438
    Luminance-gain control model of system identification, 164–167

M

Mach bands, 318–321

Mammalian feature detection, 310–325
Mead silicon retina model, 472
Mean square error, 128–133
Medulla
  fly, 428–424
  locust
    cell properties, 386–388
    physiology, 378–385
Monocular processing: size change, 298–303
Motion detection
  cat striate cortex
    discussion and conclusions, 86–87
    methods, 76–78
    results, 78–85
  directional in depth, 296–298
  fly, 432–437
  functional representations
    correlation model, 104–106
    cross-kernel design and application, 95–104
    local motion detection mechanisms, 92–93
    motion energy models, 106–107
    Volterra functionals, 93–95
  multiplicative inhibitory neural network concept, 425–444
  obliquity and, 369–370
  in robotic neural network simulation, 507–511
  speed estimation, 102–104
Movement avoidance in bees, 370
M-sequences, 25–28, 31–32, 123–125
  artifacts in, 200–202, 210–211
  fast Walsh transform, 194–197
  generation and structure, 192–194
  incomplete data sets, 197–198
  in multiinput systems
    artifact removal, 210–211
    basic technique, 205–206
    examples, 211–213
    orthogonalization procedures, 206–207
    quaternary stimulation, 208–209
    sparse stimulation, 209–210
    stimulus generation, 207–208
  orthogonalization procedures, 198–202
Multilevel analysis, 189
Multiplicative inhibitory motion detector in the fly, 425–444, see also Fly model

N

Non-Gaussian white noise, 22–23
Nonorthogonality, 167–168

O

Optomotor response in bees, 367–368
Orientation-dependent cortical nonlinearities, 334–339
Orthogonalization in binary M-sequences, 198–202

P

Parallel cascade method, 230–239, 231–233
  examples, 233–235, 236–239
  time-series analysis, 235–234
Photoreceptors
  light adapation and dynamic response range, 406–409
  optical density and light adapation, 415–417
Poisson stimuli: advantages and disadvantages, 152–153
Primal sketch concept, 310
Psychophysical models
  binocular processing of motion direction in depth, 296–298
  classic psychophysics and animal physiology, 294
  linear vs. nonlinear analysis, 295–296
  line separation discrimination, 303–305
  monocular processing of changing size, 298–303

Q

Quaternary stimuli, 189–190
  with binary M-sequences, 208–209

R

Radius of convergence, 156–157
Resistive array modeling, 472–473
Resolution
  light adaptation and, 410–411
  vs. sensitivity in compound eyes, 391–417, see also Compound eyes
Response asymmetry: frequency domain analysis, 266–280
Response variance, see Wiener kernel analysis
Robot design, 494–511, see also Figure-ground separation; INFANT controller

S

Sandwich model of system identification, 164
Sensitivity vs. resolution in compound eyes, 391–417, see also Compound eyes
Sensors: light adaptation and, 411–412
Signal-to-noise considerations, 190–191
Silicon retina model, 472
Simple cells: directional selectivity in cat striate cortex, 75–86, see also Cortical nonlinearities
Sinusoidal series analysis, 30
  discussion, 227
  examples, 227–230
  fast orthogonal search, 224–227
  prelimary computations, 223
  significant frequency identification, 223–224
  see also Sum-of-sinusoids approach
Sparse stimulation, 112–115, 162–164, 167
  with binary M-sequences, 209–210
Spatial frequency: fly visual system dependence, 441–442
Spatial frequency-dependent cortical nonlinearities, 339–342, 350–351
Spatial kernels, 482–487
Speed estimation, 102–104
Spiking systems, 132–133
Stochastic analysis, 24–25

vs. deterministic, 172
isodipole textures, 244–247
Stone-Weierstrass theorem, 230
Striate cortical cell function in the cat, 330–351, see also Cortical nonlinearities
Sum-of-two-sinusoids method, 172, 267–280, see also Frequency domain method
Sutter's Full History Walsh transform method, 120–123, 168

T

Taylor equation, 13–14
T cells of fly, 429–432
Temporal frequency-dependent cortical nonlinearities, 342–347
Temporal kernels, 476–481
Ternary stimuli: advantages and disadvantages, 153–155
Transient medullary cells, 385–388

V

Variance, 157–164
Vertical systems of fly, 430–431
Visual environment and eye design, 392
Visual evoked potentials, 162
appearance/disappearance response, 244–262
frequency domain methods
experimental results, 283–289
sum-of-two-sinusoids, 267–280
zoom-FFT and ultra-high resolution spectra, 280–282
see also Isodipole textures
Visual tracking: parallel cascade analysis, 236–239
Volterra method, 9–11
in motion detection, 93–95
vs. parallel cascade method, 230
vs. Wiener, 11–13
Volterra-Wiener series expansion
content-addressable memory in, 488–491
spatial kernels, 482–483
classification properties, 486–487
first-order, 483–484
higher-order, 485–486
second-order, 484–485
zero-order, 483
temporal, 476–479
first-order temporal kernel, 479–480
second-order temporal kernel, 481–482

W

Walsh transform, 177–182
binary M-sequences, 194–197
vs. Lee-Schetzen technique, 182–184
Sutter's Full History, 120–123, 168
White noise measurement techniques, 17–24, see also Wiener kernel analysis
Wiener kernel analysis, 11–13, 28–30, 33, 111–112
amplitude distribution and
diagonal elements, 145
exclusive expansions, 145–146
Gaussian kernels, 148–151
inclusive summations, 146–148
stimulus advantages/disadvantages
binary asymmetric (Poisson), 152–153
binary symmetric, 151–152
Gaussian stimuli, 156
ternary stimuli, 153–155
anomalies and, 124–125
conditioned kernels
example: center-surround interactions in cat ganglion cell, 136–144
general expansion, 134–135
stimulus change effects, 135
Volterra expansion, 133–134
hybrid method, 123
kernel estimation, 115–116
''full history'' expansion, 116–117
least squares method, 119–120
Lee-Schetzen method, 119
nonorthogonality due to finite duration, 117–119
Sutter's Full History Walsh method, 120–123
kernel variance
dark background adaptive systems, 160–161
diagonal elements in, 161–162
general formula, 157–159
of general kernel, 159–160
special situations, 162
M-sequence method, 25–28, 31–32, 123–124, see also M-sequences
orthogonal expansion for sparse stimulation, 112–115, 162–164
vs. parallel cascade method, 230
radius of convergence, 156–157
stimulus selection and outcome
autocorrelation properties, 126
multiple inputs, 128
stimulus sampling interval (Δ_s), 126–127
system identification
luminance-gain control model, 164–167
sandwich model, 164
units and mean square error
adaptive systems and contrast units, 129–131
flash kernels, 128–129
response variance, 131–132
for spiking systems, 132–133
vs. Volterra, 13–16

Y

Y cells
of fly, 429–432
of locust medulla, 385–387

Z

Zero-order spatial kernels, 483
Zoom FFT, 280–283

ELIHU BURRITT LIBRARY
CENTRAL CONNECTICUT STATE UNIVERSITY
NEW BRITAIN, CONNECTICUT 06050